An Open Letter Announcing

Gravity's Recipe

ISBN-13: 978- 978-1532965371
ISBN-10: 1532965370

Can you raed this?

Olny srmat poelpe can.

I Cdnuolt blveiee that I cluod aulaclty uesdnatnrd what I was rdanieg. The phaonmneal pweor of the hmuan mnid, aoccdrnig to a rscheearch at Cmabrigde Uinervtisy, it deosn't mttaer in what oredr the ltteers in a word are, the olny iprmoatnt tihng is that the frist and lsat ltteer be in the rghit pclae. The rset can be a taotl mses and you can still raed it wouthit a porbelm. This is beuseae the huamn mnid deos not raed ervey lteter by istlef, but the word as a wlohe. Amzanig huh? yach and I awlyas tghuohot slpeling was ipmorantt!

If you can raed this psas it on !!

Written in 2004 and completed beginning 2005.
Went missing in 2005 and recovered in 2016
Published 2016-04-28
Published as it was in 2004

In this four series and parts of this series about gravity there are three more books entitled:

2) An open letter Addressing Gravity's Formula

Where the author explains how the Universe came about at the first instant the Universe came about in using evidence on the four cosmic pillars and it matches the Biblical explanation in explicit detail

3) An open letter About Gravity's Prescription

Where the author explains how the Universe came about at the Solar system, as we know it took place. It explains why there are four solid planets, four gas planets and one cold structure. It also explains mathematically why all the debris is encircling the planets and where they come from. This is one part of another book entitled the Seven Days Of Creation

4) An open letter explaining Gravity's Rules

As the Author goes into detail about a new cosmos theory where the four cosmic pillars produce a cosmos everyone cam understand. It puts time in relation to space and discovers what space is in relation to time. Never yet before was either time or space understood because everyone drooled on the misconception about mass and incorrectly interoperating that mass produces gravity.

This series forms as a unit with four individual titles forms a prologue to a Thesis that introduces a whole new concept about Creation.

Matter's Time In Space : The Thesis in seven parts
ISBN 0984410-8-1
Written By Peet Schutte

In this letter I call Creation by name and prove with science that we are in Creation. I employ science to prove that that controls Creation, which is not in the Universe but is noticeably because it is not in the Universe. In the light of all proof and when facing evidence I bring I dare an atheist to prove me wrong about Creation.

In mentioning this word in a science book I break a ground rule enforced by the atheistic dominated world of science. I challenge any person to bring proof about any part where any of my theory might be incorrect and furthermore I challenge any Academic in physics to prove that Newton's mass pulling mass is anything other than fraud. I charge any one to bring proof that the cosmos is contracting by the force of mass and that mass produce gravity as Newton advocated when he committed the biggest fraud of all times.

© KOSMOLOGIESE EN ASTRONOMIESE TEGNIKA

An open letter

Announcing Gravity's Recipe

ISBN-13: 978-1532965371   ISBN-10: 532965370

Written by Peet Schutte
As part of the introducing of

KOSMOLOGIESE EN ASTRONOMIESE TEGNIKA
An open letter

Matter's Time In Space
THE THESIS

ISBN 0-620-27041-1

THAT FORMS A PROLOGUE WHICH INTRODUCES AS WELL AS IS PRESENTING A NEW COSMIC THEORY AS IS STATED IN **MATTER'S TIME IN SPACE THE THESES**, ISBN 0 – 6 2 0 – 2 7 0 4 1 – 1 which consists of the following seven books in title:

1) A Cosmic Birth…Dismissing Nothing I.S.B.N. 0-620-31609

2) AN OPEN LETTER ON: **XEPTED ASTRONOMICAL MISTAKES** ISBN 0-9584410-1-4

3) AN OPEN LETTER ON: **INTER GALACTICA SPACE TRAVEL** ISBN 0-9584410-2-2

4) AN OPEN LETTER ON: **CORRECTING COSMOLOGY** ISBN 0-9584410-5-7

5) MATTER'S TIME IN SPACE: **THE HYPOTHESIS** ISBN 0-9584410 –6-5

6) AN OPEN LETTER ON: **" STARSSTUFFN'** ISBN 0-9584410-3-0

7) " SEVEN DAYS OF CREATION" ISBN 0-9584410-4-9

There are three books entitled

An Open letter Announcing Gravity's Recipe
An Open letter Addressing Gravity's Formula
An Open letter Abiding Gravity's Prescription
Where all three as a unit explains each its own one secluded part of the manner and the means in which **gravity** as a principle and as a principal forms and rule the cosmos… and mass has no role to play in the generating of **gravity**.

All three as one unit then forms a founding part of

MATTER'S TIME IN SPACE
THE THESIS
ISBN   0-9584410-8-1

TRANSLATED BY
PEET SCHUTTE
FROM THE ORIGINAL AFRIKAANS:
"MATERIE SE TYD IN RUIMTE"
I. S. B. N.   0-620-27041-1
WRITTEN BY PEET SCHUTTE
© **KOSMOLOGIESE EN ASTRONOMIESE TEGNIKA**

TO WHOM IT MAY CONCERN,
RE: AN OPEN LETTER TO ACADEMICS ISBN 0-9584410-9-X announces

The Four Cosmic Pillars

In this letter to Academics I, the author, introduce a new method whereby the origins of the solar system can be proven, if one make use of the natural phenomena which I named **The Four Cosmic Pillars**

They are as follows:
The Titius Bode Law
The Lagrangian System
The Coanda principal
The Roche Limit.
First one has to prove how these four Pillars bring about gravity, and that I do as you are about to find out. Sit back and enjoy the ride because you are going to stray from Mainstream Physics as far as you have never strayed before.

Dear Professor,

I am Petrus Stephanus Jacobus Schutte going by the name of Peet and who is the author of the above-mentioned book(s). I hope you find your reading of this book presented as an open letter a most fruitful experience. I feel I need to warn you, whom are reading this letter that this work contained in this letter strays widely from mainstream science and for that there is a very good reason but I should add that in the least it is thought provoking. I researched the work of a man that is most exceptional and even more prominent in the history of mankind. His role in the gathering of information furthering knowledge accumulating of the human species' efforts stands second to none while most of everyone is not even aware of the full implication of his work. While recognising his work, Mainstream science bluntly ignores his work and in that they miss the full vastness of the wide influencing of his work. It is therefore almost absolutely realistic to say that what you are about to read in this open letter sent to you for your attention was never yet printed in the near or the far past although the work has been with us for about four hundred years during which time it went unnoticed. It seems to me that any research predating Newton never came into use or in practise. My investigation of Kepler's work brought about a conclusion that no one yet arrived at concerning the findings of Kepler because no one scrutinised Kepler's formula. Kepler found planets rotating around a centre but Newton saw a circle and added what is mathematically required to indicate such a circle. Newton added a mathematical $4\Pi^2$ to the formula of Kepler and removed the distance symbolising measure that Kepler introduced, using **k**. On the other side Newton changed the symbol of **k** by using the symbols G (m + m_p). This is just a longer and probably a more detailed manner of indicating **k** and a better defining of **k** but it symbolises precisely to the point what **k** stands for nonetheless.

I present you (again) with my work and this time I compiled the work into three books. In the three books, I introduce for the first time ever into science an explanation for the phenomena not yet explained. They are: 1) The Roche limit, 2) The Bode law, 3) The Lagrangian points and 3) The Coanda effect.

These above mentioned phenomena work in harmony to form the principle called gravity. They work as a whole, as an assembled unit to compile gravity. Each one has its individual role but unifies their combined principles to form gravity. Each fills a function that finds a position in the unit where it participates in forming that which we regard as keeping the Universe glued. Mass therefore, has nothing whatsoever to do with the generating of gravity. In the books I purposely explain my conclusions in minute detail, which I did not do the previous time when I offered my work for academic investigative evaluation. This time the information is now so much simplified, it is to a level where I believe it will probably be understood by any child still at school on the condition that that scholar has developed an understanding about physics to the level of a high school standard. In the previous presentations of my work that was submitted, it was rejected at the time by your institution amongst and including other institutions equal in stature to that of yours. At that time I could have been guilty of accepting that there were facts I presented that the reader was unfamiliar with, but I did not realise that because I saw those facts as general knowledge, which at the time I personally would regard as simple everyday information. In the end those facts did prove not to be general knowledge after all and that complicated some of the issues. One such an example is where to find singularity. It is so obviously allocated, that at the time I could not dream any person did not see it at first glance. The other possible mistake is not to explain how I concluded how the value of singularity came about, because that too, is so simple to see. One more example I may mention is to explain why singularity has developed space-time by specific measurements according to time and by using time. I might have failed to present the limits of time, which I clearly make an issue of indicating this time round. I took this as general knowledge being located in an obvious position where all can immediately see the presence of eternity and infinity, but it turns out not to be that obvious to others. (This is only naming a few.) However, I have corrected such mistakes and this time I have explained every aspect in such great detail that more explaining of any issue with the use of more detail would not help explaining more facts but would become tedious and monotonous. Therefore, my explaining at present is so simple a scholar should understand it.

My intension with sending you this letter is that I am sure I have discovered an enormous blunder in cosmology. I am of the most sincere belief that there is no evidence ever presented to substantiate the proof required that mass is responsible for creating gravity. Moreover is the fact that I cannot detect mass as a factor any place other than in the work of Newton. I am sure that Newton corrupted the work of Kepler in order to further his (Newton's) views on mass and by doing that he did so to intentionally mislead the entire world of science fore more than three centuries with his blatant corruption. There is no single trace of mass in all of the cosmos and I sincerely feel that I have to inform you of this fact. **What I did find is that I located singularity in both infinity and in eternity and it is the interaction between points in singularity in infinity interacting with points of singularity in eternity by providing motion, such motion is what generates gravity.** I admit that most if not all of the content is new, because I have an unusual approach to cosmology. Though my approach is unconventional, it enabled me in locating **the precise location of singularity**

that forms the connecting basis of the Universe (and this I say with some degree of confidence).

There **are two locations,** but I shall **first concentrate** my explaining effort on **the prime singularity.** I have in the past contacted various academics in South Africa but I suspect that because of my accusation of Newton, I did not generate the required enthusiasm. I have the suspicion that it is mostly because of my blaming Newton of conspiring and accusing Newton of corruption that I fall out of favour every time I send a letter to academics. I am sure that it is these views I hold on Newton that is to blame for the negative reception I received thus far in academic circles, as the response of the learned was much less than favourable. Well, be that as it may, I still hold the opinion about Newton committing fraud and Newton deceiving the world on purpose and with criminal intent. Other than the say so of Newton, as far as I can trace I found without doubt that there is no evidence of a trace of mass through out the cosmos and by that statement I stand while I write this letter. **My work is radical** and **it is unconventional**; that I admit. It seems in Africa no academic wants to be part of a radical approach. I have confidence in my work and through that, I have confidence in finding a promoter. Therefore, I thought it might be beneficial and promotional on my part in sending you a manuscript of my book(s). I am not being arrogant when I say that I uncovered a mistake in science about mass and the absence thereof in gravity.

HAS ANY PERSON READING THIS, EVER GAVE A THOUGHT ABOUT THE FACT THAT THERE IS NO SUCH A THING AS MASS? 617 DO YOU **FIND THIS STATEMENT RIDICULOUS**? I CAN ASSURE YOU THAT **THIS STATEMENT IS CORRECT** AND THE FACT OF **MASS IS RIDICULOUS!** YOU MIGHT DECIDE TO IGNORE THESE FACTS BUT THEN YOU WILL REMAIN PART OF THE LAST BASTION OF THE DARK MIDDLE AGES, OR YOU MAY READ THIS BOOK AND BECOME PART OF THE FUTURE.

Gravity is based on **Newton's presumptions**, which are **incorrect**. **My work is based on** the findings of **Galileo and Kepler,** and **their findings** are totally **inconceivable** with the **findings of Newton. I DO NOT EXPECT ANY PERSON TO BELIEVE** THIS STATEMENT OUTRIGHT, THEREFORE I INVITE YOU TO SPEND YOUR TIME ON **THE NEXT FEW PAGES** OF THE BOOK; **READ THE FOLLOWING** ARGUMENTS, THEN SEE IF YOU STILL ARE **AS CONVINCED ABOUT GRAVITY AS BEFORE.** (After saying this in person to some academics, they really got annoyed. If I did not look like the real image a Papa Bear should have, I know they would have attempted to blacken both my eyes.)

I uncovered a mistake in science. From the onset, the mistake seems as insignificant as it is small. Because the rest of the book is about the mistake, I do not intend on elaborating about the mistake itself. The mistake came about with the culture of education and the mistake in itself seems harmless. When admitting that, one must also admit that any pilgrim that got lost and died of starvation through an incorrect travelling direction, made the very first part of his ultimate mistake by looking in the wrong direction. How harmful does looking in a specific direction seem, and yet such a mistake leads to his ultimate mortality. The traveller could, when taking his first directional flaw with that the first incorrect step, only put his foot skew in avoiding a rock. Or he could have turned his face to avoid a branch and that move pointed him in a direction that led to his fatality.

It is not the mistake that becomes the penalty and it is not the origin of such a mistake that leads to the penalising, but the ignoring of accepting signs telling the wonderer of an impending error and his stubborn ignoring of such a telling sign that makes the lost party pay the ultimate price. By ignoring the mistake, for whatever reason, the ignoring of such a mistake is his undoing because the price due comes from the inability in recognising the sign indicating the presence of the mistake forming the reason for his final demise. The sooner such a person sees and admits the wrong, the less will be the consequences of his final price to pay.

I am most aware that there is this perception that Newton has never made a mistake. It is more than a perception...it is religiosity. Newton can do no wrong because Newton never did wrong. This entire concept is wrong making Newton's entire concept incorrect.

The Newton mistake is one born in culture and the penalty from this mistake is bred by arrogance. At school, minds are young and accepting, although developing. Through many tens of millennia, humans came to a habit in surviving as a specie where culture taught them that accepting the elders' advice is the same as to ensure survival of the following generations, and by such doing is also following the quickest way to an adult mind. By accepting the elders' knowledge and experience without question proves the dominance of the tribe in relation to other tribes of the same race. This is culture we cannot do without and still maintain progress. It is an inheriting method humans grew on and is the corner stone of all civilization. We cannot abandon it.

The scholar sits in class and receives from the Master information that is completely his day's breaking news. As far as his mind can tell, the news is as actual as anything else that happened at that moment, notwithstanding the fact that such news may be with the human mind for thousands of years. Whatever the teacher tells him is bound to be a first time experience, so new he has no time to digest the information. Taking into account his youthful ways (which we all had), he has little stomach to scrutinize it because learning is a painful process to all. Without pain and perseverance there can be no education of any sort. He does very willingly accept the facts as tested and correct without flaws of any kind. The scholar has to because in any education system time will not allow students to ponder about detailing information and securing a prognoses to all learnt every day. Whatever the Master tells the scholar, is taken as Biblical correct without any thought about testing the results. Where there are cases of scholars having doubt and subsequent questions, the Master takes such behaviour as being obstinate and being a reflection of the student on his (the tutor's) personal integrity and knowledge. The young mind will very soon discover that his behaviour is not tolerated by the system, and the truth is, the system cannot tolerate such behaviour for the good of the rest. Time must be spent on learning and accumulating as much information as that which the young mind can accept.

No information could be more affected by such a culture than that of Newton. You both understand Newton and are smart, or you do not understand Newton and accept that there will be very little future for you to have in the world of science. Newton is science. No Newton understanding automatically becomes "no science" -education or -learning. Without Newton, there is no other and science will be a vacuum of containing nothing. This is very unfortunate but is the ultimate truth. It is either Newton's way or no way at all. This culture also brought along the stigma that only the minds of the sharp and the sighted can accept and understand Newton and when not understanding Newton, one tends to fail your personal I.Q. test. It is a sure sign of the slow witted when the student fails to recognise what Newton said. The only way to advance in science is to understand Newton and indicate to all your pears how brilliant your mind is in accepting information.

All students have little understanding about Newton, and that I can and will prove through the next two hundred or so pages. The mistake Newton made and which I discovered is laughable small, yet it took me (not being that bright, I may add) almost one lifetime to recognise the mistake, whereas it took mankind three hundred and fifty years of research without recognising the flaw. Others in the past may have come to see what I saw, but if there were such persons, they never saw what I saw because if they saw what I saw they would also see that behind such an almost invisible puncture hole in the tube, is a reason for science to deflate and not accumulate. When "understanding" Newton, it becomes the very same as learning Newton from the heart and accept that what you memorise is what you know. The memorised knowledge is beyond question, as it has to form part of the identity of the student having secured the knowledge. It is not the hole forming the puncture and preventing inflating being as such the obstacle that is of importance, but what that hole does to the tire and the car and the travelling with the car that becomes a menace. In the extreme effort to keep the journey on course, the Academic Masters are spending an all out effort in inflating the deflating tire faster than the deflating tire can deflate. The recourses attached to this effort is enormous and without cause. It will lead to nowhere and that is where science is heading in their attempt not to head in that direction.

When students become masters, the Masters seek to break new academic ground. Masters do not ponder the ways on which they lead their students, but have an all out effort in establishing their own support of new territory that will distinct them from the rest in future to come. My uncovering of such a mistake as I did could only be from a person as ill-educated as I. I did not go through the learning process where the learning process is the very same as a brain washing and mind controlling process. This remark may seem harsh and is not intended to be such, because reality demands no other way in education. The mistake and the carrying of the mistake with the support in refusing to recognise such a mistake, is part of human training and there is very little to do, but to admit that such may occur whereby to follow in acting to correct the mistake after the discovering, rectify what ever can be salvaged and build from there.

Friend and foe alike, think of me as being slow of mind and not understanding Newton. This was what I was told on more numerous occasions than I care to remember. Pointing at the mistake I see, I am told by the wise as well as the not-that-wise such as I, that through my lack in education I cannot dream to understand Newton and therefore are committed to the position of the ILL-EDUCATED, a position I learned to accept with grace. In all cases there are more sides than one and therefore I think of myself as poorly educated, the contras to my position must be those fortunate to be SUPER-EDUCATED. In this letter there will be the addressing to you holding the position of the reader and (I hope) the un-bias judge, with me presenting my case to the un-bias (you) in concerning the third party, the SUPER-EDUCATED. Referring to the party in opposing me as the SUPER-EDUCATED, is by no means in disrespect and much to the contrary holds my whole-hearted admiration, (at times, somewhat limited, that I admit.)

The information in this book I purposely made easy to read, and easy to follow. The content however, is only a drop compared to a bucket of information contained in all seven parts of "MATTER'S TIME IN SPACE The Thesis". All information presented to you, in the first part of the book is an introduction to the second part. In the latter part of the book the conclusion comes. Without a detailed introduction in the first part, the information in the second part will be of little value, as the information is very conclusive about the first part. I have been researching cosmology since 1978, on a part time basis. The conclusions may seem simple that however, is in retrospect. Some of the arguments took me up to six months to arrive at, since I do not have all the information on hand.

When Newton formulated his theories and physics, they were the nuclear science of the day. That was a long time ago though. Science today make a great issue about how accurate their views are and how they work on facts alone. Yes, they work on facts, but the accuracy of the facts is the big question. When physics declare a fact, and the fact contradicts nature, just how accurate are such facts? A simple example we all understand but very little appreciate and even children at school find it hard to accept as they always argue with teachers about this, is Newton's statement that something going up into the air, will come down with the same force it had when going up. In Newton's time, there was no tool to measure such a claim. In the technical age we live in, everyone knows that such a claim has no foundation and even school children argue this fact. By dropping a tennis ball down a many story building, the children can see enough to make them (quite correctly) dispute such nonsense. The ball will not once reach the height it had before and blaming friction is pointless.

An object going up is fighting gravity, and the battle continues until the gravity finally wins. By winning the fight, the gravity will claim its prize and bring the object back to Earth. From a Newton stand point, this means the force that moved the object in the air was eventually equal to the force that brought the object back to Earth. The one force only cancelled the other force and equilibrium established a neutral force in the end. In Newton's day, wind was the working force of the time. Steam was a new marvel not employed on a common basis. If the wind picked an object into the air, or pushed the canvas of a ship's sail, it was a force. Anything that brought about motion meant a force was in progress.

The Newtonians still maintain that philosophy to this day. Only a force at work can bring about work. With electricity too, they declared a force at work that applied and brought about movement of free electrons. Is this view altogether inaccurate? I should say no, but with the question: "Is such a view adequate?" I should say no! Today the Newtonians will insist on your beheading if you say that the force they maintain could just as well be that of magic. This is because the Newtonians only discarded the notion about magic, but not the magic behind the force. This leaves a gaping hole in science which not one Newtonian ever explains. If we refer back to what defines mass, the definition is wide spread...yet there is one thread holding all concepts together. Mass gives distinction by allowing patrician between many. Mass brings identity to the individual and parts each from everyone else. Mass puts down what is different about the specific fitting each individually.

Galileo said everything that falls, is equal while falling. Galileo said the most massive is likely to fall as fast as the least massive when wind resistance as a factor is removed from the equation. The person of one hundred kilograms is likely to fall as fast as the object next to it with a mass of two thousand kilograms. That is Galileo's verdict. Also, it is well proved in our modern age where so many things fly and fall from aircrafts or down cliffhangers and tall buildings. There was a show on TV where three persons drop from an aircraft with a Volkswagen motorcar. While the three were falling with the car, they got out and pretended to walk around the car. On another occasion they imitated getting out of the car and pretended to push start the car. In the end, they opened their parachutes and landed softly away from the car while the car was left to fall to pieces. The descending of the car with its occupants getting in and out all the time, vindicated Galileo throughout the entire fall. It did nothing to prove Newton.

Newton saw the issue connecting to mass. He said $F = \dfrac{r^2}{M_1 M_2}$, which was changed at a later stage to $F = G\dfrac{M \times m}{r^2}$ but that was no retraction of the first statement because that was a correction to the first statement when he saw the first statement go a little bizarre. What Newton added was exactly what Galileo removed. Galileo said all objects fall equal. Newton said Mass is that, which puts differentiation between objects. If mass inspires such a falling, then the car must fall a lot faster than the persons and each of the persons must have a different rate in descending since each has its individual mass driving it towards the ground. While falling, all the objects falling have equal mass, which makes the lot have no mass at all. Newton was deliberately ignoring Galileo while Newtonians to this day are connecting what never can connect.

According to Galileo an apple will fall at the same rate as a watermelon would fall. That is accepted science. What this implies is that the apple and the watermelon share an equal mass factor of one. The apple is no less massive than the watermelon, which on the other hand implies that the watermelon is not more massive than the apple. After all, they land on the Earth simultaneously.   When putting mass into the equation there has to be differentiation, while putting Galileo's view in removes mass altogether. While falling objects have no mass it puts $F = \dfrac{r^2}{M_1 M_2}$ into dispute with Galileo. There is this choice to be made when considering the two science opinions: either mass is a factor and falling differentiates particles in such a falling, or mass is not a component of any object while falling and therefore mass cannot inspire such a fall. The fact that mass is present has to be proved mathematically or else it is not present. Newton cannot surmise its presence without proving its presence. Mass puts distinction and where there is no distinction there can be no mass. Newton (and the rest of the science world) ignored Galileo by agreeing to what Galileo said and in the same event ignored what Galileo said to acclaim what Newton said.

There is no denying the facts or getting out of the argument. If mass induces gravity it has to show in the fall differentiation. It is either mass bringing on gravity, which will put more massive object falling faster than lighter objects. Both Newton and Galileo can't be correct. It is either Newton or it is Galileo and that is where Newtonians go corrupt. They knew all along for centuries that Galileo is correct but slid passed the part that they have to choose between Galileo's proven sensibility or Newton distorting of facts to commit corruption at his pleasure.

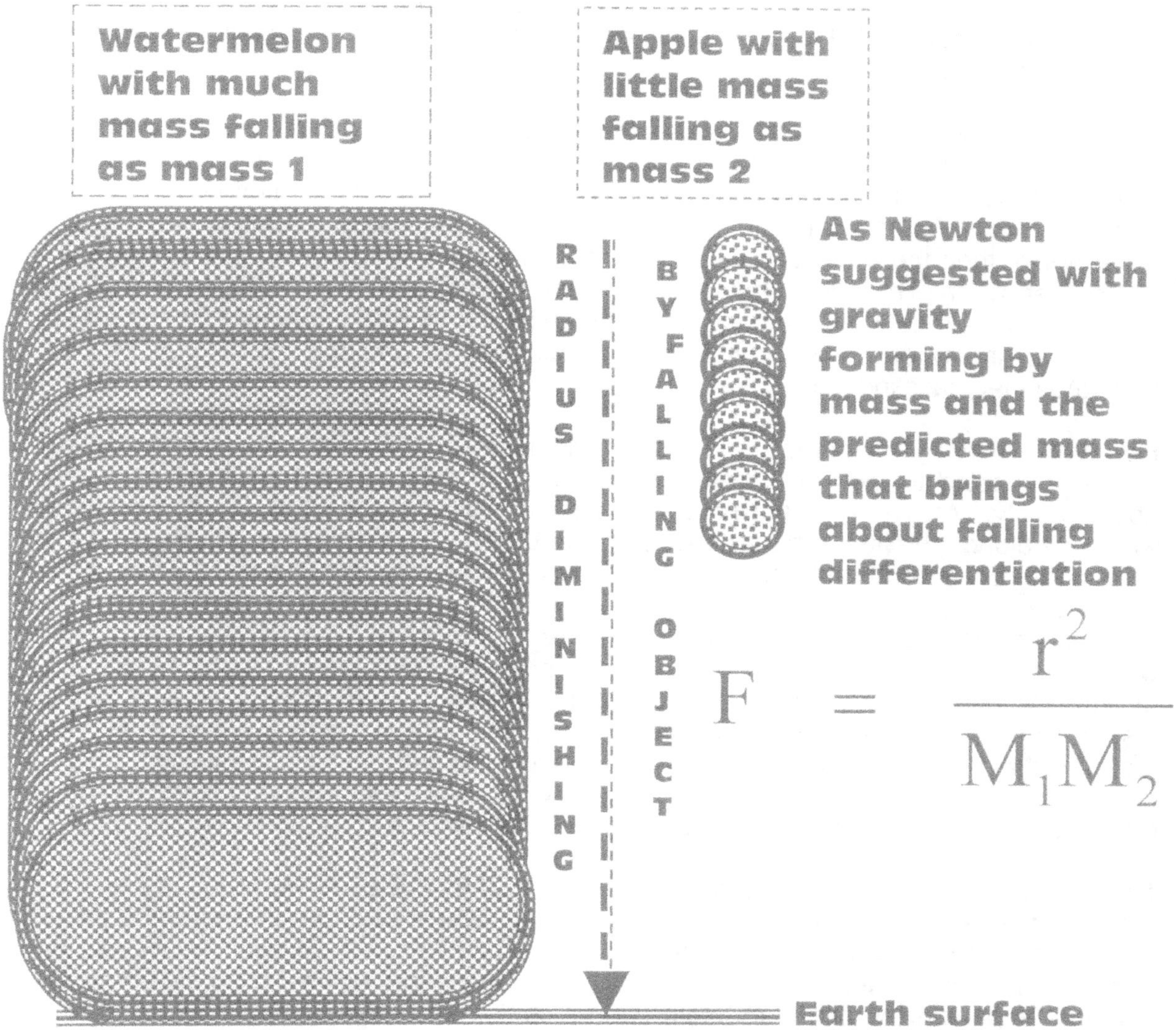

Watermelon with much mass falling as mass 1

Apple with little mass falling as mass 2

RADIUS DIMINISHING

BY FALLING OBJECT

As Newton suggested with gravity forming by mass and the predicted mass that brings about falling differentiation

$$F = \frac{r^2}{M_1 M_2}$$

Earth surface

If mass was a factor, as Newton claimed, then Newton must find that a watermelon would fall faster from his tree bearing apples than the apple fell that he saw, while it was falling. After all, the watermelon has more mass than the apple has because it is bigger than the apple. That is not what Galileo found. Galileo found that the biggest watermelon ever would fall as fast and land on the ground the very same instant that the smallest apple that there is can fall, when falling from the same height. The two would land on the Earth at the same instant, notwithstanding size or mass differences. The only way this can be accomplished is where there is no mass that is effecting the falling. That means while falling, the apple and the watermelon carries the same mass and that mass holds a factor of one. If there was mass of any description, such falling should be different and it should be according to the mass input that every object has.

Let us compare apples falling with watermelons falling and not apples falling with apples falling.

In the vision Galileo had about science, he declared that a watermelon and an apple falling will land on the surface of the ground on the condition that both start to descend the very same instant, all conditions applying are similar to the smallest detail and that the distance the two are dropped from, be precisely the same. After the death of Galileo, an Englishman by the name of Isaac Newton came along and substituted this for the idea that it is the mass that allows objects to fall. Lets put the watermelon at ten times the mass of the apple and put the argument into a sketch.

What makes this argument a big issue is that science is sheepishly running after the Newton idea that mass is producing gravity while gravity is that which will force the falling of the object. If it was mass that induced the falling and set the gravity in motion, then the falling had to go according to the mass. If the mass of the apple was one tenth of the mass of the watermelon, then the apple should be one-tenth down the drop by the time the watermelon fall to destruction. The apple should fall one tenth the distance by the time the watermelon completed its fall. Lets be adamant about one thing: Science is adamant that it is mass that is enticing gravity and it is gravity that charges the fall and the fall is provoked by the standards that mass induces.

However, if we can't detect mass in the fall and if we can't pin any differentiation on the fall that is connected to the mass that must produce the differences, then mass cannot bring about gravity. Mass is then not responsible for gravity and then Newton and science are wrong all along. Gravity is the product of some other concept that has nothing in common with mass. Ever wondered why God forgot to put wings on mountains…I would say it is because they can't fly. That could be the reason why birds are supplied with wings; it might be because they can fly. Why would birds have the ability to fly…it is because birds have wings…yes, but birds also have life. Life is not part of the Universe and that is very bad news for atheists because before they dish life out to everywhere they wish to put life, they must first find life and proves life can be anywhere but on Earth and in their imaginations. By the fact that they were incapable of showing life except in their atheistic fantasies, they must adhere to the concept that until they prove me wrong and they prove that life is ten a penny throughout the cosmos, their fantasies are cheap, mindless and childish.

The difference between the mountain without wings and the bird with wings is the fact that the mountain is incapable of moving independently. You may argue that the mountain may move with a landslide or an Earthquake. That assumption is quite wrong because as the names say, it is a landslide that is putting the mountain down as lands and it is an Earthquake putting the mountain in its correct context, it is the Earth. The motion life can produce is outside the cosmos and the motion that the cosmos produce stands outside the ability of life. The fact that mass cannot be detected in the falling of objects puts Newton and his idea of mass that is the factor, which is producing gravity in serious contention. It puts doubt on all aspects of physics and it tends to rubbish what science tried to establish for more than three hundred years.

Therefore, Newton's presumption that it was the mass of his apple that had the apple fall was way off the mark. If you have a means to find how one might go about and see the effect mass has on falling objects and can prove in what way does mass bring discrepancy as prove to the fact of mass applying, please be sure and let me know. Until such a time, I say Newton and Newtonian conceptions about mass that has the ability to produce gravity is a folly!

However, he corrected that assumption and stood by his next formula to this day.

He stood corrected by his next formula that he introduced as $F = G\dfrac{M \times m}{r^2}$.

What this says is that there are two objects in the Universe where the one object has mass while the other object must have more mass since it is bigger than the first object.

With out any apparent reason the formula is substituted with the following formula:
Gravitational force = G (M. m) /r^2 where:

G = the gravitational constant,

M = the mass of the body,

M = the mass of the lesser body

r^2 = the radius between the two bodies.

Now why would Newton find a reason to change his formula?
Let us go back to where Newton first detected mass and find how he went about in detecting mass for the very first time. We go back again to find mass where Newton saw his apple fall from his campus apple tree. In the calculation he mentioned the apple but he did not implicate how much did the gravity of the apple pull the mass of the Earth. One must see this as very important because if the two are equal partners, there can be no smaller or lesser partner and no larger or dominant partner. Where such importance arrives in the equation is to determine how much does the Sun move towards the comet and how much does the comet move toward the sun. To be able to understand such behaviour one should try to establish beforehand how much does the apple move towards the Earth and how much the Earth moves towards the Apple. I am returning in order to find the connection there was that Newton saw between his first proposal and the second proposal that one cannot blatantly find deception in. If one would put the mass of the Earth in relation to the mass of the Sun and have the two concepts pull on one another, the Big Bang would become the Big Crunch within a few minutes. The force of gravity would become so great it would have the Earth and the Sun collide with a Bang that will echo throughout the Milky Way. In that I can see that Newton saw his deception would not last and he had to change his story in order to remain convincing and have some fools believe his nonsense. The mass is destroying the radius from both ends; so how much destroying is taking place from the side that the Sun controls and how much destroying is taking place from the end where the comet is in control? There is no mention of the mass of the smaller object, **AND** there is no mention of the limit on falling objects being 203 kilometres per hour. The excuse is that it has something to do with wind resistance. That is fabricated nonsense, which, I shall prove within the next few pages.

We have the mass of the Sun going as $mass_1$, then there is the radius which is the distance between the Sun and the comet and that makes the comet $mass_2$. The comet and the Sun, is committed by the value of outer space forming G, which is the gravitational constant. That is all the substance needed to form gravity, or that then is the verdict according to Newton. The Sun has mass one and the comet has mass two and outer space being without mass has the wonder of the gravitational constant.

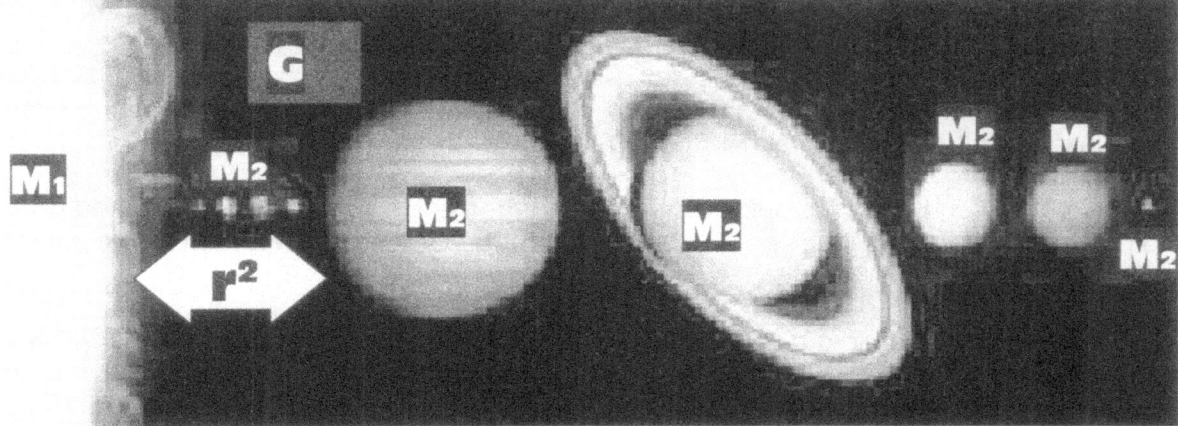

To have the gravitational constant and in order to calculate the gravitational constant one must at least have the allocated position where the gravitational constant is pulling to and that has to be the centre of the Universe. Therefore, one may presume that the first thing Newtonians did after finding gravity was to search where the centre of the Universe would be because it must be in that direction that gravity in the constant is dragging gravity in the mass. Well, be as it may, the Sun with its mass is dragging the comet with its mass across the outer space which has no mass but is blessed with the fortune of sporting the gravitational constant. The comet is coming and that is a fact.

The big object will be the Sun. Then lets assume the next impression could resemble what the idea of a comet will confirm.

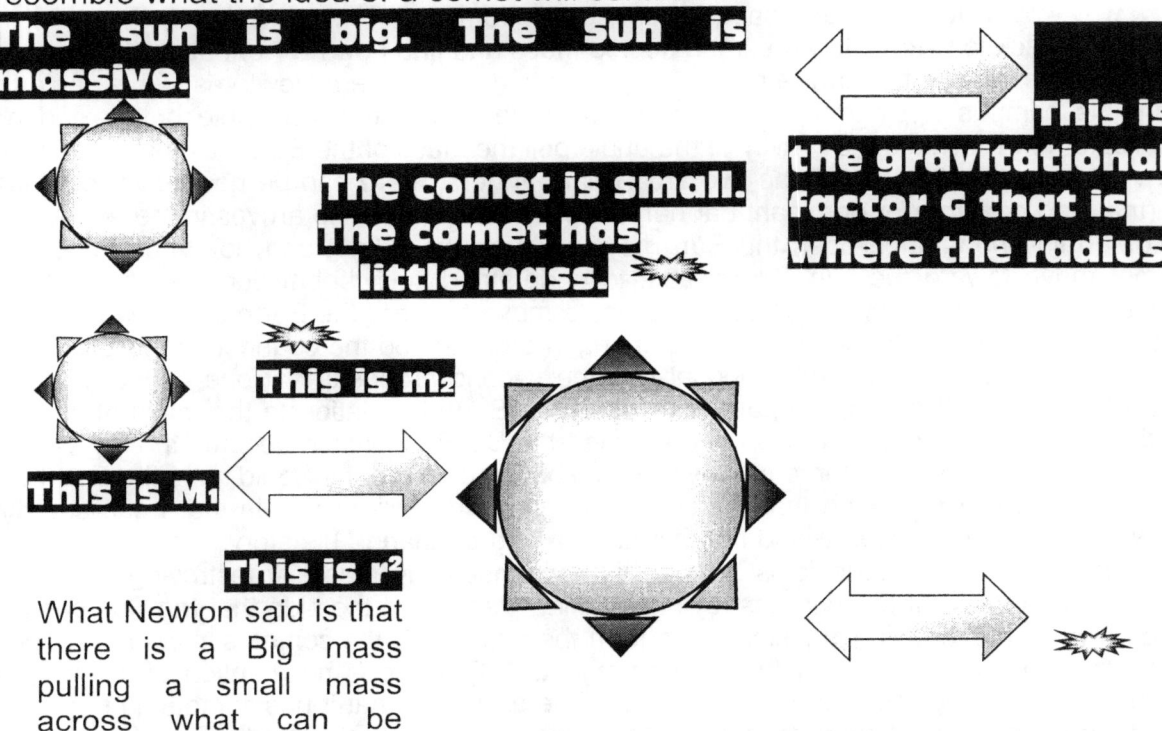

The sun is big. The Sun is massive.

The comet is small. The comet has little mass.

This is the gravitational factor G that is where the radius

This is m₂

This is M₁

This is r²

What Newton said is that there is a Big mass pulling a small mass across what can be interpreted as a gravitational constant which is what is filling the distance between the objects where we will find the radius.

The seasons remain the same because the space the planet covers differ in the time that it takes. That is what Johannes Kepler declared almost five hundred years ago. He proclaimed a formula **$a^3 = T^2 k$**. In truth it is Kepler that made the study of the solar system and it is Kepler that made discovery of gravity and it is Kepler that concluded what the cosmos was about. It was Newton that later came, that had no idea what Kepler's finding was about, that was presumptuous enough about his personal genius to the extent that he thought he could summarise Kepler. Then Newton cheated, but we will get to the cheating later on. Because the radius between the Sun and its planets changes during one orbit, the systems should draw closer, or alternatively, move further apart. However, they do not.

Gravity BASED on **Newton's presumptions** are **incorrect**. This conclusion is **based on** the findings of **Galileo and Kepler,** and **their findings** are totally **inconceivable** with the **findings of Newton.**

Nobody in the world would make me see that the FORCE holding the striped arrow will have the same effect than would the FORCE holding the dotted arrow. The length of r changes, therefore the value of blue r cannot be the same as the value of red r.

Newton stated that $\dfrac{M_1 \times M_2}{r^2} G$ = Force, because of the fact that:

1.　　　The value of M_1 in both cases is equal because that is the mass of the Earth.

2.　　　The values of r^2 are equal because the two objects have been dropped at an equal distance.

3.　　　The value of M_2 is still a mass filled with gravitons and therefore has to effect the relation in the same way as both objects consist of different compositions of materials that are used to manufacture the different objects.

4.　　　That means Force one has to have a different value to that of force two ($F_1 \neq F_2$). In contrast, time duration $P_1 = P_2$ and this cannot apply in the case of gravity because the greater the force is through more mass, the bigger the impact has to be in the time duration.

Newton stated that $\dfrac{M_1 \times M_2}{r^2} G$ = Force and the mass in each case remains the same as well as the gravitational constant. Then why would the radius change every time of each year and not draw closer (or further)? This is a result of some tiny elliptic shape that planets tend to have while orbiting the sun. It is referred to as Kepler's equation which is an eccentric anomaly of a body in an elliptic orbit to its mean anomaly and it is equated as the following E- e sin E = M where E is the eccentric anomaly, M is the mean anomaly and e is the eccentricity of the orbit.

This factor, in essence leaves a distinct problem as far as generalising mass as the generator of gravity. If mass inspired gravity, it would mean that the orbiting body or the body being in the centre of the orbit, should at one stage during the orbit lose mass and at another stage while the orbit is taking place, gain mass from some unknown source. This concept is ridiculous, but the incoherency is with the mass aspect, because the eccentricity is the fact, and the orbits do show eccentricity to a specific pre-calculated margin. The eccentricity shows cyclic recurring and mass should be a steady factor if it is any factor. Yet, the gain and the loss of the orbit is there, which indicates that the radius r does fluctuate. The factor r comes as an influence of mass from both ends and for that reason it shows to heave an influence on the force of gravity by the square. In order to scrutinise Kepler, I had to study Kepler's version on gravity again.

This is not all.
However, **this is not all**. When we regard the planets as they stand related to the sun, the effect is the same, but not as obvious. All the planets follow an oval orbit around the Sun and therefore the same factors concerning gravity applies to the letter as it does in the case of the comet. Let us now investigate the one planet we relate the best to, which of course is the Earth.

Newton saw himself fit to change Kepler's formula, and I indicate that Newton went wrong by changing Kepler's formula in the way he did. I even go further and dare to prove that **there is a difference** between findings of **Galileo and Kepler** on the one hand and the work of **Newton and Einstein on the other hand**. Kepler made no mention of mass. Kepler never referred to gravity. Kepler made a study of the behaviour of planets as it occurs in the solar system. Kepler made no wild deductions from seeing how an apple behaves. Newton stands accused of all these things and more.

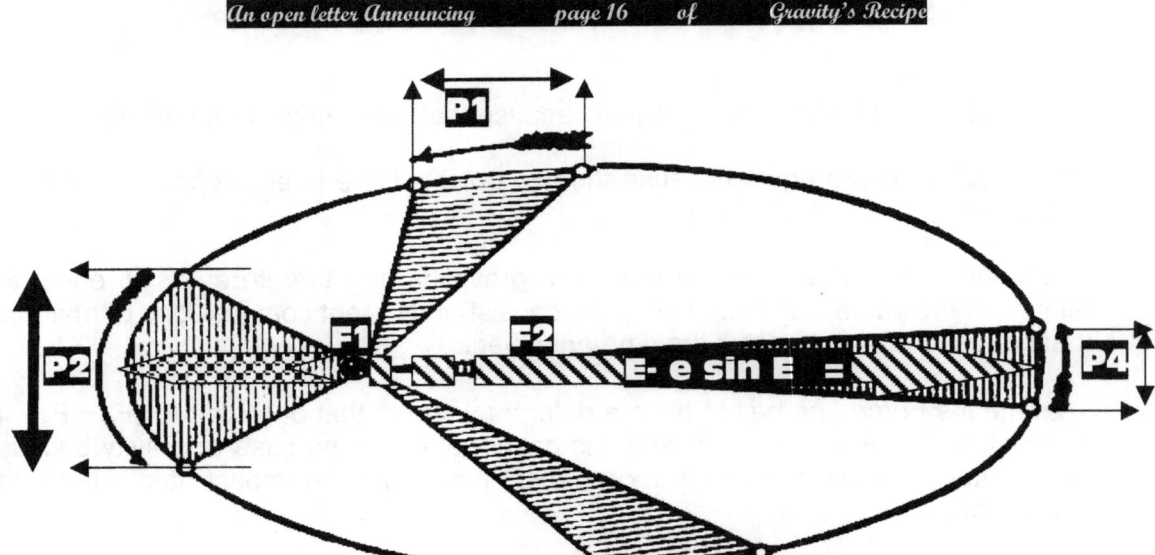

This illustration does exaggerate the radius of the Earth's orbit around the sun, but since it has taken place $4,5 \times 10^9$ times, it has no real effect on the validity of the next statement. **At one point the Earth should be drawn or pulled closer to the Sun** and **after another six months'** interval **the Earth should stand less effected by the sun's gravity.** Therefore, the earth should move away from the sun. Each cycle of twelve months would have one point where the gravity pulls the Earth closer and exactly the opposite must apply six months later when the gravity is at its least.

So, for the past 4 500 000 000 years the Earth has been re-establishing its seasonal swing towards the Sun and away from the sun, which means by now the Earth has to collide with the Sun in midsummer or escape from the Sun in midwinter, as it may then drift away into the unknown.

At one point, (R_1) the distance between the Sun and the Earth **is less than** at another point we call R_3. Let us put a value of $R_1 =$ one and $R_3 =$ three. This means that each year, for the past $4,5 \times 10^9$ years the effect of the common gravity between the Earth and the Sun has a

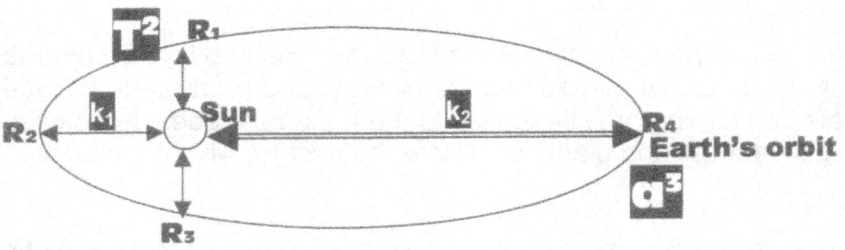

greater effect than at another point six months later. With mass generating gravity as Newton declared, then either this stands corrected, or we must find how the Earth loses mass just to gain mass a little while later, or lastly Newton stands to be corrected. There is an anomaly no one can deny and the cause of such an anomaly has in principle some connection with gravity. The gravitational constant can't be blamed because from what Newton introduced, the gravitational constant is a constant after all. The constant can't vary, otherwise it is a variable and not a constant. Therefore, it leaves just mass being the culprit for the clear anomaly there is in all planets to an equal measure. This too is hair rising because all the anomalies are working without distinction as to the size and subsequent mass differentiations occurring. The fact that Jupiter is 318 times the size of the Earth does not give Jupiter a larger anomaly than is the case with Pluto, which is .025 times the size of the Earth. How does that fit into the equation when it is the mass that is generating the gravity that is causing the anomaly, which is equal in all planetary orbits? However obvious this discrepancy is, Newtonians are very adamant that mass does produce gravity since Newton has never stood as proven to be incorrect. Again this mass generating gravity issue allows a lot of dark clouds form above the issue and questions come about that crack as if they are thunderbolts. As is the case with the falling of material, so it also proves to provide questions about mass, Newton that never addressed. Newton stated mass forms gravity by restriction $\dfrac{M_1 \times M_2}{r^2} G$ (the radius removes therefore motion becomes denied) while Kepler said **$a^3 = T^2 k$,** which is space in unrestricted motion through time**.**

Only one of these two groups' findings can be right, because there is an unmatchable difference in the concept of these two groups' opinions. Newton puts restriction as the corner stone and mass provides such restriction while Kepler says the motion of material filling space is unrestricted. Newton considers that a force existing between two bodies in space: the mass of the two bodies' product is being brought into context with the gravity constant (G). This value is then divided by the distance r calculated as a square (r^2) value. However, in this I see traces of Kepler's version of gravity in what Galileo stated, free fall comprises of. Mass is a matter of restriction and this restriction produces a free fall, which Galileo saw. The fact that all fall equal, makes all that fall, fall unrestricted and therefore equal in the fall. It is a question of perspective. Where the mass would provoke gravity, the gravity provoked must be in accordance with and must be equal to the mass that provokes the gravity and therefore the fall. However, by all falling equal, all that are falling have either equal restriction or have equal mass while falling.

How can I dare to even try to prove this? The answer is child's play. Let's take two balls of equal size where either of the two comprises of different material making up the substance. **When two balls**, one made of cast iron and the other made of wood, **but are even in size, are dropped from a building**, we all know that ***Galileo*** stated that both balls <u>**will hit the ground at precisely the same instant:**</u> a fact that has been proven even as far off as on the surface of the moon. However, being manufactured from wood and the other from steel, places much mass discrepancies between the two balls falling. They are of equal size and therefore the restraining by airflow will be equal. The mass they should have, when concerning the material from which they are made, puts great discrepancy in the mass. Notwithstanding the mass differentiation, the falling is one that proves equality amongst the falling balls. How can mass generate the fall while the mass is so much different? How can mass of inequality produce a fall of equality. The mass that drives the fall through gravity is not equal, yet the time it takes and therefore the drive that sustains the fall is absolutely equal. That makes Newton's claim on mass, which is all-important, also all incorrect. Before throwing down the book, please first explain why gasses, many times more massive than solids stay afloat, meaning they do not react to mass drawing mass as they should. If such a claim on mass was valid, then gasses should be the least massive and solids the most massive. That is not always the case because neon is twice as massive as boron but still it is a gas with boron being pretty dense for its size.

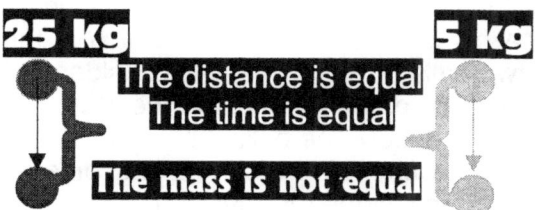

If we put a mass value of say 5kg to the wooden ball and a mass value of 25kg to the cast iron ball, and have the two become unequal even in measured size, where the size does favour the more massive cast iron ball, the fall will still prove absolute equality on both ends. As the relation in the gravity between the two bodies remain the same, this formula should still apply. If that does not put a serious question mark against the idea of mass inducement of gravity, the questioner has no idea about the issue at hand.

However, in the case of the objects that are dropped in a free fall, this formula does not apply. Why not?

The distance must erode differently when mass of different distinction drives the gravity that produces the fall. And in al this apparent reality, Newton did not abandon the idea he first had that $F = \dfrac{r^2}{M_1 M_2}$, he improvised a lot and compromised slightly $\dfrac{M_1 \times M_2}{r^2} G$ and this he did by introducing some vague description of some gravity constant.

If these two balls were thrown, using an equal force as they were thrown in a horizontal direction in relation to the Earth, the heavier ball would travel at a lesser speed and a lesser distance than in the case of the lighter ball. This comes about because a force of equal value is applied to the two balls.

When examining the case where the two balls fall vertically, gravity as the force science portrays, or as a force that Newton claimed, does not apply at all. There is no evidence seen on Earth, that suggests mass induces gravity. This comes to a conclusion where it is obvious that material without compatibility in size or mass compare equal when participating in descending or falling freely.

GRAVITY RESULTING FROM MASS, DOES NOT COME INTO EFFECT BECAUSE THERE IS NO DIFFERENCE IN SPEED OR DURATION IN THE PERIOD OBJECTS OF DISTINCTION TAKE TO COMPLETE FALLS OF SIMILAR ACTION. THERE IS NO MASS THAT ENABLES GRAVITY TO APPLY. GRAVITY THEREFORE HAS TO WORK ACCORDING TO SOME OTHER GROUND RULE.

I developed serious doubt about the correctness of mass that should supposedly be inspiring gravity and I made it a bone of contention. Even if we put fairies in that substance that holds mass and that mass is the mass that spawns gravity, and we give the fairies spawning the gravity a name such as gravitons, then the one with more mass should accommodate more fairies going by the name of gravitons than the other has fairies called gravitons.

Where any two objects with incompatible mass fall equally, the two objects should have their own value of gravity and gravitons and in comparison with the gravitons of the Earth; their value is insignificant. That is when considering Newton's idea about gravity to be correct. However, in the case of the two balls I took as an example, those two balls are in their own individual deuce to see who reaches the Earth first, and the iron ball's gravitons should give it a superior advantage. The balls fall as fast and as hard as each one can and is flat out descending to the Earth. Each one takes its fall as seriously as circumstances would allow. This comes about because the two objects are in a position where they compare in relation to one another and share a common second factor, which is the Earth. In relation to the Earth, the gravitons of the two balls do not come into consideration as placing distinction by virtue of being more in the one object than there is in the other object, but this do not play a part since the Earth is a common factor. The balls, however, are put in a situation where they stand in relation to each other and when considering a vast difference in the alleged graviton fairy supply dedicated to each by virtue of mass, the two seem to ignore the distinct differentiation in gravitons and fall as if the gravitons are not gravitons. When compared to one another, the gravitons should give the heavier ball a sizable advantage. This puts a calamity in the reality of the science paternity.

For all my crying wolf and pointing at this matter, the response that came from science thus far is one of absolutely no response. My telling them leaves them unperturbed, which leaves me thinking the matter is distinctly well known around in science circles. If my effort in raising the issue at the moment does not get them surprised by such a discovery, then when were they surprised by such a discovery? If this leaves them cold, when were they excited in finding out about this matter? When did the first realisation of this issue dawn on their consciousness? If they now are so much unsurprised by my discovery, when did they discover the issue? SCIENCE HAS IGNORED THIS MATTER EVER SINCE GRAVITY WAS "DISCOVERED". WHY DO THE MIGHTIEST OF ALL KNOWN THINKING POWER AND THE MOST BRILLIANT INTELLIGENCIA ON EARTH AVOID THIS FACT AS IF INFECTED BY RABIES? There are more to this than one would ever find them admitting to.

I think I have brought some proof to my claim that I am unable to detect the effect of mass on gravity considering Galileo's verdict and the presumptions Newton later claimed. There is no evidence of mass enticing, provoking, inducing, generating, charging or producing gravity as far as free falling is concerned by objects descending to the Earth and going as far as the

Earth. That lays waste to Newton's claim of $F = \dfrac{r^2}{M_1 M_2}$

To understand how gravity applies in outer space, it would be the best to look at how the comet and the Sun interact and then from that make certain deductions.

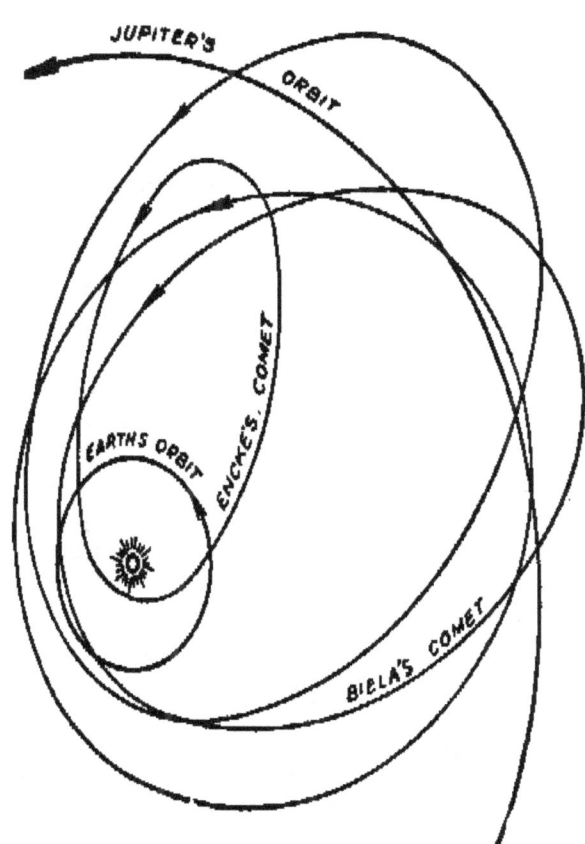

We may start at a point where I started twenty-eight years ago by determining the influence of gravity on planets as we find them in the solar system. First, let us concern ourselves with a comet.

With the comet, the Newtonians regard that there is a method in which a force has the ability to attach to the Sun in some way where this force pulls the comet towards the Sun. This is somewhat plausible since we can see that all the orbiting objects in the solar system has some connection applying that holds the planets and comets and even debris circling the Sun.

The circling can be a reason because of the fact that at the same time another force joins in that pulls the Sun closer to the comet, but such is the mass difference between the Sun and the comet, the force the comet applies never realizes. The comet is tiny and the Sun is huge and the comet is far off while the Sun still detects the comet, although it is hiding where the Sun doesn't shine anymore. In view of this, only the force the Sun applies, comes into effect because at that distance one can hardly mention the effect that the mass of the comet can produce to entice the gravity needed to commence with the pulling. The comet proves this force by speeding up its movement as it comes closer to the sun. If the force did not become greater, why would the comet gain momentum? No matter where the comet thinks to hide, it is just not far enough.

I am aware of the fact that there are comets that are not drawn by the Sun, which are hiding in Oort's belt somewhere, but in regard to gravity not being present, will prove little and we are about to prove mass. We have this big Sun detecting this little comet and with all the mass the Sun has to its disposal, it will track down any comet and drag it closer, or that is what Newton said. The mass of the Sun gets the comet anywhere any time any place.

What happens next confirms Newton as a profit. The Sun got hold of the comet it detected and with no mercy to show, the Sun dragged the comet over the mighty vast night sky towards the Sun. The Sun found the ability to act on the grounds of its enormous mass that has the ability to pull the Universe into a curve and a comet into submission. It is all down to

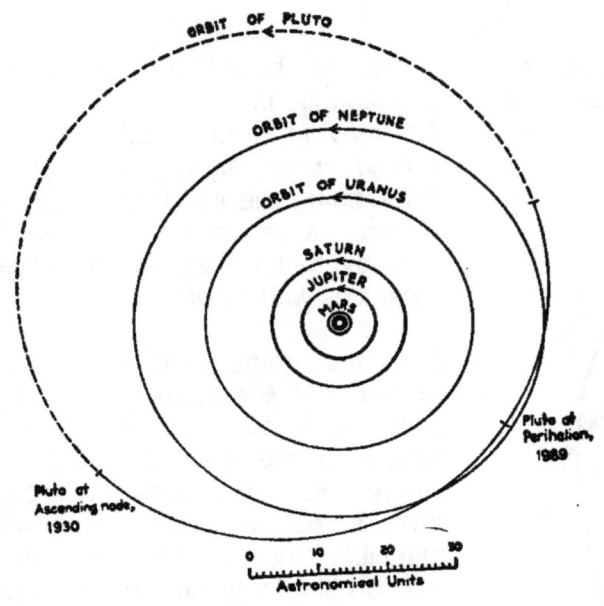

mass and it is measurable when applying the calculation by using the formula $\dfrac{M_1 \times M_2}{r^2}\,G$, which Newton derived from another formula that seems to be very doubtful as it is

$$F = \frac{r^2}{M_1 M_2}.$$

Newton says two pieces of rock will draw each other closer by reducing the distance keeping them apart. It is a very contentious statement when trying to prove this thesis of Newton on Earth.

Let's put it to the test in space. The fact of mass pulling objects, we view as proven by our response to mass, which we all can see by merely jumping in the air. No sooner have you lift off than you are back on the ground. There is a movement from space to the ground, but

is the movement mass enticed. That is what Newton said about three hundred and fifty years ago. It is mass inflicting gravity that is luring the comet to the Sun by force. By force the thought established of acting beyond will, under the spell of or being in control of something much stronger, the repudiation of a free will to enforce what cannot be otherwise helped to happen in any case or without any other outcome. With my questions never yet answered about mass, we see when falling I do have considerable doubt notwithstanding all the ensuring evidence because as Newton suggested, the comet is coming…

It is common knowledge how the comet stands related to the sun's gravity. **Firstly, picture the comet at its farthest point, away from the sun.** The **gravity** of the **Sun pulls** the **comet straight towards the sun**, this we all know. Gravity always pulls an **object directly towards** the **centre of a cosmic body**: that too is common knowledge. The comet is heading to the centre of the Sun and with all the mass at the disposal of the Sun the comet has very little say in this action. It is coming without a fight because even the mass the comet has, is helping to choreograph the demise of the comet.

It could be that the "gravity" caused by the Earth's mass is somehow connected to the radius of the Earth, and that could be a reason why Newton changed his formula. I am trying to speculate here in order to find ways in which to see what drove Newton to change his formula. Newton formed arguments to sustain and validate his force and finding those arguments will lead us to discover why mass would produce gravity. That would put the "gravity of the sun" as a valid force only at the distance of the radius of the Sun. Since, according to Newton's theory of gravity, the force of gravity is proportional to the mass, which is in turn proportional to the square of the 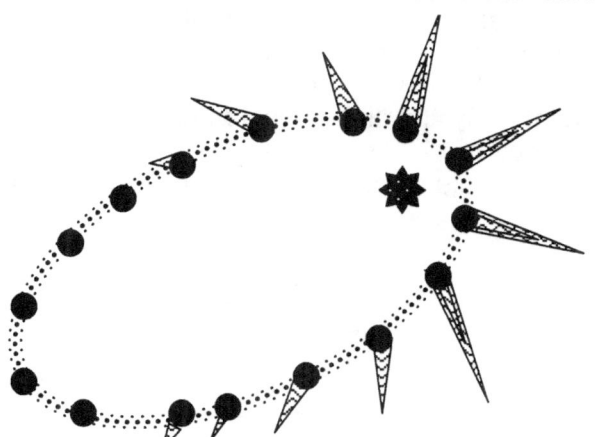 distance (in this case the radius of the body), and inversely proportional only to the square of the distance, you expect that the gravitational force/mass ratio you calculate, would increase when calculating the potential gravity of larger planets. Then, why are all the planets spinning at an equal pace, one might ask. While this might be true, it is not true because the mass of the Sun formed sufficient gravity to detect the comet and the mass did pull the comet closer. True or otherwise, it still does not explain the fact that objects can miss the Sun where it obviously has to collide. If that was the case, the Sun could not pull the comet close and that did happen.

Anyone can see a collision coming ten miles away. The Sun applied gravity, the comet applied gravity, the Sun is far too massive to fly to the comet, so the comet with much less mass does the flying on behalf of both objects. Every person with even the least of knowledge about science knows how the gravity application works. The gravity of the Sun collected the comet from no-one knows where, pulled it through billions of kilometres to the area where the Sun produces the gravity with which it pulls the comet from where the comet is, and with the pulling by mass, the intension in all of this action, is to find the comet and give the comet its last resting place. The mass of the Sun is obviously so large, it could produce gravity that can locate any comet hiding anywhere and collect it as a souvenir. Then there are those that say I don't understand Newton. To them and to those I say: What is there to understand?

The gravity of an object always points directly towards the centre of the object, the very, very middle point. Concluding from the fact that the comet is heading towards the centre of the sun, just as much as the Sun is heading towards the centre of the comet, would not be out of line. The two centre points are heading for a direct collision, the collision becomes more and more unavoidable as the radius reduces by the value of the gravity that the mass of the material produces in accordance with the gravitation constant. The comet is heading towards the sun, and by not even moving, the Sun is moving towards the comet by attaching the movement the Sun were supposed to have, on the comet. Newton's law proves to be exceptionally correct. The mass of the Sun combines with the mass of the comet to lead to the demise of the comet. As the sun/comet, radius reduces, the radius separating the mass of the Sun and comet effectively increases. This happens because as the radius decreases, such decreasing effectively increases the influence that the mass produce as a force called gravity. The relativity of the mass influence increases in force as the distance erodes and the force on each other in the form of gravity becomes increasingly bigger. The mass of the Sun and the comet increases by the factor of the reduction of the radius that is separating the two objects. That will produce a growing gravity force as the comet / Sun radius becomes smaller. By the time, the radius becomes one, the mass will grow on either side by a relevancy of 100 times what it was in the beginning, and when the radius becomes infinitely small, the relevance to the mass of both structures will raise a force with eternal power.

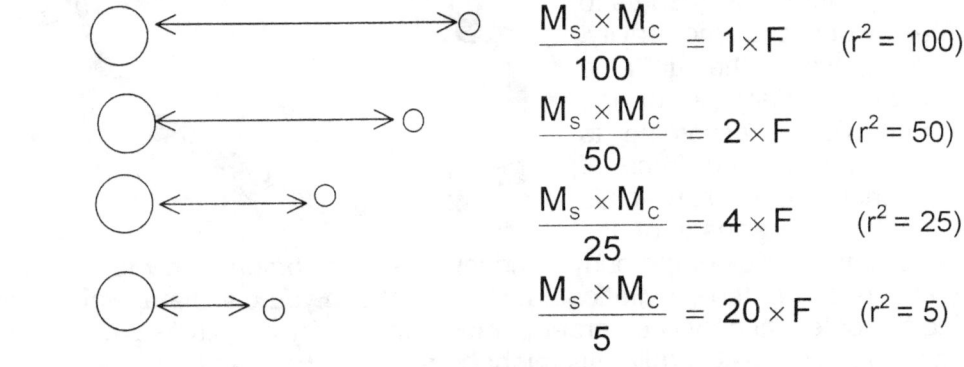

$$\frac{M_s \times M_c}{100} = 1 \times F \qquad (r^2 = 100)$$

$$\frac{M_s \times M_c}{50} = 2 \times F \qquad (r^2 = 50)$$

$$\frac{M_s \times M_c}{25} = 4 \times F \qquad (r^2 = 25)$$

$$\frac{M_s \times M_c}{5} = 20 \times F \qquad (r^2 = 5)$$

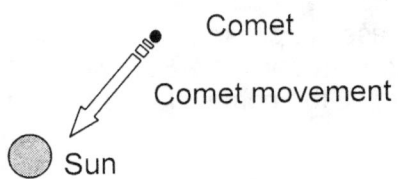

Comet

Comet movement

Sun

At a point, where the comet / Sun apply a force of immeasurable strength. Then as suddenly as it is unexpectedly, the comet breaks this immeasurable force. Remember, the direction of gravity always point to the centre of the object, and that is where the collision is heading. As the objects draw closer, the distance reduces, but in accordance to the relevance, the objects also become that much bigger in drawing power. It depends how one considers the relevancy to grow by the approaching nearness that is diminishing the distance between the objects.

Therefore, the comet is drawn directly towards the centre of the Sun and throughout its journey the comet is picking up momentum directly related to the gravity that is centred in the middle of the sun, (**gravity is always centred in the middle of a cosmic body**). As the comet is increasing its speed, the comet comes closer to the Sun and therefore the Sun's gravity pulling is simultaneously increasing as the distance between the two cosmic bodies is reducing. Each instance the comet is drawn towards the sun, the gravity that the Sun applies to the comet becomes larger progressively through the reducing of the radius keeping the objects apart. When the comet is at its **closest point to the sun**, **something odd happens which cannot be explained by Newton's gravity at all! Remember gravity should now be at its strongest point because of the proximity of the two objects.**

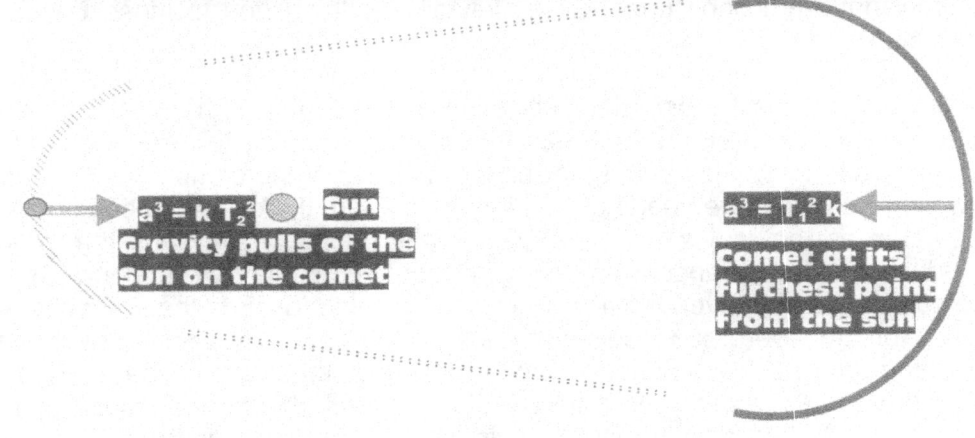

From afar the comet comes drawn to the sun. But the closer the comet comes the more it is turning away from the centre of the Sun. At one point it seems as if it is planning on missing the centre of the Sun by a country mile. That we know cannot be since Newton's $\dfrac{M_1 \times M_2}{r^2} G$ took care of that, and Newton was never yet proven wrong before. So it just cannot be that the mass of the Sun would miss.

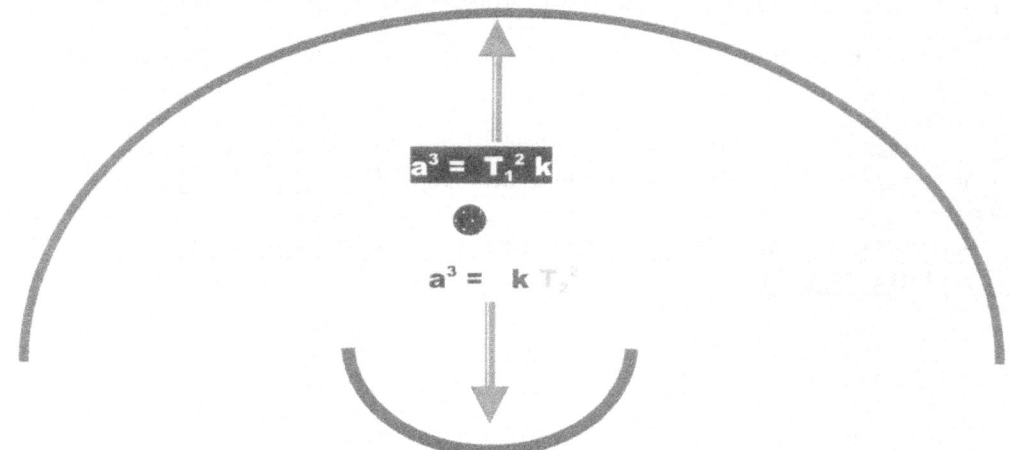

The illustration proves Kepler, but at the same time it disproves Newton's claim. You have the Sun and you have a tiny piece of rock covered by water also better known as the comet.

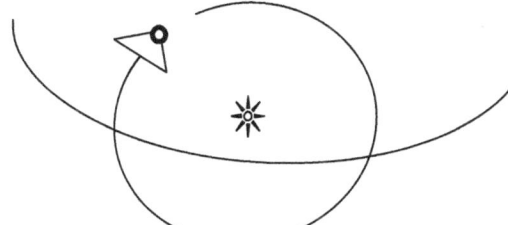

There are thousands of them flying around, but never aimless. At first Newton's formula makes pretty much sense. The Sun draws the comet towards the sun, as Newton said it does. The comet responds by speeding towards the sun, also as Newton predicted.

However, the scientifically impossible happens. That, which never could be, becomes a reality in spite of the scientific truth that it is accepted to be as true as $E = MC^2$ apparently is and it is therefore beyond a shadow of a doubt that Newton has never been proven wrong before. The comet must be mistaken because it is acting without the consent of Newton! The comet disobeys Newton's predictions!

1. The comet remains at an even distance encircling the sun.
2. No longer does the gravity of the Sun pull the comet towards the centre of the sun.

3. At this very point the gravity that the Sun applies on the comet is at its largest but it does not pull the comet towards the centre of the Sun any longer, in fact, it seems as if the effect of the gravity has been neutralized.

4. The comet stays at an even space from the Sun as it goes around to complete a half circle's orbit around the sun. It only completes a part of its rotation around the sun and then it speeds off into the distance.

5. After this, an even more peculiar event takes place. **The sun, at the point where gravity should be at its most dominant, suddenly loses its complete grip on the comet. The comet speeds away by increasing the distance parting the two objects, which before the comet escaped the Sun so vividly reduced to the smallest possible margin.**

The comet breaks the mass produced gravity of the Sun. The enormous sun lost its mass applying gravity when the gravity was at a peak. When the collision was immanent and escape was impossible, Newton became wrong. Newton, the man that never was wrong, is proven wrong by every comet that eludes the mass of the sun. The formula that physics use to base all physics on is rendered useless as it is proven incorrect by it being invalid. The comet did all this in the face of Newton that was never proven to be wrong! As the comet speeds away Newton increasingly becomes a fraud and a hoax. Where is the unshaken truth about $\frac{M_1 \times M_2}{r^2} G$ in the face of the comet eluding the mass inflicted gravity of the Sun.

Then after a pre-determinate and pre-calculated time, the Sun starts applying its gravity on the comet once more. At a point where the comet **is at its farthest point**, the gravity of the Sun becomes strong enough to bring about a complete turn around to the comet's direction of travel. **However, the gravity between the Sun and the comet is at this point, at its weakest point of influence.**

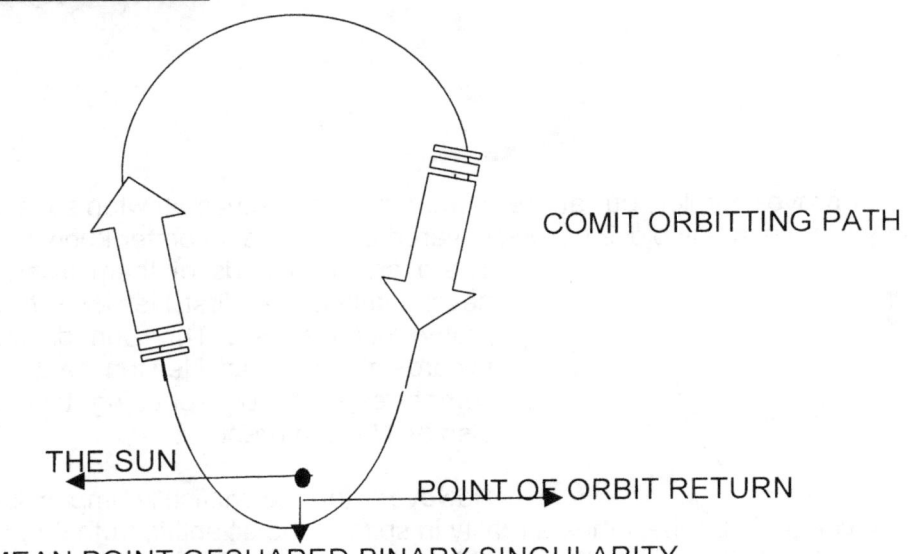

COMIT ORBITTING PATH

THE SUN

POINT OF ORBIT RETURN

MEAN POINT OFSHARED BINARY SINGULARITY.

When the sun's gravity is (according to Newtonian predictions) supposedly at its strongest, the comet manages to break loose and neutralize the sun's gravity pull in order to avoid its fatal collision with the Sun and when the Sun's gravity should be at its weakest, if mass does bring about gravity, because the comet is at its furthers point from the Sun, that gravity then performs as if it is at its strongest. When the comet escapes, the mass inflicted gravity should be at its strongest if Newton is to be believed. However, this is just when the comet cannot escape the pull of gravity and yet in total defiance of all science it does just that! Then where the mass inflicted gravity should be disappearing, the mass produced gravity comes into its full potential by drawing the comet back towards the Sun. There is definitely something very wrong, either with the comet's behaviour or the laws made up by Newton. Every Newtonian I ever came across failed to show me why the mass that draws the comet closer at the same time fails to eliminate the comet by forcing the comet to plunge into the Sun. At the same time everyone amongst the brilliant Newtonian population remains unable to show me how the formula Newton introduced entertains the comets escape into the blackness of eternity. If the mass draws close, then where is the part where that mass that does all the pulling turn around and pushes the comet away from the Sun.

NEWTON SEEMED TO BE ABLE TO CONVINCE EVERYBODY ABOUT HIS COSMIC LAWS, EXCEPT THE COSMIC BODIES THEMSELVES, WHICH PREFER TO BEHAVE AS IF NEWTON'S LAWS NEVER EXISTED. ARE THE COSMIC BODIES REBELLING

AGAINST THE LAWS OF NEWTON OR ARE THEY SIMPLY IGNORING THEM AS IF THEY NEVER EXISTED.

MY GUESS IS IT IS BECAUSE THE LAWS DO NOT EXIST, EXCEPT IN THE IMAGINATION OF ISAAC NEWTON AND HIS FOLLOWERS.

Then out of the blue, the comet finds the ability to eliminate the eternal powerful force of gravity, and keep at a safe distance around the sun. At this point, Newton goes sour. Nothing Newton predicted is happening. The comet and Sun not only stabilized the force, the force begins to decrease as the radius between the comet and the Sun is on the increase AT THE POINT WHERE THE FORCE IS THE STRONGEST, THE COMET BRAKES FREE AND SLIPS AROUND THE SUN, UNSCATHED.

UP TO THIS POINT I STILL SEE WHAT THE BRAINY BUNCH AND NEWTON SEE, BUT IT IS FROM THIS POINT ONWARDS THAT THERE COMES THE POINT THAT OUR MUTUAL POINT OF CONSENT DIVERTS POINTING OUR MUTUAL POINT ABOUT THE POINT OF AGREEMENT TO THAT OF OPPOSING POINTS WHERE OUR VIEW SEPARATE AND LEAVE BOTH HEADING IN OPPOSING DIRECTIONS ON AN ETERNAL DIVERTING PATH.

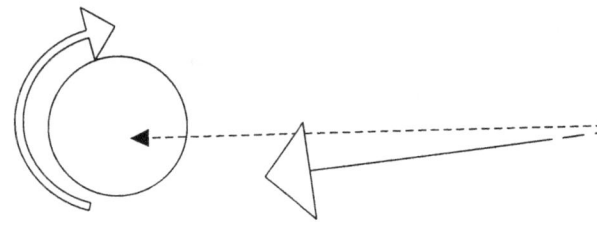

Then, in complete defiance of the Newton Law on gravity, quite the opposite too what Newton predicted applies. At the point where the radius that is separating the two cosmic objects is at its strongest, it will also bring about that the gravity force is at its weakest. At the point of almost no ability the gravity force suddenly releases enough strength to break, resulting in the ending of the immanent parting of the two structures. The force now curbs the rebel comet on its way and ends the escaping of the Sun's gravity that seemed to be for the very last time.

At the point where the force was the greatest, the comet overcame the force, but where the force was the weakest, the force overcame the comet's rebellion.
For the more mathematical minded person the argument is as follows. May I remind you, THAT NEWTON'S OWN LAWS ARE IMPLIED, and again the planets disobey these laws completely!!

THIS IS THE SUGGESTED FORMULA THAT PLANETS APPLY WHICH ENABLES THEM TO MAINTAIN THE ORBITS AROUND THE SUN, WHICH THEY DO.

A comet follows a designated route around the sun, without colliding with the sun.

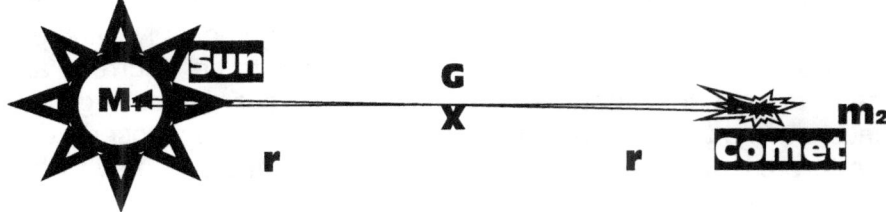

The comet travels around the sun, but it steers clear of the sun, at a time when the force of gravity should be at its strongest and force the comet in the direction that points to the centre of the Sun. Yet, somehow the gravity grip loosens and the comet finds a new direction at a point very close to the Sun but it is no longer aiming at the centre of the Sun.

When the comet reaches a centre line in relation to the sun, the direction of motion turns the comet in a new direction. The comet chooses to ignore Newton to the extent that it proves the opposite of what Newton predicted. It is as if the mass is pushing the objects away and instead of diminishing the radius, it is increasing the radius many fold. The comet is pulling away from the Sun at a pace that it was coming towards the Sun. It is now evident that the mass is producing anti gravity because if gravity is bringing the comet closer to the Sun, then it must be anti gravity that is chasing the comet away from the Sun.

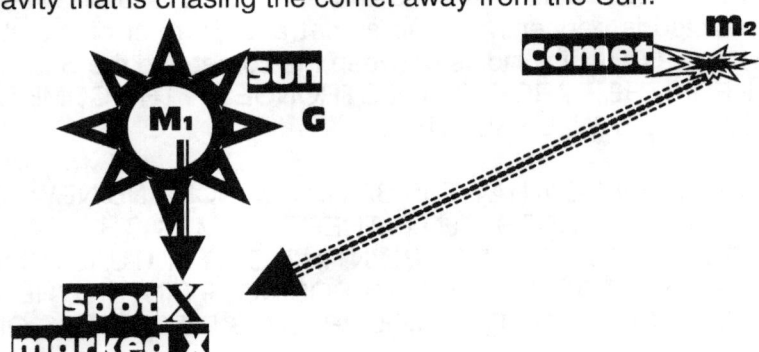

This situation carries on to a point where we bid the comet farewell and we dry a tear from the corner of our eyes, as we lose sight of the comet. The comet now again changes direction where it then finds a way to return back to the Sun. At this point again I am in odds with Newton's gravity by mass pulling over the radius theory. Just when the radius puts the mass in a dysfunctional position by rendering the force that the mass can produce across such a vast distance, the comet returns. Then quite the opposite happens when the comet is at its outward point therefore gravity should be at the weakest position of influence, but nonetheless the comet is coming back.

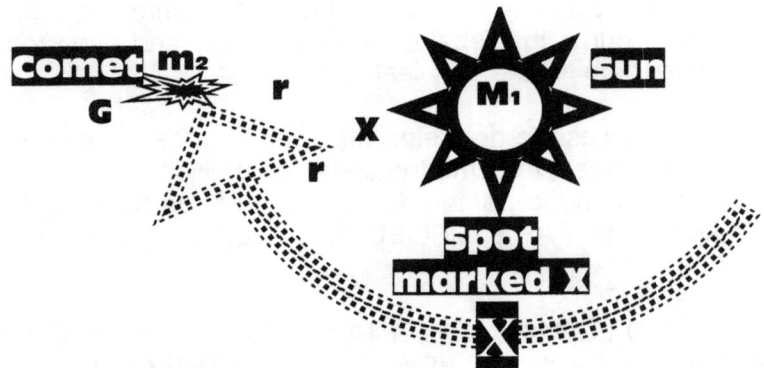

For years I tried to put this fault of not getting behind the sense about argument at my door while trying to pursuit a correct line of thinking. For decades I was trying to find what I was missing in my stupidity. It was many years on end that I blamed my insufficient schooling for the fact that I just could not see what I was missing in the arguments Newton proposed. How could I be correct and find the brightest brains in the world to be incorrect, and yet, that is what I saw when I did not see what they saw. I saw them ever increasingly being incorrect because not once did one try and propose to explain the issues I supposedly did not follow. If they are unable to present an explanation, then they are unable to prove what I wish they would prove. When contacting Academics of high standing they all suggested that I was not bright enough or educated correctly, which then was the reason for my inadequate understanding of Newton. They never said it in so many words, but I could taste it in the sympathetic manner they discussed the possibilities of such a short fall in my intellect. That is one point I stand unconvinced. I believe I can understand what the next person can understand if and when the argument is sane. Present me with facts and I shall follow the argument into the most incredible detail, in spite of my uneducated background.

The correctness of my argument is no longer the issue. It was twenty-eight years ago, when I still held the impression that I was missing some point here. I do not state this phenomenon any longer in the hope of bringing across some flaw in my understanding. The flaw in my argument is not there because the flaw is science as a whole. Newton is fatally wrong and Newtonians are disgracefully corrupt in not admitting their Master's incorrectness.

I could never understand the reason why "the ordinary", with my development level, can see what I can see, yet academics that have more brainpower in their heads, than I have life in my body, were unable to see such an obvious conclusion. You; those SUPER-EDUCATED are my superior in every sense a human can have, with the brainpower to break a wall, and yet you cannot see how far your academic tower of Pizza is leaning over.

I make the point to help you the reader to judge for yourself. If you are able to see the validity in my argument, you are not brain dead. Education has not yet bashed your thinking ability out of your scull. However, if a cloak of not understanding roll over your brain, and a numbness sets in on your ability to reason about this phenomenon, and a cloud of denial comes over you as you repudiate any such thought, then beware, you are a Newtonian. Newtonians should read this book very slowly because the effort you are about to launch, may be the most painful you shall ever experience throughout your academic career. You are going to suffer from reconditioning and Newtonian withdrawal, not that much dissimilar to that of an addict in rehabilitation. You are going to reject me, hate me, despise me, loath me as you never felt about anybody else. If you think I am sarcastic, I am not. You will reach a point where you will abandon the reading of the book. You have my sincere sympathy and with all the soothing it may bring, know that you are not the first I saw getting such painful Newtonian rejecting.

With the arrival of the comet in the Sun's domain, the Newtonians leave the argument to be, and according to me, that is where the story starts. By the order of the mass the Sun applying the force and as the mass remains the same, the Sun should remain applying the force and the force should increase all the time, accelerating the comet to the point of splash down within the Sun. The Sun gets the comet and drags the comet into arrest. It is a force just like the police are a force that can arrest and drag the arrested into custody. We must all argue that gravity is a force, which pulls an object to the centre of the larger object wherever that centre may be. It is coming from a centre and it is taking what it is dragging down into the centre.

If it is the mass inflicting the gravity that produces the force that pulls the comet closer, then what is it that pushes the comet away again? What changes the mass if the mass changes at all? Why would the characteristics of the mass change completely by establishing gravity at first and then afterwards commit the same force to quite the opposite of what science considers being gravity? Why would gravity at first pull and then afterwards push? What changes mass around to firstly become the instigator and then the decliner? Does mass disappear and if it does, then why do the Sun disappear with the mass that disappears. There are so many questions in need of answers and so many Academics unwilling to let us into their secret chamber of solutions. Then the final horrible question comes to mind: is it possible that mass is not responsible for gravity and even more horrifying is the thought that mass might just be a thought that lingers only in Newtonian minds. Is it possible that mass is not part of the cosmos but only part of Newton's imagination?

You in academic acclaimed circles are addicted to Newton. You do not understand Newton because if you did understand Newton as far as cosmology goes, then you would reject Newton. You were force-fed by your superiors on Newton at the time and during the time when you were a student and being force-fed you either had to grow addicted to Newton or die from Newton. Being where you now are it means you are not dead academically or otherwise. That means you took the second option by becoming addicted to Newton. You saw comets come and you saw comets go. You did not question the reason why the comet went because of your addiction to Newton. You accept Hubble's expanding yet you go about looking for a critical density in the hope you do not have to abandon your Newtonian addiction. You know the Big Bang cannot be possible while Newton's mass annihilates the radius between masses because then the Big Bang had to be the Big Crunch and from where we are there is only one enormous out explosion that had no Newtonian implosion. The enormous force that were suppose to contract the small radius that was in place during the Big Bang never imploded on the mass by nullifying the radius through the tremendously large mass and small radius being present at the time. All the facts modern day science accepts does not stroke with Newton. Is it because subconsciously everyone is silently admitting about Newtonian incorrectness?

Now I come and I tell you your addiction is killing you. As all addicts do, you are not willing to kill your addiction although you do realize that your addiction to Newton is senseless, stupid and all together wrong. The second choice facing all addicts including you is to hate the messenger that comes to take the drug from you. You will hate me, as the messenger. You will be willing to kill me being the messenger of evil because you see me as the evil that wished to part you from your addiction, which is Newton. You now find the bizarre feeling to put your anger distrust and vengeance on me and on the book you are reading. I have seen many academics stop reading at the part you now have arrived at. Brave yourself and throw out you fears by reading the rest. It will eventually cure you.

Even Trying to tell the SUPER-EDUCATED that Galileo said the mass of an object has nothing to do with the falling, seemed to pass the SUPER-EDUCATED'S sense of comprehension by miles. I suppose in the clear evidence of the behaviour of the comet this now is little surprising because the Sun got the comet and the comet is coming directly to the Sun. The Sun draws the comet by mass inflicted gravity. My not accepting mass was very much turned into being a dagger used against my person by those in knowledge. Whenever I said I don't see mass acting on behalf of gravity but instead I saw gravity responsible for inflicting mass, and that the two concepts stood miles apart, I was then reminded about my meagre educational upbringing. I was told on so many occasions that I did not understand Newton, but there it stopped. Never did one explain to me what I was supposed to understand to see how Galileo and Newton had the same opinion about mass acting or not acting to entice a fall of objects from the sky to the Earth. Also no one could explain to me what it was I did not understand about the comet missing the Sun by miles, where it was supposedly to hit the Sun with a dazzling impact and connecting the centre at the surface in a straight line. That the comet is missing the Sun by a distance measured in millions of miles is what is happening when the comet comes close to the Sun and believe it or not, that fact escapes young and old Academic alike. To get them to realise the escaping of the comet after the Sun caught the comet but failed to capture the comet, is beyond my abilities. The escaping act of the comet from the gravity of the sun is a point that I cannot get through to them as much as they cannot get through to me how I am miss interpreting what Newton said. Our understanding is so far apart, we do not share the same planet, and yet after all my arguments and investigation no one, and I repeat: not one could once clearly tell me what it is that I do not understand. I shall explain and then you see what it is that I, with my poor educational background miss about mass inflicting gravity in the manner the Sun draws the comet towards the sun.

Even more astonishing is facts about the Binary star system that is seldom to never mentioned.

The SUPER-EDUCATED-MASTERS-OF-FACTS never tries to explain the relation between Newton's laws and the binary star system. The binary stars are systems where two stars spin around each other while both adhere to honour a centre point somewhere between the two spinning stars and never collide. These stars are many times over the size of our sun but their mass never generates gravity to pull them into a collision. When one applies the same Newtonian formula as given above, these massive giants must crash into each other, destroying themselves in the process. The enormous mystery is not in the apparent misbehaviour of these giants, but the fact that this is known to science since the previous century. Relate the binary once again to the comet/ Sun relation and there is a distinct similarity.

I was always looking for mistakes on my part. At first, I thought there are a fifth force that I am unaware of because of my slender education, a force the academics can obviously see, but I cannot through obvious lack of education. I thought that my personal illiteracy gave me a blind spot that every non-educated has and was born with. The blind spot cleared only thorough education as education removes this blind spot in the way only education in science brings knowledge. I thought the removing of the blind spot process worked somehow similar to the way washing removes stains and spots from whites; education can remove blind spots through the process of intensive tutoring. All I wished for was some academic to help me remove my blind spot about comets and their behaviour. The comet's behaviour, I could see, was an exaggeration of orbiting patterns applied from our planets orbiting around the Sun; in the way, we observe galactica in the sky.

Then finally I came to the point of accepting defeat. It was not I, with the blind spot; it was all the academics brainwashed into a state of having such a blind spot. Science insists on repeatedly ignoring mathematical principles, because Newton had his claim to fame with one single calculation. Newton invented mass.

He made a brief calculation as a young man that saw an apple fall from a tree. Seeing this he jotted down a formula and the chucked it away. His piers and elders picked up the trashed paper with the calculation, and got all excited by the logic implication it had. It reads as $F = r^2 / (M_1 M_2)$. The mass of the two objects destroys the radius between the objects. Everyone went ballistic, proclaiming him as an instant genius, the one the world was waiting for after the crucifixion event.

I do not, for one second, deny or dispute the revelation. What I do encourage is place the event into its correct context. It was merely, and simply an apple that fell from its branch to its roots. The apple did not pretend to be a meteorite that fell from the heavens. If it were a meteorite, I am sure, with the man's genius, science would be somewhat different at this stage. However, as a young man, being very impressionable, as all young men are, and with the attention this brought about in the world of science, the matter overshadowed the fact.

I am not disputing Newton; I am disputing the relevance of Newton's scientific breakthrough. It was not two objects of cosmic proportions, colliding in a show of the spectacular. It was, after all, only an apple falling from a tree. With this miracle revealed, Newton found he was competent to improve on the work of Kepler and if I may dare say this, there must have been some political agenda behind this act and the accepting of it for Kepler was a German and which German can ever be able teach any Brit. The same politics are still applicable forming international rivalry on all fronts, even between Yanks and Brits.

The comet misses destruction by not colliding head on into the Sun. The comet finds a marker somewhere next to the Sun, heads that way, and by changing direction it speeds back into the blackness of the night sky. You might think of this as inconceivably simple to understand…yet; there are those whom have more brains than I have arrogance, who totally deny this action straight to my face! When explaining this to any child, they can immediately see that. Explain this to any Newtonian High Priest and he may have you removed forcefully from campus. I cannot find one Newtonian, large or small to accept the fact that mass does not bring the comet to meet his demise and forcefully destruct into the Sun and that the comet has no mass inflicted injuries from the force of gravity it is supposedly to have endured.

Newton cooked up some senseless argument about a wheel that never go straight and by its circling it never does any work and therefore the output of this is zero. The circle rotating motion cancels the directional motion and the result is that zero is accomplished. From that Newton takes time where time stands still. It is because he removed the one part of time being the relevance and by removing even one part, it is obvious that the entire Universe in the three dimensional concept will collapse. How bright is that? By not having a wheel rotate the rotation seizes, not the wheel. When the wheel begins to rotate, you cannot state that all things remained as it was. With the wheel in non-rotation the rotation still exists forming the infinite possibility of rotation. Then afterwards the wheel starts to rotate and by starting the rotating, the circumstances surrounding the wheel changes. The wheel standing still is rotating in relation to the rest of the Universe since the wheel is rotating with the Earth and the Earth is rotating with the Sun, which is rotating with the Milky Way and this concept goes on indefinitely. Looking at the wheel spinning or not and trying to find resolve from such a simplistic view, tends to be somewhat backwards in finding simplicity to a complex situation. A wheel in rotation is very different from a wheel not rotating and therefore cannot be the same thing. By establishing non-rotation, the wheel becomes the factor of one, and the rotating action becomes infinite because the motion allows the wheel a cyclic rotation between standing still for an infinite period of time and then continue indefinitely or eternally until the infinite standing still part of the cycle interrupts the eternal motion. But the wheel does not disappear and in the same manner does a wheel in rotation not remain still while the wheel not rotating is eternally rotating while it is being a factor in the cosmos.

In the cosmos, everything is rotating because only nothing ever stands still and "nothing" does not have any place as a factor in the Universe. The only aspect never found in the entire cosmos is nothing being anywhere or nowhere. Therefore the mean equilibrium, the common factor there is to share, has to be one, eternity, the eternal Π, because all rotating objects has Π in singularity, and sharing singularity, gives every object in space a relation with all other objects in space. After trying for many years to bring our Brainy Bunch the candle, I concluded that Newtonians are incapable of realizing that mathematical principle as a reality. They maintain they know mathematical principles far better than an illiterate such as I and yet ….

The comet rotates around the sun, and the Sun by itself has a point of singularity where Π remains without r and this I explain in due course. The comet, holding the orbit, also has a point of singularity, but since there is space separating the two objects, they cannot share a mean point of singularity, the very point of existing. Since singularity means just that, being single, there cannot be two. The comet and the Sun have a mean point of singularity but the space they occupy divides their common singularity. That is why they orbit in an oval path, a path where the one structure holds on to more space from its point of singularity towards the space it claims. Since they do not claim equal space, BY THE DENSITY they hold, the space will not be in proportion.

They do share in the common fact of singularity and singularity cannot be two, because then it will be duality or duplicity, where both find the space they occupy with the space they hold to be their individual eccentricity from singularity. The two objects are holding eccentric space around their individual but common singularity forming a point of mutual singularity in accordance with the individual singularity both claim space from. That point of singularity is

Π the circle without the radius because the singularity removes all forms or values of r, leaving Π to be singularity.

That is why Newton is corruption, and his $F = G(M_1M_2)/r^2$ is utter nonsense. The moment you say Newton or any of Newton's laws, the logic going on in the Newtonian brain stuns. For all the life in me, I could not once find one single Newtonian to see that mass has no implication on motion and gravity is motion. Mass is not having free motion while gravity is about having free and independent motion. If you say Newton is wrong, they spiral down to frenzy, and just mention gravity and they all come to have a mental seizure, have a dull stare in their eyes, seize to apply independent brainpower abilities any longer and you cannot make them say anything other than Newton is correct.

Dare you say there is no such a thing as mass and Newton is wrong, they will have you in an armed escort patrol, straight to the department of mental disabilities and psycho diseases and have you remaining there just in case you might inflict injury to yourself and others while preventing you committing acts of extremely dangerous life threatening behaviour to yourself and others. I found that to say to the Newtonian Physics Paternity that Newton is wrong is equivalent to having serious mental instability and all the while not one of those can explain why the comet escapes the mass provoked gravity the Sun condemned the comet to.

What is it the Newtonians fail to see? If an electron is orbiting around an atom, the inside of the atom must be a circle. If the atom was not a circle, it then had to be a cube. The electron cannot rotate around a cube; therefore, the inside of the atom is a circle.

If I contacted and argued with one Physics Lector or Professor about Newton, I have been in correspondence with at least a couple of hundred. What prompted me was the comet's orbit. The comet truly fascinated me from my childhood days, in the way it defies all the laws of gravity. Since my very young days, I was in search of what I at first believed to be a fifth force. I have raised this argument with just as many people that have, as I do, not being highly schooled in the art of physics and received a very different response.

Those as little educated as I am, saw what I see immediately and without effort. They also saw the comet speeding away and not being captured by the mass of the sun. The most amazing aspect was the fact that the two groups were that far polarized. The non-physics group reacted astonished, amazed, disbelieving and reserved about my view about comets at first, but with their distrust notwithstanding, everyone saw my point. The non-Educated responded in the same manner that I did at first. They argued that I was missing something of vital importance because "why do the wise not see it", was their argument. They always were of the opinion that I was too little educated to understand, while the educated was of the opinion that I was too little educated to understand. Neither party had the same view about my not understanding.

The non-Educated understood my argument, but dismissed it on the fact that it was so obvious, and therefore I missed the rest of the knowledge behind the facts that makes my arguments too difficult to understand while the educated dismissed my argument that I could not see anything they could not see because of a lack in my abilities to understand Newton. Education should bring the ability, which I am without and then that made me unable to understand Newton.

 In short, they thought I was too stupid to know the rest of the story. Polarized to the non-Academic view was the SUPER-EDUCATED where not one academic could understand my argument. The academic response was as much defending the Newtonian view as it was drawing a blank about my questions. Not once did I trace a hint of a surprise about my unveiling the misconception. They never acted astonished but were always immediately ready with a defence about the incorrectness of my seeing Newton's incorrectness. They all seemed as if their ability to understand my view completely is locked behind some wall. The non-Educated, of which I am a member, at least understood what I was saying, but dismissed the simplicity about the argument. In the corner of the SUPER-EDUCATED was no response of any kind, but to feverishly defend Newton by raising the dumbest arguments I have heard.

The arguments, even the most highly educated brought about, seemed motorized and non-responsive. I think that if they accept the points I raise with my questions, it will bring about their demise and it might disable their senses and in defence of such a possibility, they put up a block. There is a peculiar sense of numbness in the way they could not understand what I did not understand. The academics showed no signs to indicate that they could even argue my point of view, by responding that I have an argument, and from that launched a responding argument to explain how or where I made my mistake.

Their inabilities in even understanding always seemed to me as if it was rather meant to be where they found a way to hide behind a wall of not understanding the fact that someone may not approve of Newton's arguments. The problem I find is matching Newton and Kepler because what Newton saw in the work of Kepler was a mathematical circle and what I see is the division that cosmic structures hold to dividing singularity as the objects are in maintaining individual singularity.

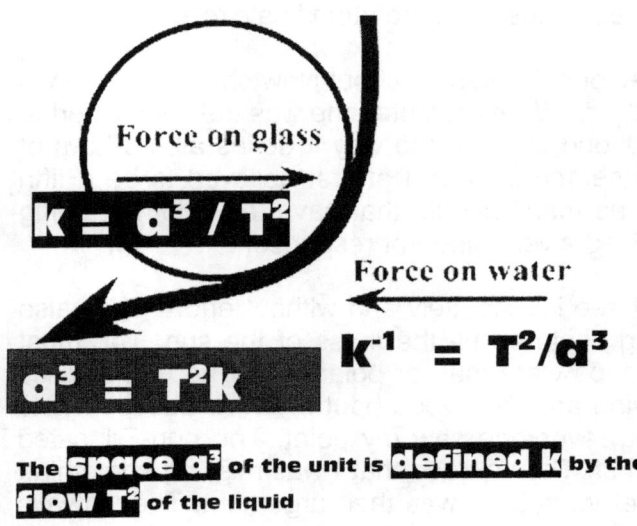

The **space a^3** of the unit is **defined k** by the **flow T^2** of the liquid

When one observes the flight path of the comet, it would much rather seem that the comet is on a designated route, one, which it followed since time began for the comet. How time began for the comet, that I do explain in another book where I introduce the beginnings of the solar system and the birth of our solar system as it is in place currently. In the Coanda effect we do not find two forces, as Mainstream science would believe, but directional flow of liquids changing direction in accordance with a centre. The flow is designated in relation to the spin and the attitude the direction of the spin will take in relation to a specific centre. The spin of a liquid forms the flow around the form of a solid and the liquid as well as the solid determines the distance such a flow would circle around a centre.

There seems to be two values distinctly present to indicate the length of the radii in both cases between the comet and the Sun, where the comet is most advanced in circling the Sun as close as it can ever get to the Sun by circling the Sun.

When circling a steady object, there are circle centres that are crossed and when such a circle centre is crossed, the direction of motion changes. The direction of the flow of the liquid follows the path of such a flow that is organised from a specific centre and that is controlled from such a specific centre. However, all the motion is in contribution to the direction of motion that the centre of the body will allow at the point where one gauge such a flow to be.

Standing in front of such a cyclic spin, there are four directions, to take into account. When the line crosses a 90^0 line directly in front of the viewer, the spin will change from coming towards to going to the right. The flow will continuously favour a change in direction until at another 90^0 point in line with the viewer, the spin direction will alternate to one still going back, back slightly favouring the correction of the moving towards the centre once more.

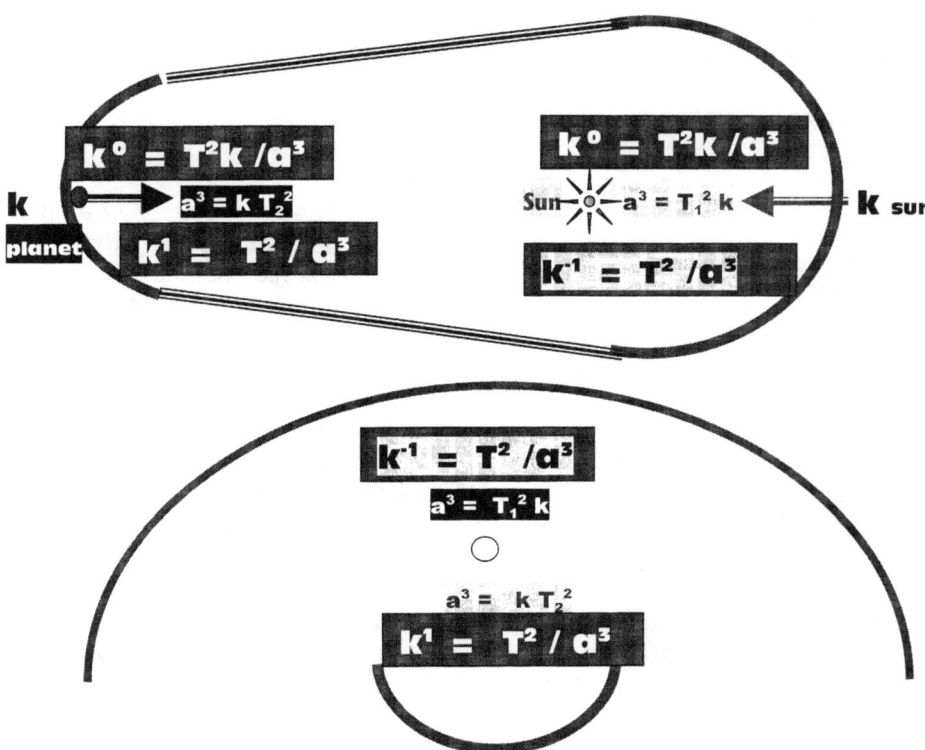

In an unbiased investigation, I again, find no evidence of Newton and his mass driven comet pulling gravity inflicting radius removing planetary behaviour. That too, just as it was in the case of the falling objects and mass infliction while they fall, so it too is a case of Newtonian misguidance with mass. Mass is not what is pulling the comet towards the Sun. It is much more another cyclic oval path the comet follows and this route has the character as if that flow around the centre of the Sun has been instituted for other reasons that has nothing to do with mass.

At the point when it reaches the circle at the back, that would be 180^0 in line with the viewer, the circle motion will make a complete turnabout in direction It will cross a point where it will turn direction from where it moved progressively to the left and to the back to where it then moves progressively forward. It will seem to still favour the moving outwards to the viewers left, but this tendency will decline as the flow progresses. Then when it crosses the centre line once more it will favour a flow changing from going away from the centre to coming towards the circle centre once more. In that entire there are Universes changing in each moment where the alternating of such circling bring complete new alliances in direction and in forming liaisons with the rest of the cosmos. In all of this, mass has no presence. It is a flow of time, in time and with time.

We think of the cosmos, with the view we inherit from the Earth. Even Newton's falling apple and the tree he enjoyed as shade, is life, not the cosmos. The force that put the apple in the tree, positioning it so that it may fall, and the fact that it fell, is all contributed to life, not the cosmos. The Newtonians are unable to differentiate between life's ability of force and the cosmos without life. Life is the only force I am aware of. Take life away and there is no force but only a flow of events brought about as a contribution of time. With the ship finding force to sail from wind, it only applies the force at work from the contribution of life placing what life produces in what life occupies. That is the result of the intervening of life by firstly establishing the presence of the ship being on the water, having a purpose for the ship being on the water and placing the sail in such a way as to find benefit from the wind. The wind is no force. We may consider wind forceful, but only as it may harm life, by blowing down trees or hurting livestock and people. When a hurricane takes a roof off, or destroys a house, it destroys only the extension of life in the form of the roof or house. Was it not for the force of life, wind would have nothing to destroy through force because the force of the wind is most unlikely to lift mountains and throw rocks around.

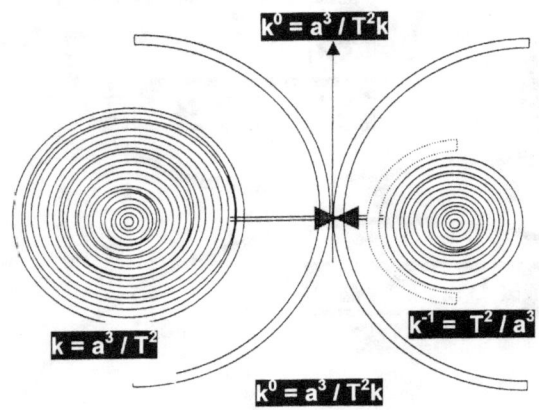

There are changing relevancies as the one cosmic object orbits the other cosmic object and there are forever on cosmic object orbiting another cosmic object except in the case of Black Holes where the cosmos circles around a centre we call a Black Hole. During the cyclic the flow of time T^2 through the four quarters or four seasons the dynamics of the relations **k** change and inter change. The location of positions and the allocation of positions place dynamics in different concepts in ratio to each other.

The Roche limit is:

The region surrounding each star in a binary system, within which any material is gravitationally bound to that particular star. The boundary of the Roche lobes is an equipotential surface, and the lobes touch at the inner Lagrangian point, L_1, through which mass transfer may occur if one of the components expands to fill its lobe. It names after the French mathematician Edouard Albert Roche (1820-83).

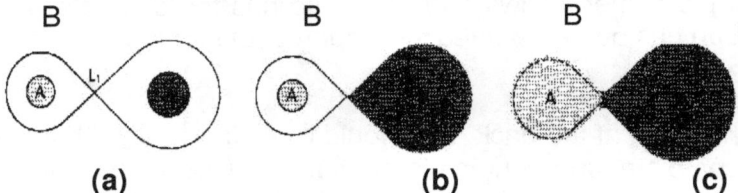

(a)   **(b)**   **(c)**

THE ROCHE LOBE: In a binary system, the Roche lobes of components A and B meet at the L_1 Lagrangian point. (a) In a detached system, neither star fills its Roche lobe. (b) In a semidetached system, one massive component, B, fills its Roche lobe. (c) In a contact binary, both components overfill their Roche lobes and share a common envelope. As with the graph I can see the two sides forming a connection therefore relevancy has to apply, all contradicting Newtonian claims of no connection but through mass attractions. The mass does not attract but one interferes with the other total influencing the space surroundings.

With this, I say there is no gravity induced by mass, no force applying gravity enticed by mass and no gravity being forceful evoked by mass. I do agree that the cosmos holds some natural phenomenon, positioning cosmic structures in relation to each other and in perspective to one another, but be that as it may, it is no force. Let us observe the force of gravity applying an influence on the approaching comet.

Once more, this phenomenon should not occur with Newton's presumptions about gravity. These bodies will and must collide and destruct, without a doubt. When the formula F = $\dfrac{M_1 M_2}{r^2}$ G applies, there should not be any force which is able to keep them apart especially when r reduces to almost infinity compared to what it is at maximum. However, they do exist and what is more, they maintain a certain distance apart.

According to Newton there cannot be the Roche limit because stars cannot face each other and not cross the threshold of 2.467 times the diameter of the star. Stars are supposed to collide with each other with the "force" of "gravity" "pulling" the stars closer, using the accumulative mass of the stars and multiplying that value with both objects by the mass component, and then this will reduce the radius r^2 progressively until r^2 reduces to zero. Seen from this view, it is little wonder that the significance of this was lost in the notion that this is yet another "mystery" of the Universe. The scientists of the day (and the past) lost the importance, which this holds for us as Earthly dwellers.

Behind this is the principle is the sound barrier and the reason why aircrafts break the sound barrier. It is all a relation in different positions of singularity. From the time I wrote the first few pages, I always was of the opinion that I might have the idea of what is missing but that I am not the person suited for the task since I am well aware of my beleaguer education. I was on a quest to find a more suitable person to take over the work, a person with much more brainpower than my abilities would allow. I, more than any one else, know my limitations and limitations they are. I knew that anyone of the more than one thousand five hundred academics I eventually contacted held more knowledge in the tip of his (or her) little finger, than I have in my entire body. I tried in desperation as I tried in vain, to convey my message to the right person that could see what I could see. It all came to nothing because the academics all sat on the same mighty Sear Tower far too high to even notice me pointing at the cracks down below. From where the Super-Educated-Masters sat, they did not even notice me. Those to whom I drew some attention be it personal, by mail or on the Internet, saw me as a nonsense proclaiming nonsense.

Every time I contacted a Super-Educated in desperation for help, they ignored me flat. Every message I sent telling whomever I contacted, that there is no such a thing as gravity, it is all a medieval hoax, Newton is altogether completely wrong with his gravitational laws and the Bible is one hundred percent correct about creation, they would not even reply or at least respond. Well… to them I say this: I still maintain that there is no such a thing as gravity: it is a hoax. Obviously, the obvious is that the Newtonian Order of High Priests would lead every body to believe that Newton and Einstein's findings are flawless. All these mentioned discrepancies are known to "Accepted science", yet they keep the charade going about other planets they are about to find, only to mislead the public and milk their tax money, in the name of research. If it were not for funding and an effort to provoke general interest in skimming tax money, then why would they deliberately spread such malice?

From where I stood, I could see the mighty tower they sat on. I could also see what construction held that mighty tower together. I could see how much that mighty tower was leaning over like the tower of Pizza. I could see how the tower will fall one day if not sooner than later, because I was at the foundation of the tower. From where I stand, at the very bottom, I see the foundation of this enormous Petranos Towers collapsing from the misconceptions it holds as a basis. Down at the very bottom where I am, I could see what no one holding a high and mighty position at the top of this tower could see. Those that are at the very top, are so high, so very secure, they would not even hear me or take notice, and yet they are the only ones that can do something about the inclining tower.

After attempting for seven years in all, years I tried in vain to make myself heard by the High and Mighty, to get the High and Mighty to notice the insignificant me down below, so small in relation to their greatness, I decided to show the world what holds them that high. I decided to show everyone how great "they" are, especially about the greatness of their misconceptions that put them at such a dizzy height.

To everyone of the Super-Educated-Mathematical-Master-Minds, I say this: Every opportunity you had, you thoroughly rubbed my nose in the knowledge that we are not in the same class. I accept that fact as I accept my academic qualifications also being my disqualifications that seem so very poor. I shall never be your next-door neighbour or the bloke living down the road from your house, because I am not in your league and I shall never be. I do not begrudge you your academic position, your mighty achievements or the height you have reached in your sphere. I shall never enjoy your company as an equal, because I can never be your equal. Your brainpower puts you light years ahead of me, and for that reason I do not even wish to have the honour of your company.

All I ask, is to listen to a mere mortal, a mindless illiterate compared to you, one sod down here at the bottom where you are at the very top and that may just see things you cannot see from the dazzling height you hold. I do not wish to join your company for I shall not fit. I never had or will have any ambition to fit either, because I am quite happy being in the sub-minor league. All I ask is to be heard. So many times you, honoured members of the clan of Newtonian High Priests, did not even attempt reading my book that I sent to you. You did not even try to pretend I had a point therefore you merely threw my book away in disgust. To you my illiterate arrogant views about science being wrong for the past few hundred years are the epitome of a mindless human. You saw me as a totally mentally underdeveloped excuse for breathing and you could not bare my company because for me to have my view such as my view is in rejecting the view of an establishment, centuries old, do fill the likes of you with disgust. I can understand your disgust, but that does not change my point and that does not change the incorrectness of Newtonian science!

Before you throw the book down in total disaproval, first answer the following argument and if you can answer it truthfully then throw the book down. If you cannot answer it, go on reading the book and you may just set your thinking mind in motion. Hear this from a mindless and you might find you still have the ability to learn and afterwards reflect on that which you learn. I say this to you, learned you might be but you cannot think, and I do not know who is the most mindless, me without education or you with education and without reason.

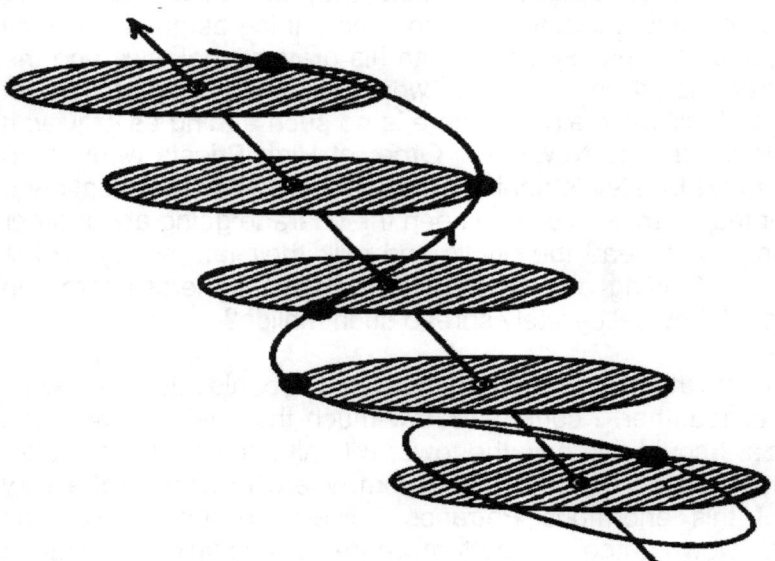

Newtonians declare that a force brought about by the content of the mass of a body will therefore pull objects closer where the pressure coming from the weight brought about by the mass of the bodies, will lead to heat.

If a cylinder is pumped with air, the pumping is a force. The force comes about because the force is that of the intentional action of the only force in the universe, the force of life. The more pumping there is and the longer the pumping will last, the hotter the air will become, and therefore the hotter the walls of the cylinder will get, as it transmits the heat from a point where the heat is most abundant to a point where the heat is least abundant. Heat flows from hot to cold or so science teaches.

After pumping stops, the heat on the inside will reduce in value up to a point where there is equilibrium between the heat on the outside of the cylinder and the heat on the inside of the cylinder. The heat reaches equilibrium with the event of time. Please for the sake of sanity, do not reply that it is the molecules in the air tank that are bumping against each other and through that collision, friction causes the heat. That is as much Newtonian rubbish as one can ever find.

Should you insist on that being your answer, then ask yourself what will calm the molecules down after the pumping stopped, which brought about a situation where they get so calm, there is no more heat in un-equilibrium? Did the molecules take drugs, or did the force calm them by telling them gentle nighttime stories? The cylinder wall is then just as hot or cold on the inside as it is on the outside. Therefore one has to presume the bumping and colliding of molecules in the cylinder stopped. Why the hell would they stop colliding when the pumping stops? The action is not with the pumping but with the overfilled cylinder that forces the molecules to collide. The cylinder is not released from its pressure, therefore all the colliding molecules remain inside the cylinder…and yet the heated wall cools down to what it is on the outside. I had so much hogwash used as arguments thrown at me wherever Newtonians defended their Master it sickens me. I may be uneducated, but I am far from mindless! Not one Newtonian gave this bumping idea a thought ever. Electrons are all negatively charged and therefore electrons repeal one another with unmatched violence.
NO MOLECULE CAN EVER, EVER TOUCH ANOTHER MOLECULE BECAUSE THE ELECTRONS GAURDING THE OUTSIDE ARE EQUALLY NEGATIVELY CHARGED, AND WILL THEREFORE REJECT ANY CONTACT OR COMING CLOSER TO ONE ANOTHER OR WITH ANOTHER MOLECULE.

Scientists know about this discrepancy in the Newtonian laws, yet there is never any mention about it.

The so-called "evidence of the existence of planets" is based on just as laughable principle. Allow me to explain: The findings which science base their proof on about the location of other planets are the gravitational pull.
At first the way in which the facts are presented does not sound that unfamiliar in argument, and one tends to accept it without a second thought. When given the second thought, the blatancy leaves one breathless.

In the evidence, the stars and "planets" are presented to be in a tug of war. This one can see in the sketch above. How this supposedly works is that the one star first pulls the other system closer with the force of gravity, and then it is the other systems turn to apply gravity and jerk the first system to its side. What they do explain in explicate detail is that they do not understand the first thing about the matters they pretend to understand. **The SUPER-EDUCATED-MASTERS-OF-FACTS knows very well that all children play this game, and therefore every one will associate this explanation with familiar events in their past, and no further questions will be asked.** Let us examine this principle with obvious general knowledge about space flight and how this applies in outer space.

I am under no illusion that you suddenly will find any inspiration to read this letter this time round, since you never made any attempt to read any of my correspondence in the past. I am under no illusion that those of you that have incisive influence about recommending what the criteria should be about printing material in major Universities or stand in as an adviser for any other Publishing Houses which may or may not form a department in your institution, those then will suddenly read my work with inspired interest and ruminate on matters I refer to. I do not envisage that you would this time round find a change of heart where you suddenly would enthusiastically devour every word in this letter, let alone get around to actively read any of my work in any of the books. I am totally aware that when you receive any letter such as this, you would immediately delegate the task of reading the letter to a lesser office since you have no time to deal with such small issues as this letter holds. I am well aware of all your dismay which I am about to release where your response is dumping this onto a much lesser desk and that is the reaction that this letter charges where it is causing the rejection that stirs in you and I am equally aware that there is a poor chance that any of your members of staff would even care to read this letter to its end. You may well wonder why I then would address this letter to your office. Your reasons for not reading my work in the past was that you were too busy and that did not allow you the time to read it, but it is more to the point of simply being because I reject Newton and you counteract that by rejecting my work from which your reason for doing the rejecting is that in justifying your action as that you simply could not generate any interest to do so. Sir, Madam, by my sending you this letter is for motivation, in order for me to find solace that you are rejecting my work because of an academic conspiracy. Reading this statement at this point might seem ridiculous and execrating, but if you think in such terms do yourself the favour and keep on reading. If you are surprised at the statement, then be even more surprised when I challenge your decency to prove there is no academic conspiracy going on to the effect of being deliberately instigated or otherwise being protective of the status quo, because you feel vulnerable about academic issues. I see no need to introduce the books in this letter because of the simplicity of the information as it is presented for reading purposes in the books. In the books I am again showing what is incorrect about Newtonian gravity. That is not my personal perception but a fact I established without a doubt, but irrespective of whatever degree of undeniable evidence I bring, notwithstanding the degree of correctness that I deliver, my statements are never good enough, not even to justify an unbiased evaluation thereof. I know from past experience that this statement where I suggested anything concerning Newton and Newtonian views as being incorrect nailed my coffin shut before you even opened the book. In the past and at present I am aware that such accusations about Newton already bring automatic disqualification to my work. It spurs immediate and total academic rejection of my work by all academic's concerned and I provoke the resentment they experience towards me in person in the most intensity any Academic can experience any resentment. They immediately feel offended by my challenging the system and I am also aware that much of the anger is as a result of their feeling personally offended in their position as guardians of physics. While they are the caretakers and Masters of the physics they guard, I come along and show flaws in what they see as being more perfect than God. Even those amongst you that declare them to be devoted Christians think more of Newton than of Christ and don't fool yourself on the matter because I have a means to prove this statement. In defence of my condemning Newtonian gravity I ask you which is more important, the ego of the Masters or the truth of the work? In the past it seemed the ego of the Masters about the stature of the absolute Master being Newton carried supremacy above the truth while overshadowing logic and reason. Therefore I say this in total confidence that I am more than aware that I am about to evoke the very same response as I have done in the past in my rejection and disputing of all the Newtonian gravity views. In my saying this which is commanding the anger that I evoke I challenge you when you go onto reject my work again, to have some degree of honesty about why you again reject / ignore or dismiss my work. Be honest to your conscience as to why you are not prepared to analyse and to scrutinize Newton with me and in the manner that I do. However, when you do go on to reject my work once more, then I charge your honesty and your sense of fairness as the bearer of an academic pillar, to go on to prove how mass does bring about gravity. After six years of continuous trying and after eight books were presented on the subject and with no clear response yet, what else must I conclude than that there is some conspiring going on to suppress my findings.

Sir, Madam, I have reached a point where I am beyond diplomacy and I have taken off the gloves of hypocritical politeness. I reached a point where I call a spade by name. I now shout fraud where I detect conspiracy. I will do so even when this is present in the highest circles of the Physics paternity. I feel tested to a point where I am no longer prepared to use cotton wool as a means to avoid confrontation with the highest in all academic intellectual circles in order to evade embarrassment by hiding honesty, decency and respect behind a cloak of hypocrisy just to save the face of a supposed respected paternity.

In the very same breath I also will admit: Sir, Madam, I feel like Satan incarnated in person while I accuse you because I do not know you in person or even as an individual person. I can only consider you as my superior intellectually and on all other levels and by all other norms and with that then what you represent what I have to regard as being above and beyond reprimand in any way. Yet when no one takes notice of what I say when I show what I say I have to come to a conclusion that there is some conspiracy going on. Why would the members of the academy ignore me rather than act surprised when I show a clear mistake that Newton made. Why trash my work regardless without reading it in serious consideration when I present a new clear definition about views concerning the basics in physics. Your ignoring comes in spite of the fact that no one knows what forms gravity and that includes even Newton in person. I bring a new suggestion to the table that holds all cosmic phenomena dearly and the phenomena are a fact whether academics attribute any reality to their being there or not. From that I define gravity in a new sense but from persons of your status I find rejection because the new definition does not involve mass. If anyone presents an answer to a problem the least one would expect, as an honest reaction is that the party which is informed would be surprised and appreciative of any new suggestion that may solve such a problem and I would think not knowing what mass is would constitute to a problem. Why would any one dismiss the problem by dismissing the suggestion and ignoring the solution that is presented? The more the lot ignores me the more my suspicions mount because I prove Newton and his mass has no validity. If you are not guilty of what I come to accuse you of when I accuse you of conspiring to suppress evidence about Newtonian incorrectness, then why ignore what I say. Why do you then not just read what I have to say? What I have to say puts the cosmos out there in a completely new perspective and if I may add, a logical and understandable perspective. I bring a perspective where one does not have to go Bohemian to show singularity and to show space-time. It is there for all to see and even a child can understand what there is to see. So far my attempt to use a respectable avenue going through the official channels and following the guidelines on offer got me precisely nowhere and it took me years to get there with all the aid I got from the Academic paternity through the years. I am now at the point where I chuck all the niceties out the window and see what I came to suspect because of the treatment I received. I made excuses on behalf of the Academics and tried to find reasons for their behaviour but the only and last conclusion I could arrive at was that I was very naïve about the honesty and unwavering integrity in which I regarded the Academic world in its complete devoted sincerity to which they strive to achieve about the factual correctness in science.

I feel like the crook in the fairy tale while I accuse you of dirty dealings but what other choice do I have. I prove that Newton does not pan out and you brush me off by telling yourself and me that I do not understand Newton. That is complete and utter rubbish. I understand Newton better than any other person I have come across. Newton does not make sense at all. Mass only has validity when gravity is restrained because gravity is the motion of space that is filled with matter in $\mathbf{a^3 = T^2k}$ and when space does not move it cannot act on gravity or motion where it becomes immobile and frustrated. Being immobile and frustrated results in that the space that is filled with matter is forming mass just because it cannot move independently any longer. However gravity is the motion of space in time and mass is the restricting of such independent motion. Read my work and you will see! If you are not prepared to read my work but always delegate the evaluation thereof to a person that is clearly not up to the task, then what must I conclude from that? If you do not even try to address the shortfall of Newton then what must I conclude? Then it becomes obviously apparent that those academics being in the status and position that you are either detest what I say or do not believe what I say. If you do not believe what I say, take my challenge serious and prove to your mind why you accept mass has gravity as a result while you scrutinize my ideas. When you detest what I say then find the reason for your detesting me including what I have to say.

Again I have to admit how embarrassed I am in my behaviour by accusing a person in your position but what else is there to conclude. The Critical Density theory makes a mockery of common human intelligence that reminds me of a bunch of drunks going on in an argument about matters they do not understand even when they are sober and in their state of drunkenness they are going mad. I just cannot believe the world's most intellectuals can reason in such an obscene manner that suits the likes of high school boys arguing about their fantasies rather that persons in such position as they have. It fits boys arguing to impress girls much more than it can be associated with the worlds supposed best minds there are. At the end of this letter when and if you have the courage to read it to the end and simply also on the condition that you will read it to the end this time round, then just be honest to your mind and think what I said and how I say what I prove in my books. I honestly do not wish to be insulting by shouting conspiracy but can you find another way that I am supposed to reason when I gauge the way you react?

The following four phenomena are real and are evidence found in the cosmos where they stand undisputed but alas also unaccepted since the phenomena does not match Newton. Newtonians rather accept Newton than would they accept the phenomena since the

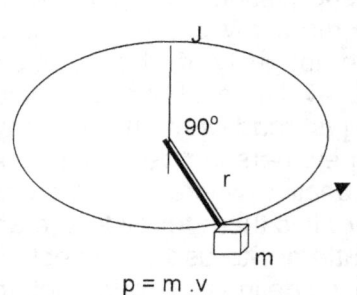

phenomena do not apply in the manner Newton explained the cosmos works. The Phenomena are as real as outer space is but because it does not match the concept of Newton and his mass by gravity, therefore the Newtonians discard them with many wide ranging excuses. The phenomena are amongst so many other fitting proof of Newtonian misconceptions and I can explain every one as much as I can fit the lot into the forming of gravity. I can ensure you they have nothing to do with mass and so has mass nothing to do with gravity where gravity has everything to do with mass since mass announces the end of gravity. Mass forms gravity's grave. The one phenomenon is called the Titius Bode law and because it is named a law most astro physicists scorn at the fact it is referred to as a law. I prove that this is not only a law but this is the Universe. This is gravity compiling a Universe. This is space-time and without this applying, stars collapse as we find Black Holes do. To this day I must still find one Newtonian that even had a glance at this proof I bring where they did not tell me to my face the phenomena are coincidental and my proof

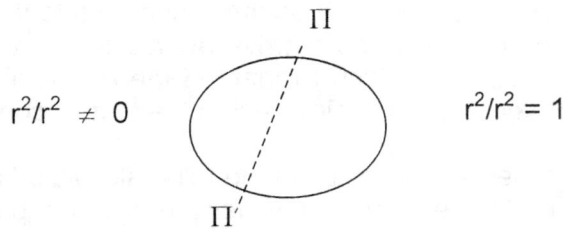

therefore is also coincidental. Should these apply then Newton cannot apply because all the planets relate to one another as well as the Sun in this precise fashion, which rebuts mass differences. The relation is there and that Newtonians do not dispute but in order to avoid the explaining they rather put it down as a coincidental insignificant phenomenon. The fact that all the planets show exactly this formation and that all the planets are distributed (all nine of them excluding not one, however Neptune does not precisely fit the order but Pluto does and I do explain why Neptune is slightly out of line) is approached in a manner that would rather 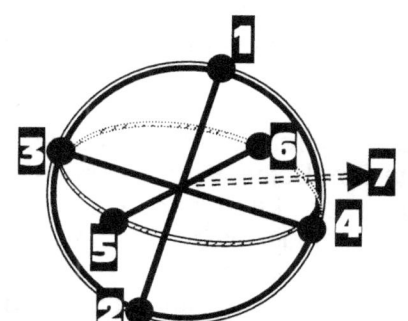 be degraded as a fanciful thought best left outside the mentioning of good conversation. They ignore a mountainous discovery because of what disgrace Newton's claims would suffer. The true disgrace is the rape Newton committed about the work of Kepler. Newton removed Kepler's formula and suggested that the rotary motion would nullify the radius since a circle does not develop any drive $\frac{dJ}{dt} = 0$. The concept is mathematical fraud as big and as wide as any scam can go. In Kepler's formula the one holding time (T^2) compliment the other factor in time (k) to allow space to duplicate as space moves through time. They are not in division as Newton claimed. Secondly there is no mass involved in outer space and in that we find the Titius Bode law that provides us with the evidence of unanimity where all planets are at a specific ratio, which removes any concept of size differentiation from the possibilities present which should be forthcoming where mass is present. One cannot divide the radius to the smallest point and find the radius would diminish to zero as Newton suggested. That is just a mathematical impossibility. The very essence of the Phenomena renders its result onto form. When a line is shortened by dividing the line will keep on reducing until the line becomes infinitely small and the line would have all point that forms the line positioned on one specific spot. That is singularity where 1^0 have no sides. If Newton did suggest that $\frac{dJ}{dt} = 1^0$ then that statement would ring true because I can prove singularity being present at that very location and that is correct! Hubble proved the universe is expanding. Then by backtracking we have to set about reducing the sphere constituting the expanding universe. If r in the circle is growing we have to reduce r to backtrack. When the circle reduces, the value located to r will become implicated because r determines specific size. Not so in the case of Π, because Π in the true sense only indicates that the circle is a square without corners and therefore Π dictates form and not size. By reducing size only r comes into contest and will point to such reduction. By reducing the circle radius r by half continuously will lead to an infinite small circle but Π will remain because the circle as a form remains even being infinitely small. In answering this question about presuming the Universe to form as a sphere, one will arrive at the point where we can begin in finding answers to the cosmos. It is the spontaneous choice the humans make by selecting a sphere when thinking about the cosmos in an overall picture that we should address. From logic comes the answer about the sphere being the form that inspires the cosmos.

To find solutions we must then investigate the cosmos in that direction. You're visiting this web page may bring along answers about the Life story of the Universe. When one draws a line through the centre of the sphere it would connect the most outward points of each opposing side. Taking such a line horizontally will draw the longest possible side from edge to edge through the very centre. Another line coming from the from to the back will run also choosing the longest line by connecting the most furthers points and the line will cross the other line at the centre at an angle of 90^0. Then a third line with the same characteristics will also cross at a centre point at 90^0 to any one of the others running from back to front. The lines will have a specific line in length as a common delineator as well the angles crossing at 90^0 and most important the lot will cross at an infinitely small precisely centred point. That point is singularity because when all the lines are evenly reduced they will all share one point in the centre that disregards size.

Finding the shortest line is also the process in which we will be able to find the beginning of the Universe. Light gives information and light travels by a line from point to point. Light uses the Universe to travel and if we wish to find where it started we must find where that which parts material, started. We must see where the line starts to find where the Universe started because the Universe is a system of combining lines that is running all over between points rotating.

In the very centre of the sphere the form dictates that the shape will relinquish all grounds in space that it can hold and the form will finally be without dimension. Being without dimension means at a point in the extreme centre of all spheres there are a point that holds singularity because this point with no space has a mathematical position although it is invisible since there is no sides to such a point to give that point any dimensions. When holding the strength of the shape of the sphere in mind as well as taking into account that all cosmos objects of importance is in the form of planets or stars and they are all in the form of a sphere, we therefore may contemplate that it is where gravity originate. We now only have to find the reason why gravity will hold a base in a space less ness as Einstein predicted. It is clear to be seen that gravity is in the centre of the sphere controlling from the centre everything that is outside the space less centre. We can reason with confidence that gravity is the strongest where space is the least. We can further reason that it is gravity that is holding the sphere in true form and since the sphere allow gravity the best working opportunity, gravity can form the sphere in as strong a shape and form as the sphere seems to have.

From every point on the surface of the sphere is a point where that point connects with the other side on the surface of the sphere by a line that runs through the spot which is space less at such a centre point in the sphere. Such a line also connects by an angle of 180^0 as but also it holds other lines 90^0 where those lines show exactly the same characteristics than the original line. These lines are between six other points forming three lines that is running from top to bottom, right to left, and back to front, where all join and cross in the centre of the sphere. There are therefore six points at 180^0 forming three lines at 90^0 crossing at one centre crossing that has all the lines at 90^0 to one another. That shows the law of Pythagoras. All those lines are connecting through a centre from any given point on the surface of the sphere where one spot on the surface will have three lines covering six points on the surface all together and cross at a seventh centre point. Such points connects in total six surface points on each side of the sphere while they all support one another through the space less centre. In that absolute space less ness in the centre holding singularity we find gravity supporting and controlling all space within the sphere as well as space connected to the sphere. That is where gravity control and guide the space, which falls in the parameters as well as under the influence of the form of the sphere. In the gravity centre space goes singular meaning space becomes space less or flat. That is where Einstein's Universe goes flat because that is where gravity is at its strongest.

Also it is true that the entire form that is the sphere has, is controlled from a centre within the sphere. That centre holds the sphere in form and shape. Therefore the strong form the sphere presents is dictated from that centre that is space less where there is no space and no form left. The natural inclining is in the form of the sphere. It is part of the roundness that the overall shape of the sphere represents and this structural strength is carrying down to the very centre. Because the circle is forever reducing that reducing which is inherently part of the form of the sphere becomes a tool in distorting of space in the sphere and is eventually removing all forms of space from within the centre of the sphere. The very centre ends up as having no space because of the reducing that continuous down to become the space less inner centre. The all roundness is the ingredient that forms the backbone of the absolute strength that the sphere has and that is the component that the sphere is so famous for. The form the sphere has allows the sphere to have a control that is coming from the centre deep inside the sphere where the space vanishes and being without space seems to keep the entire structure rigged. From the centre the sphere shape shows strength that the shape as tough as it is. How does it work in its most basic analyses? Every point on the sphere surface is in regard to every other point on the sphere surface by a space less centre.

By using mathematics the rules dictating the use of mathematics will apply which when correctly interpreted. It is about analysing Kepler's message mathematically correctly. If it is done, then such interpretation of the formula gives new meaning to claims coming as a result of analysing Kepler's formula. This is also without Newton meddling and without Newton telling Kepler what he (Kepler) should have found and whereas instead he (Newton) should have seen what there was to see when Kepler said what there was to find.

Through Kepler's investigation we now know what the cosmos tried to tell the human species if only there were some sort of skilled mathematicians since that time to decipher the true meaning of the message. Newton should have been looking at instead of telling Kepler (and in actual fact the cosmos) what, and then he (Newton) could have been able to see what gravity is. He (Kepler) said the cosmos told him that in the cosmos we would find that $a^3 = T^2 k$. He (Newton) was to be recognised as the everlasting expert on matters in motion, but his personal arrogance disallowed him to see that the cosmos told Kepler that gravity forms when space is duplicating space by the motion of space. A close study reveals that space a^3 is held in check by the motion T^2 of the space at a specific relevancy k of the motion of the space. The space a^3 will be on the move T^2 and the moving T^2 will be equal to the space a^3 in motion T^2 at a relevant distance k. There is a specific ratio of space moving in relation to a specific centre at a very pre determined distance from that centre.

The line reduces r by any point to the centre and r remains connected to the centre for all purposes that may play any role. The line will reduce until such a line is all but vanished forming singularity at 10 because that is singularity by value but the line can never become zero just because such a line always will connect material.

Taking the outlook from the point the sphere is holding from that centre out into space there

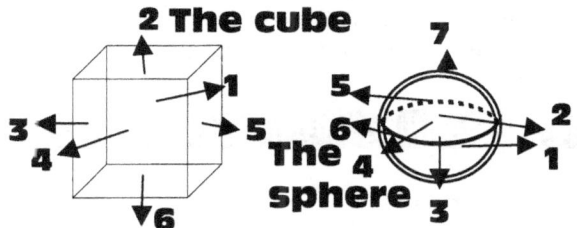

are ten points connecting to the centre. In that are the dimensions of singularity connecting to space where five connects to space in the second dimension of singularity, and five connect in the third dimension of singularity. On the other hand does the cube show a very different characteristic, which involves only six sides (at least) being connected.

$a^3 = (T^2 k) = a^{3 +2 + 1 = 6}$ with the sphere presuming the position of singularity as part of $k^0 = 1$ = **singularity**. Einstein proved that at the point where space reduces and such reducing reaches a point where space as a factor in the third dimension disappears into the single dimension (space going flat) gravity is overwhelming. Einstein interpreted this, as the complete Universe going flat while the Universe going flat, that can only be within singularity since singularity represents the Universe as flat as it can get.

Humans (including Einstein) interpretation of the Universe is faulty but the faulty aspect does not include the fact that the Universe is going flat but only which is the flat going Universe referred too. According to Einstein he proved that the Universe is alternating between going flat and holding space but his lack of studying Kepler lead to his spontaneous m'sinterpretation collected from our culture and his incorrect interpreting of what the Universe actually is. We all have a faulty perception of the Universe because not only he (Einstein) as an individual Scientist but all humans throughout has also never asked the Universe what the Universe is.

Kepler did and the Universe answered using the mathematical equation $k^0 = a^3 / T^2k,$ which when interpreted means singularity placing space-time is the Universe. No one ever thought about this statement in sincerity because from a Newtonian aspect it seems silly. But rethink the silliness presented by the Newtonian Universal centre and compare that thinking about what the Universe told Kepler then decide what is silly. Newton's never acquiring the effort to do a study of Kepler's work withheld him (Einstein) from reading his very own mathematic translation accurately because apart from Newton Einstein must be considered the second most important Newtonian ever. What Einstein saw was that space disappear and he then jumped to the conclusion that the space he saw in his mathematical equations was outer space referring to the space falling outside the parameters of the material occupied space secluded by dimensional borders. In the sphere placing the borders that the sphere holds there are deliberate and very distinctly placed edges or points forming a specific distance from the centre. The centre is also proven beyond any debating.

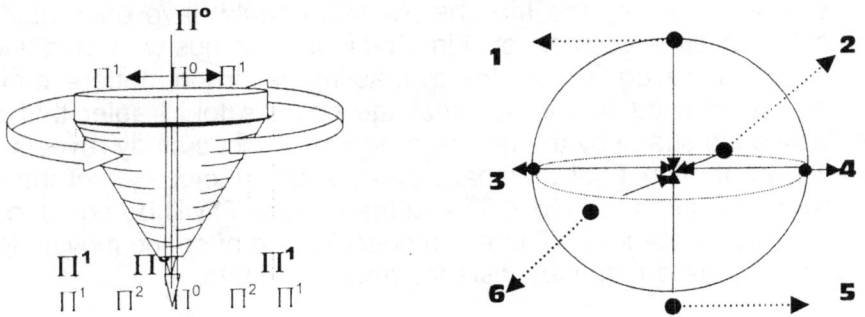

With Π^0 being a little more than the figment of an intelligent person's imagination, there are actually two values connected to singularity at Π^0 measuring Π^1 that is facing each other from both sides of the divide in relation to Π^0, and by moving from Π^0 to Π^0 by going around Π^0 the total combining value is $\Pi^1 \times \Pi^1 = \Pi^{1+1=2} = \Pi^2$, and with two sides being the very same that is opposing each other the movement will conclude as Π^2, in relation to every point that holds Π^1 and is facing Π^0. This is no more so visible when a circle is spinning which unveils the presence of a centre line that becomes a factor when an object spins in a circle.

Locating zero

Zero point — Starting point of the line — Extending the line from the start.

Zero in place

$$k^0 = a^3 / T^2k$$

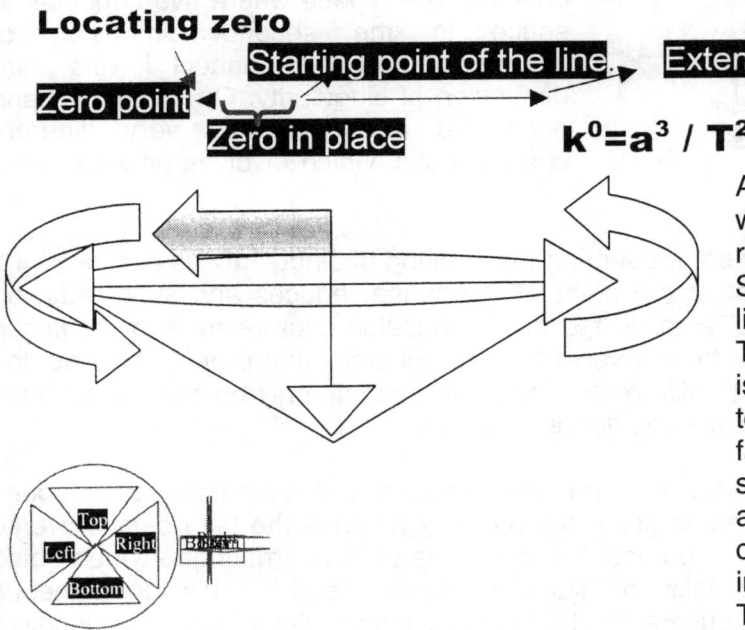

A line starting at zero has to start with zero and starting with zero removes the line altogether. Starting the line with infinity the line poses eternal possibilities. Tracing the centre of the Universe is still possible by any one wishing to find such a centre. The centre falls outside the accepted Universe since it cannot be mathematically accounted fore but that centre is in control of every thing in its influence.

The centre changes motion to gravity by diverting the straight line to an immediate circle. By tracing the line back to where the circle is no more a straight-line will uncover singularity plus one dimension. But in the entire centre forming singularity is still locatable within the Universe we have.

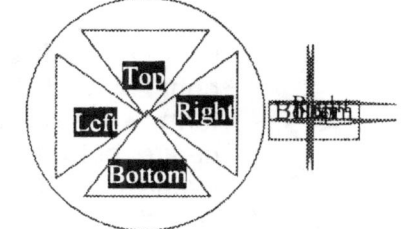

Reducing the radius r from all angles possible, throughout the circle will bring about that all possible direction will eventually land on the very same spot with no more dividing possible. Yet zero cannot be a factor since the sides still hold value. Whatever value may progress from such a point notwithstanding there are that many possibilities holding all the value that can rise from such a spot is concentrated on one point. This is at a point arrive where more reducing will land the one side on the opposite side of the line but it will not bring about zero in the equation

But keeping Π as one ($\Pi^0 = 1$) we keep the Universe in the first dimension.

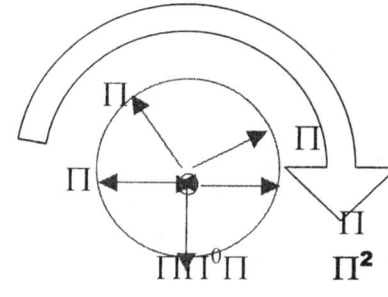

This point, which I now am referring to, is the point where Π is a fully appreciated value while the diameter D still remains a dimensional factor of one. This is the dawn of the second dimension where space was there but space was sparsely shared in some cases. It was Π^2 when Π^0 shifted to become Π for the very fist time.

What this argument further proves is that the circle reducing must then come from all points because the radius might be a line but that line represents a circle through 360^0 coming from and accounting for all possible directions. Taking that into account it is important to recognise that notwithstanding the size of a line, which any radius of any size is there is another line (or dot) eternally bigger as well as eternally smaller than the line in question. While we are in the third dimension being part of the third dimension such being in the third dimension then allows that all parts of the third dimension forever can be divided once more until the line in the third dimension is no longer part of the third dimension. When such a line leaves the third dimension it is still dividable because it might not be part of our dimension any more but it can still reduce further as part of the second dimension. By that time it has left our scope by miles but that does not mean that it end there because from our perspective that is where it ends. But our perspective does not represent reality. Yet, even then it can still reduce infinitely more until it has left the second dimension and then at last forms part of the first dimension. Only then when the line reaches the first dimension no further dividing of that line is longer possible. We can never grasp the size of a line that the first line in size that came about when the first motion broke the eternal stranglehold on space.

According to our big and small conceptions of what we perceive as large, ultra large, small and microscopic small is just mere words describing thoughts totally unrealistic in the context of what the cosmos sprang from as the cosmos moved out of the spot and formed a dot. Even by the standards of forming the dot, which was eternally bigger that the spot T, as the dot and all the many dots that came from the spot. The size differentiation only between those two exceeds all limits and divides we wish to create forming borders that we can appreciate.

Being part of the 3D we have the inclination to think of something and then we also includes in that something the space we experience. Thinking of the spinning top we will think of the edge (A) forming the end of the line, the position of seven. This concept we have is part of the 3D Universe. To understand cosmology we have to return to the beginning of cosmology. The relations then were when Π^0 formed Π.

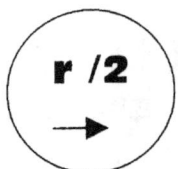

By dividing the radius r by the half of the value that then reduces r to a point where the left edge of the line reducing will be at the very same place the right hand edge of the line that is reducing will be. At one point the spots that formed the two ends of the line will be at the same spot where the original centre between the two points were. The two points would have moved evenly towards and in the direction of the centre by reducing all the space on both sides of the centre.

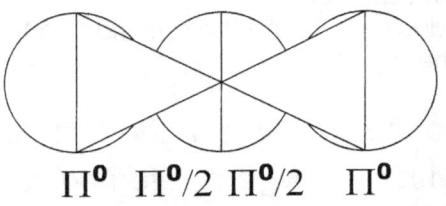

Π^0 $\Pi^0/2$ $\Pi^0/2$ Π^0

Let us start telling the story as it was.

At the very first sign of any of the sides departing from the centre shared by all, all other points must also show signs of a willingness to depart. There will be one point where r still is one coming in as a factor but pi moves out from only being a factor of $\Pi^0 = 1$ and at that point pi will become a full factor of Π.

Then by moving towards the centre they will at some point have to reach such a centre point notwithstanding cultural concepts favouring nothing to be filling that spot because reaching that centre point will land all the sides on the same side and because of the presence of all possible sides such presence of all possible sides removes nothing out of any further possibility.

Any further dividing will land the left hand spot past the right hand spot in the opposing half where it then will grow once again but in the opposing direction that the specific spot previously represented. All possible dividing then ends on one spot where such a one spot

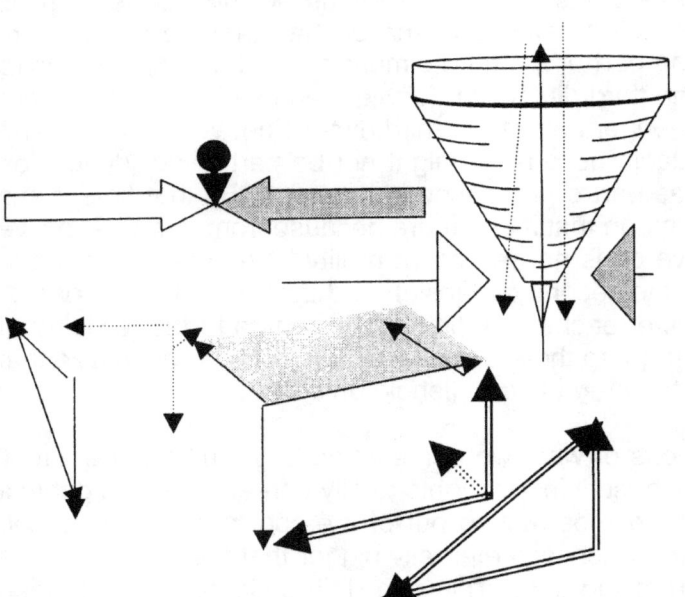

that represents the perfect centre point and that divide the left side from the right side and the top from the bottom and the front from the back will land on one spot. At that spot all the sides just mentioned shares a location with all other possible sides. The centre that then is holding all the previous points in one spot then physically is in the single dimension applying as one spot to share a location for all sides. At such a point there is no further dividing possible. That point cannot be zero because that point represents an eternity of possible growth in an infinite number of directions available to grow.

The line starts in infinity and not in zero or nothing as is taught to scholars by teachers' worldwide. Trace that centre while the top is spinning and one will find a centre that favours no side since the centre divide equally all sides while spinning. That centre proves to be no specific side because such a centre proves to be all possible sides. On several occasions in the past I have been accused of manipulating the argument to produce none-existing or overrate facts. That is not the case. I am not manipulating facts to create an argument as some intellectuals in the past accused me of.

What I am talking about is a mathematical fact that any one can prove by calculating.

At such a centre starting point all sides share one specific spot but that spot holds all further future possible growth in any direction of all sides and since everything is in there, there is no room left for zero to be there. That point is filled with all possibilities which prevent zero becoming factor since the sides share one spot and in that sharing they are present and their presence prevent zero from becoming a conclusion. While the different sides are in one place the factor and value is one to all without allowing zero any part to play.

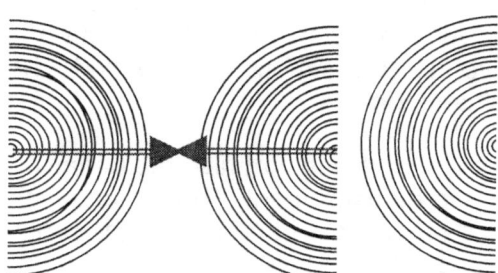 

By following a very simple procedure it is within any person's reach to detect the centre of the spinning top, which I am referring to, although there is no such a centre to detect, the centre is there for all to detect. A child is capable of using the two times table and the dividing by two every time that is the most simple form in which mathematics may be used. It is a mathematical fact that a line will reach a point where all sides are at one spot and as such the line cannot divide any more. I have been accused as being dubious about my arguments while it is Mainstream science that dubiously found a way to get to zero as a mathematical starting point of any and all lines. Then they put the double standard blame on my arguments where it is.

In the centre forms a point that connects every possible point to every other possible point in the sphere and all sides run through that centre to connect to one another. The point is no more than a dot and has the infinite value of 1^0. Motion is required to activate such a point to be more than just a spot and to evolve from the spot to the dot.

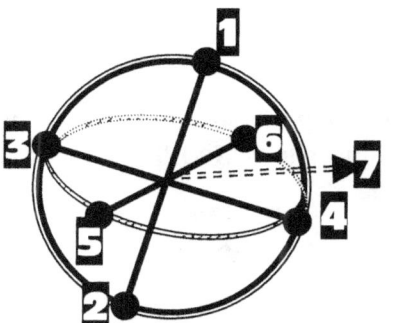

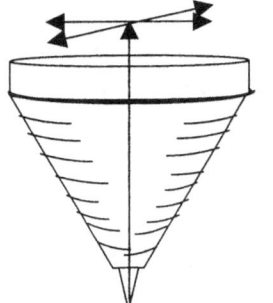

That point confirms the line that comes about when the rotating motion of the top releases the time components. Inside the top is that which has no beginning because it is the beginning that holds the entirety there are on one spot and from such a spat motion releases eternity. Eternity is the time on the outside of the too because that which is inconclusive and that which is concentrate as atmosphere but never

is on the top runs as long as it outside the top may end anywhere.

In the cosmos one finds finite space and infinite space. Anther way to put it is to say there are defined space and undefined space. There are including space and there are excluding space. All space on the inside of the atom running inside the electron is defined space. The purpose of the

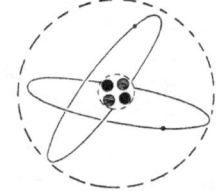

electron is to define the space. What the electron includes will be included and then be measured as the atom.

What is outside is undefined space, which has no borders but run concurrently until the space is again in touch with another atom. The border can be extended or contracted, as the atom stands related to heat. That space, which is defined, is always derived from the form a sphere holds. There are always a circle connected to the shape and depending on the cosmos development the shape can either hold relevancy equal to Π, or it can start to extend and show the presence of a line **k**. The circle is however, T^2, which is also Π^2, and the compliment of **k**, whether **k** might be Π, or running as a developed line **k**, the result is defined space being either a^3 or Π^3. In either case it prove Kepler to be the accurate scientist because that confirms the formula Kepler introduced.

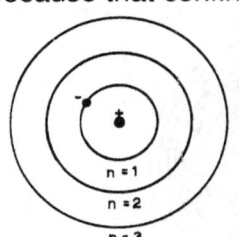

It is the electron that sets the definition to material by giving the material a limit in which to circle. Outside the sphere of the electron there is space that proves to have much less motion in relation to the space inside the sphere we call an atom.

Inside the atom there are particles that give the definition we call the atom a meaning and measurement. The proton gives the atom a defined characteristic and substance. In relation to the proton, there are (normally by equal number except in the cases where the atom goes outside the star range of space-time dismissing) neutrons. The proton indicated a definite presence of mass while the electron too, shows to have mass, but in contrast the Neutron shows an absence of mass. The proton number can add or reduce and in that is that, which underwrites the definition of the atom both in space occupied as well as in measured relation to the entirety of the Cosmos. There is always formed or defined space constructed to include and exclude that stands related to space that only includes but never excludes.

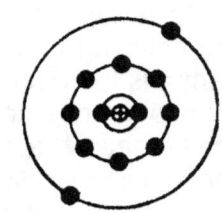

Newton said a sphere is $a^3 = 4/3 \ \Pi \ r^3$

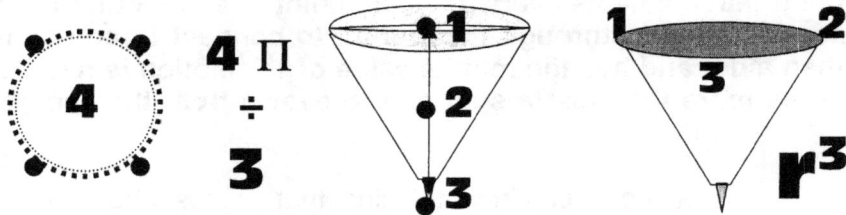

The formula to calculate a circle, where a sphere is a compliment of multiple circles, requires the square of the radius in a square multiplied by Π. That linking r to r^2 and this allow the circles form value Π to stand directly related to the square of the radius Πr^2

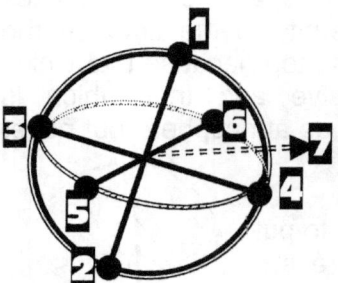

The defined space holds seven points within a specific relation while the undefined space shows six sides loosely connected and this loose connection attributes to the absence of definition. The defined space has always seven points attached by a centre and such a centre holds singularity as the attachment. There are always six points evenly aligned running 90^0 as well as 180^0 in relation with one another and the crosses at a centre were the

All connecting centre of individual connecting lines between opposing points

centre includes and presents singularity. All defined material is presented by an enclosed space that always uses Π as a defined value whereas space outside the enclosed area has little defined connection and no structure strength.

By reducing the radius in half, the size of the sphere would reduce. The reducing of the sphere must continue while the radius keeps halving every time. A point will arrive where r mathematically cannot reduce any more. The radius factor at that point then has to be in infinity $r^0 = 1^0$ since it cannot reduce more than it has reduced.

In order to build a circle, we have to increase the radius. If the Universe was the size of a neutron at one stage and the Universe grew into what we now see, then the Universe extended r quite somewhat to get to where we now are. To go back in time we have to reduce the circle by reducing the radius. Keeping these factors in mind it is clear that Π is forming the factor taken on by form and r^2 produces size. However, Π is connected in form to r^2 but is only connected.

Reducing r will reduce the circle but it will not remove the circle. However, if the mathematical proposition is correct and the circle starts with zero, then $\Pi r^2 = \Pi 0^2 = 0$, leaving no circle to grow. With that principle in mind, it would be impossible to find zero in the centre of a circle. What we must find there is $\Pi^0 r^0 = 1^0$ = singularity.

Where space comes into contact with the sphere, the cube loses one of the six dimensions it has, to the more dominating seven dimension of the sphere whereby the seven dimension in equilibrium will dominate the six dimension loosely connected by r bringing about that the cube then has 5 sides to the seven of the cube. This means that in the cube the "bottom falls out" and without a "bottom" to support objects, they fall to earth. Remember that a body "floats" in space, but at one specific point it starts to "fall" to the earth. That is gravity and it is a dimension change much more than any force. I shall explain this last remark later on.

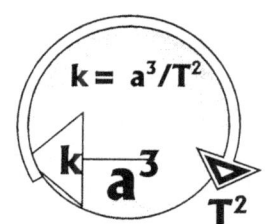

Kepler said $a^3 = T^2k$ but that could also be $k = a^3/T^2$

What this says, is that the gravity influence does not end on Earth. Although specific borders are in place in the atmosphere at various levels, the beginning and the end of such borders are in place in the atmosphere where the beginning and the end of such borders are as definite but also as vague as the gravity that forms them.

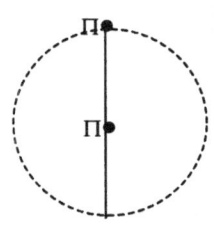

When translating Kepler's mathematical expression into a verbally spoken form of communication such as English, we can see what Kepler said also reads as $k = a^3/T^2$, where **k** is one point from a centre point that is space a^3 relating to time T^2. From a centre comes space-time.

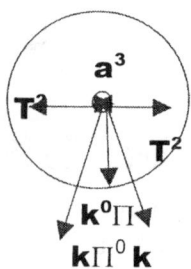

Ever considered why a water drop would freely choose to form a sphere? When a water drop is released in an astronaut's space capsule in outer space, the water drop forms a sphere as it floats in free gravity. As soon as the water drop is released from the Earth gravity and has the opportunity to float in space, it can form any shape it finds pleasing, yet it immediately turns to the shape of a sphere. It is the same reason why we would think the Universe takes on the shape of a sphere, although we know the Universe has no outside and we realize that the Universe is limitless in size. Yet, notwithstanding, we take the shape of the Universe as naturally being a sphere that formed … but why think of a sphere that the cosmos considers as the sphere

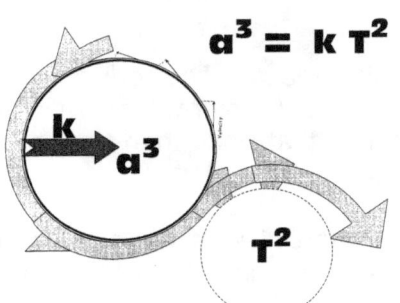
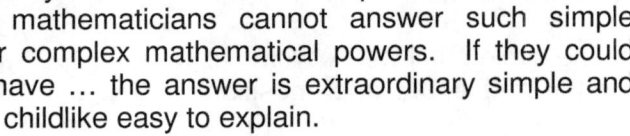

being the pre-cast shape? Well by saying it is the strongest form available, has the same argumentative potential as saying a baby is little at birth. Blaming it on no reason substantiating proper evidence that would indicate some facts more prudent than just a simple answer to dodge the question by hiding obvious incompetence. I'll give you one clue … it has something to do with finding the centre of the Universe. That brings on the next question being where one might locate the centre of the Universe.

These questions put to your are most ordinary questions … and yet the complexity in the answers rise far above the answering ability of those with the supreme mathematical skills. Those master mathematicians cannot answer such simple questions by using their complex mathematical powers. If they could have, then they would have … the answer is extraordinary simple and childlike easy to explain.

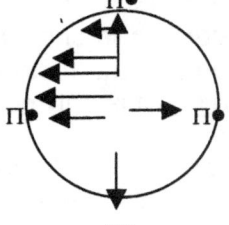

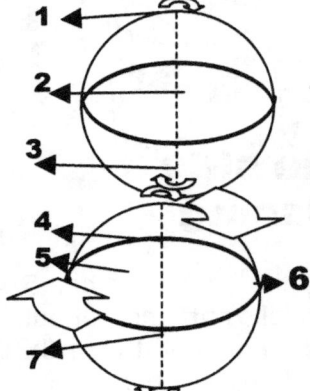

By examining the form of the sphere, we find that there are 6 points on the surface of the sphere that is holding the form at a specific and equal distance from the centre. Lines run from the centre into space at 90° and 180° angles of each other from six opposing sides. There then are six lines at 90° and 180° connecting to the centre from six points on the outside edge of the sphere. As a result of the basic shape that a sphere has, there is a spot in the extreme inner centre of the sphere where the lines in 90° relevance cross each other and others connect by 180°.

There is also at that point a spot where all space relinquishes a position and only singularity 1^0 as form remains. At such a point we find the measure of the sphere being Πr^0 with $r^0 = 1^0$. That is where the line that represents the radius as a line disappears, as it becomes singularity r^0. After more reducing continues, we get to such a point where we find only Π^0 left. At that extreme point is where space in all form disappears as the circle providing the sphere the form the sphere has, removing all possible form by going into singularity $\Pi^0 = 1^0$.

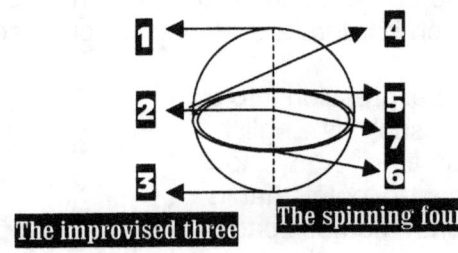
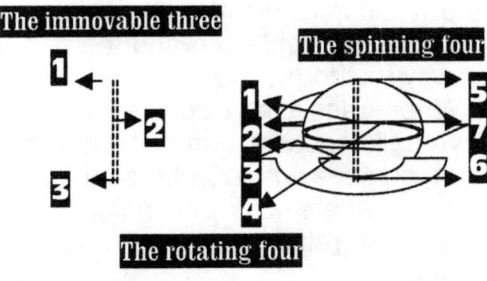

Then in that area all form of any possible space disappeared leaving only the dimensions of singularity 1^0. This too, I take much further in the book as I delved deeper into the argument and by doing that I stumbled on the ingredients forming gravity. However, from such a point there runs lines that connects to space on the outside where six points on the outside points connects to the space less point in the inside. In the book I take this argument much further, but for now, I leave the argument at that. Those lines carry the structural straight that the sphere has where the other six support every one of the six by singularity.

Where there is no space, there must be singularity 1^0 because the space is present although in singularity 1^0. If zero were a factor where all space finally halted in zero as the value, then zero would be able to remove the space from the centre and such removing would continue to remove the space until all space was removed. It will finally abolish all space in the sphere and it would remove the sphere. Zero removes all possibilities of anything coming about. Since the sphere is there, a zero factor in the centre cannot be present. Only infinity can be a factor from where space may grow because infinity can extend and grow into and up to eternity.

Where they are equal in value we must test the reason why this then is valid.

Locating and finding the presence of singularity

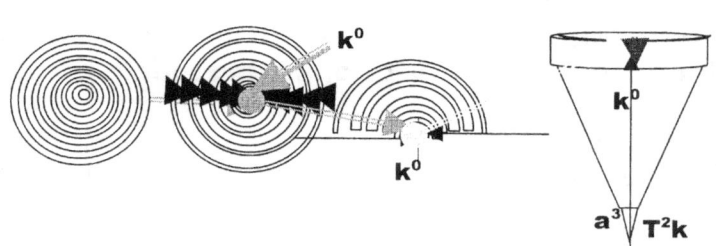

What is in the Universe is spinning. In the **precise middle** of all **objects in rotation** is a precise centre dividing the object in sectors that will **start the spinning initiation** from that centre point.

$k^0 = a^3 / T^2 k$ **states that whatever is, is also spinning in order to be present.**

Thus, the spinning object **will have a middle point,** a very specific **centre point that does not spin** and only holds Π as a specific value because no radius can apply. But also the one value such a line **cannot have is zero** because the line **is there and holds contact** to the rest of the material bringing about that **zero does not start any** line and therefore the **value of the line must be infinite,** just as described in accordance and by the definition of

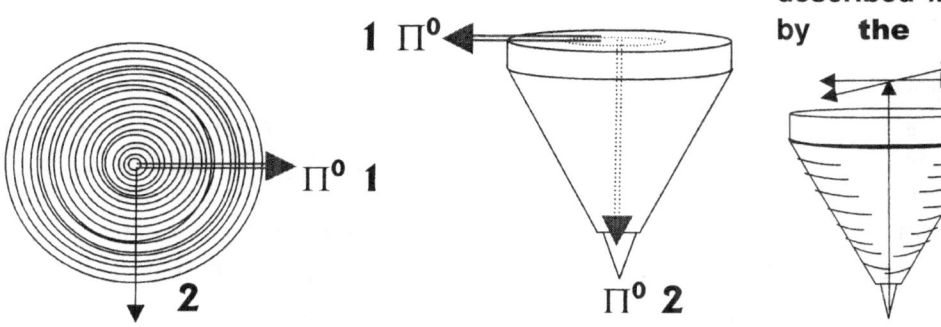

singularity. As I am introducing a very new idea, I wish to explain in better detail what I try to convey. While the top is spinning, one will find a line that formed in the centre where no line can form. It comes from spin but can never participate in spin.

That line must be singularity because if one moves any point on that line one position on, such a movement will land the point that then form on the line, on the other side of the line. The line is where the radius ends and starts because the line divides what is spinning in innumerable sectors and when reducing the radius progressively towards the centre of the spinning top at the centre where no line can be there is a line dividing the entire spinning top. At that centre point all further reducing must end because the next movement however slight will fall on the other side that is completely contradicting the one side. One movement further will change whatever is, so completely every aspect of that characteristic will contradict what it was before. There is one point that is neither left nor is it right but any point next to that point must be either left or right. The only value that point may not have is zero because albeit so small that it is not part of our Universe, still the point is there for all to witness and that point is a reality as much as the entire Universe is a reality. Whatever one attaches to the top either in the line of being material or a concept, such a concept or material has to start at the spot in the centre of the top because every aspect of the top changes in contradicting from that point onwards in all directions. **That** point albeit hypothetical, is also as much a reality none the less and is placed where that point **must be standing still** because every line **running from that point in opposing directions** is also **in opposing directional spin the other or opposing side.**

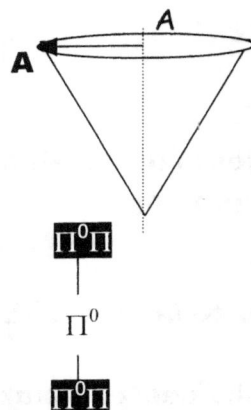

The moving of Π^0 to Π involved relegation and not motion as we consider motion. It was Π^0 getting a side and that is all. There was no true side but only a form that came into place. *Singularity (A)* received singularity (**A**) and no more of anything but the shift to comply with having a relevancy forming in relation to singularity. The dots had no sides, had no length or diameter. There was not measurable space or measurable time involved. The time could have been a micro, micro second as much a trillion millennium because time had no relevance. It was eternity interrupted by infinity, as it still is the case, however the line that eternity followed was no line because there was no space to hold the line. The line was momentarily interrupted by infinity, however with no one there, there was no one to notice. The lines were not lines but relations to sides being formed.

The relevancy that had the power to set Π apart from Π^0 is the only relevancy that still has the power, to set particles apart or join particles. It is heat in variation from cold. In order to excite singularity, singularity must establish a basis of heat that sets such a heat basis apart from cold. From there the form the atom will take on, however, the atom was still enumerable eternities to the development side.

As the rotating direction moves inwards, the rings holding Π will become smaller and smaller. The reducing of the radius r will eventually end at r^0 but the top does not end there because the top still then is Πr^0. The form we attach to the spin still applies as Π and the top finds directional contradicting change at a point that never moves because it can never move being in the centre where the spin direction ends at Π^0. **It is the only aspect in the entire Universe that can be and still be motionless because it is not within the Universe. It is the centre of the Black hole because it is the centre of the Universe. It is 1^0 and only the centre of the Universe in singularity can have 1^0** However that point where the directional spin ends is the point where the actual spin does not takes place because if its immovability. It is at a point in singularity $k^0 = a^3 / T^2k$. It is where space ends because motionless ness ends space there. The spinning is on the precise location where the point is not spinning because the Universe ends in its not spinning there.

That line running through the centre of the spinning top divides every possible side from the opposing side in innumerable points that are divided by angles and degrees. Moreover, it changes the future position of the point from the present and from the past as it redirects every point every time **k** moves to reposition in a location where T^2 ends. In the end it proves that both **k** and T^2 confirms a^3 just like Kepler stated before Newton interrupted with dishonesty.

Another huge factor that favours the use of Π as a measure in singularity progressing is that any expanding by any mathematically sympathising method will have to use Π since Π is the only route that a spot of no significance can develop into a dot that represents a Universe of development in waiting. Only by promoting through the measure of Π can all possible sides progress on equal terms in all directions simultaneously without bias. The development must progress by measure of equality to the smallest indication and that purpose only Π can serve. The progress must be generated so that it can flow equally to all sides in all directions spontaneously where not one side will favour the growing process as such. Time in it's flow does form a bias but explaining that at this stage will also involve too much other concepts which I would rather leave to the books. There is only one way to permit such a flow and

have a mathematically correct outcome and that would be using Π for such expanding. The use of Π would ensure that a dot rises from the developing that comes from the first spot. The dot would have to form a sphere and a sphere is Π in relation to seven. This bring as back to the Titius Bode concept where ten is one half of a sphere relating to seven points forming the sphere where the seven with singularity puts form to the double ten that totals (including singularity) 21.99999 and that is the overbearing dominating issue.

When the cosmos came into motion, motion was not yet defined. When the cosmos brought about motion, the first motion was relevancies. Cold parted from hot. Eternity parted from infinity. Motion parted from motion absence. Infinity broke the laboriousness of eternity for the duration of infinity. The spot became and grew into the dot.

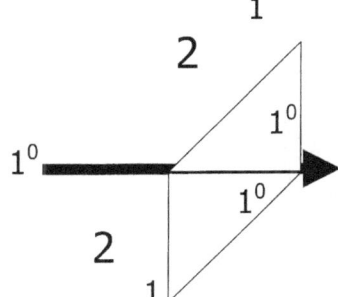

From what the spot was to what the dot now is might be just a mathematical implication of going from 1^0 to 1^1 but in reality that first motion was the creating of and establishing of an entire Universe which was with all possibilities that now is it. Never again can that much growth become a reality, although to us the growth is beyond what we ever can notice. But it is because the growth is so massive and we are so small that we are unable to notice such almighty growth. When the spot $Π^0$ became functional and established all relevancies possible, heat parted from cold as eternity parted from infinity. The expansion was not clear motion but more a parting of relevancies where a centre formed a relevancy because the centre could not provide motion. Without being capable of motion, the centre established four points, which also served singularity. From the inverse square law we know that the centre doubled by producing the four points holding singularity. We have to presume there is a time line because the Universe has this as evidence. The fact that light travel from there to here and from here to there proves of such a time line because there is no distance in outer space except in the Newtonian's misconception they have about the cosmos. Any line shows direction and the direction implicates positions according to the line having dimensions in the Universe we have in space and time. The line brings in Pythagoras and Pythagoras implicates mathematics.

1^0 going 1^1 where $1^0 \Rightarrow 1^1$

If there were progress that developed from singularity in the form of the first spot and we are the evidence of such progress, then the mathematical conclusion must be that a line formed where the line developed two sides and we have the evidence of that still present in our Universe. That brought about that three markers formed in relation to one and by admitting to the law of Pythagoras we find that what formed was 3^2 in relation to singularity 1^0 that became 10 on the one hypotenuse and 10 on the other hypotenuse which forms the square of space. Therefore mathematically space has ten positions and material has seven.

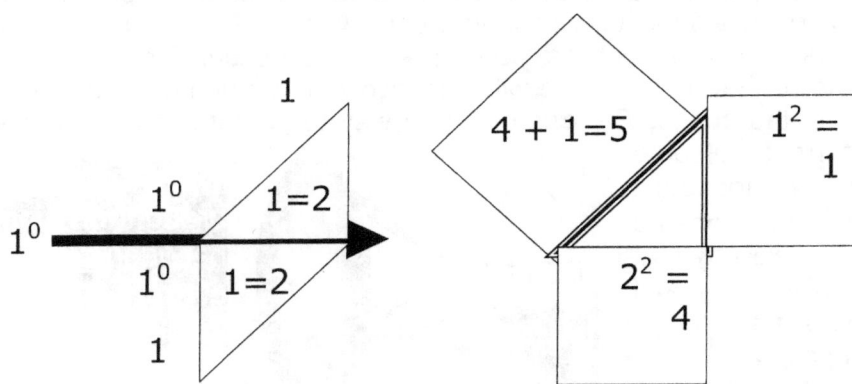

From the three the four (2 on both sides of singularity allocating the cosmic divide) the two in square developed as a mathematical consequence and that brought about the five.

The five was duplicated as a response on the other side of the divide and having five as a result of $(2^2) + (1^2) = 5 \times 2 + 10$ we find that the square of space holds the value of ten in place.

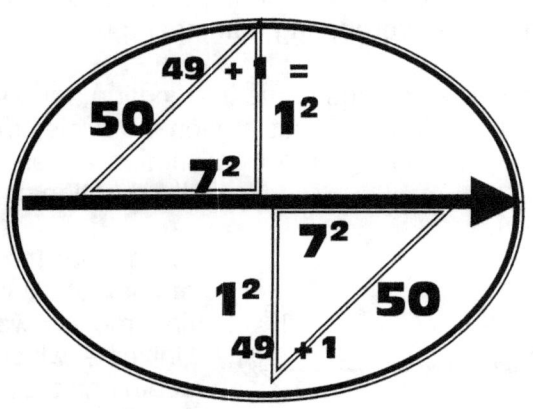

At the very beginning there was a spot. How do we know that? The spot is still with us and holds a value of 1^0. This spot is in the centre of all spinning objects. Then time came into motion and 1^0 moved to 1^1. Since from where we stand we see 1^0 and 1^1 as being the same and therefore with the moving of time in the very beginning such moving must contribute to an increase of space. Therefore 1^0 and 1^1 has to have a difference where the one is 1 and the other is one point in singularity smaller making the 1^1 coming from infinity and rising into eternity 1^0, making infinity forever one point smaller than what we find as the value one will associate with the one we find as a measure in infinity. Therefore we can judge that singularity combines to have a total of 1 + .99999999999999999999999999999999999999999999999, which then becomes 1.99999999999999999999999999999 or whatever because the one going smaller is running into infinity and since infinity is one less than eternity we are in eternity 1^1 looking at infinity 1^0 which is one point reduced in infinity. However this moving from 1^0 to 1^1 involved 1^2 as well as 1^2 on the other side of the divide. As a result of the form the sphere holds, there is a centre connecting the sides and the centre holds singularity. However, by presenting a centre where all lines cross on a point that cannot distinguish sides since that point has no individual sides, the centre holding singularity is inactive. Motion makes it active and the motion of space in time activates singularity to charge gravity that we find as a factor in the Coanda effect. That motion that establishes the purpose of space a^3 as a result of motion k

through time T^2 was what Kepler presented as a formula. Gravity is $k^0 = a^3/T^2\,k$ and to install k^0 the motion of space-time a^3/T^2 is required to complete the task.

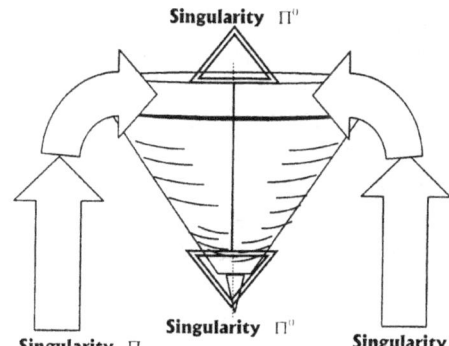

In the Roche limit the **straight line** forms part (1) and the **half circle** is part (2) and the **triangle** forms part (3) to singularity (4) Holding 5 points outside singularity

By rotational motion, the top creates a line confirming singularity running down the line and by generating the line the line charges gravity. The gravity is what drives the top as the top and as long as the top spins. There is an influence generated by the spin of the top that keeps the top upright while the top is spinning. The line is generated but the line is far from magic. The line is where the centre of the Universe is which the Universe is then that what the top filled by particles from the line to the edge of the sphere. The particles in motion generate motion by electing a centre from the centre of every particle in the spinning top. Such an elected centre becomes the centre of the Universe as far as the top relates to a Universe because all the atoms in motion elect the centre of the Universe.

When the seven occupied by material rotates, it removes the ten in contact from influencing space-time and by rotation introduces a totally new ten points in relation by the act of rotating. But such a process also comes about on the other side of the divide because it is the divide that cannot rotate and forces space-time to comply with the frequency of breaking down space-time as space-time releases from one side and if necessary shift or only apply new rotation alliances as the seven points become a part of the new ten.

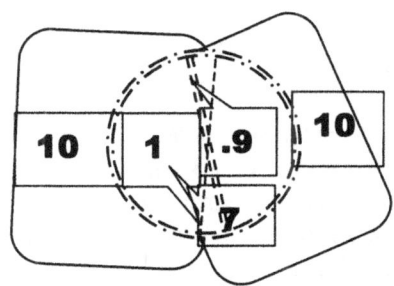

The motion duplicates space (10 + 10) by rolling over singularity (1) from the one side by reducing space (.09991) on the other side. All this motion of space 10 duplicating is directly related to matter 7. The total value established from this motion is the interaction of seven with ten producing Π^2 as the half of singularity $\Pi/2$ on the one side interact with half the singularity $\Pi/2$ on the other side and in this process Π the duplication of Π by means of matter and space produce gravity in the measure of Π^2

The conditions proving singularity is that the circular form will produce a centre point from where gravity will dictate the reproducing of space. The gravity part is the fact that motion must contradict the centre point around which the motion will produce space. The space part is proven by the motion that produces a running line of space created and followed by the liquid producing the space in the motion of the liquid. It shows that relevancy comes from space shared by space and motion separates space shared by space. It shows gravity coming from motion separating shared space.

When further consuming is not possible and when finding other singularity that challenges such consuming, the Roche limit comes into effect.

In this, it is clear why the Titius Bode ([10 + 10 + 1 + .991] / 7) and the Lagrangian 5 \\ 1 systems part their ways when applying the different processes they hold. With all the differentiating, the observer must also consider the dual message that light uses in travelling through the vastness of universal space. The thought of nothing is just what it is, a thought of nothing and although it is in the human mind common nature to present nothing as a value in the recalling of something, nothing is a presentation of the figment in the human mind. There can be no number such as nothing and that was (possibly) Newton's biggest error. Nothing represents non-existing and that is just what nothing is, it is non-existing.

When taking Kepler's formula as a valid criteria we find that singularity indicates a sphere with seven positions $k^{0(1)}=a^{3(3)}/T^{2(2)}k^{(1)}$ and $^{(1)}+^{(3)}+^{(2)}+^{(1)} = 7$, which confirms singularity within the sphere but where space-time without the aid of singularity forms a value $a^3 = T^2 k$, it also proves the absence of a charged singularity not generated into a position of control outside the sphere $a^3 = T^2 k$ and where the sphere $k^0 = a^3/ T^2 k$ meets space $a^3 = T^2 k$ the factor k becomes the common denominator that is shared which leaves a ratio of seven to five ($a^3 = T^{2 (3 + 2 = 5)}$) where through motion one part of space (5) lands in the past and five lands to the future as well as five to the past leave a distinct space square of ten. To form space-time in terms of singularity the singularity holds 1.99999 in relation to ten on the one side of the Universe and ten on the other side of the divide where the total then is 20 + 1.9999 = 21.999 in relation to 7, which gives the sphere validity. Since the Titius Bode is only proving a relation from the centre to a point and from the point to a centre and where such a point can only be at one position on one side of the divide in all events only half of the space factor takes up the given value while the object claims singularity by (1^0 = .9999 standing in for the Sun 1^0 and 1^1 standing in for the planet 1^1 making the seven the rotating action of space-time forming motion as the structure of the orbiting planet. The Titius Bode is not proof of the cosmos being in motion but the cosmos being in motion serves as being proof for the Titius bode generating gravity by the measure of Π^2

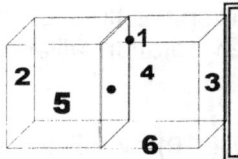

> **There is one more point in the sphere in the centre forming an addition in the sphere. That point holds gravity secure.**

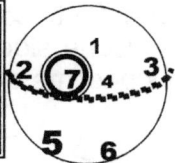

The motion that establishes gravity secures space by duplication thereof.

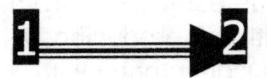

The only absolute constant we have in the Universe is that when something is part of the Universe it can go nowhere but remain in the Universe. That very first instant where the Universe came about is still applying and it is still happening because nothing can remove it from the Universe and where nothing might be a concept in Newton's head nothing is the only thing that is not anywhere in the Universe.

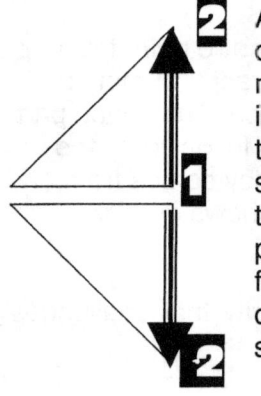

As time committed in moving along a line the rest of the motion was desiccated to produce space since a line is in one direction but that does not include the other six of the seven direction of space. One must keep in mind that singularity is the line producing the flow of time from the past through the present and into the future and therefore connects to the six sides by taking such connection in a change of position to the following time instant. On a dimensional level the flow of time progressed from a position of one to two but also it expanded into space from one to two. By flowing with singularity then it is singularity that takes the six dimensions of the sphere along and in changing newly allocated positions singularity secure the space to flow through time.

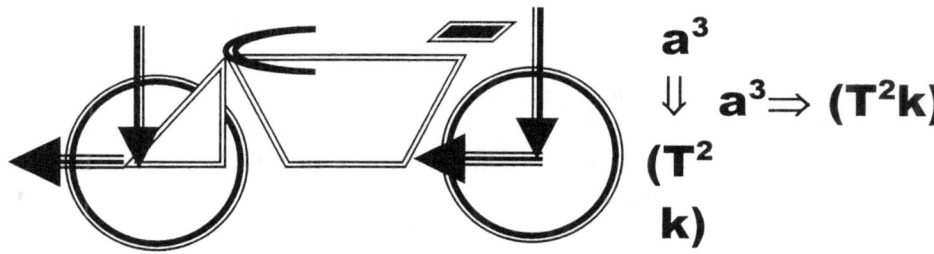

$$a^3$$
$$\Downarrow \quad a^3 \Rightarrow (T^2 k)$$
$$(T^2$$
$$k)$$

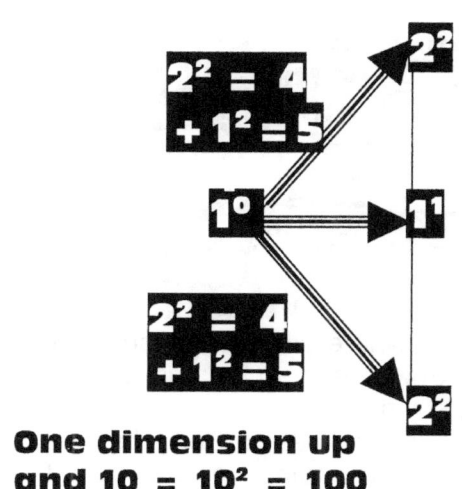

**One dimension up
and 10 = 10² = 100**

It is the movement of a^3 during the period of T^2 in relation to the relevant change in position **k** that brings about the movement. But while the bicycle has it's motion of $a^3 = T^2k$ the Earth too has its motion of $a^3 = T^2k$ it is in contrast to the bicycle's motion of $a^3 = T^2k$, which is at an angle of 90^0. In this lot mass has no part as far as cosmic relevancy goes. Yes mass is a factor but mass is a factor just because humans invented mass as a measuring unit in order to establish some common denominator to calculate but that also applies to distance as it applies to temperature as it applies to time. It is what humans may use to calculate human perception but human perception finds no valid grounds in the cosmos. It is because the bicycle time in motion exceeds the time in motion the Earth insist on that the time in motion of the bicycle find a means to keep the bicycle moving and upright. It is time in motion coming from the past, going through the present and onto the future that keeps the "momentum" higher than the moving of the Earth by duplicating the entire structure of the Earth from time in motion coming from the past, going through the present and onto the future which control what we find the Earth to be.

Newton claimed that GMm/r2 = m (ω^2 r), which I guess in itself is good and true. Then by

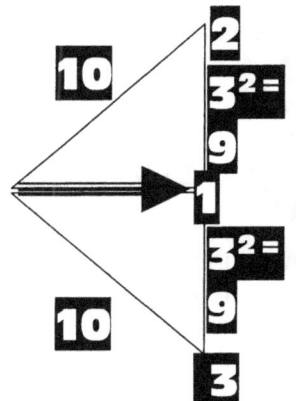

replacing (ω^2 r) with 2Π/T we get $T^2 = a^3$. That is rubbish because for one, one cannot exchange (ω^2 r) with 2Π/T because (ω^2 r) clearly is $\Pi^2\Pi$ (moving from Π to Π in relation to Π) or relevancy T^2 **k** or r where one must remember that it involves movement and movement enforces a triangle because the motion depends on a centre. The fact that Newton put ω^2, he clearly admitted a centre and the movement T^2 of a^3 is clearly by distinction of **k.** One cannot substitute that which is in a square (meaning a flat dimension) with a double Π. What was the man thinking and was the man thinking at all? He then corrupted Kepler's formula to the tune of putting $a^3 = T^2$ or $T^2 = a^3.$ That is not what Kepler said! Kepler said $a^3 = T^2$ at the rate of **k** or if you will at the distance of **k.**

If he persisted that **k** has a value of 0 then the space was zero and the time was zero because in outer space distance **k** has no relevance because **k** is a result of time and there is no distance in the Universe. Distance is a measure we on Earth attach to an unknown, which we wish to calculate. Distance is manmade **(k= a^3/T^2)** and **($k^{-1}=T^2/a^3$)** because the cosmos devised time developing instead and mass is

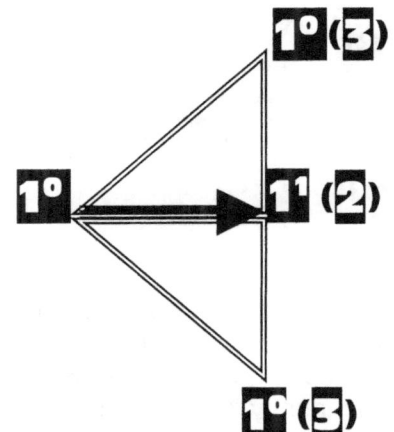

manmade $a^3 \neq T^2k$ because the cosmos devised gravity $a^3 = T^2k$ instead while temperature is manmade $T^2 = a^{3/}k$ because the cosmos devised time instead. The distance between the Moon and Earth the fluctuate while what Newton suggested was that $a^3 = T^2$ always proved that the time factor we see as distance which is using the symbol **k,** is in fact a time continuance that always remain the same. That is the measure of time because the rotation of T^2 puts the space a^3 in a relation to the duration (distance) of **k** and where the rotation of space is $T^2 \times T^2$ that motion would duplicate space in its full contingent a^3 by the measured relevancy (distance that time progressed or developed) **k,** and therefore the Big Bang came about by $T^2 \times T^2 = a^3 \times k$. In that regard we find that $(T^2 \times T^2) / k = a^3$, which is what the Lagrangian point system is. Time moves one point away from the value of $4\Pi^2$.

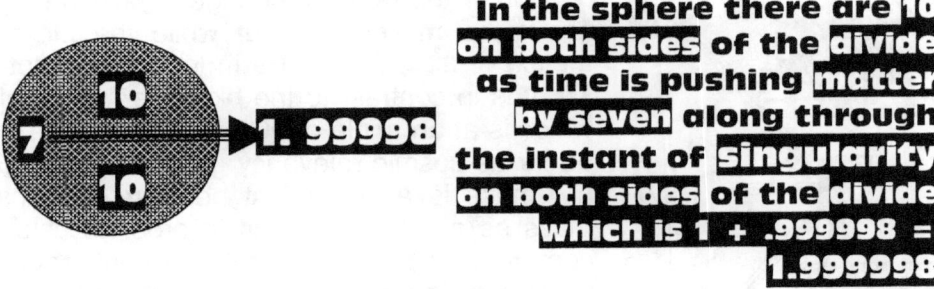

In the sphere there are 10 on both sides of the divide as time is pushing matter by seven along through the instant of singularity on both sides of the divide which is 1 + .999998 = 1.999998

However the following explaining involves the smallest area where singularity immediately commands space to motion. It is where Einstein found the Universe to go flat and where the smallest form of material functions. It is where space is time by mathematics.

We are in the zone that is still active because it is where material increases. One thing that is another part of the Newtonian dysfunctional concept about the expanding Universe is that in the Universe there can be no expanding at all.

Every person agrees that the Universe is as big as it can get and there is no limit to the Universe. The universe has no end and has no borders because the Universe is everlasting. So how can a thing that is eternally big and has no ending and presents the eternal entirety of what there is, still grow bigger and become more! How can that which has no limit still grow into something with more limitlessness? The universe is expanding by shrinking into the oblivious. That what was previously too small to be a factor becomes a factor by the shrinking of everything becoming relative to the reducing of the space. If the Universe is truly limitless on the outside then the only limit must be to the inside and it then must be in that direction that the growth of space is flowing. I explain much more of that in the book. The growth in

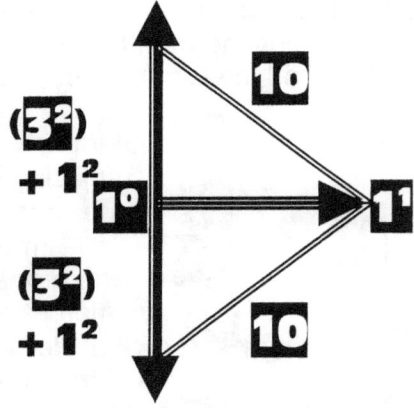

motion concerns the scope where no Newtonian can see and that is why they have no evidence of smaller aspects growing. It is because they simply can't see that. To apply motion we have to start at the smallest there is because the smallest has to shift as a unit of independent structures all working in a coherency in order to move. Motion is the most complex issue there is in the entire Universe but motion involves not the biggest as a pretty lump but the smallest acting together as a group to eventually participate as a lump of material in time.

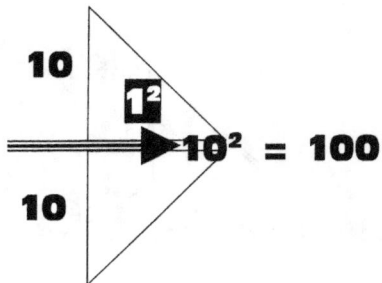

To understand the process that involves motion we have to argue about where the smallest are and that will be just outside infinity because infinity is the part of the cosmos that has no start leaving eternity to the part that has no end. Space-time is the space filled with time that parts infinity in singularity (1^1) from eternity (1^0) and where the smallest is there is that is the growth of the space-time. That is where gravity is. It is where space disappears into motionlessness and not where the Universe draws flat. Gravity is where the smallest particle meets space less ness and space disappear into singularity because singularity cannot

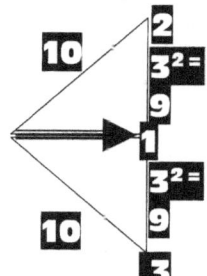

move. If one wishes to locate gravity it is there that one should be starting and not at some stupid mass idea that resonates from the dark ages. It is where space starts because it is where space disappears. When the first line moved (I choose to refer to time in beginning because although it still apply it puts a dimension to the occurrence that stands as correct because I am in time delay many moons away from where that which I refer too, is taking place at the present time) it moved by acquiring three points on one line and the one line was on one side of the Universe but since motion is the duplication of what is at present from the past into the future, it has to be on both sides of the divide in a cyclic manner.

Therefore it was three points (1^0 going 1^1 on the one side and coming from 1^1 on the other side) as it involves three 1^0 to 1^1 and 1^1 and since motion is a triangle by dimension the triangle implicates Pythagoras in the square which is $3^2 = 9 + 1^2 = 10$. As this is applying on both sides of the divide the total shift by dimension is in cyclic perspective 10 X 10 = 100 but in shifting along development it is a shift of 10 + 10 = 20. While this is in the shifting that singularity develops it is 1,999999 plus 20 and that is a sphere. This is happening beyond the dimension of the proton where the motion is still beyond space, as we know space is. On both sides of the divide we find that 10 X 10 = 100.

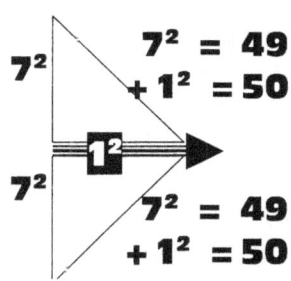

Also directly linking mathematically we have seven in the square plus singularity in the square in a Pythagoras motion association forming the triangle where then the hypotenuse is valued at fifty and on both sides of the singularity divide it totals 100. There is the connection in motion. The sphere holds seven and in the square with the adding of 1 in the square we have fifty on the one side and fifty on the other side.

In motion the square of seven fills the square of ten and in that where space has no legal status all motion is by mathematics because all motion exceeds the speed of light by a measure of many light years on end.

Material in the sphere holds the measure of seven and in seven circling the motion has to instate the triangle where the triangle enforces the use of the law of Pythagoras. That is as basic mathematics

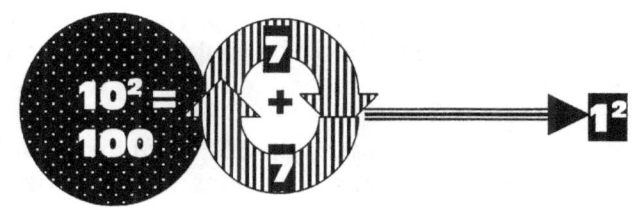

as one may get and more simplistic than this there cannot be. On the side of space the motion comes from the five sides in the past making contact with the sphere by the five sides in the present and continuing into the future by commanding another set of five sides.

In that manner we find the curvature of space-time honouring the value of Π. This is where the Lagrangian system becomes part of the space – time that generates gravity. There is four points in a circle that is completed and the fifth point is where time progresses to a new location. It is one fourth of the sphere in space where time completed

forms the other four to come to a total of 21.99999 to the seven that forms the material sector as the sphere. The motion involves the dividing of space into units comprising of six blocks where the material annexes one side leaving five sides to act on behalf of all material because where the material connected to space, that zone is not defined in connection as the material is by having specific borders and edges. The area that the five sides cover goes on indefinitely limitlessly and with no borders to form limits or endings that might remove a certain part as an excluded area. The two blocks on one side becomes a double of five, which are ten and the same value of on the other side also of applying in the same manner.

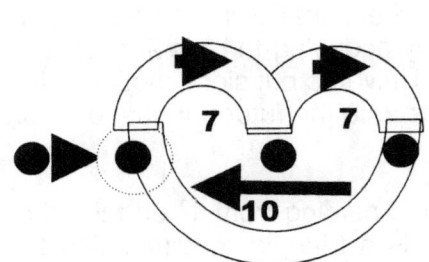

In absorbing one in four duplicated points in singularity, the centre singularity shows growth by dismissing in doubling the adsorbed heat. The centre established a density growth. One must keep in mind that space was not yet an issue there fore the half circle was equal to the straight line. Whether the growth was seen as a half circle or whether it is seen as a straight line it is the same thing at that stage.

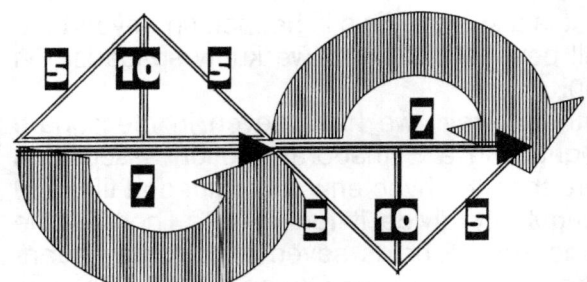

That results in the sphere and by the pin of seven points running through ten on both sides of the divide the sphere takes shape as material duplicates. Again I wish to remind that we are dealing with the very most inner space of material where space-time forms an apparition of mathematical proportions. That is where the duplication results in material and

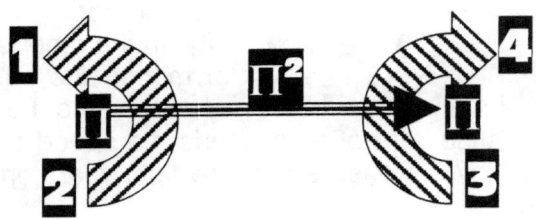

material is the result of time delay of heat that is confined to time in space. This is where the Universe draws flat. This is where time is as current as it can be. It is where gravity keeps the entirety connected to reality and the time delay is reversed.

The fact that space takes up 50 % or fifty positions in one hundred and material designated to singularity takes up 49 positions with the one in singularity and that makes the Titius Bode law the corner stone of the Universe being secured by gravity. Every aspect of space grows by the seven expanding into ten and ten reducing into seven. It puts the square of space at $10^2 = 100$ where matter that entertains infinity as it serves infinity occupies half of the square of time while time in space $10 \times 5 = 50$ holding eternity, serves material with the time within which to move. Half the Universe is dedicated to infinity while the other half is dedicated to eternity

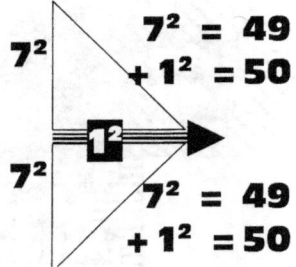

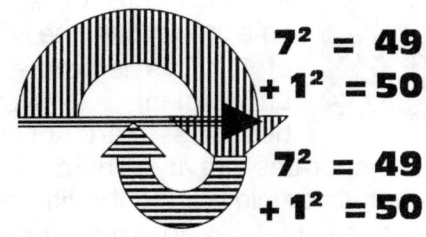

Material with seven points forms a joint value of doubling on both side of the divide relating to the filling of space. It is where the sphere fills space totally not because a square can hold a

circle and have no space left but it is where 50 + 50 can fill a hundred and have no further figures unfilled. The filling of space is a mathematical consequence of the cosmos duplicating what is within the cosmos at the time at that point in time. Yet there is no way I can describe in this little space available the full extent of what motion involves because every spot is singularity and singularity find value by performing as a point in time delay.

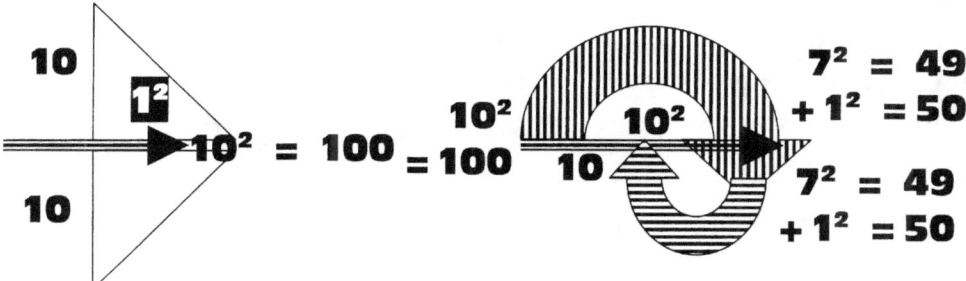

$$7^2 = 49$$
$$+ 1^2 = 50$$
$$7^2 = 49$$
$$+ 1^2 = 50$$

$$10^2 = 100 = 100$$

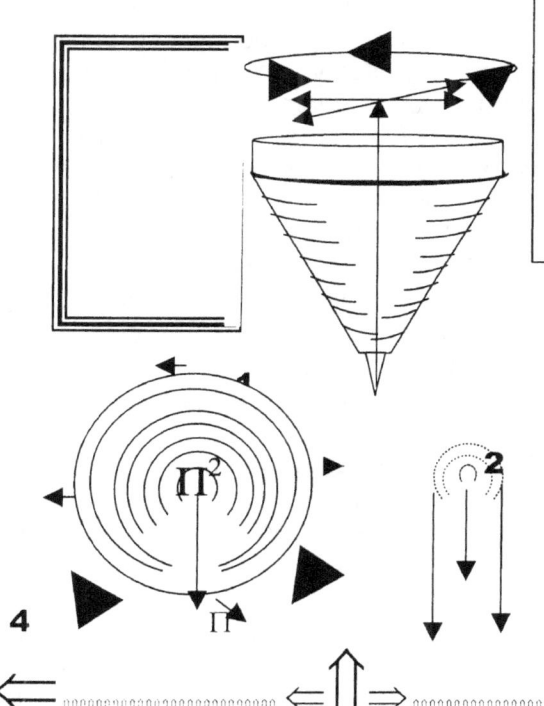

The sphere has seven points. The cube without truly being a cube but is just in consideration of having a cube in form holds five points to singularity. In the centre runs singularity to the value of Π^0, which means that which surround Π^0, holds a position of Π^2,

The spinning sphere activates the seven points, which places gravity in relation to a centre. Outside the centre there are five sides by dimension. The sphere has seven points of which four is spinning. The four spinning stands related to the gravity of spin, which are Π^2.

$$\Pi \ \Pi^0 \Pi$$

Through motion singularity forms an infinite value sustained by seven positions forming the circle in the cube through which the sphere moves.

1^0 going 1^1 1^0 ▶ 1^1

Since the square of space is directly involved with the dimensional depleting of space from ten to Π^2 and singularity divide the affect of gravity being Π^2 into two sectors forming one double of the four quadrants of time, the affect of Π^2 may not be ignored since that totally dominated gravity from even before the establishing of gravity in the form of the Roche limit.

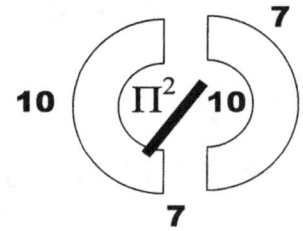

$$A1 + A2 = .49 + .49 = .98$$

$$A1 + A2 = 5.033 + 5.033 = 10.066$$

The result is the dimensional dismissing of space from a relative 10 to Π^2.

To get space to move, the time delay activates and charges a common point and the

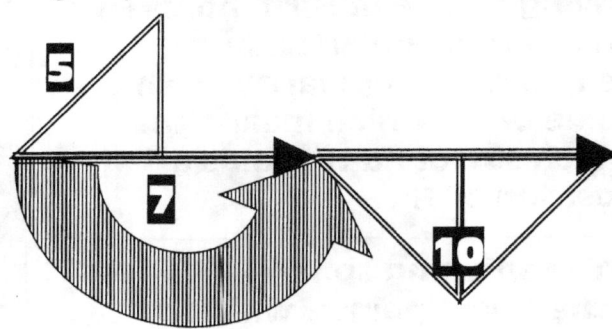

common point respond by activating time within singularity in the centre. The common point serves as the stabilizer unit around which the spin of gravity is generated that provides a spot from where space can serve material as time and material can be within space serving as time. The point holds gravity Π^2, where the gravity divides the time into four points as the four points are generating the time delay. Every point holds a spin and from the spin the duplication forms a relevancy to what is duplicated but also in as far as the total unit that is duplicated.

The unit charges a single point that serve as the divide of all the duplication that comes as a result of motion. That charging of a centre by the motion of all the atoms within the top and the centre being able through motion to generate gravity is what keeps the top in spin. That is extremely briefly explaining the essence.

It is the amount of spinning material in motion within the unit that forms the unit in motion, which is duplicating as a unit that determines the rate of duplication of the entire structure.

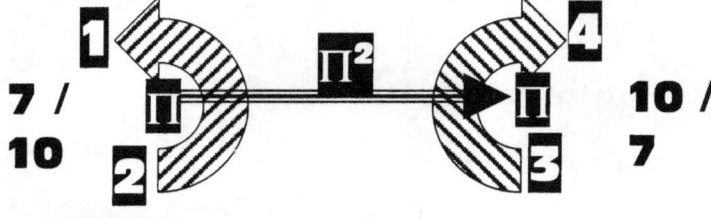

7 / 10 that is interacting with 10 / 7 across the divide or the Roche limit at Π^2 / 4

The Sun would duplicate much more frivolously than the Earth and the Earth more that the bicycle which puts the three in different time domains. The amount of atoms spinning together as a group renders the motion that is carried on to the singularity that the group is generating.

In it self the points are serving as singularity and the motion is a result of generating singularity by gravity to an elected common centre. Every point activates to fill a role as the centre is excited in being generated for the task. Every point is there as every point is time delay or time lagging behind. As space moves the singularity established in the centre is static.

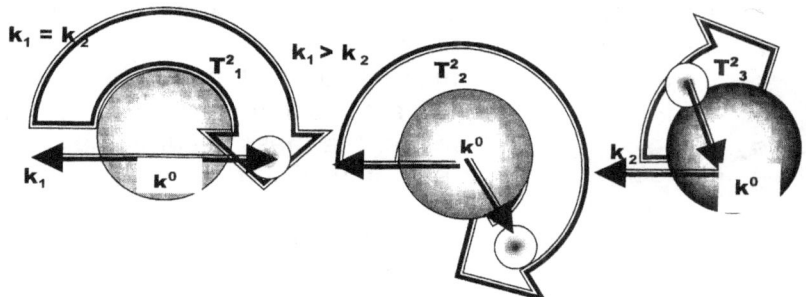

When the top starts to spin eternity is divided from infinity by such a spin. The spin establishes a cosmic identity by parting that which has no start from that which has no end and the parting of singularity in the two forms is what keeps the top in an upright stance. Through the motion the top may serve a centre in singularity that the motion charges into service. This motion identifies a Universe by attaining the standard of a Universe, as it establishes space-time by motion of space in and through time or have the entire Universe

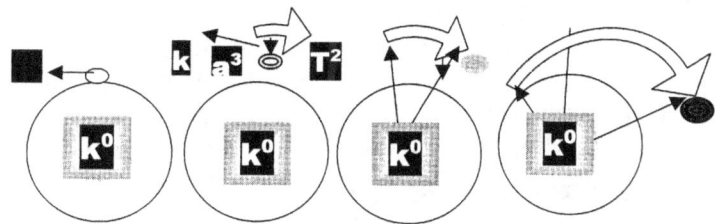

being the top in representation thereof while spinning or when not spinning have the top form a Black Hole where that Universe which was the spinning top, then by not spinning, allow the entire top Universe that was, to collapse all together.

In outer space the bicycle will willingly orbit as a satellite as long as the orbit speed supersedes that of the Earth. This has nothing to do with mass. The second the orbiting speed lags behind that of the Earth orbit velocity, the satellite bicycle will fall to the Earth and that too has nothing to do with mass. Galileo proved that when falling the bicycle would show no indication of mass influencing such a fall since all object descend at an equal rate. Again the proof is on the velocity discrepancy and has nothing to do with mass. The small object will fall as easily as a large object when speed discrepancy allows such a fall to commence. Not at any stage does mass come about to render the object with more mass superior in any way or to be at a disadvantage in any way. As one can see an object such as the Earth in close proximity has more duplication as that of what a natural bicycle in orbit would have, therefore both objects in the duplication process must set a trend to match the duplication that the Earth establish or become a part of the duplication process of the Earth in the event where such duplication of a less prominent object cannot sustain the relevance which the

$$k=k^{3-2}=k^1$$

Earth in total establishes.

By being in motion there is no evidence that such differentiation is brought about by mass

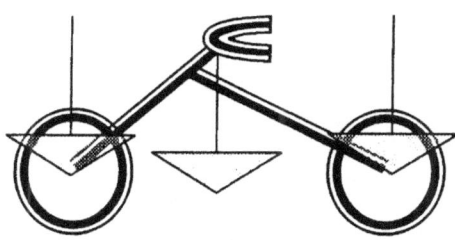

being a factor but everything depends on the relevancy of duplication or sustained orbit velocity that establishes the trend or relevancy of motion applying. Where duplication does not presume in relevance a discrepancy in relevance in terms of motion and not mass will bring about that there has to be changes in the relevancy that the motion provide. Mass is no factor to be descried anywhere in the whole assembly of motion creating space by duplication of space -time.

By accepting the mass that the Earth offers the dropped object and the relinquishing of independent motion, the object is offered a far better duplication by being part of a larger unit and the duplication as a unit is able to sustain a larger relevancy in time in the period of spin. The bicycle becomes a part of such duplication where it absorbs the duplication by being part of such duplication. The motion persists as equal to all until it hits a blockage where mass

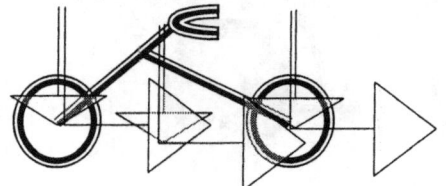

comes about and there the equality changes to identifiable differentiation as a result of mass. However the mass secures the objects new identity as part of the Earth.

Then while being on Earth and with a desire to move above and beyond that what mass killed, it then has to establish a higher rate of duplication than what the Earth provides in duplication to find

such an ability as to move with and in mass. Only life can provide that feat but life is only on Earth.

In order to move while in mass it requires a source or a supply of energy that will extend the duplication of its atoms by increasing the accumulative heat they assemble. Under

normal cosmic conditions that is not possible and the only place known where this might take place is on Earth where the qualities that life may render can accomplish that. Mass has killed individual motion but with the aid of life and where the object has such motion inherited

from the Earth life can
the bicycle time displacement
make use of the qualities of duplication that the Earth provides

and extend additional motion to the already inherited motion the bicycle receives from the Earth as being with mass part of the Earth motion. Considering the number of atoms the bicycle as a unit has to reproduce to substantiate its motion in space through time and comparing that to the space and time involved where the Earth perform the very same task one can see the relevancy that applies to the duplication of the Earth and that of the bicycle. Just duplicating shifts the Earth through time by an innumerable larger relevancy, which is completely beyond the duplicating possibility of the bicycle, where the bicycle has to rely on individual duplication and therefore it is very unlikely that the bicycle can sustain an even greater pace than the Earth does. The space a^3 that applies to the Earth demand so much more generating of gravity in the circular factor of T^2 as well the linear relevancy k and it is clear that the Earth's relation to time will overwhelm and overall the bicycle's time component many times over.

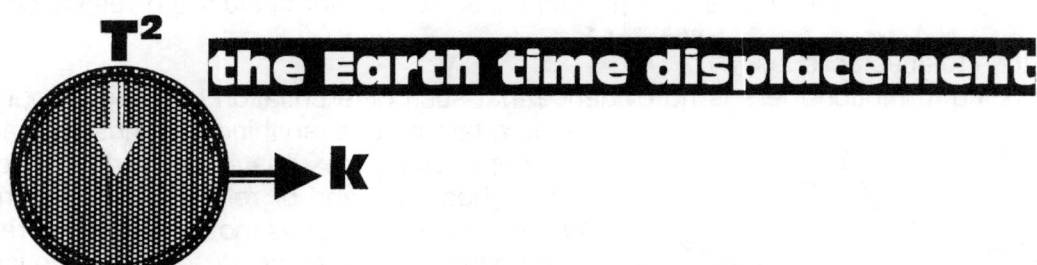

the Earth time displacement

When considering the layout of the sphere and the fact that both are just opposing factors in the same sphere the one would regard the other by 90^0. What is T^2 to the lesser will be k to the more progressive because in both cases a^3 is directly the linear factor to the other a^3. As soon as the Earth relevancy k supersedes the orbit motion of the bicycle T^2 the factor k will override the motion in the relevance factor of the bicycle and the bicycle will start to "fall".

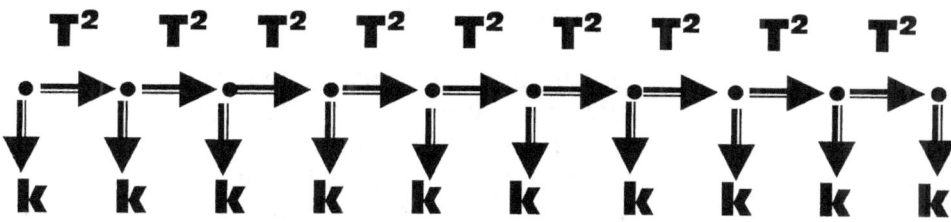

The falling is the fact that the superior relevance of the Earth by factor of **k** will overall the motion of the bicycle by T^2 whereby the Earth factor **k** will render the bicycle factor **k** of no sustainable value. At that point the Earth relevancy factor **k** annexes the bicycle motion but the bicycle still has no mass. The bicycle only starts to capitulate its time factors in favour of the total domineering Earth time factor and the Earth take position of the bicycle by supplying a time factor the bicycle in orbit cannot compete with.

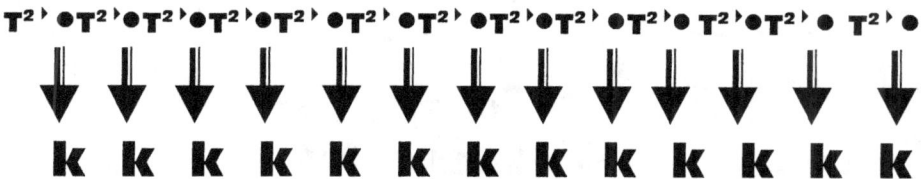

The Earth then strives to liquify the bicycle by the measure of $\Pi^2/2$ and the bicycle loses its time factor of T^2 in favour of accepting the Earth T^2. The bicycle still has no mass but the motion the Earth renders to the bicycle leaves the bicycle totally dependent on the Earth to provide motion in the time the Earth domineers.

When the bicycle capitulate it's motion and completely accepts the motion that the Earth provides, the bicycle will continue at a descending rate of $7(3\Pi^2)$ notwithstanding what mass Newtonians may connects to the structure. If that is not the case Galileo is completely wrong and Galileo is everything but wrong.

After landing in mass can the bicycle again move but only by the intervention of life will the bicycle again find motion and all motion thereafter will extend the Earth motion $7(3\Pi^2)$ by an extending factor of Π^0 to $5\Pi^0$. The motion will go from $7(3\Pi^2) \times \Pi^0$ to $7(3\Pi^2) \times 5\Pi^0$. This fits any and all criteria notwithstanding mass of what magnitude is involved.

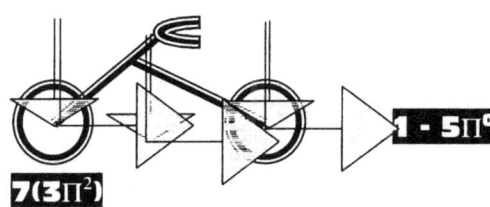

When the density caused by motion exceeds a limit of $\Pi^2/2$ in relation to the material the ratio is so thinly spread the air cannot even support sound waves travelling through the thin air. Again all of this is relying on motion, which puts a relevancy between the moving material and the space, or the time that it moves thorough.

When mass is a factor such relevancy is predetermined by the Earth and the ratio the Earth maintains with air or space or time which it really is.

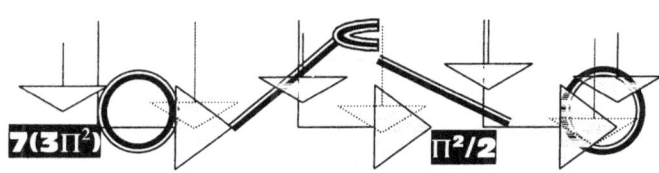

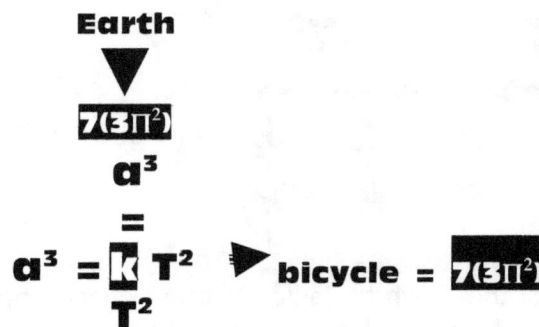

Earth

$$7(3\Pi^2)$$
$$a^3$$
$$=$$
$$a^3 = \dfrac{k}{T^2} T^2 \quad \blacktriangleright \text{bicycle} = 7(3\Pi^2)$$

However where life does create artificial motion because life is artificial motion we find the density between material and space redeploys new standards applying to mass. At a point where the bicycle has artificial motion to the value of $7(3\Pi^2)$ times $2\Pi^0$ (this figure does vary considerably in correlation with the structure profile and layout and therefore it is only a suggested figure given on my part) we find that mass will tarnish and the bicycle will at one point become airborne. This has nothing to do with wind rushing through underneath the car because to counter the air rushing through underneath the car there are even more air rushing over the top of the car so in that sense we find mechanical engineers are as Newtonian

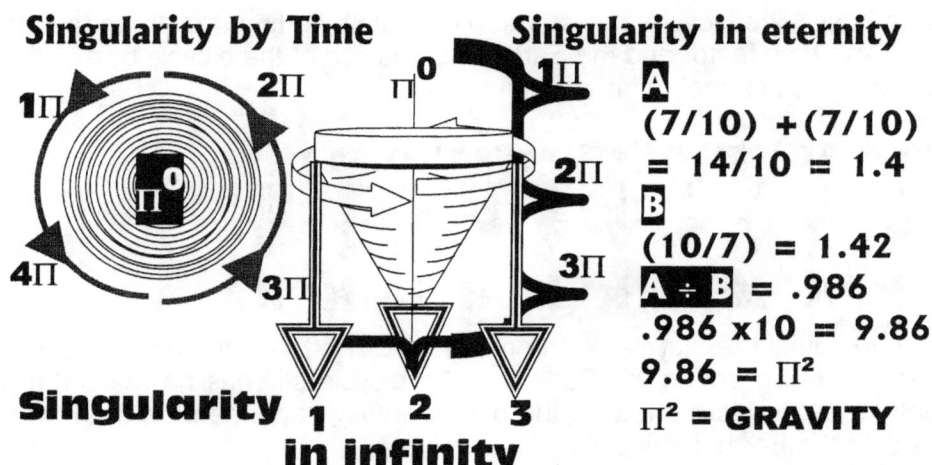

Singularity by Time

Singularity in eternity

1Π 2Π Π^0 1Π
1Π Π^0 2Π
4Π 3Π 3Π

Singularity   1   2   3
in infinity

A
$(7/10) + (7/10)$
$= 14/10 = 1.4$
B
$(10/7) = 1.42$
$A \div B = .986$
$.986 \times 10 = 9.86$
$9.86 = \Pi^2$
$\Pi^2 = \text{GRAVITY}$

as the rest are. The increasing in velocity will establish a changing density factor between the material and the air it connects to in a specific time period. More air makes contact with the same material, which makes that less material makes contact with the same space and in ratio with more air less space starts to fly. It is exactly why water skiing has a person glide on water and not sinks into water.

The moment the bicycle accepted mass it adopted the Earth motion and therefore all other motion is an extending of the Earth motion. That is why Galileo found all objects fall equal because when falling all objects fall in relation to the Earth singularity which, is $7(3\Pi^2)\Pi^0$ and mass does not yet apply.

When entering the Earth atmosphere the descending object must either adapt to the Earth motion and adopt the Earth or become liquid with a bang as Mir did. The Earth holds more

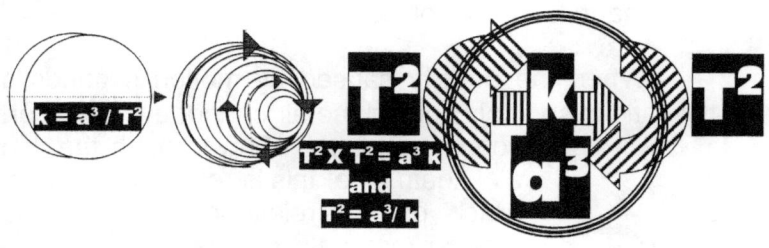

$k = a^3 / T^2$

$T^2 \times T^2 = a^3 k$
and
$T^2 = a^3 / k$

$$T^2 \quad \dfrac{k}{a^3} \quad T^2$$

atoms that has to establish more atoms from the past to the future by way of duplicating every point that is confirming the presence of material and since the task of the Earth involves more duplication than the bicycle has, the time in effect of the bicycle will have the Earth absorb such time if the bicycle of the bicycle cannot find a means to duplicate more ferociously than the Earth does. It is all a ratio with heat that science never detect and is even part of life. When we breath we breath oxygen but that also is oh so Newtonian because what do we do with the oxygen. We don't eat or drink the oxygen but we do take heat from the oxygen and it is the density of the oxygen carrying heat to our blood for use of our fibres that we need oxygen. We need the Heat the oxygen associate with to use to build our vessels and the building we call aging. Only motion can turn singularity from the sot it has in the sphere to the dot to form a line that takes charge of the independent object, which is in motion and is in relation to another centre. This confirms Kepler's $a^3 = T^2k$

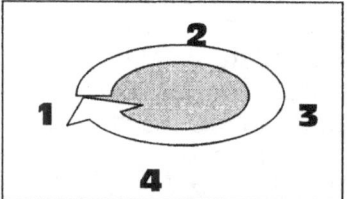

The time activates a line from which space begins. The line forms a space that turns the top, but just as important is the fact that the line extending singularity activates a dimension in time in which the top can spin.

It is the time that relates to the spin of the space filled with material that allows the position the top has when spinning. When the co ordinance between the space spinning and the timeline keeping the balance falters that the top looses independence and fall.

The issue is that the rotation activated the time component where four points in singularity moves about a centre point in singularity and the four points moving around such a centre point extends the centre point by establishing a line of three points in the centre of the top. This line can never move. This forms the basis of the Coanda principle as the rotating four points forming a circle around the top then becomes the motion that acts as liquid, although the wood is solid.

That also provides a line in the centre that can never move and that line forms the solid part in the partnership. The point around which the rotation turns becomes the solid. This is evident by the fact that the line holding three points is activated by the motion of the circle holding the four points while not participating in the spin. As the spin reduces in dynamics the turning still present a fight to the last to preserve an individual gravity that sustains a cosmic

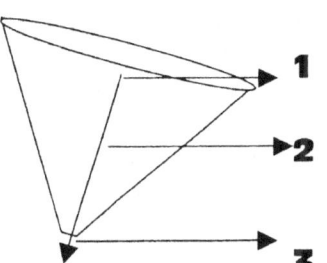

reality. The one position in time that is turning the four points around 1^0 which is the original singularity, represents eternity in time, the part that forever turns because it is time.

Infinity is the part on the inside, which represents time where time has no start since infinity is always part of eternity until space parts infinity from eternity. The instant eternity shifts from infinity through the evoking of space by motion separating eternity and infinity space-time comes about. But we are dealing with the cosmos and what was in the cosmos at the first instant is still in the cosmos at this moment because that which was part of the cosmos at the start has no where to go but to remain in the cosmos and be part of the instant of the cosmos in eternity.

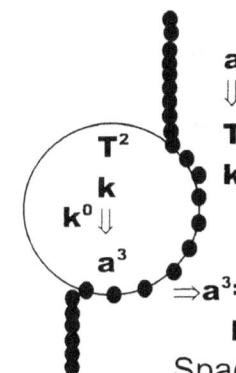

The contracting of space diminishing as a result of singularity established

a^3
$\Downarrow$ The motion provides new space and the new space
T^2 provides motion and in the midst of all of this a point forms
k through the motion where space will flow towards and
    disappear within that point of space less ness because of
    motion less ness.

$\Rightarrow a^3 = T^2k \Rightarrow \Downarrow$ k^0 $\Uparrow \Rightarrow a^3 = T^2k \Rightarrow \Downarrow$ k^0 $\Uparrow \Rightarrow a^3 = T^2k$

k^0 $\Downarrow \Rightarrow T^2k = a^3 \Rightarrow \Uparrow$ k^0 $\Downarrow \Rightarrow T^2k = a^3 \Rightarrow \Uparrow$ k^0

Space duplicating is creating motion and from the motion space is created.

The cosmos did not get bigger from the first moment. The cosmos is not expanding but is reducing. That which we cannot see where 1^0 parted from 1^1 is not to the inside of where we are but is so big it now serves as home to all there is within the cosmos and for what which was to become, to be able to be where it became, the cosmos had to reduce to accommodate that which now is where it is located in its allocated position in the cosmos. That which parted when the cosmos came about is now so big it is to big for us to see because we are too small to bear witness to something that large. The line was at one stage

one half of the entirety there is and the other half is the entirety we are able to see and the rest we are too small to see. That what now is. Was once so massive it could only be parted from that which is so small it clearly is not.

The establishing of motion is creating space and providing a centre point of singularity. This proves that the atom links with singularity and that the value of singularity extending forms a relevance with the space-time it influences up to a point that it controls the space as much as it controls the motion of the space. That proves singularity extend way beyond the space of the material it holds and establish duplicating points of singularity within singularity by merely applying motion to space within space.

The presence of the Coanda gravity principle is portrayed throughout the entirety of the Universe and every galactica holds the Coanda principle as proof that this is the way gravity applies. Yet no Newtonian ever bothered to see the indicating of the Coanda effect that

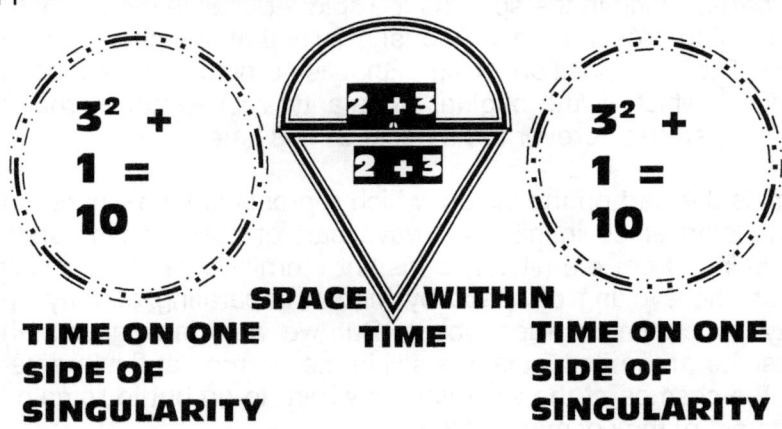

$3^2 +$ 1 =   10

2 + 3   2 + 3

SPACE WITHIN TIME

$3^2 +$ 1 =   10

TIME ON ONE SIDE OF SINGULARITY   **TIME**   **TIME ON ONE SIDE OF SINGULARITY**

produces gravity by providing motion in space and why gravity is used as the glue that is bonding the all of "the everything". But it has no mass and therefore in a thinking manner the Newtonian brain switches into a denouncing mode.

That means the second motion coming from the object travelling through space which arrange its

atoms in the same time duration as that it would have used while being stationary on Earth pushes the electron circle out of the normal sequence that electron had while being in a steady state of motion on Earth and under the control of Earth time. The principle, which I described in the motion of the spacecraft travelling through time, and space in space-time also apply in the same manner in the Coanda effect. It is the same motion producing gravity by applying a space relevancy that changes in the same duration as when not in motion but through motion the space changes to, a new motion that comes into play.

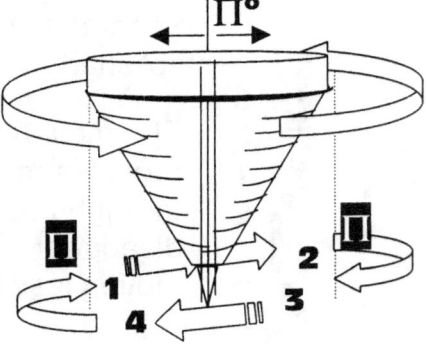

In the motion producing a relocated centre a new centre will play its part in the compensating motion that will force upon the relevancies a new space-time dispensation.

When the motion comes about we find three positions in time moving on the one side of the divide where such divide is formed by singularity and we have three positions on the other side of the same divide and because it is motion such motion transforms the line to a half circle as much as it translates to a triangle.

The motion goes around a centre that forms the half circle and is commit ting one side of the motion to oppose every value of the other side in a relation.

On the one side we have T^2 forming the space in relation to a centre and also on the other side we find the rotation committing the space defined by the half circle in relation to the centre.

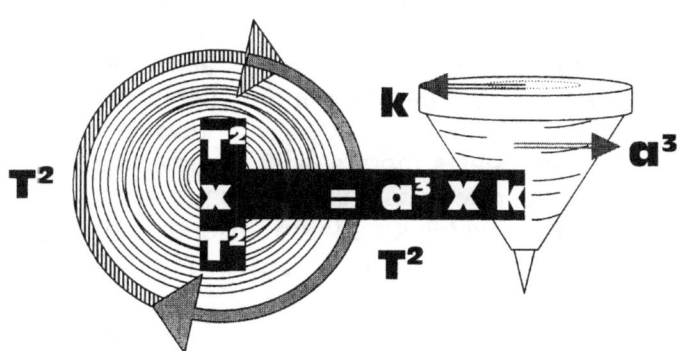

We therefore have $T^2 \times T^2 = a^3 \times k$ and that positions the completer relevancy brought on by singularity to complete what the motion provide that space to have in relation to a centre established by such motion. It is the spin on both sides of the circle that establish and define the space in regard to the dividing centre and from the centre to the outside is space in time This is what Kepler's formula confirms when one take out all the hideous misinformation Newton connected to the concept. The motion in rotation T^2 confirms the space a^3 in relation to a specific centre X k or mathematically equated as $T^2 = a^3 \times k$ that one easily derive from $a^3 = T^2 k$.

Mechanical engineers propagate that it is air flowing underneath cars that causes the lifting of the cars at high speeds.

That is as Newtonian as blaming it on insufficient mass or some other polony story they come up with. We find the top doing the same at high peed as the car or the high-speed boat does in water, where the top tries its best to lift off from the ground. If they are correct about the wind underneath

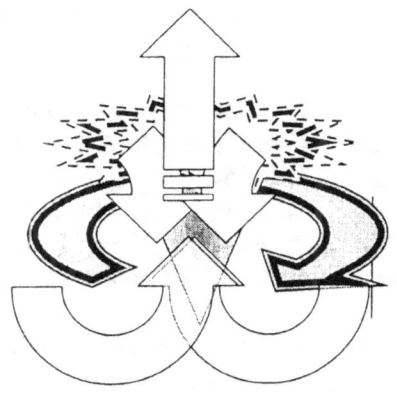

the car the next question that comes to mind is what wind is lifting the top when the top starts to move rapidly?

One should remember the wind in the case of the top must lift a needle if it is wind that is lifting the top. The process is identical and the top indicates to what incorrect surmising the engineers are capable of. The top is in motion because of it duplicating its space position in time in relation to the time it is in contact with during the instant as well as the duration of combined instants.

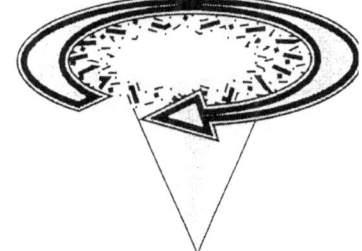

The top has the material it holds in space becoming more because of more duplicating through rapid motion and as the electron wants to extend the orbit in relation to the time factor when the electron becomes more active, the top also wants to extend the relevancy of k because the atom material becomes more in relation to the time aspect by which the duplication is controlled. With all the spinning and the commotion that the additional heat brings about by which the governing singularity extends, and by extending it is promoted to a higher time zone in the atmosphere. The singularity suddenly is in charge of more space-time by which the space is duplicated in time. This is the line that science always shows, which runs from where it grew as a motionless point in the centre's centre.

With all the excitement and nowhere to take it, the extending goes down to singularity keeping the whole top erect. That is why the top is spinning in the first place. We know the spinning is attributed to the will of life but in the cosmos where life does not apply the spinning would be a result of gravity accumulating and concentrating heat in a small area. The more assertive the spin is, the more reaction there is from the lines running towards and extending outwards. It should in real terms expand the top because the motion creates more space per time unit filled and the space has nowhere to go but to excite the newly established centre singularity. By duplicating more space a^3 the rotation of time T^2 provides for a larger extending of time k^1 to extend the boundaries of a^3, which is the result of a more ferocious spin in T^2

The spin under normal conditions can only come about as a result of heat being more concentrated by space being rapidly reduced. That in short is one part of gravity. With that aside the spin normally comes about by the exciting that brings on rapid movement, as is the case when water falls on a hot plate. This proves that material grows by removing heat from the time sector where the time component $k^{-1} = T^2 / a^3$ reduces its value and material is gaining from such motion $k^1 = a^3 / T^2$ by the measure of k as material grow.

This is consistent with the Big Bang as well as the Hubble shift because time removes its position by allowing more space to form in between particles of matter. The linear component of time k^1 removes heat from time to singularity being in $k^1 = a^3 / T^2$ by introducing $k^{-1} = T^2 / a^3$ as the flow of space-time from outside the atom to inside the atom. That means liquid heat is removing from time in the space between particles $k^{-1} = T^2 / a^3$ and is adding to material $k^1 = a^3 / T^2$. That puts the Universe in as much contraction as it is in expansion and secures the perfect balance we would expect from a harmonised Universe.

More spin increases both line directions $k^1 = k^{-1}$ that produces gravity by applying rotation T^2 where the rotation is the result of the extending of k where k^1 removes heat from time $k^{-1} = T^2 / a^3$ in order to add it to space by the measure of $k^1 = a^3 / T^2$. By heating space exceeds its boundaries which the liquidity of time allows through the Coanda effect and in the motion that results from the expanding of material by acquiring additional heat to sustain form the motion allows heat to extend by the margin that the liquid provide such extending to come about. Therefore the extending is both $k^1 = k^{-1}$.

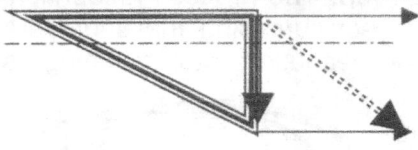

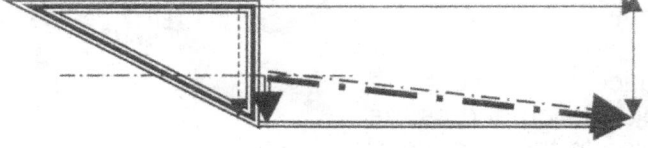

The space of the boat inside the water does not reduce. It rather distributes better over more liquid since the motion allows more liquid in terms of less solid while in overall space-time the scenario remains the same. By motion there is the very same amount of "boat" remaining in the water but because there are more space to time distributed the water represent the time the boat goes through and thus the boat seems less in the water.

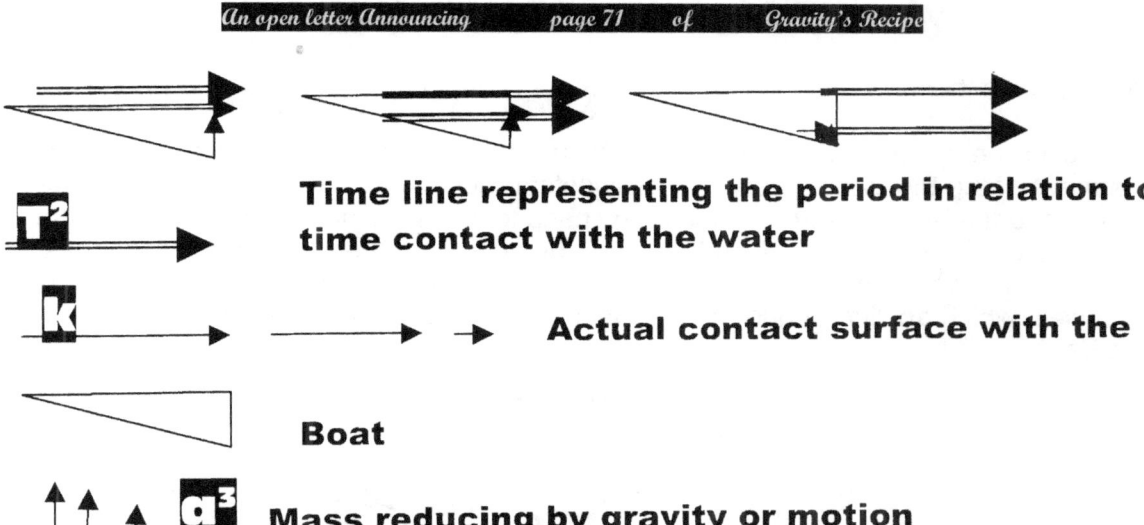

Time line representing the period in relation to time contact with the water

Actual contact surface with the

Boat

Mass reducing by gravity or motion

The racing car getting air born and the aeroplane lifting off and the water skier gliding over the water and the top spinning about its axis is all the manifestation of Kepler's formula where Kepler (wittingly or not) predicted that by moving space ^{3} through time the movement will produce space by ratio of time allowing the frequency of such duplicating $a^3 = T^2 k$.

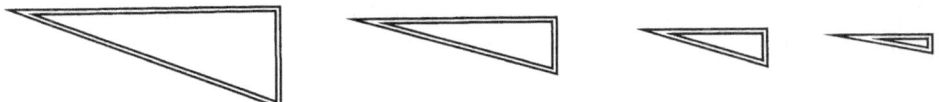

Motion is the result of space duplicating exactly as Kepler said before Newton interfered with work that was totally above his (Newton) level of comprehension. A boat lifts from the water by the increase of velocity because the increase of velocity reduces the mass the boat has.

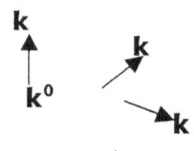

The more motion there is the less mass the boat will have because the more duplication the motion allows the less water displacement the boat will produce and the water displacement is the mass the boat delivers in the water.

By applying motion the flow of the liquid around the round solid which represents as much as rein acts singularity the motion brings about a new space-time as the factors have to compensate for the motion and the motion in itself brings about a new controlling k^0 in the position elected by all atoms forming part of the relevant motion. The motion introduces a securing of the position that is elected by all the atoms and is forming the new part that will serve as a controlling singularity by substance producing the motion or the substance of the solid securing the position of the newly elected controlling singularity.

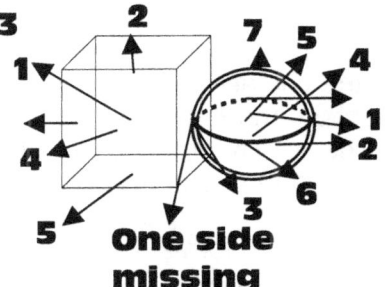

One side missing

By applying motion the Earth gravity is interrupted by the space producing motion that through motion takes charge of the space and turns the vertical flow of Earth space into a circular motion although only for a short while. This is how electricity is charged but the motion then involves the reducing that the flowing electrons undergo to compensate as much as producing speed of light. Other elements and their characteristics are also involved and in the charging of electricity the Coanda principle personifies the gravity that comes about in the condensing of space-time around singularity.

By reducing the space towards the centre in the using of the Coanda effect re-insures the correctness of the Big Bang theory. A newly established centre will create a newly controlled Universe and the motion of Iron $_{56}$ will force space towards the newly established centre. By forming the flow of space in relation to copper the space will break down and form the inside we find the space to flow on the inside of the Earth centre. We have to remember that the relevancies, which applied during the Big Bang is present in a minute capacity on the inside of the Earth.

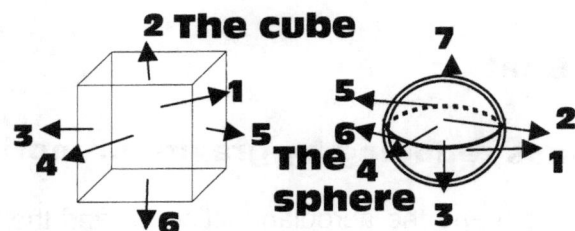

In the cosmos we find two forms available to use. The one is a six-sided cube that connects at the edges and can be found in whatever name one may attach to the angles that connect the sides but after all it still remains six sides with three sides facing three opposing sides. Then we have a sphere, also with six sides with three in opposing to three others but in this case all the sides connect via a precise centre and the centre gives the value of a seventh position that secures the precise location of all six edges and that keeps the sphere true. In the case of the sphere there are seven points where the six points representing the sides connect through the centre and the centre connecting is charging the form of the sphere to Π and with the centre forming singularity the centre is $\Pi^0 = 1^0$ that holds the form in charge and in truth.

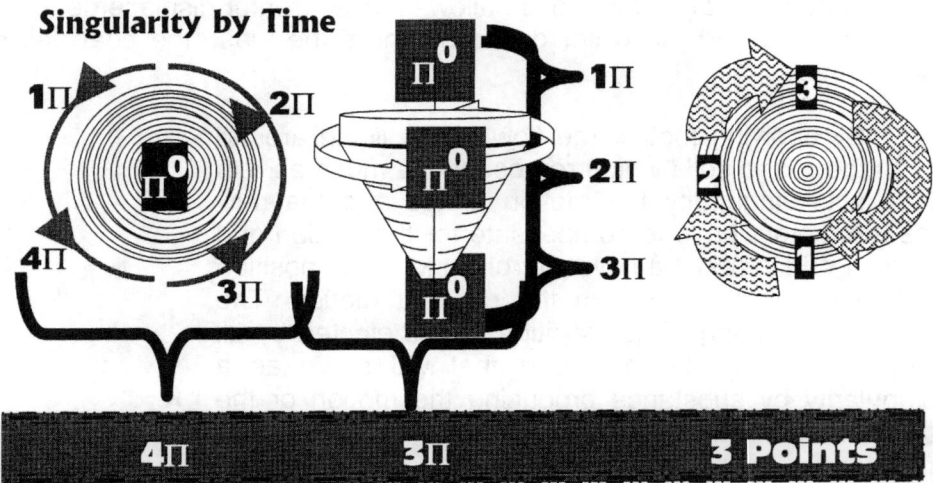

When the cube, which is having six loosely connecting sides comes into contact with the sphere with six sides but also with a centre that charges gravity by its positioning of singularity, the cube is destined to lose the one side that forms the contact with the sphere. The centre in the sphere removes the cube side that has the contact and the sphere allows that which is on the edge of the cube to become committed to the gravity of the sphere by descending to the centre. The centre will charge a reducing of space through implementing motion and the evidence of that we find in the Coanda effect which I shall explain later on. Only where such descending is no longer possible because material that forms the edge of the sphere will not allow further moving by the descending object to the centre. Then at that point where motion becomes a tendency to move, does the object receive a position of mass.

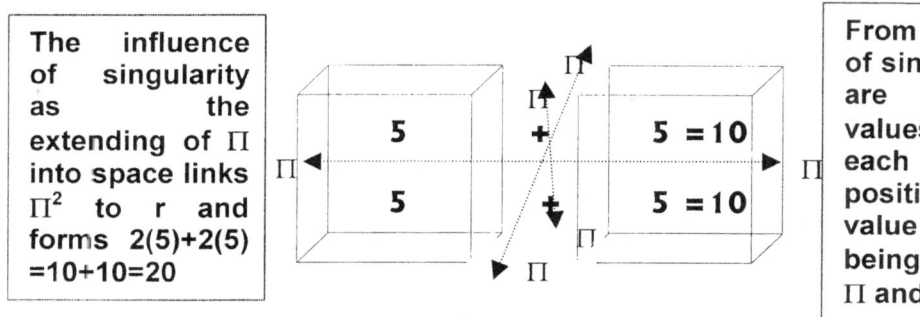

The influence of singularity as the extending of Π into space links Π^2 to r and forms 2(5)+2(5) =10+10=20

From the position of singularity there are different values in Π where each indicate a position. The value it represents being $\Pi\Pi\Pi$, Π^3, Π^2, Π and Π^0

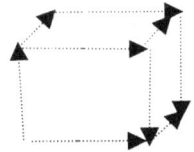 But it only becomes mass after the motion or gravity that it had became restricted and the motion becomes frustrated and confined. The motion ends where mass confines the object to the motion that the larger object confirms. As the sphere spins it have seven positions in the sphere that is eternally part of the form the sphere has, acting as one while there are six in the cube but with the demanding sturdiness of the sphere, the sphere removes the one side of the cube where contact is made. The six sides in the cube then become five sides and five in the cube then has contact with the outside leaving one point in contact with seven points in the sphere. The centre will dominate because the centre charges seven points to gravity while the cube cannot secure the position of any particle that is in the cube. The cube with five lines that loosely connect and that don't form any firm position to secure material within the cube will be conformed by the spinning centre of the sphere, which generates gravity. The fact that the cube lacks any ability to confirm material within the form it secures shows no structural bonding in the cube.

Therefore, where the sphere and the cube touches the sphere depletes the space the cube holds and remove any material to a position that relates to the centre of the sphere. The loosely connecting sides that concludes the cube as form will lose one side to the more

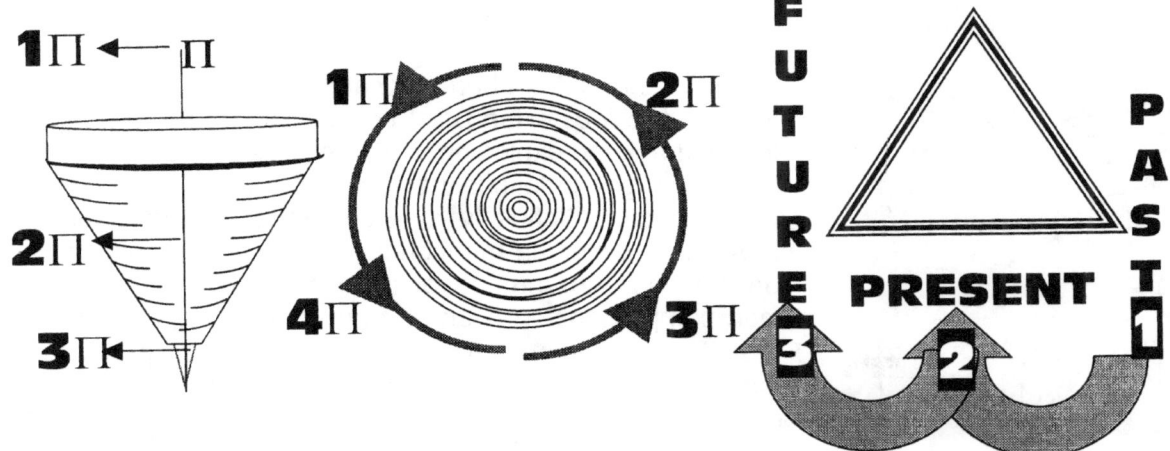

overpowering structural form the sphere has and the sphere is going to capture one side from the cube. Then with five sides standing related to six sides and a seventh in the sphere centre, which also is commanded by singularity in the centre, the pure domination is going to affect the motion of the material that is going from the cube to the sphere because of the six points that is spinning around the seventh point.

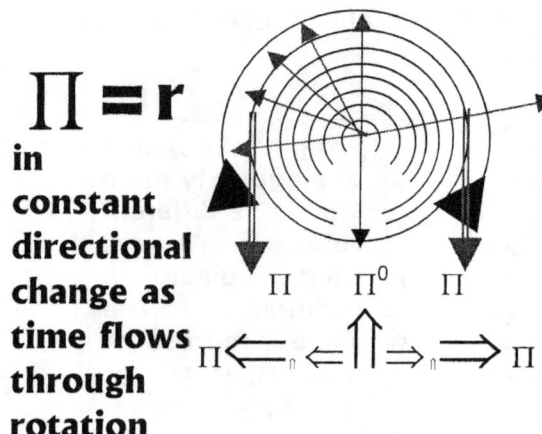

$$\prod = r$$

in constant directional change as time flows through rotation

$$\prod \Longleftarrow_n \Longleftarrow \Uparrow\Uparrow \Longrightarrow_n \Longrightarrow \prod$$

$\prod \quad \prod^0 \quad \prod$

Pinpoint positioning of singularity $\prod^0$ with $\prod$ positioning space to either side forming the border set by singularity

The new direction pointing to a new location in relation to the previous point will oppose the previous point it had in relation to direction considering the centre point.

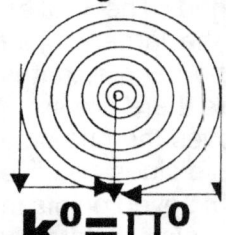

$$k^0 = \prod^0$$

The figuration found in the planetary and has taken the name of the Bode or the Titius Bode system such a system is the mould that remains in space as part of space after the concept establishes the relation and as such is then detected by observation and growth leaving the mould in space as the departing of planets from a space concept in relation with the centre. It is a manifestation of the evidence what Kepler brought from his investigation of the cosmos. Gravity is where space vanishes and as such space is in the centre of the sphere. The sphere controls not only the space included in the sphere but also the space excluded or on the outside of the sphere.

In that the cube is represented by five sides forming a form plus five on the side also forming the cube where both are constructed by the loosely connected cube in form that stands in relation to seven points connecting in the sphere and the sphere has seven to the double cube having ten and that is why the Titius Bode holds a form of ten that stands in relation to five on the one side and another five which is part of the following direction. Space –time is built in such a manner. As the movement propels it will move from a cube of five to a cube of five and the total sides being affected will be ten on the side of space.

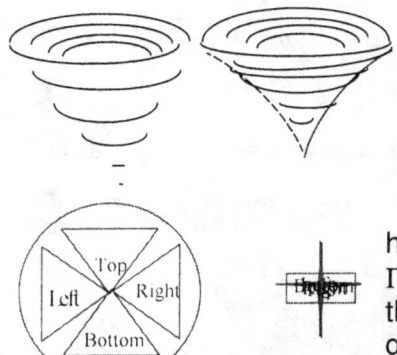

To find the invisible I had to locate singularity. I realised that my effort to locate the point holding singularity enabled me to backtrack the exploding universe to its origins

By reducing r indefinitely to the tune of half each time, r would become infinitely small, beyond human calculating means, however as mentioned in the case of the smallest dot holding one spot, r would become insignificant beyond human comprehension even, but never reaching zero and still $\prod$ would remain intact and dictating form. I am promoting a theory in which I am able to prove there is as much contraction going on in the cosmic universe as there is expansion and the contraction is as much part of the expansion. The universe rides on a balance and we have to locate such a balance. To prove my theory I firstly had to locate the centre of the universe.

Even admitting to such a notion sounds like madness, but please give me a chance to explain in more detail. If I wish to achieve success that would depend on my ability to convince all that outer space comprises of material and as such we can locate such material even if we are unable to see such material. By applying some basic effort I have located the position from where all movement came and

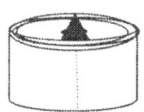
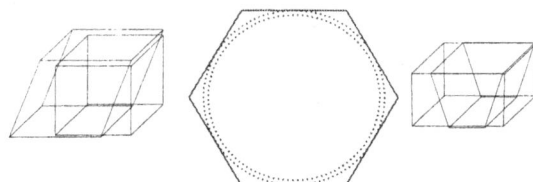

the direction it took moving forward in time.

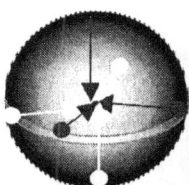

The sphere holds six sides in relation to form as unit and as does all other shapes and forms. All forms have to have at least six sides indicating different exposures to the Universe. But with gravity having a free choice, gravity always chooses the sphere. As I shall prove later on gravity is the strongest where the form produces the least evenly distributed space.

We traced the line back to the spot and so we know where it all started there was a spot. The miracle part comes in the fact that eternity shows no change. Any change, even the slightest change represents the end of eternity. With everything motionless and locked in singularity everything remained eternal because even today singularity is locked in eternity. In the centre of all objects spinning there is a space forming a divide. It divides every aspect of space spinning into sectors of space having motion as much as changing direction of initial motion. However, since it was part of the cosmos, and although it is outside the cosmos, it still is part of the cosmos.

A spot holding what ever is and can be into a dimension so small it did not even have sides. Then for a reason, which at this time I do not wish to go into some miracle happened and the spot showed motion. The motion was deliberate but the motion was eternally small. The spot had one specific value, which it clung too…it had all sides on the same side and from that rotating diverting came about. The forming of $\Pi^0 = k^0 = 1$ had all possibilities available but it was only one possibility in the end. There was no space and then space expanded producing space $\Pi^0 \Rightarrow \Pi$ and from that motion comes gravity by the value of the motion creating contracting Π^2. The motion brought along Π but even Π was subject to relevancy. From this come the most basic principles in as much as forming the ground rules of the law of Pythagoras.

When drawing a line such a line then starts of with a dot serving the spot that holds all sides

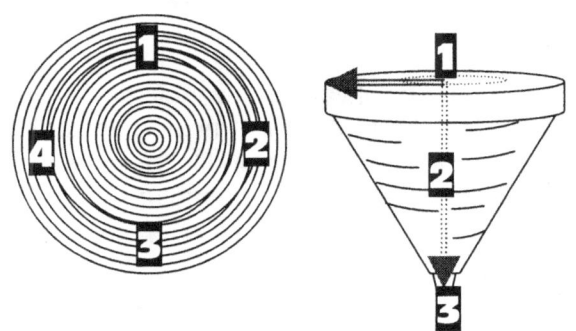

equal. That means the line serving as the future radius will be equal to the half circle which is then Π. The only aspect of the point that stands in for the end of the single line forming the radius of the circle is that we then mathematically reach the single dimension. We decreased the line to where a circle being Π formed on the single dimension. This dimension also hold the circle dividing line because from there the radius must once again generate a value and by such a gesture that the extending would form the circle that forms the sphere that eventually lead to the formation of particles.

In $4\pi^2 a^3/ T^2 = G(m+m_p)$
$a^3 = T^2 k$
$a^3 / k = T^2$ but
$k / a^3 = 1 / T^2$
$k = a^3 / T^2 =$ singularity
$a^3 / T^2 = G (m+m_p)/4\pi^2$
and $a^3 / T^2 = k$
then $k = G(m+m_p)/4 \pi^2$

All Newton's changing was possibly done with good intensions but even that I doubt. The end result however was in some cases far from good, as it does not do such great credit to Newtonian insight into cosmic affairs. Only Kepler and only Kepler unaided without the intimidation and interfering of Newton can explain the Coanda effect. I grant the fact that the Coanda effect was discovered before Newton saw himself fit to change Kepler, but only Kepler can explain the Coanda gravity effect when Kepler is without the attentions of Newton.

Where the radius reduces by a dividing of two such dividing can never remove the radius and since the radius determines the walls of the sphere the sphere may shrink but can never become zero. All lines have to start at infinity and cannot mathematically start at zero because in cosmology as in mathematics zero has no place.

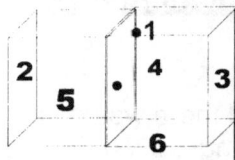

There is one more point in the sphere in the centre forming an addition in the sphere. That point holds gravity secure.

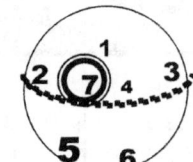

If the cosmos had any other shape the might be available contraction in the end as the final conclusion would then not be possible. Only in the sphere and more so in the circles all forming a sphere is singularity present. Singularity comes as part of the construction we find in the sphere. By singularity forming the base of the sphere the Coanda effect will forever be present and with the Coanda effect gravity applies.

As is evident from the experiment of the Coanda effect there will always be an attraction by singularity controlling the space-time from the centre of the sphere. Such control has the name of gravity, but it is no force. It is a combining of liquid (10) and solid (7) in the presence of three in time that will produce such control by motion. But why by motion?

In the sphere there are never only one direction implicated in movement. Movement are always in relation to the centre position because as a line goes up it also goes in or out. When a line goes north or south, it also comes towards the centre or going away from the centre.

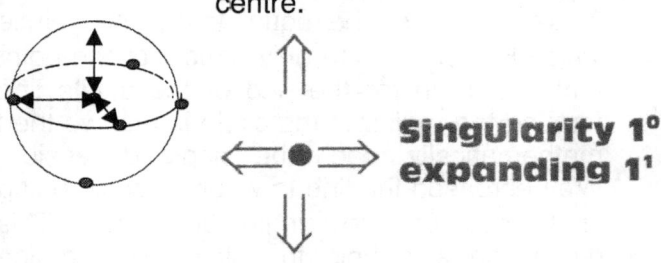

Singularity 1°
expanding 1¹

There is always relevancy present in movement. As this moving indicates direction it also apply Π^2 for indicating value forming the time factor.

Singularity 1°
expanding 1¹ by
duplicating time

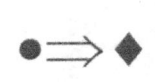

Singularity 1° expanding 1¹
by duplicating time while
shifting with time
progressing

In spite of all the overwhelming evidence to the contrary of Newton's mass performing gravity you persist that Newton is very much commendable and as sureties that the cosmos provide as clues to show the terms applying to how the cosmos truly works is swept under the carpet in favour of Newton's where it is almost overwhelming evident that the hoax Newton created is not working.

What Kepler saw was more of a dimensional nature than the practical mathematic symbols and values. On the one hand was a value to the third dimension, which equalled two-dimensional values one the second dimension, and one to the first dimension.

The spinning object has seven points serving precise equality by rotating and holds the sphere in contact with the space outside by the measure of coming from a cube that has five connecting lines while the seven points that confirm the sphere in time holds seven points in the instant of the presence and those seven points are moving onto another five points serving time in space.

The TITIUS BODE Principle Outside the sphere

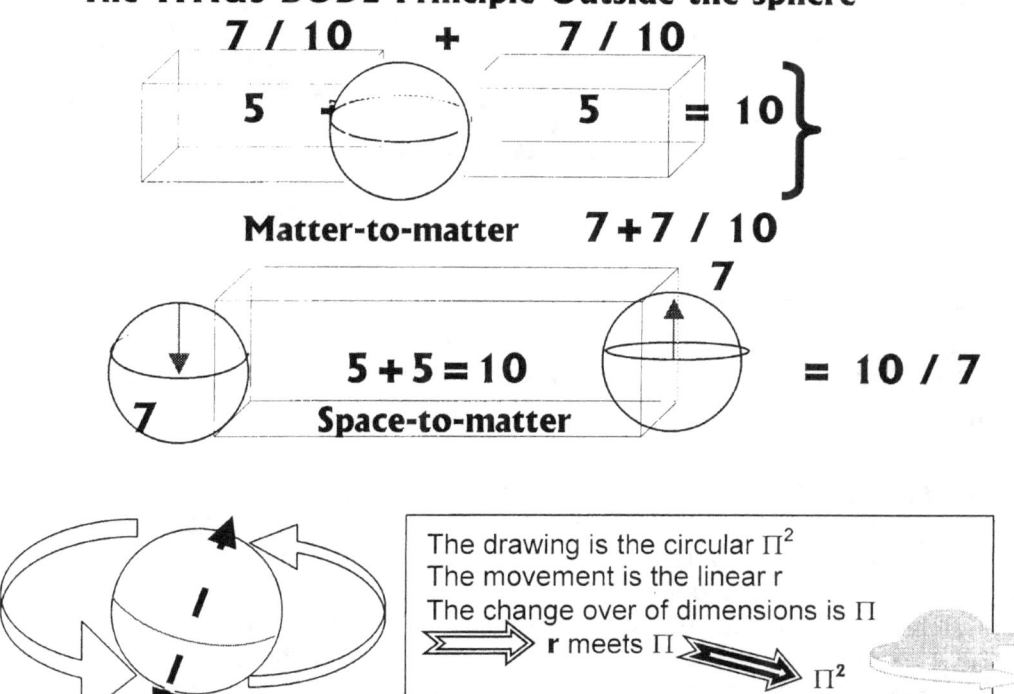

The drawing is the circular Π^2
The movement is the linear r
The change over of dimensions is Π
$\Longrightarrow$ r meets Π $\Longrightarrow$ Π^2

The result of the five to one relation comes the Titius Bode principle.

Singularity develops from 10 t0 11 to Π and from forming Π on both sides of the divide the value of such motion forms the square of singularity extending which is Π^2

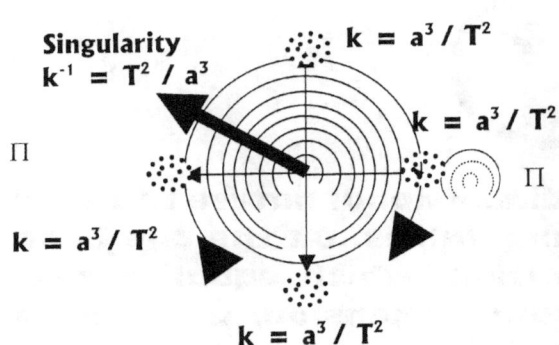

The points duplicating is four moving around a centre by the square of gravity. The motion is the sources of heating because the heat is bringing about the movement. The heat growth therefore provides the action because the action is what energises the points to provide the motion. The motion is purely is space-time duplicating and the duplicating is feeding heat to the centre from the four points overheating thus the points that shows expanding.

Let us find the smallest possible line first. We already have reached the conclusion that by reducing the line , the reduced line will eventually leave all sides on the same spot. Such a spot must be round in form. With the line being the smallest line, such a line will start off as a dot that moved away from a spot. With all possible sides being in precisely the same spot we have all possible sides onto one spot.

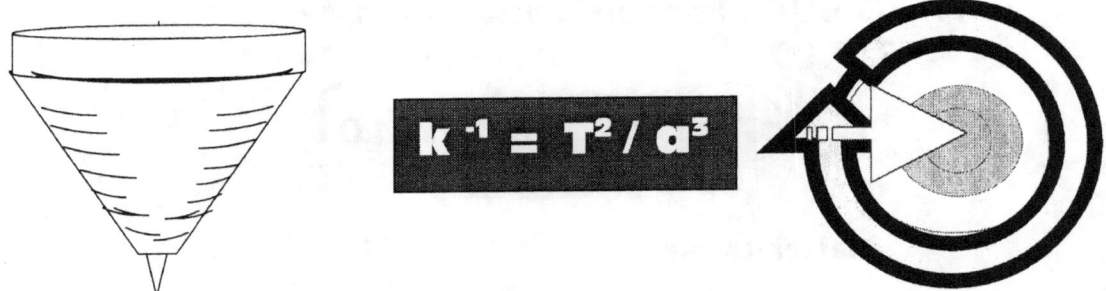

$$k^{-1} = T^2 / a^3$$

The motion established by singularity results in the implicating of the Coanda effect as much as the motion establish the Coanda effect. The spin realises the space limit while the space limits attaches the motion onto the space in the time within the time.

Mathematically the spot is in the single dimension where the space is one and exponentially zeros. There the space moved over to form the dot. We now are reaching into areas only the human mind can venture by understanding and nothing more. The understanding of this concept demands our reaching the point where the mind of the animal cannot reach. If it starts with a line that line only represents two sides being one and as such that is rather a flat Universe. The spot is not yet round because being round are requiring a shape or form and this lies beyond or before a time when any form of shape came into the cosmos scenario. It was in a period where shape and form was a part of the distant future hidden in and beyond eternity. In that time the line must have been so small it had reached a point not yet dividable in any way. If any further dividing took place such dividing would have brought growth because there then would form space between the sides going in the opposite direction. The dividing brought all there is having all sides literally on the precise same spot, and I have located singularity in just such a spot.

I came to the conclusion that the spot I found had to be singularity purely on the grounds that that spot holds only one side to serve as a start to the starting point of all directions possible.

In that side is only one spot where there is only one side applicable and one dimension present.

With all the factors given one can only come to one conclusion and that is that there can be only singularity. In such a case more dividing by two will land further positions on the other side of the divide. That point is serving as a position for all possible points and cannot allow further dividing as it is in the smallest line or spot there may ever be. This spot is the result of a most basic process of reduction as the Hubble constant is a most basic process of expanding during a matter of time. By reducing the line constantly the only value that will eventually remain without dispute from any party arguing about the facts is exponential zero. By only having exponential zero instead of a numerical zero and a radius as one in the square (the radius effectively becomes one holding any and all sides on one point) such a point might become any value of any significant measure implicating anything but zero as the radius. By expanding the line, it will be an evenly spaced structure growing into the most perfect round dot ever possible anywhere at the point when it starts to grow.

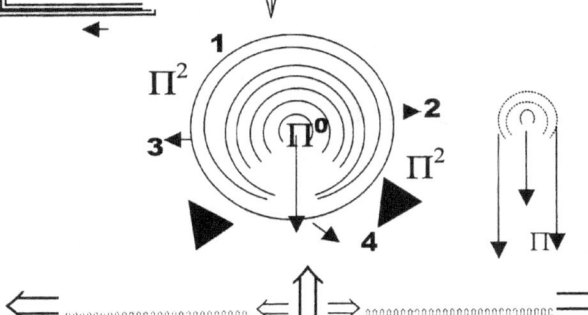

The sphere has seven points. The cube without truly being a cube but is just in consideration of having a cube in form holds five points to singularity. In the centre runs singularity to the value of Π^0, which means that which surround Π^0, holds a position of Π^2

The spinning sphere activates the seven points, which places gravity in relation to a centre. Outside the centre there are five sides by dimension. The sphere has seven points of which four is spinning. The four spinning stands related to the gravity of spin, which are Π^2.

1^0 going 1^1 1^0 ➤ 1^1 had to bring about 1^0 going 1^1 1^0 ➤ 1^1, because the eternal repeat of duplicating while contracting was not relieved from the Universe. Before the contracting was equal to the duplicating because by measure the heat was identical to the cold. It was eternity that was interrupted by one cycle of infinity and was in repeat of eternity. Once something is part of the Universe there is nowhere else to take it so it has to remain as a part of the Universe.

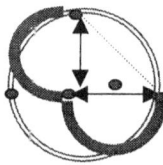

Because every moving line represents one quarter of the sphere in relation to the rest of the sphere and the line also indicate the relevant position between the point indicated and the point in the centre it is a relevancy of singularity in progress. By connecting the line, as Pythagoras will suggest the singularity within the sphere become a specific value indicated representing one half circle.

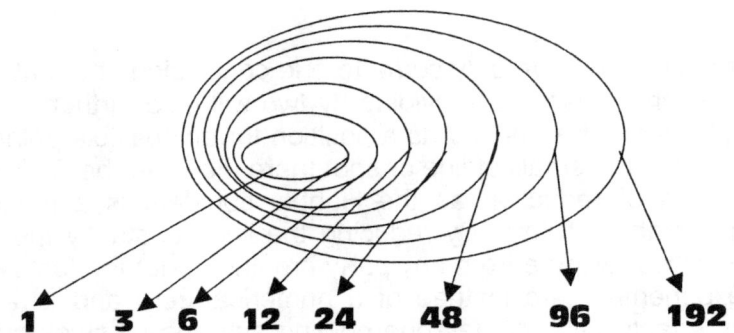

Planet	Mercury	Venus	Earth	Mars	Ceres	Jupiter	Saturn	Uranus
Bode's Law dist.	4	7	10	16	28	52	100	196
Actual dist.	3.9	7.2	10	15.2	28	52	95	192

The Titius Bode law is form where space holds a relation with material and space finds form in the cube being unconnected while material finds form in the sphere and is connected through singularity that is activated by motion. As singularity spins space converts with material and the Titius Bode law of 7 /10 and 10/7 conjuncts with the Roche limit value of $\Pi^2/4$ which finalises the value of gravity Π^2. The sphere holds seven points and the centre by dividing proves to form singularity, however the singularity can only activate into representing an active line in singularity by motion that charges singularity to take control and form the centre line. With the accepting of singularity as a point where the Universe started and that everything grew from such a point we must first allocate a value to such a point. The only value that such a spot can have while being is 1^0 because from being a spot and growing into a dot would take the development of coming from 1^0 and developing to 1^1.

Being singularity and progressing into singularity could only be at a level of one because singularity means 1. The singularity is still with us and we can see it every time a round object spins. It's a line that is but also is not in the centre of every rotating object where 1^0 are flanked on every possible side by 1^1 that has one opposing the other to every possible side. The main thing is that I can point to my claim that suits all the criteria astro physics prescribe that singularity should have but also I can say that a person is to go and look at the centre of a spinning object and find such a line there. It is a mathematical point that forms a line with the aid of spinning.

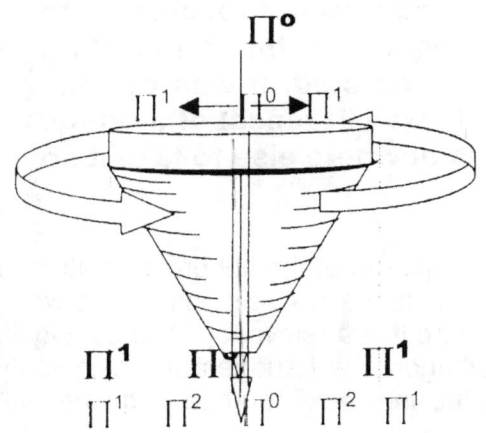

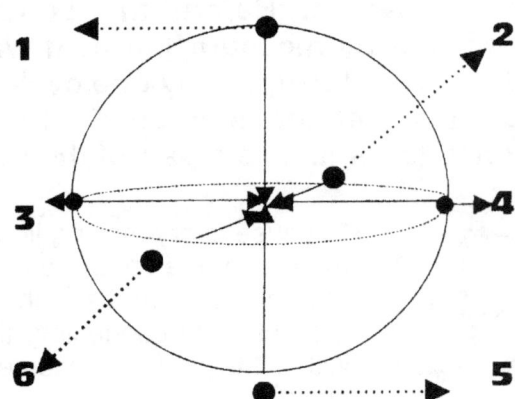

The sectors provide individual singularity a means in sustaining governing singularity by which provision comes through maintaining governing singularity the required spin in maintaining cooling. If this process did not apply, there would be no connecting individual singularity to major singularity

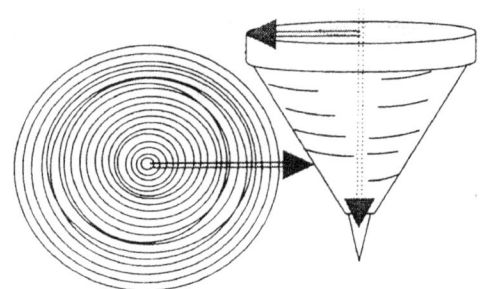

In the **precise middle** of all **objects in rotation** is a precise centre dividing the object in sectors that will **start the spinning initiation** from that centre point. As the rotating direction moves inwards, the rings will become smaller and smaller. Thus, the spinning object **will have a middle point**, a very specific **centre point that does not spin** and only holds Π as a specific value because no radius can apply. But also the one value such a line **cannot have is zero** because the line **is there and holds contact** to the rest of the material bringing about that **zero does not start any** line and therefore the **value of the line must be infinite,** just as described in **accordance** and by **the definition of singularity**. There must come a point where the ring is infinitely small, where it can reduce no more, where it reached its ultra limit, but at that point it cannot be zero, because the point is there

By the motion it gives the object spinning a freedom that the object is willing to fight for before relinquishing such a freedom. It allows the top to stand up straight and spin as an independent Universe amongst all other independent Universes there might be. It promotes singularity, which is inside every sphere to become a line, which generates independent space a^3 by the motion $T^2 k$ thereof. The space holds autonomy as the top shows while it spins and the spinning charges independence and the independence is only achieved through inflicting independent gravity onto an existing time format. The Earth is 1^0 to the Moon being 1^1. The Sun is 1^0 to the moon being 1^1.

The proton is 1^0 to the electron being 1^1. There is an eternal link in singularity and all four phenomena prove the distinct reality thereof. The centre cannot spin. The centre divides every aspect of the spin from the next and the previous where every aspect changes its relation it had with time and with the rest of the Universe in total. Everything that connects to the centre to the one side is defined by a line that is not present in our three dimensional Universe since we can't accommodate a line with the value of 1^0 going to 1^1. Being what the line is and where the line is the line has no sides but leaves all possible sides on one centre location. That is singularity, which is the smallest line or point that can ever form.

$$(\Pi^2/2)\,2$$

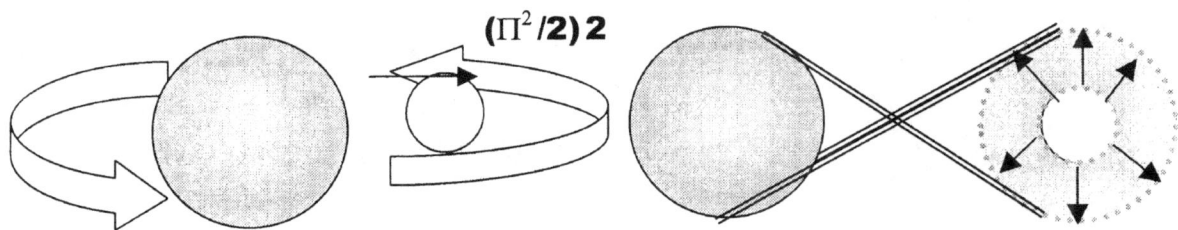

Having edges where Π^0 duplicate to present the edges singularity lost the value of Π^0 to the value of Π^1 with the same value singularity had being Π^1 to the one side and Π^1 to the other side, Π^0 must be the point splitting singularity into two parts of eternity, the eternal value of the first dimension outside eternity. It was the square of Π^1 being Π^{1+1}. That was the first dimension outside singularity Π^0 where singularity has a value of Π^1 in the form of $\Pi^{1+1=2}$. The first claim to space had a value of Π^2. This applied to both sides of the claim to space outside singularity, and the double proton became the dominant factor on matter.

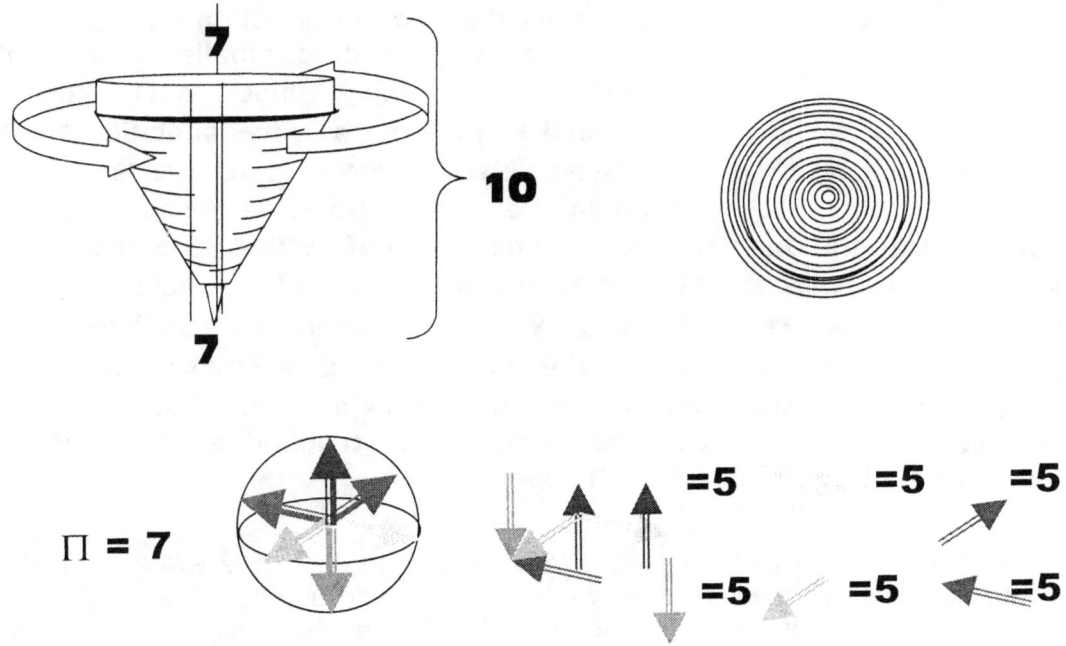

$\Pi = 7$

Singularity is always an eternal part of the sphere. Every sphere has to have a radius that formulates a distance to which Π will find a comparing value. By reducing the radius, such reducing of the line that forms the radius must eventually become a spot. Where such reducing stops is a last measure that forms singularity. For that reason alone, no line can ever start at zero because from the line come infinity that indicates eternity. Every sphere, therefore, holds singularity, both on the inside as well as on the outside. However it is the motion that activates the eternal line from the infinite spot and that parting of time is a resulting in the forming of the Universe.

Where any object is closer than $\Pi^2/4$ it becomes part of the motion of the other structure, which is reserved for liquids. The Coanda effect becomes where the material will be turned into a liquid and the liquid becomes part of the Coanda effect action.

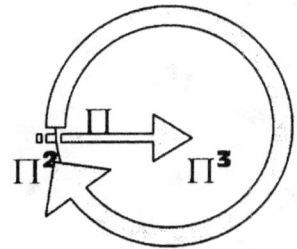

The Π^3 is matter in singularity
The Π^2 are motion or heat.
The Π is space in singularity.

The Roche limit is evidently widely applying throughout the entire cosmos but Newtonian rules do not explain the phenomenon:

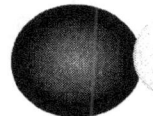

Any person taking Newton seriously should at least take on the challenge and find the comets colliding with the sun, find how much the planets moved closer to the sun since the days of Newton and indicate where there is unprecedented collision between stars. Yet the closest the Universe comes to that is to show " how stars blow bubbles" in space and that is to use the precise words.

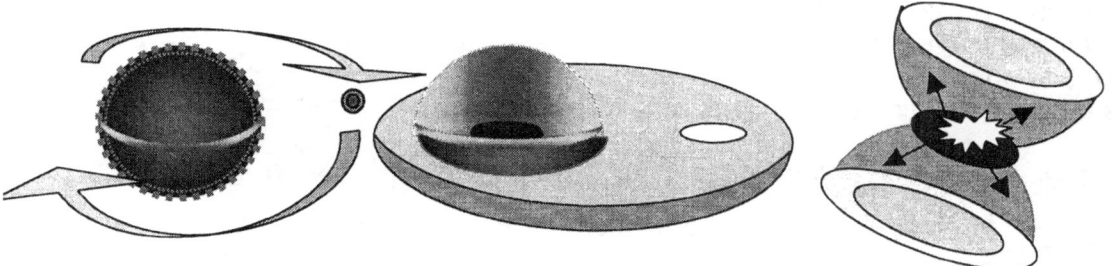

What this confirms is that no cosmic object will collide with another cosmic object and no mass draws mass closer. It is a relation where space-time relates to space- time.

5/2

Five sides divided by two spheres.

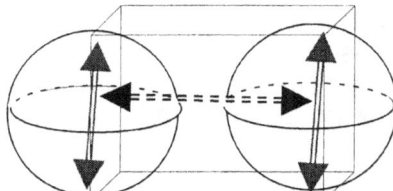

Where Π extends to lock onto the next sphere's extending indicator, Π has to connect to Π forming the square of space and translating that to the half of Π being $(\Pi/2)^2$.

The Roche limit 5/2 becoming $= (\Pi/2 \times \Pi/2) = 2.4674$ as singularity interferes

Gravity is the conforming of a centre that holds singularity and the maintaining of such a centre by rotating as well as contracting towards the centre. It is confirming the centre by motion and that forms time but the time is what keeps the space in place. It is the motion in time and the motion by time that confirms space from the past to the future through the present. I even explain what time is and that it is time that keeps space valid as Kepler said it would $a^3 = T^2 k$.

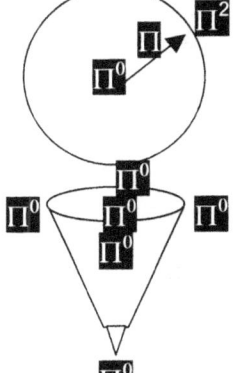

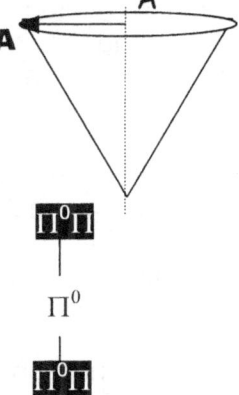

When the radius of a sphere or circle is reduced to its most infinite value it would become a factor of 1^1. That would leave the form Π as value of the sphere. In the centre where six lines cross such crossing will cut through a point where the six sides connect and that point will leaves value of singularity. Singularity can mathematically only have a value of 1 to the exponent of whatever comes as choice or of whatever symbol symbolises the radix puts the value of the exponent at a value of zero.

That point is there regardless of any other argument put to the contrary. If the is volumetric space the point in singularity is at that centre and it is a point that is so small it falls outside the parameter of measured space in a three dimensional form. It is not part of out Universe but it divides out Universe into immeasurable sectors.

That line can only be defined by the value of 1^0. Those line cuts whatever are on their outsides into sectors of infinite measurements and only by a mathematical regarding of the fact can intelligence come to be recognised. It is only the intellectual mind the will find recognising in such a point that control the Universe without being part of the Universe. Try to explain that to a gorilla and the ape will not understand the explaining. To the immediate outside of 1^0 are two (at least) two bordering values of singularity that has to stand in value as 1^1 because they still maintain singularity although they are hardly if anything bigger that 1^0.

That the lines are there is as true as the fact that the cosmos of in place but detecting the lines can only be through an intellectual apprehension of the fact that the cosmos starts at that point. Those points are there eternally in infinity and the only manner to release eternity from infinity is to provide motion to eternity outside infinity.

Because space is ten and half of the space should be five

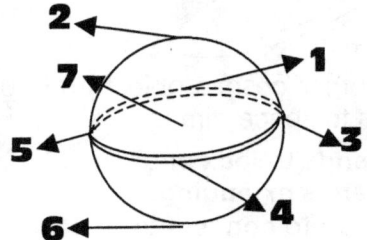

one would expect that the Roche limit would rather establish half of ten to form half of space in the square by half. Since that is not the case but gravity is involved one may presume that this value reflects of a condition that applied before all space, as we know space was a factor and everything rested in motion. The only time only motion was about and space was excluded was when the Big bang was not in place yet. By tracing the phenomena and studying the values and positions of the phenomena that will allow investigation to go into the era that is pre Big Bang.

The sphere has the seven points and the cube has the six sides. That much we can see. The sphere has seven points in the strongest form the cosmos can provide while the sphere has six sides that can be anything anywhere. Wherever the sphere makes contact with the cube, the cube loses its dominance on space at that one particular point. At the point of contact the sphere is going to take the point the cube has, away from the cube and allocate such a point in relation to the centre of the sphere. Why the sphere does that I shall explain further on but at this point I wish to explain the crucial part of this understanding.

The cube will lose one point it holds anywhere the cube has contact with the sphere…and the sphere makes innumerable points of contact as the sphere turns in the space of the cube. That contracting or diminishing space, or removing space from one cube to one sphere is gravity. That is how the seven of the sphere that turns around the square four of singularity can remove the space from the part we think of as outer space and redirect that space towards the centre of the cube.

Therefore the sphere has the strongest gravity where the sphere holds the least space because all space eventually flow to that point by contraction of the spin of the seven in relation to the ten of the cube. Wherever there is contact between the cube and the sphere, the cube will be at a disadvantage, as the form of the cube will always compromise to the strength of the form of the sphere. It is absolutely vital to realise the concept that the six sides of the cube retains five sides as the sphere removes one point wherever the sphere makes contact.

Matter in relation (part of) to the total dimension of space.

(10 / 7) \ (7/ 10) = 2.04

1.4285 / 0.7 = 2.04 Taking from both orbiting influences

SPACE DIVIDED INTO TIME

(7/10) / (10/7) = 0.49

.7 / 1.4285 = 0.49 Taking from both orbiting influences

SPACE MULTIPLIED WITH TIME

7/10 / 7/10 = 1 and 10 / 7 X 7/10 =1 Therefore not influencing change

THE PROCESS PARTED USING THE ROCHE PRINCIPLE

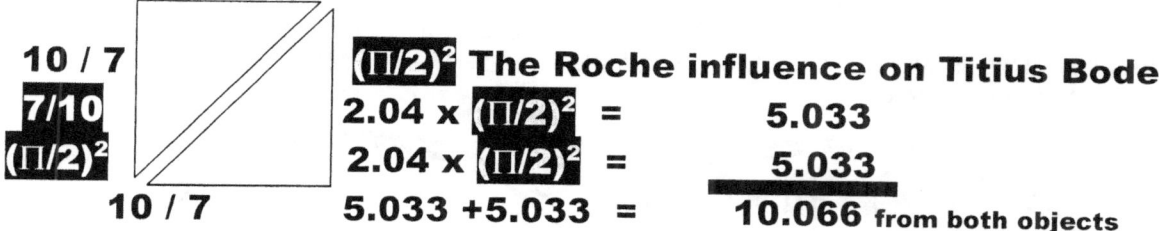

10 / 7

7/10

$(\Pi/2)^2$ **The Roche influence on Titius Bode**

$(\Pi/2)^2$

2.04 x $(\Pi/2)^2$ =   5.033

2.04 x $(\Pi/2)^2$ =   5.033

10 / 7   5.033 + 5.033 =   **10.066** from both objects

SPACE DIVIDE INTO TIME

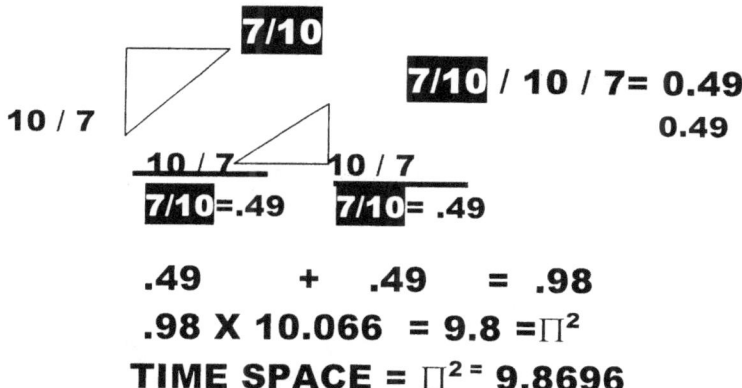

7/10

7/10 / 10 / 7= 0.49

10 / 7   0.49

10 / 7   10 / 7

7/10 =.49   7/10 = .49

.49   +   .49   =   .98

.98 X 10.066 = 9.8 =Π^2

TIME SPACE = Π^2 = 9.8696

TIME SPACE =Π^2=9.8696= Space and time in a dimensional implication.

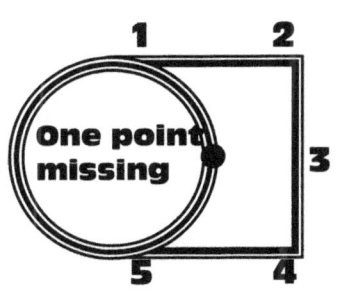

One point missing

Do not only think in terms of the macro, but rather see it in the absolute micro where every atom being in contact with space it is forcing space in the cube to lose its advantage to material. The generating of gravity is not confirmed by the Titius Bode law but the Titius bode law perform the generating of gravity in relation of bridging singularity restrictions by implementing the Roche limit at $\Pi^2/4$. It is the turning of material (7) through the dimension of space (5+5=10) and from that relation, which is motion that becomes the generating of gravity. It is true that it is immensely more complicated but how the hell would you find out how the process truly works when all the different element's functions join in the process when you never read my work!

The Titius bode is an interaction between particles in motion through time where time leaves a mark on material ($k^{-1} = T^2 / a^3$) and material leaves a vacancy within time ($k = a^3 / T^2$) and that motion of material or space inter acting with outer space or time is the legacy of gravity by motion $T^2 k$. Only motion can turn singularity from the spot it has in the sphere to the dot to form a line that takes charge of the independent object, which is in motion and is in relation to another centre. This confirms Kepler's $a^3 = T^2k$.

The interaction of the seven forming material and that concludes the sphere in relation to the ten that commits the cube is the connection between time in infinity holding the immovable part of singularity, which then forms a relation with time in eternity and that, which then forms by motion parting time, is space. That is the Titius Bode law and that is gravity, the glue that connects everything to anything.

The measure of seven standing in relation is all a forming of singularity standing relevant to singularity. In the cosmos only singularity apply. On one side of the sphere there are five plus five which becomes ten and the sphere of material holds six position on every side plus a seventh that stands in the centre and is centralising the six edge point where the point in the centre forms the point that generate gravity through singularity. The other value such as mass is a generated form of spec-time that comes about as time delay and that is what material is. This is rather slight too complicated to divulge in this short profile but is well documented in my books.

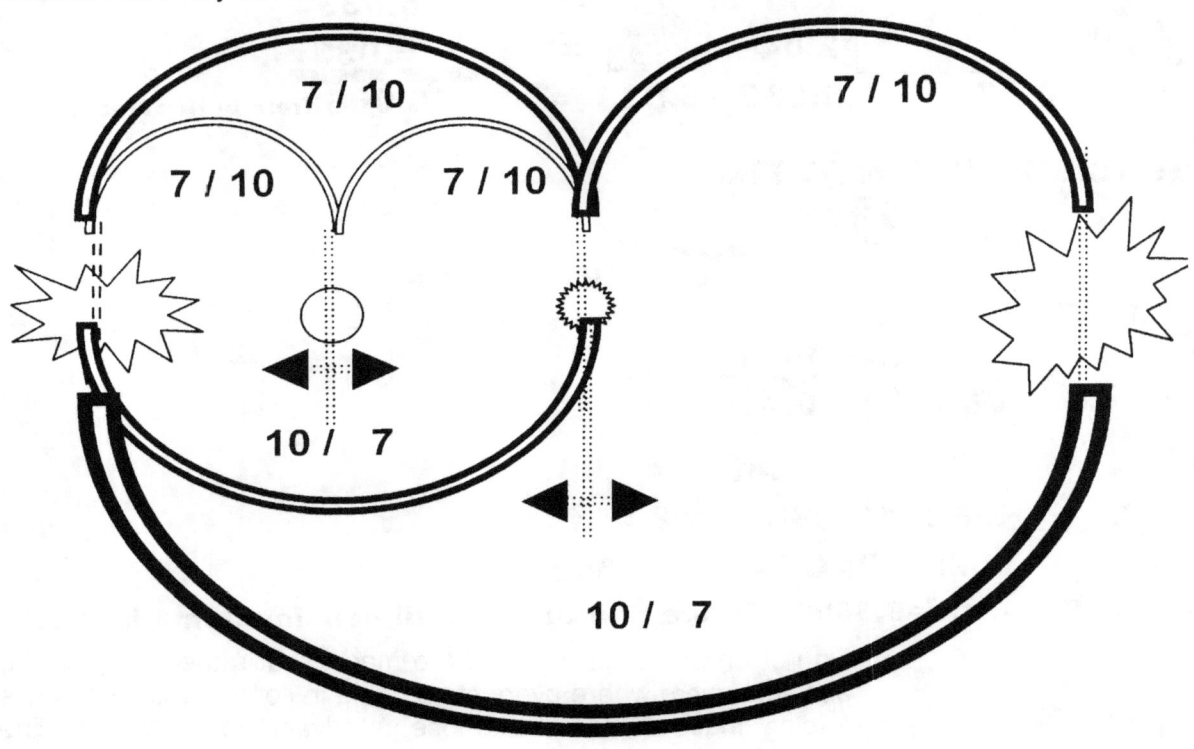

Taking the argument back to Kepler's law,

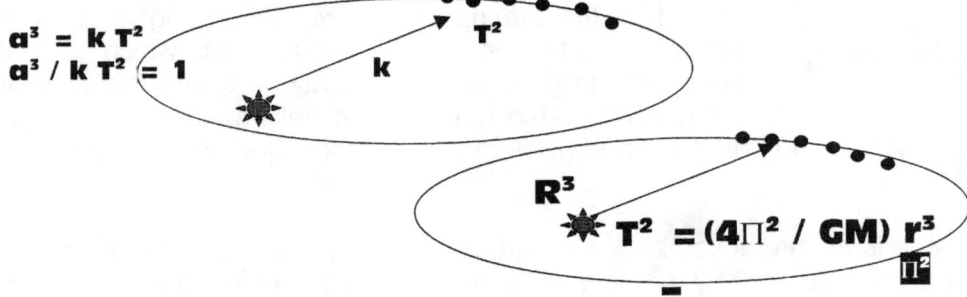

$$a^3 = k\,T^2$$
$$a^3 / k\,T^2 = 1$$

T^2

k

R^3

$$T^2 = (4\pi^2 / GM)\,r^3$$

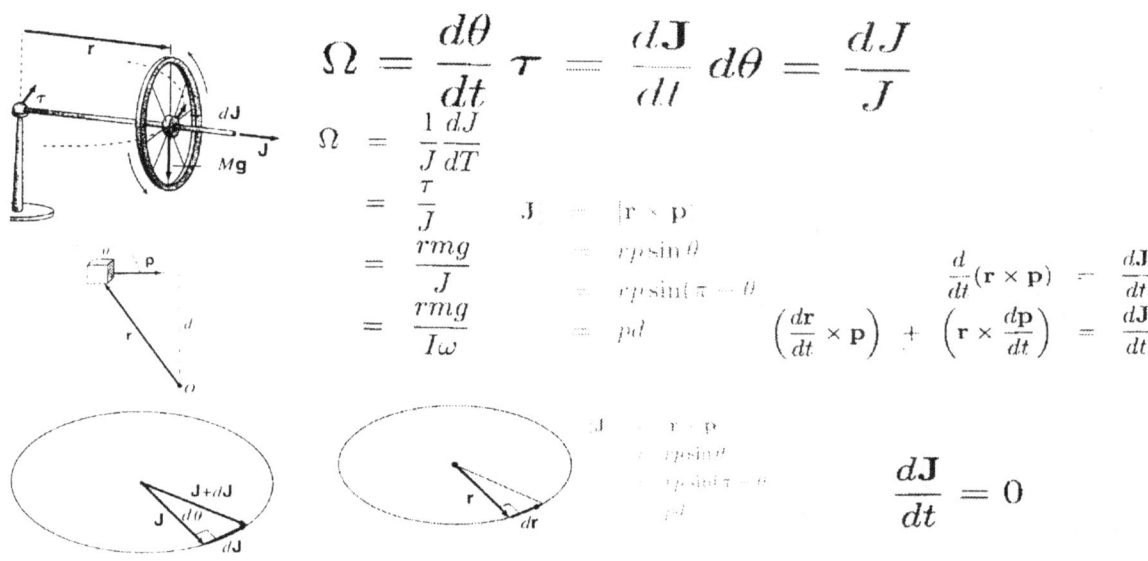

The Titius Bode law is one of the phenomena but it is also the phenomena around which gravity is centred. Alongside the Titius Bode law is equally important the other phenomena being the Roche limit wherefrom the Roche lobe derives its value and its measure, the Lagrangian points: and the Coanda effect. At present Newtonians do not even regard the Coanda effect as a cosmological function but sweeps it under the table as something aircrafts implement to fly. Not one of them went as far as linking the Coanda effect and its association with something of value in the cosmos because the Coanda effect does not apply mass as any norm by which it can appreciate the principle of flying.

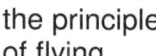 **Singularity**

 **Singularity 1° expanding in time**

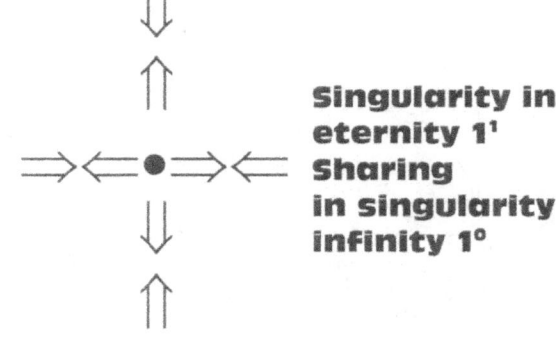

Singularity in eternity 1¹ Sharing in singularity infinity 1°

In that Newton is excluded and where Newton is excluded such things just is a hallucination of sorts. Newtonian minds cannot pass off anything that does not support Newton and mass inflicting gravity and anything that shows a lack of supporting the likes of mass they disregard with eagerness. Since there is nothing that actually supports Newtonian thinking everything of value is passed over since nothing collaborates Newton and therefore they have the cosmos being nothing. How desperately wanting and illogically incoherent can an argument get where the prospects of nothing is being offered as a valid commodity to be used by the cosmos in the cosmos where it is supporting the structure to be the ingredient that is forming the cosmos? It was the very nothing Newton introduced when he set his hoax into action and only the nothing he created finds support by the Newtonians since they have nothing other than nothing they used to form cosmic and that can at the same time be implemented to support Newton.

If this principle Newton envisage is true then the Titius Bode law of Planetary positioning cannot be true because the **k** as a factor in Kepler's formula of pace-time $a^3 = T^2 k$ has a valid position. It has and that is one reason I shout fraud and I shout conspiracy. As one can see there is a distinct relation between the planets where every planet doubles in radius by distance from the Sun. Every planet takes up a position where the planet doubles the distance that all the other planets in total have. This I explain and that I bring in relation to the charging of gravity in that manner and I include an example of my explaining.

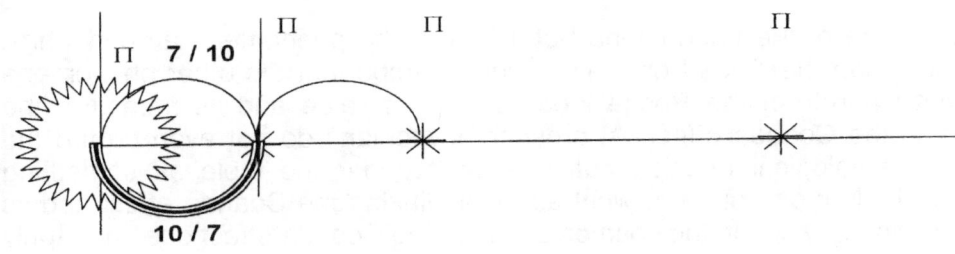

The one inner planet 1^1 next to the inside of the planet taking position forms a marker while the orbiting one positions in relation to the second 1^1 and that forms the dual singularity relation they have with the Sun centre 1^0 singularity.

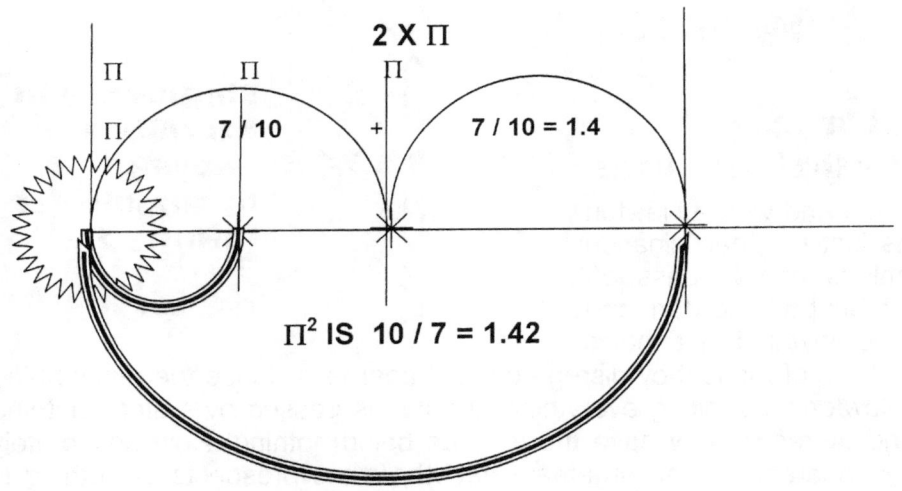

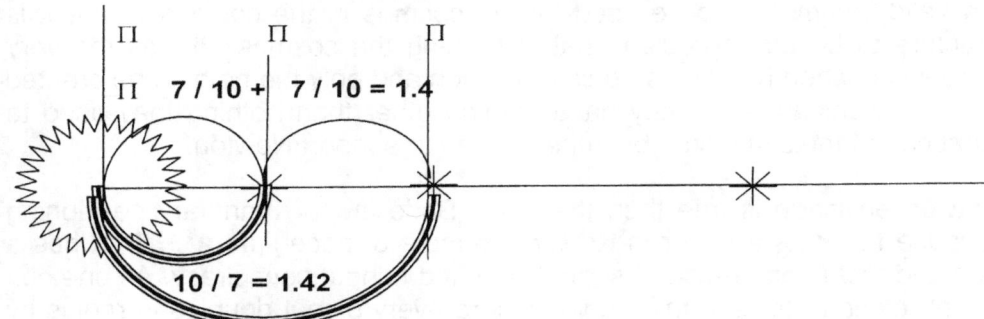

$$3 + 1 + 3 = 7$$

With the evidence of the Titius Bode law being that much apparent it should be clear to all that mass has no place in cosmology. In the cosmos there is no aspect that works by the principle of mass. The cosmos places all objects on a homogenous footing. The fact that Mercury is equally dispensed as Pluto in relation to the Sun or is as equal as Jupiter is, is undeniably clears evidence that mass has no role in the cosmos and therefore mass has no possibility of producing of gravity. We have a marker forming singularity by seven around a centre, which secures the pivot that forms a point in seven while the space in between is ten. This is the generating of gravity in the following manner: $7 + 7 = 14$ and $10 / 7 = 1.42$ where $14 / 1.4 = 9.86$ which is gravity Π^2

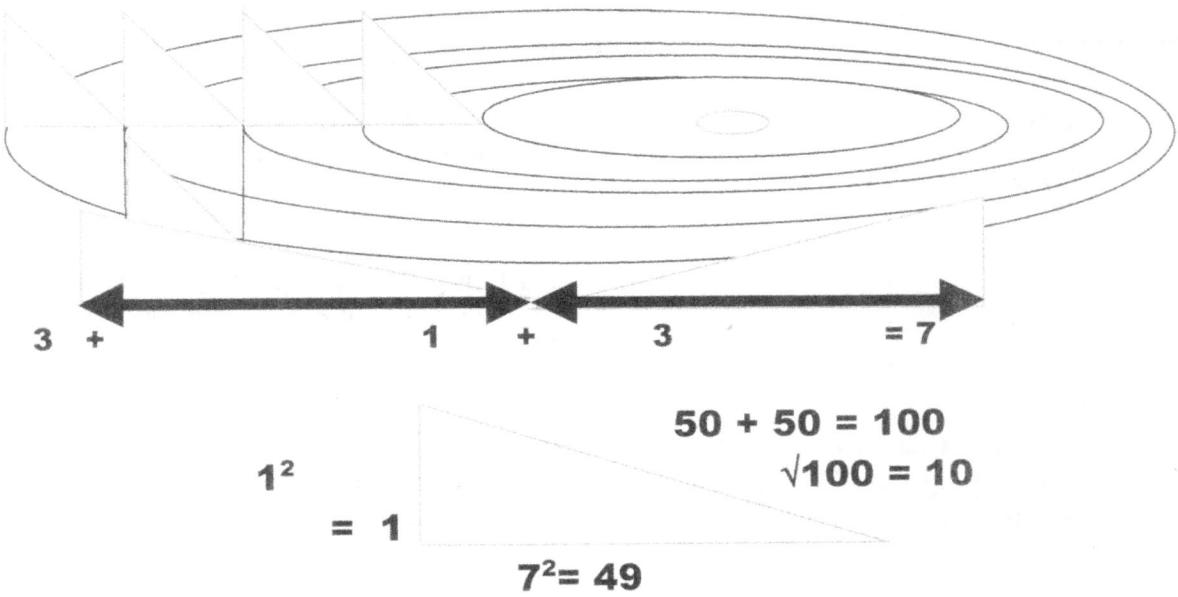

$$3 + 1 + 3 = 7$$

$$50 + 50 = 100$$
$$\sqrt{100} = 10$$
$$1^2$$
$$= 1$$
$$7^2 = 49$$

The entire truth is that there are no forces of any description but it is a flow of time coming from a liquid or less dense form of heat while it enters a controlled form of heat where spin produces the preserving of time within a sealed environment to contain heat for the accumulation of heat by singularity.

Mathematically the layout of the solar system is in terms of seven being on both sides of a divide singularity put in place and taken in accordance with the law of Pythagoras the root of distance in the square mathematical comes to seven, notwithstanding which way it is measured.

The spherical positioning layout forming the Titius Bode Principle

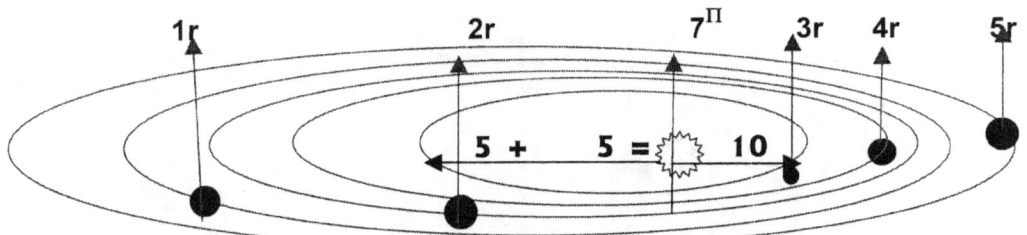

From the matter-to-matter relation in the Titius Bode configuration there are 7 / 10 + 7 / 10 = .7 + .7 = 1.4

From the space-to-matter relation in the Titius Bode configuration there is 10 / 7 = 1.42

The $5 + 5 = 10$ is a position of dimensions as space loses value to singularity. The 7 that matter diverts in points from singularity may seem, as coincidental but is valid. Still in accordance to our perception valuing the number in degrees, it seems coincidental but if it is coincidental, it is nevertheless a figure of diverting proven as accountable in all other calculations and plays a most dynamic role.

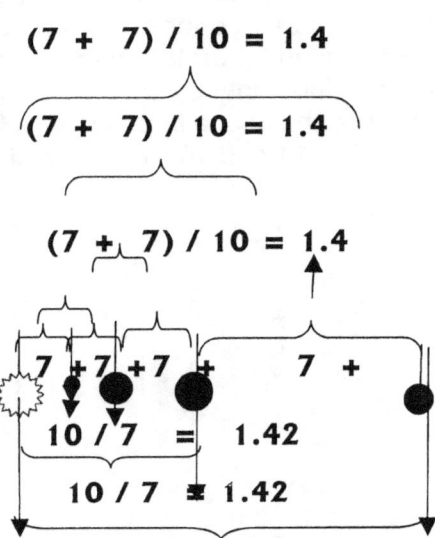

$(7 + 7) / 10 = 1.4$

$(7 + 7) / 10 = 1.4$

$(7 + 7) / 10 = 1.4$

$7 + 7 + 7 + 7 +$

$10 / 7 = 1.42$

$10 / 7 = 1.42$

$10 / 7 = 1.42$

= .7 /Λ\ 1.42

= 1.4 /Λ\ 1.42

The Lagrangian 5 point system results as much from the Curvature of space-time as does the form the Black Hole holds. The Galactica is the opposing equivalent of the Black Hole and has identical but opposing similarities being the five points positioned to singularity. The galactica is generating space and the Black hole is degenerating space.

Because the space-to-matter is in the square at 10 placing the matter-to-matter at a square of .7 + .7 = 1.4 the space-to-matter forces the matter-to-matter to double the distance by number as structures are place father from the mainΠ^0 maintaining singularity.

1 3 6   12   24

Reasons why this does not fully apply to the solar system I give in book # 7.

Gravity is generated in much different terms and does not work or rely on mass. It has all to do with the law of Pythagoras working in relation with singularity and in conjunction with the combining value of ten finding a square of hundred and seven by the square that stands in relation with the one singularity presents in the square will form o total of fifty. With fifty on both sides we have the effect that fifty plus fifty totals one hundred and the square root of one hundred is ten Therefore we find that seven is related to ten and on that is the cosmos built. From that we derive gravity by the measure of 9.8 and that is the formula whereby gravity is produced. This might sound slightly overwhelming but if one multiplies the Roche factor on relation to the Titius Bode as the Titius Bode form cross-referencing with 7/10 and

10/7 we get the value of $\Pi^2 = 9.86$ which forms the value of gravity. Mass is excluded as any cosmic reality. If gravity is what contains the Universe and gravity is responsible for the way the Universe expands, wouldn't gravity be responsible for the way the jig is set and the manner used to build the Universe…and the solar system is one part of the Universe where the Universe is one part of the solar system... being in the Titius Bode law.

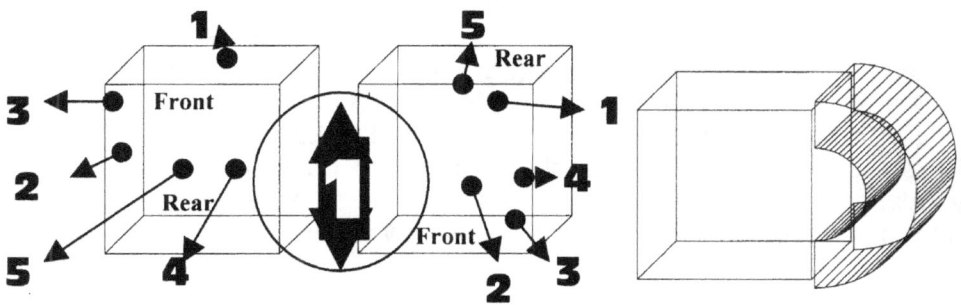

1 Relating to 5 on both sides is ten

The Roche limit is:

The region surrounding each star in a binary system, within which any material is gravitationally bound to that particular star. The boundary of the Roche lobes is an equipotential surface, and the lobes touch at the inner Lagrangian point, L_1, through which mass transfer may occur if one of the components expands to fill its lobe. It names after the French mathematician Edouard Albert Roche (1820-83).

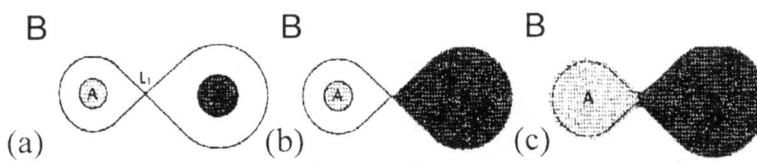

THE ROCHE LOBE: In a binary system, the Roche lobes of components A and B meet at the L_1 Lagrangian point.

(a) In a detached system, neither star fills its Roche lobe. (b) In a semidetached system, one massive component, B, fills its Roche lobe. (c) In a contact binary, both components overfill their Roche lobes and share a common envelope. Lets explain the importance of this Roche limit and how the Universe used the Roche factor to produce the Big Bang. That is where it all started…

The main thing what the Roche limit proves is that mass does not pull mass until mass devours or destroys mass by collision. The Roche limit clearly is rubbishing that and by the Roche limit that rubbishes Newtonian's claim of $F = \dfrac{r^2}{M_1 M_2}$ it puts everything about mass and the pulling thereof, in the paper basket. $F = \dfrac{r^2}{M_1 M_2}$ should mathematically represent the statement that mass pulls mass by destroying the radius, which was of course the modified blunder of $F = G\dfrac{M \times m}{r^2}$ where Newton and Newtonians after all the evidence are equally determined to convince the world that mass pulls mass towards mass. It is hogwash if ever there was hogwash because no object of cosmic proportions will come closer to another object than the value of 4.67 or then $\Pi^2/4$ and that is directly related to singularity charging gravity.

The Lagrangian point system should be very clear by now as it forms half of the half of the square that forms space in the square. The ten of space divides another halve and there gravity confirms a position allocated to material one point outside the four positions that secures time by $4\Pi^2$ It is the point where the line meets on equal terms as 180^0 with the half circle which also stands at an equal 180^0 and it relays to the triangle holding an equal 180^0. It is how the cosmos started before space became confused with the lagging of time but that I explain so specifically that no one wishes to read it. Again the Lagrangian points offer no proof of Newton's mass and the order that the Lagrangian point system holds has no indication of any relation of any sorts where mass is used to bring some alignment or manifests as a factor being present in the cosmos in any way whatsoever.

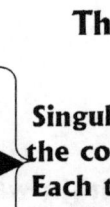

The Lagrangian System implicating the five positions extending from singularity

Triangle 1

Singularity dividing the cosmos

L_3 L_1 L_2

Each triangle claiming a side of the universe

Triangle 2

L_5

1 Half circle	$=180^0$	L_3 L_4 L_5
2 Triangle 1	$=180^0$	L_3 L_4 L_5
3 Triangle 2	$=180^0$	L_3 L_4 L_5
4 Straight Line	$=180^0$	

The half Circle $=180^0$ combining as a sphere when comprising Singularity in the matching of the value of the straight line forming the half circle and combining as the triangle and all are equal 180^0

The second one also fits in the singularity influence on the Universe.

5/2

Five sides divided by two spheres.

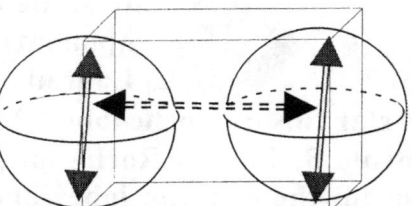

Where Π extends to lock onto the next sphere's extending indicator, Π has to connect to Π forming the square of space and translating that to the half of Π being $(\Pi/2)^2$.

The Roche limit 5/2 becoming $= (\Pi/2 X \Pi/2) = 2.4674$ as singularity interferes

LAGRANGIAN POINT:
The Lagrangian points are five equilibrium points in the orbit of one body around another, such as a planet around the Sun

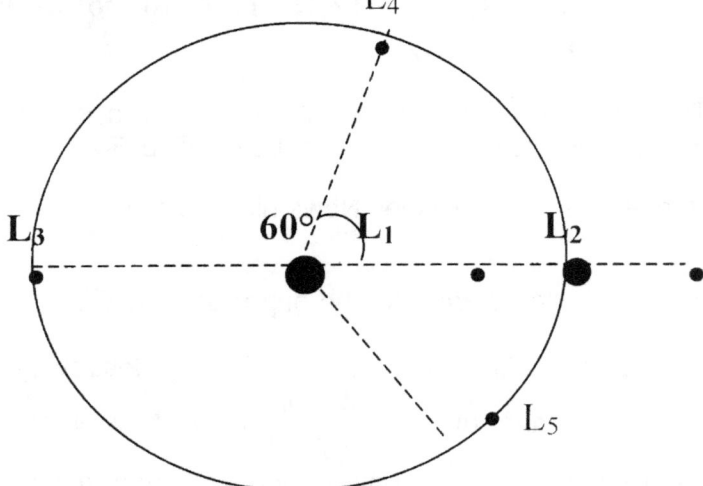

The mere fact that no planet is coming closer to the sun or to each other should be enough grounds to launch a serious investigative research into the probability of Newton's theory about mass pulling on mass in order to reduce the radius. That is lacking but moreover my saying that keeps my work unread for six years on the trod. However one can see by the size in pages of the explaining given here and space it takes when considering that the three books cover in total almost 2500 pages. With that in mind then this is nominal in context to the overall information provided in the books given to you for reading. The space it takes is such a small part even of this letter and still this explaining exceeds by far the proof of mass inflicting gravity that Newton established.

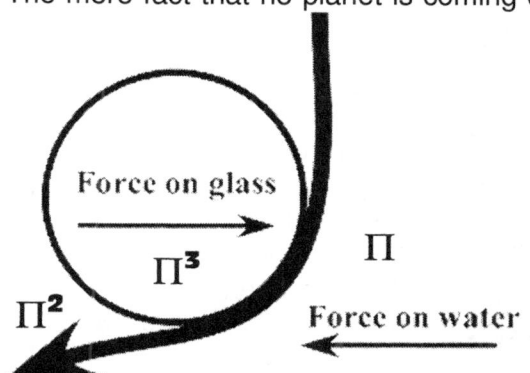

The Coanda effect is the instituting of Kepler's formula where space provides the limit of liquid and liquid confirms the limit of space. Kepler's formula states that space is equal to the flow or moving thereof $a^3 = T^2\,k$ where then gravity is the extending of space which includes the liquid forming the total space that forms $k = a^3/T^2$ and liquid is connecting to space by gravity forming the result in contraction $k^{-1} = T^2/a^3$

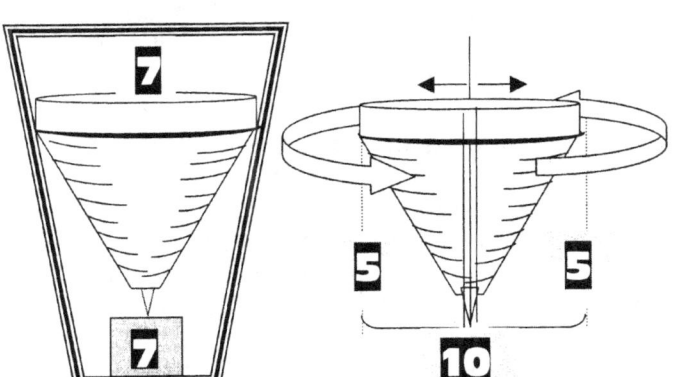

The four phenomena are inseparably one and that forms motion where motion is gravity. It is the crossing of the point that holds singularity where one then has a circle that attaches time to the circle ant the other is time attaching to the circle.

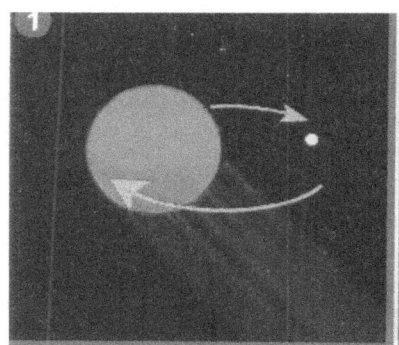
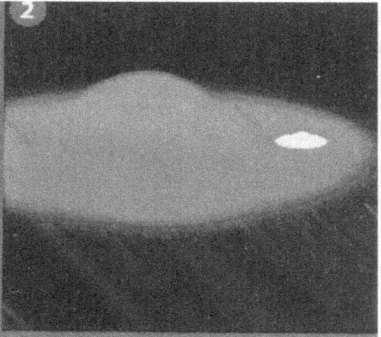

The Roche limit is to my mind the most significant evidence of proving Newton's failure. In the cosmos we have two systems that approached and became too close for comfort. The one is related to the other as a lesser relevant singularity while the other forms the conserving singularity. The two does not engage as Newton would suggest and plummet into each other's space where the mass draws the mass until the radius is no longer parting the structures. At a distance calculated to be $\Pi^2/4$ the two engage into forming a lobe that involve both spheres. The lesser star candidate overheats within this combining sphere that charges the envelope to cover both structures. Both structures losses independent identification as the Roche lobe cover both object and all the material in both objects elect a centre which provides acknowledgement to both stars. Then the

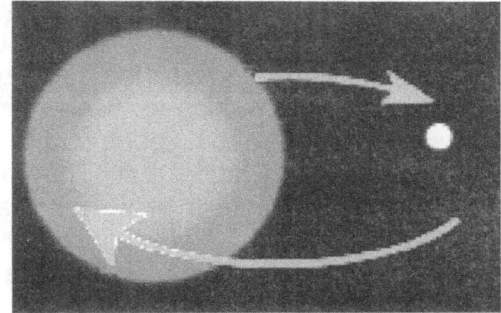

one star turns the other star to liquid where the heat forms a cloud that covers both stars.

The process that develops turns all the material into heat and it is evident that a process of

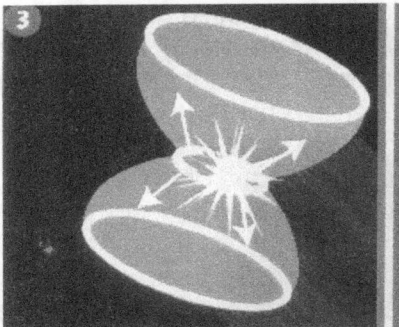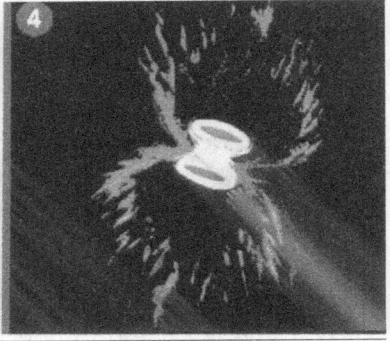

severe overheating takes charge of the proceeding that follows.

From the centre there is evidence of uncontrolled overheating that deforms the star in a circular patter where the true overheating is found right in the centre of the star. The object that supposedly retains mass holds the mass apart while both remains at a distance of $\Pi^2/4$ times the radius apart but both become liquid and both experience centres that destruct in violent heat outbursts. Newton is completely absent in this whole picture. No evidence of mass is forth coming but what is very clear is that where the control subsides, heat in an uncontrolled fashion takes over. The collision that was to come about should Newton's mass that is pulling mass close be forthcoming, was not that much forthcoming after all. Again it slurs at Newton and at Newtonian misconceptions while Newtonians hide this under a blanket of corruption.

Should Newtonians insist that their constant referring to mass is the space that material holds then the following proves just the opposite. Mass increases in stars in ratio to and where space reduces where it reaches a point where mass almost grows eternal while the space that the mass claims becomes infinitely small. Nowhere is mass a factor except when the motion of material is relinquished and mass smothers the individual purpose of the material by confining the material to a domineering structure. Why that would be is somewhat tedious to explain in this letter but since you never read my work you will never be able to see that part of the explaining. You are fixated with Newton and with mass while Newton's mass in outer space is a pipe dream. Outer space work on singularity $k^0 = a^3 / T^2k$ and space – time $a^3 = T^2k$ and mass plays no part.

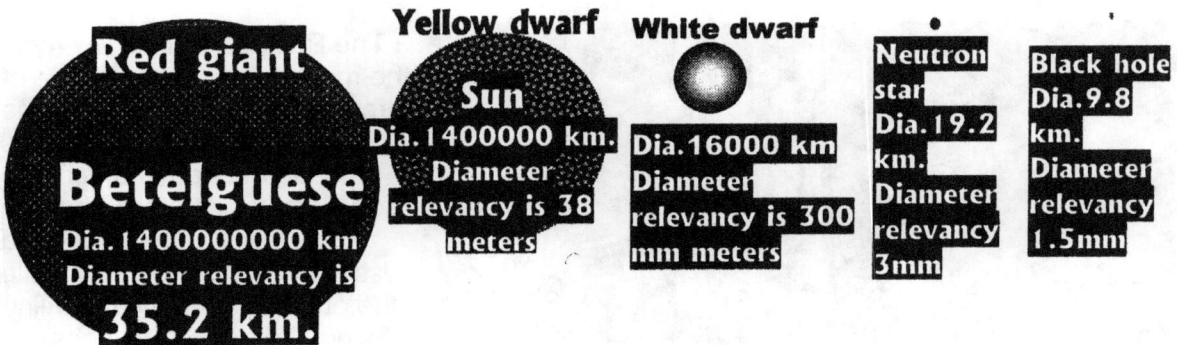

You are a Master and a Professor and you see the young minds enter their future Alma Mater in anticipation and excitement every year. The young minds are waiting for their tutorship on the condition they are accepted into the institution of learning. They come in anticipation while waiting to receive their wisdom where they would carry the flame of brilliant knowledge through their generation in the era in time where they will be entrusted to have the opportunity in doing that. They are ready to learn and accept. Some might challenge some authority about thinking but that never involves the basic arguments concerning the founding principles. Notwithstanding their seriousness it is always about matters in a superficial capacity. In reality that is a manner to channel all rebelliousness because of their eagerness to improve on that what they inherit. Before they can improve, they first have to accept what their elders confirm as the basic and the foundation values. In the case of

Physics it is accepting Newton to be better than God and my saying that is not intended as blasphemy but it is to break open a gangrenous sore. To manage the introducing of Newton as God's equal, Pavlov is introduced and not the man or his principle but his conditioning. Then comes the mind conditioning that has to be undergone by all students harbouring any wish to become a full member of the future physics paternity. The mind control is automatic in nature. The thought conditioning is a thought association where the student stands on Earth with the experiencing of mass. It uses Pavlov's theory that was instituted by the Physics department at all Universities centuries before Pavlov was instituted. The students are told that mass is the sole contributor to gravity and gravity keeps everyone solid on the Earth. They are on the Earth because they have mass that forms the gravity keeping them on Earth. In their minds no proof is required when they are experiencing the gravity about which they are taught by their tutors. They are programmed in the manner that physics is presented to their young mind to accept what they cannot refute and their standing on Earth is one fact that they cannot confute. They link their standing on Earth to the lesson about gravity and mass and immediately this presents irrefutable proof that they have mass which forms the gravity that keeps them down onto the Earth. Now their tutors have one more task to complete and although it is small, it is crucial in confirming their acceptance in their thinking patterns for the rest of their lifetime to come. They have to associate their standing on Earth while having mass which forms the result of their gravity that they are experiencing with the name Newton. Newton said this is the process where mass pulls mass to remove any distance between objects. They have mass. They can feel their mass pulling the Earth as the mass of the Earth is pulling them and that is stopping them to jump over the Moon. They fight the affect of gravity every second they are on Earth. The gravity coming from the mass is eroding their life power until they will finally perish and go to their graves. They are in gravity until they are in their grave. Gravity is pulling them to their grave. Sir, madam in your personal lives you too, were victims of this very process and your accepting of Newtonian mass that is generating gravity does not stem from your receiving irrefutable and undeniable proof when you were first introduced to Newton where you then were shown proof beyond your wildest quest for facts during your brittle youth but your sureness about the matter came from the measure of mind conditioning you were resigned to accept. If you do not believe that statement I just made, then sit back and find the day the proof about Newton was presented to you as a student. In all this, one man going by the name of Isaac Newton coined the process when Newton was also a young mind eager to prove what he was capable of. Newton formulised everything he was experiencing as mass and gravity. This concept is so easy to associate with as his frustration of having gravity, which is the process

that is sapping his strength by $F = \dfrac{r^2}{M_1 M_2}$. The formula insists that anyone has mass

while the Earth has mass and the Earth mass is pulling on all other mass which the two combine to fight the distance that might come about when any person should jump. It erodes any distance that might be in place between the mass of the Earth and the mass anyone is

experiencing. What can ever require less proof than $F = \dfrac{r^2}{M_1 M_2}$ for it involves anyone's

entire existence. It is a truth the student can never contradict and because of its familiarity and always being present throughout all person's life era. Therefore any and all further proof is unnecessary and redundant. The student has all the proof he or she can ever muster just by standing upright. Is it really that simple or does nature hold more meat that is covering the bone of contention? Galileo said no mass applies when objects fall because all objects fall equal. By Newton's very first inspiration he contemplated mass as that which is generating the force producing the energy by which the falling process of objects becomes a reality. He said mass times mass divide to destroy the distance between objects with the suggested formula of $F = \dfrac{r^2}{M_1 M_2}$. The concept that this idea promotes contradicts Galileo just because Newton introduces mass where Galileo excluded any form of differentiations. From this one would gather that Newton was incorrect even before he introduced his idea to the world, and yet not one of the bright Academics could spot this in three hundred years. If you are as convinced about Newton then explain why would Galileo have all objects fall equal without mass intervening in any way just because all objects fall equal. If mass was a factor charged with the producing of gravity, then all objects with more mass have to fall quicker than the other with lesser mass. If mass is the producing factor of gravity and mass is in place to produce more gravity, it then has more mass which then is in place to be producing more gravity that will encourage more of the falling motion which is gravity, where more mass arrests more space in quicker succession that will allow the object in falling to establish a faster rate of falling. Yet according to Galileo there is no chance that mass comes about to bring any discrepancy amongst falling objects in any way except when the motion is restricted to a stand still where the bodies come to rest against each other. Then mass becomes a factor but not while they are in the process of falling. The absolute uniformity excludes mass being a differentiating factor by elimination because mass clearly has no influence on favouring the more massive while falling. Mass only comes about as a factor when the body is in rest. In my books I explain why mass is a human instituted concept and has no cosmic validity. I prove that anything that reached a point of having mass has also by the same token reached a cosmic death because that object surrendered its cosmic independent existing which is secured in the defining of independent motion where when burdened with mass, the motion of independence becomes annexed by the motion of a larger structure and by capitulating, its motion to mass it accepts the motion that the larger object prescribes. Accepting the motion of the larger object is forming mass. By accepting mass the object becomes estranged from the cosmos and by a marriage to the larger object through instituting mass, the object divorces its independent identity it might have in cosmic motion. Through its marriage in mass it relinquishes independent motion to acquire the motion dictated by the occupying body and being part of the motion of a more developed structure it becomes something with no differentiation in cosmic terms as a distinct object. In cosmic terms any object in mass can never again move out of mass. Can any one imagine a mountain flying about even on another planet? Can any one picture two rocks chasing each other on Venus where life can gain no foothold? It is not possible because the mountain has mass and the mass makes it part of the Earth (or in the case of Venus then part of Venus) or any other planet in question. Once mass forms a part of the object instead of motion, the method of applying mass by removing independent motion is a means to secure the lesser object to the dominating object. The lesser object is connecting by mass to the larger object that captures the smaller one by granting mass and removing independent motion. The giving of mass is a way to enforce the will of the larger motion onto the lesser object by removing the ability of independent gravity from the lesser object. Then all distinguishing motion indicating independence of the lesser object becomes impossible because the motion that the domineering object takes control by measure of mass that dictates the terms of motion. Any object on Earth can only move in relation to the measure in which the Earth is moving except where life produces motion and life is consecrated to one place and that is the Earth alone. Only a paranoid fool having an eidolon would find evidence of the presence of life elsewhere or anywhere in the cosmos in any form other than what life has on Earth when life can only be on Earth as a proven reality because any other idea is merely hallucination that belongs to the insane mind. There is no evidence of life anywhere but on Earth and only rampaging insanity would deliver evidence in support of the contrary perception. Mass is the state in which the object presumes when the object in mass is not any distinct independent cosmic motion any longer. Mass is when the object surrendered its independence and with that it divorced its specific density by becoming part of that which is forming the restriction of the cosmic motion that secures the object having motion. Motion provides independence by alienating mass from forming and mass is anti motion that arrests all cosmic independence.

If a ship is buoyant on the water, it is presumed by science that the ship has mass. That cannot be the case because what then is applying when the ship sinks to the bottom of the water where it then rests without motion on the bed of the Earth? Where the ship finds a location as an individual item laying under the water and not on top of the water, then the sunken ship will have mass because it ceded independent motion where mass then enslaves the motionless object to the larger object. That then prescribes all conditions forming motion. Only when the ship sinks and rests as an individual object on the bottom of the water can the ship qualify to present mass in an individual capacity. While floating the ship takes on all the characteristics one will find when the ship is being part of the water. The ship holds the mass of the water it displaces. The ship or any floating part of the ship is for all purposes just water. The water it displaces while being buoyant is much less than the mass the ship will have when the ship is in dry dock and is fully laden. No Newtonian can ever presume that a ship fully laden while floating holds an equality in mass measured in relation to a ship fully laden in dry docks. The ship does not presume individual mass while floating above the water or even when floating in the water as submarines do. However, when it is at the bottom of the water it presumes the mass of its structure as well as all the water presiding on top of it. But floating on the water the mass of the ship might add to the mass of the water but as such the ship has no mass. If you insist that the ship can float and have mass, then please explain what you would say is the difference between the floating ship and the sunken ship, because there is one Universe of difference between the sunken ship and the floating ship. If you question my logic in this matter then go and look at the difference in the Titanic. In the one the floating ship presumes in the capacity as forming more water and in the case of the sunken ship the ship is more Earth. That is the difference but also in that is all the difference one will find in what there is between gravity and mass. However it does not stop there because relevancies change where motion is concerned. Even if you insist that what you refer to as the material filling the material-filled space of the atom or an assortment of atoms forming a structure that has mass because it is captured in a unified structure in the form it has while it has in the space it holds and being in that position it is mass then that still does not qualify to associate that with mass, because that filled space can go from presenting a few kilograms on earth to where it then becomes billions of tons in a large star. There is a reason to regard form as mass because of the increase in mass and that I explain in very detail in the books. The increase in mass is associated with the loss in form and when mass increase by relinquishing form, then form is not mass. That is not mass that you refer to when you refer to the body that forms the object because the material has the ability to increase the mass by decreasing the space that material occupies. That which you refer to is the space a^3 that is duplicating $T^2 k$ by motion into its next position in time coming from its previous position in time that it holds in the current position in time and that matches all the scenarios that one may contribute to the Coanda effect. That is space a^3 duplicating the position it holds T^2 being relative to an allocated position every time by the measure of k. Gravity is $a^3 = T^2 k$ and that Kepler proved to be gravity long before Newton admitted he had no idea what gravity is. It is the motion of space that is duplicating the space by moving through time in relevance to a measured centre point.

When a body falls into water from an extreme height the fact of having motion such motion brings gravity that changes when the falling body that is falling connects to the water rim where the motion ends by water restriction and the motion of the falling body changes its position from having gravity to become mass. While plummeting k towards the Earth T^2 where the Earth at that point holds water at the surface the water is the solid a^3 notwithstanding it being in a liquid form. The water at that point holds the mass as it forms part of the Earth and is therefore the solid factor representing a^3. While the object is falling it is not the object that falls but all the space that holds the object that moves. Galileo proved this when Galileo proved there is no size differentiation present while bodies fall. In that way the object has no mass since it travels with the space and forms part of the liquid space that moves. The body is a part of a constant waterfall by relevance of motion. The body has to be buoyant because the body has no mass and that Galileo did prove notwithstanding Newtonian denial to the contrary. It is not the body that falls down but the space the body has which is accompanying all the surrounding space as well that is falling and while all that space is falling it is taking the falling body along. That can be the only explaining of not having any distinct differentiations while all objects fall in a similar way without showing differentiation when there is such obvious and great differentiation between such bodies. The object holds buoyancy in the liquid space that carries the object while the lot of space is descending to the Earth in the form of a liquid. The object falling might be a solid but while falling, it moves and therefore it has gravity making the object being part of the liquid T^2k. The ship floating might be a solid but while it is floating it is just more water in the water. That is the only scenario that is able to explain Galileo's finding of equality of different mass in a downward motion. By touching the water the gravity goes on to form mass at the moment of impact and the object turns at the moment of impact from being part of the free falling liquid having gravity, to being blocked by a solid and at the moment of impact ending the motion and aborting gravity to accept mass where this ends the motion or gravity as this results in state of gravity ending that then turns to form mass. The relevancies at that point changes where at the point of impact the falling body will be liquid and the water will be solid. The water holds the mass of a solid and the body being part of the space, which is descending then, at the moment of impact forms the role, which all liquid has. The mass breaks the fall but the body as a factor is swapping from being part of the liquid to being part of the solid, which is the position of the water until at the moment of impact. Just after the moment of impact the water changes the position it has from first being a solid a^3 going onto presume the role of the liquid T^2k and then becomes the liquid T^2k in relation to the solid the Earth holds but only after it first presented the factor of forming a solid a^3 that breaks the falling object. That is why the water is more solid than a human body at the point of contact. The human body forms liquid while the water is solid because the water is without relative motion at first while the body shows absolute gravity and then after the fall the body becomes part of the moving water in relation to the absolutely sturdy Earth. Then, when contact is made the water becomes buoyant since it harbours the object while the object is descending further but this time in the medium of the water and the water this time has six hundred times more density than the space has just above the water. Liquid and solid are allocated to the position of motion that the one factor presents and the position of immobility that the other factor presents. First space was the density factor provider but then after impact the water took on the role as the liquid and the buoyancy moves over from the descending space in air to where it then turns the liquid state over to the more dense water that formed the solid up to such a point. Then at the point where the body comes into contact with the water the roles change where the water forms the liquid that allows the Earth to form solid. The body then will float in the water because the water has six hundred times more density than the liquid air has but is still a liquid in relation to the density the Earth presents. When the body falls on the bottom and remain motionless at the bottom of the water as part of the Earth the body then has mass since the body no longer moves on own accord. Let's investigate Newton's first concept of gravity at the point just before a hunger for personal fame got the better of his senses. Newton stated that $F = \dfrac{r^2}{M_1 M_2}$ which is mathematically expressed as a force that will destroy the square of the radius between two falling objects where the tempo of destroying of the radius by the square holds the mass of the two objects in product to achieve that. The mass factor combines by multiplication with one another in order to diminish the radius from both ends between the Earth and the object at a distance from the Earth, which then forms a square of such a distance. The mass factor makes the bodies conjunct and Newton concludes that it depends on the mass of the bodies in determining how the bodies will fall. The formula suggests this conclusion by eliminating all other possibilities $F = \dfrac{r^2}{M_1 M_2}$. That is Newton's thinking, which he unequivocally expressed in this formula. The formula is no longer in official use but the removal of the formula and the replacing thereof is not as a result of the incorrectness they established when they used the formula in the practical contexts. By using the formula in its original form indicated cosmic destruction and that they had to replace because of the unaccepted that forms part a conclusion coming from the use of the formula. Instead of revising the principle in favour of Galileo they did not revise the principle but went on to revise the formula. By the way it is revised the madness in the concept escalated just because of the protection science found a need to bolster Newtonian ambitions. By implementing Newtonian concept in a manner to determine the forces driving the cosmos it placed science on an equal footing with the abilities and brainpower of God. This is so deeply rooted that after three hundred odd years the error still goes undetected. The formula showed the madness in the matter from the start but science maintained the madness irrespective of any and all indications of such unaccepteble conclusions that stemmed from the use of the formula. They went on the revise the formula by perpetrating an even higher degree of insanity instead of questioning the regularity of the formula. Although this formula, in the way it is presented, is no longer used, it still is the essence in the concept that Newton proposed. It is indicating how he saw that the mass of the apple in conjunction with the mass of the Earth, removed the radius between the Earth and the apple. The Earth was too large to rise to where the apple was and therefore the apple did the falling. The big concern I have with the formula is the lack of concern this formula holds in regard to work that was presented and accepted at the time. At the time and long before the birth of Newton an Italian by the name of Galileo proved undoubtedly that when objects are dropped they all fall at an equal pace because after they fall freely and as they all hit the surface of the Earth the instant of contact is always the same with no regards shown to size differences. The mass amongst the falling objects could be much different but that had no effect on the time the objects travelled in concern with the actual fall while the falling was commencing as long as all prevailing conditions concerning the falling as such was the same. Galileo said $F = \dfrac{r^2}{M_1 M_2}$ was not the case because during the fall all objects fell as if on equal footing whereas mass brings differentiation. Newton came afterwards and put factors of descent on mass and that is no slip of the tongue. That was intentionally contradicting Galileo and that is intentional fraud. When I say fraud I become the dangerously insane with one sole purpose and that is to destroy all civil obedience and wreck civil standards that mankind has gained up to now. My seeing this contradiction leads to all the many times that I was rejected by institutions such as yours and persons with as admirable positions such as you have to such a degree that at present I am suffering from the "being battered yet again syndrome". I say this to indicate that I know fully well that my "repeating of past mistakes" by bringing Newton into the open will yet again bring on all the resentment from all those that are in powerful academic positions. I am fully aware of the resentment that you feel, which I am about to release and what I have already released thus far (should you still be reading this far). I say what I have to say in full realisation what resentment you as the physics administrator at this moment feel towards me. I say this remark to inform you that as much as I am aware about the damage that I am doing to my chances of having my work published and knowing full well I am shooting myself in the foot as to have my work accepted I still am adamant in continuing with such public relation damage that I evoke between yourself and I. While you are denouncing me at this moment please concern yourself with the following. When I fall down a cliff I fall at a steady pace. It is the same pace that I would have, when falling down a waterfall holding a cup in my hand. Should there be water in the cup the water in the cup will not stay behind or spill faster than any other water or myself. The water in the cup will not spill by emptying the content going up or down faster than any of the rest, which is falling. The cup will not fill with water as a result of any motion differentiation there is between the descending velocity of cup in my hand and the water that is falling with the cup located inside or outside the cup. If I had to fill the cup with water while I am falling I will have to supply additional motion contradicting the line in which I fall. I have to establish additional movement in access to the motion with which we fall. The water and I will have the same pace therefore we will fall at the same gravity. My density will not leave me superior. The water mass will not have the water fall more forceful or less forceful. The reason why a body will fall is when it has a denser personal independent density than the atmosphere surrounding the body has. Less density than the atmosphere floats irrespective of mass, which includes all gas substances and in that again it shows form is irrespective of element mass. Argon (Ar_{18}), Krypton (Kr_{36}) and Xenon (Xe_{54}) floats as gas although their equals being Aluminium (Al_{13}), Iron (Fe_{26}) and Silver (Ag_{47}) sinks. In this matter it proves that mass has little to do with falling or not falling. All objects can fall as much as all objects are able to fly. It all depends on the motion of the object in relation to the motion thereof in relation to the motion of the Earth. Floating and falling depends not on the mass but the specific density attached to the mass in question in relation to the space the material holds which forms a factor of the totality of space in motion. That is why aircrafts can fly. The forms the bodies take on as well as the fact that they travel at places where only gas is allowed by nature proves that the density of the aircraft places the flying aircraft in a different category that what it would be when it is stationary. Some has to use a specific form to carry large quantities while others go supersonic and that requires a needle shape. I explain all this when I explain the principle behind the sound barrier. The motion considering all objects is not discriminating on any basic grounds carrying specifics about size, form or weight. That is what Galileo said and Galileo is accepted as truth. There is no reason in heaven or Earth or in-between the dimensions that will establish a coherency between the truth in Galileo where Galileo says that all objects fall equally to the ground and Newton bringing in his mass to have the job of the falling done. Galileo very correctly determined that objects descend to Earth at an equal pace and although admitting to this observation of Galileo by Newtonians, science still declares that Newton was never disproved even on one single occasion. Galileo proved before Newton's birth that size and mass is irrelevant while falling takes place. Then Newton came and insisted that everything about falling while falling depends on their mass pulling the Earth mass and visa versa and in all this confusion Newtonians uphold the opinion that they are of the opinion that Newton said exactly what Galileo said. Newton said it's all down to mass while Galileo said mass has no influence and all along Newtonians see no difference between what was said. That is the result of Pavlov's mind control. In the light of this it is the truth that Galileo disproved Newton even before the birth of Newton because Newton was born at the time of Galileo's death. That makes any ignoring of Galileo by Newton not a matter of pure ignorance but it is then deliberate malice. That makes Newton's statements deliberate callousness of circumventing the truth. And the entire world not only accepts this ridiculous scamming but is also at the moment in agreement with such labyrinthic gestures of inaccuracy. Galileo proved and nature confirms that gravity has not one single common thread with mass. That is because when the object falls it has no mass (Galileo) which means that while it is true that when the object falls it has gravity without having mass. That is what Galileo proved. What is gravity? Gravity must be the part when moving is taking place because when mass is a factor there might be a tendency to move but there is no movement other than the movement the Earth dictates. Newton saw his apple fall and the falling or the movement he named as gravity. The process of falling confirms that mass does not establish or produce gravity since all objects hold gravity alike whereas mass is as individual as the fingerprints of people. If mass does establish the movement Galileo was wrong and that is not the case because modern television confirms Galileo almost on a weekly basis. Yet only Newton and Newtonian disillusions insist on mass. They maintain the factor that is producing gravity because that gives those in physics an unfair advantage in realising what others find to be senselessness. That makes Newtonians clever and the rest of us as stupid as doornails. They can see mass establishing gravity and the rest of us are too stupid to get acquainted with the total concept that the picture holds. The concept involves too much in relation to what our little minds may absorb. Now in the light of logic who seems to be stupid after all? It is shown on television on many occasions where army battle tanks are dropped from aircrafts and the tanks travel the same velocity towards the Earth as human soldiers do. The tanks do need bigger parachutes to restrain the momentum of the fall…yes that is true but the velocity of travel is the same as a bicycle would have when the bicycle is also dropped from the same aircraft as the tank is under equal terms. There was a show on television where some young persons were dropped from an aircraft while they were sitting in the car. There was no perceivable difference between the humans descending and the dropping velocity of the car. The persons got out of the car in mid air and into the car in mid air while descending at the normal rate and at one point they even pretended to get out and be next to the car where they rein acted as to push the car as if the car would not start while they were all falling with the car as a unit included in the total structure of the car or as single components falling on the way. There was no difference in the posture while they all were equally descending to the ground. The persons did not fall slower than the car did neither do the car fall faster than the people. Let your religious belief

in Newton show how that confirms the correctness of $F = \dfrac{r^2}{M_1 M_2}$ while the picture I

painted in the falling persons totally vindicate all the Galileo correctness.   Gravity might remove the radius but it does not employ mass to accomplish the destroying of any parting distance between the Earth and any falling object and that has been proven in no uncertain terms. The whole principle rests on the duplicating of the material in the space the material holds in relation to the time it takes to duplicate the material by motion of the material in relation to the time which confirms Kepler $a^3 = T^2 k$. Newton's mass has no validity in cosmic physics and falling to the Earth forms part of cosmic physics. Mass is just a man made devise or a tool with a purpose to give meaning to ideas that is inspired by man. Just as degrees measuring heat and distance measuring space, in the same sense mass too, is a method of measure. It is a scale that helps man which man uses to establish when trying to connect a sense of worth to some concepts man wishes to establish, but mass as a cosmic function is without proof. In the same sense the other mentioned measuring tools are equally meaningless in cosmic terms. **All other physics form what is a part of man-made motion and life inspired motion. In that field of physics concerning the motion that life produces, Newton was never incorrect on any point anywhere and every aspect of his work is perfect**. But it only involves the life inspired motion that is a contribution of man, produced as the labour of life where life is in control of the motion that comes about. No rock can fly and yet a manmade aircraft can. No rock can cycle in perfect balance as a bicycle can with a human in control. To qualify having motion man has to be involved in the motion, which I mention and that excludes cosmic motion from the man made motion. There is a big difference about what is life inspired and produced by life as an entity and what applies to cosmic standards because the cosmos is fatally alien to life. The cosmos can bring motion to the Earth while man with life cannot. Man with life can have structures with much smaller volume move under precise control that the cosmos is unable to achieve. The cosmos cannot take a rocket into space while man can, but man cannot influence the entire sea while the cosmos can. The cosmos take charge of a large satellite such as the Moon and never require any post- maintenance in order to secure the orbit as the Moon maintains the orbit to a precise control by naturally keeping a time component valid and ignoring the human concept of distance, which is as alien to the cosmos as the notions of temperature and mass are. Manmade satellites require frequent control or face devastation as Mir proved a while ago. What belongs to the cosmos does not fit man and what man can achieve is not present anywhere but where life is in control of the situation. Life is motion and life is the manipulation of motion that can be covered by the range of any form of motion from blowing gently on a feather to the unleashing of a rocket going to the Moon. But that is a part of life where life is confined to the Earth and life as such is alien to the cosmos. Newton motion concerns with life where Newton is alien to the cosmos. Notwithstanding my correctness, my work has been rejected because of this statement and only on the grounds of this statement and to the degree where I now have become punch drunk. Should you decide to reject my work one more time, then be my guest but before you do please take a challenge that I direct in your direction; make sure about your reasons and the accuracy of your reasons when you decide to again reject my work this time. This is no threat because I am far to small to threaten any institution as yours. Rather see it as a challenge of your spirit of fairness but please do see it in the manner that I am taking on the establishment in challenging them to prove the correctness about Newton.

Science underwrites the Big Bang, which is a science of promoting the view where the cosmos is expending and becoming bigger. They officially declare their acceptance that the distance between objects is increasing. Gravity had the best opportunity to remove the radii between all the numerous objects altogether the moment the Big Bang commenced. The influence of the mass subsides as the radius increases. The mass that is producing the force will decline its potency as the cosmos is becoming bigger from a point where the cosmos was so small then. When the cosmos was the smallest the gravity at that moment, had to be the biggest it ever could be at any point or that it can become in the future. The best chance that gravity ever had to produce a force strong enough for gravity to remove all space between particles was when being at the smallest distance apart and that then was at that moment where the radius almost did not exist, because the force will fade afterwards while the separating distance increases where it never again will find a better opportunity to endorse a stronger force. A still better moment of enforcing Newton's contraction was the chance to act by compressing the Universe that came even before the Big Bang came into progress. It should then rather be the story of the Big Crunch as a Universe imploded; that is if Newton did apply. When was mass ever that close in proximity again and when was the radius ever again that small? Take into consideration that mass times mass is supposed to remove the radius and where there was almost no radius the force was the strongest it could ever get. In the formula science officially accept as the cosmic Universal basis that reads $F = G\dfrac{M \times m}{r^2}$, mass has the purpose to the reduce the radius by unleashing a force equal to the product of the mass in relation to the compliment of the gravity constant and therefore when the radius was almost not in evidence, the mass had the best opportunity to dissolve the radius. Using the formula is a mathematical manner to express verbally that the force sets out destroying the radius by measure of the mass drawing on the mass in conjunction and with the aid of the gravity constant, which combines to form a force. Using this formula puts the mass in relevancy with the radius and the smaller the radius is the bigger in relevancy the mass product will be and in the same event the stronger will the force committing the gravity be. By reducing the radius the mass instantaneously increases, as the one stands relevant to the other as a mathematical response. If there is a partition the size of a neutron separating all objects within the cosmos such as is suggested was the case at the event of the Big Bang, then the force this state of affairs will unleash goes beyond calculations. Yet if Newton is correct his force went on to wither away because with the expanding came the increasing of the radii that is allocated between all cosmic structures. With that the opportunity of mass to increase gravity and destroy the radius altogether simply tarnished. This means the expanding went on while Newton's contraction should either have stopped this expansion or science should have gone silent about their previous notion that was promoted by Newton. In contrast to Newton's prediction they now apply the Hubble constant as the accepted norm by which science age the Universe. The Hubble constant now forms the norm whereby the progress of cosmic expansion is determined. In using the Hubble constant there is absolute proof of expanding. The results are so accurate that that any person blessed with doubt can study the expanding through a high power telescope and see the expansion personally. The Hubble constant is the confirmation of cosmic expansion. The Hubble constant has evidence, which is without dispute where the evidence insists on accepting expansion of the entirety of the cosmos. This runs directly against Newton's contracting and yet both are accepted because Newton was never contradicted. The Physics Paternity stands by its view that Newton requires no proof in spite of everything in science going on continuously to disprove Newton. Can any one find more contradiction of Newton's validity or that more blatant facts that proves more reasons in finding grounds to dispute Newton, which also at the same time can be more devastating to a ground rule of any nature or have a bigger impact that proves changes has to be implemented or that can deliver a higher degree in disproving of Newton's claims than these two principles I mention above? Yet according to the Physics Paternity that is in charge of all matters concerning physics and acts as the guardians thereof on a global scale, that body controlling academic physics maintains the opinion that Newton teaching was never yet brought in doubt. Following the evidence Hubble brought about, the response of the Physics Paternity was to create more fraud. The way the challenge of proving Newton was met was to conceive more mayhem by establishing more fabrications. In spite of such damning evidence that is contradicting Newton to its foundations, Einstein was ordered by the Physics Paternity to find how much mass was available in the cosmos in order to justify the correctness about Newton. They did not consider the connubiality of Newton or consequently proving Newton's failings that came to the open. They consecrated Newton yet again by finding a need to determine when Newton accuracy would come about. They had to find the measure of the gravity that all the cosmic mass would produce in order to determine when the mass would turn the cosmos around and vindicate Newton. In other words there is an acceptance of the incorrectness of Newton by trying to determine when Newton eventually would be right. By not admitting to Newton being wrong they found the cosmos out of order and gave the cosmos a chance to stand in and prove Newton correct. The spotlight was cast on the cosmos falling out of line by the Paternity avoiding to not admit to Newton's inaccurateness. At that point Mainstream science should have put Newton on ice holding Newton under academic arrest while waiting for his termination from the record books of physics. That is not the case because the biggest criminal act ever launched by any group of persons was set in motion on a scale no other can copy.

Einstein was ordered to find the mass deficiency. Einstein was ordered to see what in the cosmos had the audacity to not confirm Newton! The fact that this order to Einstein came about puts the Universe in dispute because Newton has never been in dispute before. The name they gave to the process used to cover such fraud and inconsistency is the Critical Density, which is the biggest smoke screen to cover fraud ever, enlisted anywhere. In comparison this cover up puts every politician and all criminals in the preschool class. Yet this is the highly respected Physics Paternity, which forms the highest level of brainpower, the world has ever seen. Of course Einstein did not find the critical density to be in order and now they use billions of dollars in tax money to defraud the unwitting public as the Physics Paternity takes the crime one step further and blow the cover on what they established in order to save Newton completely out of previous proportions by trying to find so called dark matter. The suggestion of dark matter throws Newton even further into doubt because it still suggests that Newton is incorrect. The purpose of such a suggestion is intended to cover up Newtonian inadequacy by a declaration that Newton was never yet brought into dispute. By trying to find where the cosmos went wrong (because Newton never can go wrong) they could cover the wrongness created by Newton and leave all suspicions in the basket of the cosmos which then will cover up the deceit by establishing even more deceit. All this is brought into use to cover the brainwashing and mind control they use on physics students throughout the world. Talk about shady spy institutions going unscrupulous by methods too dark to mention and at the end of the line one finds all University Physics departments located where the most extreme serves their mission of treachery. The world will never learn that the Universe is contradicting Newton because the world must only know that Newton was never in dispute. Therefore the fraud is carried on by implementing more of Pavlov's mind control about mass bringing about gravity. However this time the process is unleashed on human children showing their naive trust and is used by unscrupulous academics on unwitting students. The best is that the parents still have to pay for the malice done to their children. Because of such forceful conditioning of the mind and enforcing acceptance it shows that Pavlov's theory most of all applies to Humans with highly developed intellectual abilities. The result of this forced accepting of Newtonian beliefs is the direct result of the fact that Newton statements were never proven. Take the comet for one and find Newton in the gravity that the comet evokes.

Picture what happens when a comet is coming towards the Sun. As Newton would suggest we find the comet is heading in the direction of the Sun because we suppose Newton is correct and the Sun is pulling the comet by strength of mass towards the Sun. The gravity of the Sun is pulling the comet towards the Sun while the Sun is just too big to be pulled around by the comet so the comet is doing all the coming together on behalf of both. The gravity of the comet is focused squarely on the centre of the Sun, which is the focus point of all gravity that is producing such an enormous pulling force. This is happening while the gravity of the Sun has the comet hooked by its centre and is dragging the comet through outer space as fast as the combining mass will produce a force. This proves Newton is a genius. Then when things get critical the comet elude the Sun by missing the centre of the Sun. This proves Newton is suspect. We find the comet eluding the centre point of gravity of the Sun in spite of the enormity of the mass of the sun, which is then able to create a devastating gravity force with which the Sun finds the strength to unleash an inescapable force on the insignificant comet. In spite of the enormity of the mass of the Sun that creates a devastating gravity from which no escaping is possible the comet is defying the mass of the Sun to a certain point alongside the Sun to which it travels. When it reaches the point where the comet misses the Sun centre the comet breaks free and misses the gravity of the Sun altogether. Then the inexplicable happens where the comet defies the pulling of all the mass of the entire sun as it spins around the entire Sun by heading away from the Sun. The comet escaped from its destiny with mass forming in the Sun. The comet avoided a sure death sentence but what brought about the pardon? That small comet evaded capture by finding a means to avoid the gravity pulling force by which the Sun caught the comet. The Comet is going north and it is leaving the Sun in a position South of its Northern direction. This proves Newton is a fraud. Newton's gravity fails to explain this inexplicable action where the Sun allows the comet to defy Newton! The comet is going free in spite of all the pulling of all the mass of the entire Sun. Not once does any Newtonian care to discuss this betrayal of Newtonian principles. No mention of this is anywhere in any class where Newton is taught. That Newtonians allow to be put into the category of the wonders of the Universe and we humans find no answers to any in that category. What went wrong to the position Newton has where mass is drawing mass and there is no escaping clause permitted. The best way to go foreword is to go back to where Newton started science by applying fraud to find answers that he concocted.

Being the Professor with the status that such a position accompanies and with $F = G\dfrac{M \times m}{r^2}$

applying as the force pulling cosmic objects into a collision, can you tell anyone how many comets crash into the Sun every time the Sun's mass is pulling the comet into the Sun? Can you as a Master show how many comets destroy in relation to the number of comets escaping successfully? The Sun is pulling the comet while the comet is pulling the Sun and gravity is reducing the distance between the two. While the gravity should be pointing the centre of the comet directly into the centre of the Sun it does not happen. Every time the comet misses the Sun by a country mile where the comet makes a circle around the Sun and then heads into the darkness of outer space.

Why would the pulling gravity have the comet escape every time in the face of the enormous gravity that the Sun produces? How does a comet pass the Sun unscathed while the enormous gravity of the Sun should give the tiny gravity of the comet no chance to escape?

By using mass on mass as the force of gravity can you explain how comets get to escape from their definite encounter and avoid the pending collision they have to endure? Can you explain why the comet breaks free and travel all the way into outer space, only to return in a predictable cyclic manner? No…I know you can't explain such an obvious scenario. That is one of the numerous unexplained Newtonian misinterpretations you constantly fail to neither admit nor mention in polite conversation and is another never addressed Newtonian shortcoming in Newtonian science, which makes the pulling of mass onto mass a hoax. It is fully compatible to the likes of middle age witchcraft proposals and you keep it best wrapped where no one ever mention it to the public at large.

The gravity that Newtonians insist on where something is pulling something is a pipe dream. There is nothing that is pulling anything to bring about gravity. Gravity is charged by the combining interaction between the four cosmic principles by the manner they interact with singularity and that I prove! Gravity is the value of singularity moving in formation of space filled with material in relation with time. Gravity is the manner in which eternity parts from infinity and I prove both factors of time without a doubt. It is something that Newtonian science cannot even contemplate because the mass that is producing gravity has as much validation as a witch flying on a broomstick. There are no differences in the forces of gravity that Newton introduced and what mass unleashes by pulling power and a witch bewitching (or whatever witches do best). All four so-called forces are the very same thing except in the motion they apply and what generates to establish the forces. Gravity and electricity is the very same thing but for the scale in which they are generated. Other than that, there is no difference between gravity and electricity. How many times did I send your department these information and how many books did I send where I detailed it? Yet you choose to go the way of the criminal by proposing a concept that relies on fictive science, falsified mathematics, blatant crooked suggestions and highly suspicious arguments. I challenge you Academics in physics too show how much the Moon, is coming closer to the Earth. I challenge you to show that the product of the mass of the two objects namely the Earth and the Moon is having a mass that is pulling by force the Moon towards the Earth, and the Earth towards the Moon bilaterally. I challenge you to show how mass influences orbiting planets by contributing or increasing / decreasing or to what degree the mass is influencing the orbit velocity of any planet in any way or manner.

The orbits are in ratio but not like Newton suggested being $a^3 = T^2$ but there is a clear ratio equalising all the planetary motion. However, this means mass is not responsible for the orbiting motion or for any other form of influence there may be on the planets rotating the Sun. The mass of the Sun has no influence on the mass of the orbit of planets because neither distance nor size nor location prescribes any of the orbiting conditions applying. Something else apart from mass is dictating the terms to secure the evenness of the rotating. Notwithstanding, whatever mass differentiation each one has, the lot are still rotating very alike and in fact they are the very same. Notwithstanding the enormous mass discrepancies between the planets where Jupiter has 318 times the mass of the Earth and Pluto has only 0.0025 the mass of the Earth, they all orbit at a shared equality in ratio of about 300. The response to explain the evenness, Newtonians put the equality down to the gravitational constant that is producing the commensurate behaviour. With this in mind I challenge you to mathematically prove that there is a gravitational constant. I challenge you to prove that the gravitational constant is responsible for evening out the mass differences between orbiting planets and that this is the cause why planets all orbit in a compatible fashion using the same velocity notwithstanding mass differentiations. Remember this evidence I mention opposes Newton in no uncertain way by contradicting Newton at the very heart of physics and yet with all the contradicting evidence coming from the cosmos as such, Mainstream science remains to uphold the opinion that Newton was never yet contradicted! In my books that I present to you for your evaluation this time round as it did in the past, this repeat is being an ongoing process about my explanations that is covering this aspect. However I did explain it in the previous books where this time it also is included and is merely a better innovated repeat of that which I presented in the past. However and nevertheless you found grounds to ignore my mentioning this inconclusive assessment in science as you chose to ignore everything I said in all my work. I show that in your manner of calculations you put the gravitational constant as a contributing factor to the mass product that produces the force. The gravitational constant is increasing the force produced by the mass multiplying. This Newton invented when the formula $F = \dfrac{r^2}{M_1 M_2}$ proved to be hogwash. The force is the measure whereby the radius in the square will vanish and this argument becomes the cornerstone on which all physics hold its foundation. With the original formula in shambles Newton invented more deception by introducing one more unproven factor. He invented the gravitational constant. The gravitational constant is used as the complement of the mass multiplying the mass. In conjunction with the gravitational constant the mass multiply to produce such a force value. This then supposedly produces a force able to destroy the radius between the objects. Then Newton went even further at the time and gave the concept a name: he called it gravity. In the way the formula is presented the gravitational constant increases the effect of the mass of all the planets, which is forming the force of contraction between the various objects and the Sun. By expressing the statement in a mathematical equation it reads as follows $F = G\dfrac{M \times m}{r^2}$ F=G (M X m) / r².

The evidence that this is a fable and that mass is not pulling the objects closer was already established with Kepler introducing his work. Newton knew that the planets orbit in equilibrium by the measure if a³ / T² = 300 and all of that is in a ratio of about 300. That disputes Newton's claim on mass forming influences because it clearly shows equanimity and the proof was delivered before Newton was born.

Then when all the planets orbit at an even pace and all being the same ratio, where not one is found to come any closer to the Sun or to each other, your officially accepted reason for explaining that, is that the gravitational constant is evening out the mass differences and yet, according to the official formula you use, the gravitational constant is not evening out the mass but is complimenting the mass by improving the so called force. If the gravitational constant is evening out the mass the constant firstly has to divide into the mass to wither the effect that the mass produce and by nullifying the mass to the extent that Kepler's work proves, there is no mass factor present left to use.

The formula then should read $F = \dfrac{M \times m}{G \times r^2} = 1$ F=(M X m) / (GX r²)=1 which in normal verbal English states that the product of both masses is in equilibrium when divided by the conjunction of the gravitational constant and of the square of the distance parting the objects where the product of the gravitational constant and the square of the radius between the planets is in division with the product of the mass which then will produce the force that cancels any influence of the mass factor that both of the planet's will have on the orbit of every planet orbiting evenly around the centre structure. That is very much not what your formula you use $F = G\dfrac{M \times m}{r^2}$ suggests. Your formula shows that the gravitational constant is helping the mass product to destroy the square of the radius. In the cosmos however and in reality that destroying of the radius is the last thing that is happening. To explain this ridiculous deception Mainstream physics go on to explain that the gravitational constant which is supposedly the gravity that is keeping outer space in a unit, where that gravitational constant nullifies the mass influence on the force. Newton said mass is all-important but this explanation now suggests that the mass is nullified by the intervention of the gravitational constant.

This is a statement that is not in line with the mathematical equations Mainstream science present to serve as truth. Making such claims that is so far away from the actual mathematical statements, which came into place serving as yet another attempt to cover up Newton fraud with more deceit and becomes compatible with mafia like behaviour. This criminality is the behaviour of all physics paternity in control of all physics worldwide! This is the group of men that is in control of the thoughts in the minds of the brightest developing brains in the world.

They harvest the best intellectual group of all students this Earth can offer! To the students questioning or rejecting the Newton claims is not an option. The students has the choice to accept what is taught or face examination failure and total rejection of the institute where they are educated. If that which is presented that the statement would suggest holds water about their explaining how there is such planet mass apathy that will not influence orbit details of planets making them go faster or slower, then the entire Newton presumption is based on yet more Newtonian misinformation. If the gravitational constant were evening out the mass then it would have the gravitational constant in multiplication with the square of the radius in dividing the product of the mass. This will then be evening out the product that the multiplying of the mass will put in place. If that was said and translated in mathematical terms, then it would mathematically equate to show where the force will be in equilibrium because the radius is supported by the addition of the gravitational constant which increases the radius in the square and the total of that is in equilibrium with the product of both object's mass.

If that was true then the gravitational constant must be placed where it increases the radius in every event as well as canceling all influences mass may present and it would be stating that $(G \times r^2) = (M \times m)$ **(G X r^2) = (M X m)** is applying in all the cases so that the mass has no longer an influence on the outcome of the equation because out there in the real cosmos Kepler proved that the mass has no implication on the orbit of any planet **a^3 / T^2 = 300.** Kepler never even thought of a mass of any description and I can assure you that his not mentioning any mass was not the result of his retarded mind or not because he was mentally inferior to one Isaac Newton. With the corrected implementation of Kepler I can and I do support such statement of equality but then in that event the equation then should read that the gravitational constant is complimenting and supporting the square of the radius and mathematically such a statement is $(G \times r^2) = (M \times m)$ **(G X r^2) = (M X m).**

That removes mass as a cosmic factor. You know very well that in the planet orbit ratio the space factor **expressed as (space a^3)** divided by time factor **expressed as (space T^2)** leaves a space **(a^3) − time (T^2)** displacement factor ratio between space and time of round about three hundred in all cases, $\left(\dfrac{T^2}{a^3} = \pm 300 \right)$ **(T^2 /a^3= ± 300),** which disqualifies mass discrepancies altogether. With that the cosmos puts Newton in dispute! So why then bring in the use of mass in any event if it does not change the result but has only one purpose and that is to cover Newtonian's misconception and to hide Newton's fraud? However, the formula used by mainstream science clearly shows that the gravitational constant is in effect increasing the mass by multiplication of the mass. The formula undeniably put the gravitational constant in a multiplying position with that which the mass of both structures hold.

In the expression $F = G\dfrac{M \times m}{r^2}$ **F= G (M X m) / r²** which then in ratio is decreasing the square of the radius. That is pure rubbish and I challenge you to prove me wrong by proving your mathematical interpretation. Then all the while, while promoting such rubbish you still maintain that Newton has never been proven incorrect. When Newton is taken into the cosmos Newton physics goes bizarre. There is no evidence of mass influencing the motion of planets in any manner. Mass doesn't bring the least appreciation the orbits of planets. One can see the lack of any effect mass has on putting differentiation between planets as this table shows and it was available at the time of Newton's plotting. One can see equable motion where not one stands in disadvantage.

PLANET	SEMIMAJOR AXIS $A(10^{10}m)$	PERIOD T (y)	T^2/a^3 $(10^{-34}\ y^2/m^3)$
Mercury	5.79	0.241	2.99
Venus	10.8	0.615	3.00
Earth	15.0	1.00	2.96
Mars	22.8	1.88	2.98
Jupiter	77.8	11.9	3.01
Saturn	143	29.5	2.98
Uranus	287	84.0	2.98
Neptune	450	165	2.99
Pluto	590	248	2.99

Newton is not incorrect as it is used in normal applied physics but in cosmology there are no grounds in supporting of mass or gravity in the manner Newton saw or in the way Newton suggested mass to apply. On Earth and on the ground Newton's work is impeccably correct and that I admit without any reservations on any aspect of physics in whatever manner. However taking Newton's ideas into space becomes fraud. Mass only serves as a human concept to allocate differentiation when required when gravity is obstructed and in the cosmos gravity is not obstructed. The reason for that I have well documented in An Open Letter Announcing Gravity's Recipe.

In my books I show that mass is a resisting or obstruction by a material occupying space a^3 that prevents the flow of time T^2 to continue by extending the value of **k**. Mass is the nullifying of the directional progress **k** to obstruct time T^2 by blocking flow which otherwise would include mass a^3 to participate in such motion. When two particles are in water and are totally submerged, yet they are buoyant, the objects will flow with the water in the direction that the water flows. The flowing of the objects which are in the form using the cube a^3 will pass from a point to another point forming the square of time T^2 and the distance they flow substantiates the relevancy which **k** renders to the duration of time T^2 and the change in allocated position of a^3 lasting from point to point over the distance of **k**. Forget at this moment that water is flowing because of gravity. I only wish to present my idea. Kepler stated that $a^3 = T^2k$ where a^3 can only be space in the cube because space is only in the cube since the square is too flat to form space leaving T^2 as the motion in the flat but square motion and **k** is the relevancy of such motion completed. Also forget Newton's insane arguing about **k** being zero because of some invalid concept he proclaimed using a gyroscope he does not understand and of which leaves his entire arguments suspect since the argument has no validity. If his argument holds any water the graph has no validity and no generator can charge electricity because it is the current flowing that becomes **k** and **k** is that which he claims has a value of null, while in truth **k** is inspired by the rotation motion T^2 that is generating the electric current to a precise and pre calculated graph. His presenting of some stupid argument that one factor can divide into another factor and from such dividing it will result in zero is mathematical madness.

You as a mathematician should know that the smallest outcome of the dividing of any equation could be an infinitely small number but never can it be zero. Yet still because it is Newton you are prepared to accept what is mathematically absurd since the very idea was all part of his fraud. If **k** is zero then the entire formula becomes nullifies and we would find ourselves living in a Universe comprising of nothing because if **k = 0** then $a^3 = T^2k$ is $a^3 = T^2 X$ **0** and the entire formula Kepler presented goes away leaving the cosmos without a purpose. Therefore in order to let mathematical sanity prevail, we have to get rid of Newton's incoherency and that is as incoherent as he introduced the idea where having time, (which is the flow of material in relation to each other), find a position that would allow time to stand still. Time in the square cannot stand still because time in the square cannot be a square since only then will it be able to not show progress between two points measuring time. Time is coming from being at while going to. Time is motion from the past through the present onto the future as space moves by time $a^3 = T^2k$. The only thing that time will find impossible is to stand still because then time becomes eternity in the moment of infinity. When time stands still, time forms eternity in the moment of infinity. Time is the displacing of space in a given relative period where **k** forms the relevancy of the period taking place.

Lets return to the stream and the floating but buoyant objects. When the submerged objects float in the water somewhere in the centre of the water not being at the top or otherwise at the bottom but also not restricted in any way, then we find that the flowing body will go along with the water flowing and will go in the same relation to the water as what the flow of the water demands. Then the movement of the space a^3 would be equal to the relevancy **k** of the distance achieved of the flowing water T^2. The space a^3 in motion will be equal to the flow of the water T^2k. The motion of the space secures a free and unhindered location as the water takes the space of material as if the space of material is just more water. The space filled with material can just as well be space filled with water since the material is no different from the liquid water. When there is a retaining by blocking the flow of the space in the water as part of the water of the space changes from being part of the liquid and then becomes a separate solid that helps to restrict the flowing water. By forming restricting part of the motion the space no longer presides with the water as water but becomes a restriction of the water by forming a solid that blocks the water flow. Then only will mass form part of the particle because the resistance will be equal to the obstruction that the obstacle forms in relation the motion.

The obstruction will present the volume of the space that obstructs the flow and that volume in relation to the countering of the flow by restricting the flow in the measure of the size the object retains such restriction becomes the measure of the value the mass represents. In that case size matters because the blocking is in the full volume of the obstruction where the water gives the blocking the measure of resisting that comes about. Then that will be when the space a^3 is standing still by being retained in relation to the flow of the liquid T^2k. When objects fall they all fall alike (Galileo) and the flow of them is the same notwithstanding size or structure. One can only contribute such behaviour to the indication of the presence of specific density forming buoyancy where the buoyancy derives a value from the liquid T^2k that allows the space a^3 a medium to flow in $a^3 = T^2k$. When the retaining of motion comes in place by preventing the space to further motion then mass becomes a factor but it is not presenting the reason for the flowing of the water but is a result of that which is hindering motion and not enabling the buoyancy. The mass has no part in the motion but is a compliment in the retarding of the motion. That means instead of gravity being $a^3 = T^2k$ the application of mass would result in $a^3 \neq T^2k$. The space is not moving and the not moving while time is still moving without the corroboration of space the result then is that space is preventing a steady flow of time in relation to the object being restricted to flow with the water. What is time is rather more complicated and I leave such explaining to the books. There is a flow of liquid being time that is converting towards the Earth (or any other cosmic structure) where outer space has to be of substance to validate any distance forming between planets and the Sun. Something has to fill the space that is validating the distance between the Sun and the rest of the cosmos and that something cannot be a substitute for nothing or be substituted by nothing.

The substance filling outer space is holding the material in a state of motion and that motion of space can only be when a state of buoyancy prevails. Outer space cannot be nothing or filled with nothing and that idea is as corrupt as any idea connecting with Newtonian dogma. If outer space was filled with nothing then what would that nothing comprise of because nothing actually means it is not. It literally means there is not. It does not mean vacancy because even with vacancy there is a shortage of one part and that will leave an abundance of another. Calling outer space nothing, is not referring to outer space, as outer space is unfilled with particles but whatever should present it in composition when outer space is tagged and branded as nothing, because being nothing then in that event is a case where everything including the unfilled space has removed leaving nothing in its place and that, where I can assure you that is not what outer space is because that is idiocy and that does not represent the substance of what the Universe comprises of. The only reason why mass or size or location does not present any difference in outer space is when these features are absent. It is a sign of buoyancy. Size in cosmic terms is irrelevant and proof of that we find in the fact that a few brittle fragments that once formed a planetary structure rotates with the giant Jupiter and in the same orbit band than Jupiter. In the cosmos mass size and volumetric occupation serves little purpose but to put the unit it holds in a time development. However in cosmic location and in relation to others, the size is meaningless and the Titius Bode law serves as proof of that. The size of Jupiter is to the Sun as equal as the size that Pluto has.

Distance in between structures are a manmade concept that is meaningless in the cosmic context because all planets are at a ratio $T^2 / a^3 = 300$. Further more that is proof that time by measure of k is diminishing in density, because there are motion of k^{-1} going negative, which means in a mathematical equation $T^2 / a^3 = k^{-1}$ and in that sense time equates to $a^3 / T^2 = k$ and from that we can clearly see that all planets are at an even distance from the Sun. We might give their location purpose but that purpose we allocate for human use does not serve the Sun in the least. The very same argument goes for temperature but that is somewhat cumbersome to explain in this letter, however I go into much detail about that in the books. But if you read my books it would all be old news to you by now and most of all, I bring much more proof than the shambles Newton served science to blindfold reality. The cosmos is in two aspects that serve each other by occasion relevancy and that is why gravity is the Coanda effect. Every aspect of the cosmos is in terms of liquids $k^{-1} = T^2 / a^3$ and solids $k = a^3 / T^2$ that forms motion of space $a^3 = T^2 k$.

When that space which is holding what we perceive to be where the solid is in, is in motion then the independence of the motion that is taking the moving space along, secures the object as a cosmic independent structure. Kepler gave us $a^3 = T^2 k$ and that proves that when space is moving $T^2 = a^3 / k$ it proves that such space filled with material is allowed growth $k = a^3 / T^2$ and only penetrable substances like gas and liquid allows motion to continue by distinguishing partition through density differentiations $k^{-1} = T^2 / a^3$. We have to surrender our pretext to what is truly present in the cosmos. The question to what is liquid and what is solid is answered by what is hot and what is cold. Liquid serves as being in a higher state of heat and solid represents being in a lower state of heat. Then we should find what is heat and where it is hot.

If we fill an air balloon with air it expands the balloon by expanding the inside size of the balloon. There is a ratio in the air between the inside of the balloon and the outside of the balloon and by adding heat to the air inside the balloon the size of the balloon grows. We can vividly see from that action, that the hot air that is pumped into the balloon, the balloon becomes larger in a volumetric measure capacity. From this one can conclude and it is a logically conclusion that by introducing more heat into the balloon the level of heat to material rises in favour of the heat. The heat containing that increases puts the particles in the balloon in lesser quantities in relation to the increase in space by volumetric quantity that increased in space that the heat produce in the balloon By increasing the heat in the balloon the balloon got bigger by volumetric measure but also got less dense by atmospheric measure. The reduction in density is a result of space increasing without having any particles added to the inside of the balloon and the increase was in heat that became space where the heat was air without particles nonetheless. If one fills it with only air particles the growth in size inside the balloon would be far less substantial and the density would not produce a lifting of the balloon. The size of the balloon actually increases by receiving more heat inside the balloon and if we find that heat increases the balloon in measured volume then the air in the balloon is heat forming more of what already is present which is giving the particles of material also inside the balloon less density. If bringing in more heat allows a bigger area to fill inside the balloon then the balloon is filled with heat from the onset.

By adding more heat we promote more space and a decrease in density and that combination allows more motion and therefore the motion is equal to the heat inside the balloon that reduced the relevant density of the material inside the balloon. On the same argument we have to take the conclusion one step further. When we fill what is in the balloon with more heat then that which is in the balloon expands to entertain more of what already is present. The expanding accommodates heat by accumulating what we perceive as heat into an area that makes it after receiving the heat to be a larger area. If we allow such expanding to continue until no further expanding is possible or permitted then we have reached the maximum position heat can expand. We then are where there is no more expanding because all the heat brought about all the expanding that will ever be entertained. Then we are where there can be no more heat added that will effect any more expanding. The heat cannot grow because the expanding nullifies any increase in heat. If no more adding of heat can increase the heat level by enforcing the expanding we know that at that point we reached the position where no more adding of heat can effect the status of the heat level or the increase in volumetric space by adding heat to space. Then we are at the hottest anything can ever go. Then we are in outer space where no more adding of heat can bring any more expanding since it accommodates all the expanding all heat added can deliver. We went and added heat to the balloon to increase the space of the balloon and decrease the ratio density of material and heat inside the balloon and from that the balloon expanded. It did not expand by receiving more particles but it grew more in space and it reduced in density by expanding with accumulation of heat inside the balloon. If that is the case and that is true science then that puts outer space in the hottest position anywhere can ever get and we have to adapt our senses to accommodate reality. Outer space is not the coldest but the hottest place in the cosmos and the cosmos is the judge of that. Forget Kelvin's scale because heat as we humans would like to think of it is irrelevant and non-existing in the cosmos. It is so Newtonian to tell outer space and the cosmos what the cosmos must be because of what we think is true about the cosmos in stead of allowing the cosmos to tell us what the cosmos is. Heat and cold has no measure and the same applies to mass and distance. The entire Universe is all served where every aspect is in ratio of being in a state between what is in motion and what is deprived of motion. If one fills the balloon with hot air the heat will expand the space and put the space inside the balloon, which is in a higher level of expansion in line with a higher state of expansion in cosmic space.

By increasing the heat level inside the balloon the space inside expands and the density factor drops in relation to the conditions on the outside of the balloon where that differentiation drives the balloon to move and that moving is gravity. Reducing the heat will drive the balloon to a lower level of heat with a higher value in density. That is gravity. The adding of heat or the removing of heat, which increases or deprives the density ratio in relation to the cosmos elsewhere is what produces the motion of gravity. By adding or removing heat from the balloon that action puts gravity to the balloon where from that the balloon moves to a higher status of space that is sporting a lesser density ratio in terms of particles between particles and space. The adding of heat produces motion and that removes mass, which will expand the space. If heat expands the space then at the location where the adding of more heat cannot bring more expanding we have to accept that at that point the heat stretched the space to the limit it can go notwithstanding what Kelvin's temperature scale of any other measurement may dictate. The air outside the balloon allows the balloon to flow and such flowing of the objects in motion are unrestricted as they allow the free motion of space filled with material in space not filled with particles but will take space filled with particles along while the lot is in that space which allows the flowing object to orbit the Sun and in that the motion is secured to have gravity $a^3 = T^2 k$ or motion of particles in time to take place. That same space forms part of that space not filled with particles but that is receding to Earth nevertheless. It is that the space not filled with particles, which is receding to Earth in any event and while the space is receding it is also bringing not buoyant objects along in any case. The purpose of the gravity the Sun unleashes is to remove heat from space and not to go futile and put more heat into an area where the adding of more heat has no purpose what so ever. The balloon clearly shows why the space is floating in the air. By floating or coming to the Earth can only be if such material holding the filled space is the object that is in a state of buoyancy in the space it penetrated. Why they have buoyancy is a matter I explain when and where I explain the sound barrier. If Jupiter is "swimming " in the same "puddle" as Pluto the "puddle" will affect both planets in the same manner and Kepler did prove that statement with the expression $\left(\dfrac{T^2}{a^3} = \pm 300 \right)$ $(T^2/a^3 = \pm 300)$, being the measured flow (k^{-1}). If the "floating" was caused by a density not of the planets but of the substance in which the objects floated by buoyancy, then condition of similarity would prevail where all are effected alike and again Kepler proved that as is stated by the expression $\left(\dfrac{T^2}{a^3} = \pm 300 \right)$ $(T^2/a^3 = \pm 300)$. The fact that the planets do not show mass as an influence proves the motion is in a state of buoyancy $a^3 = T^2 k$.

The fact that Galileo saw objects fall equally notwithstanding size differences proves the objects falling equally are in buoyancy. The fact that the neutron shows no signs of mass puts the neutron in buoyancy while the electron as well as the proton shows the characteristics connected to forming restriction of a flow or depriving something its ability to flow. It is being indicative of restraining the flow of space, which provides both with mass or with space restriction. I have explained it in every book I sent you thus far in the attempt to get you to read my work. By this time I am anticipating that you are not going to read my work again because it happened for so many times in a consecutive sequence of events before on previous occasions. I have found solace in the fact that you don't care to read it. I mention this on purpose because you have the idea that you are served with a position putting you beyond any attack from other quarters and that there is no person that can rattle your cage since you have Newton to defend you. Make sure about that! Although I do realise by this time that everything I say should be old news to you since you heard it before when you evaluated my work in the past I honestly think that there is no academic that even got this far in reading this letter. But if by any chance in the slimmest of possibilities there might be one still reading and in that event if it seems all new to you it proves that in the past I am correct that no one cares to see another view except for the fraud of Newton. I challenge you to show how Newton's mass allows motion to the balloon where Newton's mass forcefully dictate mass on all that is willing and not willing to participate. If by chance you get this far in reading this letter, you do reached this point in the letter because the work interests you in some way. If you did read my work in the past you would have seen it was written in my work that you previously rejected / dismissed / ignored.

But I would be highly surprised if you are still reading at this point where I accuse the entire physics paternity and not you in person as such that you are deliberately, with malice in mind, entertaining a misconception while you are innovating mass that is responsible for establishing gravity and by that you are applauding the misgiving which is in association with the detestable corroborating of Newtonian deliberate fraud. By not investigating my accusations you then are contributing to fraud that hides any possibility of finding scientific truth. While you place mass where it does not belong and where it clearly shows having no influence at all on the working principles that guides the cosmos, you deliberately circumvent the truth by committing a heinous act of calculating deception on the part of Master Newton. Please think of my statement that there is no mass when objects fall because there is buoyancy present and see that it rings far truer than putting mass where no mass belongs just for the sake of protecting Newtonian deception. There is no mass in planets orbiting because the process works on buoyancy. While you know very well there is no mass in any of the places I mention you still ignore my work and commit to the fraud of placing mass where mass does not belong.

I challenge you to show the glorified graviton and I challenge you to prove that mass is pulling planets closer by destroying the radius distance there is between cosmic objects. You claim that this which is out there for us to see and that which we now have which is serving us as a Universe had grown from where it once had the size of a neutron as it expanded to what is now present. Just go outside at night and look at what came about since the lot out there was the size of a Neutron. On the other side there is Newton's deflating or contracting Universe. You maintain that this one principle of Newtonian mass inspired contracting is applied while at the very same time you then are in an official capacity also promoting the Big Bang development and the Hubble constant, which absolutely contradicts the contraction idea.

The Big Bang contradicts Newton in all aspects as it clearly shows beyond doubt that there is expanding going on in the cosmos. The two proven phenomena that promote expansion is acutely indicating an increase in cosmic structure expanding while Newton is projecting a contraction whereby the final conclusion will come about as all structures unite. There is no manner in which to marry Newton's contraction ideas with the Big Bang expanding and Hubble's cosmic shifting of matter notwithstanding that the latter is proven beyond doubt. There is a cosmic parting of material moving further away and that is the proven concept forming the norm in the cosmos. Mainstream Science has to cover the fact that Newton is deceptive by producing more improbity. To try and con the public in realising Newton's theoretical failure and deception Mainstream Science do not directly admit to Newton not panning out but Mainstream Science distort any such realising as they invent more misleading and scamming. Mainstream Science diverts from the truth even further with more falsified facts as they come up with more theories by which Mainstream Science puts the attention not on Newton failing but the mystery why the expanding is in progress without finding it necessary to reflect on Newton. If Newton is not, then when will Newton start? Mainstream Science do not ask the question why there is this anomaly and total contravention of the rule but would rather find no need to reflect on Newton altogether…no they do not admit to Newton's failure at all but they respond by diverting his failure as they put his success into the future. They don't admit Newton is wrong because they establish when Newton will be correct. Newton is correct, but they just have to find when he is going to become correct because they must find why the cosmos would behave in such a disgraceful manner by contradicting Master Newton! They divert all attention from Newton's failure and of course also from the proposition that science has no idea about the facts supporting physics to finding the flaw that the cosmos would present to have.

This hiding of the truth is to coward from reality and to prove the illusion Physics wished to create and substantiate. If science doesn't have Newton science in all its splendour they have nothing and science promotes the idea with atheism that science knows more about the cosmos than what a God might know. Therefore and in that light Newton may never be wrong because that would leave science as the laughing idiots that can't even explain their faculty.

With the latest admitting of The Big Bang and the accepting of the Hubble constant there now is much proof of Newton's failings because disproving Newton directly comes via the proof of the Big Bang and that is also the proof of Newton's failure. The proof that embraces the Big Bang is also the very proof of Newton failing and how much Newton was being incorrect all along. The whole concoction of the Critical Density that should prove mass deficiency in the cosmos as a whole is just a criminal smoke screen to cover up fraud. Mainstream Science puts Einstein to task to measure all material throughout the entire Universe. This idea in its individual capacity of finding the ability to calculate what there is in the Universe by own merit is outrageous and I am prepared to take on any person disregarding his or her status, to task to prove the feasibility of such an action to take place. To measure the Universe Einstein has to be in the centre of the Universe, which will enable him to have a panoramic view about what there is to measure throughout the entire Universe.

If he were slightly off centre then that would place more material on one of his sides and less on another side. How would he then be able to gauge the entire spectre when there was more material on his one side than there was then on his other side, which would then fall outside his parameter of viewing. It is madness and deception that is thought up by fools and is purposely hoaxed to serve as hogwash to all the idiots they think the public at large are. When the evidence Einstein could gather did not underwrite the fraud, another even bigger scam was suggested. Then to further exaggerate the criminality and to achieve more deception with the sole motive to cover the failing of physics, then a plan was set in action where Science was put to task with an order to start looking for some ridiculous non-detectable dark matter. If the matter is not detectable it is not contradictable which makes it automatically acceptable and overall plausible because it cannot be disproved as much as it cannot be proved and by not disproving it, the deception may just stand in for the truth. It is what Newton invented the first time round and the plan will merely follow in the Master's original guidelines because that is what Newtonians got away with for three hundred odd years. If no one can prove that mass is responsible for gravity then no one can prove mass is not responsible for gravity.

This debacle can again play into the hands of the fraudsters because if dark matter can never be proved it therefore then can never be disproved either and is very much similar to what Newton established working precisely on the lines of Newton's gravity by mass deception. If no one can detect the dark matter because it is undetectable dark, matter everyone will be hesitant to disprove the validity because it can be proven just as little as it can be disproved. The only logical reason as to why this order was given is to cover the Newtonian misconception and the hallucination of mass that is producing gravity by contracting material. Newtonian mindset dictated that after all there has to be a presence of mass should there be gravity because if mass is the producer of gravity then in that is the truth that there cannot be gravity without mass…and all along there are still those that insist I do not understand Newton! They persist that there has to be enough mass to produce enough gravity for there cannot be too little mass to charge enough gravity to pull the Universe into confinement. While there is the expanding going about, their motive is to find gravity producing mass to stop the Universe of making a fool of Newton! Therefore to protect Newton and with that to protect mainstream Science from being covered in mud and hogwash they had to find enough gravity-producing-mass, either proven or otherwise, notwithstanding whether the mass is detected or undetected.

Locating mass became a smoke screen not even Lenin and Joseph Stalin could produce which would prevent every idiot in the general public to realise the fatality suffered by science as Newton is drowned in contradiction of Hubble's Universe that insists on the moving apart of the cosmos. No method was too unscrupulous to save Newton and proof had to be found, which counteracts Hubble's disproving Newton and stands in as truth for the gravity that should start contracting imminently or otherwise science will face what was coming their way all along… In this matter as to why mainstream science committed so much treachery one can only have one conclusion and the only sensible argument is that they worked feverishly to contain the proof of Newton's irregularities by keeping the truth from the general public whereby the entire world will remain deceived as to the sorry state Newtonian science finds its position. Spelled out in more fashionable colours it would be that there has to be enough magic everywhere to establish the presence of a force that will conceal the truth and establish another populous blind spot. The deception once again became acceptable although it was undermining the truth and it was created to help with introducing more of in despicable fraud Newton started on a Global world wide academic scale. I just do not accept that Newton could mislead the world's brightest brainpower and all that brainpower was blinded by their shortsightedness and naivety. That is very unaccepteble. Not with such intelligence being in demand and in command of the subject. It was done with intent to defraud.

By duplicating Newton's first act of fraud and repeating the manner again, Physics can still pretend they are the Master in control of their faculty while no one realises the degree of deceit that is prevailing in the corridors of all the physics paternity and all is done just to give Newtonian delusions legality. In doing that the academics are defrauding the taxpayer in their falsifying facts in order to use a racketeering process whereby they steal the money they defraud the public with by promoting a falsified claim to cover Newtonian fraud. Every one concerned with physics has to realise this shortfall of theoretical consistency, which includes the solution attempt that is so detestably untrue. To falsify and defraud facts by misrepresentation in as far as presenting falsified claims as if being the truth is a way to criminally illicit money from an innocent and unsuspected public, which then is a criminal offence. It is racketeering. If there were any other convincing argument that can support honesty prevailing in the midst of madness I would love to hear about it.

The truth of the matter is that planets rotating as celestial bodies show no mass influences. The two phenomena mentioned being the Big Bang and Hubble's expanding is proof of the expanding going on in the cosmos, which by expanding and not contracting is putting distance between cosmic objects while Newton's principle promotes contraction and the reducing of radii. The two phenomena mentioned are based on highly tested and well - proven merits about facts and are not accepted on cultural merit, as is the case with Newton's claims. There was never proof about mass forming gravity. Yet while it was never proven, it still went on to become the foundation of Newtonian physics. The idea of mass that is producing gravity forms the corner stone on which science is founded and every person on earth (excluding me) accepts unconditionally that Newtonian views are accepted truth but in so many years not one shadow of evidence was once introduced to substantiate any Newtonian claims.

No one could ever produce even on just one occasion some proof that mass instigates gravity. The accepting of Newton's cosmic views comes from implementing a principle conceived through the culture of what was thought to be correct through the ages and accepting these facts as proven truths is founded by preconditioning the human brain of young students. It is founded in the culture to accept the past unconditionally and is not founded on unwavering and undisputable factual correctness born from reputable proof. The Newtonian thinking about mass forming gravity is religiosity enforced on students by conditioning their accepting the correctness thereof and linking such unconditional accepting to their failure or success. By the repeating of what is reflected on, that is coming as accepted from the past about what was taught in the past without conditioning such information to newly required proof and with no wavering of the truth about the content, is that what allows their maintaining on the course of learning about physics.

It is definitely in the case of Newton not depending on the accurateness of facts. It is instated by accepting without questions that they either gain a pass allowing them to stay on as an accepted student, where they then become part of the system or when not accepting without reservations and because of that to be shackled as a failure and then be removed from the institution. This programming of the human mind to accept facts unconditionally is called examinations and it forms the pivotal part of a system that runs through every institution of learning generation after generation. The young mind is taken in the prime time of potential learning and by teaching the culture that comes from the past generations about Newton being correct and that Newton is accepted with no questions, forms the required foundation of Newtonian science. The system allows only room for students that unconditionally accept Newton. Others not willing or not able to just accept Newton on a say so basis (be it those that require some conciliation of proof by facts being presented to fill their investigative needs to back such claims of whom I am one in that class) are implicated as being somewhat retarded and short supplied in brain matter.

Not accepting Newton without any substantiating proof classifies any person with such mentality as to be presumed being incompetent in brainpower to "understand Newton!" Those that become accepted have to take on the role of acting mindless by "understanding Newton" in their accepting what they are taught regardless of the sensibility such teachings would otherwise dictate. They stand highly apart from the crowd that wishes to be served with evidence before accepting Newton as the unqualified truth. Yet the Masters as well as the institution towards those showing total and unconditional mind control by accepting Newton regardless of any condition attached to the accepting irrespective of questioning reflect the total opposite. They are the ones tolerated to become future members because they are intellectually superior in their "understanding of Newton". In that matter of finding the acceptable candidates a climate is established wherein there are no margins of doubt allowed about any part of Newtonian issues. It is a mentality that is created by the Superiors suggesting that only those with a superior mental capacity and clarity of mind can become the chosen ones because they are intellectually much more advanced making them better suited to fill a position where they are able to see and accept Newton because they show "understanding of Newton". The others "that do not understand Newton" are detested as those who are mostly rather slow witted. Those being incapacitated by doubt because of the lack of proof and therefore are not able to see what Newton claims are the ones that is lacking the faculty to understand Newton.

They are portrayed as the dim ones and outcasts as they are slow in grasping the situation by identifying their shortcomings with having questions and doubts. It is presumed that the bright ones can clearly see what Newton claims and that makes them the cream of the harvest. That is the institution the student is exposed to with the full knowledge that the student also is compromised when "not understanding Newton". This is part of the student's life when it is also well understood that at such a vulnerable age in the era when the students are dogmatised, that in their young minds questions are not yet part of the student repertoire and their conditioning of accepting is well rehearsed, constantly repeating the idea that Newton is the unqualified and truth unconditionally. To solidify the foundation this repeating is based further founded by all the testing that goes along with the process of dogmatising, the final acceptance are entailed through examinations.

The questionnaire that is called examination papers sets the standard of measuring their conditioning. By the level of the thought conditioning relating to those standards of accepting without preconditioned wavering in doubt or insisting on proof. With the admittance accordingly come such standards of testing where on those standards which the acceptability rests to enter the academy. Admittance or rejection is subject to the level of accepting Newton notwithstanding and regardless of whatever other requirements should prevail becomes the focus point of them going further with their studies. Professor, be honest, and admit to your person in the privacy of your mind how you will react to a student that desires to carry on with physics studies while he reject Newton due to the lack of academic proof on the matter of mass producing gravity. Explain in personal honesty to your conscience what you would do to a paper wherein a student refutes Newton as I do just because there is any evidence in support of Newton's claims. How will you respond to a paper that rubbishes Newton on the grounds that Newton never brought significant proof to the investigating table? What will you do to a student that declares his rejecting Newton is because to his mind he was never presented with acceptable proof about the merits of mass forming gravity and the student therefore rubbishes such statements on those grounds? If he should dare not to complete the paper on the grounds that he has not witnessed any proof about the planets revolving around the sun by measure of mass or that he finds proof lacking to prove that objects are falling by means of mass applying because it contradicts Galileo and where he then will demand such proof before he would complete an examination paper, then I ask you what would your response be as the Headmaster of your faculty? If you are honest to yourself you will admit to your personal integrity that the student has the choice of either to accept Newton and surrender all doubt or to surrender his scholarship at your institution and in that be rejected and banished.

Yet all the doubts are present because I ask you for proof and you know there is no such proof. Students are never given proof about the matter of gravity and mass but are expected to learn Newton by over and over repeating the repetition and rehearsal of the absolute correctness one has to show which then will serve as proof to their commitment of trusting Newton. This they have to do to prove the degree of their personal ability in accepting Newton unconditionally or to use their more commonly used phrase as they put it "to understand Newton". This is so lasting even after years later no proof on the matter is ever demanded by any student of Newtonian dogma. It proves that those Masters that is in control of the future paternity in the past chose the future candidates well. When those chosen candidates are by then part of Newtonian culture, they never waver on the idea that mass provides gravity and the accepting of that fact goes without any questions asked. Any future questions are unheard of and doubt on the matter is never tolerated. I should know this well since any of this was never tolerated when I presented my case in person in previous meetings where such meetings were tolerated and such meetings were agreed on. Most Academics of importance told me to leave my books at the door of security at the gate being on the desk of a security officer at the gate because they can find no reason for us to meet in person. By instituting such forceful accepting of the very much accepted but also never yet proven, teachings is going on through generation upon generation. It is carried from the past into the future by tutors repeating their past and projecting their shortcomings onto the students which represents the position they now have somewhere into the future. It is a culture of a process where the dogma of science is unleashed onto students and without ever raising thoughts the matter becomes undisputed without any measure of proof ever served or any question raised in the line of proof.

Let any professor say why does he or she unequivocally accept that gravity is the process where material is drawn closer and that mass is responsible for such behaviour. A better name to use in describing this conditioning process will be if one uses the term "brain washing". The conditioning of the mind in accepting or face rejection by physics is the most outrageous form of mind control ever instituted by man. It is what this conditioning of student minds is boiling down to and it is echoing Pavlov's theory by confirming the total accuracy in Pavlov's conditioning theory. You may wonder what made me come to such a conclusion as the one I portrayed in the paragraph I just mentioned…well I am still at this moment subjected to just such treatment by those I accuse of deliberately practising what I accuse them of. I do not admit to Newton therefore I am subjected to rejection.

To present falsified facts as proven veracity that is equal to God given verity set without limits on the scale it is done and with the intent in which it is done is nothing less than unscrupulous plunder of the truth in unspeakable dishonesty and is an unpronounceable violation of truth with criminal intent to defraud which is comparable to delusion. It is a hoax to defraud the unsuspected by ruse and trickery in order to defraud money by means of establishing the hoax as the indisputable truth. One may not serve with chevalier in whatever capacity and no matter who does so or whom it will benefit, no matter for what reason or to what goal it eventually will achieve because nothing can merit the right or no person has the right to present falsified statements as honest truth and by doing so will be expecting to gain financially notwithstanding that the gain might merely take on the scope of claiming a monthly salary. When doing so is not only equal to swindling for it is treachery.

My work offers the solution. My work explains what applies and why there is a Big Bang while contraction is continuing. For the first time ever I am the one that is able to explain the Bode law, The Roche limit, The Lagrangian points system and I prove that the Coanda effect is gravity. The Coanda effect marries the other three principles to form gravity. Newtonian science views these phenomena in a range going from not existing to not understood to being coincidental, to being just a tad beyond explanation. Those four phenomena are what forms gravity and yet no Newtonian once had the mind to look at my work and see how the four phenomena form gravity. All investigation into my work is suspended the second I denounce Newton. Those phenomena are there and all the Newtonian denial about them or where the Newtonian science finds the explaining thereof impossible or where by using normal known methods they are not being able to explain the existence of the phenomena is casting the attention on their doubtful abilities in understanding cosmic science. While marking the phenomena off as coincidental, puts a question mark on science and not on the phenomena being or not being. But in reality the gravity generated by the four phenomena has nothing to do with mass and while that is true, such truth comes just because I can show how I am able to dispute Newton's claim on mass being the producing factor of gravity because I can conclude from what I can prove by using the four phenomena and what that idea represents where the four phenomena produces gravity, that the idea of having gravity which is founded on the basis of mass producing gravity, such a concept is utter nonsense. Mass in every purpose has nothing to do with the science of cosmology or astronomy. Therefore my work departs from that of Newton like a rocket flying away from the Earth. There is no way to console my work with the work of Newton because the work of Newton does not apply in the presence of cosmology. Because I reject the work of Newton and incriminate Newton as a falsifier of facts where he presents conditions that do not apply I get no one to evaluate my work in sincerity. I reject Newton because mass is a fabricated lie of the Newtonian imagination. Yet because of my mentioning that no one even read what I offer although I offer the solution. In spite of what I offer all in physics are fixated on the untruths that Newtonian astro physics present. The truth is that Pluto is equal to 0.0025 times the mass that the Earth has and Jupiter is 317. 8 times the mass of the Earth and all the while all three are spinning merrily at the same pace carrying on at a ratio as Kepler stated.

Incidentally Kepler was dead before mass was invented and that should also serve as a strong suggestion about all "Kepler's laws" that Newton invented. Kepler never mentioned mass but Newton defrauded the work of Kepler to fit in mass through much cunning. Now fit mass into that explanation about the mass differentiation in planets while they orbit at equal space-time (T^2/a^3) and see what I mean by misrepresentation through open fraud. My explanation does away with mass but it introduces so much more in its place…and yet you reject it because it is not in line with Newtonian thinking. Or as one Academic once put it: "it is not in line with acceptable Mainstream Physics" Newtonians do not know what time is because Newton went along and killed time by putting time equal to zero. Anything equal to zero does not exist and therefore putting time at a pace of zero kills off time. When time stands still there is no time and with Newton allowing time to stand still he removed all evidence about time from the records. He stopped time and time cannot stop because time flows. But he had to put time in at zero to find a reason for mass to apply. By taking away the value of **k** it gives validity to mass pulling mass. By saying that $\dfrac{dJ}{dt} = 0$ such a claim can give rise to the idea that **k** will diminish because the influence that **k** represents is zero. This fraud he brought about because he had to find a manner in which to incorporate the mass he was fixated about with such misleading evidence. You Newtonians talk about space-time but you know neither what space is nor do you know what time is and I mathematically prove both space as well as time.

Should I not find a publisher In South Africa willing to publish my work and one that offers me positive results this time round I am going to publish the books by way of "Print – On – Demand". Again I repeat: Please do not see this statement as a threat because from my poor academic status and low position I am far too small to be a threat to any one. But if in later development someone should find merit in my work being correct, then it will be an embarrassment to your institution when it comes to everyone's knowledge that the work has been presented once again which then will also again include those times I sent work previously which is now forming part of so many other times in the past where my work was again rejected. Then somewhere there will be some academic that would learn about the merits that was used in your rejecting / ignoring / dismissing of my work. When my book goes on sale on an international scale, there then must be some academics that will put an asserted effort in scrutinizing my work and when they do investigate my claims, they then have to learn that my presentation of cosmology is proven to the most and finest detail unlike the Newtonian principles that is furthered by the Pavlov methodology. In the books I now present I have gone into the most ridiculous small detail to present my case and the format in which they now are is down to a standard that a child can understand.

If I do not find any joy from South African institutions I am going to self publish my work by the way of another method of publishing that has much limitations but I hope this very letter addressed to your office might circumvent some of those limitations I face. The correctness of my work is beyond dispute by the uncompromising proof that I deliver. This stands as a challenge to the physics paternity while their use of mass in Newton formulae boils down to despicable fraud. Somewhere along the road someone of the right stature will read my work and will then recognise my correctness concerning this matter. It is what Max Planck did to the work of Einstein. It took one person with vision to recognise the correctness and award the claim. I do not compare my work or myself with Einstein but on the other hand I am way past being modest or being boastful and you can bet on that!

It would be to my benefit to find assessment and recognition through a South African institution, but if that cannot be done I am going to turn to the international press by way of using the system called "Print-on-Demand". Although I agree that this will not give you any sleepless nights I still am trying to secure another window to the world through which others may consider my work. I plan that in the hope that there someone will see that my arguments are correct and bring the acceptance I hope that will follow. But when that comes about, it might just bring embarrassment to those that rejected / dismissed / ignored my work in their official capacity. That is only a suggested warning of a possibility of the tiniest proportions but ask yourself if you think it is worth the chance of embarrassment when all you have to do is to consider my work without being pre-judgemental about it? In other words just go on to read my work and then try to find valid arguments when you go about dismissing / rejecting / ignoring my work. Just read it first! All I ask is when you again reject my work the following time in doing so then see to it that the reasons why you reject / dismiss/ ignore my work is based on water tight and sound principles. I show and explain the sound barrier, which is totally unexplainable when using Newtonian science. I show where one can find and physically see singularity. I show where one can see singularity by looking at the spot where singularity is not. I challenge anyone to prove that that spot is the spot that is not holding singularity. I show where time has an infinitive value and where time has an eternal value. Any one can look at the positions I allocate to where time holds eternity and time is in infinity. I show where one can see the place they both are located. Anyone can see where it is, should the person make an effort to look. I show what time is and where time is. One just has to follow my instructions to physically see it. I show by examples why time is there. I show where and how time influences the cosmos by generating gravity. That I do by mathematically proving how the Titius Bode law generates gravity. I show what gravity is and I prove that gravity is not a force in the manner that Newton saw the force.

I show physically as well as mathematically where and what space-time is. It is there in an allocated position where one may physically look at it. The information that I now present is the most basic there is. In my previous work I was under the impression that everyone could see where singularity is, or should I more correctly say where singularity is not, and why singularity has the value it has. Even a tiny child can see why singularity is allocated in the sphere because of the simplicity in the argument. I took it for granted that everyone should be aware of what time is and how time influences the cosmos. Now I do not take your realising of such matters for granted. I take the reader by the hand and guide the reader as I show where every aspect of the cosmos is. I show where eternity is. I show where infinity is. I show that gravity is motion that partitions eternity and infinity. The best of all is that I prove that gravity is the parting of infinity from eternity, which then results in the Universe we have. With using that information I then could understand how the cosmos advanced from singularity into space-time. There is no need to be a genius to understand it because even I can understand it. I explain that process how the cosmos came about in the finest detail as I take you through a mathematical tour how the cosmos came in place. All you have to do when looking for reasons to again reject / dismiss/ ignore my work is find one argument out of line or one mathematical conclusion that is irregular or one reason I give that is not conclusively proven. You can condemn my work by finding just one reason to dispute my work on. You don't need to hide behind Newton's institution status because my work offers page after page of other reasons than hiding behind the purity of Newton and if it is with reasons based on sincerity it will bring merits when you denounce my arguments. I hand you that challenge. Don't just dismiss / ignore / reject my work because I have proven Newton wrong but I challenge you to find serious grounds about the correctness in my work in order to do so. I do not have to prove Newton incorrect because nature, Galileo and Kepler was amongst the first to prove Newton was wrong and that happened even before Newton did not know what gravity is. Newton knew what Galileo said while Newton knew what Kepler said and there are a lot of others Newton also sidelined when he forced his fraud onto the world…that Newton did that intentionally is what you have to admit to that took place or otherwise prove it is not deliberate fraud!

I admit that there might have been some chance where in the past where I was guilty of cramping too much information into too small a printing space, but I am not guilty of that in the presentation I propose this time round. There are three books dealing with three different aspects of cosmology. If I could, I would suggest that the reader might start with **"an open letter Announcing Gravity's Recipe"** but that is not really important. This book I name, deals almost exclusively with the matter of gravity. It will be the book that will naturally flow out from this letter. In the event of you choosing again to reject my work, I ask you to make double sure about the foundations and the grounds on which you base your rejection. Make sure you can prove that anything on Earth can have mass when that something is displaced in a location being inside the Sun because to have mass when inside the Sun it has to be there by filling space and not be a mere suggestion. To have mass inside a Pulsar whatever is there has to be there and not be a few photons the pulsar rejects. To have mass inside a Black Hole it has to be an atom first and not a thought. When you connect a meaning to mass then define what it is that will have that mass you surrender to the object because whatever is in a massive star is very different from what it might be on Earth. Make sure about the argument and what such an argument produce in sensibility.

If my work has merit and such merit is recognised by other acclaimed academics then put yourself in a position that is ensuring that their recognition does not become an academic slap in the face for your institution. Be sure your rejecting my work again is on valid grounds that you can defend the integrity thereof about your reasons and your reasons are not just to defend the integrity of Newton. With that I then again must sincerely advise you to find grounds for your rejecting / dismissing / ignoring as it just might even by the smallest of chances otherwise lead to your institution becoming the laughing stock of the entire academic world. Make sure that when you once again reject / dismiss / ignore my work this time, that you have reasons of integrity in doing so. I do not make any presumptions on your behalf about my work, but all the reasons for rejection my work on previous occasions in the past by publishing institutions was no more than an insult to me because it proved that those responsible for assessing my work never came round to even reading my work. The comments they made had no implication on my work but only were merited to a few pages of reading, normally the first ten pages of the book with seven hundred pages to offer. Be sure this time the insult is not on your own account.

I have reasons to press the point that in the event of you deciding to reject the work, I ask you politely to make sure about the sensible validity you apply on which you base such a rejection and that your doing so is about all the aspects in my work which you took into consideration and not only the information you gathered from glancing over the first ten pages or so which was done by some pre graduate when that person had no real incentive to do so or felt any ambition to understand what I say but was doing so because the person was ordered to read my work and because the superior academics in charge of the department had neither the will to do more than the bear minimum nor did they find the academic drive to do extra work. Again I acknowledge as I repeat that this can be no threat because I am far too insignificant to be a threat to any one at this moment in time but should my work be correct and should I find such recognition internationally in the future which there is a slim chance of I admit, but in such an event ensure that your rejection could not lead to an academic embarrassment to your institution and also to you. The work is of such proportions that it either could lead to nowhere because it is never been read by any person of status that agree with my views or it is going to develop into the biggest thing to hit physics in many years. There is no middle road.

When you reject my work again then all I ask is to do so on defendable grounds that can support you in doing so later on. By your rejecting this time then make sure that your rejection is valid and that you can defend your reasons in the presence of other academics. Make sure that it will not later come to be an avoidable academic embarrassment because again you have made no effort to read when assessing the work as you did on other previous occasions when judging the validity of my views in the past. There is a slightly less than absolute chance that somewhere in the past your institution opted to use one or more of the following reasons for ignoring / dismissing / rejecting my work when they did so in the past. Choose any one or more of the reasons I supply and it will most probably reflect on a choice your institution made in the past when they chose to dismiss me. The rejections by now have almost become countless and that includes almost all institution in South Africa as well as numerous international institutions with highly admirable reputations that has as much credibility and yours have and are amongst those I implicate. Notwithstanding your institution's name you may have, it does not matter because your institutions supplied one or more of the reasons. When I inquired about why such rejection was done at the time it happened, I was given the following reasons. I was given these reasons by administrative personal about what merited the rejections in the first place and on what grounds the rejections were formed. Academics in principle motivated such rejection at the time because they did use one or more of the following reasons in the past to reject / dismiss / ignore my work. Should it please you to use any of the past reasons for rejecting my work again this time round, then please do not think of it as very original because by using it, someone already jumped you and you can therefore only repeat again what was done before. Such are the following examples. When using it, then be sure to motivate those reasons by finding reasons in my work as such, which you can state to be incorrect since that part was never yet done. These are some reasons on which you rejected my work:

1) On the idea that my English is of a poor standard to use. I am Afrikaans in every atom my body holds, which makes English not my strongest point and to that much I do admit, but to put it that my English comes across as being unintelligible is defamatory. I see this rejection resulting because everyone in South Africa would immediately see that Peet Schutte is as Afrikaans as the person with the name Jan Botha. Make sure that you do find my English to be below the level of comprehension. I am Afrikaans and I would never care to disguise or try to hide the fact. In my books however the part that I present for your attention and request, your evaluation is written in mathematics and as a Mathematician Master you Academics that I have approached on this matter, have failed miserably to read what I communicate when I am in conversation using mathematics in the past.

2) My arguments about the outer space not being nothing but is filled with substance is incoherent. If you use this line of argument again by insisting outer space does comprise of nothing between cosmic bodies then also make sure to multiply the distance between the Earth and the Sun with zero and prove the coherency about such an answer on your part because $150 \times 10^6 \times 0 \neq 150 \times 10^6$ km but is zero. If you are unable to prove how nothing can be anything else than a factor that introduce zero as a value in a calculation then you have to admit that there must be some form of substance to give meaning to the distance that is between cosmic objects.

3) My arguments on gravity hold no merit. Prove the merits of Newtonian gravity and then compare that to the proof I introduce in my presentation on gravity where the Coanda effect is gravity.

4) The publishing of my work will be too expensive to justify such a process.

5) Newton cannot be wrong because Newton was never proven as being incorrect thus far because time proved Newton correct and therefore Newton is without dispute.

6) My work does not fit the required institution criteria while no one ever says what their criteria then involves or what about my work falls outside the criteria of astronomical science.

7) The Acclaimed academic responsible for scrutinizing my work distances himself from my views as he stands by the principles Newton holds.

8) Newton requires no justification from any person.

9) The academic simply sides with Newton with no reason given for doing so.

10) My academic history and qualifications does not merit their attention or investigation into the matter.

11) A response in silence that is rather more befitting the mortuary.

In the light of all the rejection all I say is that mass forms the restriction of gravity and there is either free gravity which is unrestricted motion, which is what gravity presents without mass or there restricted gravity with the tendency to move which then is mass. However my saying that falls on incompetent ears all the time because of a Newtonian fixation on the ridiculous. When a body is in outer space the body floats above the Earth as a body would float on water. The body can only float when the body gravity can sustain a position it holds in relation to the velocity of duplication by motion that the Earth holds. In my book I present the facts why this is in the form of density that is applying to form gravity. The body maintains independent gravity but the gravity totally relies on motion that has to be in line and the rotating motion has to be superior to the orbit motion of the Earth. If the orbit motion is inferior the motion ratio will put the body in a smaller orbit and such orbit reducing will lead to the body entering the atmosphere of the Earth. It has nothing to do with mass since all bodies require a specific velocity irrespective of the size that it poses. The tarnishing of the orbit will lead to the body falling and that too has no indication of mass. When a body enters the Earth space the body will sink just like the ship sinks as the body loses buoyancy and that buoyancy stands related to the speed differentiation there is between the Earth and the independent object that enters the atmosphere at such a point. When the body falls it loses its independence in gravity since the Earth then takes command of the body's motion. The floating or not floating depends on the motion and velocity that the body can maintain when it is in orbit at a distance above the Earth in outer space and that would establish the required density attached to that. I explain this principle in very much specific detail but in short it is exactly following the rules that Kepler's prescribed that gravity should have even before gravity had a concept and that is being $a^3 = T^2k$. As is the case with the floating ship it is the case with the falling object. The only difference is when a body is no longer buoyant in the water on Earth the loss of density is because a difference in personal atomic density and the density the Earth puts down and a condition. But even that too is a case of velocity that provides the density through motion and that motion is the gravity factor.

Supplying an adding of heat gets all mass stricken objects moving in relation to the adding of the heat and in the time that the adding of heat took place. We call it either drive or exploding but in essence it is just adding heat to the relative density that the particle in focus has to endure in relation to the time the endurance is taking place. As is the case with the ship at see, once the ship sinks it becomes part of the space the Earth claims where it can no longer move except by moving with the Earth at the velocity the Earth dictates. It then has mass because it is the Earth by the extension of space that the Earth has because it then is part of the Earth. On the other hand when the body is in the water while it floats it is part of the space that the water reserves while the water presumes in the capacity of a liquid and as a fluid claims the position liquids claim because it can move independently in a variation of alternating its relevancy from the dedicated precise motion the Earth demands. The displacement is added to the value of the water making the floating ship extend the measured value of the water. The ship then takes on the mass the water provides and relinquishes any claim to independent mass it might otherwise have. It exchanges the gravity or the motion it would otherwise have for the mass when accepting all motion that connects to the Earth moving.

We that sank to the bottom of the Earth-soil have mass because the Earth sustains our motion or gravity. While we are swimming we bob as water and that motion substantiate our position as a liquid that is independent from the solidity of the soil. While in outer space we provide our independence from having mass because we hold a position where we are unattached to the Earth because we sustain our individual gravity or motion. Science call this micro gravity but it should much rather be micro mass and macro gravity. Being part of the Earth puts us in the class the Titanic at present has but being in outer space makes us part of outer space as we have independence on the precondition the independence can be sustained by a motion that match the required gravity. Then we provide our independent gravity as we supply the motion and with that establish independent mass, which Newtonians gave a name: they call it micro gravity. Being on Earth, the Earth provides the motion or gravity and sustains us in forcing us to hold the gravity of the Earth by supplying us with mass. We deprived ourselves of having independent gravity by accepting the Earth gravity on the condition that we receive mass which would make us a directly linked extension of the Earth. Now I ask you: Is what I say in my books when I say what I said above in a mathematical context so hard to grasp especially when those whom I address should be masters in mathematics? When a body takes on mass it becomes a part of the structure of the Earth and therefore loses its cosmic independence in motion as it presumes the motion of the Earth while being a part of the Earth. It then is the Earth by extending and only life can establish motion when the object in motion still has mass as it is attached to the Earth where it is the Earth that still provides independent motion but with such independence we measure that performing in the manmade measuring tool of mass. In the cosmic reality objects however, can either move as well as having gravity or it becomes part of the Earth and has mass as it is being part of the Earth in total. Being part of the Earth, the object can find mass as much as it forsakes independence and gravity. In true cosmic reality there is no chance of having both. Once an object surrenders independent gravity it will have mass because it will never have gravity afterwards except that, which the Earth provides.

The mass is the manner in which an object arrests material and claims it as confiscated possessions where by the arresting object provides the gravity the object arresting will subscribe to which the arrested material surrenders space by receiving mass. The apple that Newton gave a value of mass would remain part of the Earth in the location it fell if not for the intervention of life. Then again the reason why the apple fell was because it had life to start with. Rocks just don't drop from trees and roll around afterwards just because they can find motion independent from the Earth. Rocks have mass because rocks do not run around and relocate in new positions. Life creates gravity because life is about motion and motion provides independence by giving gravity-motion above and beyond the gravity motion the Earth provides. Life can command mass whereas in the absence of life mass destroys independent gravity.

Again you have the choice to prove to yourself and see where does mass have any prognostic influence in establishing gravity and show your honesty in the matter. I am of the opinion, which I set out and prove in my work, that mass is a man made unit of measure, just as temperature and distance forms a tool with which to measure to a specific standard applying only on Earth and has no real significance in the cosmos at large. Mass is a human invention to measure what in the cosmic context turns out to be truly beyond the measure of man. Go on and show yourself how much the Earth / Moon mass is drawing the Moon closer or how much is mass having any object fall faster or slower than the one with more or lesser mass or how does mass influence the planet Jupiter with its enormous mass to orbit faster than the individual debris in the clusters that is sharing an orbit with Jupiter. Show how mass introduces gravity other than by means of misinterpreted culture. You now have the option to side with those instating the criminal intent or you can admit to yourself what is motivating the need to find evidence that will confirm any critical density factor. With that information then bring proof that mass is producing gravity at the moment throughout the entirety of the Universe.

Make your consideration is made in a sober environment where you are detached from other influences and by standing away from performing as one that would create a screen to cover fraud when you reconsider what the critical density theory entails. Think what influences have to be subject to such a turn around where each atom is at present on route in a direction of expanding from every other particle and particles that ever formed. The idea that what is there forming what has no end can stop in its tracks is childish. Think what an effort it takes to stop a ship such as an oil tanker, which is fully laden and at full power while speeding at full capacity into a harbour. Think of what it will involve to stop the Earth from moving away from the Sun. Then stop the entire solar system, which would include every speck of material throughout in one instance. Stop the massive atoms at the same moment as one would stop the less massive atoms. Stop a star expanding while at the very same time the force will control the hydrogen on the outside with equal precision than it controls the iron in the core. Explain how the force will find a way to discriminate and apply more fiercely to the heavier atom and have them stop at the same rate as the force will apply and stop the lesser massive elements such as hydrogen and helium. Find a way to explain how the core will stay intact to the star centre and not dislodge from the star when such force of total equality enforces a will on all that is totally in inequality. How would such control on different elements with different mass deployment work? The thought must fill one with soberness. Think what it will take to stop something as gigantic as the Milky Way. Think about the impact required to turn around the expanding of every atom that there is in a body that forms something as large as the Milky Way and then multiply that with a hundred billion times the Milky Way. If that doesn't shake your foundations, you have no clear mind left. Just the consequences of what this suggests indicates the contemplating of the farce this presents being without equal of having a parallel. Think what the impact that just the thought alone embraces and the amplitude of what this entitles and what the con would suggest that such action would accomplish. What will it take to stop the entire Universe in direction and force a turnaround of the entirety that is out there? See what I mean by criminal minds blowing facts out of proportions because it is not motivated by clear minds but the motivation is to cover up deception by producing more deceit. That suggestion is that all of us (the mindless many standing in to represent the thoughtless populous) we that stand outside the academy of physics are stupid enough to swallow such slander. What on Earth would guide Academics to regard us with such dismay and have so much disrespect for our ability to judge by using sober reasoning and where we, the brainless masses, will lack any ability to also contemplate and find reason by thoughts? That entertains an insult by merely addressing such a gesture.

Prove to yourself that even when finding the so-called critical density how that then will convince the cosmos to introduce Newton's contraction and even how that will establish a start of the implementing of the system that will firstly stop the expanding. Think of it in the manner that every particle has to come to a stop. Then determine the forming of a revised working principle method in the cosmos. Show how the cosmos will stop expanding and come to a standstill by stopping all directional motion throughout the entire Universe, which includes every particle large and small and especially down to the most insignificant of particle detail. To top such madness then explain how the cosmos will then go on to start moving again from a stationary stance as the motion will start in another directional flow and where the most domineering will start to move at the very time the most insignificant is starting to move by applying the same motion velocity equally to all there are. The entire stopping and restarting must be simultaneous and if not, that will bring annihilation to an entire cosmos.

It involves the charging of the most significant as well as the minute's particle in motion as well as involving all of the entire cosmos that will change the directional motion. Go into exact detail how that will that come about where the action will involve every particle without having massive collisions and total destruction? Calculate every one of the actions of the smallest particles in an independent detail as mass will have to discriminate between large and small to equalise all. Such turnabout includes every tiny particle as well as every star in every galactica big or small throughout the entirety of the Universe everywhere. Every atom that was going the one way will be in a position where every one will have to stop at the same time, then turn hundred and eighty degrees in direction and start to follow the particle which was following the particle that then is behind the one that was behind it at first. The follow my leader has to change to lead my follower and it has to be done in a way that all possible atomic collisions are averted simultaneously. It has to stop every atom visible or otherwise, and then after that start every atom visible or otherwise not visible subatomic structure again to follow the particle that was in front of it but now finds it to behind, which was following the particle that then is behind but from then on will be in front. The one was at first behind but suddenly finds it in front of the one it followed up to that point and no atom may collide because the collision will bring nuclear destruction as atoms destroy the atoms in an escalating crash which is dooms day by the stretch of any imagination. The smallest unseen subatomic particle may become a missile, which will be destroying every atom that stops before the subatomic particle could stop. Let Newtonian gravity choreograph that one for beginners and let Newton's gravity explain how this will come about. In all of that it shows how much sensibility there is in the Intelligence of the most intellectual group on Earth where they attempt to circumvent reality and introduce total madness and all that is to cover up Newton's fraud and criminality. To those suggesting such a possibility, it is playing a game of equalling God's abilities by finding matching abilities with God on a mentally mathematical level by setting mathematical rules applied in the games in order to avoid reality by creating overbearing deception. I challenge those to come into the world of reality by leaving the world of playing mathematical games with physics. Go on and calculate the process where every particle will destroy the particle that was in front of the one that will be behind of where it was up to that point, if the force of inequality does not bring total equilibrium to everything participating in the process. The contradiction in motion will make redundant all distances achieved up to now wherefrom collisions will come about that will end the Universe then and there.

If that proves how the greatest minds in the Universe form conclusions with the worlds most intellectual group in action, then thank God I am one of those excluded and considered as being merely stupid for I am considered to be one those that is not blessed with the brains to understand Newton! After all I was accused on so many occasions by those in power of the academy with subtle suggestions where they literally suggested quite frank and to my face that I lack the insight to grasp Newton, and this they do, while they display the bravest understanding in absolute sympathy towards my insufficiencies and disabilities especially on the mental level. I was told to my face in numerous conversations when I presented my work for evaluations that I am not mentally capacitated to comprehend what Newton's work imply, but also I must add that it was done in a very sympathetic and gentle manner as not to hurt my feelings too much or degrade me very much.

Be honest and show the sensibility in such an argument and while you are about it, then show to your own view in the privacy of your own company how you envisage how gravity will overcome all the obstacles, as gravity will stop the entire cosmos when it could not even halt the progress of the expanding of the cosmos up to this point. Remember to rush and be very quick about it because with the escalating of the distance of particles the relevant gravity between particles is demising by the square so there cannot be much time wasted to prove Newton correct! The longer you linger, the weaker the force of gravity will get because the further the distance are separating structures, the less effective will the mass be in its task of creating the force. If you do not act with haste to bring a sudden solution, more deception has to follow and Newton will have to be proven yet again with another scheme of deception somewhere in the very near future! You can do that before you read my books and you can do that after you read my books. Then you can decide where you stand in the matter of the criminal defrauding of the truth forming part of my accusing Mainstream Physics of the falseness' of Newton principles. When you read my work I challenge you then also to prove me incorrect on my insisting of the relativity between particles, which is what I suggest in my work as a counter theory to Newtonian incoherency. Realising what stupidity prevail in the formulating of the critical density puts everyone of that era at the time when the formulating of the Critical Density theory was suggested, which is also including Einstein that was the one that was blessed with the task, in either a state of pitiful stupidity, or criminal conspiring. They, who were taxed with the formulating of such a theory, were the most brilliant minds that ever lived! Think in a supposition of sincerity and not a supposedly manner to cover up an hallucination that lasted since the dark ages about what was suggested by those that should know of much better and is thought to have the best minds ever, being in a position where the lot was performing in a mindless manner, acting as if thoughtless, while judging the merits of their deeds. Was that what they tried to accomplish brainless or just conniving. When using the slightest scrutiny one has to come to one of two conclusions, which are that either their intellectual status tarnished or their morality became suspect. There is another option, which is slightly slanderous but serves a better elaboration and that was that the lot that were involved in the act of conniving by creating the Critical Density fiasco intended to rape science and to rape the common courtesy of the publish at large.

I challenge you…no, I dare you to charge me with slander because if you do, that would stir the hornet's nest and bring the hidden into the light of openness. Let us then present our individual cases to the public at large and find support in the unwavering belief we have about or different interpretations that present correctness in this way. If you are of the opinion that my remarks are besmirching those viewed in admiration with intentional deformation on the characters of some of the most outstanding and remarkable persons in the world of the academy of physics then prove to your conscience how mass forces planets to go faster or slower in accordance to the mass they individually have or how planets orbit the Sun faster or slower in relation to the mass they have or how mass would have objects fall faster in relation to the mass they have. Show how I commit deformation when I accuse those deforming the truth in physics of doing just that in an unprecedented case of swindle and betrayal. Show the innocence of those with highly regarded integrity where I accuse them of being unscrupulous fraudsters. I again dare you take me to a court of law where both sides can show the justice in their views on their matter and make it a legal issue. Just go on to show where in true nature does mass play a part in anything other than being a man invented measuring unit devised to be used as a measure of calculations in a method that establishes a means of measuring.

Sir, Madam, the choice is yours to make where you may come to see the criminality in the matter and decide to investigate my work that proposes the truth of what is out there or to side with the criminality of falsifying the truth further in order to substantiate the ongoing fraud and continue to support and validate all such criminal activity. You have to agree that when one introduces falsified facts and intentionally admits to the authenticity and integrity of lies of any nature which you present as absolute unblemished truths that the doing so thereof then is criminal conduct irrespective of motive or the innocence behind such conduct or in order to pardon the bent motivation of what eventually will have a benefit from such conduct. You now have to decide which way you are taking the matter because the matter has come to your attention.

I will admit that this letter does carry another agenda than seeking your approval this time round or that this letter might find another motive than be trying to establish hopes of receiving your acceptance. I have passed that desire of finding academic approval as a motive in my effort to contact you some time ago. This is a warning that when I publish my book, I am going to do my utter most to reveal what I consider to be a case of fraud in physics as a whole. I am going to go as public as what my abilities would allow me and find as much publicity to further my conclusion about a conspiracy going on in the world of science. I will establish as big a cloud of dust as I can muster through as much press exposure as I possibly can generate with one accomplished goal and that is to portrait not only Newton as fraud but the entire Physics paternity in a state to conspire to defraud. I will do my best to show that this state of affairs has been going on for years. I will base as much as I possibly can that this is founded and partly due on their constantly ignoring me. On your part make sure that your position is as rock solid as you believe it to be and that your facts are as proven and as defendable as you might consider it because your security in your faculty rests entirely on the public's idea that are seen as unblemished. There opinions are resting on their surmising that you are being accurate beyond all compromise and in that the public belief is vested in your unshaken correctness. All about physics ride on the public opinion of never having the least bit of doubts about your accuracy.

That idea might tarnish and take with it the ultra white of your reputation. Just make sure that I cannot prove that your facts are not as unblemished as you pretend they are. On your part take the challenge and prove Newton is as defendable as you believe he is. However in all fairness, I also admit that I see this letter as being just a charade because I have to pretend to participate in this façade one more time. I have to pretend that I might still be convinced that you will be in mind to accommodate me this time when in truth you never even acknowledged receiving any of my mail in the past. If you ignored my work in the past there is no reason why you would read it at this point and if you did read it in the past but failed to act on it, then you are guilty of conspiring and in that you will not change. On the other hand I also realise that if you would acknowledge that my views are correct then you must also admit at the same time that you know less than nothing about a subject you pretend to know more about than God in person knows about science. However in the event where you go on pretending there is in science what there is not validation of and that Newton is a God instead of the hoax he is from the very start, then I guess we shall confront one another through the press. I say this as a test because no Academic would ever endure this much insulting directed at Newton and therefore I am one hundred percent sure that no Academic would read this letter this far. I don't envisage any person in academic science that would reach this very line because science never confronts the truth since they believe they and Newton sit on a pedestal from where they and Newton govern us little people.

Since no Academic will be reading this letter I therefore feel free to say that I am going to use this very letter to promote my theory to the broader public through whatever form of media a might muster and with the aid of whatever publicity I might gain on the matter I am going to call on the broader public to assemble their awareness about the matter and find their opinion in the matter. It will also then help me to sell my books without the benefiting from the benediction of science. Lastly just consider the following tomorrow where you stand in front of your classes: Sir, when you address your students about the wonders of Newton, then tell your students tomorrow how much the Earth drew closer to the Sun since the days of Kepler, and Madam, announce to the world how much the Moon came closer to the Earth since the days of Tycho Brahe while you then remain convinced that Newton is still flawless after three hundred and fifty years.

Sir, Madam, the choice is yours to make where you may come to see the criminality in the matter and decide to investigate my work that propose the truth of what is out there or to side with the criminality of falsifying the truth further in order to substantiate the ongoing fraud and continue to support and validate all such criminal activity. You have to agree that when one introduce falsified facts and intentionally admit to the authenticity and integrity of lies of any nature which you present as absolute unblemished truths that the doing so thereof then is criminal conduct irrespective of motive or the innocence behind such conduct or in order to pardon the bent motivation of what eventually will have a benefit from such conduct. You now have to decide which way you are taking the matter because the matter has come to your attention. I will admit that this letter does carry another agenda than seeking your approval this time round or that this letter might find another motive than be trying to establish your acceptance. I have passed that of finding academic approval as the divide some time ago. This is a warning that when I publish my book, I am going to do my utter most to reveal what I consider as fraud in physics as a whole. I am going to go as public as what my abilities would allow me and find as much publicity to further my conclusion about a conspiracy going on in the world of science.

My aim in writing this letter is to confess my doubt about the manner in which science regards gravity forming. I have serious reservations about the matter in as much as how science assumes that mass is the reason behind whatever is forming gravity. I equally hold strong doubts about gravity being a force of pulling or gathering. This institutionalised concept has formed the basis of science in physics during the past three hundred and fifty years but in the light of evidence that came to the attention of science in the last hundred years and more so during the past fifty years, there are reasons that puts those assumptions previously formed in the past about gravity in doubt. The new evidence points away from the previous direction and assuming. Modern evidence calls for re-examining in regards to this matter. The facts that are now available to science to form any prognoses makes the old principles that were used for such assuming much more questionable.

One only has to study aviation principles to form the doubts. Aircraft flight disproves the assumption that mass is instituting gravity and in that context the notion forming the concept of mass and gravity becomes extremely doubtful. I am about to prove in this letter that mass is motion discrepancy between two bodies and gravity is motion pure and simple not a force. Any object holding any mass of whatever description can become airborne with the correct conditions applying. It only depends on motion coming about achieving a speed over and above the earth speed that must be sustained. Mass of whatever magnitude can fly. There is no limit on the mass becoming airborne. What is required to achieve such a fete is a speed. Only that limits flying. A Jumbo jet can get into flight as easy as a micro light aircraft. The only requirements are a differentiation in velocity of air movement above and below the aircraft wings.

When a water-skier stands on the water the skier represents a certain pace time, which is in truth not magic or mystery but in relation to space used or occupied in a certain time. The skier holds its space in time $a^3 = T^2k$ in relation to the water holding its space-time $a^3 = T^2k$. At such a point where the skier is stationary the skier represents the solid and the water becomes the fluid. However not long after the skier started moving with the aid of a drawing boat, the skier represents the liquid and the water takes on the role of the solid. This becomes a reality since the movement is now on the side of the skier forming the liquid and the Earth is in relevancy solid which represents the stationary factor.

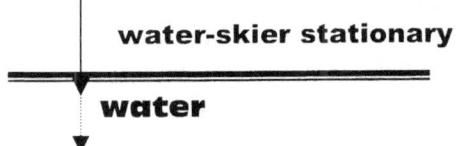

The space-time the skier represents is Π^0 in addition to the Earth holding $7(3\Pi^2)$ of space-time. The skier represents a factor of 1 and not zero.

As soon as the skier moves the skier

becomes the liquid as the roles presenting each factor changes. Since the earth motion of $7(3\Pi^2)$ also belongs to the skier as well as the Earth the Earth factor becomes 1 and all motion becomes part of the skier in $7(3\Pi^2) .25\Pi^0$. The earth then holds the factor of the stationary 1 that stands still although it is moving and the skier moves although in comparison the motion difference is minuscule.

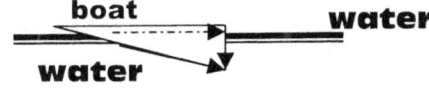

A boat floating on the water has a certain mass allocated in the proportion of the water it displaces. The mass would be considerably more if the boat were on dry land. But with the boat on the water it only represents the mass it displaces while it is stationary. The boat matches the conditions fitting a solid while the water acts as the liquid applying motion.

Then with the least of motion coming about the boat rises from the water and the depth of the previous water line reduces considerably. This can only be because the boat then represents much less mass as it displaces much less mass in the instant. Time in the square T^2 is the duration while **k** is the instant the duration ends. However in the period of time T^2 comes and goes because T^2 represents eternity while alternating with **k** where **k** represents infinity and that is ending the duration of eternity being T^2. In the duration of T^2 the boat came into contact with much more water than was the case when the boat was stationary while being in the same duration T^2. The water with which the boat is in contact, became more during the duration in time and therefore in the same context, did the water become less per instant if the relevancy was looked at from the other side. This applies because more instances fragments the same duration or the duration applies more instant endings coming as a result of the additional motion. If the water it is in contact during a period becomes more the instant of contact will have the boat reduce in mass. That is why the boat lifts from the water. It is because it shows less body in relation to more water in the time in contact with the water. The triangle in the water reduces as the motion lifts the boat from the water because the duration per time instant became less and the water in contact with the surface of the boat became more. The boat is under all conditions maintaining Kepler's true law that motion is equal to matter which is $a^3 = T^2k$

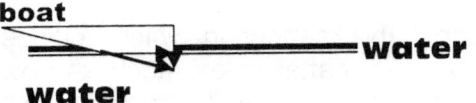

A situation arrives where the boat hardly makes contact on the surface of the water with the surface of the water. The mass of the boat per moment in the instant is so little that the boat hardly touches the water which proves that the mass of the water reduced the size of the boat in relation to the water to a fraction of what it is when the boat is stationary and bobbing on the water. The same happens to the hot air balloon when rising into the sky. Aided by more heat attributing to its motion increasing when as a result it looses mass and in the process it is receiving gravity. That also happens to the boat where the boat ultimately rises into air to become air and all is done on the balance of motion.

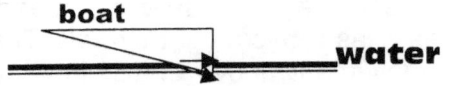

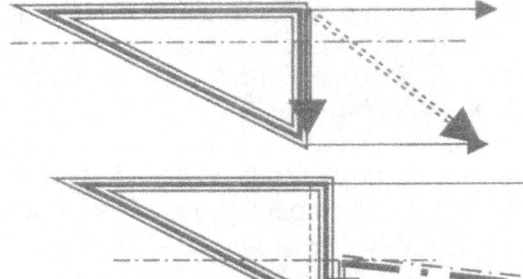

The relevancy of motion brings on the ultimate triangle of space-time. The more the Earth representing T^2 dominates the factor of motion the more the space is represented by mass a^3 where the relevancy of independent motion **k** is reduced to a factor of 1. However, as the relevance of individual and independent motion gains dominance, the factor that the earth represents as T^2 diminishes and as it diminishes so does the relevance of mass diminish as the factor of space in the instant reduce. The boat lifts from the water, which indicates the mass reduces. The boat displaces less water, which proves the space in the instant is reduced by the water taking proportionate dominance in the period. This proves the solid truth behind Kepler's findings where space a^3 is worth the motion T^2k thereof in relation to the duration T^2 as well as the relevance brought on by the instant **k** motion creates the dismissing of mass and the lack of motion increases the fact of mass. The release of the boat from the water shows clearly a release of immersing of the hull of the moving boat and the faster the motion is by velocity, the higher the boat is released from the water. It reaches a point where the boat finally leaves the water in total and no part of the hull is in contact with the surface of the water. If that is not flying, there is no flying anywhere possible on Earth, and yet Newtonians fail to recognise this by connecting it to gravity and mass.

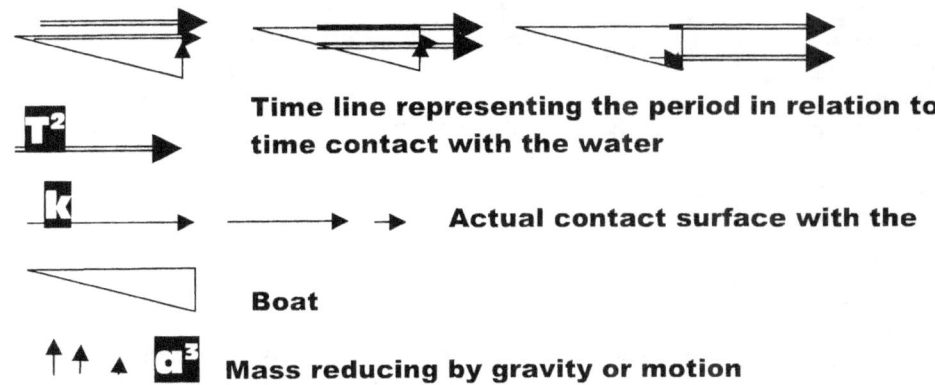

Time line representing the period in relation to time contact with the water

T^2 **Actual contact surface with the**

k **Boat**

↑↑ ▲ a^3 **Mass reducing by gravity or motion**

The lifting connects directly with the water sliding through underneath the boat in a specific duration or period in time. The more water comes past the hull through motion the less of the hull makes direct contact per time instant with the water. When the boat slows down the contact relevancy reduces as much as it increases again once the velocity increases again. The boat became a gas while the water presumed the part of the rock hard impenetrable

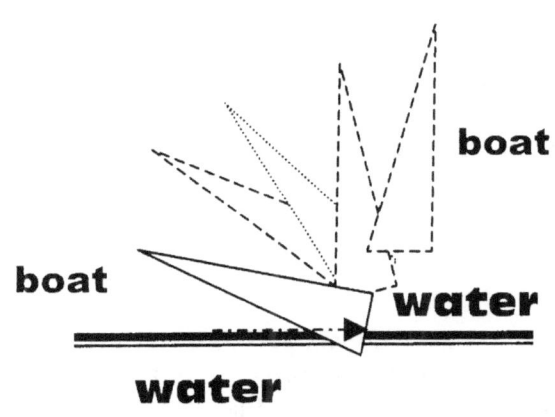

solid. The specific density of the otherwise solid boat became that of what a gas is because it floats in the air in a position of being airborne just as a gas would be while the water is so solid it will break the boat to bits at any point of contact. No matter how hard Newtonians try to allocate mass to what ever they wish to have mass, there are certain limitations that even all the falsifying of facts cannot hide the truth. The electron has mass because the electron restricts the flow of time in relation to space. The proton holds mass because the proton holds a restriction on the flow of time in relation to space to represent mass. The neutron has no mass and whatever Newtonians try to grossly falsify, it still will not establish any proof of mass. Even in their attempt to accommodate their hoax in mass, the neutron simply has no mass. That means the fact that they wish to attain mass to the fact of the presence of the particle is a contamination of the truth because the neutron simply has no mass. That means the neutron is all motion and the neutron is all gravity. The fact that it shows no indication of having mass as a clear fact represented, is not an incorrect connotation that the cosmos connects to mass but shows Newtonian incorrectness about the presumption they and their Master Newton have about mass.

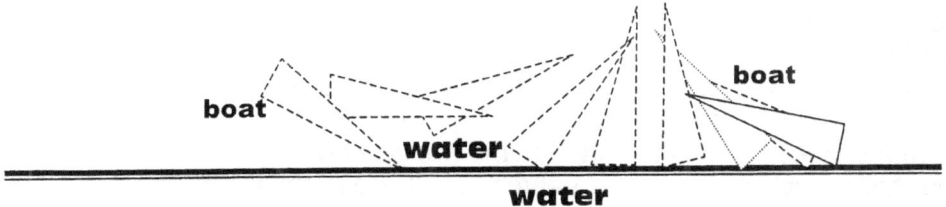

In a frame-by-frame one can see the boat starting to fly but since the drive seizes to commit to the motion we get the idea that the boat flips. It is just a case that the front part of the boat starts to end the duplicating by motion sooner than the driving rear part and the end comes charging past the front. If the drive at the back were consistent the boat would start to fly as airplanes do.

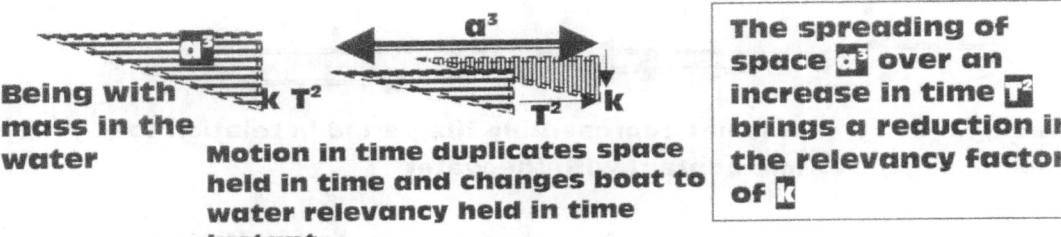

Being with mass in the water

Motion in time duplicates space held in time and changes boat to water relevancy held in time instant

The spreading of space a^3 over an increase in time T^2 brings a reduction in the relevancy factor of k

The Earth did not become bigger and the boat did not really become smaller as much as the water did not become really more but in relation to time space duplicating takes more time to produce less space. But there is a relevancy between the motion of the boat and the motion of the Earth, which the water represents in the relation that the boat holds with the Earth. As the motion of the boat accelerates the moment-by-moment contact the boat has with the water, this process allows the boat that is in space, to diminish by the same measure of water it displaces in the instant in time during the period overall as the ratio that the boat to water displacement was when the boat was stationary. By motion, the instant of displacement reduced the volume of water because more instants fill the duration of time. This comes in place to counter balance the increase in time in the instant while holding the duration in equilibrium to compensate in order to have the boat displacing the same volume of water in the total duration in time. The space of the material representing the boat to water increases on the side of the water because more water is in contact during the duration as contact in the instant reduces. The water becomes more because in the flow of time the boat makes more contact with a line of water where the water makes less contact with the space of the boat. That proves that a boat can fly and that proves that man in motion also can fly. That confirms Kepler and by the same margin it denounces Newton and Newtonian mass indicating gravity. Motion secures gravity and mass stops gravity while mass is the restriction of gravity and objects in total unrestrained motion can have all gravity that presents no motion as the neutron proves.

Mass aside, laden or not, aircraft of all description will fly. The wings determine the required velocity while supporting all mass. The aircraft supports the same mass it has after becoming airborne as it had before it was airborne and the mass remains in the air because of the speed which enables the wings to have air pushed past the surface of the wing and that is what allows gravity to be countered by flight. Science thinks of movement as momentum but in fact it is antigravity since it counteracts the containing of Earth gravity. Momentum in fact is only an increase of gravity on the object that is taxed by mass, which extends the gravity of the burdened smaller factor beyond the limitation of Earth restraining by mass. An object has mass because it moves slower than the Earth and is therefore dragged along by the Earth. Once the object moves faster than the Earth even within the Earth's atmosphere the influence of mass abates. Mass is only the friction caused by speed differentiation whereas the sound barrier is the limit on the other end of the speed differentiation scale. At the moment science differentiate in terminology used to describe what I deem to be the same issue but that puts a huge question mark over the use of terminology and the concept of mass playing any part in forming gravity. That what is considered as momentum is gravity in addition. Notwithstanding the pulling and the tugging brought about by the mass, the ultra light plane finds as much flying capabilities as does the enormous cargo jet aircraft. The pulling of the mass is many times more in the case of the heavy cargo plane than what it is in the micro airplane but still at speed the plane overcomes the pulling of the mass by only establishing motion differences between the top of the wing and the bottom of the wing. When flying, having more mass does not implicate or discriminate in any way because with the required motion both crafts keep in the air equally effortlessly. Once the craft is airborne it is a matter of maintaining speed that would secure the flying ability and not the loss in mass that would increase such flying possibilities. In the case of a satellite, such flying demonstrates my argument even much better. The space station Mir was in space for many years but it fell when the space station could no longer maintain a specific orbiting speed. While it was in space in an orbit there was no changes in mass required to maintain an orbit. The changing of the mass could not secure Mir a longer orbit life. At the point where it started to plummet to the Earth, the mass did not increase but the orbit motion decreased and that allowed the plummeting. The fuel eventually ran out and by not further being able to accelerating in order to maintain a specific required velocity Mir was destroyed but not by mass…it was destroyed by a lack of gravity. Without the speed requirements needed to match the motion equilibrium that was required to maintain a steady orbit between the Earth and Mir, differentiation set in and that caused mass, which came as a fall towards the Earth whereby Mir met its end in a fireball. It had nothing to do with mass. It had everything to do with speed being at an equal pace between the Earth dictating the orbit velocity or gravity and Mir's orbit velocity or gravity. While the satellite is able to hold a speed in a ratio with the speed the Earth manages, an orbit is secured and such securing of an orbit has no bearing on the mass that is up there flying through the sky. It has everything to do with matching a speed in relation to the distance from the centre of the Earth and in relation to the motion of the Earth that requires a specific velocity at that distance. It has everything to do with the time it takes the space to go through the motion from one point to another point. It is the required heat in the space that the space needs to move at a certain distance in a specific time of motion. Gravity is a relation between space and the heat required to provide motion which is $a^3 = T^2 k$. It is space (a^3)-time ($T^2 k$) and that brings me directly to Kepler. The only way mass becomes an issue is where a bigger object such as the Earth frustrates or hinders the natural motion of any individual particle that it its captures into its atmosphere. Then the gravity or movement results in a tendency to move but still the tendency is manifesting the motion where mass is the restraining thereof. There always has to be motion but there is not always mass. The tendency of motion, which becomes the mass when the motion is restricted, remains a part of free movement, which the frustrated object still possesses as motion. There always remains a tendency to move. The Earth set a speed of motion and in order to beat mass the object has to excel by providing amplified motion in excess of the Earth motion granted to the object and gravity extending beats the restricting effect that mass has on any body connected to the centre of the Earth. That is motion and I am about to show that gravity is all about motion.

The value of space that Kepler's indicated as a third dimension a^3 does not depend on indicating a structure forming a^3 that is in rotation T^2 but only needs one position having a constant of some sorts. There has to be a space to point to a space in defining. Any point where k as a line is ending at that point k indicates a position between the start of the line k^0 and the end k^0 of the line k where one will find a spot with the value matching a^3 and the matching location will fit T^2 at that point. The line k is coming from the centre and ends at a point a^3 but such a point establishes another space in which a^3 rotates. The line k that forms by a^3 rotating T^2 is the relation there is in the solar system between all planets and the Sun. The Sun always indicates the centre k^0 and the planets always indicate the rotation. But $a^3 = T^2 k$ is only producing a relevancy of three dimensions that is equal to two plus one dimension. The fact of a^3 being at the end k^0 of the line k is how a^3 secures k and where k then secures a^3. The line k cannot have validity without proving the validity of a^3 just as much as the rotation T^2 of the space a^3 defines k. There is no possible removing of any of the

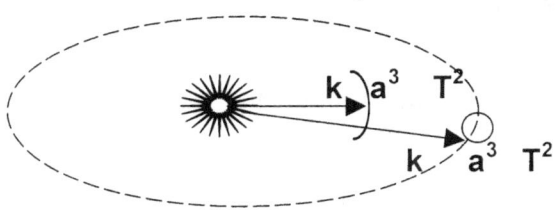

factors without removing all three factors equally. The space a^3 serves to prove T^2 that defines the line k that presents a^3 an existence. From the Sun there are three points moving between two points from one point to two other points giving the six dimensions we find in space. It is space in time or space converting space through the movement of time. It is a location of a point in the third dimension a^3 that will move according to the second dimension T^2 that will implicate **k** as a reference in the first dimension. It is about dimensions in reference to one another.

Let us take it from a point where the Sun provides a centre as one starting edge of **k** then that centre **k** will provide a line from the centre and the line **k** will provide three spots in a formation that produces a structure by the square T^2 of the dimension. Not once did Kepler indicate size as a contributing factor to a^3. That means every single point that **k** indicates, there are three positions a^3 implicating sides of a double dimension. In the same manner is **k** not limited to distance or is T^2 lesser by size $k = a^3 / T^2$. That is what Kepler said. There are three dimensions a^3 between any two points T^2 flowing as time from the centre of the Sun, which is indicated by the line **k**. The implication of the relevancy produced by the use of the formula $k = a^3 / T^2$ brings about that when dividing T^2 into a^3 there is **k** left. The fact is that a^3 is a three dimension (^{3}) of single **k** (^{1}) showing one or T^2 is two dimensions of **k** being the one dimension it means that **k** is a part of space a^3 or T^2 which is time. It is the same thing in a double dimension or space being a triple of **k** then **k** is one factor and **k** cannot show a position of zero. If **k = 0** then there is no possibility of $k = a^3 / T^2$ because **k = 0** then $0^3 / 0^2 = 0$. That does not make sense. Mathematically space cannot be zero because those being of the opinion of space being zero or nothing must first prove mathematically that space is zero. Moreover they then must prove mathematically how does zero grow through the Hubble constant. By translating Newton's vision of the circle in completing a cycle would become zero through rotation…well that does not count the use of the formula a^3. If **k** cannot be zero then **k** could not start from zero. With $k = a^3 / T^2$ no point can be zero because **k** shows space $a^3 = k T^2$ is no reference to the volumetric mathematical formula used to calculate a^3 = **4/3** Π **r³**. Nor does it show the use of the circle in the second dimension being $a^2 = \Pi r^2$. In the case of the Newton formula, the circle factor becomes the square as indicated by the duration of the time T^2. The factor standing in for the line which normally would be r and then be the square value is in the case of Kepler not the value indicating the square. That means Kepler never indicated a circle of mathematical procedure but said mathematically the distance of the planet from the Sun **k** holds space a^3 in relation to time T^2. **Lines mathematically cannot start at zero because there is no evidence of zero as a factor in mathematics. Should you disagree with my statement** the question in need of answering is this: **What will the length of the shortest hypothetical line imaginable be and moreover, what would the total overall length be in that case?**

Locating zero

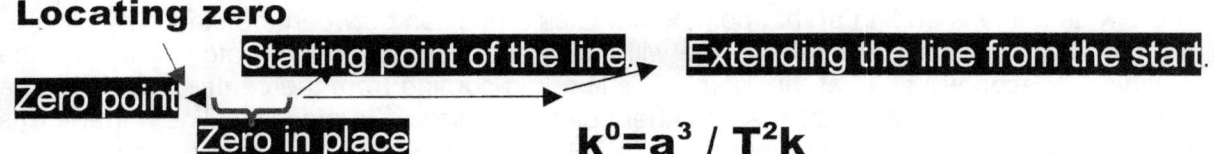

The fact of form proves that the sphere captured all sides that can possibly influence the sphere. The sphere therefore holds **k⁰ =a³ / T²k** within the boundaries designated to the sphere. When a body is placed in a location on the outside of such spherical borders, that object seems to float in any direction. There is no control one can establish which will secure movement in any specific direction of preference except by releasing heat to counter act the required motion in a specific direction of choice. We all have seen what happens to any object that comes into the border area of a sphere. The object suddenly is motivated by motion to follow a specific designated direction and the motion leads the object to move towards the centre of the sphere. It is as if the support of the six opposing sides has lost one side where the sphere took over the control and movement starts in the direction of the Earth centre. The support of one side is literally removed by the centre of the earth where Einstein claimed the strongest gravity is and the motion of the object starts in that direction. There is no pulling on the object but there is removing of space by the centre of that specific point leading the object and the space it is in as well as the space it carries to move to the centre spot. In the sphere the borders the sphere holds are deliberate and very distinctly placed edges forming a specific distance from the centre. The centre is also proven beyond any debating. The centre of any sphere has to be at the very point where space completely falls away. That will put that space at that point in the single dimension and centre is the single dimension.

The fact in the matter suggests that all three factors hold the same identifiable measure and the differentiation between the factors is in name alone. In what I explain at a later event the true value should stand connected to singularity and in singularity the value is allocated to singularity is Π. The formula should read that $\Pi^0 = \Pi^3 / \Pi^2\Pi$ and when used with the correct connotation the formula becomes more sensible. But be as it may the factors indicate the same measure carried in different dimensions and should all read the same value. The cube is a loosely connected structure with any form possible but the only precondition is that there must be at least six sides connecting. The six sides hold a relevancy or a responsibility to one another and provide a Universal accepted form maintaining the universe. From the structure one can see gravity is not strongly present. All six sides support what ever are inside evenly from all sides. The sphere is the form securing gravity. In the centre of the sphere there is a point where space vanishes. At that point where space vanishes gravity is the strongest. From the centre point where gravity is the strongest gravity hold the sphere true to form. At the edges of the sphere there are also points crossing in 90^0 and 180^0 holding relevancy and responsibility to one another but the centre spot being the gravity point positions all the points in a location that the centre point allocates. In the centre where all lines cross one will locate singularity but I am explaining that fact a little tater on. Newton changed the symbol of **k** by using the mathematical equated symbols G (m + m_p). This is just a longer and probably a more detailed manner of indicating **k** and better defining of **k** but it symbolises precisely to the point what **k** stands for nonetheless. I wish to draw your attention to the matter of Johannes Kepler's findings that Mainstream science considers as resolved and closed for many a century while it is not. My investigating Kepler helped me to resolve other unresolved matters but it was only possible by using Kepler's work. Reading Kepler with the required degree of mathematical correctness and an unbiased attitude, changed the aspect of gravity in cosmology fundamentally and as I am about to show most and totally incorrectly.

Let us for one minute leave Newton's surmising about Kepler's failure out of the picture and concern us with what Kepler found long before Newton thought about what Kepler found. Kepler said that the space a^3 is equal to the motion T^2 of the space a^3 distant from a specific centre **k**. That then is $a^3 = T^2$ **k**. Reading this mathematically encrypted coded formula of the cosmos given to Kepler and keeping it removed from Newton it reads as the space a^3 is equal to = the motion T^2 of the space a^3 in ratio to a centre **k**.
What this proves is that gravity is the motion of space provided by time being the liquid.
Please allow me to explain. In the formula $a^3 = T^2$ **k** the space forms as the space is in motion. Newton suggested that $\dfrac{dJ}{dt} = 0$ where he stopped time to have the motion of the circle demolish the work that the circle does. That means he got time standing still or being T^1 and the motion **T= 0**. Let us ponder on that thought for a while, while remaining with the formula Kepler suggested it will seem that according to Newton $a^3 = T^2k$ and in that T^2 then becomes **1**. Should that be the case then we have space going flat because $a^3 = T^2k$ where a^3 = **T X k** = forming a square instead of a cube, and the Universe we have is a three dimensional cube in every aspect there is. If time stands still then time becomes $T^0 = 1$ and **k** being in one dimension all the time it too has to become zero.
$a^3 = T^0k^0$ which makes $a^3 = T^0k^0$ = being (T^0 X k^0) = 1 X 1 = $(1)^2$.

That is mathematically incoherent and a complete fallacy. To proclaim that a cube can be a square at the same time it is a cube is ridiculous, even coming from a man that has the stature of Newton! By taking the thought further we find the same blunder in removing **k** in Kepler's formula as a factor.

$a^3 = T^2k$, then $a^3 \div k = T^2k \div k$ being $a^2 = T^2$.

That is totally going against all mathematical principles. It would be more correct when saying that $a^3 = T^2k$. Looking at the formula in this way we find that $a^3 = T^2k$ bringing about that $a^3 = T^2k$. That proves the following:

$T^2 = a^3 \div k$ or $T^{-2} = k \div a^3$. ($T^2 = a^3 / k$ or $T^{-2} = k / a^3$)

From the implementing of $a^3 = T^2k$ we can see that:

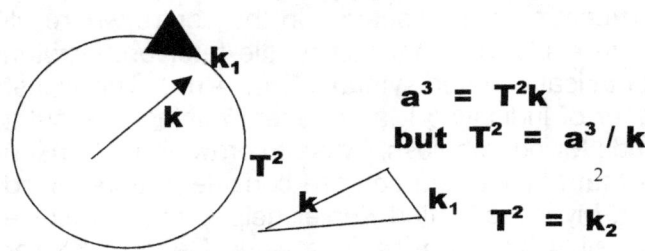

$$k = k^{3-2} = k^1$$
$$a^3 = a^{2+1} = a^3$$
$$T^2 = T^{3-1} = 2$$

$k = a^3 / T^2$	$a^3 = T^2 k$	$T^2 = a^3 / k$
$k = a^{3-2} (T^2)$	$a^3 = T^2 k^1$	$T^2 = a^3 / k^1$
$k = a^{3-2} = k^1$	$a^3 = T^{2+1} (k^1)$	$T^2 = a^{3-1} = T^2$
$k = k^{3-2} = k^1$	$a^3 = a^{2+1} = a^3$	$T^2 = T^{3-1} = 2$
is the same as	**is the same as**	**It is all the same**

That would give gravity meaning and that would explain not only gravity but also the Coanda affect which to my mind is gravity by principle.

$$a^3 = T^2k$$
$$\text{but } T^2 = a^3 / k$$
$$T^2 = k_2$$

The fact is that although the symbols Kepler used were not the same, the value they measured were very much the same. The values of the symbols were interchangeable.

From the implementing of $a^3 = T^2k$ we can see that:

$k = k^{3-2} = k^1$ is in direct relation to $a^3 = a^{2+1}$ and that is, is in direct relation to the formula $a^3 = T^2 = T^{3-1} = 2$.

With this information staring mainstream science in the pleading at them to recognise the information they turn why can man not fly off to other galactica at the speed of

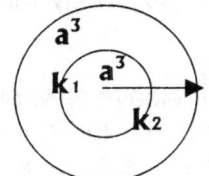

face and scream around and ask light

$k = k^{3-2} = k^1$

It takes time for space to fill **k** in the distance. In fact, it takes the distance that **k** developed since the Big Bang $k = k^{3-2} = k^1$ to fill the distance.

It also takes time $T^2 = T^{3-1} = 2$ to produce the distance forming k^2

It takes space $a^3 = a^{2+1} = a^3$ to form k^3 since coming from the Big Bang

Man could create motion but at first, such motion was far less than that motion which the Earth provides. The motion of man's ability was vested in what his muscle power could provide. But a very short while ago man grew wise to machines and the fact that machines can provide more motion much faster than could animal muscle bring about motion.

$$T^2 < a^3 / k \qquad T^2 = a^3 / k$$
$$T^2 > a^3 / k$$

By supplying machine motion it gave man extra ability whereby man extended the relation between what the object has when in normal contact with space and when extended by extra motion allowing more space to apply to the surface of the object, thus enlarging the object surface in the relevancy brought on by motion.

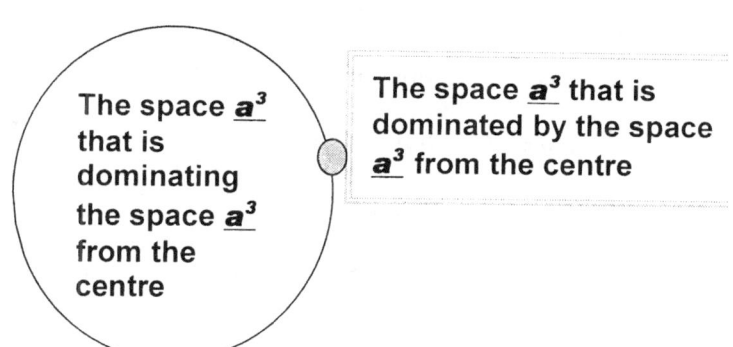

The space *a³* that is dominating the space *a³* from the centre

The space *a³* that is dominated by the space *a³* from the centre

The smaller space **a³** is distinctly distinguishing the larger space **a³** as the larger space **a³** is housing the smaller space **a³** where the factor **k** is as much indicating by length the larger space **a³** as much as it is indicating the end of the length of the larger space **a³** at the location of the smaller space **a³** by directly pointing at the position the smaller space **a³**

holds. Where the larger space **a³** ends the smaller space **a³** is. The two remain as an inseparable single unit in double motion where the motion identifies the unit as much as distinguishing the separateness in the unity and always remain in absolute relevancy.

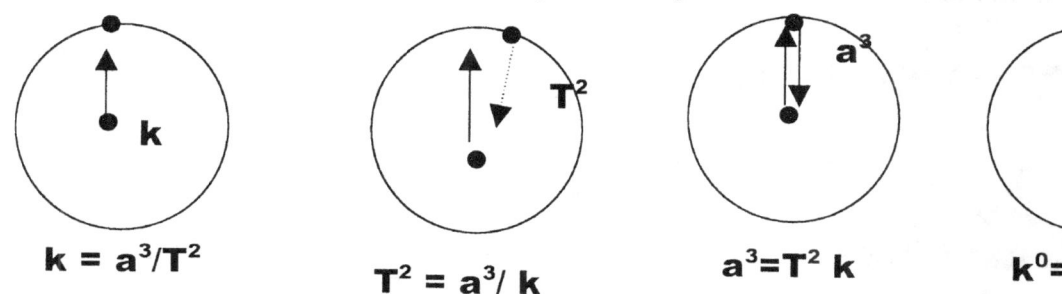

$$k = a^3/T^2$$

$$T^2 = a^3/k$$

$$a^3 = T^2 k$$

$$k^0 = a^3/T^2 k$$

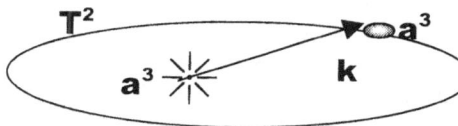

By duplicating the space of any particle sharing space within a larger cosmos structure such as an atom inside a star or a human inside the Earth there are two relations applying.

The independent object serves as the outer relevancy while the contact the Earth has with its centre forms the contact with the containing singularity. When the independent object starts moving the atmosphere allows flexibility. However, this flexibility is only limited to a point where the motion allows the object to locate a position indicated from **k** starting at…to **k** ending at the point it is at. As the speed increases the limits on the space that retains the object in motion starts to stretch and such stretching has definite limits.

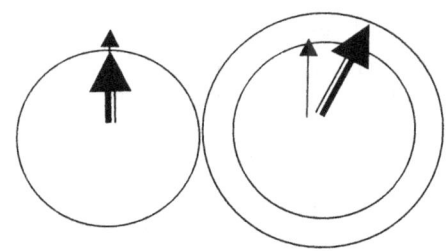

Kepler's formula is the exclusive display of *space-time* and Newton's version of Kepler's formula is plainly and commonly misleading and a deception of the truth. There is another correct connotation to Kepler's work that divert completely from Newton's vision. In the picture to the left we see the all-familiar Moon that every one blessed with the ability to see has seen at some point in his or her life.

Any person not agreeing totally with what I have to say in the next sentences knows less about mathematical principle than a newborn baby. The Moon is according to Kepler's

studies, (and discounting Newton's inconsistency if not totally flagrant rubbish) from our perspective in the space a^3 that is equal to the time T^2 it is in placed at the distance between us by the factor **k.** The Moon consists of the **space a^3** the **time T^2** and the distance **k in time not space.**

The observing "size" of what we see depends on the time factor **k** relating to the space a^3 we see in relation to the time T^2 we see it in. The bigger the space in relation to what we see gets the smaller the time factor is that allows the time frame or the motion that frames the space. From serving my cosmic position allocated to the space I claim in the time I hold Kepler stated space-time is $a^3 = k\,T^2$ and therefore mathematically it is correct to say $k = a^3 / T^2$.

When I observe the Moon, the Moon is at a distance **k** from me where I see the space a^3 in relation to the   time   or   the moving or the spinning T^2.

The figure to the left I covered the area of the Moon that I use to indicate the space a^3 is served by the sketch to the left.

The symbol to the left of this point is what I use to indicate the area time T^2 is in which the space moves or the picture to the left depicts the motion aspect of the space.

The bigger a^3 will seem the less **k** has to be in order to reduce the motion or time T^2. A large space a^3 will bring about a small time T^2 factor because of a large distance **k.** That means that $10a^3 = T^2 \times 10k$, and $a^3 / 10\,k = 10T^2$ or then the distance **k** is 10 times less. $k = a^3 / T^2$ then $k / 10 = 10a^3 / T^2$ or from another perspective it is $k/10 = 10a^3 / T^2$ that would bring about space that is 10 times bigger in the time that is relating to the space. That is space-time and that forms

gravity and that is what I am about to prove in this letter. The qualities Newton attributed to mass and the principle that Newton tried to pin to mass is just the opposite of what can be pinned to mass. That, which Newton regarded to be mass, serves the very opposite to mass because gravity is motion. Gravity is the moving of objects through a container, which every one considers being space but the truth is that the container is time.

An object coming closer will have the relative factor **k** indicating a negative indicator as a k^{-1}. In such an event the space will increase by decreasing the time aspect and this Kepler's formula most accurately show to be $k^{-1} = T^2 / a^3$. In such an event the time putting distance between the object and the space would reduce allowing the space to relate much stronger in terms of the time as part of the overall picture.

In the overall picture $a^3 = T^2 k$, the time factor, which we incorrectly see as a distance **k**, has gone very small and this is putting the space a^3 as an enormous part of the overall view and it reduced the time T^2 as a small fragment of the entire picture. The very same picture is used but by reducing a^3. We take it that the time factor **k**, which we mistakenly put as a distance and which it is not, as being huge and that large factor increase changes our relevant factors in the entire picture completely. According to our use of mathematics by increasing **k**, that

reduced a^3 because also T^2 reduces a^3 and by reducing a^3 T^2 gets a bigger share of the entire picture. That means the space a^3 stands related to the overall time being **k** as well as T^2.

That is unmistakably exactly, precisely and to the iota correctly what Kepler said when Kepler said one is observing space (a^3) –time (T^2**k**). However from that we can see time increases and when time increases it is the **k** factor in the relation of space-time (a^3)= (T^2**k**), that reduce the space (a^3) in relation to the allocated position the onlooker has. Reducing the value of **k** allows the object (a^3) to get closer to one another.

By the reduction of the time factor that **k** represents, the time between the object and the onlooker reduces, making the relevancy falling to the space a^3 larger and by the same measure does the time factor in the overall picture take a smaller part? By the same token does the time factor T^2 reduce to allow the space more prominence in the entire picture? The second picture shows where the time factor **k** reduces giving the space factor a^3 more prominence. This will indicate a much reduced time factor **k,** which reduced the time factor T^2 and this gives the space factor a^3 much more prominence. The time factor is connected to both aspects of time being (T^2**k**) and that affects the space factor.

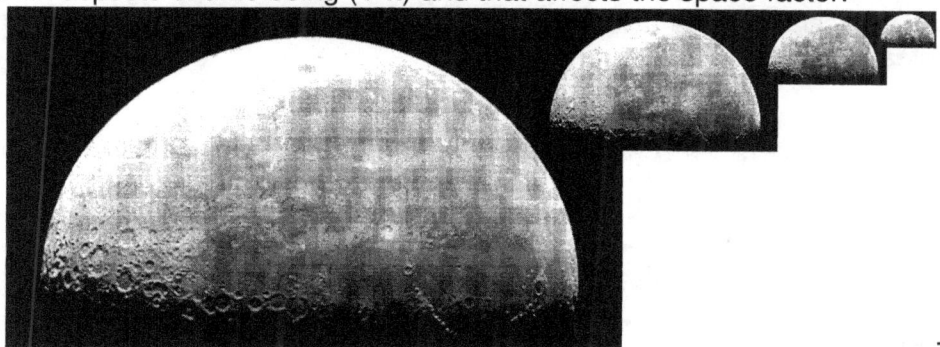

The expanding is a process where time develops space. It should be obvious to anyone concerned that as space became developed it was in line with the process of time that progressed and time progressing establish an increase in what ever is parting objects in the Universe. That is what the connecting of space and time should tell. It is space that parts time. It is the task of science to find what space parts time and how space parts time. Kepler thus gave us the answer about what Hubble found what was happening in the Universe centuries ago and centuries before Edwin Hubble's discovery. From Kepler's formula one can see that time and gravity is the same because as gravity weakens, so does time reduce and as space expands so does the influence of the gravitational factor reduce because gravity has less time per unit to control the more space per unit. Gravity is $T^2 = a^3 / k$ since the object cannot depart at any further distance between the centre and the object and is captured at that distance. Also gravity is $k = a^3/ T^2$ for the very same reason.

The circular bonding T^2 of space a^3 is enforcing an orbit T^2 to gravitationally circle around a specific centre **k**, which indicates the gravity $T^2 = a^3 / k$ in relation to the other gravity component $k = a^3/ T^2$, and it means T^2 is a circle of gravity and **k** is the straight-line distance of gravity applying motion. Still, Mainstream Academics ignore my statements that gravity is space in motion and motion of space is time: precisely as Kepler said. Any area to the cube is space a^3. However' since the Big Bang conception produced the concept that time brings development, and we know it took time to develop the moon at the position the moon is now in, that means the moon is at time **k** from the Earth. In the next 10 billion years the Moon

would be much "further" and therefore we know much more time would have developed the Moon in another allocated position.

In our minds and us being those blessed with life this picture shows the Moon closer or further away. That is the image life holds. To the cosmos there is no misplacing of displacing and every aspect comes with the development of time.

Travel is not a concept in the Universe and only Newtonians see that as an option but as all Newtonians do that also is a pipe dreams of fools. It takes time to get the Moon that much further away from the Earth. The Moon was much closer when the cosmos was younger because it started when everything was much closer. It is the time factor that is developing and that is what is pushing the Moon and the Earth apart. In that there is proof of what time is, what space is and what the concept is dividing the two cosmic differences. Space is defined while time is eternal.

That black stuff there is filling the night sky is time and not space. Since that is time and not space, that time not being space is eternal without ending. By Newton throwing the time aspect relevancy away by declaring it as a value of zero, Newton threw the baby away and kept the bathwater instead.

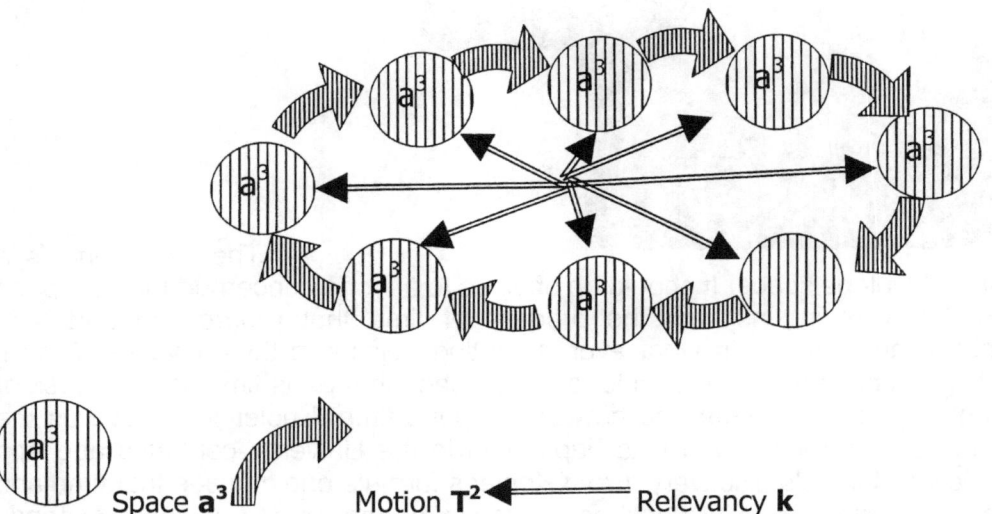

Space a^3 Motion T^2 ◄———— Relevancy k

That Sir, Madam, is mathematics on a much higher level than the mathematics that Newton could understand. That is the mathematical interpretation of words put in mathematical terms that the cosmos used to speak to us. That is a language that goes beyond the language of us, the mortals. That mathematical interpretation is a language that makes sense when one who is schooled in mathematics has a mind to read what is said. That is the language that the cosmos used to teach us what the cosmos wished us to know. That is how the cosmos told Kepler what Newton never came to understand. That is words Newtonians still cannot read because Newtonians wish to interpret into the cosmos what Newtonians want to tell the cosmos what the cosmos has to be and then humans would accept the cosmos the way they want it to be.

Time T^2 k

Space a^3

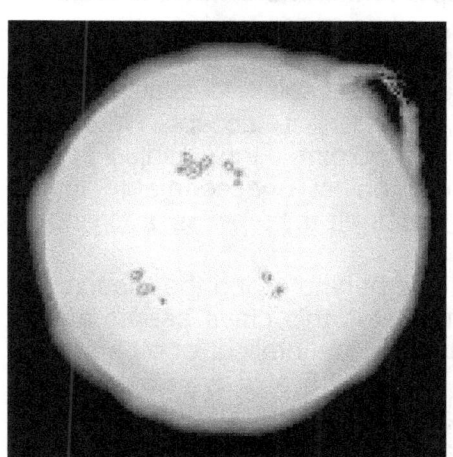

If science asserts them and truly display a need to know instead of a need to inform the cosmos in the effort they display when studying the cosmos, science would find progress. What is gravity? After three hundred and fifty years of research there is no one that vaguely knows anything about gravity. Someone came up with an idea of having a graviton but that is even more elusive than finding the gravity that the graviton supposedly is producing. The cosmos told us what gravity is but Newton knew more than the cosmos so Newton told the cosmos that the cosmos has mass. Newton told the cosmos that gravity is mass inflicted. The result is that after three hundred and more years we are as close to determining what gravity is than Newton was and Newton admitted he had no idea what so ever what gravity is! Today with all the wisdom there is going around there also still is not one Newtonian that knows more about gravity than the nothing Newton knew. That alone should guide Newtonians out of the dark ages but it does not. It only confirms the Newtonian stubborn nature to tell the cosmos what they want the cosmos to be without listening for one instant what the cosmos said. This letter is about listening to what the cosmos said on the principle Newton named gravity.

The Big Bang today is the cosmic development profile that is accepted by most and correctly so too, indeed. But in the cosmic development the Hubble constant is represented by the factor Kepler indicated as **k**, the linear time between objects. More important is that once any one accepts that this factor in time **k** grows, and then the factor cannot respond alone but has to influence the other factors it is related to by the same measure. If **k** increased, then so space a^3 had to increase also but so too included the rotating time T^2. Kepler said that there are three factors in very close relation and the three factors are the space a^3, which is directly equal to the time factor T^2 that is placed relevant to the time factor **k**.

In the days of Newton, not one of the latest concepts was realised. Since Newton, even the "bomb" was discovered and man flew to the Moon. Think what reaction this realisation would have unleashed in the days of Newton. I am of the opinion that if there were persons that had predicted at the time that man may fly to the Moon and back, and live to tell the tale, many of the Scientists of the day would have insisted that such person sporting such a thought must suffer by burning at the stake. Think of the steam that Newton admired and what steam in the latest context means. A nuclear power station is most probably the ultimate but in the end also just another type of steam generator. It is steam that drives the generators while it is not coal but nuclear fuel that turns the water to steam where the steam just turns more ordinary machines and the process is almost identical to the steam engine. It is in the fuel supply that the most significant development is.

In the world of physics there were huge development. In the world of engineering the progress was tremendous. However, compare that which was then to that which is now and spot the progress. If for example humans still had to depend on physics that was used at the time, how would that diversely effect modern society? If the mentality of the electricity department still used the basic knowledge that was used at the time of Newton, there would still be no electric departments in Universities. However, the problem is not entirely with Newton but the suckers that followed Newton into the future just like sheep follow the herdsman.

Cosmology is the only part of science that is still busy with blood letting and driving forces. Look carefully at how science explains the working principles of a star and you find the same mentality science had in Newton's day when the coal stove was burning material by letting heat. It is just as they did in Newton's day by letting blood. The sun burns so many tons of material and the star would be fully burnt out at such and such a time because the mass is destroying the…what a pack of rubbish! Cosmology stagnated somewhere between those that never think and those that could never think in any case…and if you are of a different opinion, then read this out loud to your personal senses.

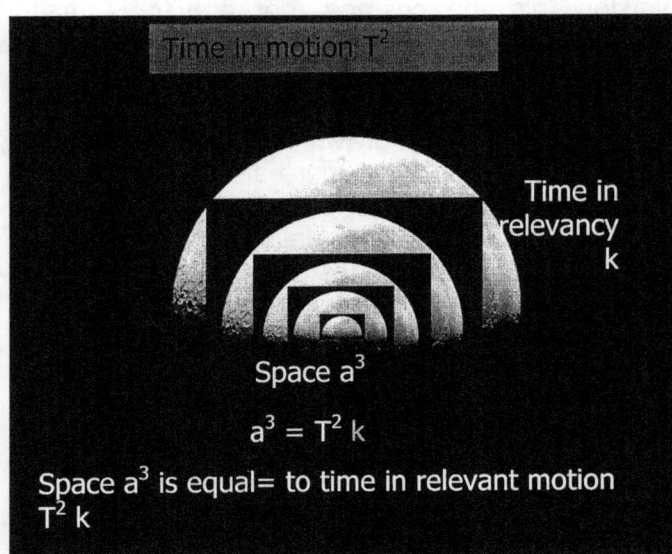

Time in motion T^2

Time in relevancy k

Space a^3

$a^3 = T^2 k$

Space a^3 is equal= to time in relevant motion $T^2 k$

All spinning matter has the point where the spin is still there but the radius is too small to measure by any means. That point is standing still in relation to the rest of the spin. In relation to that logic, I do not accept Newtonian science holding the radius of a spinning object unaccountable in the spin, whether the spin is applying or not.

The logic behind giving the reason that substantiates this claim seems almost an insult to the intellect of others. How does a Sun the size we now have fit into a Universe the size the Universe was at the event of the Big Bang when the Big Bang was the size of a Neutron?

The very accepting of the Big Bang theory insists on a revaluing of the factor **k** that influences T^2 to increase. How can **k** remain as zero if the essence of the Big Bang is the radical increase of the distance between objects in outer space? It is hard to imagine that any Newtonian never brought the concept of growth and the expanding of the factor **k** in some relation, which then will again implicate everyone connected to science in any way to be complying in the act to defraud and covering the truth to advance the untruth. As the moon advance its distance that parts the moon and the Earth, such evidence must bring a realisation of Newton being totally wrong…yet even my reminding of this goes ignored and unnoticed.

Every claim any person ever made when suggesting the Big Bang theory was that the "distance" between the objects in the Universe grew and it is still growing to this day. It is accepted as the Hubble constant. If **k** in $a^3 = T^2k$ grows, then the material a^3 must grow but also to the same measure must the circle forming time also increase to fit into the total relevancy of space-time or then $a^3 = T^2k$. It is clear that the time factor **k,** which we connect to the distance between objects are as important in the formula Kepler handed down as any other part of space-time. That completely goes against the grain of Newtonian interpretation of this significant formula. The more **k** would find significance, the more k will increase T^2 and the more **k** will reduce a^3. This conclusion is part of Kepler's formula $a^3 = T^2 k$ which is declaring that $k = a^3 / T^2$. Time decreases space when time increases relative motion. If we have a close look at Kepler's formula we find the space is in motion centred by a specific distance, which forms a conclusive part of the formula. We can see that the space as well as the time affecting the space varies in accordance to the distance **k** and it is the distance **k** that produces the time, which produces the space. How is that possible you might ask when Newton has k down to zero?

Applying Newton's second law F = ma
$$GMm / r^2 = m (\omega^2 r)$$

By replacing $(\omega^2 r)$ with $2\Pi / T$ we obtain Kepler's third law
This law predicts that $\mathbf{T^2 = a^3}$

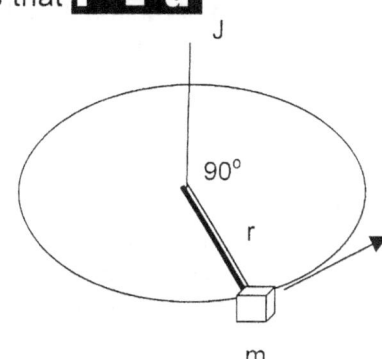

$$p = m .v$$

The mass (m) multiplying the speed (v) forms a new value J AND THEREFORE j CONTINUOUS TO IMPLY $J = I \omega$
$J = r X p$ where $p = (v = r x \omega)$
$J = r.m.v = m.r^2 .\omega = I. \omega$ and becomes interpreted as $J = I \omega$

This establishes that r = dJ / dt

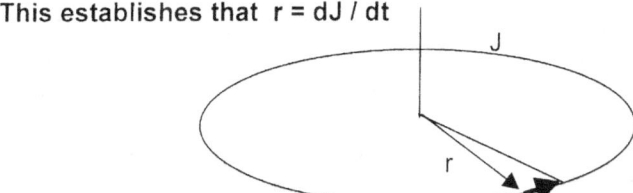

r = dJ / dt In the case of planets in orbit around the sun r forms a value of zero because dJ / dt = 0.

What this statement implies is that r does not exist. When anything has a value of zero, it is for all purposes non-existent. Only when an object is following a straight line, can the radius be non-existent because the radius alters value through time development.

A numerical sequence announced by J.E. Bode in 1772, which matches the distances from the Sun of the six planets then known. It is also known as the Titus-Bode law, as it was first pointed out by the German mathematician Johann Daniel Titius (1729-96) in 1766. It is formed from the sequence 0,3,6,12,24,48,96, and 192 by adding 4 to each number. The planets were seen to fit this sequence quite well – as did Uranus, discovered in 1781. However, Neptune and Pluto do not conform to the 'law'. Bode's Law stimulated the search for a planet orbiting between Mars and Jupiter that led to the discovery of the first asteroids. It is often said that the law has no theoretical basis, but it does show how orbital resonance can lead to commensurability. The importance that becomes known is the sequence the Titius Bode law saw in the number arrangement of 3; 6; 12; 24; 48; 96 etc.   The incorrect application of the Titus Bode law lies in subtracting the figure of 3 from 10 leaving 7. The other way of reasoning is to add four each time to the first value of three starting with 3 and so on.

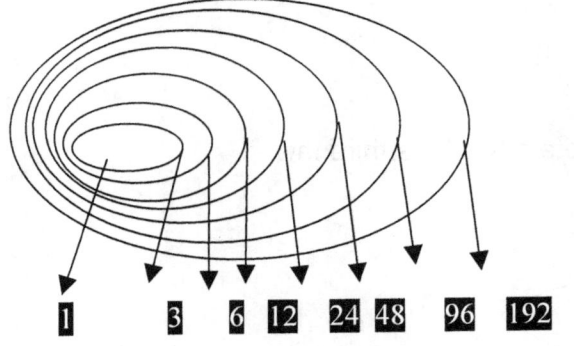

1   3   6   12   24 48   96   192

We find that the distribution of planets relating to the Sun is all in precise sequence and the name of such a sequence is the Titius Bode law, and a law it surely is! The gravity extending from the Titius Bode law forms the entirety of the building of the Universe by constructing the Universe in the using of the atoms to form the Universe in the entirety thereof.

Bode's Law:

Outer space cannot be a vacuum because if it were a vacuum, it would implode unless it was secured with very stringent outer walls. There are no such retaining walls and therefore there is no vacuum. It is again another Human myth transported into outer space that has no validity. The absence of material proves the abundance of structural composition and if matter is reduced then something else is absolutely overbearing to fill the absence of material. That, which fills outer space, is what keeps outer space rigid.

Planet	Mercury	Venus	Earth	Mars	Ceres	Jupiter	Saturn	Uranus
Bode's law distance	4	7	10	16	28	52	100	196
Actual distance	3.9	7.2	10	15.2	28	52	95	192

This puts the entire assortment of planets in neatly allocated and precise arrangements according the composition and outlay of outer space. The location of the planet does not depend on size, neither does it depend on mass. The planets are arranged irrespective of

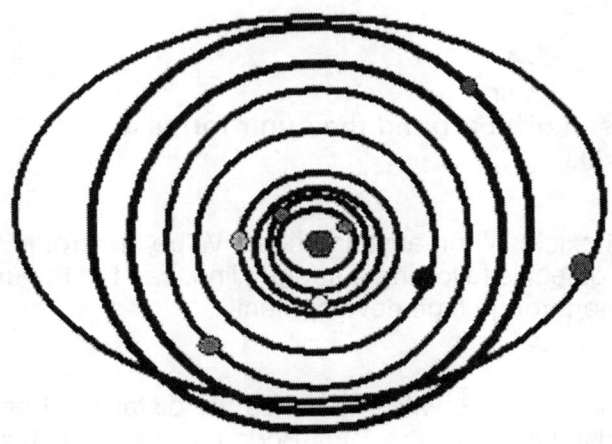

any particular factor in accordance with anomalies or other uniqueness on the side of the planet. The arrangement of the located position runs by way of number and the distance doubles every time the number is one more in position to the Sun. I know I am on record of saying distance has no value in the cosmos and it has no merit other than to be a man made device formulated by man to be used in some insignificant calculation concerning some needs man may have. This arrangement in the layout I do explain later on but at this point I wish to show that there is absolute prominence in the composition of outer space and the planets has no particular preference accept to be where they are in the cue they take from the centre of the Sun.

Let's for one minute forget the childish Newtonian argument of outer space being zero because I cannot afford to waste time in this book in disputing the validity of such an argument. I concentrate far more on this in another book about the issue where the explaining is much more extensive. In order to have a precise arrangement or pattern of positioning the planetary layout and how the planets are in location and moreover how precise the lot present an ordered arrangement, the fact of what outer space is formed by and made of, must present some merit in as much as its composition containing a notion of nothing. The planets as such can even be fragments because a structure forming a unit indicates that it does not participate in the issue. There is an arrangement of positioning in outer space, which shows that outer space holds the measure and gives the relevance.

Outer space is not just another vacuum of no importance but has absolute say when put into context with the layout is has, related to the position of the Sun. The proof is there in much more undeniable prominence than the unproven mass mesmerising nonsense Newton presents. The Titius Bode layout shows clearly no mass has any validity in the position allocated to any specific planet and therefore mass as attraction or force can be ruled out totally. It is not there while the Titius Bode law is there.

Furthermore we can find that the Kepler research proves beyond denying that **k** as a factor goes beyond dispute in contradiction to Newton's claims.

The factor **k** has a value of a^3 / T^2 and not even Newton can take such an argument away. Studying the evidence that was in hand even before Newton's birth where we find that the arrangement is such that it clearly defines the tempo in relation to the space of orbit at the location where such orbit can take place. The space has the dynamics of the other two time factors and the time factors indicate that the three factors are inter reliant on one another. The one gives the other value and the one takes value from the other. The space $a^3 = T^2 k$ to the time in motion. This means the space receives its value in relation to the value of the motion, which is in relation to the centre of the Sun. Again mass is a pipe dream.

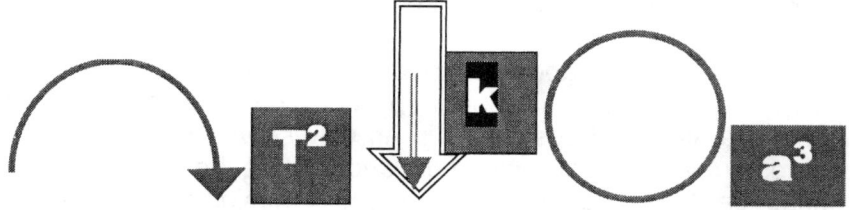

Planet	Period T years	T^2	Distance	Space a^3	Ratio
Mercury	0.241	0.058	0.39	0.059	0.983
Venus	0.615	0.378	0.728	0.381	0.992
Earth	1.000	1.000	1.000	1.000	1.000
Mars	1.881	3.54	1.524	3.54	1.000
Jupiter	11.86	140.66	5.20	140.6	1.000
Saturn	29.46	867.9	9.54	868.25	0.999
Uranus	84.008	7069	19.19	7067	1.000
Neptune	164.8	27159	30.07	27189	0.999
Pluto	248.4	61703	39.46	61443	1.004

The **space $a^3 = T^2 k$** to the time in **motion,** which would then have $k = a^3/T^2$ and in another ratio it is $T^2 = a^3/ k$. That makes mathematical sense. Compare this mathematically to what Newton suggested and compare the mathematical sanity that Newton portrays when Newton declares that $a^3 = T^2 k = a^3 = T^2 \times 0$ **(because according to Newton k=0)** $a^3 = T^2 \times 0 = a^3 = T^2$.

It is blatant corruption of mathematical law and any child that would put this down as a valid mathematical argument in any maths class will find that child has failed the examination. Yet

Newton is allowed to go insane. $a^3=T^2 k$ can only be happening if there were two relating factors in the entire issue. It can only have meaning if outer space stands in relation to the Sun and no planet has any meaning in this partnership.

	Distance (AU)	Radius (Earth's)	Mass (Earth's)	Rotation (Earth's)	# Moons	Orbital Inclination	Orbital Eccentricity	Obliquity	Density (g/cm^3)
Sun	0	109	332,800	25-36*	9	--	--	--	1.410
Mercury	0.39	0.38	0.05	58.8	0	7	0.2056	0.1°	5.43
Venus	0.72	0.95	0.89	244	0	3.394	0.0068	177.4°	5.25
Earth	1.0	1.00	1.00	1.00	1	0.000	0.0167	23.45°	5.52
Mars	1.5	0.53	0.11	1.029	2	1.850	0.0934	25.19°	3.95
Jupiter	5.2	11	318	0.411	16	1.308	0.0483	3.12°	1.33
Saturn	9.5	9	95	0.428	18	2.488	0.0560	26.73°	0.69
Uranus	19.2	4	17	0.748	15	0.774	0.0461	97.86°	1.29
Neptune	30.1	4	17	0.802	8	1.774	0.0097	29.56°	1.64
Pluto	39.5	0.18	0.002	0.267	1	17.15	0.2482	119.6°	2.03

Again there is no trace of mass of individual objects having any significance in the outlay or in influencing the time of rotation to go faster or slower on the value of the mass or that the orbit is bigger or slower in relation to the size of the planet. If one scrutinizes the whole issue as a unit, it seems as if there is no significance in the fact that there are planets participating in the orbit process. The allocated position shows order but the order concerns the planet and its location in relation to the Sun and one inner marker. I explain that later on.

The Sun, it seems, could care less about any planets being in the formation of outer space. If they were all in a specific allocated position that proves a given formula as to where they find such a position and the allocated position is in no way served by the influence of mass or size, then that proves buoyancy. Only when buoyancy enters the equation will size and mass not reflect any influence. If they rotate in precise equal fashion and the orbits in which they orbit shows identical relation to the space and the time, we can surely claim that it is more the soup that matters that the size of the crumbs in the soup.

In relation to the Sun, every planet is not only at the same timing but also at an equal relevant timing of $k = a^3/T^2$. That means in ratio to the Sun all planets are at an even distance and that could only be if the Sun regards the outer space arena as liquid where the lot is floating in a bowl of liquid with a specific density applying. That puts the arguments about gravity in a totally different principal. Where liquid stand in affect to a solid, we find the Coanda effect in place.

Newton stated that $a^3 = T^2$. That too is corrupting Kepler's fact as far as possible only to give veracity to the falsified claims Newton presented as trueness.
If that were the case then the space divided by the time would leave on $T^2/a^3 = 1$.
That is not the case because we have the tables to prove that it is a fallacy. Neither is $k = 0$ or is $T^2/a^3 = 1$ because $k^{-1} = T^2/a^3 \neq 1$. This position Newton took does not apply to outer space because in outer space the distance between two objects places the relevancy of time as the applying factor. When the Earth by shear dominance destroys the relevancy factor $k = 1$ and by mass brings about synchronised motion between the object and eliminates the spinning of motion differentiation there has to be between the two in order to establish natural gravity mass comes in because the object then becomes part of the Earth. That situation applies when objects are on earth but in outer space the objects does not have a relevancy factor of $k = 1$ and therefore mass does not apply in any shape or form. One cannot transform that which is of the Earth to other bodies just because Newton had such a desire.

The motion in value in outer space indicates a negative growth of time on an even basis towards the Earth and that concerns all material in the displacement zone the Sun claims. It is the time that moves towards the Sun just as liquid will do when flowing by gravity.

PLANET	SEMIMAJOR AXIS $A(10^{10}m)$	PERIOD $T (y)$	T^2/a^3 $(10^{-34} y^2/m^3)$
Mercury	5.79	0.241	2.99
Venus	10.8	0.615	3.00
Earth	15.0	1.00	2.96
Mars	22.8	1.88	2.98
Jupiter	77.8	11.9	3.01
Saturn	143	29.5	2.98
Uranus	287	84.0	2.98
Neptune	450	165	2.99
Pluto	590	248	2.99

From Kepler's space-time $a^3 = T^2k$ formula we find that the relevancy of all planets $k^{-1} = T^2/a^3$ in relation to the Sun is alike. This is only possible when the planets are floating in buoyancy because when in buoyancy all objects are equal in relation to the water holding them.

There is no big or small but just those having specific density in relation. If there were no buoyancy, mass or size would form some sort of resistance that would allow more and less restriction in some or other form to be present. In the sea and to the sea in that case there is no big fish or small fish but there is only fish. To us humans we think in perception of distance but in the cosmos space in the form of distance is the measure of time developed. The same goes for temperature and mass. Mass is good and mass is a product of the Human mind to put perception when it is needed but mass is dysfunctional in relation to the cosmos.

All the planets are more or less 299 in ratio from the Sun, which makes the time affecting all the planets $T^2 = k / a^3$ which is then $T^2 = a^3 X 299$ and in relation to space $a^3 = T^2 / 299$. What this does is it puts all the space in ratio at an equal distance to the centre of the Sun and it puts the time in motion rotating at an even period around the Sun. From where we stand, the planets are assorted but from the Sun it is liquid of outer space spinning in relation to the Sun that is serving as the solid. Being liquid and solid is having motion or not.

The Sun is at a level with all the planets parading past the Sun in the given ratio of 299. All the planets are precisely the same "distance " $299 = T^2/a^3$ from the Sun and has precisely the same "mass" $a^3 = T^2 / 299$ in relation to the Sun and the rest also floating about the Sun while they all travel at the same velocity $T^2 = a^3 / 299$ around the Sun.

The lot is in a bowl of liquid and it is the liquid that keeps a regard to the Sun where the Sun forms the solid as the regard to the liquid. In that regard gravity loses all it's mystery as the Coanda effect starts to explain gravity and the purpose of gravity. In accordance with and depending on what Kepler's law indicates, we find space-time is $a^3 = k T^2$. I wish to remind you that it was Newton that came up with the idea that $a^3 = T^2$ when he disclaimed Kepler's findings to validate his position on mass and the way he interpreted gravity. Let us again go back to investigate….

Applying Newton's second law F=ma
One arrive at the formula
$GMm / r^2 = m (\omega^2 r)$
By replacing $(\omega^2 r)$ with $2\Pi / T$ we obtain Kepler's third law
This law predicts that $T^2 = a^3$

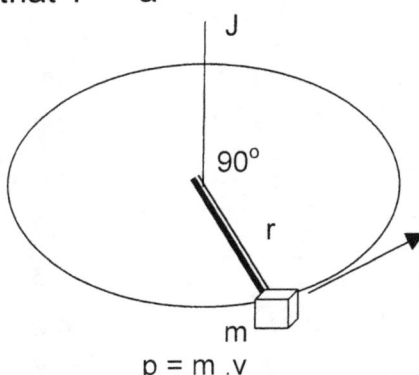

$p = m . v$

The mass (m) multiplying the speed (v) forms a new value J AND THEREFORE j CONTINUOUS TO IMPLY $J = I \omega$
$J = r \times p$ where $p = (v = r \times \omega)$
$J = r . m . v = m . r^2 . \omega = I . \omega$ and becomes interpreted as $J = I \omega$

This establishes that $r = dJ / dt$

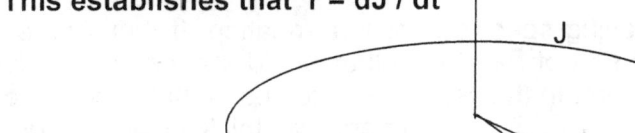

dr

$r = dJ / dt$ In the case of planets in orbit around the sun r forms a value of zero because $dJ / dt = 0$. The big incorrectness is that this presumption works on stopping time. That is impossible to achieve.
This is totally going against the grain of what Kepler said because Kepler said the motion of the space is equal to the space. On this rests the entire Universe because as soon as time stops, space collapse. The motion provides time a space to move and that is the Coanda effect.

$r = dJ /$

$$r = \frac{a^3}{T^2 k}$$

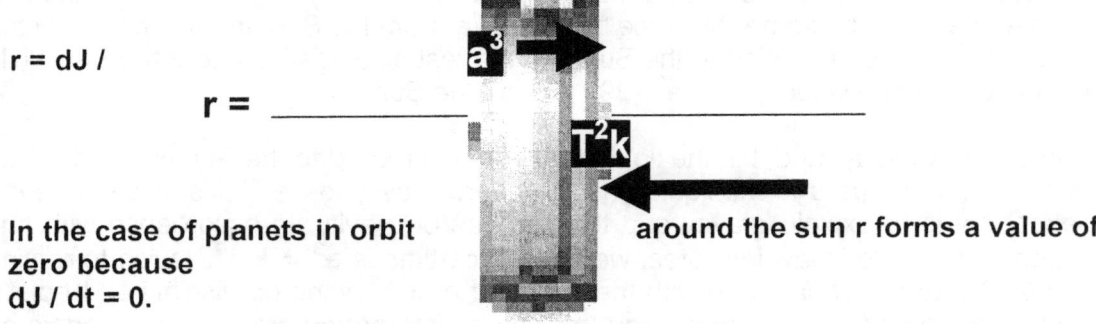

In the case of planets in orbit   around the sun r forms a value of
zero because
$dJ / dt = 0$.

That means the space will move in a straight line while it circles. That can only indicate a controlled expanding because $a^3 = T^2 k$ (also is); that the line between the centre k^0 and

a^3 / T^2 is increasing at a rate of $k = a^3 / T^2$ which indicates that expanding is happening at this very moment. The line **k** is not zero as Newtonian madness would suggest or the line is not 1 ($a^3 = T^2$) but the line is gaining by one dimension every rotation that is completed.

$T^2 = a^3 / k$ which means the time it takes to move is in ratio with the distance the space will move.

$k = a^3 / T^2$ which means the distance the space would move depends on the time allowed for moving. However if that is true and it is a mathematical statement then it also must be true that:

$k^0 = a^3 / T^2 k$ there is an appropriated centre
Then Mainstream science still has the audacity to say Newton was never proven to be incorrect...it should rather be said that Newton was never investigated and Kepler was left undiscovered because Newton raped everything about Kepler. It is as if Newton stole a vision from Kepler, which he got by studying but never saw. Let us for once seriously reflect on the matter while forgetting about Newton.

Taking the argument back to Kepler's law,

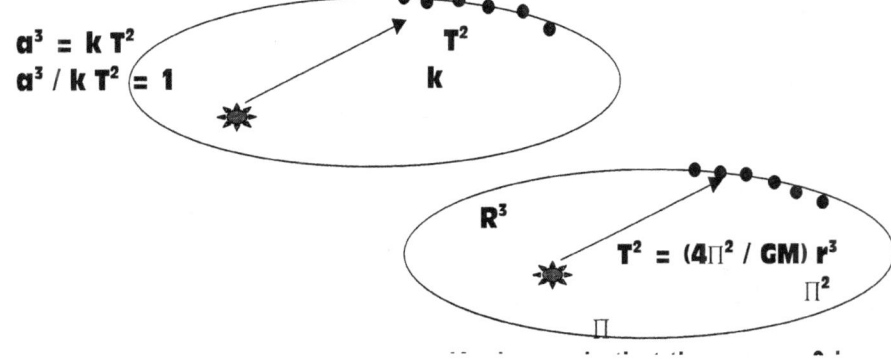

$$a^3 = k T^2$$
$$a^3 / k T^2 = 1$$

Kepler. If it is true that $a^3 = T^2 k$ and we dismiss Newton's obscenity while going back to basic mathematical principles we find that: $a^3 = T^2 k$

$T^2 = k / a^3$ Time can increase by manipulation as well as time that can reduce by manipulation. One can travel by increasing or decreasing the valid time.

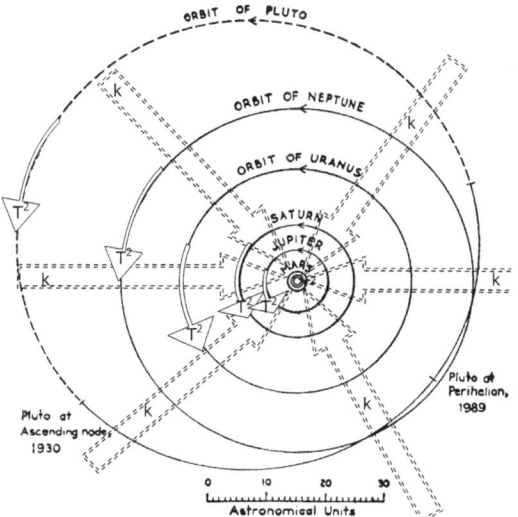

$k^{-1} = T^2 / a^3$. The distance is moving between the objects. If the objects are not coming closer it must be the substance holding the objects that is coming closer. When we have a table indicating a shift towards the centre $k^{-1} = T^2 / a^3$ that does not involve any of the planets coming towards the Sun, it would then show the space holding the planets are moving inwards because it shows definitely that something about the radius is decreasing and that means something is shifting towards the centre. The space a^3 is rotating T^2 and the only other counter action could be when the space holding the rotating space is reducing $T^2 / a^3 = 299$ by the margin space is expanding $a^3 = T^2 k = k = a^3 / T^2$ that proves the Big Bang is in progress.

Kepler told so much about so many by using so few syllabifications that a mathematician such as Newton was unable to comprehend the full implication. Allow me to explain with the aid of a manmade machine.

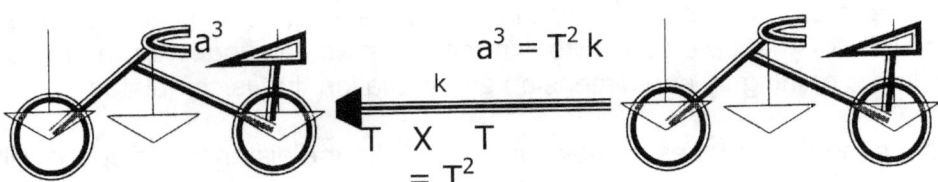

$$a^3 = T^2 k$$

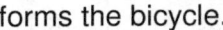

$$T \times T = T^2$$

We know time connects to the moving bicycle where the position changes from where the position allocated is putting the bicycle from the one instance to the next instance. The distance the bicycle moves is associated with time because the distance that **k** has, is the duration in time that it took the bicycle to replace **a³** during **T²** all across the length of **k**. We know the bicycle cannot jump from the one position to the next position notwithstanding whatever drive we connect to the bicycle. The bicycle constitutes of a massive number of atoms forming the structure that forms the bicycle.

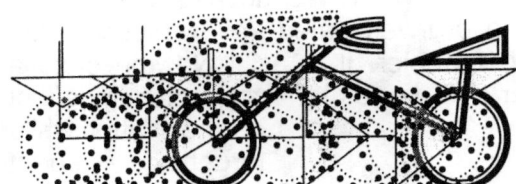

The get the bicycle from one point to another point we have to take the one lot of atoms forming the bicycle structure that is the bicycle unit to the next lot that forms the same bicycle at another location. That means to every atom there is a replacing of the one atom to the next location of the same atom at a distance of **k**, which is time during a period, which is time. Time therefore has two components being **k** and being **T²** and all the while **k** multiplied by **T²** forms space **a³**. If Newton had no regard for Kepler, then at least he should have had some regard for mathematics.

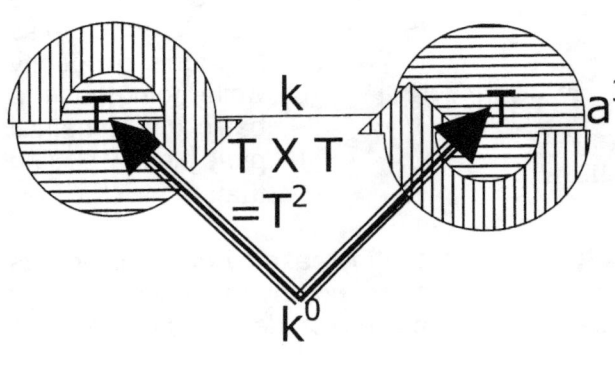

It takes the time that **k** allows to move **a³** from **T** to (X) **T** during **T²** and that is what Kepler said about space-time. Kepler mathematically said that **a³ =T²k**. Time is the motion that time allows space to move from and towards **a³ = (T** to **(X) T= T²) k** It proves that the factor **k** shares all the dynamics which the factor **T²** presents in allocating the position in time of space **a³**

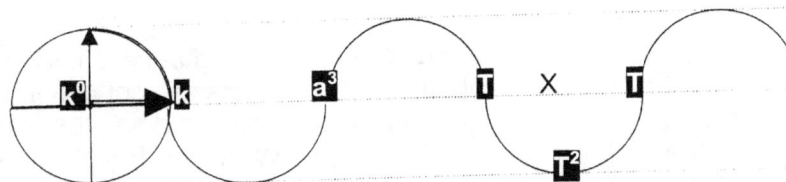

The motion of the atom follows the exact curves of the graph. Because there is motion the motion is in the square

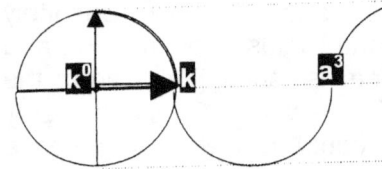

and since the square is present we can refer the result to the triangle. The motion is always 90⁰ to line of connecting the

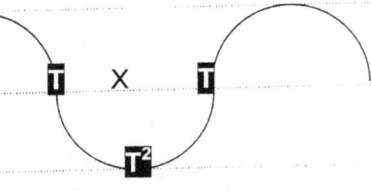

centre to the motion and where 90⁰ become present the law of Pythagoras becomes the dominant mathematical factor.

The factor **k** is always an extending line from where **k** as a straight line started to where **k** then will end. The line that forms the basis for the triangle is **k⁰** going on to **k**. If **k** is implicated as zero as Newton said then the entire Universe loses all its validity in being. That throws the entire cosmos into not being and that is the only thing that cannot ever be a factor in the cosmos since the cosmos is everything that is between infinity and eternity. The motion of a^3 is from where k^0 starts the line to where k ends the line and in between the start and the end there is a beginning and an end T to (X) T. That is also forming the other end of what totals as time T^2, but it holds time in duration T^2.

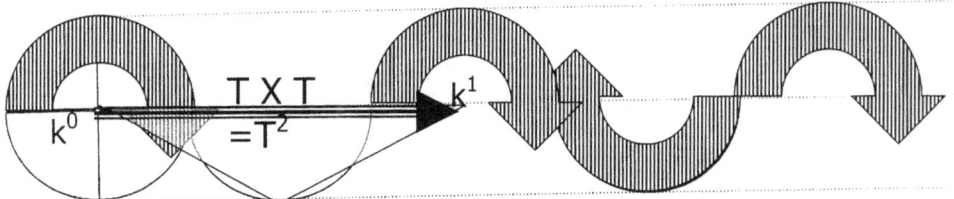

The movement of the bicycle concerns the unit as a structure that we recognise as the bicycle but the bicycle comprises of all the atoms forming the bicycle. The atom is in motion $k=3/T^2$ but also within the atom forming the atom there is motion of the atom $T^2= a^3 /$

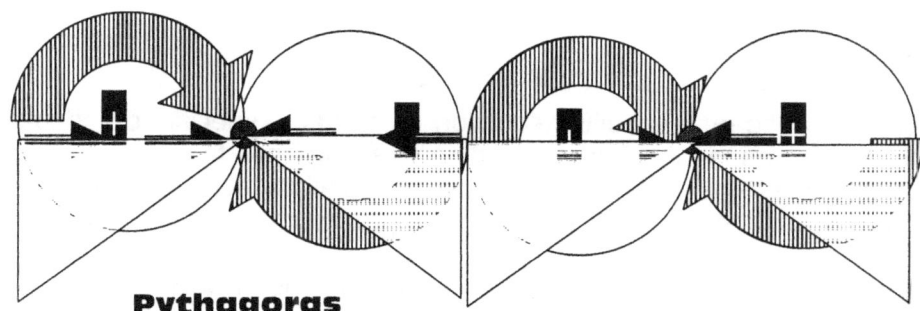

Pythagoras

k and the totality in reference form space-time $a^3 =(T^2) k.$ The formula implicates that the space $a^3 =$ is equal to the position that the space in the next movement will have $(T^2) k.$ The space becomes the motion of the space and in the space there has to be motion producing the liquid as well as rotary motion that gives the space a definition, which presents the space with motion in relation to the liquid, and then there is the space defining the purpose of the liquid.

$a^3 = (T$ to (X) $T= T^2) k$

Where there is a square present there Pythagoras is also present. I am sure Newton knew that…what is on the one side of the line forms a relation to what is on the other side of the square. The referring of motion in relation to a centre proves an establishing of a line in relation to a point. It serves a line being in relation to a specific centre and that centre projects the start of the line in a 90^0 angle with the point where the line must end. This is a mathematical reflex of Pythagoras using the line in forming a square that comes in relation to a centre.

That is the essence of Pythagoras.
The relevancy of k forms two correlating points where one relates to the centre and another relates to distance moved

The motion of the atom repositions the outside to a point in the centre of the atom and since the electron is honouring such a centre, any moving of the centre has to affect the atom in according to the dynamics Pythagoras demands.

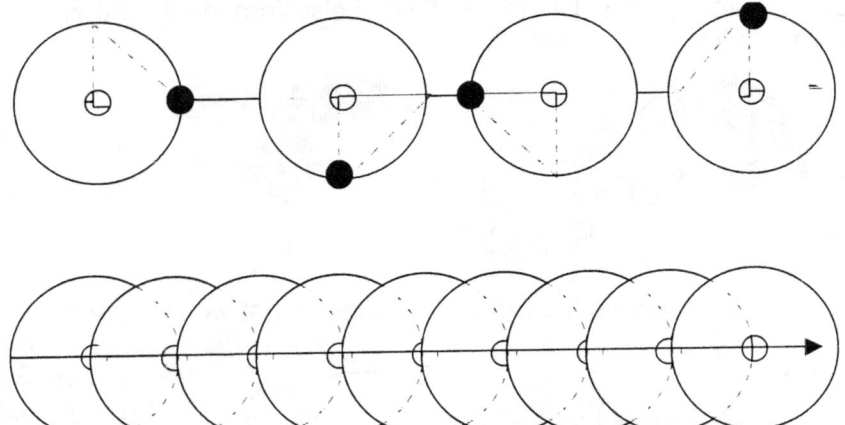

The moving of the atom is formed by a specific relocation as well as a specific reproduction of the atom in structure as well as in allocated positioning in accordance to every aspect of its surroundings. The atom must remain precisely associated with all aspects outside in order to find the precise duplicating of time in every aspect. That is gravity.

From this point the atom orbits as it honours the centre around which it focuses the spin. There is a relation by four points where the distance or the time will align with the centre as rotation focuses on such a centre. Every time the realigning forms 90^0 with the centre we will find Pythagoras applying the law aspect of the aligning.

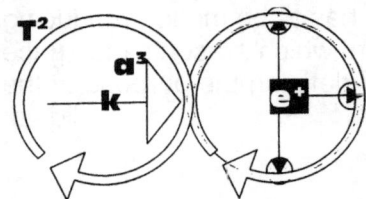 In relation to the outside there is more motion that has an effect on the atom where the centre point that the electron focuses on shifts in accordance with the atom motion as a unit. It takes the atom time in duration to remove all the material forming the atom to shift from one point to another point in order to allow motion to take place.

At the same time it takes the electron time to move around the centre and the distance the electron is from the proton serves as the measure of time that will apply when one cycle of rotation is completed. The rotation connects to time and since it takes light 3×10^5 km / sec to reposition the light photon, the speed of light is not time but is in relation to time. Einstein proved that the speed of light is time but that cannot be since it then must take light zero time to cross the entire Universe. Since light can only travel 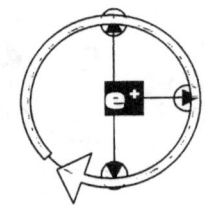 at 3×10^5 km / sec and the Universe is slightly wider than that, Einstein's argument goes up in smoke. To light, as it is the photon, the light finds time standing still but that explaining is long and is off the point as far as this discussion goes. Since light is not time but is connected to time like the rest of all material, it takes the electron time to circle the atom. One may argue it is insignificant but to the atom the cognisance is one year or one day or one period of time achievement. It is not only beauty that is in the eye of the beholder but everything about anything comes down to the measure of the beholder. To produce the space a^3 of the atom, the atom must have the electron circle T^2 once around the atom border at the relevance of **k**. The electron is focussing the relevance of the atom a^3 where it secures the border T^2 at the relevant time position of **k**. However **k** as a factor is in relation with the start and the start of any line is not zero but it is infinite. Therefore **k** comes all the way from k^0 down the length of **k** up to where **k** ends in order to position T^2, which then secures the border of the atom in a^3.

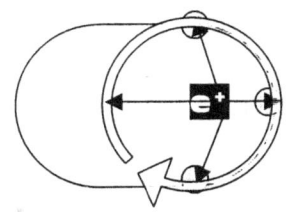

When the atom centre moves while the electron rotates such movement of the centre has a direction and this direction must influence the rotation because it will change the position of the rotation in relation to the centre of the atom. In the direction the centre moves, the space between the electron and the centre will reduce in accordance with the direction the centre moves. At 180^0 from that point, the distance between the centre and the electron has to increase by the margin the distance decreased 180^0 from there. The shift of the electron will be influenced by the intensity of the motion of the electron in the direction the electron moves. Since all the value of all the planets in the solar system mounts to $a^3/T^2 = \pm 300$, that means all the motion of all the planets related to the Sun is moving at an equal pace in time with the directional flow of time. The rotations around their individual axes might be different but the rotations they have in relation with the Sun are all equal. That proves that from the Sun we have all planets orbiting equal at $k = a^3/T^2 = \pm 300$.

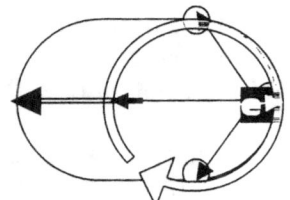

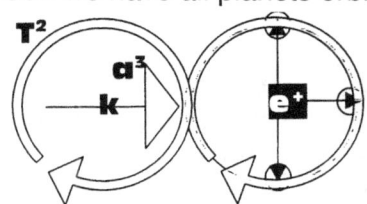

The fact of the relevancy is not merely argumentative but is a mathematical reality. Changes in the motion of the atoms will affect the atom in the rotation time it takes the electron to circle bout the centre. By shifting the centre, the time will reduce the ration period in one direction as much as it will increase the rotation in the other direction. This will have a severe timing implication on the atom as much as the time in motion increases the space it holds per time unit in rotation will decrease.

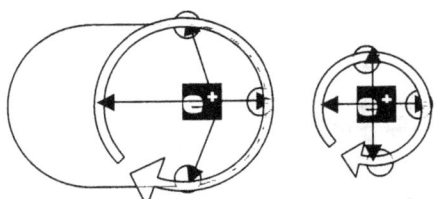

As the motion of the atom increases the electron allocates position while rotating in alignment it has to readjust with the shifting centre. If **k** shifts then **k** will insist on T^2 readjusting and with T^2 readjusting we will find the borders of a^3 reposition the circle of rotation. The more motion there is in relation to another fixed centre that represents the starting point that serves as k^0 the smaller will the allocated distance **k** be that we find between the rotating electron T^2 and the shifting proton in the centre. In this the size of the electron will reduce the size of the atom by the margin that the proton comes closer to the electron. The time it takes the electron to orbit is reduced by a ratio of four to one as the motion of the moving proton centre increases.

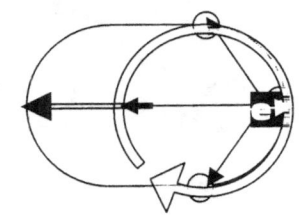

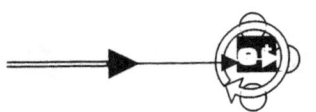

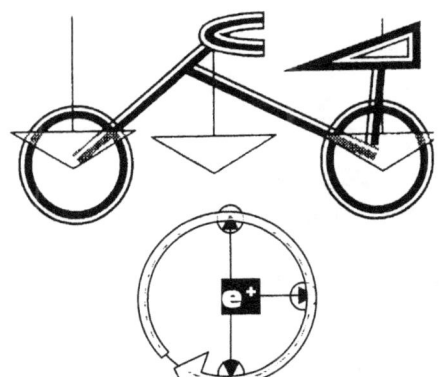

That is why gravity or motion reduces the atom as it accelerates the atom in relation to a more compact space. As the space in which the atom moves through time by duplicating increases in density that projects the ratio to reduce the rotating of the electron and with that the electron reduced the claimed space the atom holds. But it is not where the significance ends. In that way gravity increase, time as gravity reduces space where gravity is motion accelerated or decelerated by change of the density of the heat through which the space moves.

By reducing the atom, it relieves the atom of controlled heat within and such heat is ejected to the outside of the atom. The atom heat reduces by claiming less space because it holds less heat. Since the movement aligns with another centre k^0 that represents the link between k^0 and k and therefore institutes the triangle, the space held by the electron remains an

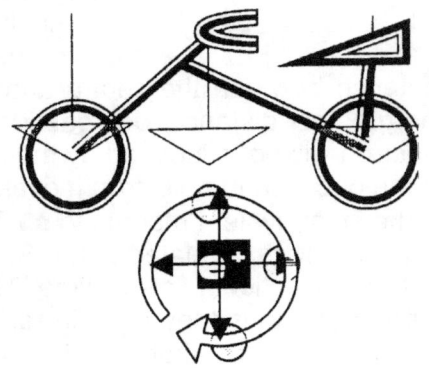

allocated value in relation to the Earth. The body shrinks because it removes heat from the atom but the atom still hold the heat in relation to what the Earth measures. That is the essence of the sound barrier. That is the essence of the heat blanket that covers the aircraft.

The atomic size of the aircraft reduces while the structure space of the aircraft remains in accordance with the size that the Earth permits and heat fills the vacancy that motion produces. The aircraft has no mass since the aircraft has motion. Newtonian misconception can never establish any reason for the sound barrier. That is what time implicates but all that is lost to Newton trying to fixate every aspect in cosmology on mass and by making the value of k redundant he removed the concept of time from the cosmos altogether. The value of k is space-time and that is $k = a^3 / T^2 = \pm 300$. **What this suggestion implicates would put the cosmos in a bowl of liquid where all swim alike**. That will explain gravity. That proves outer space has a specific density relation with the Sun and a density In as far as material within outer space goes. Every object floats just like the rest with no differences in as far as size or mass goes and outer space is liquid/gas.

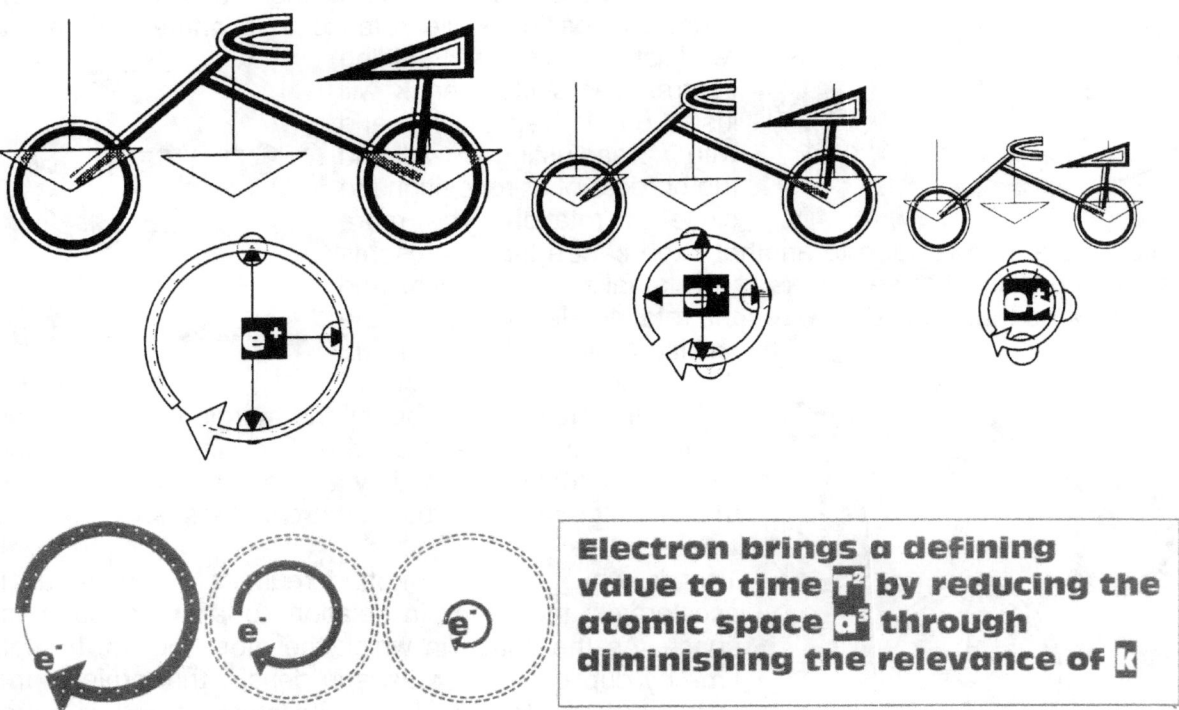

Electron brings a defining value to time T^2 by reducing the atomic space a^3 through diminishing the relevance of k

As the time intensity increases by the reducing of the factor in relevancy of k, the time factor also reduces the space a^3 the atom claims to duplicate per time T^2 unit. We still have the boat lifting out of the water as the motion in time increases ands the relevance of the space to water reduces and re establish new definitive values in Kepler's formula of $a^3 = T^2 k$.

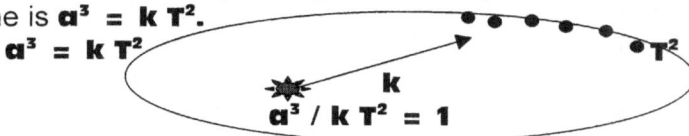

In accordance with and depending on what Kepler's law indicate, we find space-time is $a^3 = k\,T^2$.

$$a^3 = k\,T^2$$

$$a^3 / k\,T^2 = 1$$

In the relevancy we have space a^3 that is equal $=$ to the time T^2 that stand related to a specific centre k. The relation is time and space forming the interchangeable unit a^3/T^2 that stands in ratio to a centre k. The ratio between the space a^3 and the time T^2 totally depends on the measure of k.

The formula that the cosmos gave Kepler reads that the space a^3 is equal to the motion of the space in the time that it takes the space to move. That means the space is duplicating every second it moves and stopping will destroy all moving of the space. Since Newton became an institution forming the King bee of the academic cartel world wide The Brainy Bunch had Newton's vision written in the minds of the future generations almost at gunpoint...well definitely at an academic gunpoint.

$r = dJ /$

$$r = \frac{dJ}{dt}$$

In the case of planets in orbit zero because $dJ / dt = 0$.

around the sun r forms a value of

This goes directly against what the cosmos told Kepler **when the cosmos told Kepler that all space moving through time is equal to the movement of the space in the time it is moving through. That means by suggesting time may stand still such a suggestion throws the entire Universe into singularity.** You cannot stop everything that moves into a sudden halt because the momentum will destroy every aspect of such motion. Therefore one cannot remove the relevancy of motion by mathematically stopping time.

$r = dJ /$

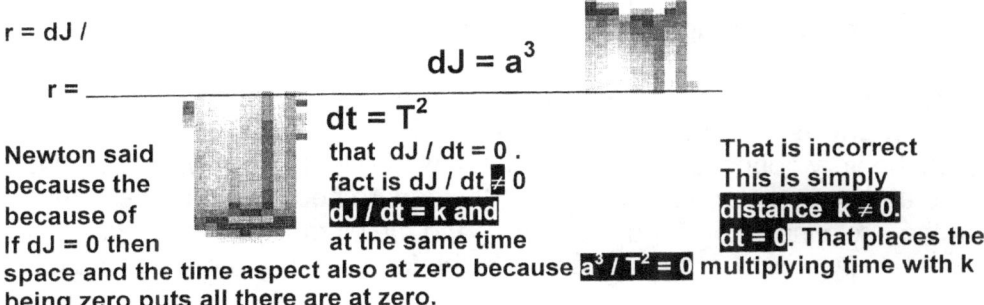

$$dJ = a^3$$

$$r = \underline{\hspace{6cm}}$$

$$dt = T^2$$

Newton said that $dJ / dt = 0$. **because the** fact is $dJ / dt \neq 0$ **because of** $dJ / dt = k$ and **If dJ = 0 then** at the same time **That is incorrect This is simply distance $k \neq 0$. $dt = 0$.** That places the space and the time aspect also at zero because $a^3 / T^2 = 0$ multiplying time with k being zero puts all there are at zero.

Gravity then is $a^3 = T^2 k$ where space is equal to the time it moves in. Space has to move or space will collapse. That we see in the case of a Black Hole. **Gravity then is also $T^2 = a^3 / k$** Time forming is valid since the object is unable to depart any further distance and is captured at that distance by the restricting of the space in motion. Gravity is keeping space relevant to determining the relevancy of motion in relation to the centre. **Gravity then is also $k = a^3/T^2$** because gravity is space in division of time which comes down to space having a velocity that time also holds space in motion. Gravity is motion of space at a specific speed or velocity, which is moving across a distance in relation the time it takes. Again that is space-time. **Gravity then is also $k^0=a^3/T^2$ k** where the Universe is being in the centre of space from where the Universe can and is claiming and controlling space- time. Translating Kepler's mathematical expression **$a^3 = T^2 k$** correctly to the verbal statement in English, Kepler said that there is a space **a^3** which is equal **=** to the motion in the time duration **T^2**, thereof where the motion of space takes up the time it uses to go between two specific points which holds a relation to a centre from where there forms a straight line k and is located on the spot where space begins the circle. Therefore that centre spot has the least space. Forget for one minute what Newton said.

Then take my challenge and prove what I said is not mathematically sane while you also know that $\dfrac{dJ}{dt} = 0$. I challenge you to prove how, where and why this can mathematically be sustained anywhere and under whatever mathematical law.

Place whatever you wish in a relevancy and have the outcome mathematically reach zero and see how mathematically appropriate that will be.

By reading what Kepler said correctly so many centuries ago the effort brings all the answers to so many unrealised questions…but it does not involve looking at what Newton said about what Kepler said… it is all about looking at what Kepler said. Not only did he introduce space-time **a^3 / T^2** but he also placed space **a^3** and time **T^2** in a relevancy long before Einstein did and placed gravity in space-time **a^3 / T^2** even before Newton named gravity. Kepler was the person who placed gravity as the ingredient in the universe that determines space **a^3** and time **T^2** and much more. Kepler was the first one that saw that gravity comprises of two factors being **k** or linear gravity and circular gravity or **T^2**.

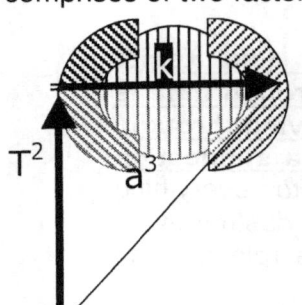

Kepler was the very first person to mathematically introduce space **a^3** in a time relevancy by **k** and related directly at all time in the universe that determines space **a^3** and time **T^2** and much more. Kepler was the first one that saw that gravity comprises of two factors being k or linear gravity and circular gravity or **T^2**.

Kepler was the one that discovered space / time as $k=a^3/T^2$

Kepler was the one that discovered gravity holding space-time relative

$k = a^3/T^2$

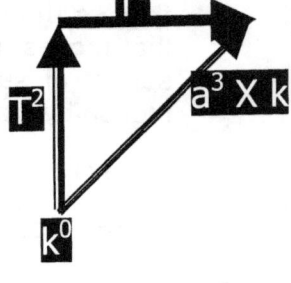

Kepler was the one that discovered there in the centre of a sphere is singularity as $k^0=a^3/T^2$ k

Kepler stated the very opposite of what Newton saw. Kepler had direct opposing ideas about the circle and what factor should be using the square because in Kepler's method of expression the circle indicator **T^2** goes square **^{2}** and the diameter indicator **k**, which replaces r remains single in the face of the volume being a cube **a^3**.

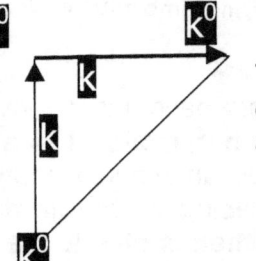

The two masters were using different dialects of the same mathematical language spoken by all.

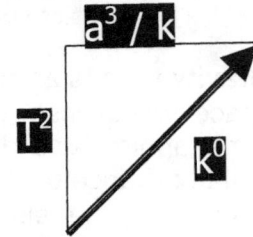

It is about translating mathematical equations and being correct in interpretations of mathematical expressions. Other factors are about certain mathematical deductions that were made in the past but were incorrectly presumed. A line cannot and therefore does not start with zero. Should you think such a statement is trivial then this book is even more especially for you because that changes where one presumes the Universe came from at the very beginning of the cosmic conception? Kepler answered all the questions we have…and much more!

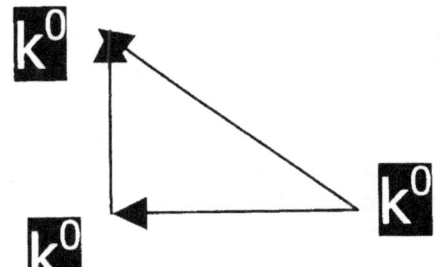

In order to have any line running there is a vertical response to the horizontal implication as much as there is a horizontal response to any vertical implication. The one response implicates the other. What ever there is there has to be another to implicate, which is. If there is **k** there has to be **k⁰** to begin **k** as much as there has to be **k⁰** to end **k**

It is a mathematical fact that where there is a line of any description, such a line will stand between two points holding infinity because a line can only start at infinity. That fact of a line puts a triangle in place because a triangle has the equal value of a straight line. If there is a straight line there has to be the presence of a triangle because both are equal to 180^0 and therefore are equal in all respects. Also, in response there will have to be a responding line of 180^0 crossing by the value of 90^0 in relation to 180^0

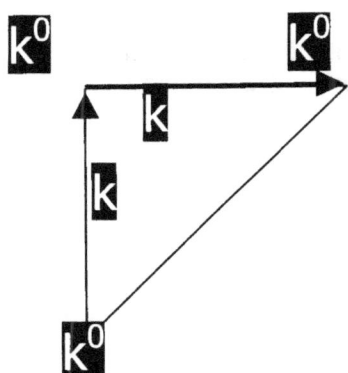

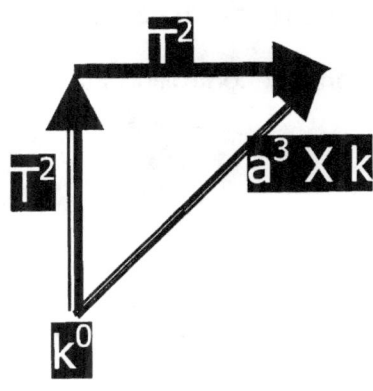

There can be no line between two zeros because only zero is between zero and another zero. All lines have to start at infinity and end at infinity and with infinity connecting lines all lines stand related to each other at the least of the value infinity. If that was not the case then the Universe had no validity. The first line assembled a Universe of possibilities. The universe is possibilities waiting for a chance to be whatever it can hold. Where you now are could and will become the centre of a raging star at a time when life is not even history. For all we know the Sun could have been a planet teeming with life and as the Sun developed into a star, life was desecrated by development. The Universe is not nothing but is every possibility waiting to happen.

The moving of one line establishes another line because by departing from singularity such motion establishes two dimensions. Motion goes by the square, which means that no matter how long the line is, the line also has to be wide to be a line. Being wide in the face of being long puts the line in a square and with the line in the square the line immediately represents the other factor of time where time is in the square. Therefore, drawing a

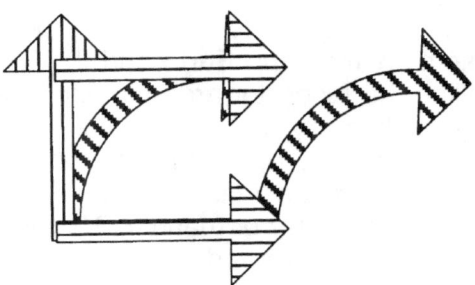

line has the same implication than enticing Pythagoras because every aspect of the cosmos is the cosmos because of the law of Pythagoras. The cosmos is Pythagoras but that is not what the cosmos introduced when the cosmos introduced it to Kepler. The cosmos presented gravity, the glue that keeps the cosmos together.

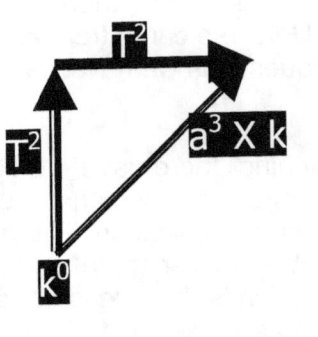

Pythagoras and the cosmos is $a^3 \times k = T^2 \times T^2$ and that is putting the three dimensions of space in motion that is in relation to time moving from where it was (T^2) and towards where it will be (T^2) while it is at where it is at that moment ($a^3 X \quad k$). However in this book we do not enter that arena because this book as a letter is dedicated to collectively indicate Newton misgivings. Yet that is the Coanda where the double square of time instates the space by three connected by the line to the centre holding singularity. All of this is meaningless if it is not connected with a line to the space that is defined by time That is the atom and that is time and that is gravity that is forming material in time.

Planets orbiting a given centre move by a line that measures by a triangle, which forms a circle. That is the conclusion of Pythagoras but that is far better explained in another book.

The gravity controlling stars works different from gravity working planets but gravity, is the

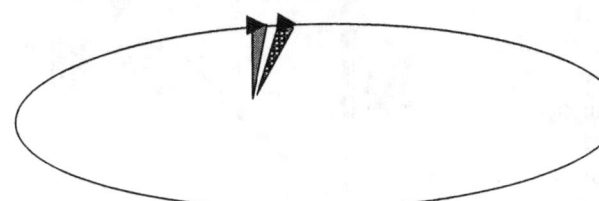

motion of space in time in relation to time. That puts all motion in context to the triangle and that puts all motion in relation to Pythagoras. The gravity, which was referred to when Kepler's studies revealed gravity was the motion of space a^3 through time T^2 in relation the a centre. The line running from the centre confirms the square of time by the distance. This is what Newton did not see.

If we investigate the Coanda effect, it is clear from what effect the Coanda principle institutes that the space of the circle depends in size on the distance that **k** forms but that distance reverses with the action of the motion, which the liquid produces.

I challenge you Academics in physics to show how much the Moon, is coming closer as the product of the mass of the two objects namely the Earth and the Moon having a mass that is pulling the Moon towards the Earth, and the Earth towards the Moon bilaterally. I challenge you to show how mass influences orbiting planets by contributing or increasing / decreasing or to what degree the mass is influencing the orbit velocity of any planet in any way or manner. The orbits are in ratio. Notwithstanding whatever mass, the lot are still very alike and in fact they are the very same. Notwithstanding the enormous mass discrepancies between the planets where Jupiter is 318 times the mass of the Earth and Pluto has only 0.0025 the mass of the Earth, they all orbit at a shared equality in ratio of about 300. I challenge you to mathematically prove that there is a gravitational constant. I challenge you to prove that the gravitational constant is responsible for evening out the mass differences between orbiting planets and that this is the cause why planets all orbit in a compatible fashion using the same velocity notwithstanding mass differentiations. Remember, this evidence I mention opposes Newton in no uncertain way by contradicting Newton at the very heart of physics. Yet, with all the evidence coming from the cosmos as such, that is contradicting Mainstream science directly, those in charge of Mainstream science stubbornly and sheepishly remain of the opinion that Newton was never yet contradicted!

Mercury

Venus

Earth

Mars

° From what we can see the Sun holds the space a^3 equal to the time T^2 at any given distance from the Sun. The ratio of space in relation to the motion in time is at an equal footing. That means to the Sun the only thing that is out there is outer space, which is

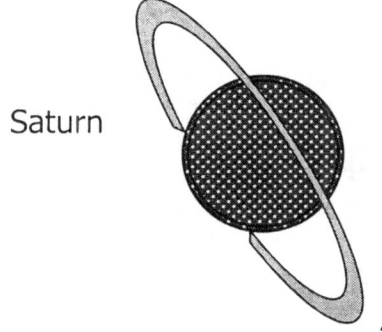

Jupiter

representing a unit in relation to the Sun forming

Saturn

another unit. That is exactly what Kepler inferred with

Neptune

his formula $a^3 = T^2k$. From the space the Sun holds, the Sun is

Uranus

equal to the motion of the space and the Sun forms the one

Pluto   ° factor of space a^3 where the other equal factor is then that, which is moving. Every planet is in ratio as far as any other planet, which puts all planets on equal footing. From the time of relation to the size in equality to the distance they are in relation to the Sun all proves a marked pattern of evenness.

The force is the measure whereby the radius in the square will vanish and this argument becomes the cornerstone on which all physics hold its foundation. The gravitational constant is used as the complement of the mass in conjunction with the gravitational constant to produce such a force value, which then supposedly produces a force able to destroy the radius between the objects. Then Newton went even further at the time and gave the concept a name: he called it gravity. In the way the formula is presented the gravitational constant increases the effect of the mass of all the planets, which is forming the force of contraction between the various objects and the Sun.

The only way this can be present is when outer space is equal in relation to the Sun while to the Sun planets are just floating in outer space and is the same as outer space is. The Sun holds no planet in any different way that it holds outer space in regard to the Sun. All planets are at an equal distance where the planets are point in an area where no point has special value. There is no mass in any consideration that the Sun gives credence to the planets.

Considering that neither distance nor size proves to be any different in relation to the Sun the only conclusion that can come from this is that the part outer space must perform as a liquid or a gas of sorts giving the planets a buoyancy regard and puts them all in equalisation. That can only be true if there is buoyancy in outer space, which makes the one part perform as a liquid T^2k while the other part serves as space a^3. This brings the Coanda effect to mind. It clearly shows a density factors as all the evidence of mass or size disappears. The only way that can happen is when the Sun becomes the space a^3 part and outer space becomes the liquid T^2k.

By expressing the statement in a mathematical equation it reads as follows $F = G\dfrac{M \times m}{r^2}$

$F=G (M \times m) / r^2$. The evidence that this is a fable and that mass is not pulling the objects closer was already established with Kepler introducing his work. Newton knew that the planets orbit in equilibrium by the measure if $T^2/a^3 = 300$ and all of that is in a ratio of about 300. That disputes Newton's claim on mass forming influences because it clearly shows equanimity.

Then when all the planets orbit at an even pace and all being the same ratio, where not one is found to come any closer to the Sun or to each other, your officially accepted reason for explaining that, is that the gravitational constant is evening out the mass differences and yet, according to the official formula you use, the gravitational constant is not evening out the mass but is complimenting the mass by improving the so called force. If the gravitational constant is evening out the mass the constant firstly has to divide into the mass to wither the effect that the mass produce and by nullifying the mass to the extent that Kepler's work proves, there is no mass factor present left to use. The formula then should read

$F = \dfrac{M \times m}{G \times r^2} = 1$ $F=(M \times m) / (G \times r^2)=1$ which in normal verbal English states that the product of both masses is in equilibrium when divided by the conjunction of the gravitational constant and of the square of the distance parting the objects where the product of the gravitational constant and the square of the radius between the planets is in division with the product of the mass which then will produce the force that cancels any influence of the mass factor that both of the planet's will have on the orbit of every planet orbiting evenly around the centre structure.

That is very much not what your formula you use $F = G\dfrac{M \times m}{r^2}$ suggests. Your formula

shows that the gravitational constant is helping the mass product to destroy the square of the radius. In the cosmos however and in reality that destroying of the radius is the last thing that is happening. To explain this ridiculous deception, Mainstream physics go on to explain that the gravitational constant which is supposedly the gravity is keeping outer space in a unit, where that gravitational constant nullifies the mass influence on the force. Newton said mass is all-important but this explanation now suggests that the mass is nullified by the intervention of the gravitational constant.

This is a statement that is not in line with the mathematical equations Mainstream science presents to serve as truth. Making such claims that is so far away from the actual mathematical statements, which came into place serving as yet another attempt to cover up Newton fraud with more deceit and becomes compatible with mafia like behaviour. This criminality is the behaviour of all physics paternity in control of all physics worldwide! This is the group of men that is in control of the thoughts in the minds of the brightest developing brains in the world. They harvest the best intellectual group of all students this Earth can offer! To the students questioning or rejecting the Newton claims is not an option.

The students accept what is taught or face examination failure and total rejection of the institute where they are educated. If that which is presented that the statement would suggest about their explaining how there is such planet mass apathy that will not influence orbit details of planets making them go faster or slower, it is based on yet another Newtonian lie. If the gravitational constant were evening out the mass, then it would have the gravitational constant in multiplication with the square of the radius in dividing the product of the mass. This will then be evening out the product that the multiplying of the mass puts in place. If that was said and translated in mathematical terms, then it would mathematically equate to showing where the force will be in equilibrium because the radius is supported by the addition of the gravitational constant which increases the radius in the square and the total of that is in equilibrium with the product of both object's mass. If that was true then the gravitational constant must be placed where it increases the radius in every event and it would be stating that $(G \times r^2) = (M \times m)$ **(G X r²) = (M X m)** is applying in all the cases so that the mass has no longer an influence on the outcome of the equation because out there in the real cosmos Kepler proved that the mass has no implication on the orbit of any planet **T² / a³ = 300**.

Kepler never even thought of a mass of any description and I can assure you that his not mentioning any mass was not the result of his retarded mind or not because he was mentally inferior to one Isaac Newton. With the corrected implementation of Kepler I can and I do support such statement of equality but then in that event the equation then should read that the gravitational constant is complimenting and supporting the square of the radius and mathematically such a statement is $(G \times r^2) = (M \times m)$ **(G X r²) = (M X m).** That removes mass as a cosmic factor. There is nothing wrong with this supposition but the way it is presented. The way it is presented kills Newton's claim on mass, which then should run through to the abolishing of his formula and the discarding of his theory. It shows that firstly k is not insignificant as zero but presents much significant proof that the Sun places all planets on an equal footing of 10 / 7. I shall explain that statement in a short while. Every distance of every planet in relation to the Sun is equal and every orbit of every planet is in harmony. Mass does not implicate or profess to have any influence which is strongly suggesting the characteristics of a density factor as particles would have when they display buoyancy characteristics in water.

You know very well that in the planet orbit ratio, the space factor (space **a³**) divided by time **T²** leaves a space **(a³)** – time **(T²)** displacement factor ratio between space and time of round about three hundred in all cases, $\left(\dfrac{T^2}{a^3} = \pm 300 \right)$ **(T²/a³ = ± 300),** which disqualifies mass discrepancies altogether. With that, the cosmos put Newton in dispute! So why then bring in the use of mass in any event if it does not change the result but has only one purpose and that is to cover Newtonian's misconception and to hide Newton's fraud? However, the formula used by mainstream science clearly shows that the gravitational constant is in effect increasing the mass by multiplication of the mass. The formula undeniably puts the gravitational constant in a multiplying position with that which the mass of both structures holds.

$$F = G \frac{M \times m}{r^2}$$ **F= G (M X m) / r²** which then in ratio is decreasing the square of the radius.

That is pure rubbish and I challenge you to prove me wrong by proving your mathematical interpretation. Then all the while, while promoting such rubbish you still maintain that Newton has never been proven incorrect. When Newton is taken into the cosmos, Newton physics goes bizarre. There is no evidence of mass influencing the motion of planets in any manner.

On the contrary all evidence shows a clear equilibrium present in the relations with the Sun and the Sun's planets.

Newton is not incorrect as it is used in normal applied physics but in cosmology there are no grounds in supporting of mass or gravity in the manner Newton saw or in the way Newton suggested mass to apply. On Earth and on the ground Newton's work is impeccably correct and that I admit without any reservations on any aspect of physics in whatever manner. However taking Newton's ideas into space becomes fraud. Mass only serves as a human concept to allocate differentiation when required when gravity is obstructed and in the cosmos gravity is not obstructed.

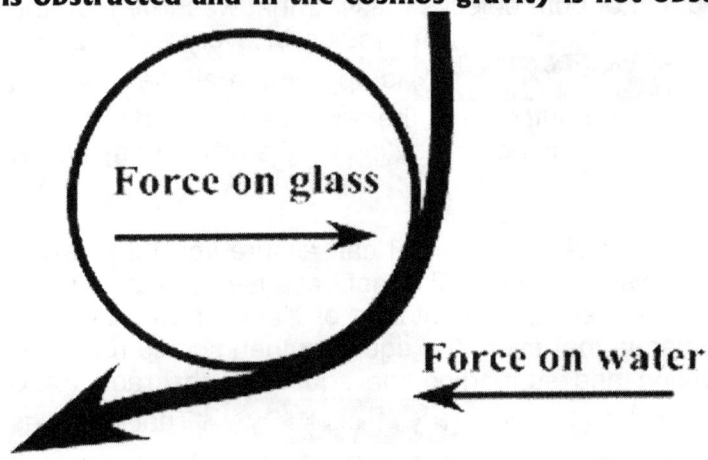

In the Coanda effect we find that when a liquid such as water runs down the side of a round shaped surface, which is a glass amongst others, the water follows the contour of the glass. Where water normally is used, the water would fall directly to the ground as gravity effect the water and only allows the water to run in a straight line towards the Earth. When the water is poured over the round surface that the glass provides, the water follows the contour of the roundness of the glass.

This part Newtonian physics admit to and they put two forces to explain the action but fail to give any further explaining as to why the forces come about or what would initiate the forces to counteract and disturb the gravity pull that should have the water running around the surface of the glass at the top end and then as soon as the water would find the opportunity, it will get a gravity pull and run straight down ignoring the interrupting of the glass contour since it then no longer blocks the path of the running water.

If we return to Kepler's formula we find the answer, but the answer again totally contradicts what Newton claimed about a circle having no effect on motion. According to Newton the rotation of an object is not time related and has no influence on the surroundings of the object.

When a circle of an object rotating produces the motion there are four points on each side that plays a part. Each point produces a direct relation as to the coming towards / the going away, in passing the front and in crossing the back marker.

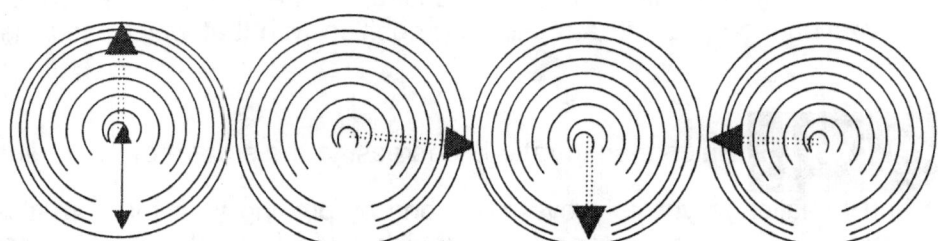

The Coanda effect proves otherwise when studied closer.

 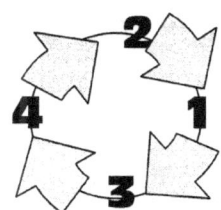 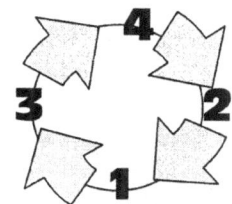

We can see that there are some coherency between the time and the rotation of objects.

That coherency we think of as time. The rotation of the object finds new alliances with the surrounding and if that was not the case then no fluid drive of objects was possible. You may discard this remark as not applying to the cosmos, but later in the book I am about to prove that outer space is a lesser-concentrated form of heat and the atmosphere of the Earth is just a higher concentration of heat.

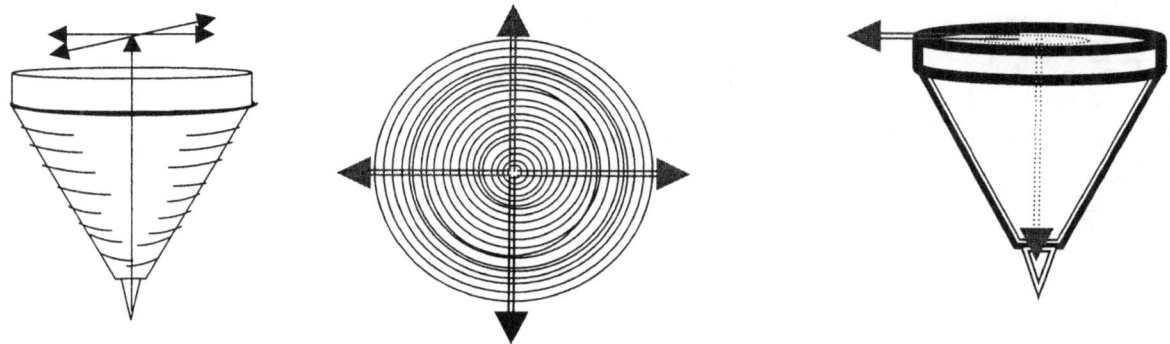

Please allow me to explain the relation between time and rotation by a simple experiment. Let's take a top and think of the same top as having two ends in opposition to each other.

When we take an object in rotation as having one side running left which it has, and another side running right, which it also naturally has. The side heading left is totally opposing the side running right because of conflicting directions in the spin.

If we put the same object's two sides in position, as they are when they rotate, we find the tow sides are contradicting each other by not being the same as one another. The one is a clone of the other but the clone is an opposing duplication of the other. This is part of the characteristics of all rotation and in this rotation it entirely depends on a liquid state that limits the space where the space includes the liquid $a^3 = T^2k$. **<u>Gravity is merely the motion of particles in space.</u>**

When any and all objects spin the spin is in opposing relevancy notwithstanding being part of

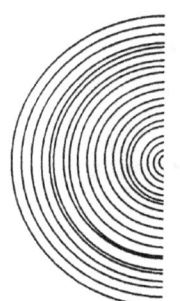 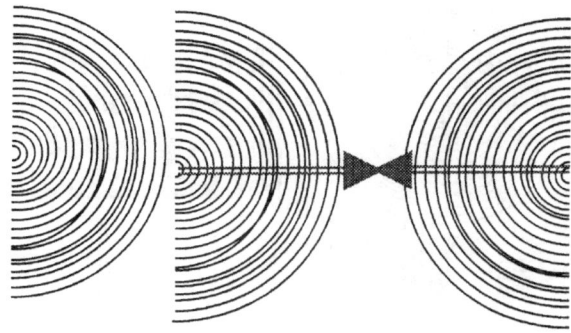

the same object or being the in the location where the next allocated position is precisely where the

opposing direction at that point forms the opposite of what it will be as the next part of the same but ongoing rotation produces the rotation graph. That is because no object can be twice in the same Universe and every object is in two parts of the same Universe in different locations. No object share Universes but forms duplication by motion of the same Universe on two sides of the same and one unifying Universe.

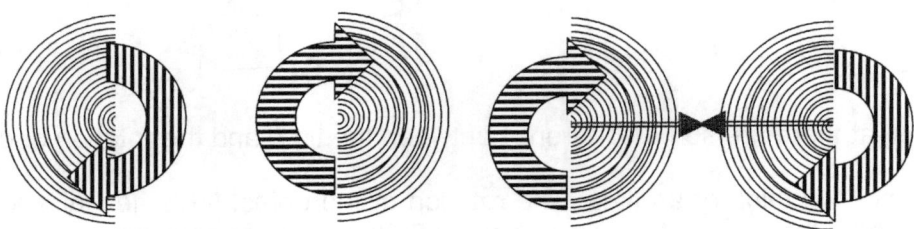

Taking this one step further in order to explain gravity there is two factors found in sharing the same unit. The one factor is the liquid to motion in relation to the other substance forming in the relation the part of the solid or the stable not moving part.

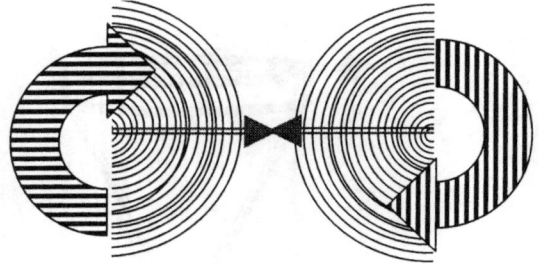

Taking a considered look at the spin procedure we find not forces but a rotation ratio at work which produces the opposing of two sides in exact opposition of each other where the sides in opposition will cyclic interchange and become the other party as the direction of the spinning objects change. In the spin we find a divide running as a line in the centre of the spin and is the point on which the ration is grounded.

From that line of division two sides oppose each other all the time. On the one side we find **k** being the distance having a relation with time in division of space and on the other side we find space being in division with time.

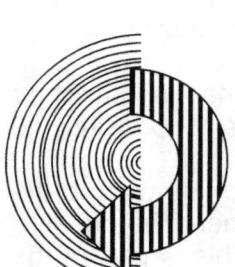

If we take Kepler's formula on individual merit, the cosmos told Kepler that space is equal to time by motion from a specific centre. It is well known as $a^3 = T^2k$, but at the same time very badly understood by the science paternity.

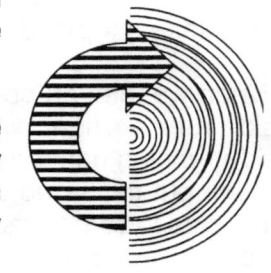

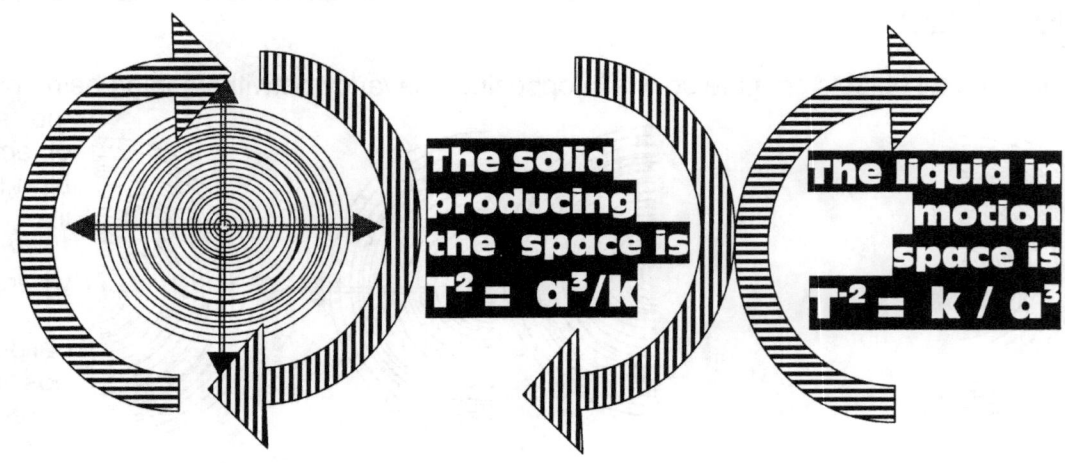

The solid producing the space is $T^2 = a^3/k$

The liquid in motion space is $T^{-2} = k / a^3$

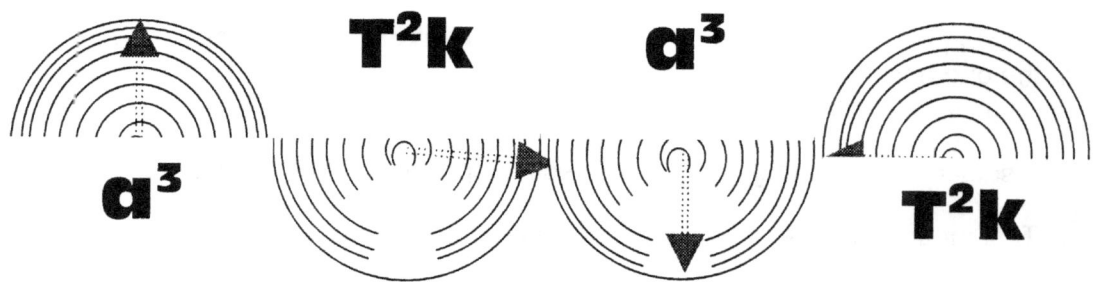

Reading the mathematical statement correctly we find that there are two opposing parts of the same unit contradicting each other by being on two sides in a total equal balance. On the one side is space a^3, **which is equal= to the motion thereof** T^2k. It is documented at this point that any object spinning has two sides directionally opposing one another and we have the cosmos telling Kepler that space divides in opposing sides as space a^3 and T^2k motion which forms the time aspect. We even use that side as time by indicating a cyclic year from the completion of one rotation and monthly and seasonal positional points indicating the different periods.

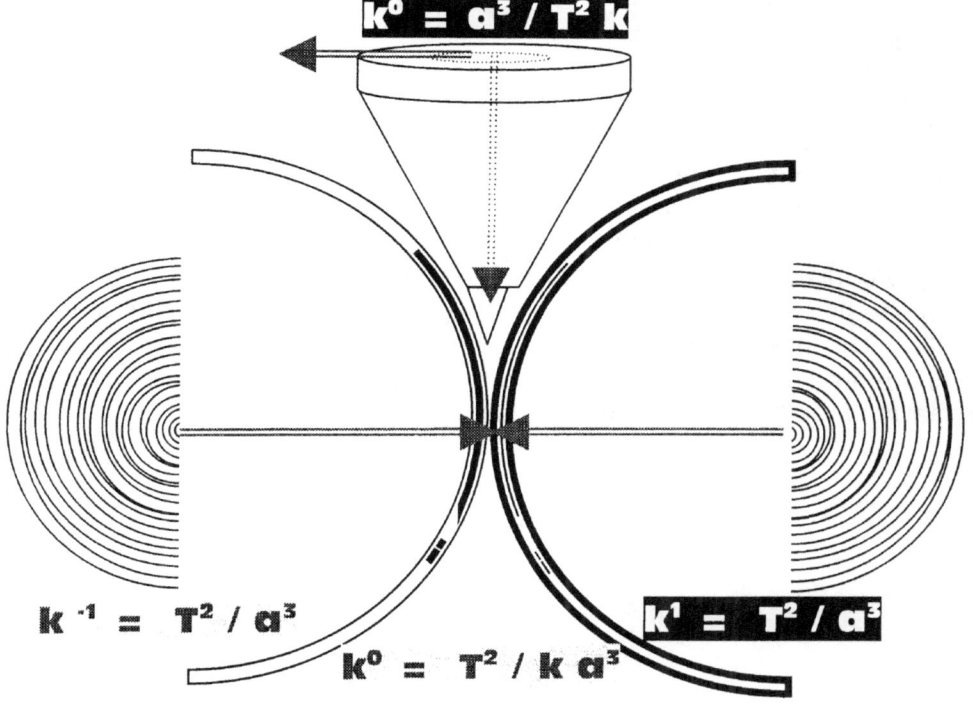

In this we have the one side holding **k** related to space a^3 in division of time T^2 being the relevancy of $k = a^3 / T^2$. There then is an opposing relevancy to this because of the opposing we find in the direction of the same object and being in opposition it then must be $k^{-1} = T^2 / a^3$. However with that being the case we then find objects are in opposing relations to time as well with time going positive and negative or going bigger in relevancy or smaller in relevancy. $T^{-2} = k / a^3$ and $T^2 = a^3 / k$. This proves that the object in rotation around a specific centre will reduce the relevancy or distance by reducing or increasing the time and space relation. That is the reason why comets do not slam into the Sun by force instigated on grounds of mass as Newton would suggest but rather cyclically circling the Sun in rotation by going away further and afterwards coming back to come much closer. Guarding this very jealously is the balance of the line being $k^0 = a^3 / T^2 k$ and every time the rotating object crosses this dividing line the relevancy changes to the opposite because the direction changes to the opposite.

This we find as a principle in the orbit patter of the Sun and all planets where the relevancy shifts from the one onto the other. But mass has no role to play in this because the structures do not have mass, they have gravity.

What Kepler said in his formula $\mathbf{a^3 = T^2\,k}$ is directly opposite to what Newton suggested with $\dfrac{dJ}{dt} = 0$. This means that the spin is uninfluenced by that which is forming the surrounding of that which is spinning.

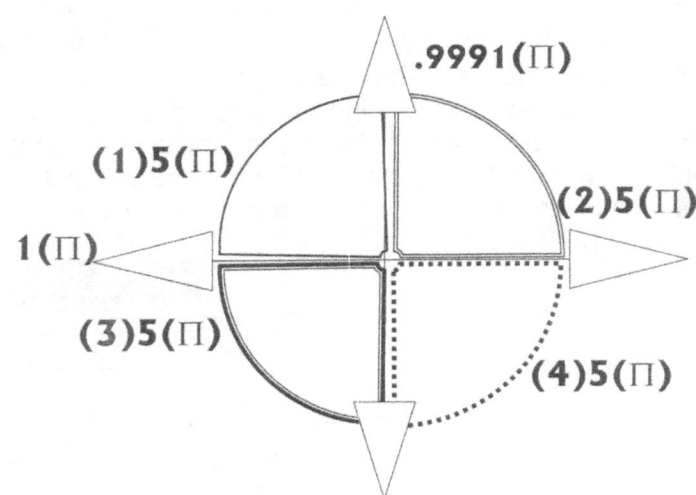

The fact that there is motion in converting 5 to a ratio suiting the time value places 5 in relation to the space factor held by time. The motion then is Π^2 on one side of the Universe plus Π^2 on the other side of the Universe putting the two in Π^2 in relation to the five it moves through. This will bring about $5(\Pi^2+\Pi^2)$ bringing matter and space to singularity. There are Π to the number of 5. From this comes that 4 (the four Π in the double proton in relation to the proton $(\Pi^2+\Pi^2)$ will always form the time value $(4(\Pi^2+\Pi^2)$. From this comes the fact that $3(\Pi^2+\Pi^2)$ will hold the space component, as that in fact is half space-time. Space-time being $(\Pi \times \Pi)(\Pi)(\Pi \times \Pi)$ and half of that to any direction is $(\Pi \times \Pi)\,(\Pi)$. This is why space-time in the geodesic sense has the value of 10Π and that 10Π in a sphere are Π^3. It is both triangles relating to one half sphere which is both triangles (7) in the totality of the half sphere (10).

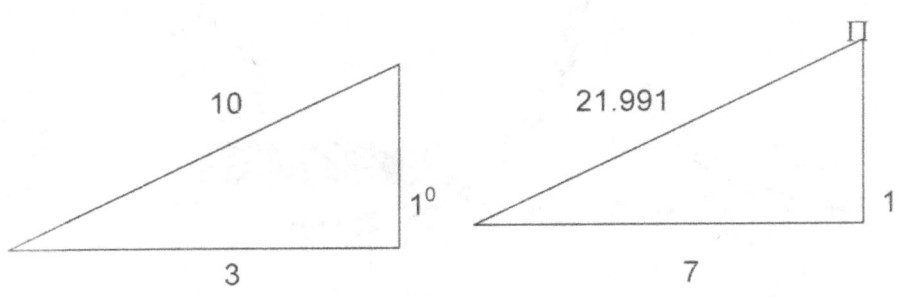

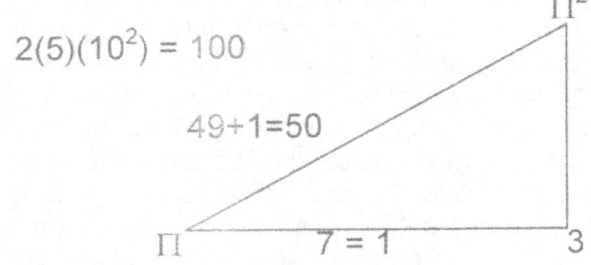

$2(5)(10^2) = 100$

$49+1=50$

Π^2

$50 = 5\,(10^2)$ where the complete is Pythagoras be comes a factor $\sqrt{100} = 10$ the value of space. $T=1$

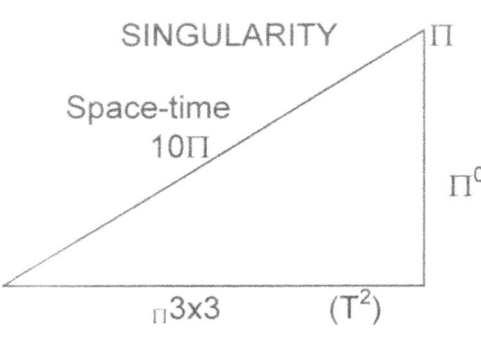

SINGULARITY

Space-time
10Π

Π

Π^0

$_\Pi 3\times 3$ (T^2)

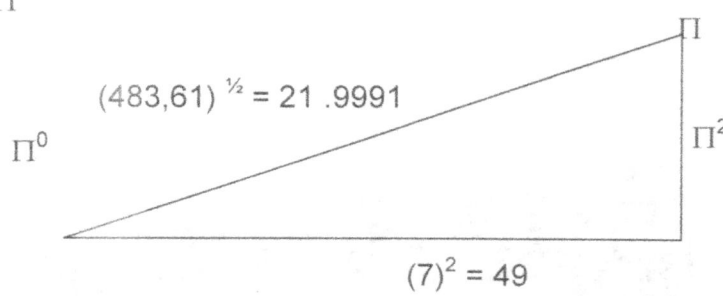

$(483,61)^{\frac{1}{2}} = 21.9991$

Π

Π^2

$(7)^2 = 49$

**10 /7
A / B**

B

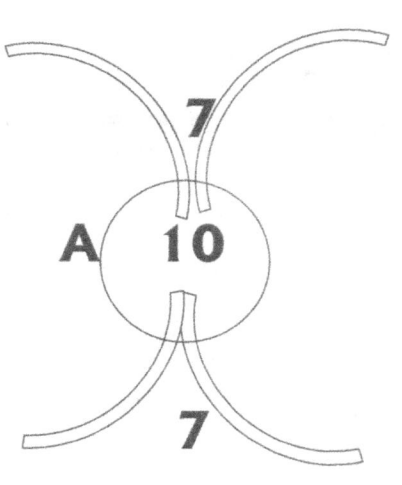

7

A   **10**

7

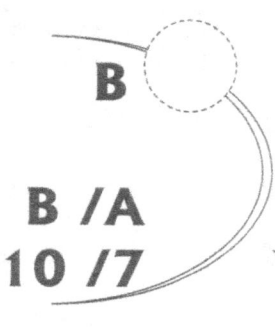

B

**B /A
10 /7**

$$T^{-2} = k / a^3$$

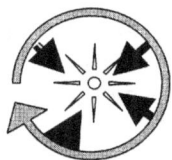

The sun is in a position of contracting by preserving the limits of the borders of the duplication limits.

$$T^2 = a^3/k$$

The comet is in a position of expanding by setting the duplication limits.

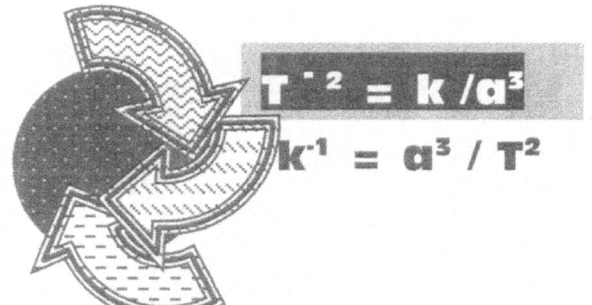

$$T^{-2} = k /a^3$$
$$k^{-1} = a^3 / T^2$$

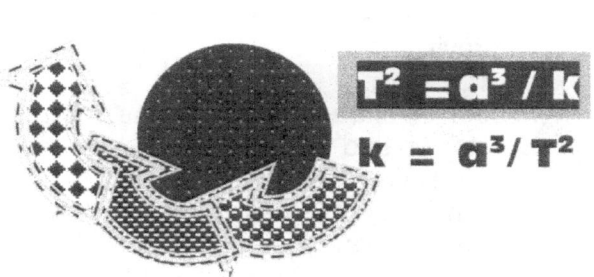

$$T^2 = a^3 / k$$
$$k = a^3/ T^2$$

From this then go on and try to match the comets true behaviour as seen how Kepler's formula interprets such cyclic repeat to that of what Newton falsifies as he introduced mass.

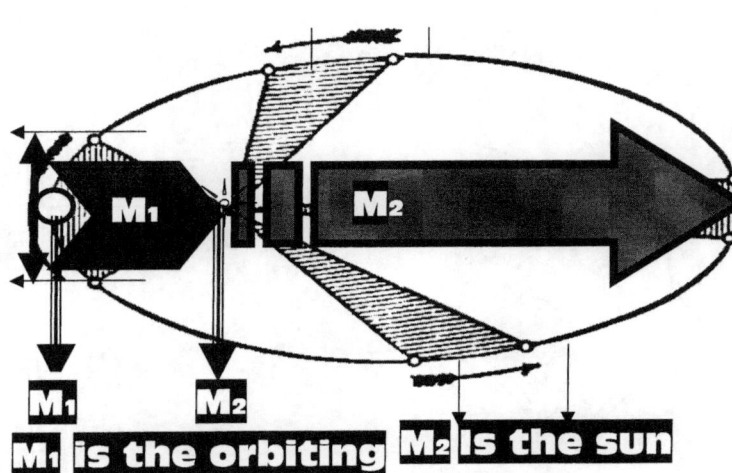

M₁ is the orbiting **M₂ Is the sun**

Newton stated that $\dfrac{M_1 \times M_2}{r^2} G$ = Force and the mass in each case remains the same as well as the gravitational constant, then why would the radius change every time of each year and not draw closer (or further).

F = G(M . m) /r² where:

G = the gravitational constant,

M = the mass of the body,

M = the mass of the lesser body

r² = the radius between the two bodies.

Newton stated that $\dfrac{M_1 \times M_2}{r^2} G$ = Force, because of the fact that:

1. The value of M_1 in both cases is equal because that is the mass of the earth.

2. The value of r^2 is equal because the two objects have been dropped at an equal distance.

3. The value of M_2 is still a mass filled with gravitons and therefore has to effect the relation in the same way as both objects consist of different compositions of materials that are used to manufacture the different objects.

4. That means Force one has to have a different value to that of force two ($F_1 \neq F_2$). In contrast, time duration $P_1 = P_2$ and this cannot apply in the case of gravity because the greater the force are through more mass, the bigger the impact has to be on the time duration. I think in all fairness it now is rather obvious even to children that orbiting comets do not destruct as they collide with the Sun by the pulling force of gravity.

Except in extra ordinary cases where some interference changed the orbit cycle of comets, we find that comets pass by the Sun unscathed to go into another cycle. In such manner Halley was able to predict the cyclic return of the comet named after him. As far as cosmology physics is concerned, little can be further from correct than the idea that it is just the mass bringing destruction by the pulling power which it unleashes instigated as a force by the mass in relation to the product that the mass on both ends exchange for becoming the force of gravity. The comet misses the Sun every time without failing. That means if it is occurring in repetition there is a balance in place.

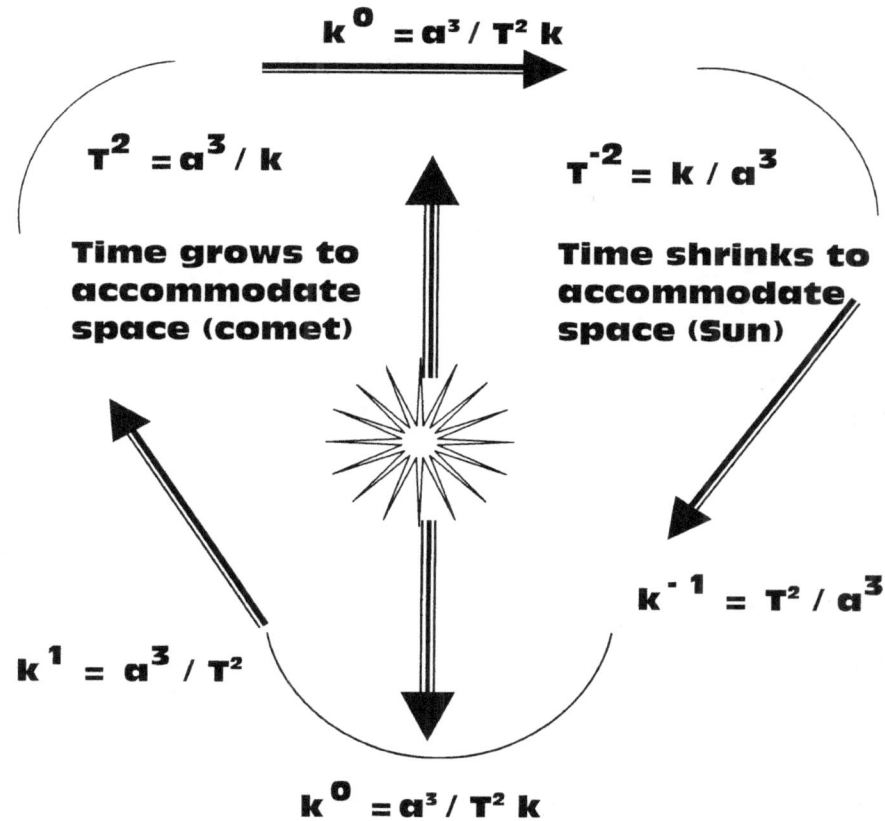

$$k^0 = a^3 / T^2\, k$$

$$T^2 = a^3 / k$$

$$T^{-2} = k / a^3$$

Time grows to accommodate space (comet)

Time shrinks to accommodate space (Sun)

$$k^{-1} = T^2 / a^3$$

$$k^1 = a^3 / T^2$$

$$k^0 = a^3 / T^2\, k$$

On the one side we find that the structure controlling the contracting aspect holds dominance in providing the pivotal circle relevance $k^0 = a^3/T^2\, k$, then this switches around where the other plays the key factor providing the gravity as $k^0 = a^3/T^2\, k$. Seen from the Sun the one side of the rotation would be time in progress of becoming more by applying $T^{-2} = k/a^3$ and in this the other orbiting structure would provide the relevance. The factor then would be in relation $k^{-1} = T^2/a^3$. As the orbiting object crosses the centre divide, the complete opposite comes into action where the time factor once more grows and $T^2 = a^3/k$ applies. However at that point from the other side we find that the factor also grows in $k = a^3/T^2$. Since the factors inter act by taking on different roles through the orbit we do not have collisions as Newton suggested but controlled orbit interacting as Kepler suggested.

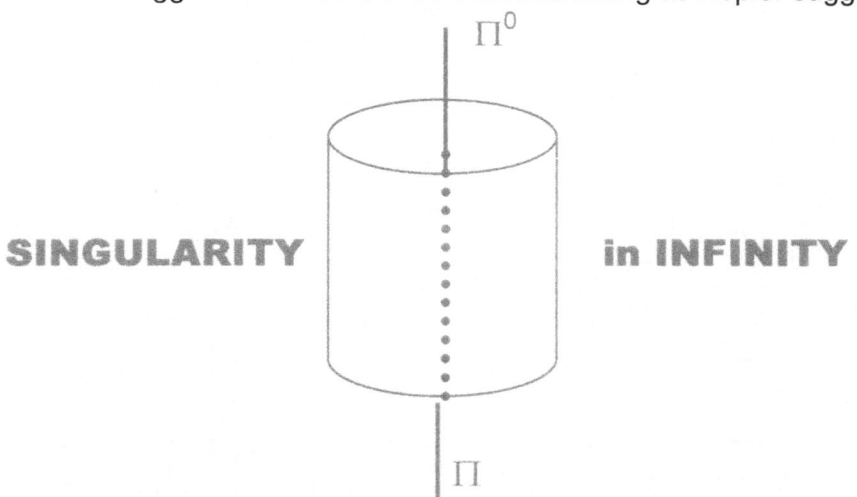

Π^0

SINGULARITY | in INFINITY

Π

SINGULARITY IN AN INFINITVE POSITION

Between the relevancy of infinity and eternity is space-time. Infinity is that which cannot move because it represents the part in the cosmos that is eternally frozen. Eternity o the other hand is the apart of the cosmos that can never stop because it represents eternal motion. Time in delay forming heat parted the two concepts in time and from that space-time came about. Space -time is the liquid representing time in eternity and space is the time delay covering that part of the Universe that can never have space because it never can start. In the centre joining this space and this time ids gravity where gravity is the motion of space in time and through time. All this culminates in the concept we named the Coanda effect. This truth is one eternity away from the nonsense of Newton.

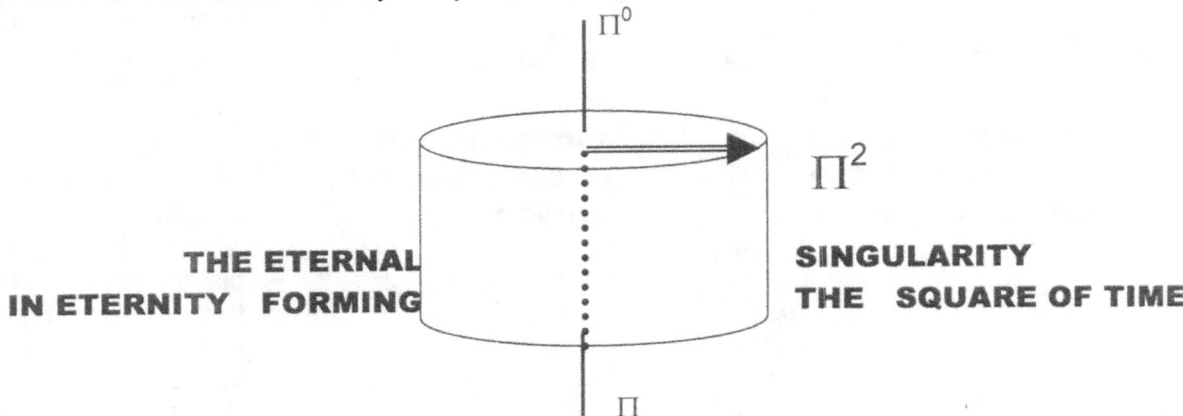

By applying the Coanda principle there are two relevancies at work and the motion as well as the direction time takes the motion will place either one of the two relevancies in prominence at any particular given time. There is absolutely no chance that both the factors may apply simultaneously and either of the relevancies has an opportunity to dominate but with motion applying stronger, there is a situation that presents the contracting to overshadow the expanding but that I explain in a later stage of this letter.

From this comes the working principle we at present give the name to as the Coanda effect. In the Coanda effect we have full gravity working as a compliment when the liquid provides motion to implement the gravity action.

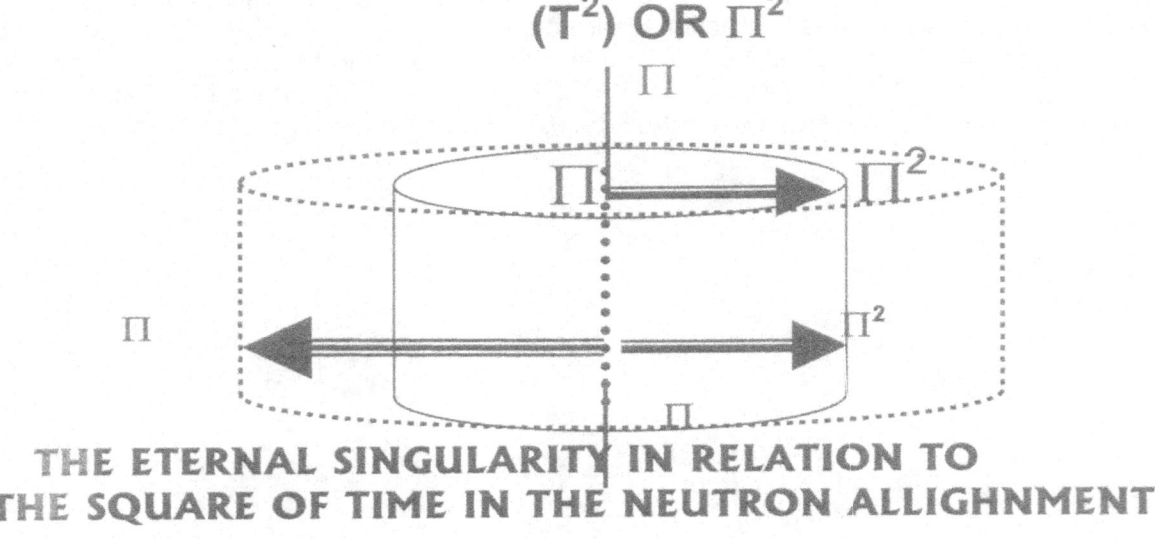

THE ETERNAL SINGULARITY IN RELATION TO THE SQUARE OF TIME IN THE NEUTRON ALLIGHNMENT

It is gravity finding motion through material making contact with time. It is the motion of space in time as Kepler said where material a^3 is in motion or in a neutron state T^2k or $\Pi^2\Pi$. The Universe is the neutron and where the Universe expels the neutron space-time collapses entirely. But the most important function of the neutron is the guard the relevance factor **k** in spite of Newton's contradiction of the fact. In the more advanced books such as ***Starstuffin'*** I in fact go even more literal with Kepler's formula and prove that the Universe a^3 is the neutron T^2k or better said that singularity forming space Π^3 is the neutron $\Pi^2\Pi$ and when a star discards the Neutron T^2k or in singularity terms $\Pi^2\Pi$ the star discards space a^3 or Π^3.

 This goes directly against what the cosmos told Kepler when the cosmos told Kepler that all space moving through time is equal to the movement of the space in the time it is moving through. That means by suggesting time may stand still such a suggestion throws the entire Universe into singularity. You cannot stop everything that moves into a sudden halt because the momentum will destroy every aspect of such motion. Therefore one cannot remove the relevancy of motion by mathematically stopping time.

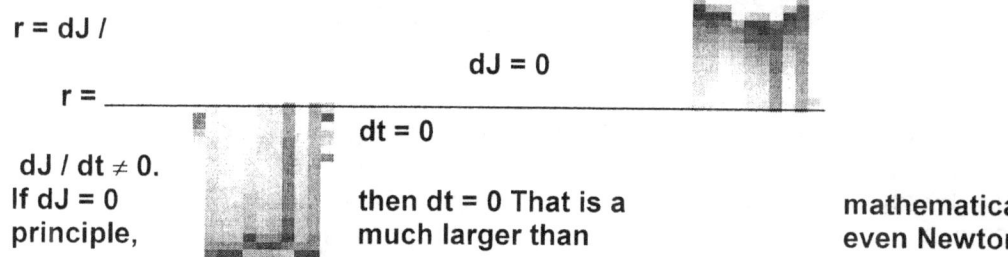

r = dJ /

dJ = 0

r = _____

dt = 0

dJ / dt ≠ 0.
If dJ = 0
principle,

then dt = 0 That is a
much larger than

mathematical
even Newton

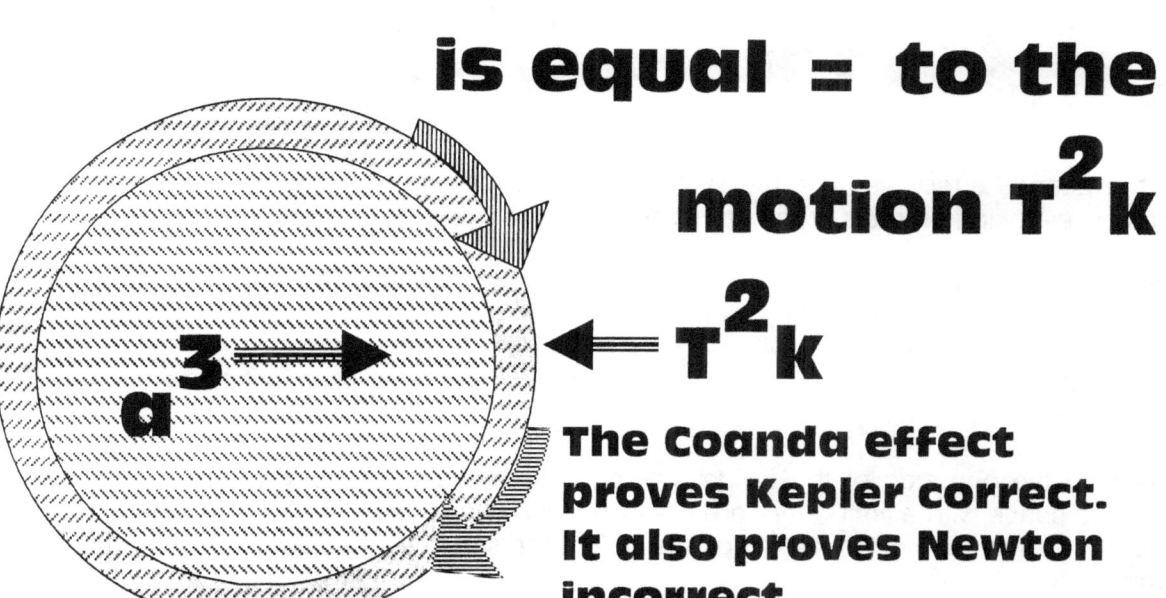

The space a³

is equal = to the

motion T²k

← T²k

The Coanda effect proves Kepler correct. It also proves Newton incorrect.

Let us argue what Kepler said mathematically and without Newton trying to convince every one about his discovery of mass.

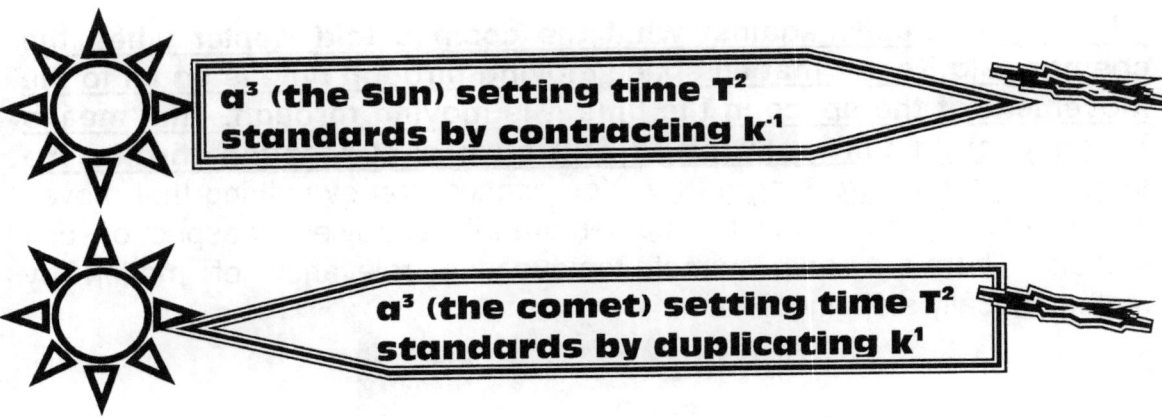

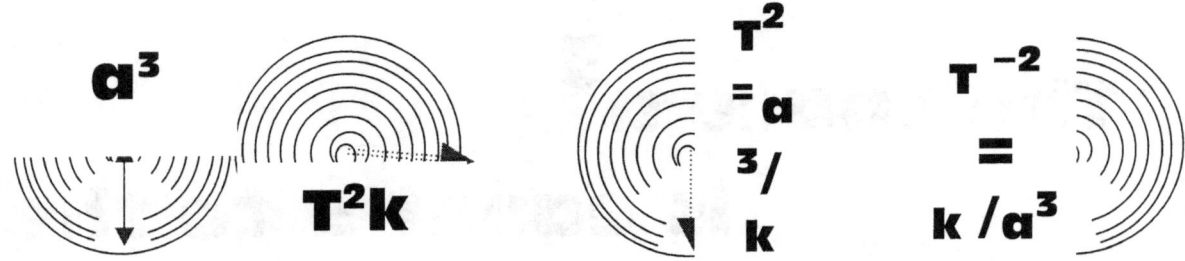

It would be far more prudent if we left Newtonian forces (and this I say in spite of notwithstanding all the rejection I thus far encountered from the celebrated mainstream physics paternity by my criticizing Newton on cosmology) and witchcraft to the Middle Ages from where it comes and see what gravity is by standards applying in the modern age.

In the Coanda effect we have the same opposition forming the same unit but two different substances taking on roles in the opposing sides and from this springs gravity.

Going back to the spin action we saw that one side is opposing the other side while in rotation and while forming a single unit. This we take back to gravity and in that we find what forms the motion we find in gravity.

As I explained previously there are two actions forming part of four factors bringing about a rotation action.

While on the one side of the divide $k^0 = a^3/T^2\ k$ we have $k = a^3/\ T$ where the time factor provides space with a limit, there are the other side where the time factor being on the other side of the divide $k^0 = a^3/T^2\ k$ extends the limit of the space by the liquid to provide the limit to space at $k^{-1} = T^2/\ a^3$.

In all of this the lot that I mentioned prove Kepler different to what Newton portrayed Kepler to be and also it proves Kepler correct in gravity whereas Newton is absolutely incorrect in his surmise on cosmic gravity. The liquid is part of the spinning circle but is also part of the other side of the divide $k^0 = a^3/T^2\ k$ and on that side the flow or motion $T^2\ k$ establishes the limit to space while the solid forms the space.

On the other hand the space a^3 finds an extending of k by the liquid that forms a part of the unit as the motion is part in Kepler's formula $a^3 = T^2\ k$.

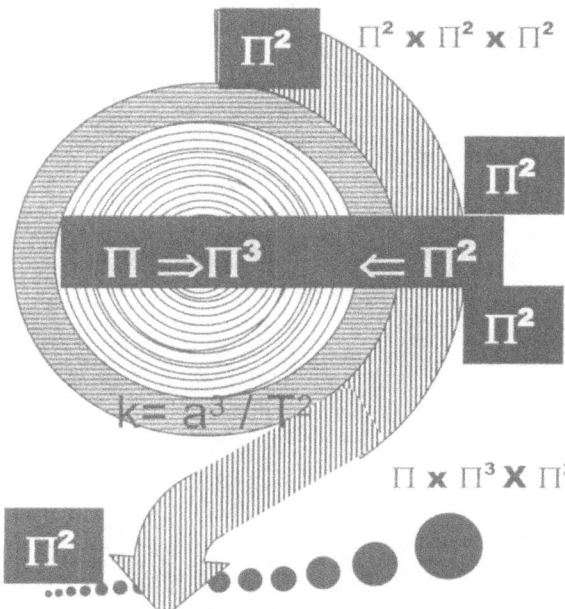

Π^2 $\Pi^2 \times \Pi^2 \times \Pi^2$

Π^2

$\Pi \Rightarrow \Pi^3$ $\Leftarrow \Pi^2$

Π^2

$k = a^3 / T^2$

$\Pi \times \Pi^3 \times \Pi$

Π^2

brings on.

Because the liquid seems to run in the opposing direction while it is running in the same unit and therefore in the same direction it is being influenced by the same relevancy factor **k** but in the other opposing direction **k** $^{-1}$.

Although both the solid and the liquid form part of the same circle and are in one unit the liquid as one part serves one side of the Universe or the divide while the solid serves another part of the same unit but on the other side of the Universe or the divide.

That is gravity. The rotation forms the gravity by forming a divide that initiates the start of a Universe. There is no possibility that time can ever stand still or that motion of rotation has no influence on the gravity bonding that forms because the gravity is the result of the divide that the rotation

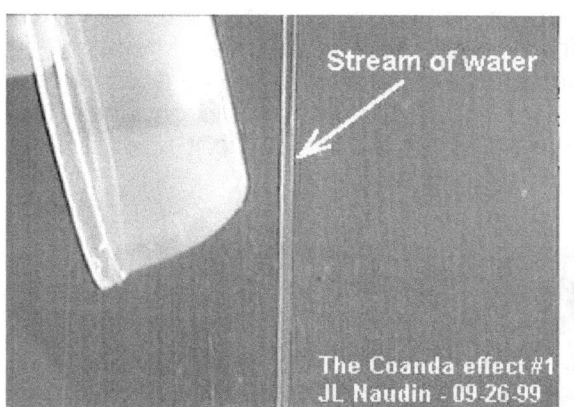

Stream of water

The Coanda effect #1
JL Naudin - 09-26-99

<u>**Sir / Madam, the fact that Newton presents k as zero, is quite impossible because of what Kepler said. Kepler said a^3 = T^2 k.**</u>
When motion produced space and motion came because of singularity that is unable to produce any sort of motion of any nature whatsoever, motion appointed three positions holding singularity that formed a relevancy with other positions that produced the space in relevance to singularity as a factor.

The reduction of space will bring about heat. Injecting fuel in a place where such reducing of space already exists it increases the heat level further and the fuel will "spontaneously" ignite. The igniting creates a rise in the heat level to a point where such a level that one can only find that level in the stars. Fuel will establish conditions that are in accordance with the laws of cosmology. Such heat will only apply to stars where such heat levels will indicate the enormity of the gravity present that is generating the massive gravity accumulated in the absence of space. Gravity is the concentration of heat in the utmost reducing and concentration of space. Thus a star is born in the gravity on Earth. In this example, **k**

Stream of water

The Coanda effect #2
JL Naudin - 09-26-99

increases as thrust pushing space **a**^{3}, which the **k** creates to a new **T**^{2}. That started the Coanda effect but that also started the Universe before the Universe and space – time. The Universe then was spinning at speeds faster that the speed of light and everything in the Universe only had form. Later on with the event of the Big Bang the Universe received dimensions but form remains the template of dimensions and even today that is how we interpret timesaving three positions to a centre. We still have the evidence, which is even more obvious than most other certainties we uncover in the Universe.

The Coanda effect is gravity and moreover it is the movement of a liquid in relation to a solid that forms gravity. In the phenomenon we named the Coanda effect, we find that water runs down in a straight line seemingly following a direct route to the centre of the Earth or to where it will contact the soil.

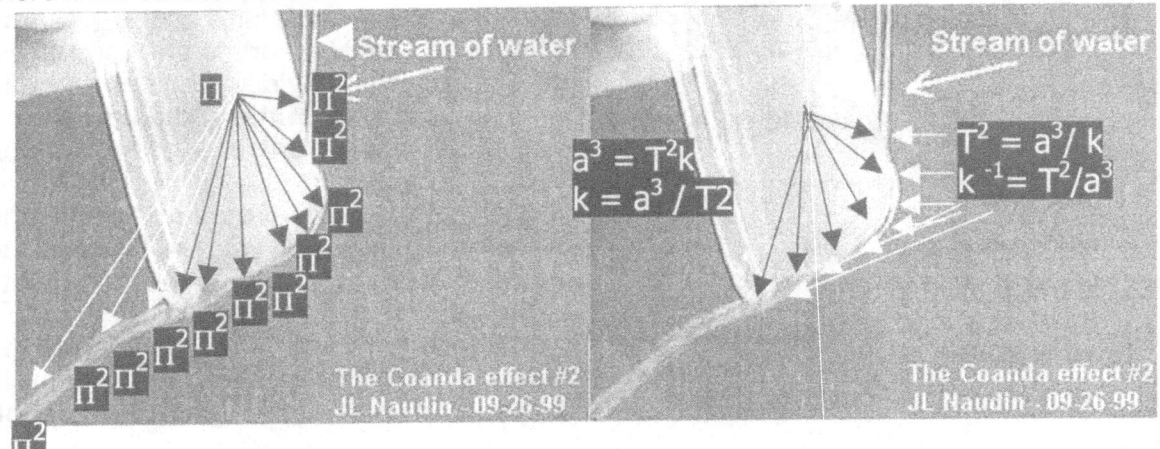

The function of the Coanda effect is well known but not that well understood and has never yet before been connected in terms of physics forming gravity. No one ever attempted to connect the Coanda principle in terms of cosmology or in the manner as to how the cosmos came about. Yet the evidence is there that the Coanda effect is the mould on which the entirety of the Universe formed from the very first inclination of motion right down to the understanding of the Black Hole.

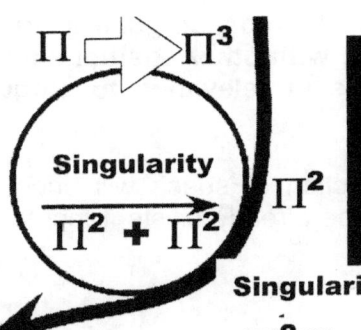

The Coanda effect is creating gravity. It is not replacing gravity. It is not recreating gravity. It is not substituting gravity. What the Coanda effect is is what gravity is. The Coanda effect is gravity.

Singularity extending the influence on flowing

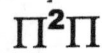

The Coanda effect is also the perfect example of the curvature of space-time brought about by the extending of singularity, influencing due to the shape that imitates or duplicates the value of singularity and again conforms Π. By establishing a new value of singularity as Π, singularity can once again take control and establish a new Π^2 as gravity in the new Π^3 forming space.

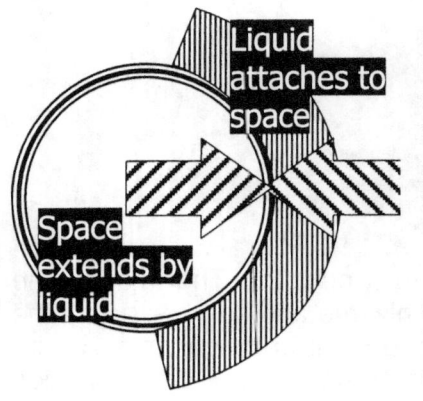

Looking at the Coanda principle, we find two distinctly separate parts forming one Unit. Looking at Kepler's formula, there too we find two distinct sides forming the same unit. In the Coanda principal there is a space **a³** forming a basis to which a moving **T²k** attaches. The **T²k** extends the space limit while the liquid provides the motion, which the space uses to become larger. The space a^3 is one side with the motion of the liquid T^2k being the other side. As I shall indicate later, Kepler did not introduce a mathematical calculating measuring formula as Newton would have suggested, but it is a cosmic principle on which the entirety of the Universe is built. It is gravity by all measures there can be. It is does not mean that the one side is in precise measure to the other side but it states a principle on which all aspects of the cosmos rests.

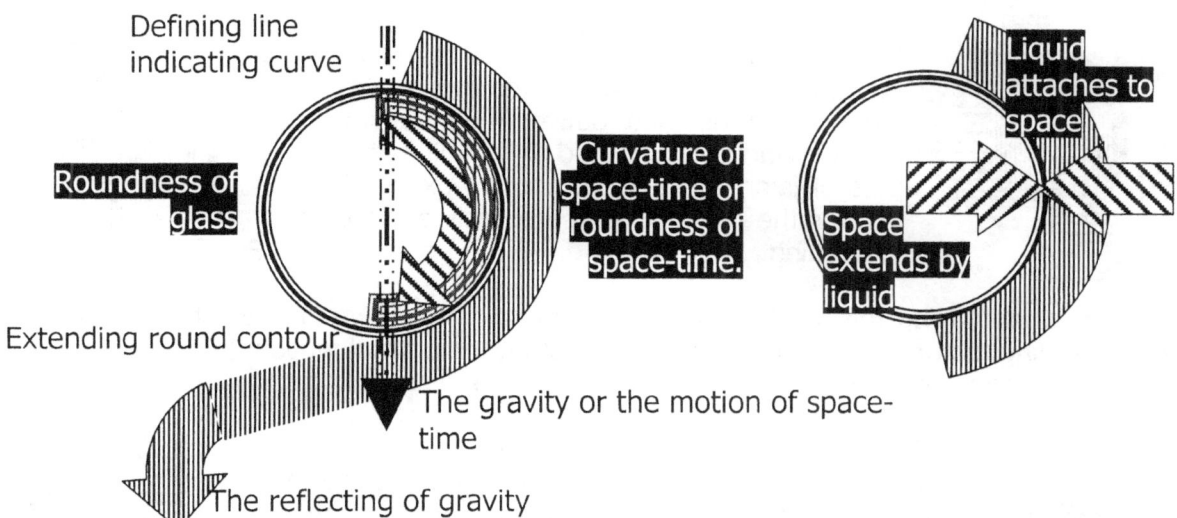

Defining line indicating curve

Roundness of glass

Curvature of space-time or roundness of space-time.

Liquid attaches to space

Space extends by liquid

Extending round contour

The gravity or the motion of space-time

The reflecting of gravity

$$k = a^3 / T^2$$

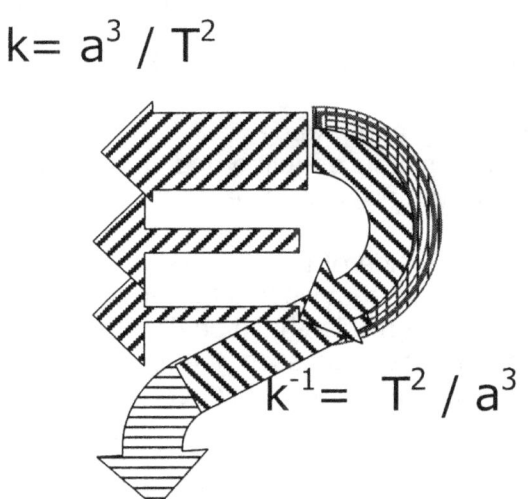

$$k^{-1} = T^2 / a^3$$

We find a line running through the middle of the round glass, which is dissecting the two parts in two directions. The line is $k^0 = a^3 / T^2k$ where there is space a^3 on one side of the divide k^0 and a substance providing motion T^2k on the other side of the divide. On the one side there is an attraction $k^{-1} = T^2 / a^3$ where the liquid is controlling the space border and on the other side there is $k = a^3 / T^2$ where the space is identifying the position the liquid has as the liquid then becomes a part of the solid space.

As we can witness in the behaviour of the comet and all other bodies orbiting a centre structure, the concept of spinning is bringing about changes in the alliances between the two opposing orbiting sides. That comes towards the centre on the one side then departs as it moves away from the centre on the other and where one side bring about contracting the other side flings motion away from a centre. This is all part of singularity demanding motion in relation to a specific centre. It has nothing to do with mass but every aspect concerned all depends on extensive control of the centre governing singularity.

Looking closely at the principle there are two principles where the

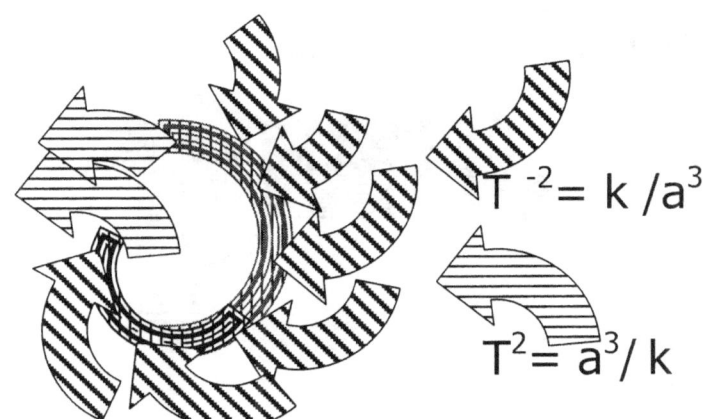

$$T^{-2} = k / a^3$$

$$T^2 = a^3 / k$$

one involves a contracting to the centre and the other form an expanding or a rejecting of the centre. There is one side that openly and exclusively favours contracting while there is the other side that provides expansion of the possibilities to be. In the two contradictions there normally is a fair balance but as life interferes with physics, we find a shift in favour of contraction as the motion life provides throws the balance towards the motion part.

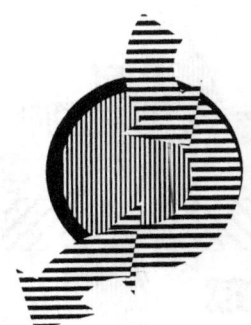

The contracting and expanding balance is totally in line with the motion of the liquid that provides the balance. In normal gravity motion, the balance between contracting and expanding is divided in the line that puts division in place. As motion excels, it favours the contracting to the decline in the expanding.

However, when saying this we have to realize the symptoms we find are in relation to the Earth providing the gravity conditions and by excelling the normal motion, the conditions then would amplify the contraction.

The one part is forming a motion and the motion is attributed to the fact that the part is a liquid. By improving the motion the contraction benefits but that is because the motion that favours contraction turns the balance in the favour of motion. The higher the motion, the more the motion will overburden the expanding and reduce the solid factor. With us completely engulfed and in the control of the atmosphere we experience a complete contracting with no possible expanding.

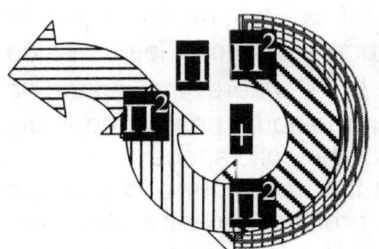

In all rotation we find that one side of the Universe contradicts the other side of the Universe. In that we find the workings and the influence the Pythagoras principles have on the foundation of the Universe. When the flow crosses the balancing of Singularity it crosses the entirety of the Universe. Then all of the principles change which makes contraction expand and expanding contract.

Only by regarding Kepler's work in its full extends can this behaviour be understood. That is the fundamental of gravity. Gravity stands controlled by singularity, which is not part of our Universe. Crossing the divide bring alternating because of Pythagoras.

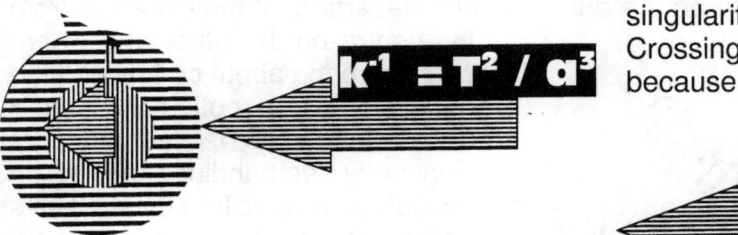

$$k^{-1} = T^2 / a^3$$

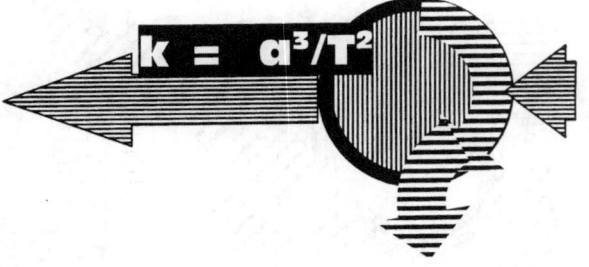

$$k = a^3 / T^2$$

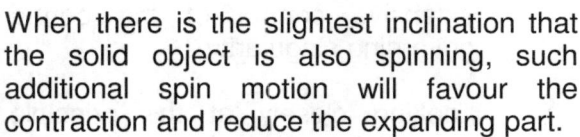

When there is the slightest inclination that the solid object is also spinning, such additional spin motion will favour the contraction and reduce the expanding part.

This is a fundamental part of the nature of motion as Kepler's formula introduces such a principle.

It is the motion of the spinning that produces and substantiates the attracting or the expanding.

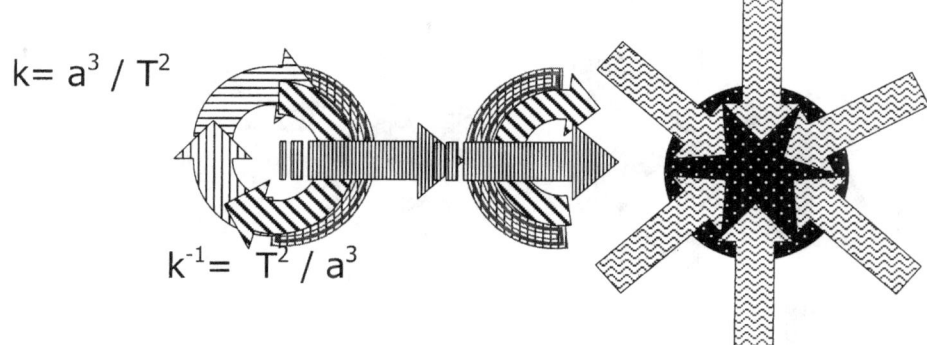

$$k = a^3 / T^2$$

$$k^{-1} = T^2 / a^3$$

With only gravity working in the motion sector will reduce the contraction to only one sector while moving the solid as well as the liquid will bring altogether favour to the contracting part.

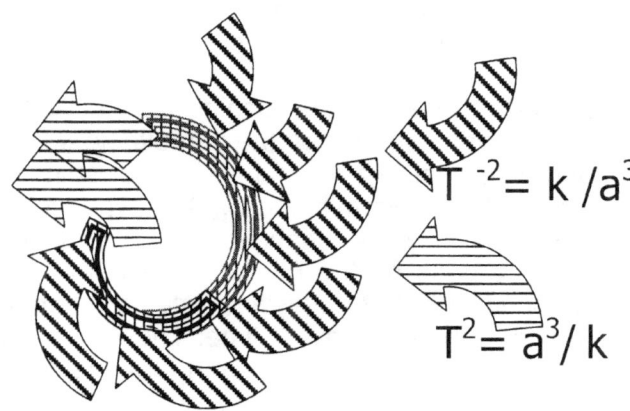

$$T^{-2} = k / a^3$$

$$T^2 = a^3 / k$$

That is gravity and that is motion. There is no force or else whatever there is then is a force. It is directional motion of a circular nature that govern the relevancy in strength as well as direction of inclining. There is an expanding to the contracting and the expanding is in direct relation to the motion of the liquid, which brings about the contraction

Even the direction of motion needs not to change because it is the relevancy when the motion crosses the divide that sets the motion apart from what it previously favoured. It is always a liquid that brings on the motion when the liquid in motion is connected to a solid that gives the liquid stability and the liquid provides the space in question with a limited definition of securing a precise border to end the space. It is a liquid in defining a solid that produces the influence of the contraction in relation to the expanding and in that we find gravity. It has no bearing on mass what so ever. It is motion of a liquid relating to the solidness securing the position of a solid.

$$k = a^3 / T^2$$

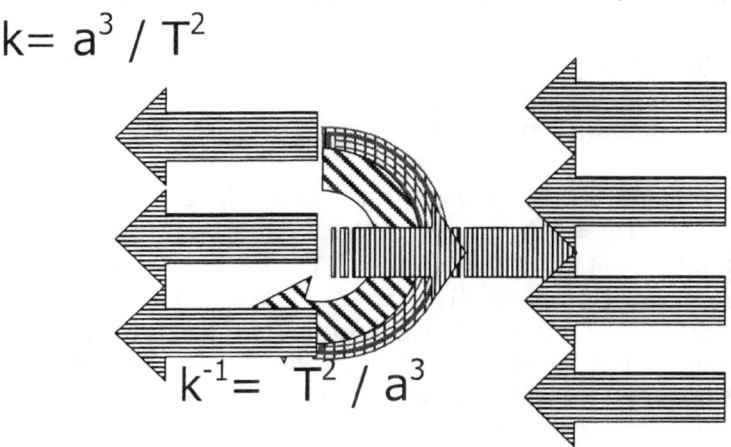

$$k^{-1} = T^2 / a^3$$

At this stage it is not important to study the direction of motion but it is vital to acknowledge that contraction comes about from one side where contraction does not totally dominate, we find expansion also part of the equation. From Newton's time we have observed the gravity from the position we hold as trapped belongings within the Earth's liquid motion.

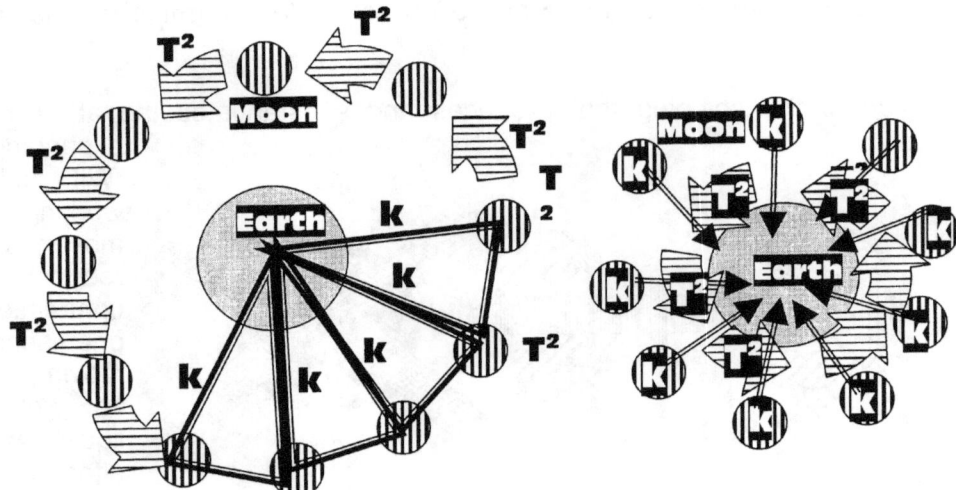

The Moon and the Earth is in a Roche connection and are sharing some minor sort of a Roche lobe with the two being slightly outside the limit. What connects the Moon and the Earth is inverse relevancies from either side.

Seen from the Earth there is a line running from the Earth to the edge of the Earth. Seen from the immovable centre, the Earth cannot spin just like the centre cannot spin and the centre connects to the Earth's edge without variation of any sort that would implicate changes in position. The centre points at one point on the Earth's horizon and that line secures a connection that never fluctuates or changes on the same point on the edge every time. The connection proves to the centre as a solid according to the Earth centre. The fact that the Earth spins around the axis is to

the axis neither here nor there because the line **k** connects to one secured point on the Earth edge at all times. While the Earth rotates, the Earth seems from the centre to be the solid stable partner, which puts the part of being in motion or liquid to the other partner that is in outer space. In relation to what applies to the earth, the moon is outer space and the Moon holds the earth in the same regard or disregard.

The Earth faces the Moon in a new allocated position every time and seen from the centre of the Earth the Moon is a liquid that is moving in ratio with the Earth time. It is holding a position of time as $4\Pi^2$, which represents time in a complete circle in relation to seven degrees of change per rotating interval. However, we are on Earth and we look at matters from such a vantage point. From the centre we find that the Moon rotates in the boundaries of time (7) $4(\Pi^0)^2$ because it is the singularity Π^0 at the centre that holds the prominence from which the cyclic days arrive. From the Moon the cycle lasts one day and from the Earth the cycle lasts 28 days. On the other hand since the moon is still part of the atmosphere of the Earth, therefore, according to the Earth we find the Earth applies another type of Roche limit being $\Pi^2/2$ in relation to the Moon and that means the Earth day gets one cycle $(\Pi^0)^2/2$ from the centre singularity divided into two entities of one cycle considering the Moon.

From the centre of the Earth the Moon is as much liquid as the space it is in and by rotating, all the motion belongs in the factor, which the Moon represents. The Earth holds the Moon in a Roche lobe and the Moon is to the Earth, just more liquid rotating. It is important to note that the Earth regards the Moon as liquid and with the Moon already being liquid the Roche limit does not come into effect. It is also very important to look at the Moon and find that the Moon is where it represents the Earth limit at the end of its atmosphere because the Moon is still in a Roche limit with the Earth while the Moon is very much acting as liquid would that carries all the moving abilities.

Seen in perspective, one arrives when taking a vantage point from the centre of the Moon, the Moon is as solid as the motionless of a solid needs to be. The Earth represents all the liquid and all the motion since the Earth is representing another point towards the Moon every time the Earth rotates.

The Moon is standing dead still as any solid requires to do because the Moon does not rotate around its axis. All the rotating belongs to the Earth, therefore the Earth represents all the motion and therefore the Earth is the liquid while the Moon represents the centre of the Universe being as steady and as sturdy as it seems to be from the Moon's centre of the Universe. The one to the other is in liquid and the one to the other is liquid. The liquid it represents is outer space it is within.

The centre of the Earth is the immobile unmoving sturdy dead solid part whereas we know the Earth is spinning like the timing gear of a Swiss clock. Again what the Earth uses to focus when it takes regard of the Moon, applies in the very same manner when regarding outer space.

The line coming from the centre of the Earth and which is going to the edge of the Earth is always connecting to one spot, which also is always the same spot. From the centre the line never shifts and therefore the line is always consequent.

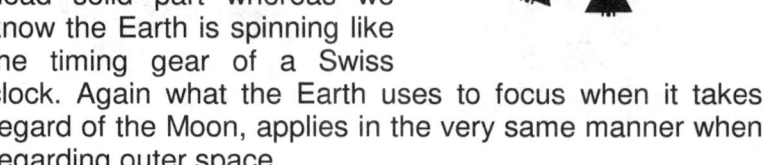

Therefore even though the outer space does move in relation to the Sun but maintains equilibrium in relation to the planets, outer space does in affect move according to the motion of the Sun but is completely motionless in regard to the position the planets have within outer space and outer space and the planets are one thing. All the planets are going as much straight ahead in relation to the Sun as they are rotating around the Sun and therefore in relation to the Sun outer space is coming towards the Sun at a pace of approximately 299 in ratio, but in relation to the planets, outer space is standing still.

From the centre of the Earth there is a line **k** that holds a position of 1^0 to 1^1 in all possible directions. However, as said the points never reaffirm new allocated positions in relation to previous positions and no change comes into place as result of change. However that part represents eternity and we know the Universe is about patricians forming in between time and moreover between eternity and infinity.

From the perspective of the planet such as the Earth, the rotating of the Earth does not implicate change in relation to the Earth centre. However change does come about and the Earth is rotating, which places a movement between the Earth and outer space. The new alliances with changing relations between the earth and outer space find focus on the part of

outer space where outer space is focussed on then being the liquid in relation to the moving solid being the planet. The planet is moving but represents the solid while outer space in accordance with the Earth is standing still but is focused as being a liquid representing the motion. The changes coming from the rotating of the Earth in relation to outer space puts the liquid emphasis on outer space because it translates singularity from the edge of the planet to the connecting point between the planet and outer space.

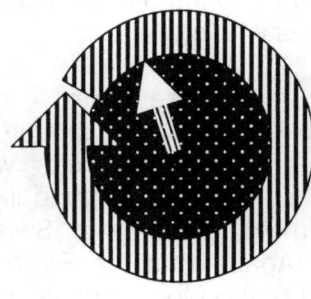

Considering that outer space then represents 1^1 in relation to the centre that is holding 1^0 the change diverts to the changing position that outer space shows in the partnership. The connecting is always 1^0 extending space-time to 1^1 and therefore the Earth or all other planets and structures for that matter hold the solid position. The planets providing motion onto outer space where the motion is projected from that which moves to that which in relation does not move transfer the solid position that the Sun gives to outer space for doing its job in housing the planets.

Therefore outer space is granted a steady position that it maintains in its relation with the planets when the Sun grants the security of a centre or performs the task of the solid. Outer space should by any stretch serve as a solid in regard to the planets.

To outer space the planets are in equilibrium by motion, therefore the partnership should not reveal a definite solid and a specific liquid. But by rotating the planet should be the factor that supports the liquid part. However in the sturdiness of the lien that comes from the centre of the planet to the edge such a solid line gives the impression of immobility and transmits such mobility to outer space, which is motionless in relation to the planet. In the relevancy there is between the Moon and the Earth the Earth moves in a circle and the moon moves in an orbit. This is very significant and is important to use as an example to understand how the cosmos functions. In relation to the centre of the Earth the Moon is changing positions all the time while the Earth is locked in one position that does not change.

The Earth stands steady while the moon is orbiting around the Earth. The Earth is the solid

while the Moon is a part of the liquid. Then from the vantage that the Moon centre singularity holds, the Moon is standing still because the Moon is not rotating around a personal axis. That puts the Moon in relation to the Earth as being away from the Earth while directly still linked to the Earth and although the Moon is many miles from the Earth the moon is in cosmic terms part of the Earth. What this implies is that Man has never yet left the Earth because the singularity which man is attaching to has never crossed any divide. This is a very serious obstacle in the road those anticipating future ventures to Mars or outrageous planetary visits. Going to Mars is not just crossing the time line from Africa to America, but poses serious questions of safety and human sanity and cannot be classed in the same category as getting a bit of jet lag. It is a life threatening issue, which should seriously be studied with animal experiments. I am seriously of the opinion that by going out of the atmosphere interferes with human thinking and behaviour and those that are in the future planning on trips for other adventurers and explorers to Mars will play with the sanity of those on voyage to explore. Man thinks by using the measure of Earth gravity.

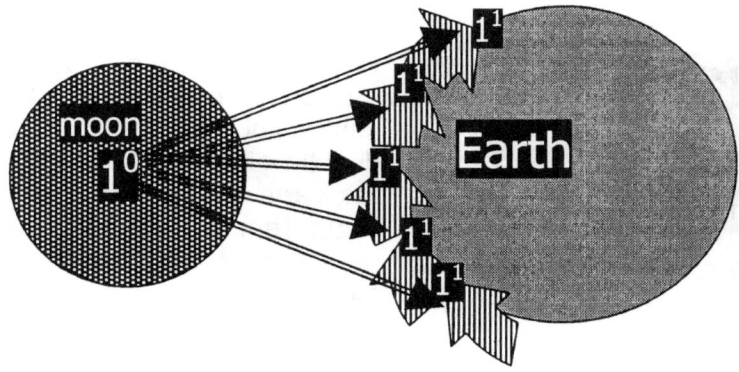

The Moon is relatively standing still 1^0 while the Earth is rotating or applying motion 1^1 The Moon is solid 1^0 while the Earth is liquid 1^1

The Earth however, is rearranging its position constantly by providing motion, which is realigning with the Moon centre by every rotation motion in the minutes of moving. In this instance again the Moon is the solid while the Earth forms the liquid by securing a motion free centre in relation to the Moon and the Earth motion.

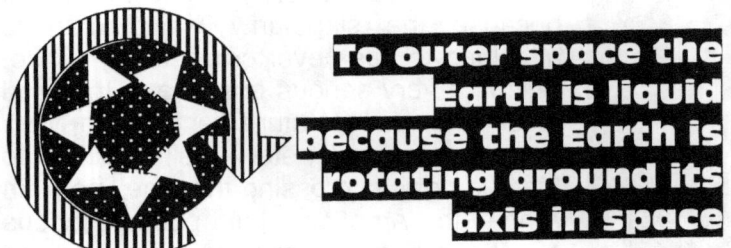

The Earth is relatively standing still 1^0 while the Moon is rotating or applying motion 1^1 The Earth is solid 1^0 while the Moon is liquid 1^1

Being a liquid or a solid depends on what provides the anchor or pivotal role and what provides the motion within the relation. The planet moves in orbit as well as around its axis. This duel capacity is all motion while it is also all solidity.

To outer space the Earth is liquid because the Earth is rotating around its axis in space

The cyclic rotation forms a liquid in relation to the point where the Earth meets outer space, but since the Earth at that point is connecting to singularity by seven, it is outer space that then carries the motion at the point. The contact point at the earth's end remains the same although it moves because it is solid, but the outer space are show a new point in relation every time although it is steady.

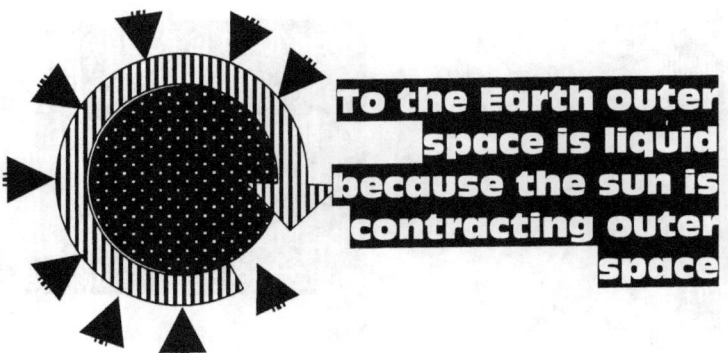

To the Earth outer space is liquid because the sun is contracting outer space

To the Earth it is outer space that shifts while in fact it is the Earth that rotates but that rotation does not affect the centre since the centre remains directly aligned with the edge of the Earth solidity. However, the Earth renews a contact point with outer space, which is the motionless part in the relation at that point but serves as the changing factor in the relation.

With the Earth rotating while the Earth considered its position as stable and solid motion is reflected to outer space, which at that point is solid. Outer space, which is stable, is facing a new position in relation to the Earth but since it is not part of the solid structure of the Earth, outer space shows changes and diverts its position in ratio to the centre of the Earth. This ratio gives outer space the liquid partnership.

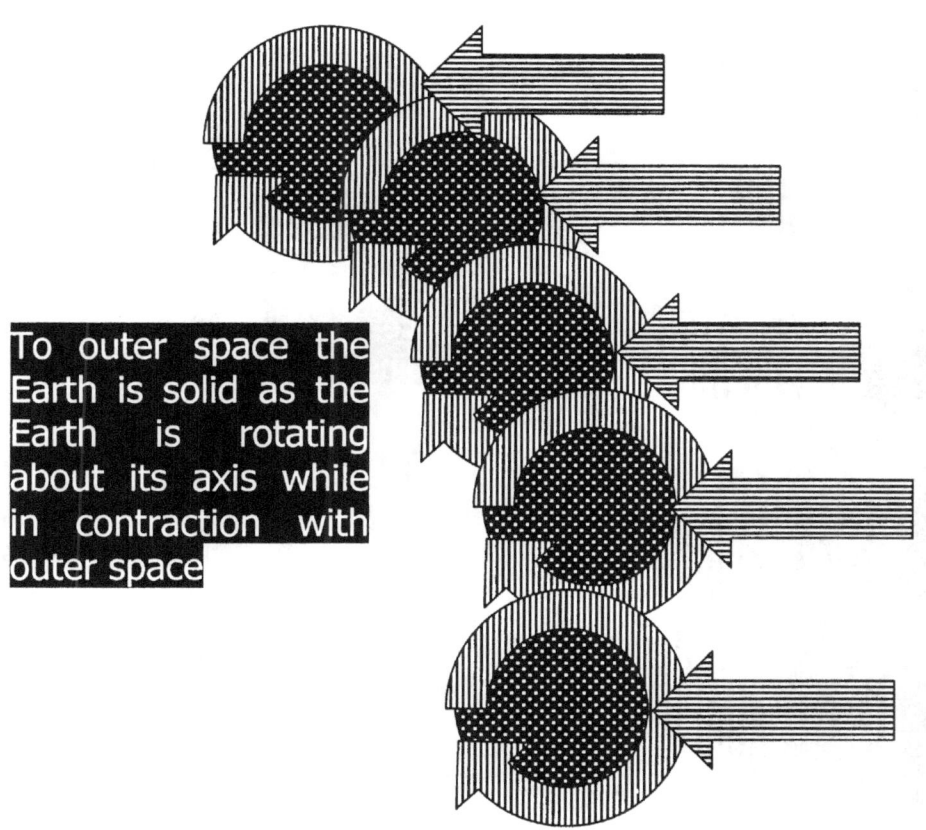

To outer space the Earth is solid as the Earth is rotating about its axis while in contraction with outer space

But then the Earth shows another side in the affair where the Earth in orbit tears through outer space being without motion. This brings about that outer space will allow the Earth to move and while the Earth is moving, the Earth is aligning with the centre of the Sun. To outer space the Sun represents all that can be stable and in that regard it takes the Earth as another solid principle. In that way the Earth again proceeds as the solid or stable factor while outer space holds the motion.

The Earth centre holds the surface we walk on, which is the edge of the material that connects the inner singularity to time, as being solid. We know the Earth rotates but from the vantage that singularity governing forms, the Earth is not moving, but is constantly being solid. As the Earth rotates, there remains a fixed position that finds connection with the centre singularity. That point remains locked to singularity. In relation to the point in singularity that connects to the point on the surface, there is no movement detected or any variation of whatever means. In relation to the centre of the Earth, the Earth surface remains immovable and solid. However, one micro millimetre away from the surface the singularity centre detects a different allocated space forming time. That point always shows a different position in relation to the rest of the cosmos.

This point that is not connected to the surface of the Earth by a fixture, is different every time. Therefore, according to the centre of the Earth singularity, space is moving and from that it takes space to form the liquid. Although we know it is the Earth that is spinning, in relation to outer space it is the space on the outside of the Earth surface being past the edge of the soil that is the movable part and singularity treats that part as such. With time outside the solid that is always shifting, the gravity we find applies in the part the Earth holds as the moving part. The gravity we have does not apply to an object that is within the solidity of the Earth's structure and therefore tombs like in mummifications for instance degenerate by many times over and many times faster when opened after centuries of being covered. The area where Galileo saw all objects fall at an equal pace, is the area that the Earth applies gravity, because although the object is falling straight down to one particular point in relation to our observation, the falling object relocates every instant to another position in relation to the centre of the Earth. That makes the water that is a liquid act as if it is a solid and the solid object that is falling into the liquid water being treated as if the solid falling object is a liquid.

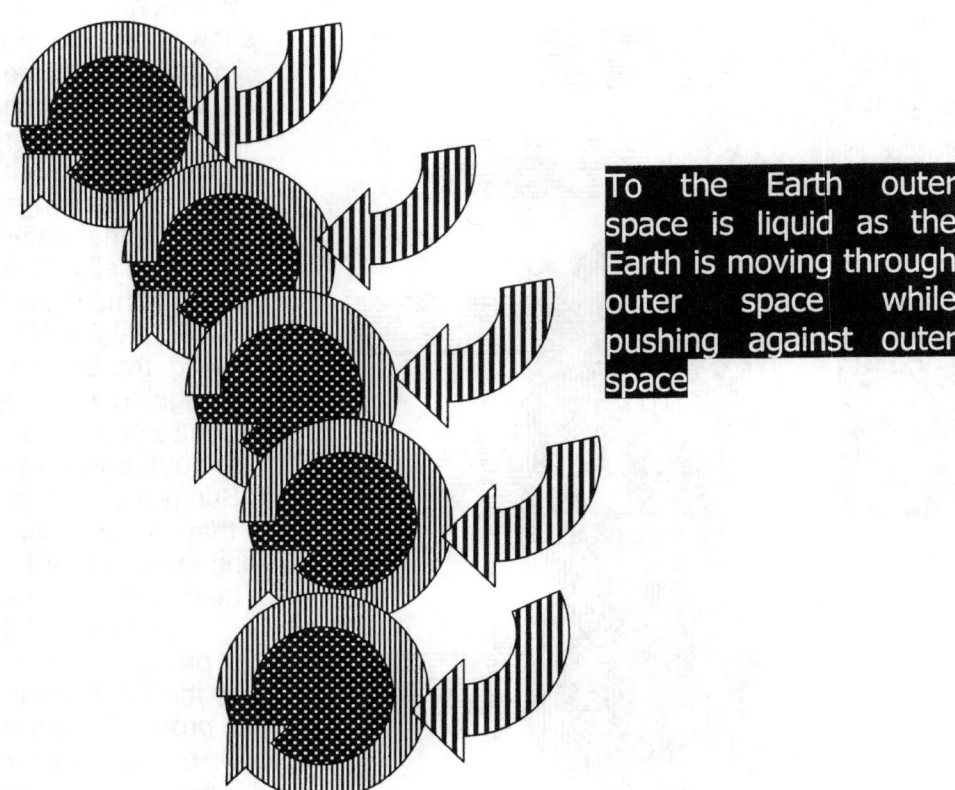

To the Earth outer space is liquid as the Earth is moving through outer space while pushing against outer space

That is not yet the end of the Coanda partnership. The Earth is pushing against the outer space and while outer space is inherently a liquid, outer space gives road to the moving Earth. Therefore while it is the Earth that pushes against outer space it is outer space that is moving away to allow the Earth the motion it insists to have. Therefore again outer space is moving in relation to the Earth which is pretending to be solid and the total result from all the activity is that outer space seems to move at a rate of 10 X 10 X gravity which is Π^2 and that gives space-time a value of space (a^3) / time (T^2) is (=) 298 but I shall come back to this when much more information is exchanged.

The whole debate now in this part resonates around the fact that mass never comes into the argument and mass is no factor in the Universe. It is all about motion and being solid (not moving) while the other party in the equation is moving (being the liquid). It is as Kepler said $a^3 = T^2 k$ or then we find that $k = a^3 / T^2$ and $k^{-1} = T^2 / a^3$.

My first nut I cracked in cosmology as an individual standing apart from what I was reading about cosmology through the avenue of mainstream science, was concerning motion in relation to the speed of light. Today the fact that it took me a full six months to solve is a joke. That it took me so long to get to such a simple answer is in hindsight, not very complimentary but please keep in mind that at the time I had a blank paper in front of me, which was blank in more than one way. However, that was what set me on the way to be able to crack the first code. I must admit if I did not break the seal I would be totally lost in cosmology but still such a simple solution took me six months of head breaking arguments with myself. Einstein said that if light were travelling at the speed of light for one year it would be away from the source of origin by a distance of one year. That is acceptable even to a person with my mental capacity. Then came the jawbreaker.

The two photons travelling in opposing directions will also at that instant be at a distance of one year apart. It takes one light year to go in one direction while it takes the other photon one year to move in the other and opposite direction and yet the two are one year from the light source it left while also being one year apart from each other. That baffled me into almost madness. It was just way above what I could mentally cope with. The two photons opposed each other while travelling but at the same time moved apart only by one light year of total motion. The total that should add to a double was the same as the single, which was the same as three totally different points. Something told me in this was the key to understanding cosmology.

Then one day the simplicity about the whole argument hit me between the eyes like a ton of bricks. I was staring at outer space while viewing outer space as a distance. Outer space is time and not distance in as much as forming space. It takes space a^3 time T^2 to bridge time k^1 and reach space a^3. The time it takes a^3 is in relation to move at time T^2 to cross time k^1. We all confuse space and time. Time is what is between the Moon and the Earth and not the distance. The time lapse was one year and therefore the time that parted the objects and the source of origin was one year but also the time that parted the two photons travelling merrily was one year.

Light travelled in a straight line as well as a half circle and where light is connecting to time the half circles 180^0 is equal to the straight lines 180^0 which is equal to the triangles 180^0 that means outer space has nothing to do with space but is all about time while time is all about motion. That is the key to solving the riddle we call cosmology and by using that key I found a way to unlock so many answers. The light flowed in a straight line, which is 180^0 while they move apart by a half circle also to the value of 180^0 and while being in a triangle position in relation to the point wherefrom they came which was also a value of 180^0.

Time was moving and if time was moving then time has to be liquid. Since time and space is not the same space then space has to be a solid holding time as a liquid in relevance and knowing that, it is time that is between the Earth and the Moon it made that which is between the Earth and the Moon the liquid and that made the Earth and the Moon solids. How simple can everything be if one takes the correct line of arguing? Even I, being who I am, could start to understand what everyone should understand. I feel obligated to explain my referring to myself in the position I have. When saying this about myself I must request that you should never forget while reading, that I am only a motor mechanic and that is all I can ever be.

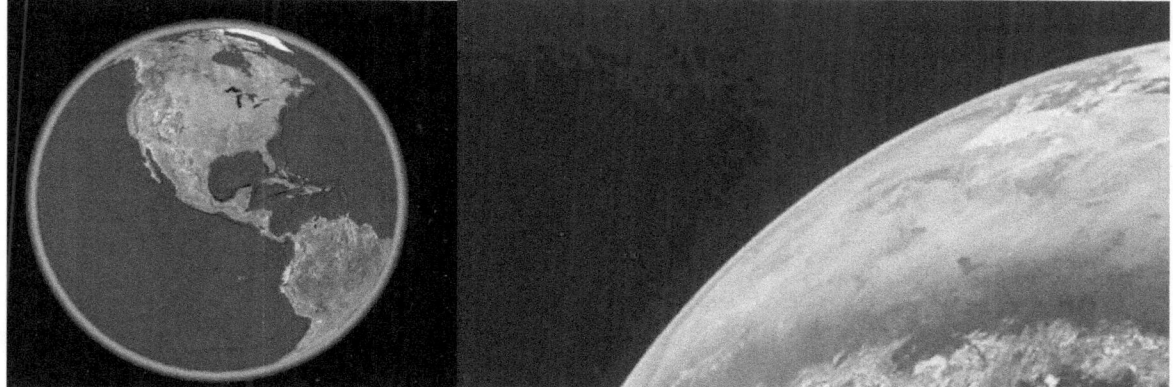

We are part of a thin top layer, the part that is contracted without really being part but only part as an extended part of the Earth. That is part of Newton's gravity where the liquid we are in and the liquid that we are becomes part of the Earth as it is confined to the Earth by the atmosphere of the Earth that forms the liquid restricting us to the Earth. That is why Newton and his apple were on the ground very much secured. The intensity of the increased motion that the Earth provides when cooling the time expanded region we call outer space brings about that the intensity increases as time shifts in relation to what the earth cools with motion and shut cooling is the result we find in the space we refer to as the atmosphere. However the region called the atmosphere does not heat up as science wish to see it but the heat reduces. This simply becomes a factor because anything that decrease the space it holds does cool down and anything that heats up increase the volumetric space it holds. Hot and cold has nothing to do with temperature and al to do with volumetric space held in relevant measure.

We are in the motion, which forms a liquid. We are part of the liquid because we are part of what moves. Every aspect surrounding us brings proof that we are contained in liquid and we are preserves as being part of the liquid. Where the wind blows, it indicates a wave pattern and a liquid leaves a wave pattern. When looking at a mirage, we distinctly see that where the atmosphere becomes more dense as the heat at that point becomes more intense, that what we see and which we call a mirage is water, is a liquid substance floating in waves where the concentration in density varies. The liquid engulfs us and in being part of the liquid we are experiencing only the contracting aspect since we are totally secured by and in the motion the Earth provides. The earth atmosphere puts us on the ground as the atmosphere secures our positions on Earth. We are part of the $k^{-1} = T^2 / a^3$ while the earth being in motion and providing the motion forms $k = a^3 / T^2$. That is the time aspect, which the Earth provides and that time aspect enable us to read the time by the measure of Galileo's pendulum.

$$k = a^3 / T^2$$

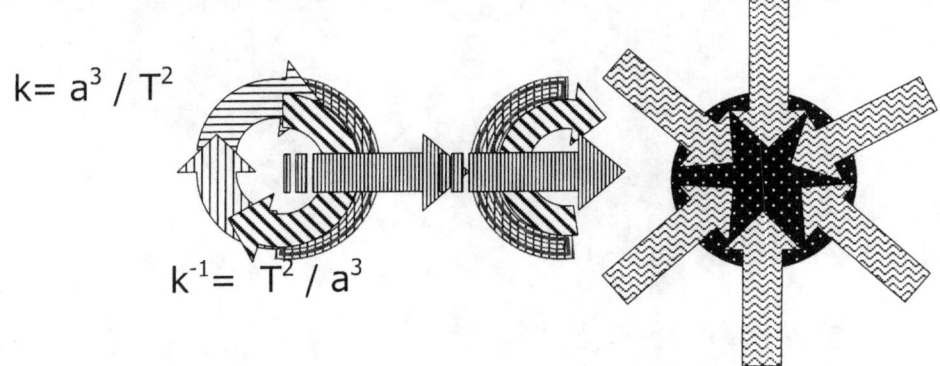

$$k^{-1} = T^2 / a^3$$

By the motion in rotation as well as the motion in the lateral, we find that the liquid being air confines us. It is not the particles in the air that is the liquid, but the substance separating the particles in the sir that form the liquid. The Sun is, by moving around its axis, confirming the singularity governing and while the planet axis is moving around the Sun, this process is putting everything confined to the Earth as liquid relating to the Sun. However, in the process of forming a liquid by motion, that puts every aspect connected by mass to the Earth as a solid part of the Earth because it accepts the Earth motion, and by being mass, the mass is confining the structure to a solid, as it is forming part of the solid Earth and it is that, which confines us to a position that only Newtonian gravity, apply. There is only one direction flowing towards the centre of the **Earth. That flow is the space a^3 that the liquid T^2 secures k to the space $k = a^3 / T^2$.** In this however there are no grounds that support the suggestion that mass is producing the motion or that mass is indicating the flow.

By observing a fire we can see where and how the flames bring intensity to the heat in the air. The flames are the densest form that heat can have while being in a liquid form. The flames are so dense a liquid it provides light and light is pure heat in minute quantities of liquid space. In the picture we see three forms of liquid heat where in each case the element responsible for producing the liquid heat contains the heat in a different form. We see the flames souring and that is the responsibility of the nitrogen where the nitrogen being$_7$ expands heat into space. Then there is the oxygen that contains the heat in a dense material we think of as smoke. The next we can see is the wood or carbon$_6$ that keeps the heat contained in the particles. Every layer in a star has this duty that the substance hold in providing and managing the heat within the stars structure. Taking the idea of air (not particles in the air but atmospheric substance containing the particles in the air) to further proof of finding a liquid we again have to go to the Coanda principal.

This distinct relation between two forms of space-time being material and heat has gone unnoticed in principle by science because they are totally mesmerised by Newton's mass fantasy. In the Coanda effect two forms oppose one another in order to join the cosmos by motion $a^3 = T^2k$. It is not only the aircraft using the principle to lift off and fly, which allows the aircraft to fly as it applies to airflow over the wing, but it also applies with the rotational turning of the impeller and the turbine engine that uses this principle in order to generate motion to fly.

The creating of singularity by infinite dimensions and then putting material between infinity and eternity is how gravity is generated throughout the cosmos. But all this has gone undetected because all Newtonians have this Newton mass blind spot through which they only can see nothing. Again they have come up with the most elaborate names, which is another way that shows how far Newtonians miss the point. The thrust the engine generates is just the way the turbine transform heat concentration to space expanding and in the conversion of heat to space the translation promotes the gravity of the craft to a level higher and much higher than what the Earth gravity perform. It still uses the Earth gravity, but goes above that ability and that is what gives the turbine engine "thrust" to perform and sustain motion on a level above of what the Earth's performance delivers.

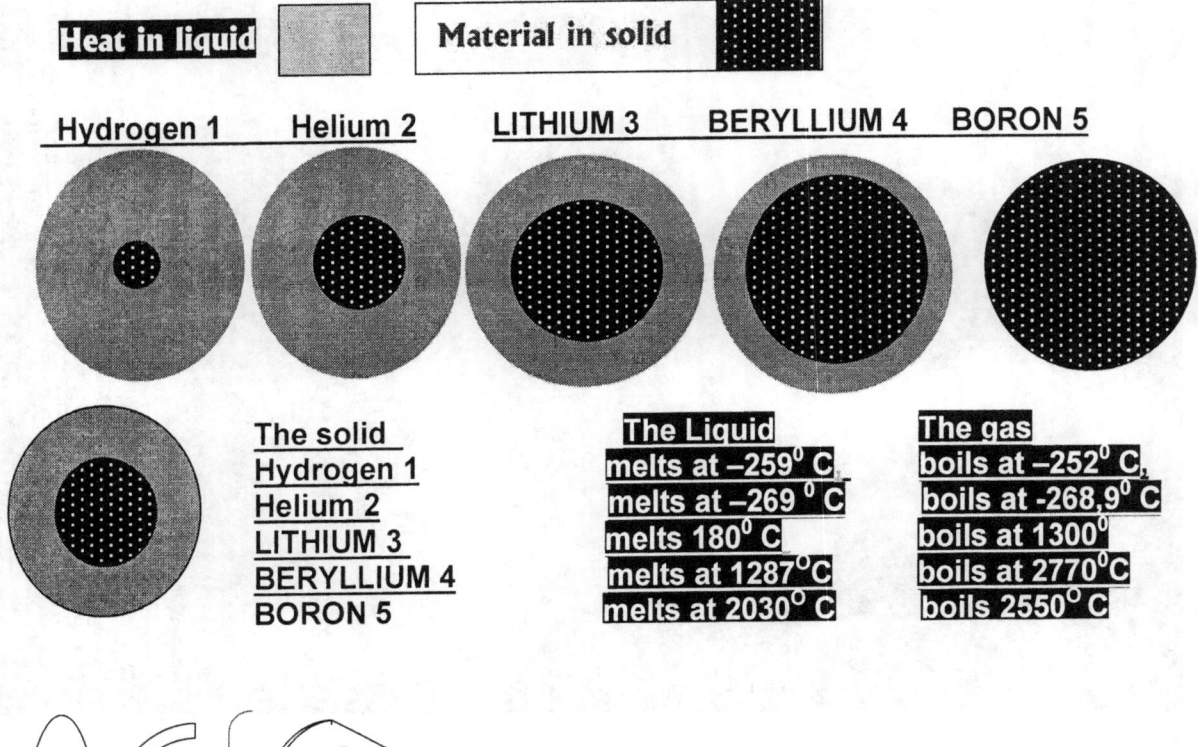

Heat in liquid   **Material in solid**

Hydrogen 1   Helium 2   LITHIUM 3   BERYLLIUM 4   BORON 5

The solid	**The Liquid**	**The gas**
Hydrogen 1	**melts at −259° C**	**boils at −252° C**
Helium 2	**melts at −269° C**	**boils at -268,9° C**
LITHIUM 3	**melts 180° C**	**boils at 1300°**
BERYLLIUM 4	**melts at 1287°C**	**boils at 2770°C**
BORON 5	**melts at 2030° C**	**boils 2550° C**

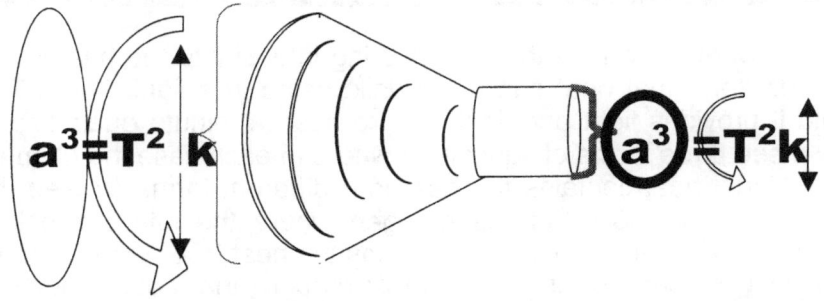

$$a^3 = T^2 k \qquad a^3 = T^2 k$$

The particle (atom) secures the solid basis that provides the motion, which enables us to categorise elements according to our perception. It is not the truth or cosmic reality but it is our perception through culture that teach us about gasses, liquids and solids, about noble gasses and heavy metals and non ferromagnetic or good conducting and immeasurable other characteristics we attach to elements except what truly is important as far as

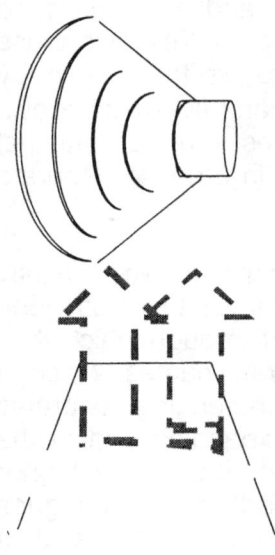

cosmology goes. All material are solids as much as they are gasses or liquids. It depend on the concentration of the heat surrounding the element in that particular element that turns the element at that point and time (temperature) to be either a solid, a liquid or a gas. The state that the element is in is a response to the conditions, which the heat levels bring on, and that is a response to the Coanda gravity motion that serves the atom as a liquid at the moment of response.

However when saying that it is a liquid in motion around a solid, we also find that the principle drives turbine engines and the air compresses within the turbine to cause heat. It is the air that flows when driven by the turbine rotor, which provides the solid. Then from that we must deduct that the air producing the flames forms the liquid, which provides the contraction when the solid spins the contracting turbine and the spinning turbine serves as the solid part.

The Coanda principle works in two parts where there has to be a solid securing space and there has to be a liquid performing the motion that results in contraction. The ingredient is about a liquid in motion T^2 capping a limit or an end k or k^{-1} where the space $k = a^3/T^2$ holding the solid extends to appreciate space $k^{-1} = T^2/a^3$ that holds the liquid. It reduces space by motion providing contraction to a point where the space goes liquid in finding flames. That is what Henri Coanda first saw when he tested his new enclosed propeller. It is all so exciting but what the Newtonians of the day and those today completely misses is that for the Coanda effect to be functional, one needs a solid and a liquid. The solid part the rotary propeller provides and with such motion it contracts the atmosphere into a compressed flaming liquid. The atmosphere (not the particles in the atmosphere) compresses to become a liquid.

That is the conditions applying to the atom, which provides the conditions applying in the atom. We differentiate the particles by giving every part a name and try to find meaning in the particle combination. In a hundred years from now we are about to "discover" another million smaller particles and in a thousand years from now there waits another billion more, which are all smaller to be discovered. We will go on discovering until eternity once more meets infinity and still we would not trace and name them all. It is what the combination provides each other in supplying space-time that finds importance and not naming the lot individually.

The major function that the combination of the electron provides is the expanding and contracting of solids in relation to time being a liquid that has a flow and a direction of flow. The atom fills a solid as the atom forms from seven points (the sphere in the past) through seven points (the sphere in the present) towards and into seven points (the sphere in the future) Three times seven plus singularity in relation to seven is the ultimate circle.

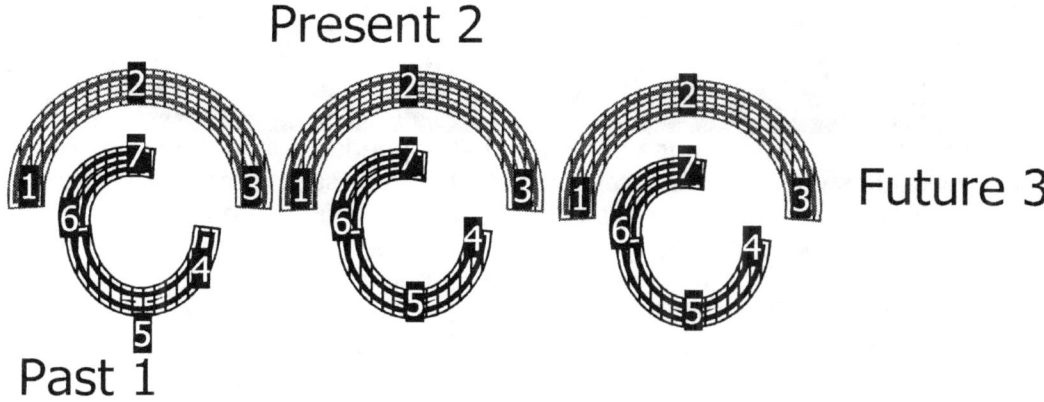

Present 2

Future 3

Past 1

It is the everlasting flow of time from the sphere in seven points in the sphere of seven points and into the sphere of seven points that establish the electron time frame of three. It is time duplicating the seven of the sphere by the repeating of three times. The proton is $(\Pi^2+\Pi^2)$ which is working as a solid in relation to the liquid $(\Pi^2\Pi)$ that the neutron provides and this flowing of seven Π through the repeat of three in time duplicating the atom leaves the distinct displacement relation of 1836.

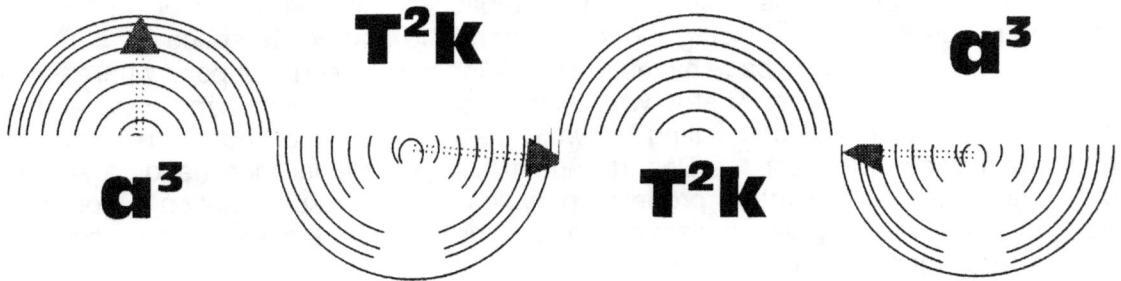

It then becomes pertinent to find what is liquid and what is a solid in cosmology. The first aspect we have to abandon is our preconceived notion about what liquids are and what solids are. Everything that moves or forms part of that which moves or may move while another in relation is standing still becomes that which is a liquid while the part forming the motionless factor then becomes the solid. From that gravity becomes reality and gravity is motion that has no bearing on mass in any way possible.

Looking at the night sky I see images of events that happened a long time ago. To my vision it happens now at this minute in the present. Yet what I am seeing most probably happened before there was an Earth or a Solar system we can use. The picture I see is in my past and was part of a time that went by a long time ago. It is there in my present but it is not in my present because that which is in my present while being where the light originates would ultimately be very different from what there now is to what I am seeing while I am looking at it as if it is the present. The light connecting to me is in my present and is reminding me of what was, but that has validity too me where I now am in the location I now fill while I fill the present space in which I am. The light came all the way representing what is afar as it was when the light left, while what there is over there in the immediate the reality changed a lot while what was left over there remained where they were. That tells me there are two times in relation but also opposing each other. There are allocated time that represents the past and there are the present that confirms the past. The fact that the light is there confirms my future because the future represents the fact that the light is there. The light of the Sun is confirming the Sun but the light that hits me was confirmed as a reality most probably 10 million years ago. Then the fact that there is light confirms my future for at least the next 10 million years because if the light that hits me is 10 million years old, the light entering the Universe now confirms time for the next ten million years. The light developing within the Sun in my present second I now witness is the light hitting me in a time span of ten million years from now. The light is not only in my present but on another place, but also is my future at the place I now am coming to me through time that will last the next ten million years while the light will be a constant present in my future for at least the next ten million years. The Sun being there confirms my future as it represents my past for at least the last ten million years. The fact that the light is there confirms time spinning while the light hitting me confirms time contracting and all in between confirms my present during the past ten million years while also standing in as my future. I can see my past coming towards me in my present that never stops in the form of light travelling to me through time and announcing what my past and my future represents in my present. It is my past. Let us give it a value of 1^0 and it is my future where the future value is 1^1. I can look at it ands say I see my past or I can look at it and see my future. If I look at it as my past, it is sturdy and can never change 1^0. I can also look at it as 1^1 and see as an ever-moving ever-changing future while I am in my present. The connotation I have to it in the time frame depends on the motion I attach to the idea I have about what I see. That which is secured and does not move allows that which always move to come to what never moves. That what always move represent my constant past present and future by confirming time in motion and what time is drawing towards me represents my future being destroyed by my present as my past. The time can be the solid in the Coanda effect that never moves and time can be in the Coanda effect that for ever moves towards me as well as changing all the time in relation to me. This rules physics to the extreme but Newtonian physics have no room to accommodate this thought. This is the ultimate measure in physics of physics.

The Coanda effect is the motion of liquid in relation to solids. That it is but also it is so much more because it is the relevance of the present being in motion where liquid is moving through time in relation to solids. Allow me to explain in this way. If I look outside I see the night sky. That which I see is the past. The light tells about some event, which is what happened during my past. It brings to my present the past events and while it happened in my past the learning about it is occurring during my present putting the past of that which happened to my immediate present. That past is also my continuing present because the light that tells the story is oncoming from that which is my future. If the light stops coming my way I have no future so the oncoming light represents my future of what is the past. That past is coming to my present forming my future. While being in the past it is in my present that forms my future that is ahead of me. That which the light tells me happened at the point it happened when other things happened where I am. My planet might have been here but was in a total different state. That makes what I see my past and while it is my past I am witnessing my past as my continuous present. While the light is on its way coming towards me that light might be in my continuous present coming from my past as it is going into my future where being in my future it remains in my continuous

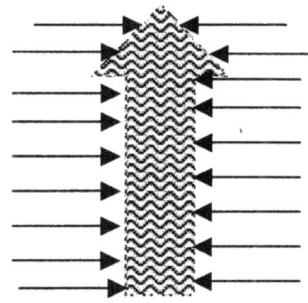

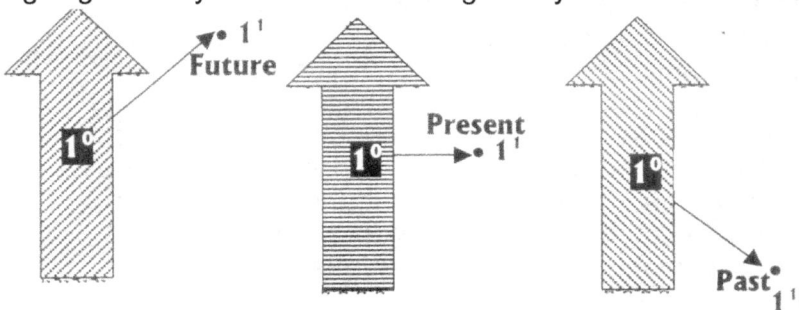

present.

This sounds a lot like high philosophy but while it is philosophy it is also exactly what forms the core of mathematical physics. When an object moves the object is having a contact point holding relevance to a solid while the liquid moves. Being in motion puts a future holding a point whereto the motion is heading and a past where the point discontinued. The light moving through outer space heads my way. I am the light's future and the light heading my way is my future. The light travelling towards me puts points in a line-up that serve as my continuous present running from my past as my future. The light coming towards me becomes my past while in the present the light must be absorbed by my singularity and being absorbed and accommodated by my singularity it then forms part of my infinite past. Then that reference point placed the future within the present that forms the eternal past in infinity. It then becomes the end of the line for the future eternal present. Time caught up with time as time then contains time.

Being the solid presents a point where the motion will come and take the point from the past through the present past the future. Those are time points we observe as light but with light reflecting from the object, the object as a cosmic structure presents a reality that forms the centre of the Universe on own merit. As a point of reference and being an isolated Universe the structure remains part of the continuous present moving by motion onto our past, while it is forming part of the continuous present. It remains part of the Universe as a Universe.

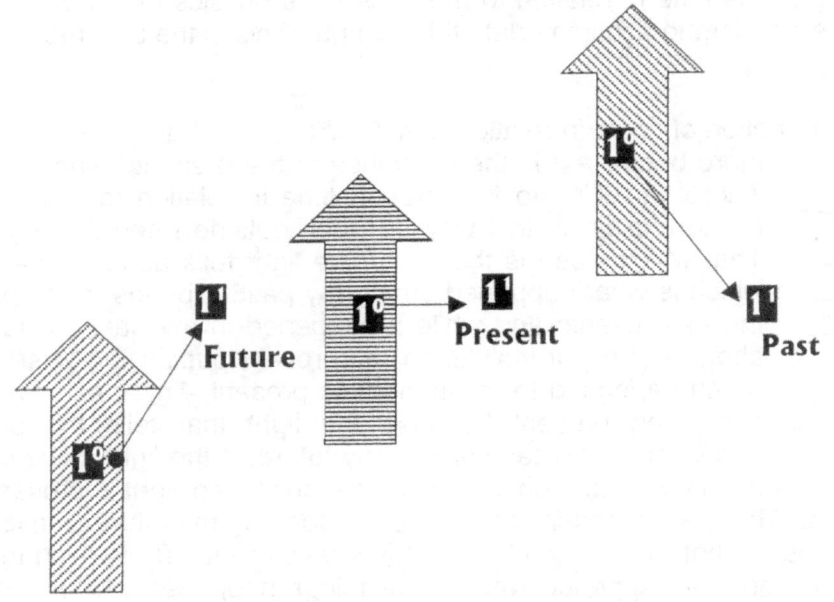

By moving the object is travelling both in time and also through time. Space is splitting time while the space is pure time delay being in between time in eternity and time in infinity. It contracts time into infinity while space expands into eternity. That is why it is space-time having time on both sides of space. Space is that which parts that which cannot part because that which space parts has no inside and it part that from that which has no outside and therefore space has two time components. The movement of space in time through time constitutes to the position space has in relation to time. Space forms as a back log of time placing eternity apart from infinity ($a^3 = T^2k$) but in order to achieve that it has to move to connect time by parting time. In the Black Hole there is no more space that is parting time from time and without space

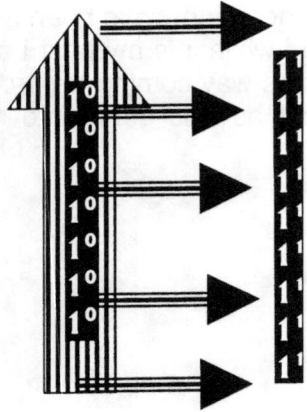

moving eternity is catching infinity.

By moving through time in time we have a relation forming between time in infinity and time in eternity. Space has to duplicate through eternity while maintaining to contract and sustain infinity as not to allow infinity to overheat. To stop infinity overheating it has to allow that which has no start to feed off that, which has no end and allow that with no end to slowly unite with that which has no start ($k^0 = a^3 / T^2k$) When we sit in a moving car the car represents the solid that is moving by never moving at all and the space outside the car is the time that is always moving by remaining stationary.

I have been through the part of life not forming part of the cosmos and I shall repeat it further on but at this point I again have to reiterate about life. Only life allows motion that is apart from cosmic motion.

Only life can extend gravity beyond the space in time retarding. It makes life special and scares, so scares it is only part of Earth. Because of lies and dogma can science portrait life as a part of the cosmos, but to do that they have to rubbish reality and drive the lie into a farce to do so. Only when they can show planets holding life in other regions do they have grounds to put life as a cosmic reality anywhere else but on Earth. Until then their portraying of life as a cosmic substance is as mad as their view about gravity and mass being all over. The following scenario is a cosmic reality, but in the manner we find and use it, is much more part of life than when it is applying in the cosmos. The motion of a car is part of life and therefore the next example has to be seen in such a context. The flow of a river holds life and although we may see cosmos behaviour in it, it only applies on such a small scale as we find on earth where life also is part of Earth and water. There is no river flowing on any planet and just because there is no liquid of sorts on any planet. Therefore there is no defining motion other than that the planet moves through space.

There is motion on solid planets as there is motion on liquid planets or Micro Stars. The motion we find on this structure contributes to liquids acting in relation to solids where the wind blowing or the gas flowing forms part of the concentrated time and that stands correlated to the solid not moving. Where wind blows dust and the dust moves the moving of the dust is not deemed as solid particles but all the particles form part of the liquid substance. The dust therefore is not solid but is liquid. Winds blow but while dust moves the dust is not solid but liquid.

In the cosmos we are not a reality but just more time delay. Life is the ability to manipulate space-time and it does so by motion. Life holds a body intact but the body in a form which life is holding is not a reality. As soon as life as motion departs from the body the body distorts and decay sets in where the decay is returning the material to a cosmic form of different atoms. After life departs the body disintegrates to atoms. Therefore the body is not reality but a function that life contributes. Therefore once the body leaves the unreality of a form it held in tact and in a combined form, life moves into reality where time is eternal and infinite. That time is indestructible so therefore that form of time is reality. The fact that life has the ability to compile and use that, which is not reality, makes that, which forms the drive, called life an eternal reality. Moving from point to point and without planned reference only belongs to life but life also leaves the body liquid. After death rigor mortis sets in and the body stiffens to become solid. The solid state that the body forms after death shows that the liquid of life being that which contributes to the motion which life in essences is, has vacated and the body became just another normal cosmic object with no special features. This part about life requires a lot more discussion in order to define the argument and therefore there are other books in which I do extensively elaborate on this view in order to cover the entire complexity of the issue. Those atheists wishing to distort what I say with such a brief introduction, to those I give the challenge; take me on about the information and arguments I make in books I wrote with such intentions in mind.

The point I wish to make is about life and while it is also

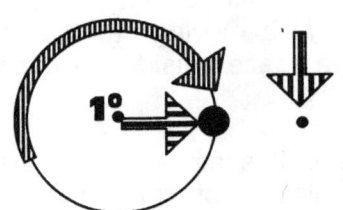 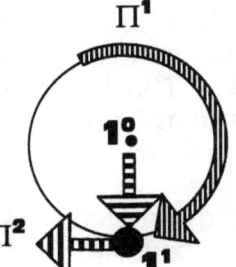 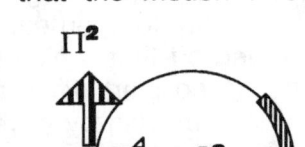

that the motion I refer to is an action part of the larger cosmic concept, the way we find and use it regards life in particular. The whole debate now in this part resonates around the fact that mass

never comes into the argument and mass is no factor in the Universe.

It is all about motion and being solid (not moving) while the other party in the equation is moving (being the liquid). It is as Kepler said $a^3 = T^2 k$ or then we find that $k = a^3 / T^2$ and $k^{-1} = T^2 / a^3$.

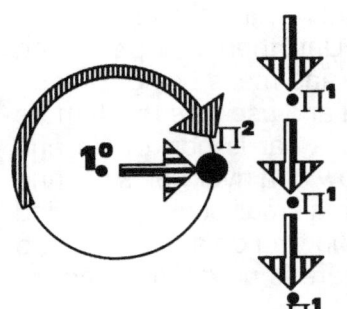

That is why in all cases the ratio of orbit in space-time is the same. The planet rotating outer space is in relation to the liquid Sun gives outer space. vantage point that outer the orbiting object is in a while performing as the the partnership in relation

in the zone of movement in position the From the space holds, space ratio liquid partner in to the motion

the Sun. This is because the Sun is holding the position of the solid and that allows outer space with all in outer space to be the liquid. The motion is a result of holding singularity within the Sun steady as a solid while all in outer space is constantly repositioning and moving through outer space which then allows outer space to hold the object singularity as being a liquid to the Sun having the solid reference point while placing the liquid part over to outer space.

The Sun is moving but this movement the Sun reflects onto outer space while the Sun affirms its position as a solid. This is how it is done. We know that it is the Earth and the objects secured to the Earth that performs in the role as solid that is in motion. The Earth with all it has as mass that is connected to the Earth is in motion. The Earth spins around its axis while the Earth also spins around the Sun by using the Sun as another axis. On the other hand outer space seems rather unmoving and stationary because the only motion one may detect that outer space has is $k = a^3 / T^2$, which is the small part it expands in density loss as it contracts $k^{-1} = T^2 / a^3$ and flows increasingly towards the Sun by density loss

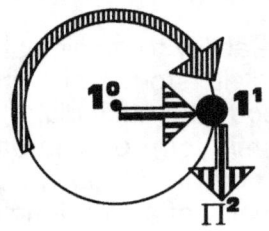 

While we know and we can see that it is the solid structure (the Earth) that is the prime moving factor in the relation, it still is the centre or the governing singularity that rotates while it cannot move and having a fixed unit of reference with the expanding singularity that forms a point on the edge of the Earth surface. It is that fact that totally contradicts the issue. Any piece which has mass is occupying a position on the surface of the solid (Earth) and the position that the occupying spot has in relation to the governing singularity 1^0 is forming the expanding singularity 1^1, which never moves but stays connected to the centre singularity. All this applies while the centre singularity can't move.

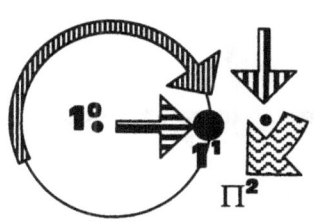

However, the solid may move as we may witness, the expanding spot of the solid is always in a firm position in relation to the centre. The position that the occupying spot has on the edge of the Earth surface and thus forming the end of the Earth or 1^1 in relation to the centre 1^0, is one of forming part of the what the solid entertains as an immovable factor and being part of the solid, it is therefore blessed with having mass. Granting the object mass is a deed by which the Earth entertains the object as forming part of that which cannot move.

In contrast to such a perception, cosmic law forms another reality that is not the way we normally see the relation. Since the spot is constantly in a firm relation to singularity governing at the centre, it is outer space that is constantly showing a new point in the relation.

As the solid turns, the solid part is in contact with the centre and thus it is unmovable, but outer space shows a constant change in reference point to singularity and therefore singularity is always changing at points holding outer space. In that way outer space forms the liquid in the relation and then has

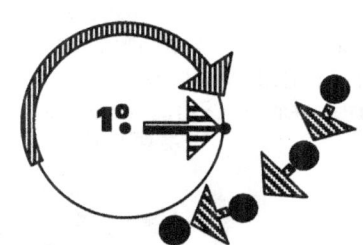

the four cubes with five sides that is in relation to singularity 1^0 = .99999999 and singularity expanding 1^1 = 1. In that the liquid that is the moving part forms the 21.9999 that stands related to singularity expanding and then is seven, which in own merit becomes the unseen solid. On the other hand the seven forming the solid is the factor, which is constantly changing and forever moving. Although to everything else the solid holding seven moves, in terms of saliencies to infinity it does not move. As far as infinity goes the solid of seven holds relevance without altering location. It moves except in the manner that singularity in infinity experiences. The solid holds a constant point of reference to singularity where this puts material in terms of time in the time past = 7, in time present in the sphere = 7, and in the future going into the sphere = 7. There is a body (7) duplicating in time (3), which is 3 X 7 = 21 points forming the sphere by motion is the solid by never moving 10 = .99991 and when moving through the seven that the liquid represent is therefore singularity expanding = Π.

This puts all material in two distinct possible relations where it can be part of a solid and

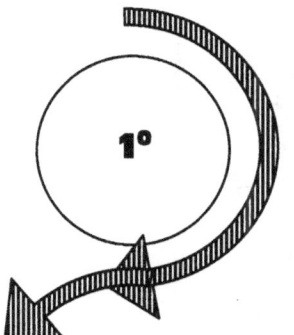

carry the distinction of forming part of an immovable solid object or form part of the liquid moving although the object might be a solid. It can be part of the ten in terms of another seven or it can be seven in terms of another ten.

In terms of human thinking an object might have mass. In such an event the object surrendered its individual

gravity to be adopted by the Earth as part of what forms the larger construction, and adapt to the motion or if it pleases more, the gravity that the Earth holds the object to confinement.

The object is in all senses immovable and part of the bigger structure as much as being the bigger structure. On Earth it might have a body holding life, or it might be an object that is formed as result of the intervention of man, or it could be a lifeless object used as an extension of life, in which case it may form mass only because the body can move after the body is cosmically dead and removed from the rest of the cosmos.

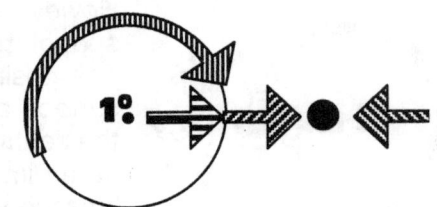

When being cosmic removed it forms mass because it shares the cosmos with the mass of the entire Earth body. The other possibility is where the object has micro mass being away from the Earth in outer space and then has absolute gravity because the object sustains its viable independence by motion or gravity individual motion apart from the Earth. The object is in a position where it has motion in relation to what has no motion although that which has motion is moving in relation to that which in affect should be standing still. The body can have mass and form a unit with the Earth as it adopts the Earth or it can have gravity while being independent and have micro ability to form an independent structural gravity. The object is either in the relation forming part of an object, which in that case it can have mass and is therefore part of the solid factor forming one part of the seven points. Then the other case is where the body can have gravity which it then is part of the ten that forms the liquid and is the part that should be in motion in relation to the part that is solid and therefore immovable.

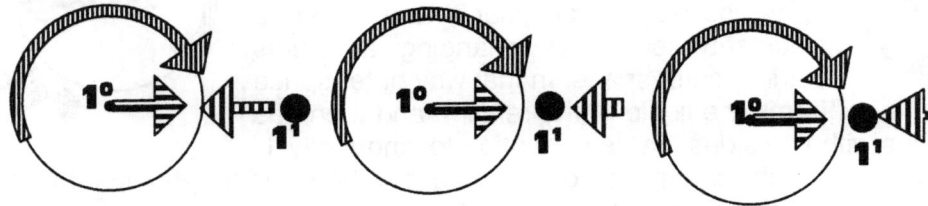

This is where Newton's deliberate fraud comes in. Correctly Galileo dismissed mass while an object is falling because from a Universal stance the body that falls shows gravity before mass becomes a part or a factor. As the body descends towards the Earth the body alters its relative allocated position in terms of the solid reference points the Earth holds. By altering and changing such reference points the falling structure is a liquid with gravity and in motion.

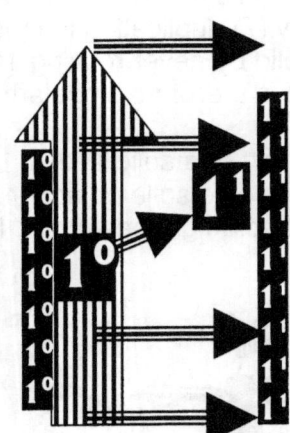

As the body moves through the sky, the body is part of the motion of the liquid, which becomes denser as it becomes known as the Earth atmosphere. However denser it might be, it still is a denser part of the outer space. Since it is part of the liquid it is detached to any individual value the body may have and is just more liquid in the cosmic sea that is holding liquid as the first layer that all cosmic objects have. This first layer being the liquid of outer space is what connects all objects in the Universe because this outer star layer being the outer

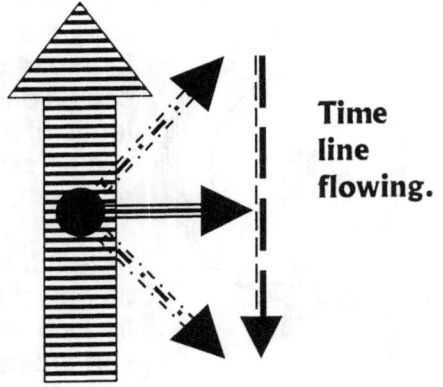

Time line flowing.

space is the time in which the current Universe finds the past moving into the present and onto the future. It is time connecting whatever to everything else. This outer layer being the liquid is time relating to everything that is related to a point securing a solid singularity and thus it is time confirming the Universe. It is the first layer of all stars in whatever size or form. All atoms are stars because all stars are atoms and this time then places all stars with atoms in the same time range.

As is the case with the body holding life while life occupies that space the space as a cosmic factor also does not really exist but is a result of Creation where heat or light parted time from infinite and time from eternal. In the relevancy there are two factors applying where time in the infinite stands related to time in the eternal and I saw fit to give the one a name holding its value as 1^0 and the other carries the main and value 1^1. It is space being time retarded that holds the two apart but also in relation.

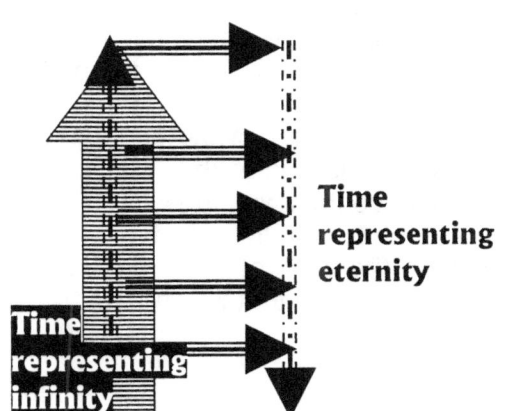

Time representing eternity

Time representing infinity

Therefore material applying as a particular function on earth will not apply nearly the same on other structures. There is no mass attracting mass and those still believing it must enjoy their soggy insight. It is a tool man created to use to the benefit of man where man can establish on Earth, the work that is a product of man's manipulation ability. Since man cannot be on Jupiter man can manipulate nothing on Jupiter and therefore man's creative tools such as temperature, pressure, mass and work does not apply on Jupiter or any structure that is not viable to life's occupation.

When we drive as we sit in the car everything in the car as well as the car forms a secluded excluding unit with what is outside being outside and what is inside being inside. We humans have the concept that it is the car that is moving while it is that which is outside the car that is standing still.

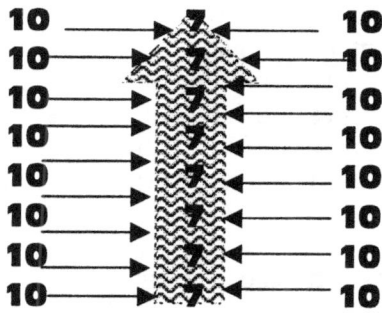

The mountain we pass is standing still because we will find the mountain on the same spot the very next day. We know it is the car that will be located at a different destination. To correlate singularity with the concept singularity forms quite a different view. Everything in the car is standing still while everything outside the car is moving. This is because as time flows the structure in the car does not change relevancies as time flows. What's on the outside though is moving and is changing constantly therefore seen from the immovable time in infinity, it is time in eternity on the outside that moves.

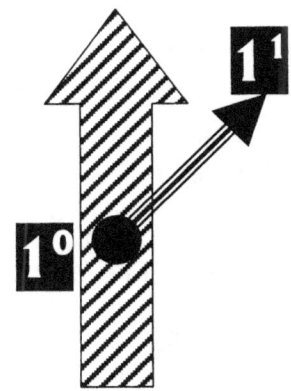

Life only has the ability to use what is available in the cosmos for the purpose and the benefit of life. The motion life exerts by travelling in time through time in a vehicle is cosmic principle dedicated to man's benefit and becomes man's contribution as he has the ability to manipulate. The car secures the material to the value of seven or of confining. The space through which the car travels stands in regard to the ten and therefore relates to time in eternity or excluded. The car secludes time in eternity from time retarding that is serving time in infinity. There is a relevance being deflected onto that, which is not moving by relating to it as moving. From what we can see inside the car, the outside of the car with everything in it is moving because the relevancies change as time flows. That, which is inside the car remains related to each other without changing positions and that represents allocated position in relation holding some order.

All contained within the car represents and confirms 1^0, which is 1^0 and that which is on the outside of the car distinctly refers to 1^1. Then we find a loose paper that starts floating inside the car. The small paper playfully frolics around on a breeze within the car.

Everyone inside the car knows to grab the paper immediately to secure the paper if the paper is judged to be important because the wind will blow the paper out. The paper darts to the front and to the rear and while it is the wind that is blowing the paper around the wind is not to blame for the paper leaving the car.

The paper is boisterous and skittish and while making the playful gestures on the whisking of the wind inside the car, the motion will bring about a change of legion. This makes the paper moving and motion becomes part of 10. At a point the paper will float in the air but will remain still in the air. Then in a whisk the paper flies out the open window. The flying cannot be as a result of wind blowing out because the speed the paper moves exceeds any blowing in the vehicle. The wind in the vehicle is in such case not in contact with the wind outside because if it was, the paper would blow out the window without frisking inside the car first. But even in the case of the wind blowing out the result remains the same but when using that which I am trying to explain becomes less obvious to understand.

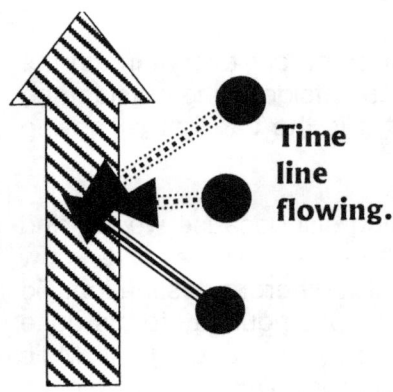

Time line flowing.

As the frolicking paper becomes stationary without obtaining mass, which means lying on something else, the relevancy of attachment changes. The outside becomes the stationary while the paper becomes the moving factor. The outside now correctly aligns as the stationary since the paper lost its allegiance with the inside of the vehicle. Now it is the paper that forms the moving part 1^1 and it is outside that becomes 1^0 while holding the relevance of not moving. It might not seem much prudent except that motion establishes attachment with time. In time we have the future, the present and the past and having that in motion puts a centre to the Universe from where singularity in infinite responds. This is where we return to the reason why we can see stars.

Newtonian say that because the photons are so flat that makes it possible to see the star. The travelling light run from the star to my eye and because of photons running all the way I can see the star. This expression is so vague, totally undisruptive and unimpressive as saying the wind blew the paper out of the car. It shows incredible small insight into physics. How does light travel to get from over there and then get to over here because the issue is truly utterly complex?

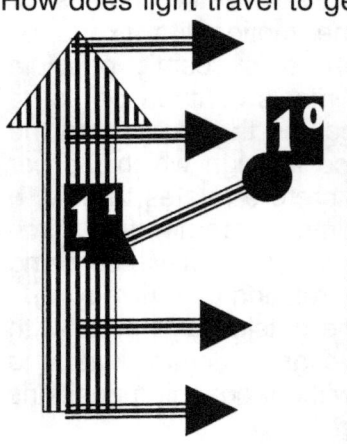

There is time in infinity, which is forming in the centre of the seven represented by material and is even inside of material that holds singularity in the centre of the seven. This seven is expanding by duplication rotation and contracting by restoring form within the limits the ten allows. This is the flow of time. Time is the uninterrupted flow of eternity towards infinity, which allows a relation between eternity that forever flows and infinity that never flows. Every second time marches on the Universe will never be exactly the same as it was the very previous instant. Some would change by contraction while the rest will rearrange locations by expanding and plain normal shifting. The Universe is never the same as it was or as it is going to be and that rearranging and relocating by contraction and expanding is what time is. Time is the flow of singularity in relation to singularity being absolutely stationary. What is between the two realities is space forming an unreal location. There is time delay we call empty space and then there is time delay we see as filled space or material. Between time and time there is time delay forming space and time delay forming material. Time moves from afar onto me and into me. This is present past and future and not specifically in that order.

That is the time in relation to contraction but then there is time flowing in the present continuous that I observe as my past coming to me as my future in the present I am experiencing. That is the continuous flow of time in duplication relating to contraction. The time in continuous motion is relevant to 1^1 and time in contraction is 1^0. The flow and direction of flow gives the legality of relevancy applying. When the paper is floating through the car interior the paper holds relevancy to all in the car as forming part of the contraction while the space outside is that which allows the car to move by duplication.

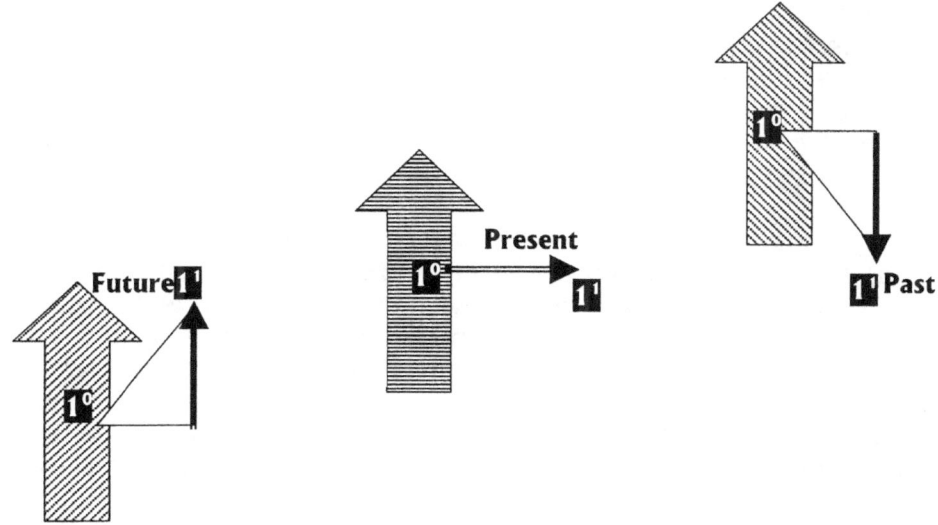

As soon as the paper stands still while not being connected to any object that is part of the stationary in the car, the paper becomes that which moves and the outside time markers become the stationary or solids. With no material that is between the paper and the outside parting the floating paper from time outside, time outside then forms the solid and the floating paper transforms its position to the liquid. It then is the paper being at task to move to the outside because the outside has the solidness of standing still without presenting motion and the paper with no connection or attachment to anything inside the car becomes the liquid. It is clear that the speed in which the paper leaves the car is relevant to the motion of the car and not any wind blowing inside the car. If it was a wind inside the car that blew the paper out, every lose object including every ones hair must be blown out of the window or blown in the direction of the window. The leaving of the paper stands relevant to the time that the motion of the car applies to the parting of the time factor.

The paper moves to the outside by time motion. The paper swapped attachments in the brief time that it was in the air and motionless. The relevancy of liquid to solid as far as the unattached paper is concerned changed its alliance and it became the liquid with outside being the solid. The paper now became part of that which brings us light and therefore has to move towards us by means of contraction. The paper got a time marker into its future and went on and attached to the marker in the fleeting moment it hovered in the air. Then as the marker formed time and move towards the paper, the paper aligned and as the moment of the future moved into the present, the paper moved to its time marker. This was done by the speed in which the car moved.   By the time the marker passed the car, the point of attachment went past, but the paper met its time destiny and was outside the car. Now you may ask what has this insignificant paper got to do with the importance of physics? Newtonians are over the edge with pressure, mass, heat variation and all the manufactured concepts they create and therefore they control that which puts man in control where man has very little control in reality.

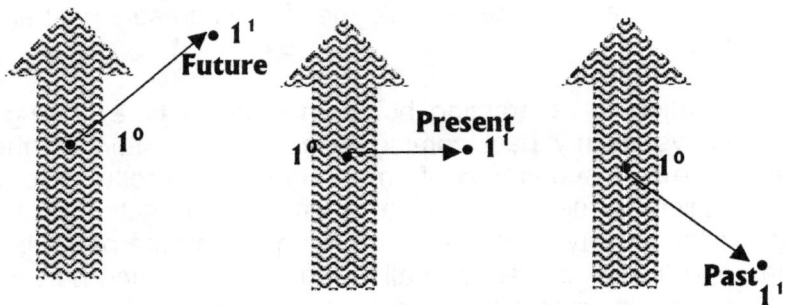

When water flows in a rapid stream we find that objects on the side get washed into the flowing centre. There is no pressure from the side to the centre therefore it must be the same time activation. Let's presume it is a flood release of flowing water and there are cars that are on the side of the flooding river stranded on the bank of the river while the centre of the stream is rushing by. We first find the cars start getting buoyant and floating as the river rises.

As the water raging water rushes past in the centre of the river the flood level rises and the cars on the side start to turn and move as if not knowing why they move or where too they are moving as they turn aimlessly. With the water rushing by the water speeds off in the centre it brings about that the increasing flow rate will get even more rapid and this will allow the relevancy of the banks to increase to accommodate the increase in water. Te flowing in the centre will increase and that will put the rate of flood flowing on the banks even to be much slower than what the pace is of the flowing of water that is flowing in the centre. The speed increase in the centre will let the object on the banks seem even slower than what they are in reality. The river rages at the centre with water piling on the flowing water coming from the back to increase the flow rate in front because there is more water in progress than the water currently flowing. The water coming behind is pushing the current water and that gives rise to the water level overall in the river. That makes the flow of water becoming more than what it was. If it was pressure it should then secure the cars being on the bank because the water mass in the centre will disallow any moving of mass from the bank to the overfilled centre. That contradicts reality because in real physics we find the water is rising while the centre is drawing water and mass from the banks.

The water should be pushing from behind representing the river's ever- increasing and that is due to the fact that the water level rises. That means the mass of the water from behind must fill the mass of water in front since the mass of water is filling the vacancy the water, which is running away in front left. There is more water coming than what the water is that already came. The water coming must therefore push the water that is not in the immediate flow more to the side as it pushes all things forward and sideways. That is if only mass and volume contributed to the proceedings. We will have everything floating on the bank moving towards the outer edges of the river and nothing that is in the direst line of flow will have a chance to come into the main stream flowing at an increasing speed and this increase will push more to the banks while holding what is on the bank, on the bank. Time in relevance of motion brings a complete different scenario to everything.

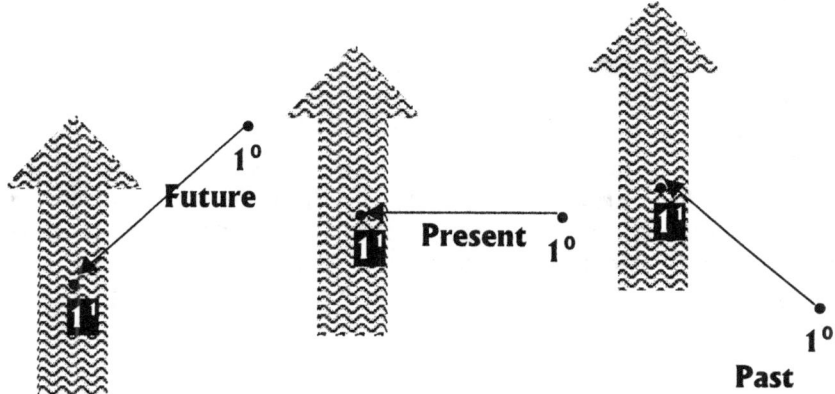

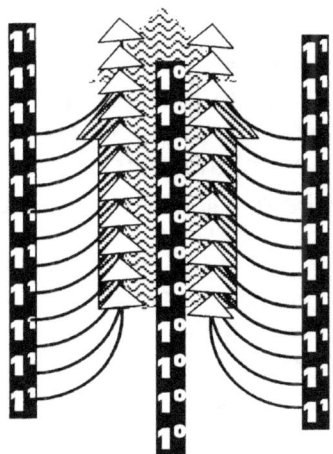 If the cars are mounted to ground on the bank or show signs of mass as they connect to the Earth, the motion that the river has in the centre is not going to affect the cars in any way. If the cars on the bank start to whirl and circle in the water they seem to find a personal centre because due to buoyancy they lose the attachment they had to the Earth, and with that then, relevancies begin to change. Then the motion reverts to the side as the line of motionlessness in the centre gathers that which does not have attachments on the bank. It is so easy and so Newtonian to declare the stream is drawing the objects in and then start to calculate the "force" with which it is done, well knowing there is no manner in heaven or hell to control such calculations. Why would it be drawn? That is physics and that is beyond Newtonian thinking because if it were at a Newtonian level of thinking someone somewhere would have drawn the correct conclusion. This, which I now refer to is the 3 used in motion presenting time. The speed of light at $3\Pi^2$ or the atom having the electron value at three where the Neutron is $\Pi^2\Pi$ and the proton is at a relevant value of $\Pi^2+\Pi^2$ making the repeat of duplication 3. That gives the atom a relevant measure and that adding of a time factor completes the atom value at $(\Pi^2+\Pi^2)(\Pi^2\Pi)\,3$

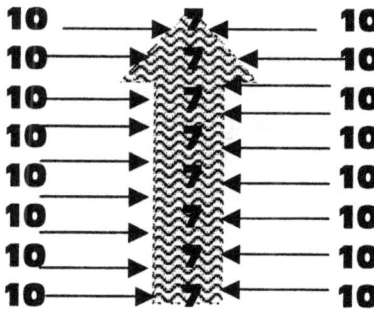 There is a secluding of time on containment in relation to time allowing duplication and however this is normally fixed; gravity can change the direction of time forming motion. Where this becomes most apparent is where one can see an object flying from the airplane fuselage. What is in the fuselage is contained by the fuselage and is excluded from outside. It may be pumped to air pressure but in itself that has no importance. It may help to instigate immediate motion direction but it is not the reason why objects fly out of a broken or punctured fuselage.

The object inside the fuselage is contained as mass within the aircraft and forms a secure part of the moving aircraft. Everything said about the Earth moving but deemed as solid or the paper darting inside the car applies to all within the aircraft. At this point I wish to get back to life and the body housing life in order to bring to light Newtonian fallacies about science. Life contains and creates the body. The body does not contain and establish life. The body cannot place life back into the body because life has left the body. The body disintegrates and dissolve as a compiled structure as soon as life leaves.

The first thing the body loses is its ability to move and then to be a liquid with a solid density. That the body holds life is just more inconsiderate rambling of Newtonian proportions and mindless Newtonian madness. If life contain the body one may replace the body with other life and the body will revive. When life starts out life starts by building a body to the requirement life sets out to have and fore fill. The body is not there and then the body first being lifeless starts filling the body with life. There is first life being small and then life grows a body. Life puts control to the body as the body becomes body and the way the body performs is an attempt as well as the result of what that specific form of life is able to produce. Humans should start to recognise how relevancies apply in the correct order.

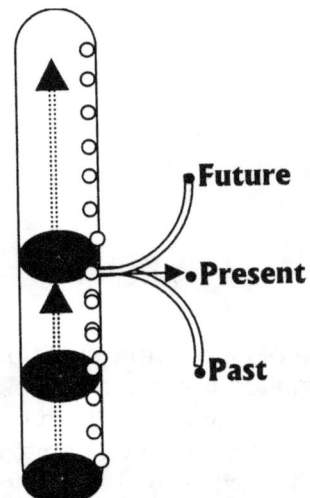

This argument I introduce at this point is part of an argument I make in another book where the argument I make in the other book is more than five hundred pages in all so do not think for one minute it is this simple. The body secures time within while time without a body allows the control of the body to take place. The same applies to the aircraft. Life produced the aircraft and life secludes that which is within from time on the outside. The

pressure is to benefit motion may be a result extension of influence uses applying to the aircraft time flies by in presenting the future continuing present past is the future in the and motion in particular correct context would be filled with because the Newtonian understand reality. The one example although it

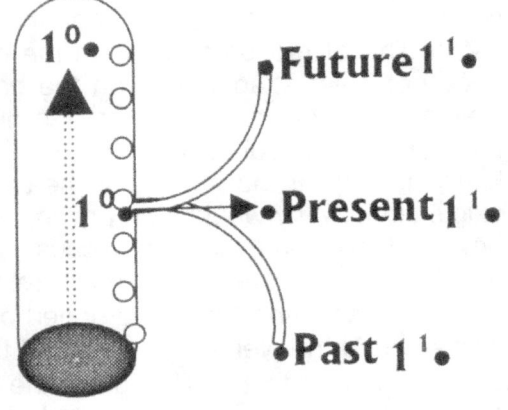

the existing of life but the of what life finds an but in reality it is what life cosmos. Outside the the form of space present and past in the continuous where the present. The fuselage must be seen in its otherwise the cosmos unsolved riddles mind is too simple to aircraft body serves as is not the example. It

merely serves to highlight one side of time while the outside highlights the other form of time.

The fuselage forms the part where time is continuing in the present continuous while outside

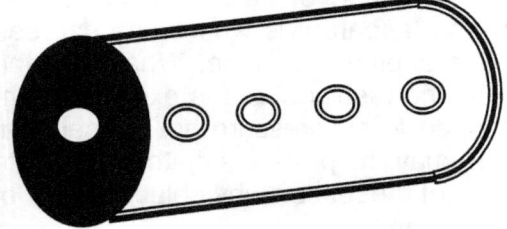

time allows duplication and present the opportunity to have time flowing from the future through the present onto the past that flies past as other objects holding individual centres of the Universe takes on allocated relevancies, applies such relevancies and then loses such relevancies. On the inside of the fuselage all components are secured and attached to the structure. On the outside time is moving at a pace bordering the Roche limit of $\Pi^2/2$. The relevancy of time borders any limits the earth will allow. The structure holds every one in position and secluded from the fierce time displacement going on outside. The structure shield the passengers like a star shield the interior layers from space on the inside by applying motion as well as securing the relevance of time in progress on the outside. It is the perfect motion of material holding 7 in relation to gravity Π^2 that is holding time towards the outside at 3. That is time in duplication that puts a relevancy in action to duplicate the earth's gravity relevance of gravity at $7(3\Pi^2)$. On the outside the inside worth $7(3\Pi^2)$ is duplicating at a rate of about $4.5\Pi^0$ giving a displacement speed of $7(3\Pi^2)\ 4{,}5\Pi^0 = 932.6$ km/ h.

Time is going by at a rate of the earth plus $4.5\Pi^0$, which is kept nicely contained and apart from bothering the well-to-do paying customers inside the aircraft. The pressurised capsule has only one function and that is to see to the comfort of the customers. On the outside it is time whisking by at a pace of $7(3\Pi^2)\ 4,5\ \Pi^0 = 932.6$ km / h while inside everything is at a steady motionless continuing of time by presenting the future in the present as the past.

While flying that factor that jetlag presents is not affecting any of the passengers because the aircraft fuselage is keeping the customers flying in a nice paid for Universe where everything is attached to each other while nothing is attached to the Earth. The flying passengers are in a world that is made to suit them. Only when they land do they have to take stock and adapt to a new Universe on Earth, which they have to adopt because of their new location. This is where jetlag hits them as they surrender the comfort of the previous Universe from where they came and found security in the Universe they bought with their flight ticket.

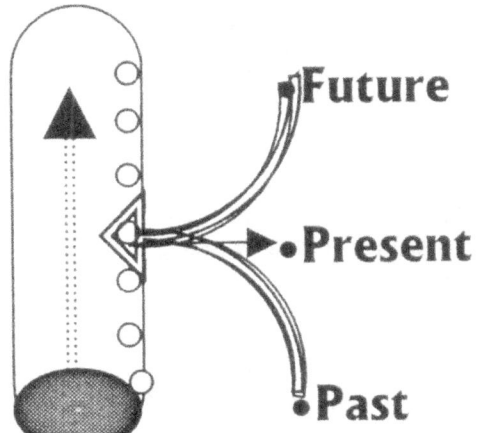

But once in a while things go wrong and the paid for Universe ruptures. When a puncture comes about in the container it is not the release of the compressed air that pushes the object in the containing capsule out. The difference in air pressure would not amount to much significance and it is there that Newtonian science leaves crash inspectors in dismay.

Recently I have been a witness to such an investigation, not me in person but I saw how the procedure of tests was performed. The conclusion in the end was that the pressure was not responsible for the body to fly out any window and therefore the body flying out the window of the aircraft is a fable. It is Newtonian science that is the wanting fable. As soon as the isolation raptures and a whole appears, there is contact

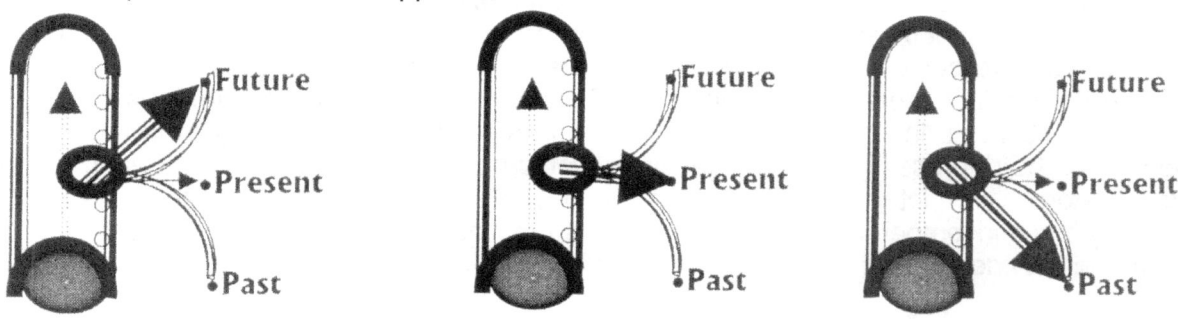

Align   **Connect**   **Attach**

The body raptures and a direct link come about between time outside and contraction inside. Only where such a direct link can attach does the focus of time change relevancy. Only at the moment when such a moment represents the opportunity does such a swap in relevance take place. It is the instant in the moment that the changing of time by motion swaps from that which truly moves to that which truly is stationary. After the event has passed and as the occasion slipped away in time does the status quo remain once more. The gravity on the inside is not sufficient to respond to the time in the instant when the rupture occurs.

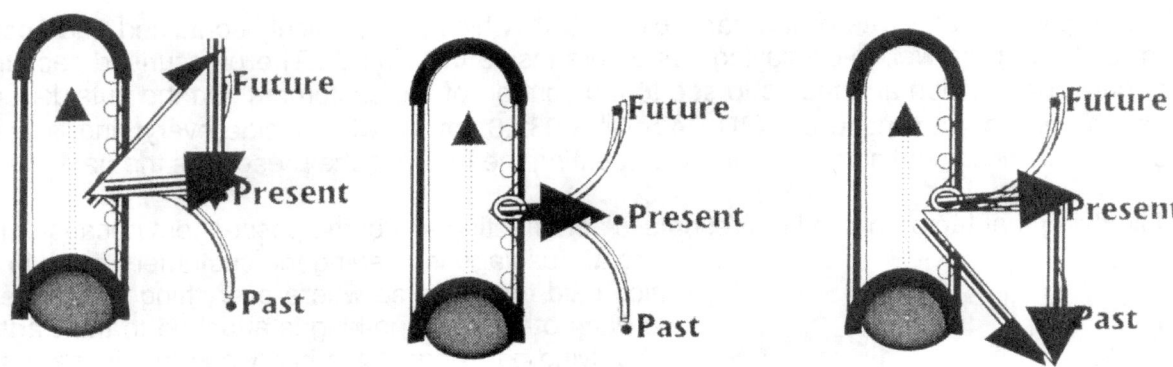

In the instant of connecting, the gravity that the cabin represent as a secluded Universe, is far too inadequate to leave the responding object secured in the face of time in motion on the outside of the fuselage. The responding object can be anything from a walking hostess to a piece of unsecured paper to a man in a chair being secured by the strongest bolts to the floor, it doesn't matter. It is not the mass of the object or the security the object has or the size or the volume that is of any magnitude. It is the contact that is made between that making the contact and that swapping focus. I do not picture for one second the Hollywood rubbish presented as a show where the person slowly pushes through an open window while shouting in terrifying agony. There also will not be any person fighting the movement towards the open window while shouting orders to other less intelligent passengers. The person or object will leave the fuselage faster than what the eye can witness.

The one instant the object will be inside and the next instant the object will be outside with no one being able to witness the event. The time attached to the process responds on the motion of the aircraft as allegiance in time having motion changes from inside to outside. If the swapping involves a living being then I think death will occur before the motion takes place. Life will not be able to participate in such cosmic relation changes and the person's body might still have a heartbeat for another split second or maybe two, but the mind would have died as soon as the cosmos changed relevant motion from inside to outside the aircraft. Death will come about as the relevancy swaps and that is before any motion of any sort takes place. The person most definitely will surrender his or her consciousness and will be unaware of the last seconds on Earth being in and with life. The air pressure blow out might have a function, as initiating the starting of the movement, but in actually participating the blowing of the depressing of such a fuselage would hold no significance in the least. In that the Newtonian verdict is correct, but trying to apply the Newtonian pressure release is showing much incompetent understanding about the whole issue. Mass plays no part. It is all about motion and relative motion.

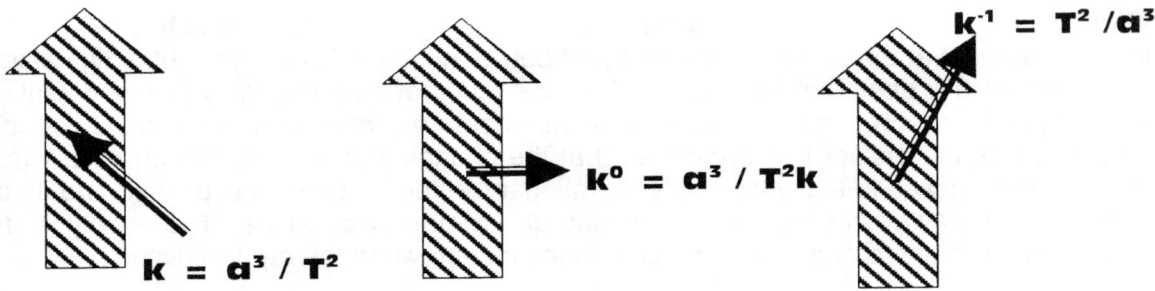

$$k = a^3 / T^2$$

$$k^0 = a^3 / T^2k$$

$$k^{-1} = T^2 / a^3$$

What rules this principle is the same law that rules the principle that allow the comet to cycle around the Sun and that keeps liquid motion in relation to solids. It is still Kepler's formula we find and the Roche limit either applying not applying that allows the change from one part of the Universe to the next part of the Universe.

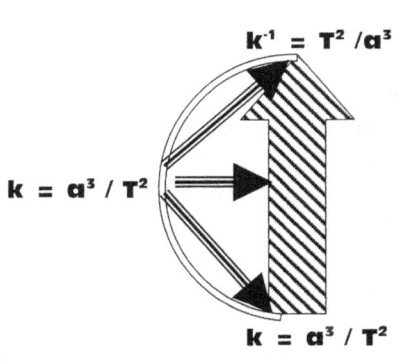

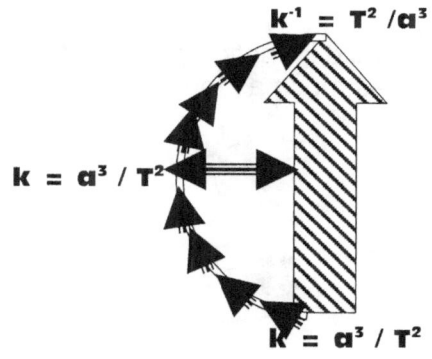

As the water is coming towards the object that is drifting on the bank while spinning around an independent axle, the alignment holds the water coming closer to the object as the active motion. It is then $k = a^3 / T^2$ applying while the distance between the floating object (the solid) and the flowing river (the liquid moving) reduces.

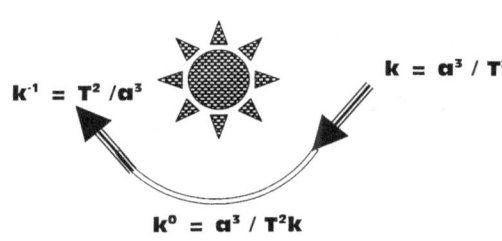

This is also what is happening between the Sun and the comet while the comet holds a position within the liquid and the Sun is steady as a solid. However, in the case of the Sun and the comet the relevancies also change in favour of the one and then when crossing the divide it favours the other side of the Universe.

As the water is coming closer the relevancy factor readjusts as the motion puts the material in a growing stance, which allows the time reduce by growing larger. In this we find that the time factor diminishes as it comes closet to the point of singularity at $k^0 = a^3 / T^2k$ Then singularity places every aspect on the other side of the Universe and all factors change position.

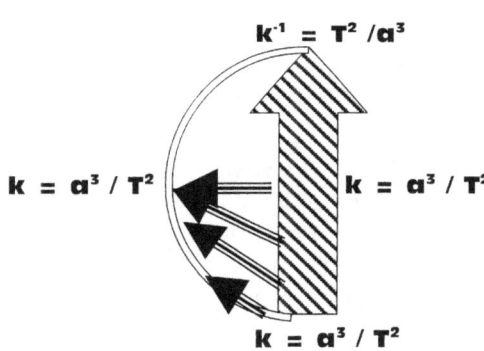

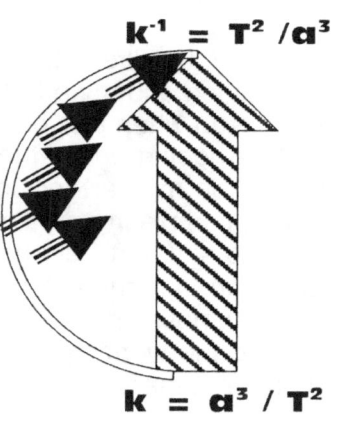

It is the restriction of mass that prevents gravity from becoming a reality and because the Earth restricts our motion as we find motion that the Earth prescribes, we have mass and not gravity. An object may have some part vested in mass while having another part acting in gravity but mass does not bring about gravity. In gravity there has to be motion of space in time and when time ends, space collapses. In this letter I am about to introduce you to gravity being the motion around a solid that provides the contraction. Gravity has two parts where the one part is expanding by motion and the other part is controlling the expanding by contracting the motion. It is $k^{-1}= T^2/ a^3$ and it is $k= a^3/ T^2$ but above all to find gravity we must find the centre of the Universe being singularity at the centre **k**

$^{0}= a^3/T^2 \, k.$

The liquid in motion provides the gravity and by initiating the motion the gravity contracts the solid as much as the solid is extended by the motion, which the liquid provides and the liquid

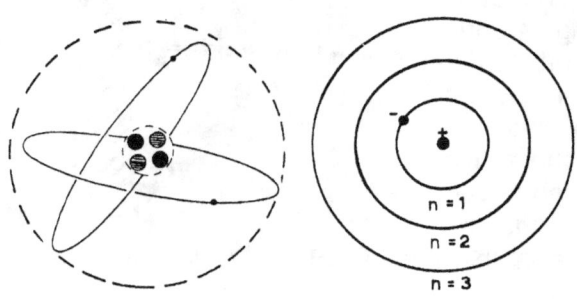

is the gravity or the motion of the solid where the solid provides restriction or mass to the liquid. It is in this manner that the entire Universe operates because from the spinning atom to the most prolific galactica and to the other end of the spectrum where Back Holes destroy time, the entirety out there works in this manner.

I first started my studies in the field of Cosmology as a spontaneous development of my natural curiosity spawned from childhood interests in the field of cosmology, which I developed even before I went to school. The studies were a reaction (I would imagine) that was part of my personal childhood development in how I was forming a personal concept of a lifelong interest that followed me into my future. At first I conducted all my earlier studying mostly on the basis that inspired me to find out more about what made the Universe tick, with no intention ever on my part to reach a point where I would be writing books on the subject. At first I was investigating cosmology on a part time basis. This went on, periodically, for the best part of twenty odd years (*as* time and *when* time would permit).

Then in later life with my health deteriorating, I committed myself to more intense investigation and my effort developed onto involving a study using time that is only permitted by a person when that person is involved in such a quest on a full time basis. That quest has now been going on for the last seven years in full devotion and if one includes all the years invested on my part including the twenty odd years before, part time, then the time I have spent in completing my theory when adding all it comes down to almost twenty eight years. This is to say that I did not come to realise what I am about to introduce on a light-hearted conclusion. I mention this because I wish to ensure the reader that he should have no doubt about my most sincere commitment in producing a cosmic theory on matters concerning the start and the working of the Universe during and before the Planck era. At first I began by arguing that there is something that is blocking our progress.

There is some barrier preventing humans passing a threshold whereby our understanding will pass such an obstacle. If there were any way that any one may break through that barrier is there and that is preventing normal research to go pre-Big Bang, it would be accomplished by finding the barrier. Then the vision we use to focus would pass such a limit. If we wished on progress in our pursuit of the very first cosmic moment then we have to find and cross the barrier that blocks our view. We have to look deeper and in another direction should the desire driving us be strong enough to commit us to reach into the very birth of the cosmos. We have to rethink the strategy that we use. Max Planck was one of the most brilliant men of all times and even he, notwithstanding all his personal brilliance, accomplished little.

There are parts missing in what we have and that which we have at our disposal to use, because if there was no such an obvious barrier then the Wise-Men involved in science would by now have found the way to break through the seal that is locking us out of the critical past which will uncover the origin of the Universe's infancy stage. I went about trying to find what everyone since Adam, (meaning all of the rest of mankind and myself) were missing throughout the ages of speculating and interpreting while philosophising about whatever we find inspirational. The obvious we saw; that was clear. Therefore I had to find a route that would lead into the not so obvious that all of us were missing, notwithstanding the best efforts of the best qualified to accomplish such a breakthrough. My effort involved trying to accommodate that which was in the cosmos available to use by the cosmos in all phases of developing. If I had any hope of finding the answer, such an answer had to be simple because I am not very inclined to unravel what is deemed as complicated. The simplicity had to be locked in what was not yet understood about that which was in the cosmos as it formed part of the process used in forming the cosmos. My realising this brought me to focus not on that which we understand.

There is not a lot we actually understand because even gravity is very poorly understood. In fact, gravity is so poorly understood that there is not one person alive that can claim the prestige of understanding gravity and among the dead there is even less that can make such a claim. There are several phenomena that are presented in nature and acknowledged by science but also discounted by science and therefore not presented as accepted science. By admitting that, that which we have available to us to use concerning our research of cosmology in an attempt to better our understanding of cosmology, is useless to use, then one realises that not having what there might be makes what we already have useless. It then is useless to use what there is as part of the big picture we are trying to paint because what we use is not really part of the picture. This leads one to believe that the picture of the cosmos Mainstream science is painting, is being painted without painting a full picture.

In my first attempt to understand the full picture of what science was painting, I found so many colours missing, that the result was that there was no picture painted that anyone could appreciate. This is what made me decide to go on researching the 'unknown' in the hope it might clarify the 'known' and as the book unfolds, you as the reader may agree that I was correct in pursuing the misunderstood and rejected phenomena. Finding the missing phenomena helped me to place the phenomena mentioned above in a theory where the principles also mentioned above form a part of the overall gravity used in binding the Universe. I believe what is in the Universe is not able to be coincidental because of too many influences contributing to what there is - notwithstanding the fact that that is the manner which science uses when they refer to the Bode law. What is in the Universe has a role as it had a role, which is the same role that phenomena has had and in future will have. This is establishing a very new idea about the working relationship between particles and in explaining it by using Kepler's studies. Redefining the work of Kepler's views brings a new Universe to light involving new concepts that are based on old principles. The old principles serve to update man's view about cosmology and therefore are new in that capacity. Through that new vision I was able to come to realise what the reasons might be why Kepler never saw it fitting to include the measure of Π in his formula. I do not suggest his neglect thereof was intentional, nevertheless the formula he devised without using Π proved that there was no need for the inclusion of Π since his figures brought about a correct answer in the final end result leaving a well concluded fitting answer.

The numbers he produced brought about a specific space $\mathbf{a}^3$ contained in a circle $\mathbf{T}^2$ at the distance of $\mathbf{k}$ from a defining centre, thus the calculations did not require the use of Π to find a meaning. In that Kepler did not see a need to include Π. I would not go as far as declaring with absolute certainty on his behalf that he did it deliberately, however there never arrived such a necessity. It is prudent to agree on whether or not such a need is necessary, because if one is agreeing about such changing not being required, a new Universe emerges. The circle that Kepler discovered came about without ever forcing Π into the frame because it is clear that the circle formation came about as a natural consequence and came spontaneously delivering an equation while he was working. In this book I prove that the reason for adding Π to the rest of Kepler's formula is unnecessary. This unnecessary addition is because when going one step further in the investigation one will find that $\mathbf{k}$ and $\mathbf{a}$

and **T** are symbolising the same value with the only difference being that each one represents a different dimension to our six dimensional or six sided Universe we enjoy.

In fact I shall show that Π replaces **"a"** and **"k"** and **"T"** and that Π is the true value that should be replacing each factor as to indicate the correct value to the sides nominating Π. We humans work on a numerical basis using ten as a basis where we count to nine and re-establish a new decimal numbering line by adding a nought behind the number in value. This is using the numerical basis of ten, which I suspect we took from ancient knowledge about cosmology and not from using our fingers and toes as the earliest calculating processors. In this letter there is unfortunately no room to explain my suspicion but another fact I do prove is that the cosmos uses Π in the cosmic numerical basis as a means to measure and quantify. Therefore in fact the Kepler formula should read instead of $a^3 = T^2\,k$ as it does, it must be $\Pi^3 = \Pi^2\,\Pi$ where I shall show that Π represents singularity wherefrom the entire Universe sprang from Π and by forming as $\Pi^3 = \Pi^2\,\Pi$ it is confirming that space is equal to the motion thereof. Kepler's greatest achievement was showing that the cosmos is space –time $a^3 = T^2\,k$ while time is the motion of space in space. The value of Π is the primeval and most basic of measures applying as an accepted cosmic legal value that the cosmos used exclusively in the very beginning and as it does today. The measure of Π in the Universe, values particle development that brought about all development ever conducted in the Universe. Only after this stage did the rest come including mathematics and went on to freeze spilled singularity into frozen material. Reading this statement may sound suspiciously senseless but as the book unfolds the sensibility will become apparent.

The full implication of such a statement will become clear when one dissects different facts coming from studying Kepler. My discovery of this fundamental basis of legal valuing ensured me again that there was no need for someone the likes of Newton to add Π in any form to the work of Kepler because Kepler discovered the ultimate Π in the Universe, the Π giving the Universe form and gravity. The concept of Π that is the only single form of all other forms available that can by duplication of Π's assemble the value of gravity. When replacing the symbols with Π the facts of the Universe become self-explanatory because the most basic form that forms the cosmos has a definitive and uncompromising value.

But getting this far took me down roads overgrown by ignorance and which I had to uncover myself as if hacking away miles of overgrowth with a machete chopper. All of the disbelief science showed to my work in the past and their refusal to see past Newton made any and all attempts on my part as bad as they could be, strangling and smothering my attempts to announce my uncovering of the newly found insight on my part.

For decades I tried to come to terms with the inability there is in science to explain the cosmos in real terms, when using the science of official reputation. That which there is makes a mockery of science because the undisputable clues left in the cosmos makes what little correct explaining there is available, seem like a comedy of errors, when it is mixed in with all the other near Dark Age errors we still use after so many centuries that provided countless opportunities to revise the old muck. By applying current accepted Astronomy as such, the phenomenon found all over the cosmos is still beyond the explaining ability of Mainstream science. This is true and it is a shame because it also is an undeniable fact in spite of the vast knowledge and progress in other forms of science taken in the manner science uses when it approaches cosmology. Cosmology truly lagged behind while the understanding and advancing of physics, mathematics and chemistry as subjects were flourishing. By comparison I saw how little there was available in explaining cosmic phenomena and how much improvement in understanding the other departments such as chemistry, electronics, medicine etc. could offer as results were coming about from research. Even where there is little explaining available in cosmology, it turns out that such explaining is confusing to say the least and at best it highlights the manner in which science is applying double standards. For decades photographs were the only progress forthcoming as an addition to improve the meagre field in cosmology and that improvement was artificially stimulating cosmology.

By providing a false impression of advancement, everyone missed what and how much was missing…To the connoisseur desperately looking for more than the obvious stirred in with some out-dated misinformation dating back to the Middle Ages, it all seemed as if it was a picture portraying the ridiculous to make the sublime look good. The pictures only proved the opposite of what progress in cosmology will represent. In truth and as such in cosmology the cover up that was hiding the lack of progress about the science of true cosmology was only forthcoming in the improving of electronic optical telescopic advances and spectroscopic progress. There were only photographs carrying beautiful pictures which pleased the less informed except the photographs did not bring progress to cosmology at any intellectual level by promoting insight. The explaining that the photos demanded about the subject had the opposite effect of installing hope because what it did do was underline what lack in any notable progress there truly is in our understanding of cosmology and laws in the cosmos.

While such Hubble telescopic images might seem to be as clear as daylight it was more than clear that there was little academic value to them. To the person in need of more stimulation than being impressed with pictures of God's marvellous Creation and the sightseeing that always accompanies such pictures, such persons always felt very disappointed. The pictures did give satisfaction to those more easily impressed, but the rest of us seeking knowledge accompanied by understanding the images left us despondent. Although they leave the vast majority in total amazement, there are those amongst us being less impressed about not knowing the 'why' and the 'how' that we wish to see in such amazing pictures. I know the group I fall into may be the greater minority and the majority may only demand the portraying of the images, which is what that easily satisfied group demands. The rest of us rouse with anguish at the lack of information about what is known and what lies behind what those pretty pictures are conveying.

Nevertheless there can be no real progress in scientific understanding about the images portrayed by the Hubble telescope, and others, if no one is able to show the slightest clue of a deeper understanding of what is going on in the Universe. Everyone is almost breathlessly waiting the commentating by the most informed, which accompanies the magnificent cosmic portraying of God's Creation. When we are portraying the new images, we should also be investigating what the cosmos is portraying through the lenses. The lack of actual believable explanation coming from investigating by means of telescopic imaging should impress one and all, but the impression must not be based on the colours in the images but the sensible information attached to the image investigated. It is ***that***, which we wish to see. What we wish to see must at least be accompanied by scientifically backed information, which provides the proven understanding coming from science. When science is employing new explanations with such photos it should also be discarding senseless baggage carried over from the past. Most images contradicted Newton and for saying that, every Academic I ever came across in the past ostracized me.

That bothers me little! I know I cannot possibly be the only person absolutely discontented with what Mainstream science accepts as science. Here I refer to the out of date theorising Mainstream science still accepts amongst many others as how they suggest stars and planets are forming. One cannot promote cosmology in honesty and advocate scientific fact whilst dishing up such fairy-tale nonsense to students. Moreover I hold the opinion that amongst Academics in particular there must be many if not most that share my personal serious doubts or have an inclination to share some of them. This I say when considering the overall doubtful picture painted about what there is and what one believes there should be. I just cannot believe those forming the most intellectual group of mankind are unaware of the mismatching facts seen over the broader picture because the contradiction and lack of a plan, makes what there is so very doubt provoking. Newton dismissed the formula Kepler presented as all factors forming motion. That is where the apple cart derailed.

In honesty we have to realise that we cannot dismiss the whole formula that Kepler produced as being motion. It is so much more than just motion. It is $a^3 = T^2\,k$:

That is what Kepler brought into civilization for all time to come. He saw space a^3 being in isolation due to the time it uses to move T^2 claiming such space forming independence according to the lines k indicate.
Let us look at the factors in more detail before we proceed with the rest of the book.

a^3 symbolises a mathematical interpretation of implicating the three-dimensional space.

T^2 is representing the period or time that Kepler suggested we should use to calculate time that holds the orbiting planet in direct contact with the space in relation to a very specific centre.

k is the space taken from the centre to the end of the line from which the planets must have grown if one accepts the Big Bang growth of particles and the affect of the Hubble constant on all cosmos material. The specific value about the centre is most important because from the specific centre gravity always applies the strongest influence.

One cannot justify Newton's dismissing of Kepler's formula as that all factors only contribute to the motion indicated because that is misleading. We all accept that the true cosmic form *would be* and most probably *is* a sphere. Everyone accepts the Universe on a whole as a sphere…but why would the sphere form? What would be the reason why the original form that we devote to the Universe would take on a sphere as a natural form? Apparently our imagination grabs the sphere as form. In all natural events the gravity in that space which stands apart and independent from all other space takes on by cosmic pre-casting the sphere as form of shape … **it is because gravity chooses the smallest space to hold the strongest force**.

I am of the opinion that gravity is about dismissing space to the advance of heat increasing in such a specific and concentrated space using the concentration as measure for the heat as well as the space holding the heat in space. According to Kepler that is what he found to be true. Space a^3 will always be circling space around as T^2 in any position from the centre k. That is what Kepler said when he said $a^3 = T^2\,k$.

Kepler indicated space a^3 would forever fight for independence and show separate individuality in remaining apart as identifiable cosmic components by means of motion. Every space will cling to independence indicated by k through fighting off the integrating of another coverall unifying unit by applying the motion of T^2! The problem we have to solve in this letter is what will the cosmos use to secure such independence between all particles? What sets space apart from the rest of space? First we have to admit that Kepler was the one that introduced the following.

Kepler gave us the answer to the following but no one ever took notice!

Kepler was the one that discovered **space / time** as $k = a^3/T^2$

Kepler was the one that discovered **singularity** as $k^0 = a^3/T^2k$

Kepler was the one that discovered **gravity** is holding **space-time** relative by the measure of distancing **k** as $k = a^3/T^2$ and $k^{-1} = T^2/a^3$

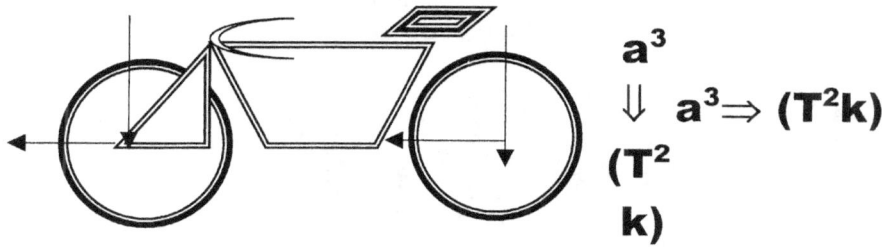

$$a^3$$
$$\Downarrow \quad a^3 \Rightarrow (T^2 k)$$
$$(T^2$$
$$k)$$

Everyone able to read mathematics has to realise that Newton suggested collisions between cosmic structures must eventually come about as gravity erodes the distance separating the cosmic structures multiplied by the product of the mass of both structures from both ends. Newton said the multiplying mass of both structures destroys the distance between the structures by using the eroding force of gravity in the square. The cosmos then must end in a Big Crunch with all material joining together but that joining is not forthcoming at all…and that only indicates how much insufficient understanding there is on offer in cosmology by the educated–to-be-wise-about-these-matters. There is precious little available to explain about their field of cosmology amongst the ranks of Astronomers. So…let us return to the beginning of cosmology before every one became oh so wise and see what there is to see. Let us see how the humble bicycle can teach the so wise about what gravity is because it is easy to demote the prominence of riding a bicycle, when it can be so easily explained as just being a balance.

A person that acquired the skills of peddling while staying upright on the bicycle has achieved the method of rearranging gravity within singularity. Without motion the bicycle falls on the spot it holds. When the bicycle is put in motion the bicycle can maintain the upright stance as long as the motion applies. When the motion stops the bicycle drops. To introduce motion to the bicycle, the motion brings about a stable unsupported upright stance where balance can result from the motion the Earth enforces to the balance coming about by the bicycle using independence gained from motion of the space holding the bicycle. The space that the stationary bicycle holds is the direct result of the Earth providing the motion.

That is why it then will adhere to the gravity or motion that the Earth will enforce. The motion restricts the static bicycle to one allocated position that the Earth supplies to the bicycle. When the bicycle starts to move the bicycle gains a cosmic independence. The gravity effecting the redirecting of the Earth gravity response comes about as the result of additional motion that is introduced to the bicycle. This is the very same process that the aircraft needs to get airborne because it replaces or repositions the singularity the Earth holds to the singularity the bicycle develops in motion. The aircraft only takes the change in direction of what the gravity is insisting on through changing direction in motion through phase one and into phase two. It all is still part of the Coanda effect. With more motion contributing to acceleration, the bicycle will become airborne on condition that it is also given the advantage of a set of wings to increase the effect of creating space-time to the advantage of the motion requiring the change in singularity direction.

I specifically chose to use a bicycle in my explaining because the bicycle is the object that relies the most on singularity achieving the required balance in which to operate. It is singularity, which puts space in balance of the time the space uses to duplicate. The singularity creates space-time and such space-time results in a balance of space and time $a^3 = T^2 k$. It is through the Coanda effect that marries the motion to the space that gives the balance that keeps the bicycle up right while it is singularity that allows the bicycle to move or duplicate the material by relocation through time. It is an act of balancing singularity that gets the bicycle as a machine working properly. It is also the next best thing to illustrate how singularity by motion provides gravity in addition to that which the Earth already produces. In the Coanda principle there are two factors where one is motion and the other is space and the two provide both duplication as well as contraction of space-time.

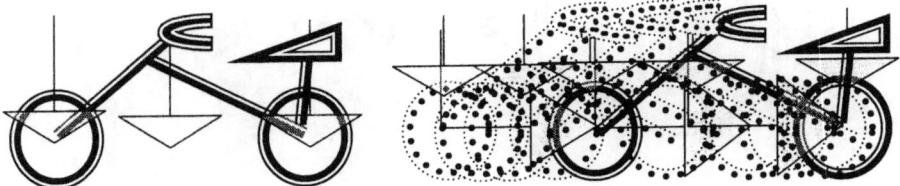

In the normally applying of gravity, we find contracting lines running vertically as the lines connect with the Earth centre. There are numerable lines running from the outer regions to the centre in diverting by 7^0. Motion provides extending of the 7^0 establishing the centre connecting points to the Earth to which it connects. When it is only the Earth providing the motion there is only one spot of space allocated to the bicycle in the time frame applying at the time. The Earth takes on a specific size by which it duplicates the space it holds in relation to what it renders the bicycle also sharing the space, which the Earth has. By that standard the bicycle also holds an exact relation of space and volumetric size related to its position within the Earth. **This is where motion through duplicating changes the dimension in equation. Motion is the duplicating of existing space from time in the past through time in the present towards in the future to time.**

The instant the bicycle starts to move independently it increases its share of space it holds within the Earth. By duplicating the space allocated to the bicycle at a higher premium than the Earth does with the motion the Earth provides, the space the bicycle charge increases in ratio to that which the Earth charges because the bicycle maintain the duplication that the Earth grants but then still adds to that space by enforcing more space provided by the motion

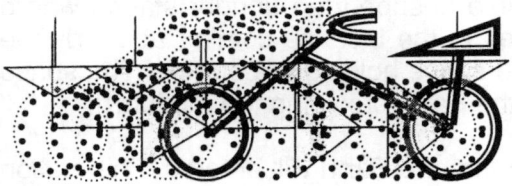

addition. The motion of the bicycle not only extends the vertical connecting lines and not only changes the direction of the vertical connecting lines, but does both. The value added and the change in direction contributed is what brings about flying and moreover is the cause of the sound barrier.

The Earth takes the position that was before the motion of the independent object came about previously and held by the Sun. By establishing the directional motion in accordance with the k^0, which the Earth then provides instead of as previously provided by the Sun, relevancies replace previous ones. In normally applying gravity, we find contracting lines running vertically as the lines connect with the Earth centre, which is the gravity we confuse with mass. It is a state of contraction and is the result of space being confirmed in relation to the motion of liquid time. Motion provides extending of the 7^0 establishing the centre connecting points to the Earth to which it connects.

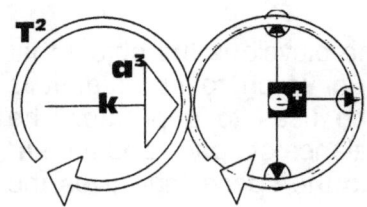

With The bicycle moving, the bicycle changes associations and where **k** was the factor securing the bicycle as part of the Earth, that factor **k** then goes 90^0 in relation as the bicycle in motion then is in association with motion putting it then in terms of what forms a part of the liquid. The atom moves and as the atom moves the atom spin. The atom spinning has to deduct the linear motion of the bicycle from the circular spinning motion since the linear motion does subtract time from the spinning. The faster the linear motion is the more will be deductible from the spin. In the end it is the atoms forming the bicycle that moves and not the bicycle.

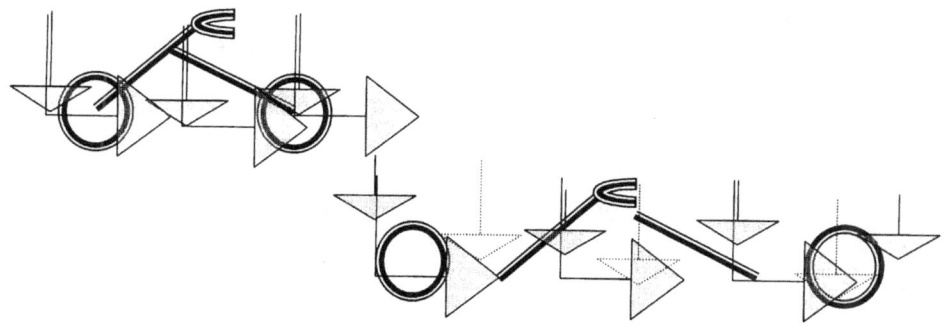

The motion provides the bicycle that is already confined to share space with the Earth its due in motion. The bicycle can only have independent motion if and when the bicycle has the correct number of atoms filling the space the bicycle holds that will grant the bicycle independent motion. Without the unit being able to concentrate the correct amount of heat that will enable the space to generate motion, such motion under cosmos standards does not exist. In fact under cosmos standards the entirety of the bicycle unit does not exist. Only when the unit forming the bicycle can hold the number of atoms, which will produce the amount of motion through which the required heat will be subtracted from time in order to generate the gravity or motion needed, will the bicycle under cosmic standards gain motion. Before that the cosmos does regard the bicycle as the Earth and only the atoms standing independent from the Earth seems to regard the bicycle not being part of the Earth. As the bicycle confirms the required and the displacement to launch an independent duplication wills the cosmos regard the bicycle in such a manner. When the atoms forming the bicycle unit are able to condense from time enough heat to give drive in order to provoke independent motion does the cosmos put the emphasis on the unit as a star. But in order to be the carrier of such independence the bicycle will have to grow. On its own as it is there is no way in hell that the bicycle can get fired up and start going as a star, cosmos style. If not for life it will go nowhere but be consumed by the Earth to become a part of the Earth in time. In the cosmos the atoms forming the unit provide the motion and only when the total effort of the combined unit atoms manage to move abruptly and with the required confidence can the gravity be generated where the gravity generated is the motion of the entire unit. That means by having motion the fact of having motion grants the bicycle unduly respect from the Earth. The bicycle having motion of any status enlarges the bicycle in atomic space in ratio to what it has when without motion in relation to what the Earth has when only the Earth provides motion and when the Earth stands in size in relation to a static bicycle.

When the bicycle is motionless, the bicycle is part of the Earth by gravity applied. As soon as life steps in and brings about separate and artificial motion but still uses the support of the motion that the Earth provides, it will inevitably do better than the Earth as long as the motion that life provides is not in conflict with the motion the Earth provides. The bicycle becomes an object with the ability to transform the direction of the Earth's domineering motion by redirecting gravity there and find the ability to change the direction of gravity. **When gauging what happens we must also admit that it is highly unlikely to find running rocks on**

Venus or moving craters on Mars. The motion that applies to the bicycle is an extension of the second force in the Universe, which is not part of the Universe and only affects a very small part of the Universe within the Universe. Life giving motion is an alien product and gives an unrealistic adding to the Universe. The motion however has nothing to do with mass but is only extending what was to what will be through what is. It is refurnishing what will be with what was before it now is present.

When the motion of the bicycle accelerates such points that are forming by the increase in motion, the forming connections extend to match to motion putting a standard of duplication per time unit extra to what previously was the norm. More space fills in the same period or the period reduce to match the filled space. The motion then contributes by increasing the space factor to keep the commitment with gravity valid. The bicycle breaks its form but because it is structurally bonded, other aspects concerning gravity have to commit to the breaking of space. When expressed extremely crudely it is put as follows but is very bluntly stated.

Yet, it still is the best way to explain the basics of the sound barrier. The bicycle is the compiling of the independent space within the atmosphere space or time concentrated to be more exact, that is holding the motion in duplication where the motion is continuing from a facet going to the next facet by duplication what was through what is to what is going to be. While the bicycle is filling the space, in motion that is part of the space holding all aspects within by the atmosphere and the atmosphere is holding all that is in it together in the atmosphere of the Earth. That is time performing in ratio to space filled by material. Because the conflict the gravity experiences by having motion within motion, gravity first tries to break the object that is in independent motion.

As the motion continuous and as motion extends, the reflex to the situation is then to contain by the breaking to the connecting devices such as the sound waves in the adjoining space. The atmosphere does the breaking of the relevant **k** on behalf of the object in motion since the moving space holding the object in motion as a unit shows much stronger bonding in structure unifying. We experience such breaking of space as the breaking of sound, which is showing motion or gravity differentiation.

The wing holding singularity while maintaining singularity puts the object in a Universe apart from other Universes. The aircraft becomes a separate identity maintaining a singularity and as all singularity does, such a singularity insist on two factors. The one factor is duplicating by motion the production of space-time while the other is the dismissing space-time, by controlling of space-time. The contact with space-time allows the motion to present the wing (and the aircraft) as being much bigger than in reality because it is not only the size of the space-time that maintains the singularity but it is also the contact or relevance which holds the dimensional area of control which stands as a controlling factor.

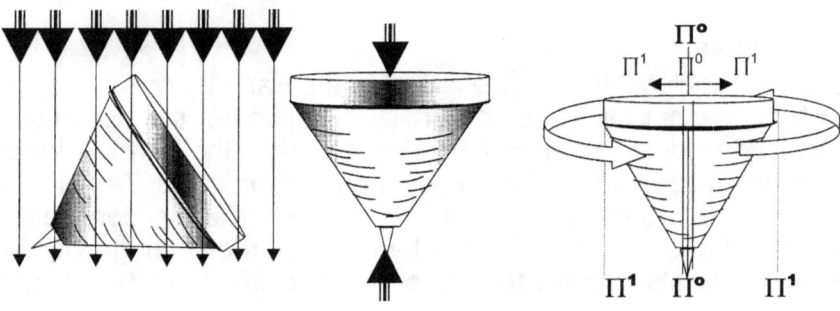

These factors combined make the aircraft become bigger as the capability of motion suddenly allows the relevant size of the aircraft to grow in stature. The contact that the air or space has on one side which is more than the contact that the wing has on the other side where airflow is restricted, makes the side having more airflow larger by motion than the other side has in size by restricting motion. The motion enlarges or restriction reduces the contact and therefore the size per time unit. This alters the size as much as it redefines the balance and that makes the craft fly.

From the allocated position we hold the bicycle seems to be stationary when it is not moving. When we stand still it seems to us that we are standing still when we are not moving. However that is a human conception like mass is and is far adrift from a cosmic reality. We move as fast as the Earth spins.

We move as fast as the Earth rotates around the Sun. We move as fast as the Sun rotates around the Milky Way. We move as fast as the Milky Way is rotating around another common centre because there shall be such a common centre that is allocated to order another common centre and this role diversification goes on running up to eternity.

When any object is not moving, the object forms a part of the object, which holds the first object, captured. The cosmos disregards the existing of the first object in the event of it being stationary. However as soon as motion applies the cosmos grant the object having motion a position of existing by recognising it as an independent Universe that is entitled to all the privileges granted to a Universe.

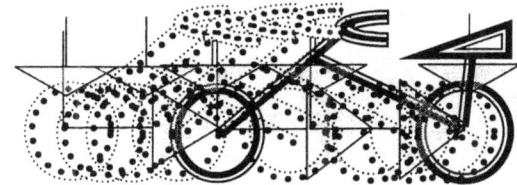

As soon as the motion enters the equation the bicycle gets cosmic status because the bicycle generates time in the manner of parting singularity with time. The bicycle achieves cosmic independence and Universal recognition as an independent cosmic Universe. The bicycle is keeping upright because it forms a relation between time and singularity. The position is changing because time is changing singularity by the movement through time that gives the bicycle the opportunity to remain upright.

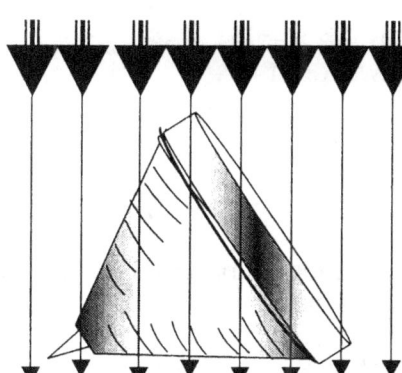

The position the bicycle had in the past taking the bicycle through the present into the future is

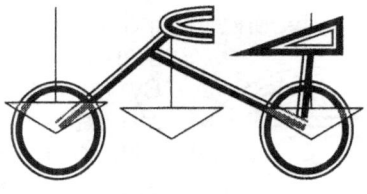

what is keeping the bicycle in balance because the one position in time is supporting the next position, which is supporting the next position. It is the fact that time has one position, which supports the next position by supporting the next position, and this support forms a line, which brings about the balance by which the bicycle stays in line. Once the motion is no longer present, the bicycle – Universe collapses.

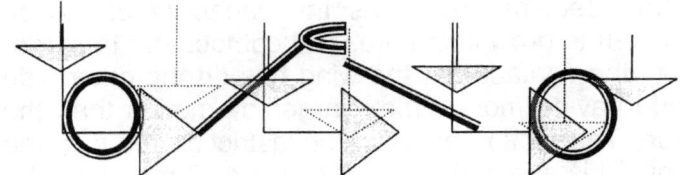

As the bicycle is standing still it is still in motion but holds a position to time where it fills a certain volume of space in that given time in motion. It is Kepler's $a^3 = T^2k$. Let us now forget the fact that life is responsible for the motion and pretend it is all a cosmic affair. Since the bicycle moved faster than what it did when it was within the Earth's motion it now fill more space than what it did before the individual motion commenced. In having individual motion on top of the motion the Earth supplies, the bicycle is filling more space than it did before when it was stationary.

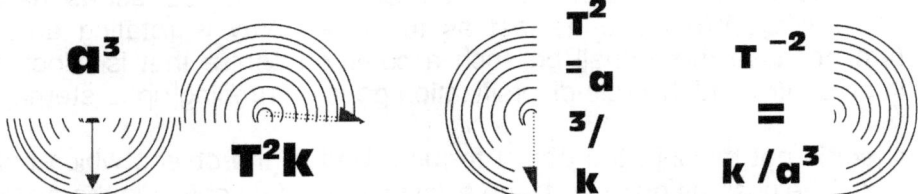

$$a^3 \quad T^2k \qquad \frac{T^2}{=a^3/k} \qquad T^{-2} = k/a^3$$

As Kepler stated there are two positions to the Universe unit. There is always space confirmed by the motion thereof in relation to the motion thereof.

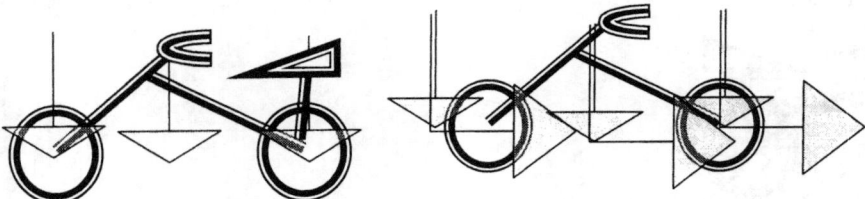

The motion fills space the Bicycle fills but also the bicycle fills space the Earth provides. When moving, it is filling more space than what it was filling when it was stationary and with mass and was just more Earth within the Earth. When it is with mass, it is in a position where it fills space as if it is Earth, which it then is being space within the Earth and of the Earth. By moving however, it is standing in as more space being filled, than it did when only the Earth provided such space through motion that only the Earth provided.

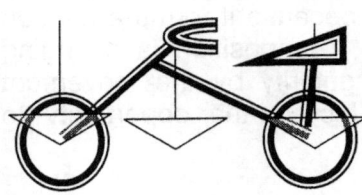

When moving, it takes up more space being what it then is, as an independent and individual particle within the Earth's confinement than when it is in mass and being part of the Earth. By having more duplicating motion, the object is more than what the Earth provides when it stands still. When moving it is bigger in it's filling of the Earth space than it did before it started to move independently. It has more of the gravity going around than it had before when it was only dependant on the Earth to allow gravity going around. The bicycle grew bigger by the same margin that the Earth grew smaller in ratio to each other.

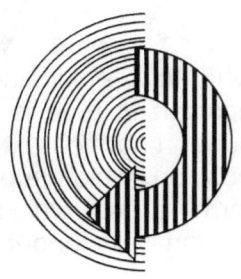

The bicycle without motion forms part of the space, which the liquid space confirms as part of the space. The bicycle then forms part of the other side of the Universe that is part of the solid.

In the motion we have, there are two contributing factors that play a crucial part in the dynamics as to how the material in forming the atom construction reacts on such motion. The atom comprises of spin around the proton by the electron.

Any motion exerts an influence by all factors on one another and the dynamics is irrefutably connected. Yet at the same time the factors are so small it can never be detected and yet it is so huge it spans across an entire Universe. Remember every atom is a Universe in its individual making.

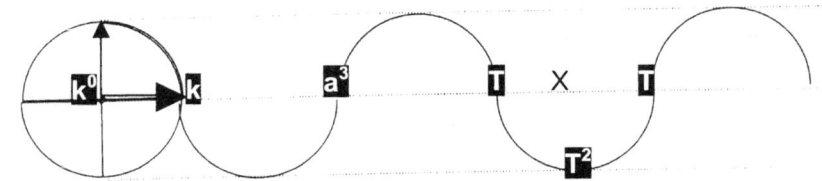

The moment the bicycle moves, the bicycle switches sides and switches allegiances. The bicycle then becomes part of the liquid that the space confirms not as space but as the liquid that extends the space. It then forms the gravity adding space instead of the gravity confirming space. It becomes $k^{-1} = T^2 / a^3$ instead of $k = a^3 / T^2$. What applied before then does not apply any longer because the bicycle is then on the other side of the Universe. Yet it is much more complicated than that because when stationary, the bicycle was part of the Earth extending solid space which is gravity in relation to contracting being $k = a^3 / T^2$. As soon as it moved it became $T^2 = a^3 / k$ which is relative to the old position $T^{-2} = k / a^3$. **This too has no principles that it can share in the mass applying gravity idea. It is about being stationary in relation to moving.**

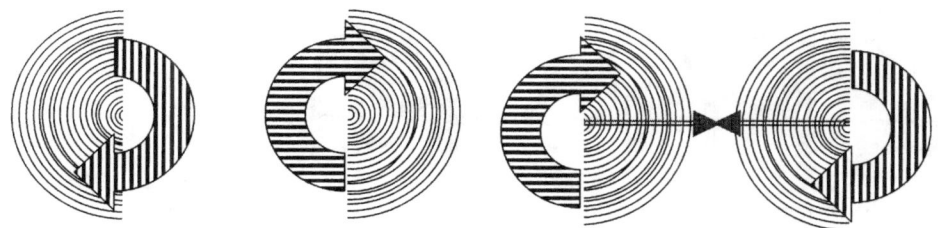

Taken from the Earth's perspective there are lines (**k**) running at 90^0 to the rotation of the Earth T^2.

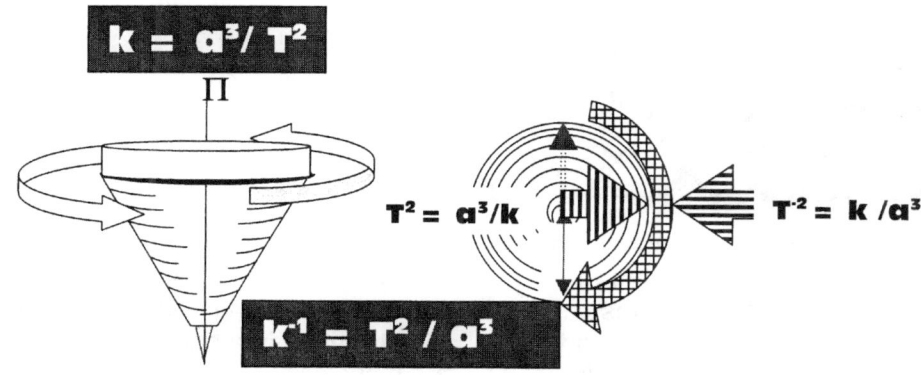

$$k = a^3 / T^2$$
$$\Pi$$
$$T^2 = a^3/k$$
$$T^{-2} = k /a^3$$
$$k^{-1} = T^2 / a^3$$

The Earth gives the bicycle one line in singularity in which to confirm its position as far as the space the Earth grant the bicycle to manage. That gives the bicycle a specific space to hold in relation to what the Earth has and in relation to what the Earth offers the bicycle. That is gravity... however, as far as the evidence at hand shows, we can see that it has no obvious principles that it shares with mass pulling mass. It is being on the one side of the Universe which space holds motion in relevance $a^3 = T^2k$. When moving, every independent part of motion finds the alliances switching, where the bicycle forms legions with motion by duplicating the space it holds in relation to that which the Earth grants $k = a^3 / T^2$ and therefore the pace it holds becomes part of having more space because of motion than that space it relies on when it only depends on the Earth's motion where motion forms part of time being $T^2 = a^3 / k$.

There are always two sides of the same Universe forming one Universe. There is the space extending to confirm the liquid by producing a solid and there is the liquid attaching to the space to extend the space by motion that secures the space. The one stands related to the other by opposing the other.

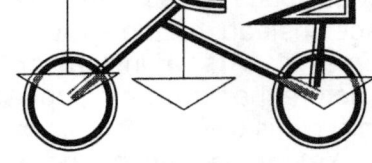

When the bicycle is part of space, the motion confirms it by implicating it as space in the motion, which forms the extending of space. In that case the motion forms T^2 that produces the contracting lines in singularity that runs a reducing and reclining formation into the centre. When the bicycle is a part of the space the line attributes to its space disposition and the space the bicycle holds allocated the bicycle the position it has.

There are always two opposing time lines forming one united space. The one is the line k and the other is the half circle T^2 where from those perspectives there then forms the triangle a^3. By moving the bicycle then forms the 90^0 cross reference to the allocated position it had before. It is $k = a^3 / T^2$ or it is $T^2 = a^3 / k$.

The moving of the bicycle involves duplicating the space and the position the bicycle has in relation to the space the Earth allocates and the position the Earth allocates to the bicycle. The faster the bicycle goes is actually the number of times such repositioning of the space the bicycle holds are in response to what the Earth allocate and what the Earth takes in a specific period of motion of the Earth. When faced with the question of how the bicycle manages to stay upright it always comes down to charging the achievement to a balance…but a balance of what? What goes into balance to achieve the upright position? The bicycle repeats its position in relation to the position the Earth grants and when the repositioning of the bicycle is faster than the re-allocating of the earth, the duplicating of the bicycle from one position to the next will sustain that the bicycle can cross the vertical lines faster than the vertical motion will effect the stance of the bicycle. The bicycle firstly crossed its allegiance by no longer forming a partnership with space but becomes a factor of liquid presenting motion.

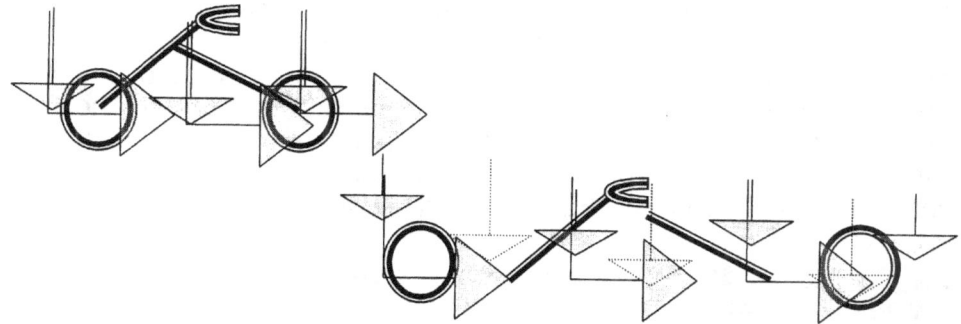

With the motion the bicycle is duplicating in ratio more space than what it had while being

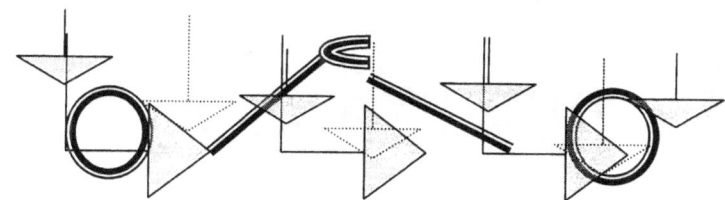

part of the Earth. The motion holds more space because it holds less space per time unit and there are more units of space per time unit. Since the bicycle is propelled at a faster pace than what it was when the earth alone supplied the space and forced the time by establishing the duplication tempo in the time contracted. By supplying more motion the bicycle grew larger in ratio to that which it had when the Earth was the sole space-time provider. After accelerating the bicycle then has more space in relation to what the true status is. The bicycle holds a larger part of the Earth by which the Earth has to reduce the space it offers the bicycle. The earth had to shrink in order to provide the bicycle with more space.

In the normal relation that the bicycle has with the earth when the bicycle is motionless (or in the manner we think about the status of the bicycle in terms of only its position within the scope the Earth provides) being motionless and then started moving, the space increased rapidly as the Earth space decreased rapidly.

With the new motion the bicycle finds much more duplicated space and that disturbs the ratio of space shared by the bicycle within the confinement of the Earth. The earth now presumes the bicycle to be much bigger than what it was. By moving the bicycle physically got larger and this is a fact not only by relevance but also by actual annexing and capturing of space in any given period of equal ness. The earth provides a certain value but as the bicycle moves faster the bicycle annexes more of the Earth space and that improves the size of the bicycle in a volumetric and physical measurement. It is a^3 growing bigger, therefore the earth a^3 has

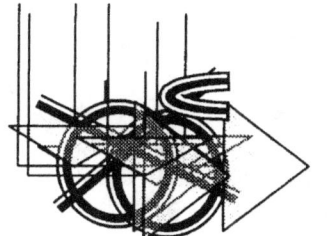

to compensate by reducing that much actual space. In that a problem arises.

The bicycle does not even contribute a morsel of space when compared to what the Earth delivers. The earth that actually became smaller resents it becoming smaller by the demand of such an outrageous exploiter. To the Earth the bicycle is motion and the motion is liquid and therefore taking space contravenes the being liquid part. Being liquid is also being heat but being heat means becoming hotter when becoming more. To the earth the moving bicycle is liquid motion and being more makes it being hotter and being hotter is therefore the same as being more volatile. Therefore the Earth refuses to become smaller and the bicycle being space claimed cannot become hotter without destroying its independent molecule unit.

By the size the Earth holds the Earth will not allow such a renegade to grow bigger and the Earth sanctions the time in space in accordance with the Earth generated governing singularity. Since the generated singularity the bicycle has still adheres to the dominance that the Earth generates, the bicycle reduces its proportional space in the face of the Earth showing such a strong reluctance to abide by the will of the smaller bicycle. The Earth crushes the bicycle in response to the bicycle growing and when the response is more than what the bicycle can withstand, the bicycle crushes buy reducing space. That is what happened to Challenger when it entered the Earth and was liquefied by turning into gas in time.

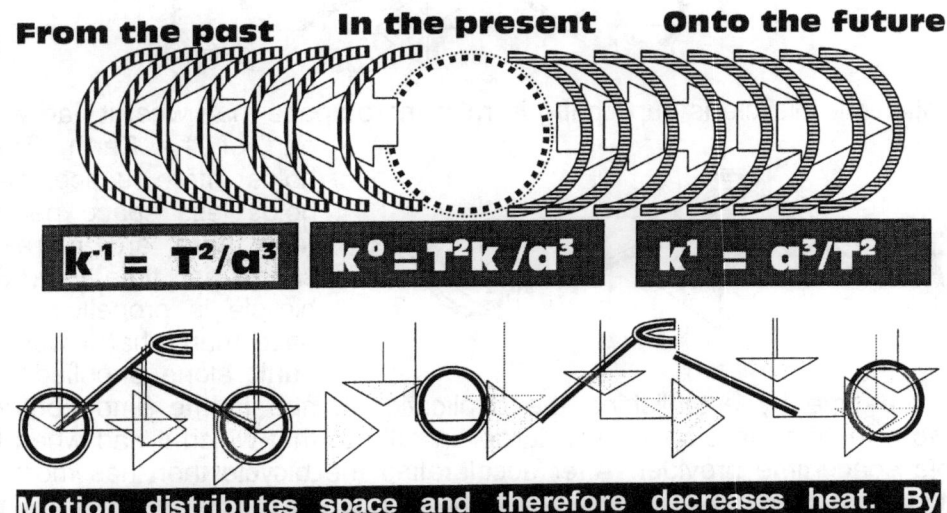

From the past   **In the present**   **Onto the future**

$$k^{-1} = T^2/a^3 \qquad k^0 = T^2 k /a^3 \qquad k^1 = a^3/T^2$$

Motion distributes space and therefore decreases heat. By spreading the space over a larger area the heat in the space is reduced because the density of the heat allocated to the space reduces. Motion decreases the heat and therefore the Sun is the coldest place in the solar system while outer space is the hottest part of the Solar system

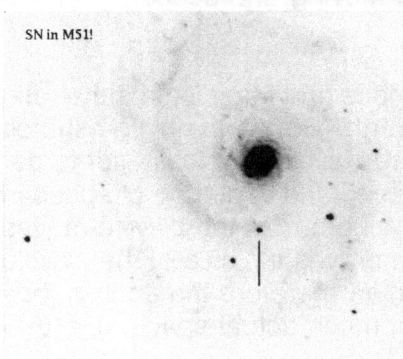

SN in M51!

The ratio between space and heat goes into an imbalance where the liquid that is standing as ($k^{-1} = T^2/a^3$) totally dominates the space factor $k = a^3 / T^2$. The cosmos informed Kepler of another gravity, which the cosmos applies much more widely and is used by nature all over the Universe. Being with life and being part of life, we humans take motion in content and as part of our daily existing because life is independent motion that is not part of the cosmos. We regard motion as nature because to life's abilities, motion is nature. That is not the case with the rest of the cosmos. Life is the manipulation of space-time by motion and since we can move because life is about motion we

generalize motion with contempt. Motion is the most complicated process in the Universe because the Universe is motion. But why would motion occur because the fact of motion proves the presence of a Universe.

Supernova stars are stars that overheated. The overheating came as a result of the motion in duplication that was not in ratio with the control in contraction. The expanding of the space the supernova held before, had nothing to do with mass. The expanding had nothing to do with Newtonian gravity. It is all a ratio that puts a relation

between the space that is taken by the material in concentrated heat (or time) and the time in ratio to the duplicating material. If the material forming the motion is not very highly dense the motion is poor. By the same token is the material density high when the motion is volatile. It all depends on the motion that

the atoms forming the Unit generates and that motion is forming the gravity. There is no mention of mass except in the imagination of the Newtonian mind.

There is no pulling between particles compressing them into plums of pressure. That is nonsense because a star has no containing wall on the outside to capture the pressure on the inside. It all depends on the duplication of material confirming the containing of the unit in relation to the distributing of the material in ratio with the liquid time with which it is in partnership. It is all about containing heat and distributing heat at the same time. The more motion that is present, the more heat is condensed and is contained.

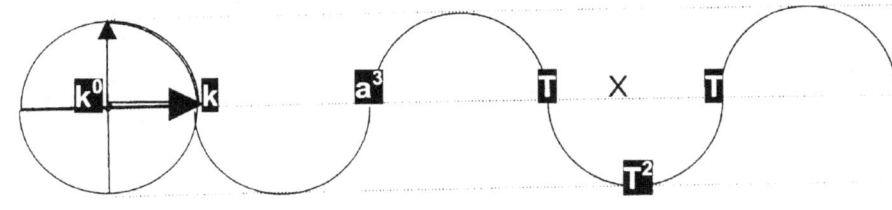

By duplicating at a higher ratio the heat contained is spread over a greater part of time, which reduces the

heat contained in the material as it is distributed in a larger ratio to the liquid heat that is maintaining the balance. This again represents the boat rushing through the water and the water holding a relation with the surface of the hull it is in contact with.

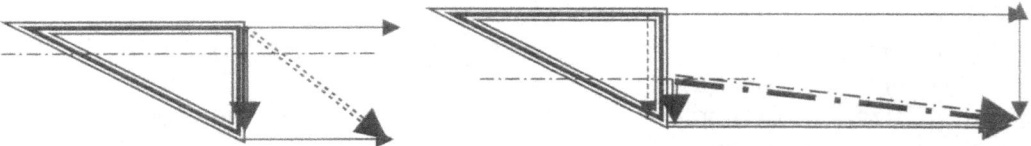

The water the boat moves through is more per time period, which makes the hull moving through the water being less during the same time period. That places a cap on the speed that boats can move through water and Donald Campbell the ex world land and water speed record holder reached such a cap when the met his demise during his fatal last attempt to improve the water speed record. I go extensively into this matter in the book *Inter Galactica Space Travel*. We find the same being even more prevailing where it and solids meet.

There is a defined ratio between liquid heat that is the basis for time and solid heat, which forms the norm for space. The higher the ratio favours the liquid, the colder the solid will be 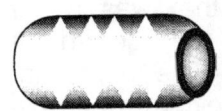 and the lesser the solid is in contact with the liquid, the hotter the solid will be. Stars are ice-cold ice cones floating in outer space because of the motion they generate. If we pump air into a compressor the air gets more inside the compressor. The compressor gets hot while the air gets more. The size or the compressor remains the same while the air gets more inside the container and that means the compressor is shrinking while the air is remaining the same because the air cannot get more while the compressor is unaffected.

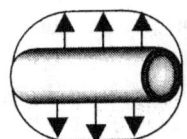 In relation to the heat, the heat gets more because the surface of the compressor remains the same while the size of the compressor within the relation shrinks. Because the size of the compressor that shrinks, the space outside the compressor has to accommodate the flow of heat because equilibrium has to be re installed. The size will remain reducing up to a point where the compressor is just too small to accommodate the air. The air then will expand. However the air was always expanding from the pumping started because the compressor inside became smaller as the heat level rises to expand the size of the inside of the compressor to match the outside of the air.

The same goes for material blown by wind to reduce heat. The object has an indicial size to 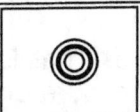 start with. Then we put heat to the object and the heat makes the object increase in size. That is hardly the increase worth noting because the relevancy of heat in the air to the heat in the heating object goes array. The heat has to increase the size of the object in relation to the match it has to find in the space it is within.

With the heat coming into the object, the relation the object has with the heat or air outside makes the object that many times bigger because the ratio in the heat balances is disturbed. If we blow air over the object we increase the size of the object by allowing the surface of the object to make contact with much more air in the same period of time, which will bring the size of the object back to normal because in relation and considering the contact with air, the object expanded by the motion of the air in contact with the object.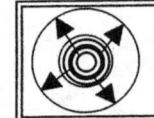

In the normal flow of time the object has a heat to space relation set by the time the Earth dictates. Then we go and increase the heat on the object and in that event we actually 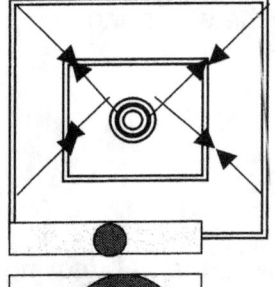 increase the size the body has in relation to the heat in the air. By blowing air over the body we increase the air and therefore we increase the size of the body in the same period of time.

There is now a dispensation of many times the body carrying more heat and contacting many times the heat or air which brings the equilibrium back to normal. There was a body size and by applying heat the balance shifted to the reducing of the body size in relation to the heat. The body then had to expand in heat because the body was too small to incorporate the large heat. Blowing the air over the body increases the size of the body and heating the body decreases the size of the body in comparison with the air it comes in contact with. The body is either expanding or the body is redefining and the balance in heat places the body in relation to either gravity cooling by contraction or expanding by overheating. The very same principle applies in the sound barrier.

In spite of outer space being as expanded as anything can get, we still regard outer space as being incredibly cold. Anything expanded to its limits and which can heat no more, is as hot

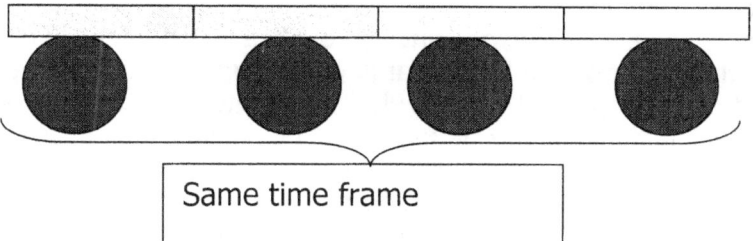

Same time frame

as anything will ever get. Outer space is the very edge of expanding of space where heat cannot expand into space any more. On the other hand we find that concentrating heat is producing cold making anything in the atom in stars as cold as it can be anywhere in the cosmos.

That heat filling the outer space lacks motion and is therefore space in another form of material that could conduce by diverting from space to constrain further expanding through motion and therefore was unable to marry the union of space by becoming more space. One must look at outer space and judge outer space from the findings only considering outer space. This suggests strongly that we'd better be getting very suspicious about the idea of mass contributing to gravity. But in contrast to this, science is unshaken about their confidence in the perfection about facts they use in terms of correctness. It is well known amongst all persons that science only uses dependable and ultra reliable facts coming from sources beyond doubt.

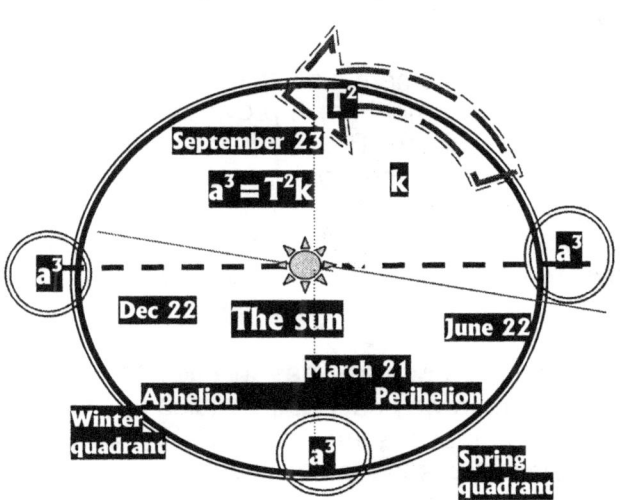

Referring to any work done by any scientist will find a remark about science only accepting facts they use to work with. It is accepted overall by all communities that in science those in science use one hundred percent accurate facts or they use no facts at all. If our view was as perfect as science would lead us to believe it then must be the Universe that is imperfect as it otherwise would not behave so mystifyingly. The unshaken confidence science uses has us believing at first consideration that the drawing of gravity should produce an even diameter positioned between the Sun and the planets because of ever dependable evenly distributed gravity... but I believe there is a perfect Universe and our understanding carries the doubtful suspicions.

Delving deeper uncovers even more contradictions and the level of accuracy contained by our scientific understanding then arouses more suspicion about the correctness of science. Remember Newton changed what the cosmos told Kepler leaving much suspicion as to how far the misdirection takes science. We have to correct the facts we doubt because when correcting the facts they use in science concerning our view about science, such correcting brings along a better understanding and then the Universe has to become ever more perfect as one learns to understand the perfect Universe even better. But it does require an open and clear mind and it needs no culture driven preconception that should confirm interpretations about facts surmised even before they are carefully studied. It becomes obvious that Newton never gave careful attention to Kepler's findings because if he did, he would have seen what gravity was. Kepler described gravity without using the name that later was given as 'gravity'.

By motion, space duplicates and by space halving, it removes heat in space as well as by dismissing space. In the case of material the electron is spinning at the speed of light to contain the heat inside the atom at a higher rate than the speed of light. It is containing heat at a greater motion than the velocity that light can travel. The atom by motion is the condensing of heat by contraction in relation to expanding or duplicating of heat. The concentration or release of space with heat or space from heat is a direct contribution of the singularity in control of the space-time. The regard of the singularity stipulates the conducing of heat in space or the release of heat to form space by means of bisecting the occupied space. By applying motion the space duplicates what it is from the past through the present and into the future. This is no hypothetical suggestion but is the actual flow of time coming from outer space as a liquid and is incorporated into the spinning atom on the way to confirm singularity.

While we are in gravity the manner in which gravity applies in our use of gravity makes us part of the Earth by mass forcing us onto the Earth as a semi unit with all other Earth belongings. Is that which we have truly gravity? By using mathematics, the cosmos spoke to Kepler personally and by the use of mathematics as the medium, it provided Kepler with information about the cosmos coming directly from the cosmos. The pictures we see coming from the Hubble telescope shows why, in the perfect Universe…but can the Universe be perfect when... we see a radius between the Sun and individual planets is not using a regular distance as one would expect of gravity in being a force driven by the mass and in that sense the mass is producing the gravity that always remains even because the mass doesn't alternate.

As the mass is never changing on either side, that steady mass has to keep the gravity steady. But in our imperfect understanding of the Universe we find that the radius that should be constant varies considerably proving either that mass somehow adds by measure unnoticed while the structure is in orbit and later allows the same amount of mass to escape undetected; or it's the seasons adding and removing mass at will. This is an absolute contradiction to reality if mass was the factor determining the radius we find between the Sun and the planets.

Kepler did not give the name gravity, but Kepler's studies gave Kepler the insight to coin the concept of gravity. Nevertheless it was a name and not the concept that was later named by Newton. The naming was the contribution of the Englishman. The concept that Newton later introduced is totally incompatible to the concept that Kepler introduced. What he (Newton) introduced as the force of gravity, he connected to mass, which diverts totally from Kepler's findings. With giving a name, the Englishman also changed the concept that Kepler introduced. Kepler made no mention of size or mass as part of the phenomenon that later was named as gravity, yet it must be gravity that holds the Universe together.

The Englishman changed the concept simultaneously with the introduction of the name he gave to the concept. That what he introduced, he corrupted beyond recognition. The concept that accompanied his new name strayed completely from what Kepler introduced. Newton brought in something that was mismatching what Kepler saw in Kepler's view of the phenomenon that holds the Universe true to form.

The name was dominant but even more dominant and totally inaccurate was the other concept Newton introduced. In truth Newton only gave the world a name of an idea, which he then corrupted as far as cosmic physics are concerned. It is important to admit that as far as cosmology is concerned Newton gave the concept the name but *only* the name and not the concept of gravity. Newton's persuasion on matters of gravity as gravity functions between cosmic structures orbiting one another as we find in outer space is inaccurate.

What Kepler saw Newton saw differently and used the opportunity that Kepler left by not giving any name to the process he (Kepler) and Tycho Brahe worked on for two life spans. Newton did seize the opportunity to name what he, Newton, saw but that what Newton saw did not include that which Kepler uncovered. In Kepler's era the name or title was lacking but Kepler established the concept of gravity and the formulation thereof. The concept came from Kepler even before the name gravity was used by Newton to describe in the concept of whatever we today (after Newton) became accustomed to believing what the concept of gravity is about. With the help of Newton everyone since Newton confused Kepler and Newton on the issue of gravity and this confusion even begins with Newton. Gravity might not have been named but became a proven concept and factor after Kepler formulised it, which is before Newton named it. The concept of gravity that Kepler saw is about the manner in which the structures orbit because there is a space that circles around a centre and this process has kept planets secured, connected and rotating around the Sun, which is the same concept that is keeping the Universe secure and comes about with a process Newton later named as 'gravity'.

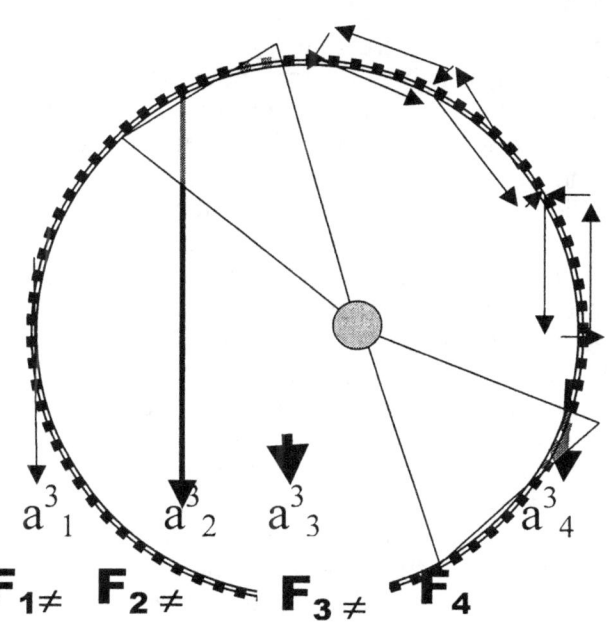

$$a^3_1 \qquad a^3_2 \qquad a^3_3 \qquad a^3_4$$

$$F_1 \neq \quad F_2 \neq \quad F_3 \neq \quad F_4$$

What Kepler saw is not the same as what Newton saw when he saw two objects drawing closer by pulling on each others mass. Then later on Newton named, what he thought he saw as the force that Kepler saw but introduced another completely different concept. Kepler saw cyclic formations keeping the Universe together and never approaching each other. Newton ignored what he wished not to see but he changed it as he saw fit and what he thought that should be. His experience as a young man drove him to establish a process he formulated as the process that is keeping the Universe together. In that act he corrupted as much as ignored the work of Kepler, which he also named as the same gravity that he saw as a young man. Why he chose to ignore Kepler's findings on gravity we shall never know but why the world still chooses to ignore Kepler's findings about gravity almost four hundred years after the fact, I shall never know. My saying this has literally made Academics avoid me as they would avoid the plague. I am not pretending nor do I exaggerate when I say there were those in Academic institutions that questioned my mental development. Some went as far as seeing me as a joker of sorts and I have correspondence to show evidence to that fact. I know by now while Newtonians are reading this letter I have aroused the tempers of every Academic reading this far, therefore let's see what is being ignored by the Academics which I blame to do just that.

We live through seasons which comes from being that at one point, (a^3_1) the distance between the **Sun** and the Earth **is less than** at another point we call a^3_3

Let us put a value of $a^3_1 =$ one and $a^3_3 =$ three. This means that each year, for the past 4 500 000 000 years the effect of the common gravity between the earth and the Sun has a greater effect than at another point six months later.

That means at one point the earth should be drawn or pulled closer to the Sun and after another six month's interval the earth should stand less effected by the Sun 's gravity, therefore it should move away from the Sun.

Kepler said gravity in space is about the area a^3 that would always keep equilibrium with the time T^2 it takes to travel the distance of the full circle position placed by the indicator **k**, therefore adjusting **k** as the need arrives. With **k** shifting in length a^3 will have to readjust and therefore T^2 will find a new relating value each time. This was the findings of Kepler and came after his intense study of orbiting planets.

Before I attempt any investigation into this matter, there must be coherence in our agreeing about what gravity is. If you, the reader, insist that the falling of objects is the only gravity found, any further reading will convince you of little else. Anything we do decide upon must support the fact that it is gravity that prevents planets from dislodging from the grip the Sun has on them. Gravity is not about the Sun trying to catch the Earth by attracting the Earth…no, there is so much more to gravity. We must be under no illusions about what gravity is and that being the focus of our discussion and where that gravity is because we have to identify and not confuse the gravity we are looking at. We are now discussing the gravity, which is keeping planets circling around the Sun, and stars around a specific galactica centre. In that we do not find one example to use as proof in connection to stars coming tumbling down on galactica centres and crushing into galactica centres. If that is gravity keeping structures in orbit around specific centres we must look at the behaviour of the structures in gravity. We have to find a reason why the planets do not reduce the radius between them as Newton suggested but we must trace the reason why it is gravity, which is keeping them apart because if anything, they are departing as they are extending the radius connecting them to the Sun. That is gravity because it applies throughout the Universe. The gravity Kepler found is the general gravity that is keeping structures from colliding and in that the principles are avoiding collision or on the other hand avoiding abandoning each other. It is about confirming respect for one another's independence and clearly staying at a predetermined distance while at the same time both are sharing a common space unit. That then must be the defining of gravity we have to study to find the Universal enticing gravity holding the Universe together. By close investigation one will find three factors in urgent need of investigation. There is firstly a centre that draws the object closer. This gravity is clearly a synonym to what Newton saw as gravity. If it were not drawing the object closer the object would not be orbiting around the centre and applying motion. It will draw and absorb all rotating things in its field of gravity.

The fact that it does not draw the object into its ranks is because there is another gravity standing alongside this first mentioned gravity. Our recognising the first gravity forces us to accept the presence of another part of gravity. This forces us to recognise the second gravity. When saying this we are not using Newton's cosmic formula concept $F = G\,(M.m)/r^2$ because that can barely be what is out there happening. What Newton saw was falling. If that which Newton saw is the only gravity then whatever Kepler saw including all other parts of everything out there that are spinning around some centre must come closer to one another and connect in collisions. While that is not happening we must start to look past Newton to new grounds we can investigate. We have to go beyond Newton and admit there is more than that which Newton led us to believe because it is clear that what Newton had us believe…is not happening. That confirms the presence of the second gravity. The fact proves that everything is departing and not arriving. Even the moon is drifting away from the Earth and this information comes about from the most advanced investigation up to date, including a moon visit and the placing of measuring devices there.

Looking at the gravity intensely we find the roving structure travels in a straight line, which repeats another circle around another centre but because of the influence of a centre keeping the roving structure attached to such a centre the motion allows a circle to form by reforming motion from the original straight line to that of a partial circle. There is a centre; a connecting line travelling between the two points that form a line when the points are connected, which establish the specifics of a centre within a circle and the end of the circle.

According to Newtonians the centre supposedly draws the rotating object closer. That is half the story.

I suggest we do some deliberation and in deliberating may I remind you THAT NEWTON'S OWN LAWS ARE IMPLIED, and again the planets disobey these laws completely!! In the modern age all evidence points away from contracting and favours eternal expanding.

The latest news confirms that the lot is apparently not coming any closer!

Newton saw his apple fall and then went on to blame everything on mass... and you think it is all that plain and simple? Kepler on the other hand said (and let's forget what Newton said

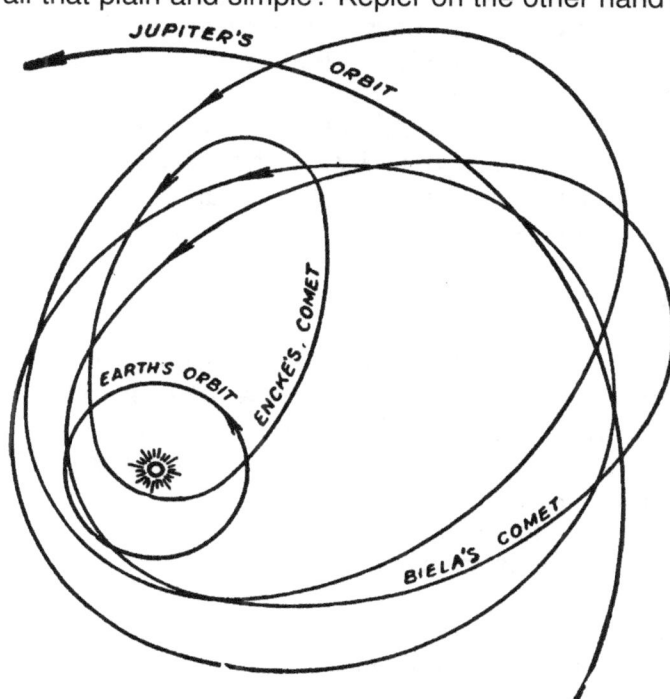

about what Kepler said for a bit) that a space of cubic proportions a^3 that will always keep equilibrium with the time T^2 it takes to travel the distance of the full circle at the distance (or relevancy) in ratio or relevance k with an indicator pointing the distance the circle is from a specific centre. That means k is as crucial as T^2 in positioning a^3. In placing the allocated position a^3 requires in determining the sectors (we think of it as seasons) a need arrives to predetermine k in order to measure T^2. Every spot a^3 fills is located at T^2 and is allocated where k indicates. When k shifts relevancy (from Earth to mars or even from season to season) the space in a cube a^3 will have to relocate as well as readjust and with that T^2 will be redefined. This was the findings of Kepler after completing the work started by Tycho Brahe and then later himself. The line forming looks straight from the onset but the line never moves straight but goes bended in relation to the relevancy that k indicates. In what k contributes as a factor it introduce T^2 and from the alliance of the product that $T^2 k$ delivers we find that a^3 forms an eternal circle about an eternal centre.

That is gravity if you wish to call it gravity by name. That is the force that is no force that prevents planets escaping from the grip the Sun has on every planet. Gravity is what puts

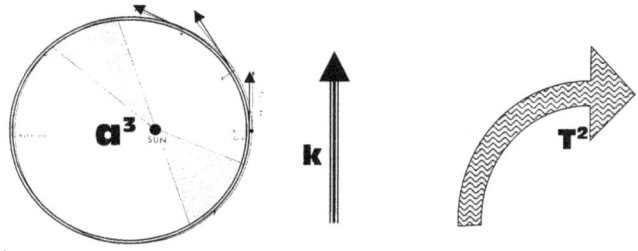

order to what would be the most chaotic arena there can ever be. That gravity has no bearing on mass because with mass pulling that order which gravity then would bring will result in complete destruction. That is not happening. On the other hand if we don't do what Newton did by putting words in Kepler's mouth we find that Kepler gave gravity a completely other meaning. Kepler said (when ignoring Newton's uncalled for interfering and meddling) that the motion T^2 puts the space a^3 at a distance from a centre and the relevance factor k prevents a^3 to come close or drift further away but stay in the allocated position where T^2 locates a^3. The factor **k** has it at task to prevent a^3 coming closer and orders T^2 complying with a measured value. By denouncing **k** everything Kepler said goes array.

> $a^3 = T^2 k$ **then** $k^3 = k^2 k$ **and this is showing that the space** k^3 **is equal** $=$ **to the motion** $k^2 k$ **of the space** k^3 **seen form one specific point.**

Every object in outer space holds as much turning as it commences its run in a straight line. It turns as much as it goes straight. The comet coming towards is no different from the comet going away and by passing a dividing line, they change direction from coming towards to going away but in that they remain equal. A space remains between the comet and the Sun that defines the line it crosses when the comet changes from coming to going. That gives the comet a cyclic approach and departure and does not put the comet on any collision course with the mass of the Sun. There is no death defying all destructive route calling for a disastrous end with no chance of avoiding the immanent disaster colliding that is unavoidable predestined to happen as Newton's formula would suggest. As the comet approach it is the relevancy brought on by the changing of **k** that puts a^3 in a ratio with T^2. It is k that sets the approaching limit as it sets the departing limit and it is k that prevents a collision as it prevents an escaping departure. That is what Kepler said...and that is not what Newton said Kepler said. What Kepler said is quite a different story from what Newton said Kepler said...

In Kepler's formula $k = a^3/T^2$ the smaller **k** becomes the smaller a^3 becomes and the bigger T^2 then gets. T^2 represents the gravity that positions the space a^3 at distance **k** from the centre capturing the structure through gravity applying T^2. We all are very aware which star is the mighty gravity producer. So has mass the least say when gravity is generated? It seems most likely to be true.

$k = a^3/T^2$; that which we see as distance depends on the position that the orbiting holds while the object develops space-time

$a^3 = k T^2$; the space depends on the distance the space develops from the centre and the speed the space moves around the centre.

$T^2 = a^3/k$; the speed the space orbits around the centre depends on the distance of development and the size into which the space developed.

Gravity has two factors influencing space, which are a straight-line **k** and a circle going around the centre T^2. $a^3 = k T^2$

Translating Kepler's mathematical expression $a^3 = T^2 k$ correctly to the verbal statement in English, Kepler said that there is a space a^3 which is equal $=$ to the motion in the time duration T^2 thereof between two specific points which holds a relation to a centre wherefrom there forms a straight line **k** and is located on the spot where space begins the circle. Therefore that spot has the least space.

The value of Kepler's space he indicated as a third dimension a^3 does not depend on indicating a structure a^3 that is in rotation T^2 but only needs one position having a constant of some sorts. Any point where **k** may indicate a position one will find a value matching a^3 and the matching location will fit T^2 at that point. That is the relation there is in the solar system between all planets and the Sun . The Sun always indicates the centre and the planets always indicate the rotation. But $a^3 = T^2 k$ is only producing a relevancy of three dimensions that is equal to two plus one dimension.

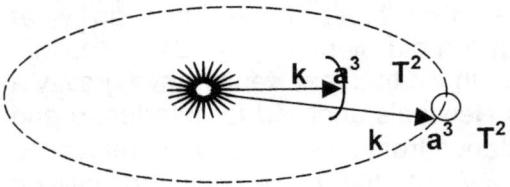

From the Sun there are three points moving between two points from one point to two other points giving six dimensions we find that is forming space. It is space in time or space converting space through the moving of space in time. It is

locating a point in the third dimension a^3 that will move according to the second dimension T^2 that will implicate **k** as a reference in the first dimension. It is the duplicating of space by time providing the dimensions to do so.

Let us take it from a point where the Sun provides a centre as one starting edge of **k** then that centre **k** will provide a line from the centre and the line **k** will provide three spots in a formation that produces a structure by the square T^2 of the dimension. Not once did Kepler indicate size as a contributing factor to a^3. That means every single point that **k** indicates, there are three positions a^3 implicating sides of a double dimension. In the same manner is **k** not limited to distance or is T^2 lesser by size. $k = a^3 / T^2$

There are infinitely more implications in the statement Kepler delivered than what is merely a contribution to motion and only motion as Newton was of the opinion. What is there mathematically not correct in my interpretation of Kepler's manner of translating mathematics to English and why is any changing thereof by Newton or any other person necessary in any way?

We can test any of the following symbolic values in the mathematical expression and also test the principals behind the expression in which Kepler stated them. By such testing we will find that time after time there were never any corrections in the translations required since the translation thereof was never incorrectly presented and in that a case asked for no alterations to secure the correct reporting of the cosmic information being translated. By taking the formula on face value it can change as follows: $a^3 = T^2 k$ can become $k = a^3 / T^2$

When translating Kepler's mathematical expression into English we can see what Kepler said also reads as $k = a^3 / T^2$ where **k** is one point from a centre point that is space a^3 relating to time T^2. From a centre comes space-time. The centre **k** brings space a^3 in ratio to time T^2, which is space / time a^3 / T^2. Reading this correctly cannot bring any dispute…yet it does…and it's been doing it for centuries on end!

The cosmos spoke to Kepler about space-time coming from singularity. Kepler gave us his findings. Any discomfort that may come when we read what is revealed must be set aside, because we must remember it is not me, or Kepler, but the Cosmos that is doing the revealing and lending us the tools we can use to decipher what the cosmos is trying to make us understand. Kepler translated what the cosmos told him (Kepler) as $a^3 = T^2 k$. Translating Kepler's mathematical expression $a^3 = T^2 k$ correctly to the verbal statement in English Kepler said that there is a space a^3 which is equal $=$ to the motion in the time duration T^2 thereof between two specific points which holds a relation to a centre where from there forms a straight line **k**.

What is there mathematically not correct in Kepler's expression and why is any changing thereof necessary in any way? It says where there is space such space has to move. Test the following symbolic values in the mathematical expression and test the principal behind the expression in which Kepler stated them. Convince yourself about the evidence that Newton saw what Kepler saw where the translation thereof that was done by Kepler is mathematically incorrectly translated by Kepler's interpretation from mathematics to English:

a^3 The fact that any symbol uses a value to the third power indicates space or a volumetric established and separate unit which is serving an under dividable dynamically separate space being within a space. Although being apart the two in space sharing a unit can never be apart but serves as a unit by division of motion. It is space because it is volume using the third dimension. But since the space is smaller than the Universe, it must be space being within space, which is within space. There is relevancy between factors forever present.

T^2 Is an indication of space apart from the surrounding space by granting the independent space by establishing borders through motion, an ability of moving from one point to another point or following a flat distance between two points. It is motion that is taking time in the second dimension.

k^1 Is the symbol used to indicate a straight line between two points with a definite beginning and a specific end position. The two points are valid only by re-aligning an eternal straight line bringing into relation also the figuration of a circle through alternating as well as recognising the control coming from such a centre. It is Pythagoras by the triangle, half the square and the straight line sharing value in the 180^0 they represent. Kepler introduced this absolute basic mathematical principle.

That is what Kepler said. There are three dimensions a^3 between any two points T^2 flowing as time from the centre of the Sun, which is indicated by the line k.
The implication of the relevancy produced by the use of the formula $k = a^3 / T^2$ brings about that when dividing T^2 into a^3 there is k left. The fact is that a^3 is a three dimension (^{3}) of single k (^{1}) showing one or T^2, which are a two dimensional duplicating of k being the one dimension. It means that k is a part of space a^3 or T^2, which is time. It is the same thing in a double dimension or space being a triple of k then k is one factor and k cannot show a position of zero.

If $k = 0$ then there is no possibility of $k = a^3 / T^2$ because $k = 0$ then $0^3/ 0^2 = 0$. That does not make sense. Mathematically space cannot be zero because those being of the opinion of space being zero or nothing must first prove mathematically that space is zero. Moreover they then must prove mathematically how does zero grow through the Hubble constant. By translating Newton's vision of the circle in completing a cycle would become zero through rotation…well that does not count the use of the formula a^3. If k cannot be zero then k could not start from zero. With $k = a^3 / T^2$ no point can be zero because k shows space $a^3 = k T^2$ is no reference to the volumetric mathematical formula used to calculate $a^3 = 4/3 \Pi r^3$. Nor does it show the use of the circle in the second dimension being $a^2 =\Pi r^2$.

In the case of the Newton formula, the circle factor becomes the square as indicated by the duration of the time T^2. The factor standing in for the line which normally would be r and then be the square value is in the case of Kepler not the value indicating the square. That means Kepler never indicated a circle of mathematical procedure but said mathematically the distance of the planet from the Sun k holds space a^3 in relation to time T^2. What Newton saw was the mathematically measured space of a particle having mass and by having mass as it is filling space in the Earth's realm. But in cosmic terms the particle was doing the impossible in a cosmic relation. The particle was standing still while the particle was moving with the Earth. That is why in cosmic terms any object having mass is not being in the cosmos. Having mass means the object is filling space while the object shows no independent movement except as being part of the Earth. The object then has $a^3 = 4/3 \Pi r^3$. However, in cosmic terms all objects must move and that are what Kepler's formula realises. Space moving through time $a^3 = T^2k$ is not the same as that which Newton saw as $a^3 = 4/3 \Pi r^3$, which is an object filling space as part of another objects time. Newton just was far too arrogant to read Kepler and thought of himself in terms of as being a much better mathematician as the likes of a lesser person such as Kepler.

Newton's mathematical vision was the way to calculate the space by using a mathematical formula used as

$$a^3 = 4\pi \times (r^3/3)$$

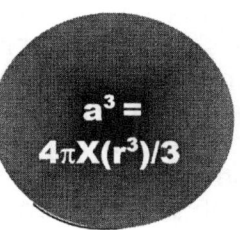

Kepler's cosmic vision was that in the formula $a^3 = k\ T^2$ **the space** a^3 **is equal to the movement** $T^2 k$ **of the space, which comes about as time** T^2 **in relation to a distance** k

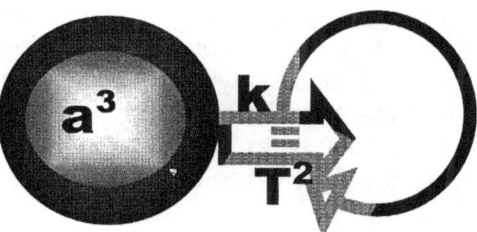

May I remind you THAT NEWTON'S OWN LAWS ARE IMPLIED, and again the planets disobey these laws completely!! **In the modern age all evidence point away from Newton's vision**
The lot is more likely moving away from the Sun. The lot is not coming closer!

$F = \dfrac{M_1 M}{r^2} G$ This is the suggested formula confirming the behaviour of planets used by Newtonian scholars underlining the argument that contraction is coming about between all cosmic objects. What Newton witnessed, if my memory serves me correctly was an apple falling from a tree where both the apple and the tree were part of the Earth and this did not constitute - or lead to - or come as a result of a catastrophic cosmic event happening. In the mathematical sense it does not make sense when Newton's argument is taken out and used in outer space. What Newton saw with his falling apple was a mass influencing another mass to reduce the distance as the influencing involved motion that came about. In outer space there is another gravity where in the case of those cosmic structures in outer space there is no mass pulling each other about or pulling one another onto each other. In the case where there are particles falling from space onto the Earth, that falling also results from gravity, as much as it varies from the cosmic gravity

The lot is more evidently moving further apart

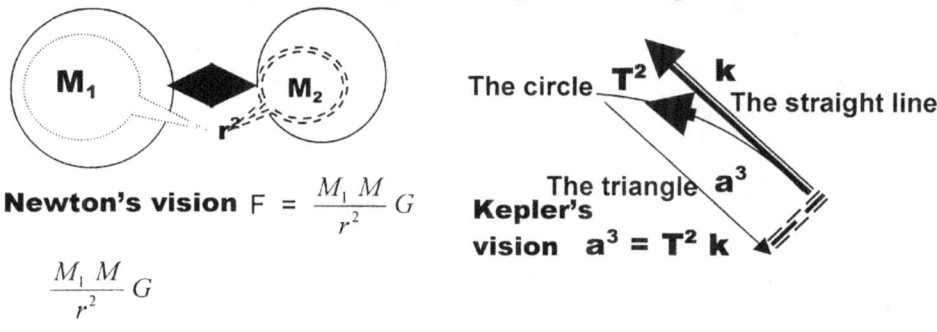

Newton's vision $F = \dfrac{M_1 M}{r^2} G$

$\dfrac{M_1 M}{r^2} G$

The circle T^2 k The straight line

The triangle a^3

Kepler's vision $a^3 = T^2 k$

. There is another type or form of gravity different to the concept Newton introduced. The concept Newton introduced is not the cosmic gravity Kepler formulised. What Kepler introduced is a duel where both objects are clearly in an eternal compromise therefore neither party relents its position. Newton saw just the opposite...Newton saw both compromising their individual as well as each other's position. But since the mass in both cases is unchanged and the mass is the factor that is establishing the force that is used by the circle to hold the radius steady and in place, these facts point to a balance that formed bringing about the above-mentioned steadiness. In the view of science however, it is the mass that either draws the orbiting objects closer or is keeping them apart. The mass does not change and since that mass of both produces the radius between both, the logic is that there has to be an even and steady radius that develops. The radius has to be equal all the time since the mass never changed throughout the rotation. The radius must be the same from any and all given points that form the rotating circle which must keep the radius equal from every angle…yet we know that Kepler proved this not to be the case even before Newton's naming and changing of Kepler's work came about. What we see is that there is one factor that is trying to run away being a lesser space within the pulling powers of a larger space (the second factor) trying to capture and control and a referee (the third factor) is seeing to it that the even-handedness is at all times applying in the fight. (That gravity which is what I am familiar with and what I know is there). In some part but not in all out representing all the gravity there might be because I cannot see the jerking, as much as I do not feel it.

That is then most probably another gravity I can see and which is Kepler's gravity which $a^3=T^2k$ represents. We have a motion of pulling…yes and that is what Newton saw…but then there is another motion of establishing a motion trying to depart, leaving the centre by tearing away from the centre and thirdly there is a motion that sees to it that the balance evolves as rotation. That is what Kepler said when he saw all three factors whereas Newton saw but one of the three. The one space is filling the next space as the space duplicates the position it had in the next moving moment that brings about the next position through motion. This eventually will have confined the next point by using a circle motion, which at first was intended to be a straight line, which is stopped by another straight line. The quest in this book is to find out why the other two factors apply in outer space as only one of the factors comes about on Earth under normal applying conditions.

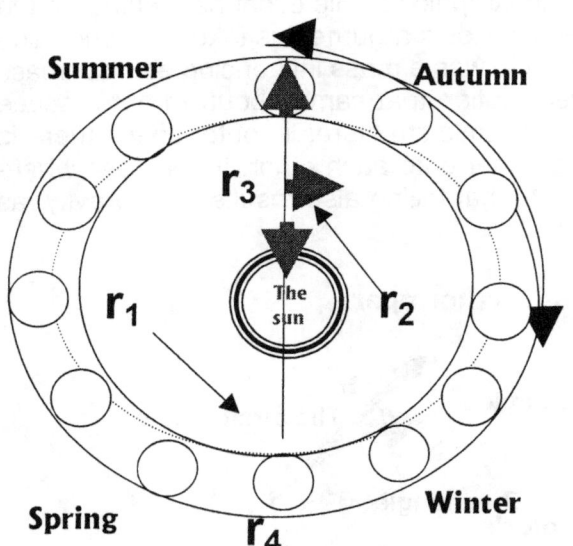

As the two factors are in a motion directional dispute, there is obviously one of the two factors or strengths fighting to cut loose from the other one's grip and run off. If there were not such a force trying to escape, the first force would have a quick and decisive victory by reeling in the loser just as Newton predicted. The fleeing object and its matching fighting partner has a third party referee that allows the fight to go in a specific direction as long as there is no decisive victor.

This book, which I produced in the form of an open letter, is on a quest to find the missing two factors and I can declare with some delight and with even more certainty that I found the missing factors. By Newton's introducing gravity as a force with the formula $F = G \frac{M_1 . M_2}{r^2}$ a precedent was set in relation to the concept of gravity being a contracting force forcing distances supposedly to grow smaller. Apply Newton's view to a comet's behaviour. Newton insists that the Sun has gravity reducing distance between the objects and while lecturers are teaching this during the day, at night they all witness how the comet follows this principle in detail showing Newton as a prophet. No sooner does the final conclusion draw near by orchestrating the final demise of the distance separating the two cosmic components when the opposite changes all concepts taught by institutions of science, the next minute out of the blue with no pre warning of the comet changing its mind, the comet defies all logic in scientific circles that apparently even included defying Newton and his logic. Because at the very point you'd think there is no chance of any return where gravity supposedly should peak because the comet is so close to the Sun and due to that fact makes the collision unavoidable…then the comet chooses that very point to dart away into the blackness of outer space, missing the definite collision by miles. By the time the collision is truly unavoidable with the radius between the Sun and the comet being as small as it realistically can be, the comet starts gaining on the radius distance in spite of Newtonian denial of any possibility that such an event can in fact take place. The radius that should be shrinking further is instead enlarging. The radius that now begins to stretch proves Newton incorrect and it even depicts Newton as possibly being a fraud. The gravity applied that focussed on the comet reducing the radius between it and the Sun was not acting predictably by maintaining the reducing of the distance until collisions come about as Newton insisted on. In our reading the Newton formula in English it says that $F = G\,(M_1.M_2)/r^2$ which when one translates that which is said in mathematics to a verbally spoken linguistic dialect, the translation then suggests that a force is committing the material that forms the factors involved, and forcing the material into a path that is leading to a collision. It says that the two will eventually collide because of the non-retractable mass inside each one that enforces the pulling by which the mass in each case is creating the force. The unchangeable ability of the mass and the unavoidable pulling each mass creates, would bring about such a collision. The mass contributes a force making a collision imminently unavoidable. The collision is beyond any attempts of diverting any oncoming objects away from the inevitable possibility of contact. The force that mass contributes is ruling out all possible evading each other or avoiding the destruction. (By enforcing a mass, it created force that removes all chances from diverting away from the collision that is about to occur). Such a force then removes all possibilities of avoiding the oncoming collision. The force will not allow any attempt to try and bring into the equation other possibilities in as much as rerouting the approaching object and changing the course in the imminent collision that is due and in due course will come about between the comet and the Sun . That which I explained is what Newton mathematically suggested with the formula. That is not what Kepler said notwithstanding so many arguments with Academics that I had in the past who tried to prove to me that the two visionaries views were equal and the same. Well…it's not the same because when we go onto translate Kepler to the verbal English the letters that come out do not even spell the same words.

It is conducive to remember that there is another part of the two relevancies applying where one is a^3 that is relevant to k but also there is the point where k has a duty to place a relation to a^3

The correct Translation of Kepler's mathematical expression will be $a^3 = T^2k$ which proves that Kepler said that there is a space a^3 which is equal $=$ to the motion in the **time duration T^2** thereof between two specific points, which is a straight line k that holds a relation from a centre to an end where the two ends run from the beginning of k to connect at the end of k. I might not be the smartest boy on the block but I'm not that stupid either. I know how to translate… and I translate as follows:

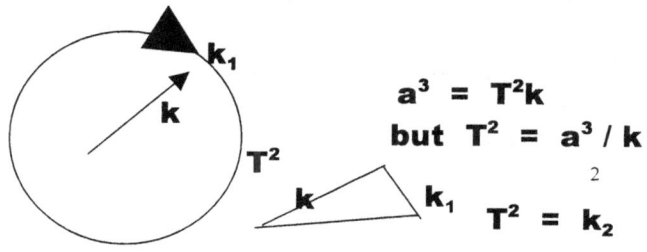

$$a^3 = T^2k$$
$$\text{but } T^2 = a^3/k$$
$$T^2 = k_2$$

a^3 must have a volumetric interpretation because the third dimension is sure evidence of multiple conjunctions of dimensions put together in three sides opposing three sides having the third dimension in place. The fact that any symbol uses a value to the **third power a^3** indicates **space** or a volumetric established and separate unit. Using a cube by three dimensions symbolises a cube, a room, a space to be filled, a unit able to hold other ingredients on the inside when empty or partly filled. It is space because it is volume using the third dimension.

T^2 is an indication of something having a cubic nature other than the square forming motion that is provided by the motion the square indicates, which is where the moving object is representing a third dimensional object that is moving from point to point and it is this point to point that multiplies into the square. The space is moving as a unit from one point to another point and the moving between the points are represented by a flat square or following a flat distance between two points. The cubic space was in one instant in one place and then the second instant in the other and because time can never stand still or become single dimensional (this I am about to prove as the letter unfolds) insisting that time must always support the motion it consists of or time cannot be. It is motion that is taking time, which is motion in the second dimension moving the space in the cube.

k^1 is the symbol used to indicate a straight line between two points with a definite beginning and a specific end position. It is the location where the cube is holding space and where the space was and where the cube in space is going to be in very the next split instant that follows. That will then in multiplying form the square that indicates the time the journey took to move the cube of space from one point where k is indicating the location of the space to where the next indicating of k will shift the space being the cube pointing at the end of k. Since time represents the square and with k being the distance that proves that the k represents the distance the space representing the cube went to take the time represented by the square through the motion. It is the distance moving space in the cube to complete time in duration in the square of motion; therefore k is permitted to be in the single dimension.

There are infinitely more implications in the statement Kepler delivered than what is merely a contribution to motion and only motion as Newton was of the opinion. What is there mathematically not correct in my interpretation of Kepler's manner of translating mathematics to English and why is any changing thereof by Newton or any other person necessary in any way?

We can test any of the following symbolic values in the mathematical expression and also test the principals behind the expression in which Kepler stated them. By such testing we will find that time after time there were never any corrections in the translations required since the translation thereof was never incorrectly presented and in that a case asked for no alterations to secure the correct reporting of the cosmic information being translated. By taking the formula on face value it can change as follows: $a^3 = T^2 k$ can become $k = a^3 / T^2$

That proves that the establishing of distance k will produce space a^3 and set space a^3 in motion T^2 where such motion is in opposition to singularity, which means gravity or contraction is the deliberate opposite of expanding $a^3 / k = T^2$. In the beginning the expanding then also involved three more points all just outside the border of singularity but within the atom exclusivity. It extends k while it introduce a returning relevancy back to singularity k^0 by creating motion in spin and duplicating space by reducing space.

With this mathematical reality what then later formed the grounds for any individual to develop any need to change Kepler's translations from the cosmic given to mathematics and then from mathematics to English while the guilty party is renowned for his superior skills in mathematics?

Kepler translated what he found to be the cosmic given to mathematics which we humans are able to interpret from the mathematical expressed to the verbally pronounced and written but Newton still saw a need to change what the cosmos said about how the cosmos is presented and by no one less than by its own interpretation of its self structured composition.

When viewing my interpreting of what Kepler said I might have asked myself countless times what did I not translate correctly from the mathematical expressed to English after encountering a battery of Academic onslaught and resentment on my Newtonian views because after all it is directly diverting strongly from the teachings presented by Mainstream science and the diverting is not coming in a small way.

In truth from my diverting I came across very new ideas I am able to prove. By my translating Kepler's work correctly I came upon answers not yet uncovered by Mainstream Science

Kepler gave the World mathematically translated cosmic answers he received from the cosmos that Kepler uncovered long before Newton, Einstein and others got wise about cosmology...and later the wise came up with old news (old views as far as Kepler expressed their views before they, the wise were born with the purpose of coming to the conclusion that those wise men eventually did) and where the conclusions that the wise concluded brought much surprise to the world with the originality of the later Masters' initiative while Kepler said the same thing ages before...!)

Such is the advantage of recollecting Kepler facts that it does answer many questions, which went unnoticed and therefore not spoken about up to now and some were previously never even thought about.

Newton said a sphere is a^3 = 4/3 Π r^3, which is mathematically correct, however

Kepler said the cosmos told him a cosmic sphere is a^3 = k T^2 There are the two distinct possibilities which Newton saw and which Kepler saw and both are most valid. Between the two concepts there is literally one Universal difference and the two can never be mistaken as promoting the same principles. 'Ever try to answer facts about the Universe in as much as...what brings about the expanding? Kepler said the Universe plus its entire content is expanding centuries before Edwin Hubble realised what he was seeing through his telescope.

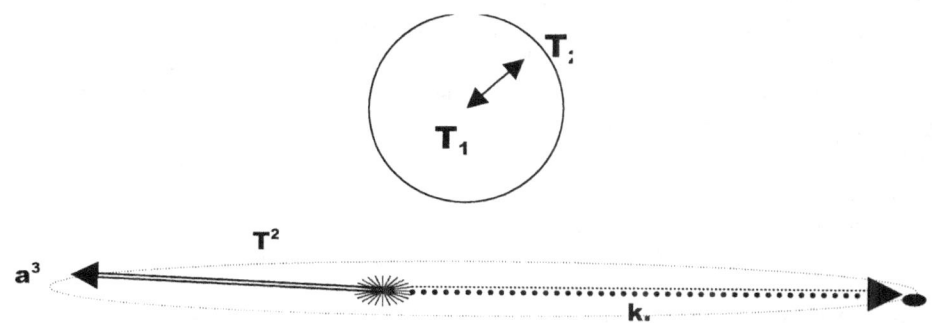

Kepler was the very first person to mathematically introduce **space a^3 centre k** and **time T^2**. Not only did he introduce **space-time a^3 / T^2** but he also placed **space a^3** and **time T^2** in a relevancy long before Einstein did and placed **gravity in space-time a^3 / T^2** even before Newton named gravity. He showed that space **k** is growing in the measure of what means the Universe attend to by promoting space-time as **a^3 / T^2 = k^1**. Kepler was the person who placed gravity as the ingredient in the Universe that determines **space a^3** and **time T^2** and much more. Kepler was the first one that said that gravity comprises of two factors being **k** or linear gravity and **circular gravity or T²** as gravity keeps space in form while all is staying together.

Although not one Academic has ever openly admitted to me that they as members and part of Mainstream science are more aware than I am of all the facts and doubts I point out to them, such evidence then becomes clear whenever I mention the matter to them I get more than the impression it does not come as a surprise to them and hit them like a brick between the eyes. The lack of surprise and initial doubt they should show at first when they discover the incorrectness of evidence in their theory is a telltale sign confirming my suspicions about their evidently knowing all this information all along. They clearly seem very agitated about every detail I show when I bring the mistakes and double talk to their attention in the hope that they may confirm my doubts.

Never is there a whisper of a surprise or a hint of a suggestion that would initiate an argument carried on by the bewilderment or the astonishing surprise they should feel confirming my arguments because there is a mild complacency in their voices. My jumping them total unexpectedly about matters they never contemplated in the least leaves them unturned. The rush in blood pressure that should be a factor on their part and part of the instant where total surprise will bring about some confusing thoughts that will inspire the unleashing of an argument in defending their holy grail should at least carry a surprise in an attempt to save what they believe as being the Gospel in science and with that defending their honour. They lack embarrassment, which they should have in their disputing of my claim as they fight off my allegations with a countering of denial claiming foul on my part as they are in shock when finding out about any doubts. A lack of true emotion on their part is a telling sign that they also may have some serious thoughts on the quiet about any inclination presenting a flawed view about what they always thought they knew to be true.

There is only that eerie dismissing of the seriousness and the lack they show in excitement that would deny or support my credibility as I present my findings. If they know about the inconsequential facts in science why is it not generally acknowledged and pronounced as a matter of fact? Why is there the covering up and hiding facts that we associate with some professional criminals such as politicians. The fact that Academics are aware of this evidence even in general terms, where time cannot prove Newton and about the subsequent misinformation and doubtful evidence about Newton's cosmic vision but moreover underlying this, is their total denial of knowing about it and that is what is so seriously unforgivable. The fact that all Academics are aware of my evidence even before my presenting them with such evidence is beyond doubt. If that is the case then why are they forever trying to kill my viewpoint and forever try to silence me where I am only the messenger because I bring the solution and the answer? Please note that the answer and the solution are unbelievably simple and unsophisticated. It lacks all the splendour and grandeur expected by all Academics concerned. It is because it is so simple that it went amiss for four hundred years. It is because it is so simple that it misses the grandeur that will entice them. Instead, every academic accuses me of not understanding Newton while they can't show me what part it is that I can't understand and I on the other hand can't see what there is not to understand.

Newton said that it is the reducing of the distance between the objects that would bring about the un-reversible reducing that will end in a total demolishing of the radius that is between the cosmic structures, but instead we find the gravity applying in outer space is one of the instances where gravity provides an orbit circle that gravity seems never to completed as the orbiting objects follow from closing any circle that is leading into a following circle up to where the circle is completed in cyclic precision. That is not the gravity that Newton identified although Newton admitted that there is a presence of a centre forming a point in the middle between the two objects. He was unable to know what caused or even the presence of the

Coanda principle, which forms so critical a part of my theory. The formula concerning cosmic balanced gravity however leaves no room for the admitting of such a point and by not leaving a possible inclusion of such a point in his formula Newton did by such gesture in principle repeal his admission of such a centre. This had me cast doubt on what is taught at institutions of learning. It motivated me to venture back to an era before Newton came to influence science. I came to acknowledge Kepler as I came to understand Kepler. The accepting of that which I understand from Kepler involves much more reading into what Kepler said by finding what Kepler did not say in the way that he did say what he said than the reading about what Kepler said as it is written in the precise detail and to the letter used in his statements. He never directly stated what I say he said. Again I must stress this point: when I refer to what Kepler said, it then most likely means reading into the part that is being a part of the part that he did not say when he was saying what I presume he said but also, which I accept that it forms part of the part that he meant to say and whether he said it intentionally or not, my repeating what he said makes his not saying so not the least less valid about that and him not saying so intentionally or not, is the part on which I cannot comment with absolute certainty. However, I am very certain about the accuracy of my deductions. My deductions from what Kepler said when Kepler did not say what I say he said, are far more accurate than the miserable corrupted failure Newton came up with as his version of the deduction he made about the work Kepler introduced. That which I say is what I am reading into what Kepler was told by the cosmos and translating that from what Kepler then said as part of what he did not say but meant to say. I have to read more with my mind than with my eyes. This comes as a result of interpreting Mathematics to the verbally expressed. I had to learn to read with my mind and not my eyes and I found that that is the manner in which one has to approach cosmology. From the first time I discovered what manner one should use if one wished to read into Kepler's findings, I saw Kepler was all about uncovering the unknown. Realising that, the conclusions I drew by reading in such a way cemented my better understanding of Kepler's work, which then helped me improve my insight into Kepler's work as it increased my understanding about cosmology several fold. This helped me to realise what implications were to be found underneath Kepler's discoveries. From my realising what approach I should use, it helped me to improve my cosmic realising by using the method of reading Kepler and from that I could come to appreciate what Kepler introduced.

Only then did it bring insight and proof to me as a student of Kepler and this proof I found by dissecting what Kepler **did not** say instead of what he **did** say, which I now present to you with this letter, you being a superior intellectual person. Kepler said $a^3 = T^2 k$ and that correctly translates to a mathematical expression $k^0 = a^3 / T^2 k$ which in the verbal statement in English translates that Kepler said that there is a **space a^3** which is **equal =** to the motion in **the time duration T^2** thereof between two specific points which holds a relation onto a centre k^0 where from there forms **a straight line k** that is centred on the spot where space begins from k^0 **that produces k** as well as producing the circle therefore that spot $k^0 = a^3 / T^2 k$ has hold k^0 at a value of having the least space. The line **k** is centred onto a spot where space begins specifically at k^0. This point not only produces the line k^0 but represents also the space that forms the eventual circle T^2. Therefore from the centre holding k^0, k^0 leads to **k** that forms the roving space a^3, which is rotating at a distance **k** where T^2 forms the outer limit of k^0. Mathematically $a^3 = T^2 k$ will be $k^0 = a^3 / (T^2 k)$ because $k^0 = 1$. But $k^0 = 1$ also presents the single dimension where all factors are a product of one. If one can locate k^0 one will find singularity. That is where gravity is because gravity is strongest where space is least. This suggests that gravity is strongest at k^0 because space is least. That is gravity because that is what keeps the orbiting object in orbit but also that is what Newton completely missed when he changed Kepler's work. Newton failed to recognise gravity as the only ingredient in Kepler's formula. He admitted he missed this because he admitted he did not know what gravity is while Kepler explicitly showed what gravity is. Gravity is what keeps the orbiting object orbiting. **k $= a^3 / T^2$** is **distance[1] = space [3]/ time[2]** forming from a pivoting centre k^0.

That is a cycle and moreover it is a cycle formed **by space/time**. What Kepler said is that space is a^3 **in motion** T^2 **k.**

That says **space^3 (a^3)** relates directly to **time^2** that uses the symbol **T^2**. This is also what I refer to when I say one has to read what Kepler did ***not*** say when one wishes to see what he ***meant*** to say. Kepler introduced space^3 –time^2 long before Einstein's date of birth appeared on any calendar although Einstein is credited with the formulating of the concept of space-time and giving it a name. Going even further Kepler stated that the space a^3 is on the move T^2 around in a circle at a distance **k.** That is what that comet we are discussing is doing. The space^3 (Comet) is circling the Sun using a radius **k** to establish the cyclic time^2 as a period of continuous motion and continuous motion is gravity. That reads much more correctly and closer to the truth than what Newton predicted what according to him (Newton) was happening in space. Remember in this statement I am separating cosmic principles applying from the way that gravitational principles apply on Earth. I distinguish that which is the rule in the cosmos from what we find ourselves trapped in on Earth. The two just don't mix. I am removing cosmic physics from normally accepted physics because the gravity concerned is not the same.

The proof I bring is real however simple it may seem. It has none of the mind-blowing complexities normally associated in the presentation of investigative analyses of Astronomy. I realise the information in this book carries the arguments in a childlike manner which are very simple to follow, and for that in the past I have been blamed over and over again as being unprofessional. In my answer to that I can only reply by using another question: Are only professionals adequately equipped with minds that make them (the professionals) the only ones able to think? We being part of the human race are all thinkers. Everyone as a human being can think. Every person on Earth is a thinking thinker that uses his brainpower by exploring thoughts mainly and normally to his or her personal benefit. It is what we think about that produces the results of our efforts by which we accomplish what ever we are thinking about. I have met professional Academics whom I found foolish as much as there are other cases where the so-called amateurs can credit themselves with much more wisdom and insight. Albert Einstein as a patent clerk was that much but to name one. Please understand that I do not compare my achievements or myself in any way, shape or form with the likes of a Master such as Einstein although I speak my mind when not being totally in agreement with some of his or other views. My unsophisticated retracing of Mainstream physics concerning the Big Bang in detail helps to reinvestigate established principles and moreover investigate proof in the light of modern evidence. In principle I distinguish between Kepler and Newton in that Newton is one hundred percent correct concerning gravity on Earth but as far as outer space forms gravity the conclusions of Kepler and Newton do not match and they had totally different ideas about what they saw in gravity. I am in disagreement with some basic principles that science acknowledges and I divert strongly from all accepted roads Mainstream physics follow. My doing that prompted those who are considered and accepted as self-proclaimed members of Mainstream Physics that in their eternal wisdom, have categorised my views as incoherent in the past. That I do not accept. I admit that my line of thought is extraordinary and controversial but only to Mainstream science and not to the standards lay down by nature. Since the concepts I follow start at the beginning, and I take Kepler forming the start of such a beginning, Kepler (and not Newton) is at that point where modern cosmology began. It then is with that in mind and that mindset which puts me in, that I re-evaluate Kepler's work. I start by tracing a new approach as to what I see Kepler found. The main aim in my investigation is to establish a divorce between what Kepler said and what Newton thought to add to what Kepler said. It is this divorce I create that Mainstream science finds repugnant or even in some persons' opinion repulsive. I believe the repugnancy does not come from or is not manifested in any part of my work to the letter as such, but rather what my work suggests and who is doing the suggesting. To my view in cosmology such adding to Kepler by Newton was unnecessary and it diverts Kepler's work away from cosmology. But as the generations moved on Newton became religiosity in the mind of science wherever science was taught. To students there is little or no choice in the matter since the only choice left to them is one of understanding by forcefully accepting or die an academic death since Newton is academically accepted without asking questions or raising an opinion. For the second choice, the less accepting students are greeted with a Dear John good-bye letter sending them off into the unknown Sun set that such a future outside physics will bring them. That is brain washing.

From studying Kepler I saw that we have to gauge what we find in the Universe. What we find is not that what we realise with our eyes but that what we observe by using our minds to translate from visions coming from our eyes to our minds. We have to test the part that we are seeing much more than merely accept what there is to see on face value. We have to not only see what other life beings blessed with much less insight most probably also should see. We must stop using our eyes in the same manner as animals do and start seeing with our mind, as humans should do. By being the superior evolved species that we are, it gives us the ability to read into that which only we can see and that we only can see by using our intellectual mindset. By seeing with an intellectual understanding what there is to see when we see what we can observe, we should therefore have the ability to be in understanding by looking at what we can see but moreover understand that which we cannot see. It is the same as playing chess. See what there should be moved instead of noticing an object not having an ability to move by own initiative. This I first found to be true about Kepler's work and when I started projecting this method of observing what the Universe is, as it scattered most previous perceptions I found that using the new method brought along answers so fast I could sometimes hardly keep up with the interpretation thereof. But as is the case with Kepler so is the case with the entire study of cosmology: One should see what there is about the cosmos, which is unseen to us and then we may find so much more in the cosmos unseen to us representing that which we cannot see and that which we cannot read because we have to learn to read what is not written in light. Armed with this realisation I then proceeded from that point by further arguing and debating the full implication of Kepler's contribution. Kepler placed cosmic structures in relevance to one another and so does the Big Bang Theory. The backbone of the Big Bang is that relevancies apply in dynamics and such dynamics are placing all structures without any reservations independent from each other. As the Big Bang progresses all filling the Universe to the inside is in the same Universe that was at the time of the Big bang just as much as it will always be and the lot remain the same, however the relations that the elements comply to bring across new relevancies with new positions to fill. The father of the Big Bang concept is a person by the name of Father LE MAÎTRE, GEORGE ÉDOUARD (1894-1966) who was a Belgian priest and cosmologist. He was the first person to embrace the fact that the Universe expanded from an infant stage. His model of an expanding Universe (1927) was superior to that of W. de Sitter in that it took into account mass, gravitation and the curvature of space. Similar models were proposed in the early 1920s by the Russian mathematician Alexander Alexandrovich Friedmann (1888-1925) but Friedman compiled various such possibilities. Lemaître argued further (1931) that the quantum theory supported an origin in the explosion of a 'primeval atom' or 'cosmic egg' into which was originally concentrated all mass and energy. As modified by A.S. Eddington, Lemaître's model provided the springboard for G. Gamow's Big Bang theory. In the wider picture of science in general a lot changed to just allow such turnabout in thought since the day of Isaac Newton. From Newton's attraction and contraction many things came into place that allowed change in the most hardened minds. Accepting facts about the Big Bang concept is quite radical. By promoting expansion the Big Bang theory contradicts gravity and our accepting of the Big Bang has to change all other concepts. By accepting the Big Bang other changes are also involved.

KEPLER, JOHANNES (1571-1630) is a name never mentioned alone but is always attached to another name being Isaac Newton. This is because every Newtonian holds the impression that only Newton gave Kepler's work substance while in fact Newton had to degenerate and compromise the validity of the work of Kepler to give substance to the work of Newton.

The German mathematician and astronomer KEPLER, JOHANNES (1571-1630) became Tycho Brahe's assistant in Prague in 1600 A. D. where he undertook to complete the tables of planetary motion Tycho had begun. Kepler first calculated the orbit of Mars. He spent much time trying to reconcile Tycho' s accurate observations of the planet with a circular orbit, but concluded (in Astronomia nova, published in 1609) that Mars moved instead in an elliptical orbit. Thus, he established the first of his laws of planetary motion. A theory that the Sun controlled the planets by a magnetic force led him to the second and third of his laws, which were published as part of his treatise on theoretical astronomy, Epitome astronomiae Coernicanae (1618-21). The Rudolphine Tables (named after Tycho's patron, the Holy Roman Emperor Rudolph II) of planetary motion appeared in 1627 and were still in use in the 18th century. Kepler also wrote De Stella nova, on the supernova of 1604 and Diptirce on optics and the theory of the telescope. The overall view followed in this book **_An Open Letter announcing Gravity's Recipe_** places the true significance of his work in true contents. In KEPLER'S EQUATION is the equation that relates the eccentric anomaly of a body in an elliptical orbit to its mean anomaly. The equation is $E - e \sin E = M$., where E is the eccentric anomaly, M the mean anomaly, and e the eccentricity of the orbit. It is important as one of the mathematical relations enabling the position of a planet about the Sun, or a satellite about is planet, to be calculated from the orbital elements for any time. However this only relates to the solar system, and KEPLER'S LAWS only apply in the contents of the solar system. The three laws governing the orbital motions of the planets, discovered by J. Kepler is as follows: The first law states that the orbit of a planet is an ellipse with the Sun at one focus of the ellipse. The second law states that the radius vector joining planets to the Sun sweeps out equal areas in equal times. The third law states that the square of the orbital period of each planet in years is proportional to the cube of the semi major axis of the planet's orbit. The first law gives the shape of the planet's orbit; the second describes how the planet must continuously vary its speed as it follows its orbit, moving fastest at perihelion and slowest at aphelion. The third law gives the relationship between the planets' average distances from the Sun and their periods of revolution.

Instead of studying the true value and contribution of Kepler's laws an Englishman going by the name of I. Newton placed his own interpretation to Kepler's laws, and in doing this, he wilfully destroyed the principle working of the Creation. Saying this, I hear the alarming hooters announce Newtonian dismay. In the past my experience was that all the revered Academics lost their appetite for any further investigation of my work. That is sad as much as it is regrettable. Through Newton's tunnel vision, he applied his own misinterpretations to the correct presumptions of Kepler and through the Newtonian tunnel vision Academics did not move an inch away from repeating the same procedure. In the past it was this that had Academics shying away from me because at the point where I raise criticism of the Newtonian viewpoint I am rejected. The point where I declare my suspicions concerning their accuracy and the correctness about their theories, which is where I should then be raising their doubts about their way of thinking is the point where in stead I raise their suspicions about my way of thinking. That is what caused the rejection of my criticism about Academic Newtonian science and evoked their criticism in the past about my views instead of them following the logic by investigating what I said. Their rejection of self-investigating got me and my work rejected to a point where the applecart lost its wheels on every occasion. When Academics read my remarks, they instantly retaliate with wrath. I say this because I realise that reading my remarks or hearing me remarking about this notion brought much resentment on their part and if the reader at the present moment is a Newtonian, boiling his/her blood. It is blood boiling because I believe they see my remarks as belittling that which they feel they have accomplished. This is not the case but still my remarks have the same effect on the Academic as pouring icy cold water down the back of his shirt. I mention this because I know it has happened many times before and if possible I wish to avoid this response. Therefore I ask you kindly to please be warned about the negativity you must feel towards me where you are the Newtonian and I am not. Before you lose interest in reading this letter any further, please allow me to finish. In the past Academics thought me to be presumptuous and that normally became the point where all the Academics found their interest vanished. That should not be, because if Newton's work is as utterly accurate as those with faith in his work believe it is, then every aspect about Newton should stand above any and all reprimanding or any form of doubt causing a notion to reprimand. The testing of Newton's work should withstand all testing notwithstanding the person or the prominence of such a person's social or academic standing in the Academic society or even the prominence that such testing will deliver. From what I see about Kepler's work, it is a flow of circumstances that lead to Academics neglecting Kepler's work and the realising of the theory I suggest is not forthcoming due to my personal brilliance. I do not consider myself to be the brilliant in any way as to be the one that can remove the verbal splinter from the eye of the Academic. Yet…if there is a splinter what else should I then do…Newton reduced the implication that Kepler findings hold by introducing to the law of gravitation. He, (Newton), then went about and changed it to three laws of motion. It is clear that while he formulated the laws on motion he missed the way Kepler introduced gravity as space a^3 coming about through motion T^2 and that gravity is space a^3 within space k within motion T^2. Newton also missed the fact that gravity is at its strongest where motion and space cease to be. It is of the utmost importance to recognise the two interrelated forms of gravity. . I. Newton generalized Kepler's first law, verified the second law, and showed that the third law should be amended to the form; $4 \pi^2 a^3 / T^2 = G (m + m_p)$. In this, the value of "T" and "a" are the period of revolution and semi major axis of the orbit of a planet of mass m_p about the Sun of mass m, and G is the gravitational constant.

It should be clear to any person investigating Johannes Kepler and his work that Isaac Newton hijacked Kepler's work and any time there is the slightest referring to Kepler about the research Tycho Brahe and Johannes Kepler did, such referring to Kepler always lead to and always include the mentioning of Isaac Newton changing the work of Johannes Kepler. It is as if the World never could acknowledge Johannes Kepler because the work of Johannes Kepler would be completely wrong and misleading if it were not for the intervention of Isaac Newton saving the skin of the less admirable Johannes Kepler. This comes in the midst of every one realising that Kepler used the information he received directly from the cosmos. I do stress this on many occasions throughout the letter because the embarrassing part is that Newton changed the work of The Universe and not of the man called Kepler. Should you, reading the letter, entertain the opinion of Newton and feel any urge to defend Newton you should ask the question as to whom is standing corrected, is it Kepler or is it the cosmos that gave Kepler the information he concluded? The cosmos supplied all the information by using mathematics, which Kepler then had to translate. But Newton destroyed the accuracy by, altering what the cosmos said and directly by adding to that what he (Newton by name) thought that the cosmos left out. This set a precedent by Newton in cosmology and also set a trend, which was retained in all future cosmological development and it lasted in cosmology for three hundred and fifty years. In this book you are reading, I am about to show that such practise should no longer be accepted in cosmology. In the process the world of Mathematics developed by the world of cosmology stood still for almost four hundred years. Faculties contributing to cosmology and feeding off cosmology improved as much as they developed, but when cosmologists see the Roche limit in action in the lens of the Hubble telescope and refer to the event as "stars blowing bubbles" being the ultimate response coming from those persons who are supposedly the Masters of cosmology affairs, then the truth of what I just said comes down on you like a ton of bricks. Everyone having any remote interest in cosmology will find they are being very disillusioned by such "official" testimony about the evidence the Ultra Wise report about. This book is about showing how great Johannes Kepler was and how enormous his work was. It will show he preceded all ideas of everyone that came later and officially introduced the novelty of such ideas. During the time Kepler introduced his work, the stature and the magnitude of his work was beyond any person's understanding (including Isaac Newton) and this prevailed for most of half a millennium. I do not say I am the brilliant one to uncover Kepler in the face of everyone failing that came before me, but as I am not a Newtonian such bias was not part of my repertoire and denying me the fortune of being a Newtonian added to my fortune of realising Kepler. Yet as you will notice, the work I contribute is much below the sophisticated norm of modern investigative research and the levels that modern research accomplishment demands to better the effort of the understanding ability in the splendour that investigative research work should deliver in view of our modern times. It is only pure neglect in science circles that moved science past Kepler. Not seeing and therefore not investigating through almost half a millennium has paved a road past the inferior levels that the researching of Kepler's work holds because it was rocket science four centuries ago but the brilliance of it has faded since then. My contribution holds no astonishing flair that may add to science in general. Only failure to notice what I see on the part of those truly brilliant can explain my being able to present my contribution about my work in investigating Kepler. Only by their passing such degrading levels of the Academic establishment in the past and the present can bring the blame for such an obvious discrepancy because any involvement in the work at such an inferior level as that which I bring cannot interest and excite a salted Academic and when thinking about it, the idea is totally unthinkable. This letter (although it is on this inferior level of cosmic importance) is about correcting this tendency and has in mind the effort to put in writing that would place Kepler in the greatness and glory he deserves.

As I already said, if Kepler was wrong then the cosmos was wrong about facts and applying relevancies and tendencies in the cosmos. I yet again wish to reiterate. We should never for one moment forgets that Kepler received his information directly from studying the cosmos so how could the cosmos stand corrected? In spite of all the brilliance attributed to Newton nonetheless if Newton had the mind to change Kepler's work and my saying this includes all persons agreeing with such changing by Newton of the work of Kepler, those persons admit that he or she or Newton never took any time to really and truly investigate what the cosmos told Kepler. Through understanding the work of Kepler I prove gravity, the Titius Bode law, singularity, space-time, space-time relevancy, the Lagrangian system, the Coanda effect and the Roche principle, the sound barrier and the principle behind the Black hole. The precondition for my ability in doing so is that I have to remove Newton's opinion about Kepler's work from Kepler's work. Whenever cosmology comes into question and all the phenomena, which I mentioned just now remains unexplained and by that token alone it shows to what degree did cosmology remain undeveloped. Whenever there is any mention of Newton, Kepler is never mentioned. But the reverse is always applying. Mainstream physics holds the opinion that Kepler may only have an opinion if Newton can change the opinion. Kepler (wittingly or not) gave the world space-time, gave the world gravity, gave the world singularity, gave the world the Plank theory, gave the world the theory on relativity but no one ever found Kepler's work deserving enough to launch any investigation such as I did. I belabour this because of what revulsion my rejection of Newton unleashed. That is one barrier much unnecessary but it has been an insurmountable barrier this far.

NEWTON, ISAAC (1642-1727) and NEWTON'S LAWS OF MOTION
An English physicist and mathematician who developed his principal theories about gravitation, optics and mathematics between 1665 and 1666. In 1668, he made the first working reflecting telescope. Most of his work remained unpublished for long periods, partly because of criticisms by c. Huygens and the English scientist Robert Hooke (1635-1703) of his early work on the corpuscular theory of light. However, in 1684 E. Halley persuaded him to organize his work on the celestial mechanics of the Solar System, which was published as the Principia. Newton's other major work, Opticks, was not published until 1704. It contains his corpuscular theory of light, and the theory of the telescope. His greatest mathematical

achievement was his invention of calculus, independently of the German mathematician Gottfried Wilhelm Leibniz (1646-1716). His profound influence on physics and astronomy is reflected in the phrase 'Newtonian revolution'. Three laws published in 1687 by I. Newton concerning the motion of bodies.

1.   A body continues in a state of uniform rest of motion unless acted upon by an external force.

2.   The acceleration produced when a force acts is directly proportional to the force and takes place in the direction in which the force acts.

3.   To every action there is an equal and opposite reaction.

4.   However, there is one more law on motion that went undetected by Newton…This book is not about trying to disprove Newton…it is about adding to science more than there now is available without removing anything that science already accumulated.

In this book I use Kepler's formula to either prove or to disprove the following accepted principals in cosmology and if any person in the past gave only the slightest attention to Kepler's work, many statements would have come much sooner delivered by someone else or may never have come at all. By applying Kepler's formula correctly in this letter I can either agree with or in other cases deny the following principles.

It began with NICOLAUS COPERNICUS who changed the status quo. COPERNICUS, NICOLAUS (1473-1543) was, according to the Anglo Americans, a Polish churchman and astronomer although this is just more politically inspired propaganda because his parents were both German (in Polish, Mikolaj Kopernigk). While he was completing his studies, he had realized that the Earth revolves around the Sun and not vice versa. Such a view was in that time, held to be heretical. As I pointed out in the first few articles, the Church regarded the geocentric world-view of Ptolemy as consistent with its doctrines. Copernicus set down his basic ideas around 1510 in the Commentariolus, which he circulated anonymously, because of the Islam link. In 1512-- 29 he conducted his study and concluded the observations that he needed to support his theory, while carrying out ecclesiastic and local administrative duties. In this time, he had to defend his mother in court on charges of witchcraft. In 1539, the Austrian astronomer and mathematician Georg Joachim von Lauchen (1514-74), known as Rheticus, became a pupil of Copernicus and began to spread his ideas. The published work was openly spread as the Copernican system, in spite of the life-threatening dangers connected with such a "crime", in 1543 in the book De revolutionibus orbium coelestium. However, the reality of a heliocentric Solar System was only commonly accepted, after the work of Galileo and J. Kepler. The ideas introduced developed along and proved to be correct until such a time it met a solid wall with the investigation of Max Planck.

PLANCK CONSTANT
(Symbol h) A constant that relates the energy of a photon to its frequency. It has the value 6.62076×10^{-34} Js. It is named after the German physicist Max Karl Ernst Ludwig Planck (1858 – 1947). PLANCK ERA. In the Big Bang theory, the fleeting period between the Big Bang itself and the so-called Planck time when the Universe was 10^{-43} s old and the temperature were 10^{34}K. In this period, quantum gravitational effects are thought to have dominated. Theoretical understanding of this phase is virtually non-existent. It is named after Max Planck (1858-1947).

PLANCK'S LAW

A mathematical description of the energy radiated at different wavelengths by a black body: $E = hf$, where E is the energy of a photon and f its frequency. It was formulated in 1900 by Max Planck (1858-1947), who realized that energy is radiated in discrete packets, which he called quanta, and it formed the basis of quantum theory. The quantum of light is a photon, the energy of which depends on its wavelength.

There is one rule which is well established and which Mainstream science agrees about. It is one aspect, which forms the very principle that holds the theory about the cosmic starting together under the covering of a verbal blanket. All in science agree that it all started with singularity but I manage to go one step further where I prove that it is also where it ends, as singularity reunites space-time, which is from where Creation split in the very beginning.

Singularity is as follows: Singularity: a mathematical point at which certain physical quantities reach infinite values, for example, according to the general relativity, the curvature of space-time becomes infinite in a black hole. In the big bang theory the Universe was born from singularity in which the density and temperature of matter were infinite. From singularity flows space-time.

Space-time is as follows: Space-time is a four dimensional position of the Universe where the position of an object is specified by three coordinates in space and one position in time. According to the theory of special relativity there is no absolute time, which can be measured independently of the observer, so events that are simultaneous as seen from one observer occur at different times when seen from a different place. Time must therefore be measured in a relative manner as are positions in three-dimensional Euclidean space, and this is achieved through the concept of space-time. The trajectory of an object in space-time is called world line. General relativity relates to curvature of space-time to the positions and motions of particles of matter.

SPECIAL THEORY ON RELATIVITY
A theory proposed by A. Einstein in 1905, based on the proposition that the speed of light in a vacuum is constant throughout the Universe, and is independent of the motion of the observer and the emitting body. A consequence of this proposition is that three things happen as an object's velocity approaches the speed of light: its mass goes up, its length shortens in the direction of motion, and time slows down.

Hence, according to special relativity, no object can ever reach the speed of light because its mass would then become infinite, its length would become zero, and time would stand still. In addition, Einstein concluded that the mass of a body is a measure of its energy content, according to the famous equation $E = MC^2$, where c is the speed of light. This equation describes the conversion of mass into energy in nuclear reactions within stars.

GRAVITATIONAL COLLAPSE
The collapse of a body that is unable to support itself against its own gravity. Gaseous bodies undergo such collapse if they are not hot enough for their gas pressure to balance gravity. This can happen in the early stages of star formation, or when nuclear burning ceases in a star's core. The time taken for such collapse decreases rapidly with increasing density, varying from about 100 000 years for the birth of a new star to less than a second for the formation of a neutron star. Star clusters may undergo a similar collapse if the random motions of their constituent stars are insufficient to offset gravitational effects, either during their formation or at an advanced stage of their evolution.

GRAVITON
A hypothetical particle or quantum of gravitational energy, predicted by the general theory of relativity.   gravity - motions have not been observed but are predicted to travel at the speed

of light and to have zero rest mass and charge. A graviton is the gravitational equivalent of a photon. It is this anti-photon-being-a- gravity - motion by just merely swapping direction and all is proved that I find not very indigestible in modern science. One of the main issues that I wish to protest by my writing of this is my argument that if the Universe can be compressed back to the size it had at the point of 10^{-38} seconds after the Big Bang the daily outdoor temperatures of 10^{27} K will also come about once more. The expansion was the result of compressed space, which then formed into heat and in turn resulted in finding a Universe with all the insufficiency of space less ness prevailing throughout and wherever space was needed. By that it forced space-time to come into being. Space-time came about at the time of endless time duration without space availability, which brought about the period of the Big Bang wherein space growth was the converting of such heat to space. If the Universe was in a vacuum as big as being available now then what was the temperature of the vacuum while it was empty before material filled it later. Then I presume the vacuum was there present as it is now in this present day. If the Universe then employed the space of say one atom, the impression comes through that from edge to edge and from Universal border to border the space occupied was the same as one atom will claim in our present day and age. Normal gravity started at 10^{-43} seconds. The Universe was the size of a neutron or somewhere in that vicinity. The Big Bang began and GUT, (or the grand unified theory), produced the attempt to describe the strong and weak nuclear forces and electromagnetism in one single mathematical theory. Somewhere before 10^{-12} seconds of counting the Universe cooled to about 10^{15} K, from where the electromagnetic and the weak interactions acted as one single physical force. Science reckons that unification may come about at temperatures of 10^{27} K, which was the temperature of the day at 10^{-38} seconds after the Big Bang. This statement echoes my viewpoint but one has to look carefully for that to surface.

In the suggestion the presumption is claimed that all the space that the Universe made available at that time was the total space one atom might take up today. If that might be the case then where was the rest of the space that now fills the Universe? Or was the rest of the space we now find in the Universe and what is now explained away as the vacuum, also available back then. Did the Universe only have that one tiny hot spot it filled with huge volumes of heat? Was the rest of the space vacant being out there all along during all the time running to the present date but filled with emptiness standing around as a big vacuum with no better to do than sucking on the Universe while the Universe was exploding at the speed of light. Then that statement suggests that in this hot Universe there were light-years upon light-years of vacuum waiting to be filled by the intense heat soaring in the smallest spot. If that is the case then why did the vacuum not fill in the blink of an eye by all the exploding expanding material growing at the speed of light? Was the Universe overall bitterly cold where the vacant space was locked in with one spot of the vacuum filled with temperatures so hot we can only produce it in numbers suggesting a value but never claim to be able to digest the reality thereof in the human mind? If so what happened to the natural consequence that heat flows in the direction of cold and equalises between hot and cold. Was the space being available at present available then or was the hot space the only space available at the time. If so what prevented the heat from instantaneously filling the eternally cold vacuum because with the rules controlling vacuum in affect, it should have filled in such a manner in less than a heartbeat? I believe that singularity formed space-time and space-time developed from the overflowing thereof at the time it was extending.

With time marching onwards and outwards to this day, space-time developed. Space-time developed another product that everything in the cosmos has to have. It must be in such large quantities everything imaginable in the Universe has to have it and that is space using time to move about. I suggest that it is space that is holding heat in a quantity providing density and ratio to space available and in relevance to the space being available to quantify the presence of the heat and which then proves to form the time factor. The container and contained all together mixed by motion. From that very first separation of heat and space, which is what formed from singularity to produce space-time. The Universe was full... It was overflowing by the speed of light in the beginning...so where and when did vacuum or nothing enter the Universe as a factor if and when the Universe was so full?

The answer to that is absolutely crucial because how did the Universe decide to fill some parts with a variety of something and decide to fill some parts within the in-between with nothing? If that is true, why did gravity not prevent the vacuum filling because no gravity that came about since, can beat the force that gravity had back then? This leads to another question following the previous one in asking why did gravity at the time when it was so strong with r^2 so much compromised not fill the nothing immediately as it entered with something that could absorb the nothing. At the very beginning the mass that was pulling on the mass by force was immeasurable and none quantifiable. Even more to the point is the question to be asked in how big was the radius between the materials with the immeasurable mass placed in such a little space. This is all the more important when considering that the smaller the radius is, the bigger the force will become from the immeasurable mass pulling.

With the immeasurable mass that was producing the first gravity between the particles divided by an almost non-existing radius, the gravity produced had to be in gigantic proportional quantities and with the separation of the radii being in the infinite measure that it was at that point, then how did the Universe establish the chance to expand? It did expand, as we all are witnesses to in spite of this contraction of gravity that had to have been compromising the expanding factors. Still the expanding filled the unknown part of the unoccupied Universe, which at the time was there or was not there and where it was not there it was then filled with nothing. If the nothing was not "nothing" then the nothing that was not being nothing was also filling the rest of the vacant Universe that was or that was not because if it was, it was filled with nothing and if it was not then it was nothing.

This is then taking into account that then all the reducing that is resulting from Newtonian contraction that was going about in the space available at that time was something filled with nothing and surrounded by more nothing. With everything in the Universe being that much crowded and crammed, where and how did nothing enter the Universe and fill the rest that was unfilled? What factors introduced nothing into the picture since the entire Newtonian concept finds its base on the principle that matter reduces using gravity by force which then brings about reducing or the removing of the many nothing between particles, which will then lead to nothing that has to vanish even before nothing can enter the space. This question may seem small-minded belonging to the mentality of a child or to that of the mentally impaired with not much factual appreciation developed yet. Please do not see it that way. If you think in those lines it will be because you do not have an answer to challenge these silly questions. Beware silly as they are they represent official backing by the Wise-and-Informed. If the space is nothing and if the space was as large as it is at present then there was no need for such a small area to fill with something leaving only the rest filled with nothing at first since all the space we know about was there present and by being present it was there then for the taking. What ever filled the Universe had to start at the centre of the Universe and fill the entire Universe all over from a centre as it moved outwards filling from the inside outwards. This is a natural human instinctive realisation but is beyond proving by using Accepted Scientific policy. But that leaves Newtonian science with a massive unsolved problem: where is such a centre at the present time and where does the centre produce the limits or border it apparently has to form as it expands?

By expanding there is an additional contribution to that which was when that was. It was receiving more than there was before the addition increased that which was and by then becoming more than there previously was, it had to be improving the border from where it must have been before the adding took place to where it was after that, what was added, was added. I cannot name that, which was added at the time because that which was added, is still being added on the other side of the divide and by coming across from the one side to the other side of the Universe divide, makes that which was added and still is added, everything there was, there is and that there ever can be. That makes the task in naming such a concept going beyond my capabilities.

When that was less than it became when it was added, it was at the limit that was there before it was added too and that limit there was, was a limit that is the limit that I am referring to as a border being there. The cosmos is filled with unrecognised borders. The expanding has to be an ongoing filling that is at the same time expanding from the inside towards the outer limits of the Universe. Since nothing can enter from the outside where nothing is, the filling of nothing as a substance that would take up vast quantities of room had to fill from the very centre spot where all other filling came from. This filling of nothing with material has to be well mixed.

The truth about cosmology is that space forms no borders but by using any Newtonian centre from where mass is attracting we must find a point where there has to be the ultimate Universal centre which is the cardinal point in the entire Universe and it is the first, the prime position to locate coming before any other concept one wishes to put forward because all concepts have to start with locating that cardinal centre. There has to be the ultimate r^2 radii located precisely between the ultimate mass drawing the other ultimate mass closer. If there was a Big Bang then there has to be the spot, from where the Big Bang developed therefore there has to be such a centre connecting the past to that ultimate centre with the line of development flowing onwards to this day.

The fact that science is Newtonian proves that in the meantime Mainstream science is still of the opinion that there was the specific centre in the Universe that is nowhere to be found as it was filling the unknown with nothing coming from nowhere, but which somehow is still somewhere in the centre of all of that which is something. On the opposite side of nowhere there is an outer border in space producing a limit to nothing and serves nothing with a specific point to stop being nothing because that point is precisely where nothing ends and forms a beginning of a Universal border or a Universal end. How one will stop a vacuum being no longer nothing was a question everyone comfortably missed to ask therefore no one ever seemed to deliver any form of answer.

One night some years ago I enjoyed a meal with a very close friend of mine that is going by the name of Johan Boonzaier. It was at his restaurant that we started (or stumbled would be more accurate) on the matter of space and the fact that he could not see that outer space was never ending. As the conversation progressed he asked me about my opinion as to what space is and where it must end. I tried to explain to him what I believed in comparing my view to what Mainstream physics believes, but soon saw I was not gaining grounds on the matter of his understanding my concepts I tried to project to him. Then I decided to jot it down on paper and he could read it at his leisure as he saw fit. That led to the first book written by me (in Afrikaans my native language).

What I tried to explain to Johan Boonzaier that night, is that if the Universe was the size of say even a tennis ball with only the size of a tennis ball being the very all of space there is available, then yes, it must take time to expand from that having the excessive heat there was back then in all the space we have at present. It then is converting heat into space bringing about the expansion. But one will find most expanding within the atoms, as the atom must grow since the Universe in all was the size of what one atom is today. The space in the atom pushed the space outside the atom but there must be plenty more to the growth. Something outside the atom contributed in it own right because there is more expanding than there can be blamed on coming from the atom.

But the space then also developed as the Universe developed and if space developed then it cannot be a total vacuum filled with nothing because "nothing" cannot develop. You the reader must judge whom is correct between my view that space developed with the Universe as part of the Universe and reject the official view about space being nothing or otherwise you the reader must then decide that I am wrong, but should you do that, then find a reason why the Big Bang started out small and filled all the available vacuum or what is contemplated as a vacuum that we have with the motion of time.

When Mainstream science accepted the Big Bang as the principle that will take science into the future, the view about such a Big Bang concept unlocks a different door to another view on the cosmos from birth to end. It calls for the revision of all aspects of the entire history on cosmology and changes that what is dead wood and that, which needs to be chucked out. Most of all it was my following the lead I got from Kepler that unlocked the doors I now present to you. I claim there is no gravity - motion as there is no gravity forming weight or forming mass. I hope the sketch contributes to my explaining effort:

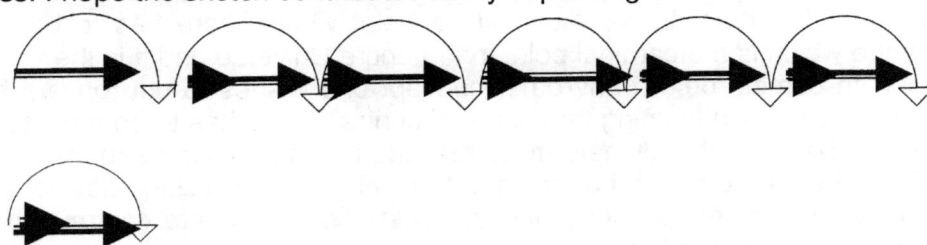

The duplicating frequency the Earth shows as k_1

The frequency of motion duplicating my body maintains as k_2.   **k_1 minus k_2**

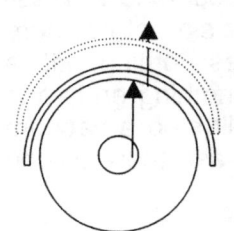

The frequency of motion difference my body has minus to what the Earth has where that difference in motion becomes my mass. It is the sum total of the reducing of the motion that my body has in comparison with the motion capability of the Earth that is the mass value.

In k_1 as well as k_2 the symbol represents motion, however in the case where **k_1 minus k_2** that shows an incapability of motion, which is motion, frustrated or a more commonly used name would be <u>mass</u> is created

The new **k** that is applying the relevance, must link the space a^3 being equal to the motion T^2 to singularity k^0 in order satisfy k^0. The flying of the aircraft is then unequal to the motion in the previous relation that was in place where it was part of the Sun and the Earth motion relation and the new motion will bring a correcting in the relevant distance **k** to put the motion in balance with space. Space will always demand a correct establishing of the miss-interpretation of the equilibrium that is needed to sustain the effected singularity because of the space-time factor.

There is a point where the two points forming the relevancy unite in shared singularity. It comes because of shared motion. In all space-time, one finds at least two relevancies where one is at the centre.
That part Newton saw and formulised.

He missed another part. Crossing a limit of inclusion is the limit of division and such limits are in distinction by motion producing the gravity, which is parting the two objects. Motion brings about a relevancy where two positions no longer share a common point in singularity. That is what Newton missed. That is the gravity aspect Newton and all other Newtonians miss.

Two objects of substantial size differences are travelling at the same time but one has a space, which it has to move when it travels that is considerably different from the larger space. The larger space will produce an extending line equal to the space it moves while the smaller space will also produce a line in ratio to fit the space it holds relevant and which it has to move

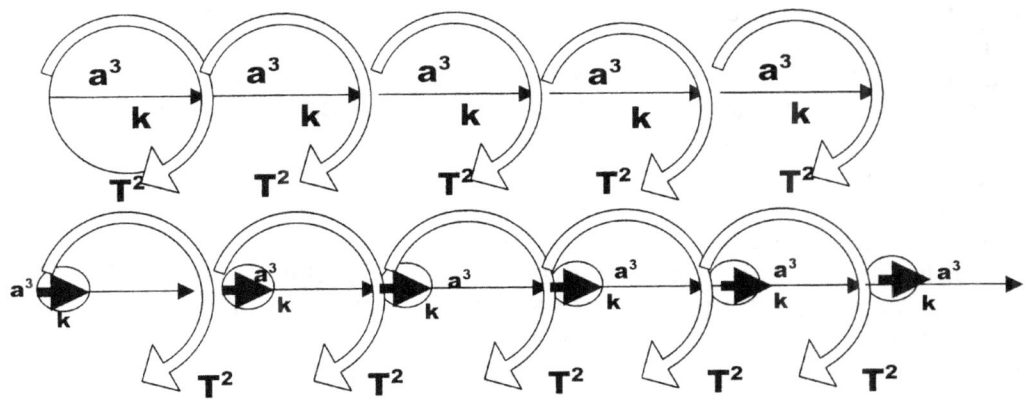

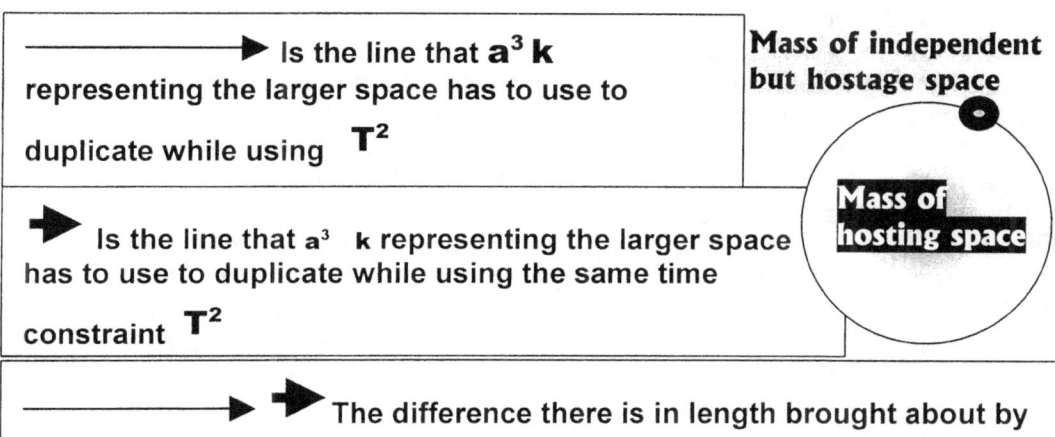

—————► Is the line that $a^3\,k$ representing the larger space has to use to duplicate while using T^2

► Is the line that $a^3\;k$ representing the larger space has to use to duplicate while using the same time constraint T^2

Mass of independent but hostage space

Mass of hosting space

—————► The difference there is in length brought about by moving the larger space in the same time T^2 as the smaller space is what brings about mass. There are other factors too which I shall touch on as the book develop.

Mass has precious little to do with the whole affair except to be an obstacle intended to restrain the motion of the hosting space. The difference in size between the one in circular motion and the space in contracting motion must bring about that the smaller object has to move about a circle much closer to the centre because the larger space form the centre hosts it. However there is no large or small in the cosmos but only those better developed or those poorer developed. By duplicating there is more to duplicate in the better developed than in the lesser developed. When the lesser-developed space is duplicating the less developed space would hold a lesser extending from point to point forming a shortfall by distance in comparison. The motion being extended needs less extending and should therefore be closer to the centre in relation to what the better developed space would need in extending by a duplicating effort. This is the principle we find behind the sound barrier. The motion the aircraft produces forms an increase in the duplication of the aircraft, which extends the duplication of the aircraft splitting the Earth and the duplication that is producing an extension of the aircraft. The splitting does not align gravity lines with the Earth as it did before. The aircraft is reproducing more in much shorter time duration by duplicating and extending space filled by material that goes beyond the attempt of the Earth's extending through duplication by such motion.

I know this may sound barely believable but please hear me out. While we use gravity, the use of gravity as such makes us part of the Earth. We see gravity as some influence or force producing mass and that mass is forcing us down on the solid ness and onto the Earth. By having the mass we become a semi unit with the Earth. That is how we on Earth see gravity but when investigating gravity in outer space we must come to a basic question: Is that what we experience as gravity on Earth truly gravity? Much of the proof about gravity is part of our perception about gravity because we experience certain conditions with gravity while we find ourselves bogged down on Mother Earth. But are our perceptions about gravity truly correct? We experience mass but are the mass the result of gravity or are the mass the product of gravity. We only experience gravity, as a factor from the position we have on Earth and the conclusions we form is a product of a perception we formed while we are being forced to be part of Earth. It's as if we are upside down and have to decide on which route we should follow. I want to make a suggestion, which I aim to prove in the following pages. My personal being on the ground and having mass that is keeping me on the ground comes about because of the speed that I travel through space being the very same as that which the Earth has.

By me not applying a speed difference, I then inherit the speed the Earth places on me. But my space which I use $a^3 = T^2k$ to travel and the space which I use tot travel through is much smaller than that which the Earth burdens with to move and to move through. By me having a smaller space to move $a^3 = T^2k$ the space a^3 being moved k in the time it would take to move T^2 will produce less space a^3 to shift k and therefore a smaller distance k to replace all the space a^3 that is moved in the time T^2 the space a^3 needs to enable it to move k. To duplicate by motion, the smaller space requires a smaller distance to shift the space but the motion will take up as much time to complete than would the larger space take to complete though the space the larger space has to duplicate will require a longer distance to complete the total duplication of the larger space. A large space a^3 will produce a large extending k when using a^3 the same time durationT^2 when using the same time factor as that which the smaller space is required to use when under obligation to use the same time constraint. Behind this is the most basic principle hiding which allows us the fortune to be able to fly using a flying machine. It is all about motion supplying relevance and enforcing time constraints.

Because my body that I have is travelling so much slower than the Earth is travelling due to my size in relation to the size the Earth has and although I am using the same time as the Earth does to move, such a speed difference is not in the time differences it takes to complete but in the space differences that has to be completed in the same time but is unable to fill and the space is trying to crush me into the Earth where I am forced toward the centre. If I were able to penetrate the soil solidness I would reach a point where my speed as zero would equal the space I occupy.

The space I duplicate by moving from one position and placing the space I hold in the next position while keeping my space, I move, as it is identical in the next spot but locate in the next position. Such moving by duplicating takes a certain time to move from one spot to the following spot and it will use a certain frequency that will have the same ratio in bridging the gap from one point to the next point as that which the Earth has. My speed of duplicating by motion has to be even in frequency because I am within the duplicating space, which the Earth is duplicating and being where I am, I am part of the space that the Earth is duplicating. But in size there is a massive difference between the space I hold and the space the Earth holds but to duplicate will take me as long as it takes the Earth. Notwithstanding this common factor the Earth has to use equal time in duplicating its massive space, as I have to duplicate my small space when we both have to share a frequency that will keep us duplicating evenly. Therefore the frequency of duplicating using the same time period will be a lot different to my much shorter frequency of duplicating space.

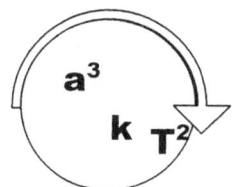

The difference is between me being in mass and me being in the correct position in the space-line the Earth has, which will place me in the correct position but the heat that then will surround me will fry me into non-existence. Fortunately, for life, the soil forms a barrier through which I cannot fall any further as to correct my location. Being where my position would have no mass, would allow me to float there in that location in the same manner as I would float in water. I would be buoyant. It is because I do not harmonise the displacing frequency that I should, that I have mass.

My having weight is what Mainstream Physics use to give me my gravity. Science purposely switches my having mass and confusing my mass with my having weight to explain what is beyond explaining. It is said that while I float in outer space in a state of suspension hanging above the Earth in the weightlessness, I still have all the mass that I had on Earth. But in order to prove that those in science will give me a mass even in outer space whether I deserve it or not. By that token science first has to cheat all logic by reasoning in some bizarre way that I take my mass up there to where there is only micro gravity. They firstly claim that all of a sudden I take my mass to outer space and in their next argument they say I have micro gravity in outer space since my body is floating as if it is in the sea. But if I stop floating and start falling to the Earth my body and I did not gain any mass. My falling then comes as the result of my motion being much smaller in relation to the space I claim and my motion then is being less than what is required to keep me in the position I have in which I maintain my orbit up there. By moving too slow I fall. I do not fall because my mass grew. But science has been proven wrong by their work without any of them ever admitting to such a defeat. All the satellites fall if the satellite motions are not reset. The satellites do not gain or lose mass. They gain or lose motion. By amplifying either my space (using a Hot Air balloon) or by accelerating my motion that I have in relation to that which the Earth forces me to have, I will break free from my weight or mass. I shall become airborne and float as if I am in outer space, by pretending my mass can be multiplied many times over in using a process, which then is called not gravity but momentum. But motion and gravity is all the same because motion is gravity that is redirected, which then forms another part of gravity where gravity again is also only motion applying. Science maintains the argument that when I am in outer space and am no longer part of the Earth I then will only have mass. But since there is only micro gravity I will be in a state of weightlessness. My mass is what gives me gravity and while being up there I take my mass along with me. But with my mass up there I will only have micro gravity. I am floating with my mass and it is my mass that is responsible for my gravity and I am floating above the mass of the Earth, which is right down below me, but still I have micro gravity. That is true if I wish to incorporate the dubious use of double standards by separating mass from weight. The mass my body will have in a Black hole will be a billion times (at least) more than what it is on Earth. With that the Black hole destroys the fact propagated in science that my mass will be the same everywhere. That is more than permitting double standards. Because our motion is much slower than the Earth is spinning, we place a breaking effort on the velocity the Earth has and that breaking effort we accept as the mass we have. The truth is that my mass comes about from the lack of motion I have in relation to the space I occupy and has nothing to do with any gravity - motions pulling me down. If I increase the motion I have there shall come a point where my motion will be sufficient to pull me into the air, as I then will have the required velocity to lift from the ground. That motion being in excess of what I have and is complimenting the motion that I receive from the Earth counteracts the motion of gravity that is containing me. The motion I adopt then release me from the motion containing me and if motion can release me by only becoming more, then gravity is my motion not being enough in the first place to keep me onto the Earth. Nowhere and at no time does my mass ever gain by having more protons that will get me back to the ground as if I am bigger or carrying more material or does my mass reduce to get me into the air as if I am smaller or carrying less material. Please note that this is my way of explaining to you about the fact of bodies having weight or mass. It is not mass or the lack thereof or any means to measure occupied space within the atmosphere of a larger body that pins me onto the ground. My body is claiming space by motion in space. Gravity is the result of motion because it is in the motion that bodies have that gravity affects them. This is proved because by adding motion the mass does get more but the body never gets bigger or hold more material, and in defiance of that statements, by increasing the motion my body lifts and flies. The reality is that my body in motion has more mass being momentum but still my body lifts when motion allows my body to lift. This statement confirms Kepler that a^3 becomes more (massive) when motion T^2k becomes more (moving).

Mainstream physics admits all along that nobody, human or otherwise knows what gravity is. While investigating Kepler's work with employing much motivation and detail in order to give his work the much duly credit it deserves it will also serve a valiant purpose when by the same token we try to establish what gravity is, because I believe Kepler possibly answered that mystery. We have to start with the person that introduced gravity or so does everybody acknowledge. Newton saw an apple fall from a tree and he subsequently realised there is some force pulling the apple to the Earth.

Although he still was a student he announced his findings and became a genius on the spot. The concept he introduced as gravity gave him instant admiration from which he became the legend he is today and that reputation he gained there at that moment would last him from that day he instantaneously unveiled his mastermind, and that same genius still serves him in his honour to this day long after his death. He found that this force has to have some thing to do with the weight and the mass of that particular object and the mass of the Earth. There is some force pulling that apple as much as the force is pushing the apple and the same goes for the Earth because the mass the Earth has is doing the same to the apple. Between the two objects facing gravity there is a force that develops where such a force is pulling the apple on a constant basis towards the Earth even after the apple is already in a steady state on the Earth. That forms the mass and the mass forms gravity. He concluded that the mass is responsible for the pulling. Remember this observation came three point five centuries ago when knowledge and brilliance carried a much different defining than what such defining of brilliance is worth today. He realised the pulling on that apple brings about weight that brings about mass because the apple departs from its location and arrives at its end location when the falling is completed. Then he went out on a mission to convince all those that were standing in line with a desire of finding the conviction needed to be convinced because no body before Newton thought of what Newton thought quite in the way that Newton thought about gravity.

Newton succeeded because he found a way in presenting science with the fact that objects move closer because of some force. He went one step further and named the force he fathered as gravity. But there it stopped! Any and all other further defining the matter or going into any possible observations of whatever magnitude concerning the topic never realised any motive to go further. Inspiration to further commitment just flew out the window as the essence to do so immediately expired as far as the rest of science is concerned. What might he have missed if he missed anything? We all fall down when we are unstable and out of balance. He never realised that balance is more crucial than brutal gravity because that part is the defining part about gravity. No one ever gave a thought about the balance part even centuries later even as we grew into all the sophistication we now enjoy. What brought about the balance that secured objects in an upright stance and supplied some form of control over the managing of a position? Any other position than being flat on the floor would have a better defining than being just at the mercy of the force gravity. Standing tall is a stance that defies gravity so there is another force other than the pulling of gravity. Admit tingly the force would first and foremost have to aspire to the rules of gravity and then comply with other demands. True enough is the fact that that position would ultimately and firstly by all accounts have to satisfy gravity before any further motion could commence. Yes but then by balance motion defies gravity by changing gravity's force of pulling everything straight down towards a visionary centre between the objects. In affect this means somehow there is control over gravity and gravity does not leave objects beyond outside control. Gravity is manageable and can be controlled; we just have to find a way…

Years later some one came up with the novelty of hot air ballooning. Ballooning proved that there is antigravity but that part was missed by all even to this day. Some people speak of antigravity as if that is some mystifying mysterious concept that is so well hidden in the secret annals of the hidden Universe that only Ali Baba and his magic words can reach it. Please consider the following statement. If gravity was bringing the object down, because of the affect of gravity which is that which we experience as the gravitational sensation and that is what we interpreted as gravity by our sensation and observation, then that is only coming about by our bodies that is in a state of being dragged down. The dragging down of the body is in the direction of the Earth's centre. That sensation of being firmly locked onto the ground constitutes to what we believe we experience as gravity.

When some influence brings about the very opposite effect, which then results in establishing the opposite result, it deserves to be anti. In example we feel dragged down but anti will be the lifting of the body into the air. Anti will be going in an opposing direction of the motion that gravity inflicts. It will counter the influence that gravity applies. Such motion has to indicate antigravity. The counter acting of the mass dragging us down must be anti gravity pulling us up into the air above the ground. Antigravity must come from such an opposing influence that will bring about the lifting of my body. If hot air ballooning gave the object an opportunity to lift, then ballooning must be antigravity. The balloonist and the entire balloon found a manner to counteract the pulling of gravity enforcing weight. The balloon can lift what gravity depresses and if Newton said gravity is the falling then later Newtonians must agree that the opposite of falling is flying or lifting. A balloon is lifting-and- flying. If gravity is pulling down objects in the direction of the centre of the Earth then flying is antigravity. Moving away from the Earth by means of motion and in particular flying is using whatever means to defy gravity where the lifting can also be the hoisting of a body by a crane. Lifting by ballooning in a hot air, such balloons escape from gravity where the balloon constitutes to bring about the effect of establishing antigravity. Climbing up mountains must fall into the antigravity department because parachuting down the mountain definitely falls in the gravity department. Nevertheless it still does not answer the question of what gravity is.

Let us look at antigravity because the antigravity is releasing the object from the gravity that controls the object by an Earth fed force. The balloon starts flying when the confined space of the balloon is ferociously and violently heated in access. The balloonist shows us that in order to overcome gravity we have to introduce heat. That is the only manner in which we can defeat gravity. Even by an engine driving an aeroplane such flying can only result if an engine combust solid fuel by creating motion as the fuel mixture is turned into heat. It is heat that makes the difference. That is the very thing that Kepler said. Expand the space a^3 and the motion T^2 will move further increasing k. Blowing hot air into the balloon is increasing space within the balloon a^3 which then results in providing the balloon with a larger distance k from the Earth centre k^0 that still holds time with in the Earth atmosphere with the Earth T^2 within the space of the Earth k. Using Kepler provides us with insight and the ability to see what gravity is by showing us what antigravity is (a^3 gets bigger and that will bring in a larger k). But moreover the larger space in enough compensation to bring about extra motion that will defeat gravity by the extending of **k**. If that is not antigravity, then we can forget about Ali Baba and his magic rhymes too.

The balloon assists us to escape the Earth's hold on our body, because there has to be the force producing motion countering the motion of the Earth's gravity. The balloon shows that releasing enormous quantities of heat into an inclusive area excluding space such as that which the balloon canvas provides, which is establishing the release from the gravitated containing force on the body giving the body a means to escape by floating about above the ground. The motion is at that point breaking free from the containing gravity by moving in a specific direction, other than the direction the Earth gravity inclines the body to travel. By concentrating the releasing of heat into the balloon, the direction of motion starts to contradict the enlisting of the Earth gravity and the heat breaks the balloon's confining properties while the balloon is released from the Earth as the balloon and us lift up into the air and away from our confining to the Earth.

At the point of explaining we arrive at the point where we can say what we think the difference is between the balloon floating in the air above the Earth and a body suspended in outer space floating above the Earth's atmosphere. The difference is the heat that is in the confined air per volumetric ratio favouring the heat being more in the space than what the heat is outside the confined space. If we had any method to put the required heat we need to escape from the limits of the Earth to outer space into the canvass of the balloon, there would be no canvass left to contain the heat. The heat is available to do the job but the means to do the job with the tools in hand is unavailable as far as we can use the balloon. By having more heat in the one area than there is in the other area beats of the pulling off gravity. Obviously it is antigravity that keeps the balloon in the air and what keeps the balloon in the air is having a larger volume of heat per space unit than what is in the atmosphere. The balloonist shows us that by applying more heat we can defeat gravity more. Someone took the advice, because the next minute the Germans had rockets. The launching of rockets brought about the ultimate defeat of gravity but it involves almost the ultimate releasing of heat.

In antigravity we find heat more concentrated in one definitive area than the heat concentration is elsewhere. The more the heat is that we release into space, the more the antigravity is that we achieve and the more release such antigravity can produce. But what connection can gravity have with heat and if there were any connection between heat concentrated and gravity, what would such connection be? The history behind Carl Benz should bring the answer but more so would be the story behind James Watt and steam although the James Watt story may not be that thought provoking because it is much less filled with the ever popular cheap thrill only sensational gossip can provide…Still both stories cover the same principles. In the Carl Benz story a housewife leaves a pot of benzene fuel on a coal stove. The pot with benzene heats up where the pot with benzene becomes hot and under pressure. This performance of heat increase, such increasing expands as space and releases the heat as newly creates space, which then removes the housewife with her house from the neighbourhood she used to regularly frequent as her residential address. Afterwards almost the entire neighbourhood is not there to tell the tale or ask why...

It was a stupid tragedy that brought about the end of steam and the rise of the internal combustion engine and on Earth billions on billions of human souls are in torment not to please or suffer for the advantage of coal Barons any longer but now they are dying and suffering in agony to please the wishes and desires of oil Barons. How much did the world not change…While it is no longer the coal Barons shackling us in chains and telling us democracy broke our burden of slavery, we have now the pleasure of the oil Barons enslaving us with democracy and telling us to be happy as we are the fortunate slaves because there are others circumstances in which they can enslave us that will leave us worse off. All this came just because the pot of fuel created a houseful of space that was enough to remove the house from the address the house previously enjoyed. But Mainstream science neglects to appreciate this. They see the heat, they see the antigravity but they fail to add the heat, the anti gravity and the space that no longer housed the house of the naive and rather impractical thoughtless housewife. They call the tragedy an explosion but then again everything that expands while using a noise during the expanding is an explosion. Adding of new space to the space holding the house at first altered everything that was previously proportional positioned in the space where the house was. Such exchanging of heat to accumulate and introduce more space in the process referred to as an explosion was bringing in more space that came directly as a consequence from the explosion which was producing more space where the increase in space brought disorder because the well organised material distribution and placing was before the event filling just enough of the required space arrangement that was holding every object in a prearranged order of tidiness.

Then suddenly out of the blue the space, which held the house in a tidy arrangement had to accommodate more space therefore the ratio of material per space volume increased dramatically many times over in the favour of the space in the balance. That part no one ever acknowledges. However the losing of the house was not much surprising to Mainstream science back then and even today because who cares about old news. All of Mainstream science was at the time, as they are today, very familiar with all explosions because of wars and bombing that leads to maiming and killing and all the unspeakable monstrosities we associate with war so that the dirt poor can suffer and die to leave the disgustingly rich even richer. The poor has not the means to pay science to be clever and devise methods to save their lives, so the rich does the poor the favour of paying science to find methods whereby more poor could be killed as long as the rich saw it as a good investment with great capital gain on the part of the rich. Therefore science is well established in the method of creating more elaborate and destructive explosions that the rich pay them to invent. In the explosion caused by our housewife no one put up money to investigate what happed during the explosion but money went to why the explosion happened.

That inspired an investigation in connection with the fact of finding out more about what takes place during the carnage as more money goes to finding means to create more carnage per money unit spent. At least that is why the poor were invented and that is why wars are invented. It is invented so that no money goes wasted on saving the poor people except if the poor has the money ready and available to pay the rich for medicine to enable the poor to stay alive. So science goes out and develops more fuel for carnage but fails to find out why the housewife and her house are no longer part of the neighbourhood she used to frequent. The loss of the presence of the ignorant housewife with her house her neighbourhood and all were a normal way of leaving us with a new way of tapping and harvesting energy and so untold riches which was born with the death of the absent minded housewife. But according to the mindset of science they saw not what the incident presented in space producing for to their view nothing new came about since it was just another exploding of fuel...so no body bothered as to finding out how. What they missed was the part the coal stove played in the whole tragedy. The coal stove intervened by producing the heat that turned the liquid fuel to liquid heat and in the process went on to liquefy the space that turned the liquid space into a gaseous space where the liquid space revealed its true incentive in nature by turning out as space and the newly created space that was in fact liquid space that went on to become more space, well that space was providing the one main factor in space-time relevancy. The stove's heat was producing space by transferring the heat from the stove to the pot filled with volatile fuel and the transfer of the heat to the fuel brought about the expanding of the space that the fuel claim to need in the pot filled with the heating and volatile fuel. The fuel space requirements became more as the heat filled the space that was filled with fuel and that took up more space, which the pot could not cope with since the pot had no room to allow such an increase on the demand for more space. The fuel expanding as such was claiming new space to sustain and accommodate the growing requirement for more space to be created in order so that the volumetric increase of heat added to the fuel could be accommodated. At one point the asking for space became a claim on new space, which we see as an explosion. The heat transformed to new space by an exhilaration of breathtaking increase in heat forming space and this increase of newly formed space was transforming all other surrounding space within the room, the house and the neighbourhood in general. By increasing the volumetric quantity of the space it rearranged all space, which included some of the space held by material that was in a solid form and scattered the rearrangements as fragments in all other designated places far away from each other. This meant there was excessive and all around rearranging and relocating as well as re-allocating of space in general. Science sees this as shock waves resulting from an explosion but it is merely heat expanding space to set new required standards. It is rearranging every aspect that contains space or that space contains. It will bring a much different looking end. Everything about this concept is missing from Newtonian science because Newtonian science failed to investigate Kepler. Kepler said space a^3 is equal to the motion T^2k thereof and then that says without Kepler directly saying it, it says that if space a^3 goes bigger as a result of the explosion then

such increasing in space will constitute to more space a^3 which has to produce an increase in motion T^2k where more motion T^2k will bring about faster displacing space. This is one small fact that Newton robbed the world of realising with his ignoring of Kepler's work.

From every point there may form on the outer circle line of every part of the circle structure and all structural positions of the circle in all circles, all circles refer to the centre in perfect alignment. Every point wherever located on the sphere has a matching and equal but an opposing point on the other side of the circle but in equal position on the other side of the circle. Between the two controlling points runs a precise straight line connecting the two opposing points in counter balance. When drawing the connecting line between the two controlling points and connecting such points on further edges of the circle by lines formed, the lines will all cross the centre. From wherever a line may cross and from every point forming a line to the other side of the circle rim holding the connecting points there has to be a counter point located on the very opposing side that when connected by a line, such a line

All connecting centre of individual connecting lines between opposing points

crosses in the centre. In the middle the centre spot bonds all sides coming from any and every direction there can possibly be. The line will run to an equal point on the other side across the same distance from such a centre and that then has to be where the strongest gravity can be located.

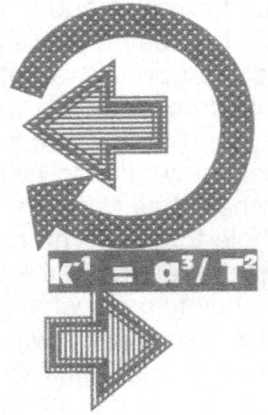

$$k = a^3 / T^2$$

$$k^{-1} = a^3 / T^2$$

We are now serving time in the twenty first century. One Professor once told me I must realise that Newtonian science took man to the moon and back several times and in such a view I am rather annoyingly presumptuous to criticize Newton. The Professor missed the point. I criticize Newton on what he did not give us, which he gave us as incorrect by his own admission that it is mostly guesswork on his (Newton's) part. His guessing about the facts later became institutionalised facts believed by all concerned to be correct and to be proven to a degree of correctness that is far beyond doubt. Newton gave us gravity but Newton never gave us the explanation about gravity. At the time Newton met strict opposition from his colleges and piers because others felt his introducing of an unexplained force was taking Science back in time, which of course it did. Many scientists at the time accused Newton by name of dragging science back in the wrong direction of progress by introducing unexplained forces acting in a superstitious and mediaeval manner. I went one step further by asking myself the question: If space becomes more when heat becomes uncontrolled why can space not become heat when space is under control? If space becomes more as we see with every explosion of every kind and such heat forming space releases energy, then why would space being managed not form heat being under control and produce energy. We only have to see what Kepler said gravity is. Motion gives us substance in which to be.

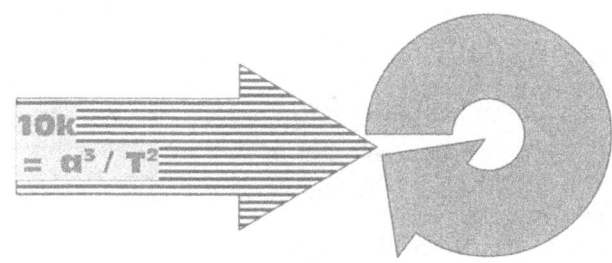

In the normal flow of events there is a certain ratio of liquid time that stands related to a specific gravity, which is generated by the motion of the liquid in relation to the solid that binds the liquid to the solid. That side of the relevant motion uses symbol representing the factor **k** in the Kepler formula $a^3 = T^2 k$. This symbol is standard on both sides but due to the inherent nature of rotation qualities, it opposes that what it was by crossing the divide there is. The liquid

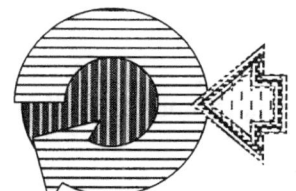

attaches to the solid by diminishing the relative factor bringing about that **k** goes negative $k^{-1} = T^2/a^3$.

The solid on the other hand confirms the liquid attaching to the space by the motion thereof and with that the solid recognises

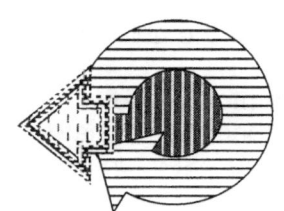

the shift in limits where there forms a new boundary and a new location where space that include the attaching liquid then ends.

When an explosion occurs, the solid space melts and the lot in the space becomes liquid. That is a direct result from the expanding of the limiting or relative factor **k** that reposition the boundaries of the liquid turning space into an altogether new dynamic. As the solid melts, it

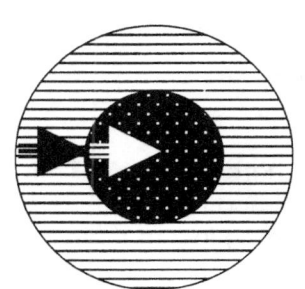

amplifies the space held by the heat many fold by tarnishing the spin T^2 that increases the space a^3 and reduces the density of the heat in the space a^3 where the velocity of the motion T^2 is reduced as the factor **k** is increased in distance. By increasing the claim on space as the spin velocity reduces and the density intensity decreases, the liquid part becomes more since the overheating destroyed the solid part and that which we know as gravity reduces because that we know as concentrated space reduces in intensity. The expanding is due to more overheating than what the governing singularity can control. Gravity is a product of overheating. The overheating brings about expanding while the expanding brings about cooling. Since there is an excessive demand on space through the overheating that leads to expanding into more space, the claim to space is that which reduces the density of the solid and turns the solid into liquid. By overheating there is a reduction in heat density and by that expanding there follows a lost of singularity control. The increase in liquids demands more space in the light of the decrease in liquid density and the motion reduces as the density diminishes. It will take years of increasing the controlling ability of the singularity governing space-time and a rebuilding of the solid under the control of the motion that increases while the space reduces, which then confirms the increase in density and the improvement of the control of space-time.

Where space is the least, which is in the centre of the circle, gravity is the strongest. The gravity located in the circle's space less centre holds not only the sphere together but all that is in the surrounding of the sphere outside the sphere as well. It is from there in a giro action that gravity bonds all atoms forming the structure of the sphere as one unit together in a unit as well as distributing a specific alliance in shape and form. How the atoms manage that, we will get to in a while, but there is a law allowing for that to take place.

Gravity is the strongest in all cosmic structures holding the form of the sphere and gravity controls all around from that very centre where space is the least, therefore the more material there is to generate motion within a star, the more secure would the generated centre be in any star where such centre produces gravity. It is not the material but the motion the material accomplishes that becomes the factor of gravity. The smaller the star is as far as volumetric occupation goes, the stronger the gravity is that is coming from such a centre. The less the space there is the less the motion is and therefore the stronger and more deliberate the motion is evoking gravity. From the centre in the middle where space is absolutely at a premium the gravity grows stronger as it draws all material.

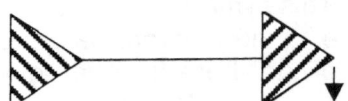 If outer space was nothing then the crossing of light in outer space was impossible because of the total destruction nothing leaves when contacting anything (1X0=0). Light crosses space at so many points holding infinity that it becomes the speed of light or the ability of light to convey space by motion thereof.

The motion is one of confining the space to a centre by the moving or trying to move the flow of space and whatever is in the space into the centre where the space is least. Take the Neutron star and the Black Hole as an example and compare that with the Sun and the answer is simple. I claim that gravity is all about reducing space and not attracting matter but that I explain a little later on. Therefore the matrix of gravity must be permanently located in the location where space is the least. Looking at a sphere, we find that what holds the sphere true to form is placed in the centre of the sphere, which then has to be the most intense point of gravity. Gravity is confirming the round shape without favouring any specific point. Such evenness of gravity comes from what is applying at such a centre and is in control of the surroundings. The centre that secures all of the space and material in the space holding the specific form has to be round if it is anything. That shows that in the sphere one can see that the sphere as a form is dominated or controlled from one specific location in the centre. The explanation about the reason there is control coming from the centre, has a very childlike simple answer.

The Big Bang was where gravity held the Universe in the least space there ever was. To find the original gravity we therefore have to reduce the sphere to the circle and reduce the circle from there narrowing the circle down to as far as one can go. The Universe is a magnitude of spheres constructed by a complexity of circles. This is because everything sprouted from singularity. To narrow any circle down will be the same as narrowing down the Universe. In our reducing of the Universe we must first acknowledge that the Universe constitutes many spheres, which is giving the Universe gravity as a combining unifying part which is the part of the sphere giving the sphere form (or gravity) and that confirms that the sphere is a circle in many times over multiplying the positions from where gravity secures form. If we wish to go back in time by taking the Universe back down the same route and at the same time maintain some coherency, we must concentrate on a single circle because a sphere is a circle by millions of possibilities linked together by just a name that changes the concept. When one takes this accepted route in thinking that by reducing the connecting line to the connecting circle point in the centre of the lot, it must take us back in time at the same time as the circle reduces to the time during the Big Bang.

During the Big Bang where all circles were as small as they can get, we run into an unknown substance we came to know as antimatter. This theory is propagated according to Mainstream science but what is most surprisingly is that I do agree with this part of the statement. All material produces gravity. I go one step further and say all material applies motion where some motion may be to contain by using gravity attributing to the contracting that leads to the reducing of their space. Then as everything in Creation has an opposing to restore and maintain balance, there had to form another or other material that did not by our lamentable standards produce gravity because those material produce antigravity, a concept beyond human discernment. Antigravity must be the expanding in counteracting contracting. A counter action to contracting is where expanding provides growth of space. By growing, the heat distributes over a larger area and there is then the same heat in a larger area making the heat less in a specific area.

By reducing, the cooling contracts into less space to that which was used when the overheating was in progress and the cooling then reduces the space which results in both the gravity effects. Forming pappy provides more space by losing density to the advancing of their space. Material either have gravity by solidifying or concentrating the space they hold in ratio to the material within the space they hold whereas others lose their solidness by entertaining more space within the ratio of material to space where such material becomes liquid and in more extreme cases they become gas. Being a gas they float which gives that material a high degree of antigravity being airborne. It is however not clear if antimatter produced gravity as it did when it went to lunch on and ate up all material in the immediate surrounding. It was cannibalistic but the unanswered question is this: was it a gravity producing predator or a non gravity-producing carnivore. Did material find a comrade in their gravity forming of form or did the gravity it produced bring on the demise that subsequently followed the event as is reported by the highly informed.

The Accepted statement on antimatter reads that matter composed of anti particles where such subatomic particles that have identical rest mass to corresponding particles of ordinary matter but opposing charge and are opposing in other fundamental properties. One example given is that an electron would have a positron, which then functions as the anti particle and has a positive charge compared to the electron's negative charge. That is put bluntly in its utmost simplistic form. Unanswered and tough questions arise from such a statement. What kept the electron bonded to the atom since the protons must by implication produce expanding or by definition be repelling the atom and surroundings instead of the normal contracting or confirming of form?

What is a positive compared to a negative charge, because it is human concepts that put the directional qualities of material into a positive or a negative contexts as we did with hot and cold. It is human standards that humans brought about to make all human inadequacy by lamented human understanding better but it is not applied cosmos principles. If there are extracting electrons performing in the capacity as antimatter, then there better be protons by another name in service to the anti electrons, which then of course serves the anti electron in the capacity of an anti proton with an equal but negative charge to that of the proton. When matter and anti matter meet, the two opposing particles annihilate each other until one vanishes from the Universe. I have to add that at the time this theory was devised the first computer games became a crazy fashion played by young and old, those wise and those foolish all alike. This game was called the packman and the packman ate up all the skulls and after eating left nothing as evidence. The theory about antimatter has some very striking similarities to that packman game. It still does not answer the most ardent questions: What makes a positive electron different from an electron in the working place each has and can any person show such an object found in nature?

Can people take a positive electron to an investigative bureau and are acclaimed for bringing about such evidence? It is unwise to substitute nature with human concepts just to further mathematical equations. This was apparently presented as normal as nature was when nature developed with the Big Bang and nature then did behave this oddly just after the Big Bang came about. But one huge misgiving in this argument is declaring that everything the antimatter had as a meal vanished and even moreover then antimatter went and vanished too. Where could the combination that was produced when the matter and antimatter collided go after it disappeared and did it form the by-product of antimatter science is talking about, which since then apparently vanished too? What a bloody non-intellectual fairytale that is.

This does not even make sense to a child still in school. It is so obvious the person that came up with the fairy tale was one of those science fiction writers that NASA apparently employed to serve as visionaries while they are paid to concoct made–up-as-they-go-along stories, which are told by persons that supposedly should know better. Since there is no place other to find a location to be within than being in a place inside the Universe it is hardly possible to vanish from the Universe except in fairy tales because for one simple fact: there is no other home to have but the home we have which we call by the name the Universe and we have no where to escape to but within the walls that the Universe provides for such a purpose.

There is one Universe containing all and preserving the lot. Mainstream physics is accepting this fact. But then by the same margin they accept a principle that allows property that once was part of the Universe to leave the Universe and go somewhere outside the only Universe. They create a loophole whenever it suits them to misplace what they cannot explain readily and logically. According to my view there can't be any hiding of anything any place of whatever description in the Created Universe. This they admit and confirm although with the same breath those very same intellectuals also admit that there is another place outside of what we are able to find in the Universe. When someone comes up with the marvel where such a person can declare in all honesty that the product of antimatter or singularity escaped from the Universe to God knows where, that person should leave the field of science and go for fantasy writing such as fairy tales or reporting about politicians inner deepest chastity and integrity. That is what we can find outside the spectrum of what the Universe can deliver. With such a statement on anti matter or the loss of any Universal product disappearing from the Universe then alarm bells should go off in the mind of the trained and professional Scientist working with such matters.

Yet those in charge do not once belabour the question on the validity of a statement that involves stating losses occurring of substance being in the Universe going lost or being removed from the Universe. They seem surprised when stating they detect the loss of factors and when they are declaring the possibility about such losses, they forget to follow the argument into a sensible conclusion. They never pronounce where those factors at present are, since it then has to be allocated in space that is somewhere outside the Universe, and the only place anything can ever be, is inside the Universe. Go back and read the definition as to what the Universe is and how the Universe is defined. They never even give the notion a thought as to how that, which once was part of the only place there ever can be, came into a place that is no longer part of the only place there ever can be. It seems the mathematical awe of their abilities to fantasize with the aid of mathematics douses their connection with reality. Their personal fantasies mesmerises their clarity in thinking. They can read mathematical calculations and agree on an outside the Universe without stating it in an explanation as to (in their mathematical concluded opinion) what happened to the lost and found or then to what happened to their ability of introducing the concept as a reality, which they claim it is. It seems to miss their minds that it is quite impossible for whatever there are in the Universe, to go outside the Universe and leave the Universe by causing a Houdini vanishing act of a never–to-be-repeated-again status. It is totally inconceivable that the most intellectuals there supposedly are, those in astro physics Science, might think us of the lesser fortunate in intellect as to having us believe that antimatter went into hiding in a place through a manner where such a manner then, is out of the Universe.

They applaud this thoughtless presumption while fully knowing that there is no position outside time to place such vanished substance as anti matter. They present the argument while being fully aware that there is no other place for anything that is wishing for a place to be within, then to be in any other place than what is allocated to a position inside the Universe. If it ever was anywhere, then it still is within the Universe merely because there is no other place to go than to be inside and part of the known Universe! It seems to be beyond their realising that there cannot be some factor that can be misplaced in a location outside the Universe. If anything was a valid factor inside the Universe, no calculation of any distinction or of any value can prove the disappearance of such a factor. No mathematics has the means to prove that by disappearing it no longer is. If it was in the Universe, it must still be in the Universe somewhere and as a science it is the duty of those in charge of that science to better start looking for it.

Another big issue is that what ever the Big Bang produced must be in equal terms everywhere. The Big Bang was a process that had the Universe act as a high-speed cocktail mixer of no repeating ever again. Whatever the Big Bang was of all that it was, the most it was in the beginning was that it was one massive mixer mixing everything in it at the speed of light. The relevancies might change slightly and balances may change favouring opposing ends…yes and known appearances did change…yes. But in the end all the factors must always be present everywhere through out the Universe. By this lacking of a fundamental explanation about what antimatter will look like when found Mainstream is incredibly poorly judged by scientific standards. Those mathematicians calculating physics suggest that science should take antimatter as a cosmic fact and then in disregard of other realities they dispose the truth by discarding its properties onto the unknown.

That hardly suggests plausible science by any one's admission. In the cosmos is, was and will be all the material there can ever possibly be. The concepts we put forward can be faulty but nature cannot ever be at fault. Our arrangement of our ideas can be at fault, but we cannot pull a vanishing act on certain cosmic products and in doing that, then dismiss the existing of such a factor or factors, which we then claim, have vanished in the further developed Universe. Our concepts of what became of them may be at fault and by changing some basic principals; such changing may produce a better understanding about what we think we read into mathematics. Mathematics is purely a language and mathematicians are purely translators. Mathematicians translate from the language they read to the verbal equivalent they speak. As in all translations made, certain concepts may become misinterpreted. The terminology used to explain this is "lost in translation" Mathematicians must see what there is in the translation and try to incorporate what there is available in the cosmos to what the Mathematician sees in his mathematical calculations. The Universe was full of heat and it was full of material but it was not full of free space. If that is the case then where did the heat come from and where did the heat go? Hiroshima and Nagasaki taught us many things about the horror of human nature but most of all it taught us that material is heat secured in atoms and atoms are heat tightly wrapped in a cocoon, which we named the atom. Heat in any form cannot have anti in another form. The package holding heat wrapped can unwrap as it does with nuclear atomic demise. But the anti to heat is cold and cold is space.

The undeniable fact about the Big Bang theory is the accepting of a growing state in which the entire cosmos seems to be in. With all the expansion that went on we came to the point where we now are at and in such growth all aspects in the Universe must grow in relation to quantifiable progress in all different aspects, which takes us to that which is seen and that, which is unseen and which came along as products in the Universe where everything took everything on a growing spruce by unveiling space. That is where we now are. Such expansion includes all there is including everything and not just with outer space growing. The dynamics of outer space alone cannot grow by leaving the growth of material behind. Should we wish to see where we came from, we have to reduce that which we now see in our surroundings to apply to the measures that once applied in all aspects of the cosmos. Mainstream physics is over pronouncing the growth of space and with that suppresses the part matter must play in such growth by simply ignoring the issue. That is the reason why they prefer to ignore the evidence that material is growing notwithstanding that material is growing or that their disbelief about the matter of material growing do not change that material is growing in any case. Because they cannot find any reason why material should grow, they refuse to admit that material does grow.

This is hiding from the truth by hiding the truth. If space grows and the Universe is getting bigger, then all space grows to allow the Universe to get bigger. That includes matter and space not in matter. Space can only grow if materials that also hold space also grow within the space that is growing with the growing space. It means that stars get bigger by the cosmos growing from the Big Bang onwards and outwards to the moment in which we are at the present. But if stars grow then the atoms forming the stars are doing all the growing as they secure more space within the space they claim. If Hubble saw space grow, the growth of space must include the growth of space holding material as well. In studying the Hubble's expanding theory we come across evidence that makes it clear that all material expand in a manner as if the expanding comes from the centre of each and all particles within the expanding space and the expanding grows outwards from every particle centre. It is using every star's centre to grow from in all directions proportionally in all directions evenly. This leads one to believe that gravity is this securing of space in the material just as Kepler showed it to the world. It proves a connection with deliberate implications coming from every as well as in every specific centre. It proves that the centre $k^0 = a^3 / T^2 k$. It becomes apparent if and when separating Kepler from what Newton thought about the work of Kepler which Newton accepted as being inferior and all incorrect.

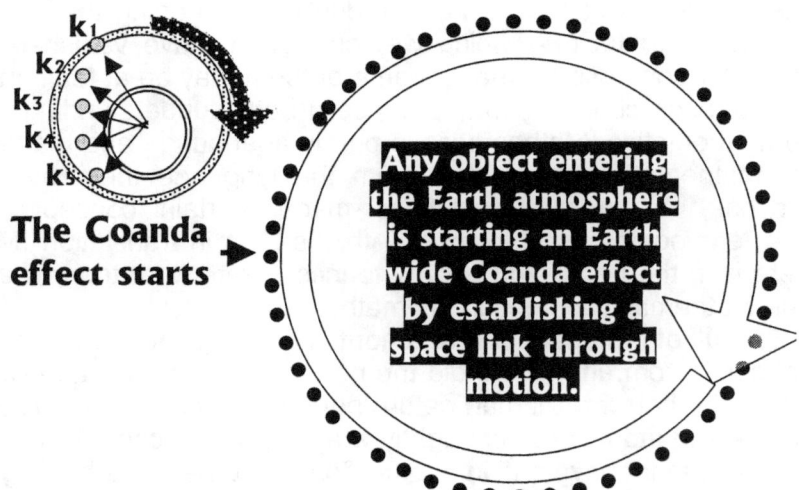

The Coanda effect starts →

Any object entering the Earth atmosphere is starting an Earth wide Coanda effect by establishing a space link through motion.

That is making the Universe small and as man grows man allows the Universe also to grow in relation and corresponding to man's ability to comprehend. We see the cosmos as a circle and we accept the circle because the circle is what gravity implement when the choice of form is coming from material that has all options to freely choose from. By taking the circle back will trace the rout of the cosmos to where it then started. All stars are many circles in many dimensions, which form when all circles join into what we call a sphere, but that leaves us only with the circles in the plural. Taking the cosmos back can only lead to one point and there Kepler told us we will find singularity $a^3=T^2k$ which is $k^0 = a^3/T^2 k$. We can only reach $k^0 = a^3/T^2k$ if we repeat $1/k = T^2/a^3$ in a continuing manner indefinitely. When one does the effort of reading this correctly, it says that when distance **k** breaks from singularity $1=k^0$ that is then $(k^0 =1) /k = T^2/a^3$ where the space a^3 produced a time T^2 equal to singularity k^0 and singularity k^0 is equal to eternity which was where all was equal to a never changing cosmos that was holding the single form into one dimensional space that included all the filled and vacant material filling in from all sides.

This is one way of looking at the issue and by doing that I am about to prove that singularity is Π. I am about to prove that not only are the planets adhering to the Titius Bode rule of seven over ten and ten over seven in relation to the Roche limit, but that the Roche limit explains the very, very first instant the Universe experienced outside eternity. The atoms relate to space in the very same manner of seven singularity positions to ten points and from this motion of material interacting with space material is secured on the inside as well as on the outside. By that motion gravity comes about finding the value of Π^2. Gravity uses the relation of the Titius Bode seven on ten and ten on seven as well as the Roche factor to form gravity and gravity is always Π^2.

This I see by reading Kepler's work as Kepler produced the work and introduced the work as

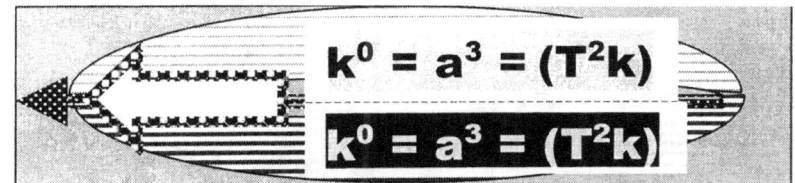

$a^3 = T^2 k$. With this formula $k^0 =a^3/ (T^2k)$ must also be true because $a^3 = T^2 k$ is a relevancy that has to be in relation to singularity and therefore singularity must be $k^0 = 1$. Where will we find $k^0 = 1$?

When an object is within the singularity alignment of the Earth, meaning it is either stationary, or free falling on pure "gravitational" momentum, it holds the $4 \times \Pi^2$ in relevancy to Π. What I try to say by that is that the Earth holds all objects within the atmosphere to a displacement value of $3\Pi^2$ within the seven are where it is using 10 as the liquid in space. Because the aircraft does not leave the Earth's atmosphere the 6^2 and the 10^0 does not come into effect.

Mass is the refusing of any object to dismiss the form it has and to join the Earth solid

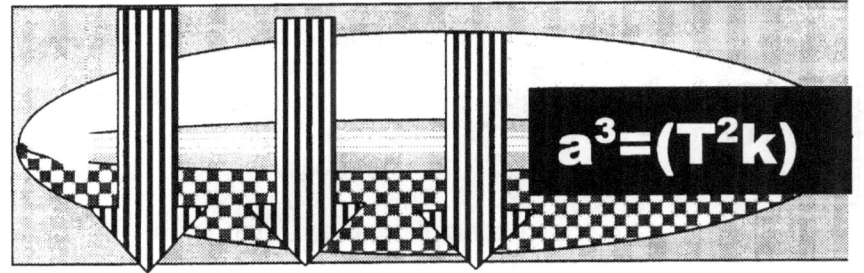

structure. Mass cannot and does not contribute to the establishing of gravity except by depleting space through motion and such numbers of the protons in a space forming an exclusive unit. It is when the motion exceeds the mass that the aircraft has the ability to break the sound barrier. Galileo proved that no mass is present in falling, which is also matter in the process of flight and because of that fact the sound barrier can become some form of constant. The establishing of independent motion of the craft secures an individual gravity and such individuality leads to the breaking of the sound barrier because the one gravity can no longer subdue the smaller motion, which is producing gravity.

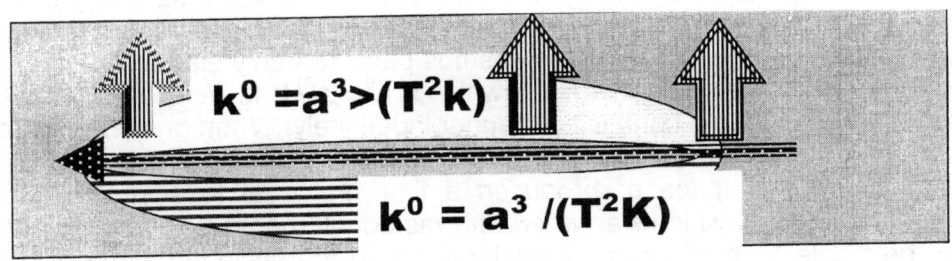

The Earth
the mass of the wing becomes compensated only by a motion of a relevancy that comes about at 2500 km per hour. In that case the craft has to apply motion at a rate of 2500 km / hour just to create the required velocity to keep the aircraft in motion in the sky. Motion creates gravity just as Kepler said when he said gravity is about space moving through time $a^3 = T^2 k$. Also it translates to the dismissing of space through the motion, where the duplication establishes a centre that controls the balance that the newly allocated singularity that secures the motion will provide in such a secured centre. When the aircraft stands, the aircraft holds mass and with mass the aircraft is cosmically dead. To the cosmos the aircraft is the Earth. It is the earth and being the Earth it forms a centre in the Sun where the Sun provides such a pivoting centre, because the Earth including the aircraft spin around the sun, but when the aircraft asserts independent motion, the differentiation of a time centre within the Earth comes about where the point shifts from the Sun to the Earth centre.

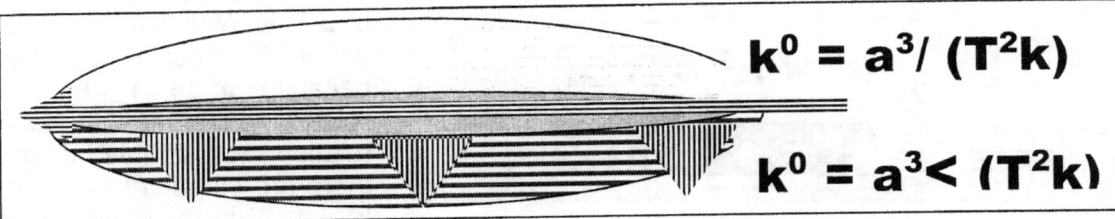

The aircraft's independence establishes a new centre focus where there is a line in contact between the points in singularity that the Earth holds. The independent motion then forms a new relation in respect to the singularity activated by the independent motion of the moving body, where the aircraft reallocates, while it takes on a trip in motion. With it forming a position that the relevant singularity is claiming, it then claims the minor relevancy and it grabs the lesser position as part of forming the allocated place as the minor partner.

Everything that is in the Universe is heat. There is liquid heat that can flow and there is solid
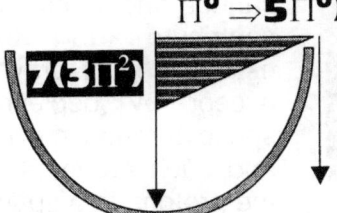
heat that cannot flow. The solid state or liquid state has no bearing on human perception but is in motion. When the body moved in relation to the body that does not move, the body that moves is liquid and the body that is relevantly stationary is a solid. Movement is liquid and immobility is solid. '

The sound barrier is in principle Galileo's pendulum arm

The sound barrier is directly related to the swing of Galileo's pendulum arm. By standing still the arm points directly down to a centre point within the Earth. As the motion becomes a factor in the aircraft or any moving object for that matter, the motion will push a diverting of the stationary line as the moving object changes relation to the lines, which the Earth dictates.

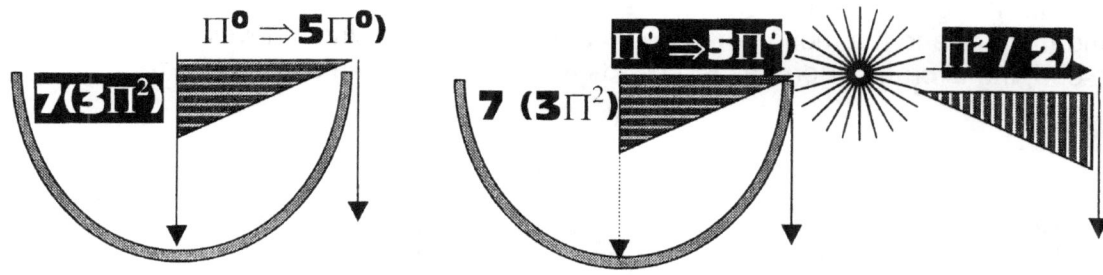

The line holds a velocity of $7(3\Pi^2)$ and that is the motion the space has in relation to the time the Earth dictates by contracting motion of time to the centre of the Earth. Any more motion that will underline the independence of the object in motion will show a diverting from the Earth singularity line by measure of Π^0.

Motion is connected to heat in every principle applying. It takes heat to drive motion and it takes drive to establish independence of singularity. By additional motion brought on by adding heat, the motion will show how far the independence diverts from the Earth's prescribed singularity. The sound barrier comes about when the second form of the Roche limit is crossed at $\Pi^2/2$

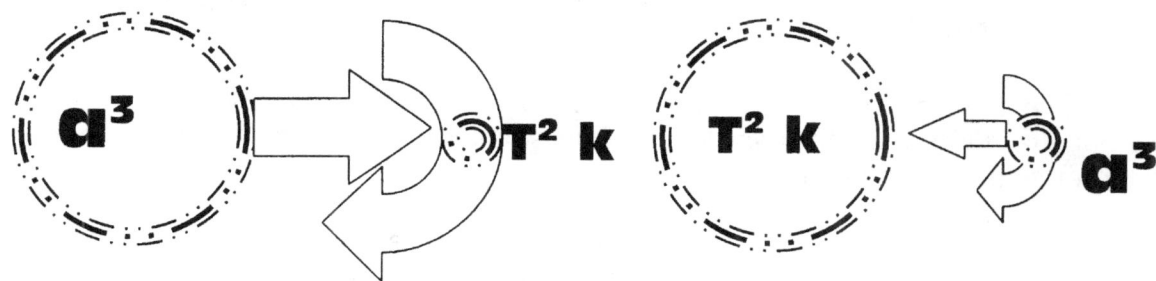

Every aspect however small or however large is serving as an atom that is maintaining singularity. Every atom is a galactica and every galactica is a star. Everything moving is containing heat and everything containing heat is preserving singularity where singularity is the Universe. Gravity's limit according to the Earth singularity is $7(3\Pi^2)$ where 7 is material (Π^2) is singularity displacing and three is the time factor as singularity displaces space-time. Moving beyond $7(3\Pi^2)$ $5\Pi^0$ the Roche factor comes into effect and this is all in the space-time depleting zone established by the neutron atmosphere of the earth.

That means the earth still holds its own $21,991/7$ neutron base, but the object breaking the Titius Bode law bring in its own value of $\Pi/2$. This value of $\Pi/2$ is n the neutron parameters of the earth where the removes the linear factor value of the neutron from 1 (Π^0) to Π. That means there is a dual for supremacy of a neutron position is as much as ($\Pi \times \Pi/2$) becomes $\Pi^2/2$. That places the sound barrier at $7(3\Pi^2)$ $(\Pi^2/2)$.

At the value of $7(3\Pi^2)$ $3(\Pi^2/2)$ more implications will come about, because the object will establish an own space within the space-time of the earth and this will lead to the object wither joining the time of the earth $4\Pi^2$ or securing itself an individual space and time. Obviously the heat supply will be insufficient to bring about the value the object needs to place it in outer space $(4\Pi^2(7(3\Pi^2) / (360° \times 10))$ so the aircraft will forcibly join the core of the earth, crashing on the ground.

In the same manner we find every layer of all stars to be another star contained by motion differentiation. The one layer is not just a group of elements within the part of the star, but assembles a unit, which forms an era of performing of the star. The motion that the layer allows puts the gravity of the star at a specific development. In all it is an array of atoms forming the star and the motion of the atoms places an era related development on the star. In the end, the aircraft perform as is it needs another layer in which to perform.

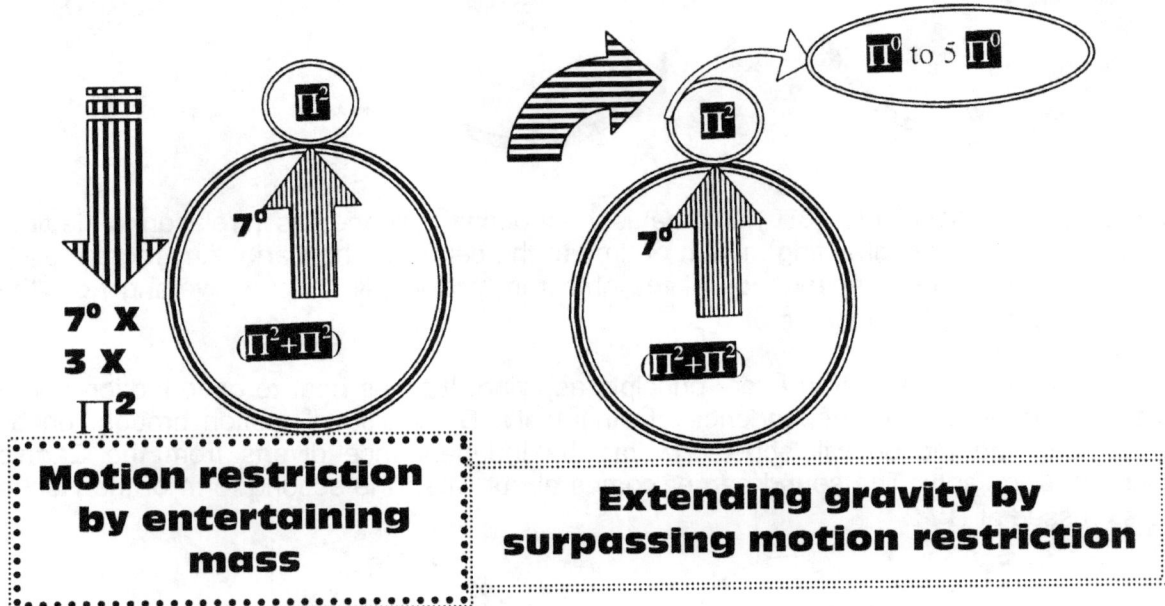

Motion restriction by entertaining mass

Extending gravity by surpassing motion restriction

The statements taken from the pictures to the top and to the left is indicating that science does not attach size with mass any longer. The smaller the diameter of the structure is, the larger science thinks the mass is.

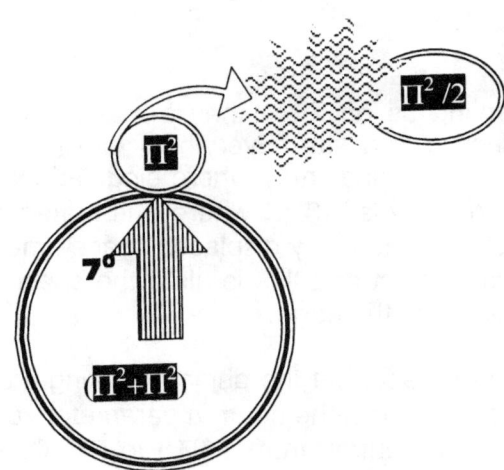

That is where the Mach principle finds a solid foundation. What separate time and space in joining singularity is the matter and matter is heat. However Mach placed all emphasis on mass and mass is only the resistance of matter in parting with own space to join another object's space. Part of the resistance of mass or parting with individual space is to apply the Roche limit that will remove the neutron value of the object from $\Pi^2\Pi$ to a value of $(\Pi/2)^2$.

This will improve density of the proton considerably because it removes the Titius Bode principle of matter in space as it eradicates 7 to 10. I have shown how 7 to 10 are the development of 21,991/7 and this of course is Π. By removing Π from the equation there no longer exists space in time but a value of half time to time $(\Pi/2)^2$. The influence that this brings along is a huge increase in the time value of occupation of space-time that the object initiates. Where the object normally holds a relevancy of space to time in as much as we find that the atom represents, $(\Pi^2+\Pi^2)(\Pi^2\Pi)(3) = 1836$, the atom forming by relation of the Earth adopts a new relevancy of $(\Pi^2+\Pi^2)(\Pi/2)^2 = 97,4$.

The flying craft now assimilates an atom that formed where the electron stretches the atomic boundaries. The atmosphere acts as the neutron and with the neutron stretched beyond its limit the aircraft that simulates the role of the electron puts the atmosphere in a relation of) $(\Pi/2)^2$. All this happens within the atmosphere of the Earth. The structure holds a normal value of $T = 7 (3\Pi^2)/10^2$ under normal Earth movement and if one wishes to replace the Earth gravitational confinement of matter to the outside one has to add to the normal linear value of Π^0 by exceeding this linear gravity displacement. This again shows the Lagrangian position shift five places before the Roche limit sets in. This $T=7(3\Pi^2)(\Pi^2/2) = 1036.3$ km / h then forms Mach 1. At mach 3 the space-time demands will become $T=7(3\Pi^2)(\Pi^2/2)\times3 = 3108.9$ km / h. This places the relevancy where that particle can demand own space in own time due to the density it acquires from space. It has all the properties that a star in space requires, but it will still be confined to the earth. In that way the object then must acquire a relevancy of the full Roche limit, without the protection of the Titius Bode law.

Moving beyond $7(3\Pi^2)$ $5\Pi^0$ the Roche factor comes into effect and this is all in the space-time depleting zone established by the neutron atmosphere of the earth. That means the earth still holds its own 21,991/7 neutron base, but the object breaking the Titius Bode law bring in its own value of $\Pi/2$. This value of $\Pi/2$ is n the neutron parameters of the earth where the removes the linear factor value of the neutron from 1 (Π^0) to Π. That means there is a dual for supremacy of a neutron position is as much as ($\Pi \times \Pi/2$) becomes $\Pi^2/2$. That places the sound barrier at $7(3\Pi^2)$ ($\Pi^2/2$).

At the value of $7(3\Pi^2)$ $3(\Pi^2/2)$ more implications will come about, because the object will establish an own space within the space-time of the earth and this will lead to the object wither joining the time of the earth $4\Pi^2$ or securing itself an individual space and time. Obviously the heat supply will be insufficient to bring about the value the object needs to place it in outer space ($4\Pi^2(7(3\Pi^2) / (360° \times 10)$)) so the aircraft will forcibly join the core of the earth, crashing on the ground. From the sound barrier and the effect the sound barrier has on matter, one can measure galactica and how that "growth of space" seems to become a reality as the Hubble Constant indicate.

When the object holds a position above Mach 1 and Mach 2, the temperature to the outside of the aircraft rises dramatically. This is just the same what we can see in the core part of the galactica. The structure of the craft becomes a solid or a proton. The craft holds the value of ($\Pi^2+\Pi^2$) in relation to the earth because it maintains an own space value in the related time of the earth. The craft applies a demand for additional heat, because to all cosmic purposes, it shows a higher resistance in sharing space and time. Place this in a human context and the aircraft has a higher mass. Should the aircraft wish to break free from the atmosphere while applying a linear displacing position, the craft will need a velocity of $7(3\Pi^2)$ ($\Pi^2/2$) (10Π) x 3 (Mach 3) leaving it with no less than a speed of just meter 1 00 000 km/h required and even at that velocity I am not sure it will pass the Π influence of 21,991/7 barrier that shield the earth remaining totally intact and unscathed. I wish the alien hunters that claim alien can come and go as they please will use their brainpower in the direction of common sense and not common fantasizing. No way in heaven or hell can there be a craft that will slip in and out of the atmosphere at leisure to the demand of its alien pilot. The aircraft has to have a structure built at least from Californium 98. That shows how ridiculous the alien detectors are and they do not even know their own stupidity.

Science holds the movement of the structure in the category of momentum, but momentum is only linear "gravity" and mass is circular "gravity". In order to claim the space-time occupation that the craft demands, will mean an excess of protons in order to bring about the space-time occupation required for that strong relevancy to heat. The cosmic law does not allow for artificial heat production and life has no normal place in the cosmos. To the cosmos it is all a relevancy of supply versus demand.

The sound barrier shows that the earth is an atom containing what is inside by the motion of what is inside. When the motion of some particles inside this Earth sized atom moves to fast in relation to the rest, the faster moving particle forms an electron which then becomes the craft flying. When the relevancy of the "electron" flying becomes larger that what the Roche limit within the Earth structure can contain7 $(3\Pi^2)\Pi$, the atom forms space larger than what the Earth allows contain $(\Pi^2/2)$.

$7 (3\Pi^2)\Pi^0 = 207.2$ km/h. $\mathbf{R^2/T} = 7 (3\Pi^2)\Pi$ and $\mathbf{R/T} = \Pi^0$

The Roche limit also changes to accommodate this change and becomes either Π^0, Π, or $(\Pi^2/2)$. Falling "free" will then mean that the object holds a position of $7 (3\Pi^2)\Pi^0$ to singularity. In the half circle applying two of the quarters in time, the position is in the triangle of singularity placing the half circle in the first quarter. Any further linear movement will follow the triangle, the second line by multiplying the line by the value translated as a number of Π^0.

$7 (3\Pi^2)\Pi^0 = 207.2$ km/h or $(3\Pi^2) 2\Pi^0 = 414.5$ km/h.

$\mathbf{R^2/T} = 7 (3\Pi^2)$ and $\mathbf{R/T} = \mathbf{2\Pi^0}$. This is the first barrier but we only see the linear part as a value and peaks at $5\Pi^0$

$7 (3\Pi^2)(\Pi) = 651.13$. In this $\mathbf{R^2/T} = 7 (3\Pi^2)$ and $\mathbf{R/T} = \Pi$ where then the linear component Π becomes the second line in the triangle of singularity. The linear component or negative displacement can go as high as 2Π but then another barrier would come about and in this we start to locate the principle behind the breaking of the sound barrier. There are definitely TWO barriers to comply with when breaking the sound barrier

$(\Pi^2/2)$

$7 (3\Pi^2)(\Pi^2/2) = 1022.79$. In this $\mathbf{R^2/T} = 7 (3\Pi^2)$ and $\mathbf{R/T} = \mathbf{2\Pi}$ where then the linear component Π becomes the second line in the triangle of singularity.

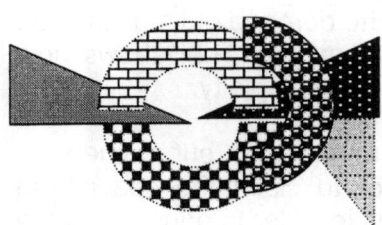

$2(\Pi^2/2)$

$7 (3\Pi^2)2(\Pi^2/2) = 2045.59$.

From this point on and above there are a boundary no one can cross because singularity will not allow the crossing of the third quarter by any object.

The spinning top is all the evidence any one needs to come to such a conclusion. In the past when I have acknowledged the fact that I have no academic background in the field of cosmology, this admission brought about discontent, or rather dismay and I might even add some scorn from academics. In my opinion this trend of behaviour is uncalled for as every

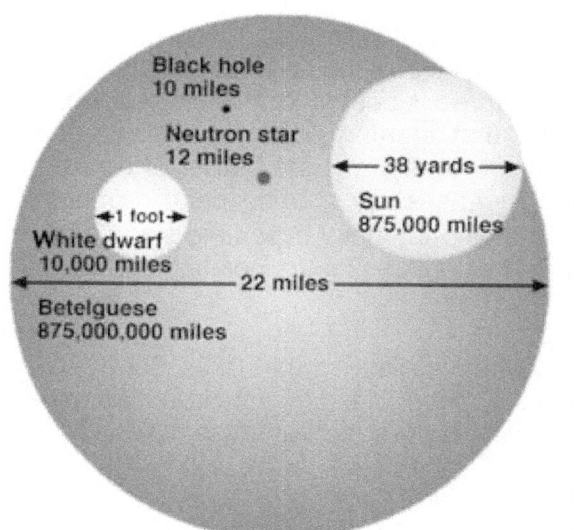

Black hole
10 miles

Neutron star
12 miles   ← 38 yards →

←1 foot→   Sun
White dwarf   875,000 miles
10,000 miles   ← 22 miles →

Betelguese
875,000,000 miles

person is entitled to be opinionated. Every person has an opinion and it is the opinion that has validity, not the persons social standing. It is the way one absorbs and reflects a personal view about matters that is important and not the manner in which the person arrived at such conclusions but it is the validity of the conclusions that should carry importance. The cosmos contains matter, space and time in heavens or dimensions standing in relevancy to one another. Whether you refer to dimensions as dimensions or heavens it is no different because it is the same thing. Singularity brings about heavens as it brings about dimensions and singularity is a mathematical fact. A straight line cannot start at zero and still be a straight line because zero extending to wherever brings about a full zero. A straight line starts at the point where the pen point meets paper. That point may be any distance from infinity to a measurable dot, but it cannot be zero.

Any straight line is also half a square because the line forming the square cannot start at zero for the reasons I just mentioned. That is singularity pointing an eternal direction from a point of infinity and that, is the basis of the cosmos as much as that is the basis of mathematics. To escape from nothing one has to become something and by doing that one could not have been in nothing in the first place. If one holds a point in nothing, one cannot become something because of the nothing value.

To back this argument that no line can ever start at zero is to ask the simple question: what will the length of the shortest possible line be. It must be a line where the starting point is so close to the ending point the distance parting the two is incalculable yet there is a line therefore the end and the start is apart still sharing the same spot.

To this end any shortest line will start running and ending in infinity Should a straight line start from infinity and never be able to reach a point of zero (because that will bring in the factor of no line (0 X 1 = 0)), that means the line dips into infinity and it has to come out on the other side leading in the opposite direction of the first line in an attempt of avoiding zero.

The length cannot be zero because zero means no line. The starting point and the ending point may be inseparably the same point with virtually no space between the two points, but neither of the three points can be zero simply because there is a line (be it infinitely small but it is there all the same). If the point is zero, then the line will be shorter than the shortest possible line. If there is a line and the two points starting the shortest possible line and being the continuing of the line, it may still be the same point and even by sharing the spot as the point ending, the shortest line possible the line must be there and the line holding the start and finish is next to zero but can never be zero otherwise there is no line. If that is the case then the one side of the square also presents the point of singularity to the other line holding the other dimension. That then makes the straight line not 180^0 because it is half that of the complete straight line and indicating the point of singularity of the other line and that results in any form holding only one aspect of the full form. But because the two lines are in a relevancy to a common starting point, both will enjoy the remaining distance of flow still holding the relevancy in comparison to each other and that brings about a triangle that also has 180^0 in number. That means any straight line is also a half square which is a triangle bringing about 180^0.

The co-ordinates of referral will only hold three references as surface contact areas pointing in one direction of the possible six surface areas available. That is mathematics and mathematics cannot be bias or lie. Every object in the Universe holds three points of face value to any direction possible. You can only see three sides of any six-sided object. Because the lines have a starting point in infinity, that starting point has to represent singularity outside the sphere because at some point the lines cross the border of being the shortest lines possible to being normal lines. That means every starting point represents singularity by measure of r instead of Π. Every point is also the point indicating singularity at another point pointing in the direction of a point holding singularity as the pointing line as well as the supporting line.

The prominence of this will become most apparent when explaining the Titius Bode law in relation to matter relating to space and matter relating to matter claiming space, but as usual, I am getting ahead of myself. What I have just indicated is pure mathematics moreover the most basic mathematics there are.

The circle to the left would come about from a straight line r growing, influencing the 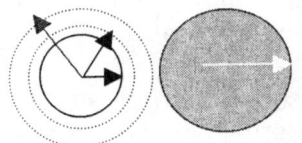 appreciation of Π, but to influence Π would lead to a breakdown in r as Π and r are different entities. The circle to the right shows a continuous growth by extending Π every time and since Π is the same part as the previous Π, only extending that billionth of a millimetre each time, the circle will be truly continuous without any signs of a break

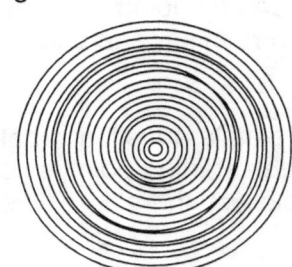

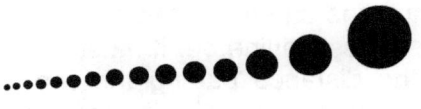

In the circle using $r^2\Pi$ the r has to have distinctive qualities placing it as a factor apart from Π. Where the growth shows no separate distinction but a continuous flow from the precise centre to the precise edge, the flow would become in relation with Π depicting the circle and Π replacing r as reference to any point on the circle. By using r distinction in the circle is possible but by using Π there is no distinction possible.

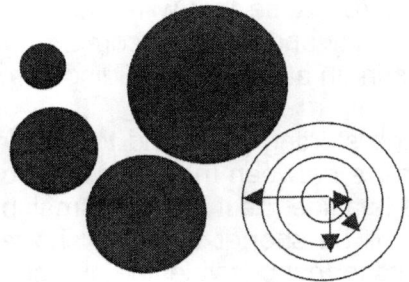

When working with concrete and heavy metalled solid objects r would show as a crack distinctly parting solid structures, while Π indicates a continuous flow of solidness giving the material an overall and continuous structural strength, yet engineering never recognised this difference. By confirming Π, the circle employs singularity in all components and therefore proves to be a much stronger support as building choice than other shapes.

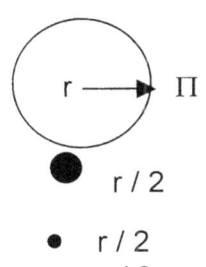

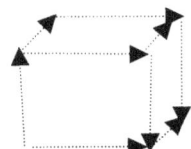

The shortest line in the realm of possibilities must have a start and finish holding one spot and such a line will also be a dot or a circle. Not favouring one direction puts all directions at equilibrium meaning that any form of what ever might develop from such a spot with the end and the start being in the same position also has to be a sphere because the flow outward will be equal in all directions. This reasoning prompted me to look for singularity in such a spot because if the prime spot from which all came was a spot holding all, then the spot must hold the shortest line but more prominent it will hold the smallest form including the smallest circle or for that matter the smallest sphere. One possibility that the shortest line or smallest spot can never have is having a starting point on the zero mark. If the mark of zero holds the start it must also hold the end because the end and the beginning has the same position. If the position of zero then is the beginning, the end will also be zero leaving the line or spot without an end as well as without a beginning. Such a spot will constitute all of nothing.

There are always two lines running in supporting but opposing directions claiming space from that one point of singularity and a third line confirming (controlling) space in a third direction.

Every line is as much part of a square than it is pointing in another direction relating to a position holding singularity. That brings about my conclusion in changing Kepler's formula of $a^3 = T^2 k$. The square of outer space does not resemble motion and one therefore must deduct that motion is only contributed by the sphere where space-time is confined to the motion of $a^3 = T^2 k$

Squares rotating will find a point where it goes into self-destruction through faster motion. This action is imbalance but in principle it has a point where Π forms a meeting with r and Π at that point forms r.

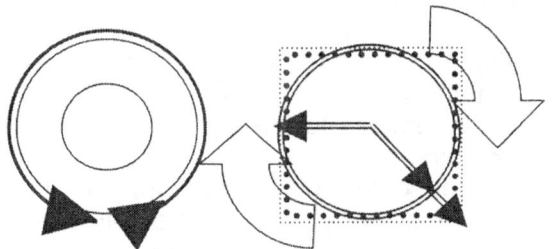

That is space-time holding one square of space in time and directing time in space in a specific direction of flow. You may not agree with me on this issue at this point but my reasons for such a claim come in other parts of the book and there I substantiate the claim with much more explicate mathematical detail. Space-time is matter-claiming space in time while directing space in that same duration of time from any point, forming singularity. Now one arrives at the question of positioning singularity. Singularity cannot be in space because space is claimed from the point of singularity and space is confirmed from that point of singularity. If space is claimed and controlled, Einstein must be at fault looking for singularity in outer space.

To find such a point one should once again look at the possible dimensions available. A square is six sides holding three to any direction. A circle is a square without edges. That is what we are looking for. A square without edges can only be a circle with a definite invariable point of singularity indicating a specific location. A square can place singularity at any position because space holds no specifics where as a circle holds specifics at a definite point. Space provides no specifics because an astronaut can and does drift in any direction when not being attached to a sizable object. That throws outer space out as a possibility.

Take the top with some astronaut to outer space and ask him to spin the top in outer space. We do not even have to bother such a busy person because we know the answer. The spinning top will not be able to turn because there is no stability factor in outer space. The top needs support from the needle running upwards, therefore the top will then have thrust running downwards, and the accumulative effort will bring about a point where the opposing points will allow a spin to occur. In outer space there is no such a point therefore the point we seek is within matter. A top cannot spin in outer space. We already established the fact about singularity being the position where two points originate in a square and the third will show direction. The top is a circle.

The difference between the circle and the square is in the direction the indicator follows and a square cannot spin, but through contact with a sphere as a circle cannot be motionless The factor of Π indicates eternal motion and NOT zero motion. There is a massive difference in that concept. If no line can have a zero point to start with, where will the circle get the zero to indicate motion! This principle is the most basic mathematic rule and even I the ILL EDUCATED can see this. When the end of the rotation arrives, the end rotation also announces the beginning of another rotation and not nullifying the previous rotation because the rotation will have a line showing the effort it made and as it forms a wave, the wave will be there forever. The pitch may decline to a straight line, but the line remains. When calculating the motion, a triangle does the honours.

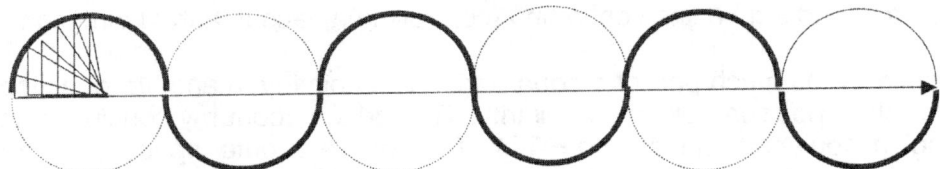

The wave confirms rotating directions followed by the circle as it spins. By stating that a wheel has a relevancy of zero by completion of a rotation, such a claim denies the wave its entitlement of existing. The wave going flat, as it becomes a straight line also has an indication to singularity. All spinning matter has the point where the spin is still there but the radius is to small to measure by any means. That point is standing still in relation to the rest of the spin. In relation to that logic I do not accept that science holds the radius of a spinning object unaccountable in the spin, whether the spin is applying or not. It is in the very fact that

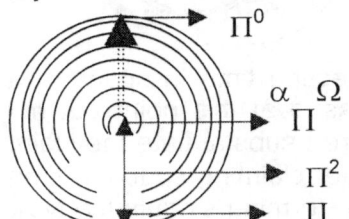

the spin places the reference factor in the graph that makes the graph that astonishingly accurate. It is mainly mathematics through the graph, which affects day and night and not the Earths standing in relation to the Sun that brings about climatically and weather changes and Earth developmental conditions.

In the centre there is the singularity from which all stems. That is the centre of the Universe carrying the value of Π^0. By rotation at Π^2 a line forms Π to Π with the value of 2Π. The forming of the line by initiating Π^2 is the realisation of gravity and the constituting of the Universe, which releases the Universe from Π^0 to Π by Π^2. The motion spawns the gravity because the motion is the gravity. To block the motion is to resist the gravity and such resisting forms the mass aspect that counteracts gravity.

Nothing said so far is high tech or mind bending complicated. All the above arguments from the first page to this point reached are simple and there's only ordinary primary school mathematics involved that every scholar should know. One does not need a brain fitting Einstein to come to these conclusions but just thinking about everyday issues.

All circles have something in common with squares. It has a surface but where the square holds a surface pointy, the circle holds no points. To calculate a square there has to be a point where the line starts and that line will hold a value of at least infinity running from that point in two opposing directions. Arithmetic presents the possibility of zero and mathematics excludes such possibility. That is the difference there is between arithmetic and mathematical science. It is where mathematics departs from arithmetic and is as basic as counting is in arithmetic. There is nothing outrageous about that which I have mentioned.

Neither does one need the brainpower of a person like Einstein to come to such basic conclusions.

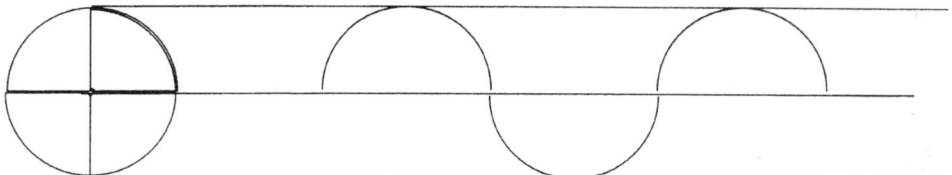

From the graph one can establish the link in the circle's rotation around a conforming unit being singularity.

Being a circle means the thing must be round and spinning. In that case, let us take an example well known to all, the spinning top. The top spins on the thinnest of points, and still maintains a balance.

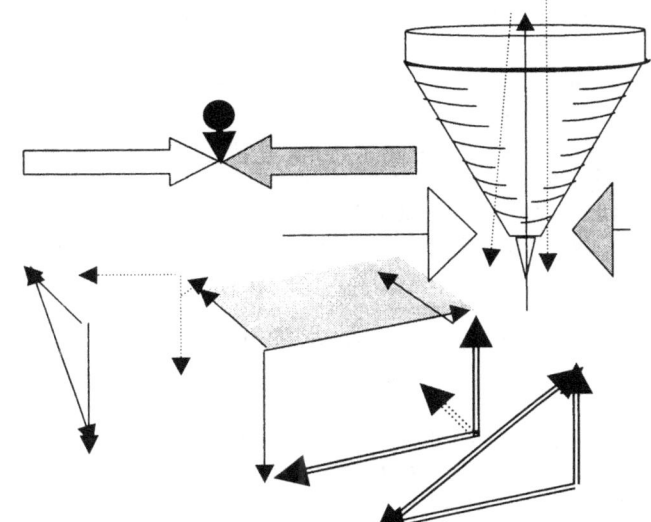

All circles have something in common with squares. It has a surface but where the square holds a surface pointy the circle holds no points. To calculate a square there has to be a point where the line starts and that line will hold a value of at least infinity running from that point in two opposing directions.

In the scenario depicting the square there is always some three pointing triangle everywhere and the common denominating figure is therefore three.

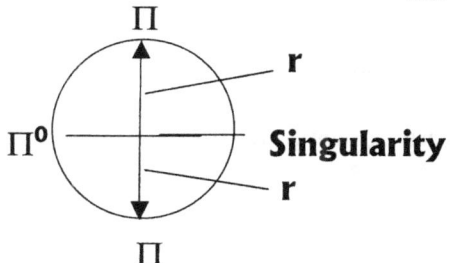

In the circle, there are also two lines where each line holds one point to singularity. Splitting the line in the two opposing directions, so in that way it is the same as a square, but the third line indicating direction brings about a difference that distinguishes the circle from the square. The circle direction indicator is always Π placing the pointers at $r \times r = r^2$

Applying the second law $F = m\,a$, one arrives at the formula $G(Mxm) / r^2 = m\,(\omega^2 r)$ claiming a zero influence between the radius and the orbit of planets. I shall later again return to this issue but firstly I would like to take your mind to one other thought that seems to have escaped every body. It concerns the medium science relies most on for the gain of facts and information about the Universe. It is the influence of light.

When realising the error of science in accepting a value as zero to be legitimate in mathematics, one can establish from that, that the circle does not employ zero as a value after the completion of one rotation. Therefore $F = G\,(M_1 \times M_2)/r^2$ is invalid, one has to return to Kepler's $a^3 = T^2 k$ and establish a value from that.

From the graph one can establish the link in the circle's rotation around a conforming unit being singularity.

Saying that, one therefore has to admit that the smallest spot has to hold space because the most insignificant dot can transmit light and being able to accomplish that, one must accept it to carry a value of something. If that spot had the value of nothing, it means that spot was not there to begin with. Holding space-time one should return to the original formula indicating space-time in as much as $a^3 = T^2 k$ where a = R and T = T. Being time it has to alternate positions and that can therefore only apply to **k** where **k** will indicate a relation to the space-time in question or the relevancy to singularity being $k^0 = 1$. This reality has serious implications on the speed of light we take as a constant.

Have you as you sit reading this part at this minute ever sat back and gave a thought about the light enabling you to read? Such a thought brings to mind the most simplistic answer one can imagine. The light hits the page bounces from the page and contact the lens of my eye where the lens conveys the photons becoming electricity to a part of the brain that translate the electricity to an understandable message and that makes one read. It is as simple as that! Ever gave a broader thought about light streaming across the night sky, coming from the ends of the Universe we do not even realise is there? How does the photons manage to convey one complete picture coming from as far apart and as wide an area as it does? With a few photons connecting the eye or lens no one ever noticed the wonder of light. The photons reflect a view that seems as if coming from all the billions upon billions of stars. But most is coming from darkness covering an area no man can measure. Yet how many photons can actually connect to the lens of the camera or the eye? Still a few photons coming from a single direction directly ahead eventually tell the entire storey. It is very simple to take the process of seeing by means of photon conducting very lightly and I have never heard one of the Brainy Bunch really in sincerity dissect the process to its potential. It is impossible that light from such an array of assorted sources can simply come together at the eye lens and show a picture of objects spanning across a Universe as wide as our mind can receive where the objects they reflect is beyond human measurement and the quantity is inconceivable many.

Light is much more than the medium science takes it to be. Light connects the Universe in a way we cannot contemplate. Light being far apart originating from regions not in the same time or universal space connects in a way that presents us with a picture holding the Universe in an understandable content. From the point we stand and we watch the Universe the significance of what we see surpasses the sense of understanding of what we are experiencing. How can the few photons that our lenses catch coming from such a vast area covered by the night sky, transmit the complete picture of what we see. Take a few seconds and inspect the picture of the night sky then rethink the picture applying the full content in the picture to that of the size of your eyes. Think how big the picture is that your eyes take in and translate that area to the size of your eyeball in an effort to determine a ratio. One will be forgiven if one thinks of the ratio as eternal to nothing. Yet a few pages back I showed that according to mathematics there couldn't be anything as nothing. Consider the path the light followed from the source connecting to light from all other sources where all particles of the other light may come from and bringing a full picture to the lens one use to look through. In

your mind connect a line from every atom producing light and connect the lines to your eyeball and see how you can manage to fit all the lines, as small as the lines may be.

Scientists think of outer space as geodesic zero, with nothing in outer space but space. Geodesic zero means the light travels in a straight line from where it originates unhindered all across space to where the light connects the eye. Such an idea by itself is outrageous because the stream of photons reduce in space to such a minute quantity, that taken the area the photons travel and the space in vastness it covers, the chances of one photon coming across many hundreds of light years through billions upon trillions of cubic kilometres of space and selecting my eye to convey the electricity is less than infinite. Yet such conveying takes place every second of every minute. The position of the location of the second singularity, which is the precise duplication of the first singularity but in a diminished capacity, is obvious to miss when one is not applying a detective mentality, as one should in scrutinizing the cosmos. Culture will have us believe that when one sees a colour shining from an object the colour is associated with the object. Logic tells a different tale. A yellow dot is all the colours in the spectrum but yellow because it is disassociating with the yellow. That goes for red blue and all other colours we may visualise. I think the norm accepts this as scientific fact with very little argument or substantiating proof.

If light came as individual streams of photon flurries our visage would translate that as such shown in the fragmented picture above. It would be a picture unconnected bringing across some photons in the manner where every object stands apart not being related in any way and that will be what we see, if it is anything that we see. That we know is not the case, but that means geodesic zero is as much rubbish as anything Newtonians regard with simplicity and with careless thought. Geodesic zero means nothing and how can I see nothing as darkness because "nothing" is not darkness, nothing is "nothing" and the darkness I see is darkness showing the darkness as something.

What then about colours that are technically not colours as is the case with black and white? White is simple. By spinning all the colours in the spectrum, the colour white shines through. Black is quite another matter. A friend of mine whom is one of the best painters I have ever come across told me that one couldn't paint black but have to make black a dark blue to show shade on the canvass. That apparently is his success in achieving the realism.

He also went on to explain how many variations of dark blue form the shadows in one simple tree. This remark set my mind in motion. One cannot see black because black has no colour to show, but black is the colour most prevalent in the universe. One can see only by colour and since black is not a colour we should not see black, but we do.

If the darkness was the representation of "nothing", then that should be exactly what we must see, nothing but the stars. Taken from the top picture some stars and leaving the rest to nothing is what we see in the picture below. A blind person sees nothing but when we look at space, we see something that we think nothing of which we see as space. One cannot have the ability of sight and see nothing except by closing your eyelids and then you see nothing. But in that case you do not see "nothing" in contrast of "something" you see "nothing" without it contrasting to "something".

Nothing is all about not being and not "not seeing".

By the ability to see the darkness, renders the darkness something other than nothing and that changes the acquired value of the darkness from nothing to something. There is an eternal difference between something in infinity and nothing.

The arguments introduced up to this part of the introduction prologue only touches the most basic aspects of my work and by no means can such an introduction secure an opinion. Yet, not once through all my long investigation in the past thirty or more years have I found any other person claiming such views that I have brought about even in this skimpy way as I do in the prologue.

The arguments introduced up to this point of the introduction prologue only touches the most basic aspects of my work and by no means can such an introduction secure an opinion. Yet, not once through all my long investigations in the past thirty or more years have I found any other person claiming such views that I have brought about even in this skimpy way as I do in the prologue. As it applies with all things, so it does in this case as well that when delving deeper into any issue. The complexity of the issues truly comes to the fore ground when analysed in more detail. I wish to advise the reader to treat the seven books as seven different works and in that light I have separated each work in volumes of seven separate books with individual I.S.B.N. numbers with adding one part, the one you are reading, with one sole purpose and that is to bring about an academic introduction to clarify a quick perspective. Then the next three parts being of a general introductory nature, there are overlapping in some sense but each highlighting issues in a different manner as to clarify facts used in the last three parts bringing conclusion to different cosmic perspectives. Yet the work is seven parts of one thesis and as such it serves.

I WISH TO DEFINE THE CATAGORISING I USE AS PART OF THE BOOK.

I have the utmost admiration for Scientists and I shall never dream of placing me in the same category as academics mainly because of their intellect and achievements. To substantiate this segregation I use some referring to place distinction between the highly schooled super trained academics that spent most if not all of their lives in preparing to further their minds. Because I tip the opposite of the scale and spent as little time in an official capacity on learning and education, I have to be on the "other end". From where I stand and admire you, I can only see intellect: being the academic's common denominator. If that is the common denominator on the one side, the joining factor on the other side *"my side"* must then be the class of stupidity. To you and your class such a remark would be an insult but to me and therefore my class it rings truth and that makes it not an insult but a norm we should accept and learn to live with. It is rather a pity that while the SUPER CLASS will never say it to our faces; the SUPER CLASS is strongly of such opinion that we on the one side of the Universe have no minds to think in any way, and it is our duty as much as privilege to accept what you the ones occupying the other side of the Universe, inform us to accept and you live by that idea. As I said, I have to live with it too and if I am the illiterate, then the SUPER CLASS must be the SUPER–EDUCATED; where I am the class amounting to stupidity the SUPER CLASS must be the Brainy Bunch. It all comes from the fact that there is such a huge differentiation between us. You consider one with merely one Baccalaureate degree a stable boy; think how low your opinion must then be of my type including me that never even came that far. To distinctly point to grouping or class or whatever you wish to consider the division there is between you and me, I refer to your side of the Universe by the names I use above. Further more when I refer to mistakes that I do prove to be mistakes in the book as we go along I refer to it as Xepted mistakes to clear another distinction of necessity.

Introducing the book **Matter's Time In Space** written with facts about the creation in mind produces a problem because the complete picture that I introduce has nothing in common with current accepted science. The issue remains comprehensive even by using it in a very simple form. In spite of this, I shall explain three of the four unrecognised phenomena I use in proving my statements in this very book aiming at a theme of simplistic introduction being "an Academic Letter". Then in the following books I go into extensive detail proving all of the Cosmic Pillars. With my introduction of the phenomena, which I named the four cosmic pillars, you will find it obvious why science do not accept them even if it is documented throughout the Universe and is quite commonly found. Applying them totally annihilates Academic's formula of the basis on which science rests in the formula being $F = G (M_1 M_2)/r^2$. The four cosmic pillars are the following:

1.   Roche-Lobe
2.   Titius Bode principal
3.   The Lagrangian principle
4.   The Academics Gravitational Concept forming the atomic relevancy.

The Universe is a combination of many material formations holding positions in space. Some of such material was covered in the blanket of heat, distributing into more spacious surroundings as the material expanded from the centre flowing outwards. Hubble's constant is proof that the space between cosmic structures are departing from many centred positions between such objects and this is a trend being located between all the objects throughout space but also indicating a definite growth in the radius and such radius growth follows a pattern where the growth seems to flow from any such a centre point away from the centre. With out the absolute and undeniable proof coming from the Hubble constant that brings proof beyond any possible doubt in any one's mind we can visually see that the expanding is very much and a very big part of all Cosmic activity, the accepting of the Big Bang would not be in place. $F = G (M_1 \times m_2) /r^2$ is in essence a big issue about contraction while Hubble showed the space was not dividing. The space was multiplying. The stars are growing apart and so is the galactica. This then brings in the question of space available.

I do not whish to go into the formation of structures at this point in the book, but I would like to point to the fact that my following referring to the solar system is actually referring to a similar solar system that is somewhere and is now a part of a galactica we do not know about. I bring this in to disqualify any academic loophole that may come about from an argument about the solar system coming into place at a later stage of the cosmic development and therefore the argument I am about to present is that such an argument does not apply. To avoid such a loophole we now use a hypothetical but real solar system in space, which formed as the Big Bang took place. There are those who avoid admitting to inconsistencies by arguing that my argument about growth is invalid because the solar system was not in place at the Big Bang. To them we now present a solar system that is identical to the one we know in precise duplication thereof. But it represents a precise duplication of our solar system and was in place ever since the Big Bang. That means with the solar systems being apart for millions of kilometres there was a time the planets and the Sun were apart by the measure of kilometres. The Big Bang shows a growth in space. Then there must have been a time when the planets were between fifty-nine and fife hundred and ninety kilometres, away from the Sun. How did the planets being the size they are at present fit into such a space and still be apart? Material too must be part of the growing. This line of arguing I suppose is much below the Academics pursuit of matters, but since I am much lesser in mental standards of development than they are, such reasoning prompted me to go on some investigative journey. Light journeys through out the cosmos and it will be sensible to follow the travelling of lights.

With objects being apart at some distances and light flowing in straight lines between them, it must take light a straight line to travel between cosmic objects. The distance the light has to cover depends on the radius there is between the objects and as such the Universe is then about structures claiming space and space setting objects apart being the radius standing between those objects. The objects are circles by dimensions and the space is also dimensions that are crossed by lines travelling through the dimension. With light being a line and the Big Bang coming from a situation that was a lot more cramped for space than at present, the correct path to follow if I wish to trace the steps of the Cosmos back to the Big Bang is to reduce the straight line between the structures and find where such a line will no longer be a line.

The same procedure will apply to the material structures all being in a sphere form. A sphere is a lot of circles forming a unit but not repeating any occupation of the space, which any one particular one claimed. Such a circle also applies a straight line only known by another name but still serves the same purpose. Reducing the line will lead us to the beginning of time.

By dividing the radius r by the half of the value that then reduces r to a point where the left  edge of the line reducing will be at the very same place the right hand edge of the line that is reducing will be. At one point the spots that formed the two ends of the line will be at the same spot. Any further dividing will land the left hand spot past the right hand spot in the opposing half where it then will grow once again but in the opposing direction. All possible dividing then ends on one spot where such a one spot shares a location with all other possible sides.

The centre then physically is in the single dimension applying as one spot to share a location for all sides. At such a point there is no further dividing possible. On several occasions in the past I have been accused of manipulating the argument to produce none-existing or overrate facts. That is not the case. I am not manipulating facts to create an argument as so many accuse me of. What I am talking about is a mathematical fact that any one can prove by calculating following a very simple procedure. A child is capable of using the two times table and dividing by two every time is the most simple form that mathematics may be used. It is a mathematical fact that a line will reach a point where all sides are at one spot and as such cannot divide any more. At that point all sides share but all sides prevent zero becoming a factor since the sides share one spot. While the different sides are in one place the factor and value is one to all.

Reducing the radius r from all angles possible throughout the circle will bring about that all possible direction will eventually land on the very same spot with no more dividing possible. Yet zero cannot be a factor since the sides still hold value. A point arrive where more reducing will land the one side on the opposite side of the line but it will not bring about zero in the equation

What this argument further proves, is that the circle reducing must then come from all points because the radius might be a line but that line represents a circle through 360^0. Taking that into account it is important to recognise that notwithstanding the size of a line, which any 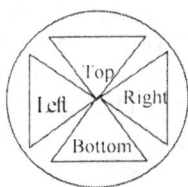 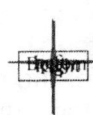 radius of any size is there is another line (or dot) eternally bigger as well as eternally smaller than the line in question. While we are in the third dimension being part of the third dimension then allows that all parts of the third dimension forever can be divided once more until the line in the third dimension is no longer part of the third dimension. When such a line leaves the third dimension it is still dividable because it might not be part of our dimension any more but it can still reduce further as part of the second dimension. By that time it has left our scope by miles. It does not mean that it end there because from our perspective that is where it ends. Yet it can still reduce infinitely more until it has left the second dimension and then at last forms part of the first dimension. Only then, when the line reaches the first dimension, no further dividing of that line is longer possible.

We can never grasp the size of a line that forms the utmost or the least of possibilities and therefore size belongs to the human mind forming conceptions of big and small, but it has no place in the cosmos at large. This concept not only applies to size, but to all limits and divides we wish to create forming borders that we can appreciate. When looking at the circle in the conventional manner, we persist with errors brought about in culture and not by applying some significant modern logic. Take a circle and reduce such a circle constantly to where it no longer can reduce. Reduce it to a point where only form remains part of the circle because the radius has gone beyond human measure and becomes so small it is not noticeable with what ever tools man may use, then what remains is pi since pi does not indicate size but indicates form, and form is all that then will remain. In any circle or sphere the size only depends on the fluctuation of r, as a component to the circle or sphere but that does not affect the form by indication of Π in any way there may be. The conclusion I drew from following this process is that from this line no start can be at zero because that will be a mathematical impossibility since no line can ever reduce to zero.

A line will forever be able to reduce further becoming smaller but it can never reach zero because zero is not on the scale of lines. If a line cannot reduce to zero it then cannot start at zero. A line or spot starting at zero would therefore be shorter than the shortest line possible. For obvious reasons can no line, or any line grow or extend from zero because such a line must then quit zero and become something, thus abandoning its original value. That would mean the start of the line has a different value to the end and a line holds conformity through out. When any line is starting from point zero, which it can never leave zero because of the influence of being zero disqualifies any possibility of growth. If the line then had to grow in all directions at the same pace, the line must then become a circle or being three-dimensional, then forms a multi circle we name a sphere. Since the Universe is about circles and lines connecting circles, I came to conclude that flowing from this fact is that in the Universe there can be no zero improvising as a filling ingredient for the space of a point or be unfilled space. In the case of the growing sphere the value of the circle is Π, and that is where creation must have started. That gave me the clue where to start looking for singularity. One would find singularity in the value Π and the value Π will be in all things rotating in a circle. As usual I am again shooting the gun before the hunt started. Lines in mathematics do not start from zero and that is no discovery on my part that was a realisation I came to.

UNIVERSE
Everything that exists, including space, time, and matter. The study of the Universe is known as cosmology. Cosmologists distinguish between the Universe with a capital 'U', meaning the cosmos and all its contents, and Universe with a small 'u' which is usually a mathematical model derived from some physical theory. The real Universe consists mostly of apparently empty space, with matter concentrated into galaxies consisting of stars and gas. The Universe is expanding, so the space between galaxies is gradually stretching, causing a cosmological redshift in the light from distant objects. There is growing evidence that space may be filled with unseen dark matter that may have many times the total mass of the visible galaxies. The most favoured concept of the origin of the Universe is the Big Bang theory, according to which the Universe came into being in a hot, dense fireball about 10-20 billion years ago.

UNIVERSAL TIME (UT)
A worldwide standard time-scale, the same as Greenwich Mean Time. Universal Time is the mean solar time on the meridian of Greenwich. It is defined as the Greenwich hour angle of the mean Sun plus 12 hours, so that the day begins at midnight rather than noon. It is closely linked to Greenwich Mean Sidereal Time (GMST), since the mean sidereal day is a precisely known fraction of the mean solar day. In practice, UT is determined by a formula from GMST, which in turn is derived directly from such observations of the meridian transits of stars. The version of UT derived directly form such observations is designated UTO, which is slightly dependent on the observing site. When UTO is corrected for the variation in longitude due to the Chandler wobble, a version of Universal Time, UT1, is derived which has genuine worldwide application. When UT1 is compared with International Atomic Time (TAI), it is found to be losing approximately a second a year against TAI. Broadcast time signals use the time-scale known as Coordinated Universal time (UTC). This is TAI with an offset of a whole number of seconds. The offset is adjusted when necessary by the introduction of a leap second, and UTC is always kept within 0.9 s of UT1. On this issue there is much more to explore than the meagrely mentioned. Time stands related to the

position an object holds to a centre such an object refers to while in rotation. Kepler found for instance that T^2, which holds the orbit to a rotation specific, is directly dependent on **k** to value the space a^3.

Einstein proved that in the presence of a strong gravity, time slows down. Surprisingly, with that evidence being around this long, nobody since then in science took those statements and made any further progress from there. It was left in some drawer to dry. Science still sticks to its change that time did not change slightly since the beginning of the time and holds the same pace ever since. With the entire Universe including all the gravity now present and not excluding one Black hole or dust speck pressed in an area possibly the size of a lepton the gravity extending from that must have been beyond what words can ever describe. If the gravity was that high and Einstein already proved gravity slows time down, then there is one logical conclusion and that is that time was in fact standing still. Mathematically it is incorrect to allow gravity to compress the Universe into a spot smaller that an atom and exclude any other factors and relevancies to change. But before coming to the mathematics, I would first like to bring your attention to the practical side. I am promoting a theory in which I am able to prove there is as much contraction (moving in the direction of the Big Crunch) taking the cosmic Universe back to the size it had during the Big Bang as there is expansion (moving apart by Hubble's Constant) and the contraction is as much part of the expansion. By contracting the Universe is expanding and everything is based on gravity providing both actions. The Universe rides on a balance and we have to locate such a balance. To prove my theory I firstly had to locate the centre of the Universe. Even admitting to such a notion sounds like madness or in the least a tasteless joke, but please give me a chance to explain in more detail. I realised that my effort to locate the point holding singularity only stood any chance of success if the reducing of the line enabled me to backtrack the exploding Universe to its origins. By applying some basic effort, I have located the position from where all movement came and the direction it took moving forward in time…and yes, during my search in locating the centre of the Universe I also stumbled on time as such.

Let us find the smallest possible line first. Reducing the line will eventually leave all sides on the same spot. Such a spot must be round in form. The line being the smallest line will start off as a dot. A line so small it has reached a point not dividable any more, will have all sides literally on the precise, same spot, and I have located singularity in just such a spot. I came to the conclusion that the spot I found had to be singularity purely on the grounds that, that spot holds only one side to serve as a start to the starting point of all directions possible. There in that side is only one spot is only one side applicable and one dimension present. With all the factors given, one can only come to one conclusion and that is that there can be only singularity. In such a case more dividing by two will land further positions on the other side of the divide. That point serving as a position for all points that cannot allow further dividing, is the smallest line or spot there may ever be. This spot is the result of a most basic process of reduction as the Hubble constant is a most basic process of doubling up during a matter of time. By reducing the line constantly the only value that will eventually remain without dispute from any party arguing about the facts is Π. By only having Π and a radius as one square (the radius effectively becomes one holding any and all sides on one point) of any significant measure, as the radius it will be an evenly spaced dot. From the smallest ever possible dot will grow a line in every imaginable direction relating to a prospect of Π not favouring one direction that puts all directions at equilibrium meaning that any form of what ever might develop from such a spot will have the end and the start being in the same position, which will also have to be a sphere as the flow outward will be equal in all directions. Please think clearly, is that not precisely the commitment we find in gravity, where gravity is flowing from singularity outwards but never favouring any side? This reasoning prompted me to look for singularity in such a spot because if the prime spot from which all came was a spot holding all, then the spot must hold the shortest line but more prominent, it will hold the smallest form including the smallest circle or for that matter the smallest sphere.

With gravity always being in the centre of a sphere where the space is least available in the entire structure (there is not even space left to fill) one finds a flow of gravity from that centre spot outwards in all possible directions even-handedly. The fact that the original gravity will begin as a circle or will be a circle, is the direction it will take when being the first spot created. All progress will be evenly in all direction because no direction will stand out or be in favour above any other direction at first.

The spot forms a full circle, but the line running through the circle is forever present because that is the future radius of the circle that will one day develop the circle, which is equal to the present diameter. The fact of the presence of such a possible line in such a possible circle dividing the possible circle into two parts makes the centre line equal to the half circle. The line forms the half circle but not only that the line presents the half circle as much as the line is the half circle. The line then is 180^0 and the half circle is 180^0 because in singularity the two factors are the same. The same value is of course $\Pi^0 = 1$.

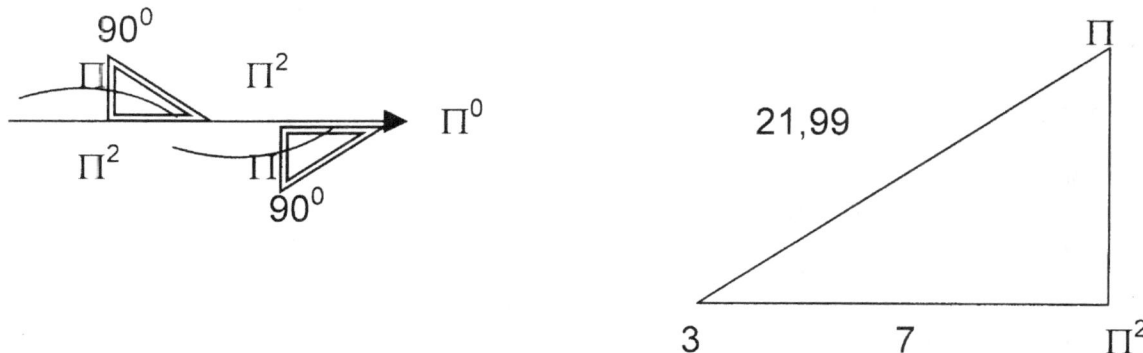

In this half circle of the future, which is no half circle as yet because of a lack of space, there are three future points indicating the space less ness that will go on to become space filled with something. On top of such a circle to form must be a marker indicating an awaiting boundary or future border and at the bottom of the future circle there also must be a similar marker that is no marker as yet. Between the two possible points that are not there yet is a future line running that is not there yet. Then indicating the possibility of a position to come that will bring about the half circle being a future distance apart from the future line indicating a diameter that will one day be, there a third such a marker must be established for the future. That forms a triangle with two more sides being connected by either a line being one or half pi being one. From singularity comes about that the line is the same as the half circle is the same as the triangle and they all have one value being 180^0. From this come the most basic principles in as much as forming the ground rules of the law of Pythagoras.

When drawing a line, such a line then starts off with a dot serving the spot that holds all sides equal. That means the line serving as the future radius will be equal to the half circle which is then Π. The only aspect of the point that stands in for the end of the single line forming the radius of the circle is that we then mathematically reach the single dimension. We decreased the line to where a circle being Π formed on the single dimension. This dimension also holds the circle dividing line because from there the radius must once again generate a value and by such a gesture that the extending would form the circle that forms the sphere that eventually leads to the formation of particles. This leaves a problem to investigate.

With no line possible, there had to be another dot that formed since the Universe has many dots that formed lines. But let us not to get confused and lost in the range of possible diversions but let us stick to two dots. One dot was next to the dot next to the dot, but as I said we stick to one dot next to the second dot. M X M / r^2 is the first step gravity began with. That leaves us with a huge problem in as much as when r = 0 then $r^0 = 0$ and 0 dividing any value will leave 0 as the answer. If the particles were inseparable at the start it must bring about that gravity would not be forming since the distance will not permit any dividing. By allowing the distance separating the particles to be zero, the particles melt into a unit.

Again this is Mathematics and not my incoherency as some Academics dismissed my work. Let me run through the argument one more time because I have been insulted by Academics in the past telling me I am bending mathematic rules with my applying double values to try and produce some argument. The two particles formed by an inseparable unit separated by a sharing of a spot. We know that at least two spots formed because there are many more than just two that remained to become part of the visual Universe. Let us name the spots because that is what humans do best if they do not know what to do with what they have to do. Let us call the one em and the other one spot next to em we then call emtoo. Between em and emtoo there were nothing because em and emtoo were inseparable. By they're being inseparable we would naturally be inclined to think that the separation value should be nothing or at least zero. But putting zero in that place is a mathematical excluding procedure leaving future mathematics excluded. With m multiplying m_2 and then dividing $\div$ r with zero (r=0) such a procedure will leave the lot at zero and with that, nothing is going nowhere. That means although we think the space between the two parts are nothing the non-existing space has to be at least one to be a future factor.

Every part of the argument is sound but was never yet used. I repeat once more, if my argument reflects on inconsistencies, those inconsistencies are not about my work. In order to disprove my argument, replace Mass one and Mass two with any number possible, and then divide such a number with the square being zero. If there was no space then the value of the particles had to be one. If there was no space between the particles the particles then had to form a unit. But if there is a mathematical possibility of reducing a line to the single dimension then there had to be a factor representing r as a factor of one. Take $(M_1 \times m_2) / r^2$ and substitute any of the factors with zero and the result coming about has to be zero. The factors in the equation have to have any and all the elements at a value of at least one. Only if r was a factor of one can gravity bring about any mathematical equation developing from this argument.

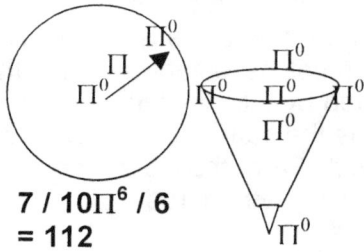

$7 / 10\Pi^6 / 6$
$= 112$

That means the mass on both sides must have a factor of one being a limit, which does not allow such further reduction of r and any further reducing of r beyond the limit will not be tolerated. Only if r = 1 then r^2 can be 1 and mass can be apart. Like it or not but believing in the Big Bang must also bring about the acceptance that the cosmos moved apart somewhat. The fact that r brought increase in the space separating the mass, produces a problem that was solved already. About a century and a half ago Roche found just such a limit. Once again I were confronted by zero becoming growth.

There is a huge hole that needs filling when bringing into a relation any forming of an alliance between a cosmos coming from nothing and filling with nothing and a cosmos growing spontaneously through balance shifting prominence. Mathematically the fact of applying nothing serving as a factor applying in the cosmos is not a strong and convincing argument. The minute one brings in zero as a multiplying factor forming a definite value working into the calculations of the cosmos, growth disappears. If growth was not a factor, the zero factors could be involved with some form of maintaining stability and where then further growth will accept the responsibility of zero

The region surrounding each star in a binary system, within which any material is gravitationally bound to that particular star. The boundary of the Roche lobes is an equipotential surface, and the lobes touch at the inner Lagrangian point, L_1, through which mass transfer may occur if one of the components expands to fill its lobe. It names after the French mathematician Edouard Albert Roche (1820-83).

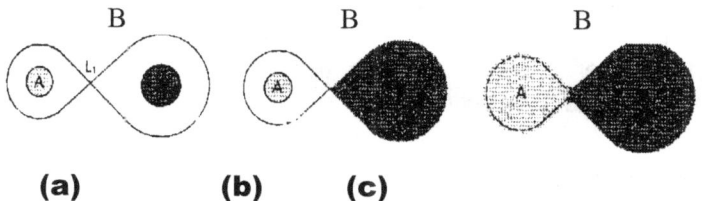

(a)   (b)   (c)

THE ROCHE LOBE: In a binary system, the Roche lobes of components A and B meet at the L$_1$ Lagrangian point. (a) In a detached system, neither star fills its Roche lobe. (b) In a semidetached system, one massive component, B, fills its Roche lobe. (c) In a contact binary, both components overfill their Roche lobes and share a common envelope. As with the graph, I can see the two sides forming a connection therefore relevancy has to apply, all contradicting Newtonian claims of no connection but through mass attractions. The mass does not attract but one interferes with the other total influencing the space surroundings.

The closest encounter worth noting we ever had with this law in the modern age of news and Television was the Shoemaker-Levy 9 incident during the previous decade. At the time and even in the present, no one drew any similarities but after completing this book the reader should find why I could draw such a similarity, which there is between this incident and the Roche limit. Even the phenomenon called the Sound Barrier became clear when applying the Roche factor with the laws governing the influence of singularity.

The gravity Newton suggested and the gravity Mainstream science is in search of is not there. In fact there is no gravity if we consider what science would wish to locate as gravity. Mass does no instigate, initiate or create gravity by any means or measure. Gravity is the independence of matter in relation to its depending on matter in order to realise its independence. The motion that produces gravity is what inspires a new Universe into being

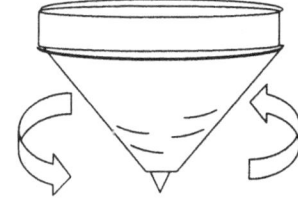

Every time a top or anything else starts to spin the spin creates a newly established Universe, which fills that Universe with the surrounding time in that Universe with a new centre and the centre is the gravity that rule and dominates that Universe.

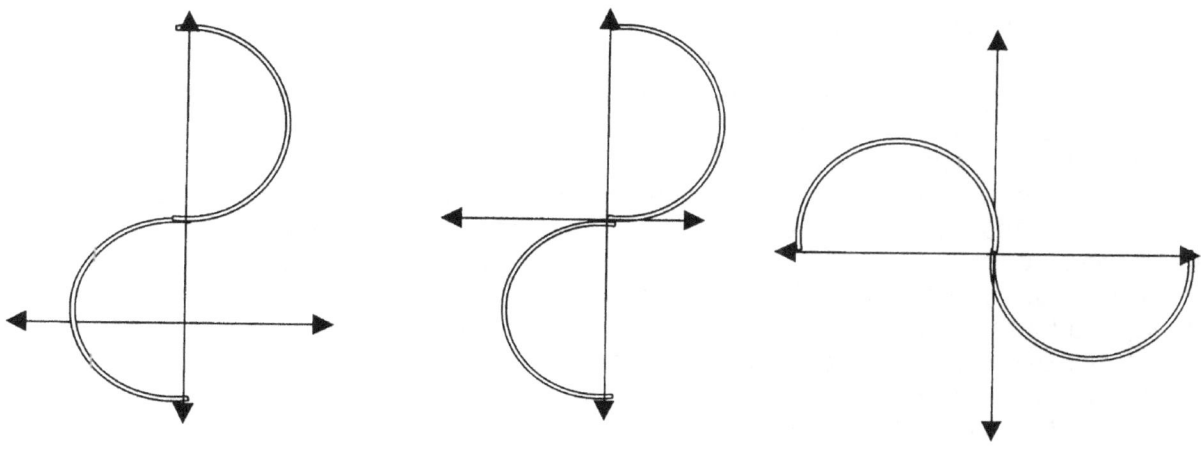

The graph is the result of motion where singularity forms a centre but the centre formed is not zero. The centre formed is Π^0 and from such a centre space-time forms by expanding **space a^3** to the value of **Time T^2k. In the centre of the graph is Π^0.** This point, which I now am referring to, is the point where Π is a fully appreciated value while the diameter D still remains a dimensional factor of one. This is the dawn of the second dimension where space was there but space was sparsely shared in some cases.

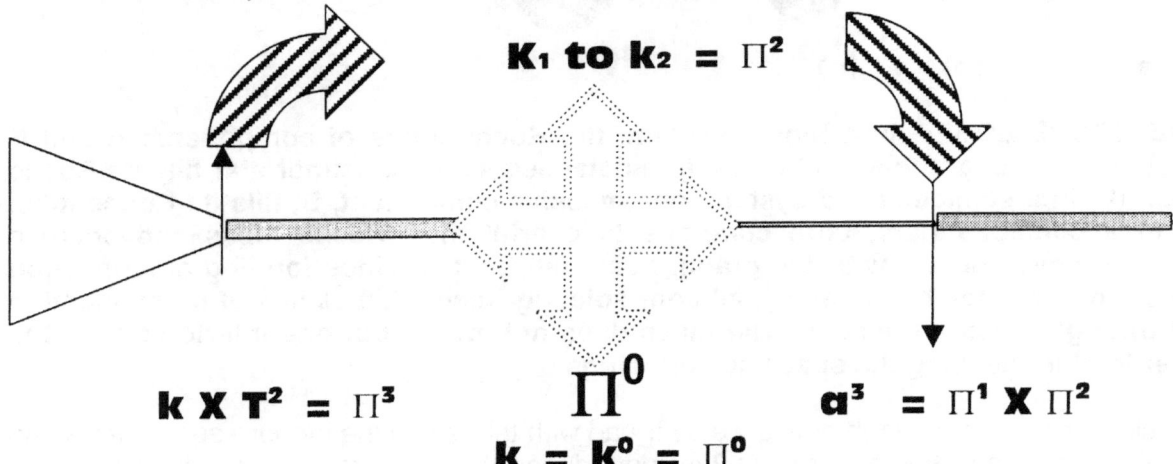

$$K_1 \text{ to } k_2 = \Pi^2$$

$$k \times T^2 = \Pi^3 \qquad \Pi^0 \qquad a^3 = \Pi^1 \times \Pi^2$$

$$k = k^0 = \Pi^0$$

It is when Π^0 shifted to become Π for the very fist time. The point without movement, the point holding singularity must have a value of Π being the eternal dot but since the dot has no dimension in having form the Π that indicates the dot must be Π^0. From such a point there has to be to the side of the centre point be a point where space do start. That point will then receive a diameter but that point will have form only in being a circle. In that point there is a shift from in relevance from Π to the centre Π^0 and for the first time it brings about two separate values for Π.

Because the three points existed on equal terms in singularity sharing a same spot the coming out of singularity will enforce that equal value comes to all. That means the circle gets to become Π, the diameter becomes Π and the distance setting the structures apart will also become Π. This is what the coming from one point brings along. Only when being part of the second dimension can separate values come into being.

While the form was still being in the single dimension from the one side of the form the dots had to establish identities apart but not separated yet. The one circle had a factor of $\Pi^0 = 1$ and the centre had to have a value of $(\Pi^0 / 2)$ extending past the very next object but also cutting such an object into a square double half value that was going to come about as soon as the other dimensions came into form.

The only definite place one will locate zero is in between the starting point of the lines going in opposing directions in the position the lines hold before there was the least of directions applied, but that is only because there is no such a position, not because any line is coming from there. The two lines are still one holding the opportunity of parting as an option but have not yet parted and therefore are on the very precise same spot. The line coming from there is already there because it already has the choice of going in any and all opposing directions and when it starts running it will place filled space in that location because the space was already filled with a line starting and not with a line not there at all. When reversing a line we might find a better idea of what is in place and where it is in place. Gravity is officially a force without limits going past and through borders and has an unlimited reach. It seems to remain even and this is conflicting with the flow of perceptions about mathematics.

The formula **F = G (M₁.m₂)/ r²** is unable to explain the principle discovered by Titius and later by Bode and in contrary to all statements to that effect made by Accepted Science policy makers, the Titius Bode principle is not coincidental. In fact it is one of the four most adhered and important cosmic pillars holding the cosmos structure in place. From the two above comes gravity. In a few pages I prove how one can arrive at the facts that prove how the Titius Bode Principle leads us in the direction of the origins of the solar system. But before we can accept the influence of the Titius Bode Principle, we have to deal with "Nothing" and as such dismiss nothing from science. "Nothing" in the Universe is coincidental; "nothing" in the Universe does not apply. Where mathematics connects to lines nothing disappears. Should any principle not match an accepted theory or change the accepted theory, the theory does not apply.

The content of my work holds a new view about Cosmology, which I have been working on for the past twenty-seven years and exclusively for the past six years. I always had a problem with the idea that space constituted of nothing, while I came to realise that lines mathematically couldn't start at zero because there is no evidence of zero as a factor in mathematics. Should you disagree with my statement the question in need of answering is this: What will the length of the shortest hypothetical line imaginable be and moreover, what would the total overall length be in that case? The shortest possible line (hypothetically) must be so short it must have an initial and ultimate point sharing the same spot. The two points must be one and only then can further reducing of any line not occur. If it used zero as a start, the zero part would not count, because the line will only start at a point past zero where the line then will start forming an infinitely small dot. I press this point to urge understanding because there is such a point, but in an attempt to recognise the point, I have to convince the reader to abolish four or five thousand years of accepted and practised mathematical culture and that is no easy feat. Taking the line back as far as possible brings a dot because of the equilibrium that will stem from such a position. The dot is in infinity, however small, it is not zero. Zero ultimately means not existing and then that point, as a start does not exist. The smallest line has a beginning and an end at the very same spot located in infinity, and infinity may be beyond human scope, though infinity is still not zero. Infinity may constitute of something we do not yet understand, but we may not define our human misunderstanding by that, which is not present in our minds and therefore formulate a concept by the degree we have the understanding as nothing. In this aspect lies the difference there is between arithmetic and mathematical science where arithmetic can have position such as zero since arithmetic excludes the cosmos calculating numbers only. Cosmology is not about numbers because no one can calculate the number of stars. Cosmology is all about lines and angles positioning objects, and in those there features no zero. No line can be zero long and forming a position of zero degrees in relation to another object.

A man may have that many oxen or so many sheep and even this amount of wives, (in Africa) or not have any therefore having then a total of nothing, but there cannot be nothing between the Sun and its orbiting structures. The having and have-nots are part of arithmetic. Light will indicate a line flowing between the Sun and whatever planet, following dot after dot thereby proving the existing of the possibility of something going about by a straight line, and any straight line in relation to other straight lines will be under the law of Pythagoras in as much as obeying the rules of trigonometry. There is no possibility of a straight line not forming in space. If there is space, there can be a straight line. The mere fact of two spots having different positions in space gives the two dots different values. If the line has the length of zero it is not present. If the triangle has one angle at a value of zero it is no triangle because the zero would dismiss all the other angles. Mathematics converts the values of integrating lines according to Pythagoras and arithmetic is about numbers to be added or subtracted. By mathematically excluding zero from cosmology a new Universe opens to the human mind. With the distance between the Sun and Pluto being roughly one hundred times more than the distance between Mercury and the Sun, the distance must hold something more than a pure vacuum filled with nothing except one atom here and there occupying the vacuum between them and the Sun. If space supposedly comprises of nothing, how can nothing then become plural forming more or be multiplied by a number as to indicate a growth in something not even existing. As the one becomes one hundred the one cannot substitute a value of nothing but then must be part of something. If the one substituted the nothing, all laws of mathematics will go in disarray because when one multiplies any number by zero it becomes zero placing both planets in the Sun. If Pluto was one hundred times closer than it is at present, was it one hundred times nothing closer? $100 \times 0 + 0 = 0$. That is mathematics!

By allowing the three hundred a value, the nothing must form one making that which is between Pluto and the Sun not to be nothing but there has to be something. This argument follows mathematics to the letter and in precise detail. That which is keeping the Sun and Pluto apart at a certain distance has to have an ingredient of something in place that is forming the substance, which is responsible for the distance between them with specified cosmic positions. The same substance is forming the specified space between Mercury and the Sun, only it is ten times more of that which is between Mercury and the Sun. Therefore it is one holding time multiplied by the ten times the time we find in that space standing relative to other space regarding whatever the space becomes what is between the Sun and Pluto. That factor cannot stand in for being not one (even expressing this incoherent concept verbally shows how ludicrous the reasoning are of those thought to be the best in the business), which is the same as nothing is, that is being there by not being there, because one cannot take or fill the allocated value of the place in the position that zero secures. By excluding nothing from the equation, space becomes something bringing in a value lying inside the realms of the infinite that must form singularity. As the zero becomes a dot, something else becomes clear about the dot. Looking at the night sky we find darkness overwhelming the space in relation to the stars bringing across light.

My approach to cosmology shall prove to be somewhat unconventional but through the abandoning of the accepted, it enabled me in locating the precise location of singularity that forms the connecting basis of the Universe (and this I say with some degree of confidence). There **are two locations** but I shall **first concentrate** my explaining effort on **the prime singularity**. Singularity did not vanish into the unknown after the completion of the Big Bang development but is in a place science incorrectly valued and classified incorrectly and in that, there is something hiding the truth. If singularity was or is where the beginning is. We have to go back and see just where such a beginning was. I cannot accept that the Universe started at zero and neither does anything else in the Universe start at zero. My excluding the possibility of zero includes that the Universe is not filled to the top with nothing and neither is nothing part of outer space. The Universe is about lines allowing light to flow from one point to another point and in following that line it has to continue in the line as the line has to represent something. The Universe is all in relation about lines indicating distances between cosmic structures. The cosmos is in short about lines connecting points in space being apart. It is about a line starting and continuing from such a start. But science advocates their opinion that such a start of a line flowing between any and all objects can hold zero because according to them the Universe is full of nothing. If the Universe in as much as outer space is a container filled with nothing at the present moment, and there is no place anything that was part of outer space previously could release to and there was no emptying of what ever filled it before, then it could not get rid of what was in the outer space when it started with what it started off with. We must then accept from what is not in the Universe was not in the Universe at the time during the start and what at present is present according to science because it then still must contain the same nothing and must have that same filling from the start to the present. If it was nothing it still must be nothing and that same substance being nothing is what it also used to grow using it as it grew because it filled outer space with nothing growing from and growing to nothing. Is that true? The filling of the Universe could not go anywhere, so one has to presume it started off from nothing and from there it kept filling with nothing since what ever was in the Universe at the start had no place to escape to or no place through which to escape. That is only applying if it is nothing filling the Universe at large. Can nothing grow as much as a line is growing from a start of nothing? The answer is that such lines not only indicate a distance but since the Universe came from such a small space as science propagates with the theory of the Big Bang then all particles in the Big Bang Universe were rather cramped for space when the Universe started from that small line between particles and is now the same line but is now so big. In the past it seemed being so small and showing the space between particles to be awfully short at one time. It was short but how short was it? Did it start off as nothing? Is the line starting at nothing as science wishes us to believe? If it does then all lines must start from nothing so we better investigate this trend with the start of a line. With the following work, I show my argument with which I hope to prove the counter part of what science believes. I am about to prove that which science sees as nothing in space and in material is the very location of singularity.

Lines mathematically cannot start at zero because there is no evidence of zero as a factor in mathematics. Should you disagree with my statement the question in need of answering is this: **What will the length of the shortest hypothetical line imaginable be and moreover, what would the total overall length be in that case?**

Locating zero

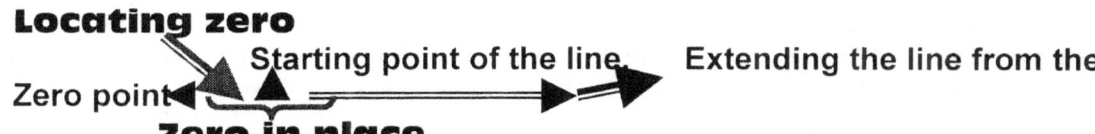

Zero point◄ — Starting point of the line. — Extending the line from the start.

Zero in place

Let us duly test my statement by taking the line back as far as possible. The shortest possible line (hypothetically) must be so short it must have **an initial and ultimate point sharing the same spot.** The line that **cannot reduce** any **further** must be **so short** that **directions flowing away** from each other **are located** in the **same position.** Any theoretical line being the shortest possible line cannot have the line holding the initial starting point at point zero and advance from there. Mathematics simply will not allow it. If the point had zero as all it had to offer, such a point is not present. The zero means there is no such a position. If it used zero as a start, the zero part would not count, because the line will only start at a point past zero where the line then will start. Zero ultimately means not existing and then that point, as a start does not exist and where the line then starts is a point in existing. When the line **has a beginning and an end at the very same spot** and it wishes to extend the position as to further the possibility it has, which direction should it favour? Extending the line in any one direction will favour one direction without any clear reason not extending in other directions. The fact of direction being present only proves and is proved by another point established, which is placed in relevance to such a second pointing a position already established by the relevancy of two points located in a direction opposing one another. But if one point starts one line there is no favour of direction since there is no established direction yet. The only mathematically sensible option about extending any line starting at a pre-designated point without any other point to establish a pre determined direction will be non-bias progress in all directions equally in order to give a meaningful flow of mathematical equilibrium. Not one direction stands superior to other directions and all directions are equal with no bias anywhere. Of this statement the Pythagoras mathematical principle is proof of and that I explain later.

Let us dissect nothing, as we find nothing in the presence of the cosmos. The distance between the Sun and Pluto is roughly one hundred times more than what the distance is between Mercury and the Sun, and that indicates to form the square of space (10^2) multiplied by the distance between Mercury and the Sun, but both has nothing between them and the Sun. The space filling the distance from the Sun to Mercury has nothing more than the space between Pluto and the Sun. That means the distance between the Sun and Pluto is as equal in relevancy as the distance is from the Sun all the way to Pluto, since both is the measure of nothing multiplied by millions, which in the end, still assembles nothing. If the one substituted the nothing, all laws of mathematics will go in disarray because when one multiplies any number by zero it becomes zero placing both planets in the Sun. The distance between the Sun and Pluto **is Pluto is 5900 X 10^6** kilometres of space, but in that statement we take it that the one of a kilometre is present in such a multiplication. The one constitutes the presence of fact being a statement of a value. By saying the distance constitutes of nothing we have to substitute the one factor with a factor of zero. Then the calculation must read **Pluto is 5900 X 10^6 X 0 = 0.** Including nothing as to state the presence of that part contained by the calculation delivers the total of zero. By excluding nothing from the equation space becomes something bringing in a value lying inside the realms of the infinite that must form singularity. Applying this logic to the Lagrangian system and interpreting that information to the law of Pythagoras, a clear pattern comes about.

The reaction responding from my argument is that it is silly, but should that be your personal opinion too, then do a test to find where the silly part applies. Bring the zero into the calculation, the zero that science so eagerly places in outer space and see the mathematical result. By applying the distance, one accepts automatically that the figure become calculated with one as it represents one in being a calculating part of the cosmos. The calculation as all calculations normally are, is in order to calculate something and the something will at least stand in as one in relation to the rest being part of the calculation. But saying that, the factor of one in fact represents nothing since nothing is so much the part in the calculation being calculated, and then the zero has to replace the one as the fact of being calculated.

The claim becomes obvious when observing the connection between the half circle, the straight line and the triangle, which could also promote all the qualities lurking behind the pyramid. Consider the connection between 180^0 sharing three different forms all part of mathematics where each is different in form, but equal in value and then one may realise in considering the very basic in mathematics being the Law of Pythagoras on which all mathematics are focused. The triangle stands in for one factor represented by one at a value of 180^0. So does the straight line become a factor of one and the half circle also becomes one where the factor of one equals all 180^0. All three are most seriously part of shapes in the cosmos. Revalue any one form to zero and the rest too must follow and share the same value. The Law of Pythagoras is about angles in relation to lines and not one angle about representing zero, because that will reduce all the lines also to zero. The measure of angles between stars at a distance uses parsec as the indicator, but the parsec between the stars indicating an angle has to represent an angle whereby one may measure distance and such a distance cannot be zero because then the parsec will be equal to zero. Again it is multiplying the factor with the measure but if the measure is about a factor of zero, then the factor too becomes zero. That is as basic mathematics as I can present.

If the argument seems ridiculous it is not my mentioning such a fact that is ridiculous but the mere fact of the reasoning also becoming a recognised argument accepted by science, making it as such ridiculous. If space is nothing then it has a number to use indicating just that value being zero or the capitol O indicating zero. Try and indicate what is measured and calculated in space, but not by simply not thinking about the fact and therefore simply ignoring that which is measured forming the sole value of space, but put the value of nothing as part of the distance in calculation because that is what is measured. When stating the distance between the Earth and the Sun, place on paper what will allow the kilometres measured to represent the factor that is being measured. If represented by one being the total of one by hundred and forty nine million kilometres of nothing put that language in the International language of mathematics that spans all dialects spoken on Earth. Put it in mathematical terminology by saying there are 149 000 000 X 1 (multiplied by the kilometres) multiplied by what it is being measured which is 0 and what will the total come to… a full zero.

149 000 000 X 1 (km) x 0 (indicating what the km are made of) = 0

Mathematics says it. If there is something to be measured then the least value the measurement can have in relation to what is used in the measuring has to be one. It cannot be zero and be measured…and we do measure outer space! It sounds as if something here is at fault. It is not with my mentioning the inconsistency one should find fault but the fault is with the fact that it is there and no one noticed! I am not to blame just because I am mentioning it, but the blame must go where it belongs.

I think it is by now little understood although I imagine not nearly accepted that by adding a million of nothing to one nothing there will remain one nothing and that is still nothing. Nothing cannot accumulate therefore I cannot accept anything holding the vastness of space being able to constitute nothing as the major component.

When reducing the circle in size one have to reduce the radius or the diameter because the pi is the indicator of the form as being a circle. Divide the r until there can be no dividing any further and that cannot in the end indicate zero because no matter how small, in that will forever be a value in place.

There will forever be subatomic particles building atomic particles because the reducing of space goes down as far and to infinity. There will always be some material forming a part of more material that builds into something which ultimately becomes the atom as a unit. In every centre of every subatomic particle running down into infinity there will be a centre composing singularity and the group will establish a centre governing singularity.

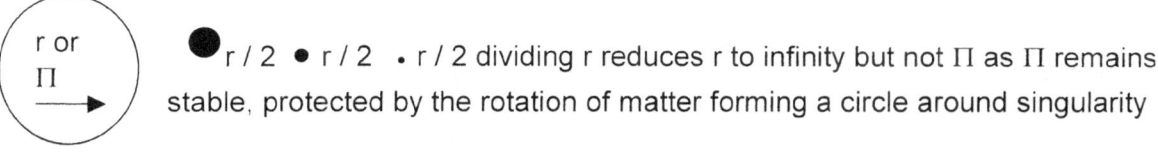

r or Π → ● r / 2 ● r / 2 ● r / 2 dividing r reduces r to infinity but not Π as Π remains stable, protected by the rotation of matter forming a circle around singularity

Taking that into account it is important to recognise that notwithstanding the size of a line, there is another line (or dot) eternally bigger as well as eternally smaller than the line in question. We can never grasp the size of a line that forms the utmost or the least of possibilities and therefore size belongs to the human mind forming conceptions of big and small, but it has no place in the cosmos at large. This concept not only applies to size, but to all limits and divides we wish to create forming borders we can appreciate. When looking at the circle in the conventional manner, we persist with errors brought about in culture and not by applying some significant modern logic.

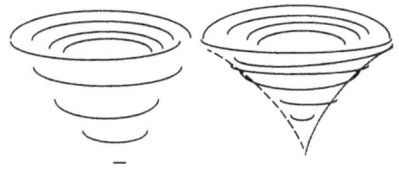

By reducing r indefinitely to the tune of half each time, r would become infinitely small, beyond human calculating means, however as mentioned in the case of the smallest dot holding one spot, r would become insignificant beyond human comprehension even, but never reaching zero and still Π would remain intact and dictating form. I believe one can begin to see where my suspicions were heading the first time I went looking for singularity and then subsequently found it exactly where I suspected it was. It was where

Kepler said it was. It is at $k^0 = a^3 / T^2k$. It is where a line truly starts and not where it is presumed to start. The line cannot start at zero because the flaw comes about in the manner mathematics are practised for thousands of years. But before coming to the mathematics I would first like to bring your attention to the practical side. I am promoting a theory in which I am able to prove there is as much contraction going on in the cosmic Universe as there is expansion and the contraction is as much part of the expansion. The Universe rides on a balance and we have to locate such a balance. To prove my theory I firstly had to locate the centre of the Universe. Even admitting to such a notion sounds like madness, but please give me a chance to explain in more detail. If I wish to achieve success that would depend on my ability to convince all that outer space comprises of material and as such we can locate such material even if we are unable to see such material.

To find the invisible I had to locate singularity. I realised that my effort to locate the point holding singularity enabled me to backtrack the exploding Universe to its origins.

By applying some basic effort I have located the position from where all movement came and the direction it took moving forward in time

The reversing of the circle radius is not alien to nature at all. An observation coming instinctively to mind that one may recognise, is that the form reminds rather explicitly of natural phenomena such as hurricanes, water whirls and even the shape most commonly favoured to express the cosmic object referred too as a Black Hole. The similarity may be more than coincidental. Let us consider the statement in the reverse. In our calculating of a circle we apply two formula methods. The one uses an r to indicate the radius and the other use a D to indicate the diameter, which is double the radius and therefore needs to be divided by a four to eliminate the Newtonian inverse square law amounting to the difference there will be between the two. The one using the radius is Πr^2 and the other formula using the diameter is $\Pi D^2 / 4$.

In any circle or sphere the size only depends on the fluctuation of r in the square as a component to the circle or sphere but that does not affect the form by indication of Π in any way there may be. The conclusion from this is that no line can start at zero because that will be a mathematical impossibility. A line or spot starting at zero would therefore be shorter than the shortest line possible.

For obvious reasons can no line, or any line grow or extend from zero because such a line must then quit zero and become something, thus abandon its original value. That would mean the start of the line has a different value as the end of the line and a line holds conformity through out. When any line is starting from point zero it can never leave zero because the influence of being zero disqualifies any possibility of growth. If the line then had to grow in all directions at the same pace, the line must therefore be a circle or being three-dimensional, a sphere. Flowing from this fact is that in the Universe there can be no zero point or unfilled space. In the case of the growing sphere the value of the circle is Π, and that is where creation started. That gave me the clue where to start looking for singularity or the centre of the Universe.

One would find singularity in the value Π and the value Π will be in all things rotating in a circle. You might wonder how that applies to the cosmos and moreover to gravity? In my search I stumbled on two accepted but not intergraded laws and when I found and located singularity, the two laws became very much plausible and factual. Take a circle and reduce such a circle constantly to where it no longer can reduce. Reduce it to a point where only form remains part of the circle because the radius has gone beyond human measure and becomes so small it is not noticeable with what ever tools man may use, then what remains is pi since pi does not indicate size but indicates form, and form is all that then will remain.

I realise that my effort to locate the point holding singularity enabled me to backtrack the exploding Universe to its origins. By applying some basic effort I have located the position from where all movement came and the direction it took moving forward in time...and yes, even time as such.

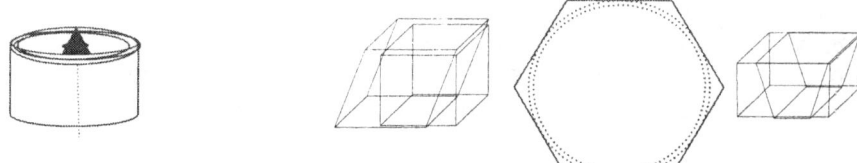

Anything occupying space in the cube will apply r and by r, I mean just a distance not using Π because Π serves as a form indication while the collective product of r will determine form as well as accumulative dimension total. Notwithstanding the name used confirming the shape, or r named as length width or height, it is all just a straight line bringing about the cube with all its other names that may find attachment to specific form but nevertheless still remains only a six-sided cube with connecting lines applying different angles changing in some cases. The normal perception is that any circle growing spontaneous would grow by the radius, which is r. In mathematics that may be true but it is not true in nature. In nature that cannot be the case because r is an indication of a straight line. By growing with the aid of a straight line from the centre to circle the influence that that would have on the circle would result in many circles following one another and not a continuous growth.

Gravity is the dimensional changing of space holding r as reference in the cube as to the sphere holding Π as the reference. In order to generate spin, producing time in matter occupying space, therefore creating dimensional change, Π has to be a factor indicating the possibility of spin because implementing Π, the circle sides will follow one another without establishing separation. The answer must be in finding Π, and thereby locating singularity. If singularity is in affect the original point of the cosmos birth, the reducing path we should follow will indicate the whereabouts such a point must be.

There are two standard formulas used to calculate a circle. The one uses an r to indicate the radius and the other uses a D to indicate the diameter, which is double the radius and therefore needs to be divided by a four to eliminate the Newtonian inverse square law amounting to the difference there will be between the two. The one using the radius is Πr^2 and the other formula using the diameter is $\Pi D^2 / 4$. However one looks at the mathematical expressions and Kepler's formulating of space-time, there is an exceptional difference between the two scientific uses. When investigating Kepler's formula one do find it appreciably differs from the normal Mathematical equation like $a^2 = r^2\Pi$ and $a^3 = 4/3\ \Pi r^3$. In the normally used mathematical expressions such equations tend to concentrate on the volumetric aspect. In the case of Kepler's expression it is something else that wants to surface. It is another idea that is coming to mind. In Kepler's formula $\mathbf{a^3}$ stands to symbolise the third dimension and such a third dimension becomes equal to two other dimensions grouping and sharing value to equal $\mathbf{a^3}$ efforts. It is not the circle of the rotation because with such a normal circle the radius is in the square and Π evaluates form. Here there is no mention of a factor Π, which one would suspect to be somewhere applying since the circle is Π and Π is the circle and the two are inseparable, but not in Kepler's $\mathbf{a^3}$, where there is no mention of Π at all. The fact that there is a radius of some sorts used to indicate a position cannot hold the square as it normally does in the case of the normal equations. In the mathematical equation, the factor indicating the position of the circle edge has the square value being called the radius or in some cases the radius doubles and which then is the diameter, and the circle indicator is Π. But in this event the formula value will bring about a square value to the answer one receives. It will bring a value to the surface of the circle. In Kepler's formula it specifically does not.

I realised before starting my quest that one possibility that the shortest line or smallest spot can never have is having a starting point on the zero mark. If the mark of zero holds the start it must also hold the end because the end and the beginning has the same position. If the position of zero then is the beginning, the end will also be zero leaving the line or spot without an end as well as without a beginning. Such a spot will constitute all of nothing. Any line starting from zero would inevitably start from a point where it ignores the zero mark because the fact of zero does not implicate a start or a size of value, but only the not being there of that position.

All lines would form a duplication of another line sharing value since there will always be a possibility of yet another line in the realms of singularity lying between the two lines in question reducing the size infinitely to either side of the divide we humans create. Boundaries therefore are human and as man made substances it does not belong to the cosmos outside the influence of man and must be discarded. No mathematics will ever measure the thickness, because as the line that is standing still, it cannot have a width at all. The moment a width appears, which one can measure or calculate, the line will become part of the factor forming the divided and not the divide. The instant when space connects, the spin direction will produce the partisanship of space and spin. Any form of space (even in the most minute) will expand as it favours a direction but changing the direction is by rotary motion.

The moment there is an area there is a measurable rotating brought about and no longer a non-interfering divide. Such a line holds space in a position that runs far beyond the boundaries and limits of the three-dimensional. Another factor of such a line would be that the radius (let us substitute the radius r with the using of Kepler's k), k would be immeasurably small. The factor k cannot be zero because infinitely close to that first k is the start of the third dimension where time plays the part as the fourth quarter. The presence of k is undeniable and recognisable yet it is not visible. The fact that k is there albeit stripped of any influence, disqualifies it from being zero and therefore not being there. With k already beyond any measurable space, leaving a^3 as a factor of one and not being able to pin any volume measure to that one k will have to be to the power of 0 being k^0. In Kepler's formula $a^3 = T^2 k$ the area a^3 would be one because of the dimensional non-existing of measured sides in any direction. If $k^0 = 1$ and $a^3 = 1$ the only alternative T^2 could possibly have is also one. The factor of T^2 identifies the time in the formula and when the formula indicates time as one, the time component must therefore be eternal. Only time in eternity does not change

The real formula applying when the calculation of the sphere volume is calculated is $a^3 = 4/3$ Πr^3 where it places one third of the dimensional (but lesser) factor in direct relation to another third dimensional relation held by the radius and all aspects about the factors being in relation is to acknowledge the form that is applying and serving as a sphere. However there is no criss-cross matching of dimensional accumulating. It places time in the square directly in relation to space in the cube in association where time shows two distinct qualities. The one factor is time in the circle rotating while the other is in the linear or the straight line implicating the position that the other would have. In all instances of measuring the distance the orbit travels around the Sun, as the space displaces or space covered by travelling in the time it is covered and dividing such a ratio, one finds the distance of the orbiting object from the Sun in relation to the other factors form one or very close to one. It is relevancies carried from the Sun and the Sun is the governing singularity representative for the entire solar system. This is about relevancies applying throughout the Universe. This balance is much, much more than what the figures say. It underlines and it explains gravity as a life form in the cosmos other than what we consider our life to be.

The German mathematician and astronomer KEPLER, JOHANNES (1571-1630) became Tycho Brahe's assistant in Prague in 1600 A. D. where he undertook to complete the tables of planetary motion Tycho had begun. Kepler first calculated the orbit of Mars. He spent much time trying to reconcile Tycho' s accurate observations of the planet with a circular orbit, but concluded (in Astronomia nova, published in 1609) that Mars moved instead in an elliptical orbit. Thus, he established the first of his laws of planetary motion. A theory that the Sun controlled the planets by a magnetic force led him to the second and third of his laws, which were published as part of his treatise on theoretical astronomy, Epitome astronomiae Coernicanae (1618-21). The Rudolphine Tables (named after Tycho's patron, the Holy Roman Emperor Rudolph II) of planetary motion appeared in 1627 and were still in use in the 18th century. Kepler also wrote De Stella nova, on the supernova of 1604 and Diptirce on optics and the theory of the telescope. The overall view followed in this book **Matter's Time in Space** places the true significance of his work in true contents. In KEPLER'S EQUATION is the equation that relates the eccentric anomaly of a body in an elliptical orbit to its mean anomaly.

The equation is E − e sin E = M., where E is the eccentric anomaly, M the mean anomaly, and e the eccentricity of the orbit. It is important as one of the mathematical relations enabling the position of a planet about the Sun , or a satellite about its planet, to be calculated from the orbital elements for any time. However, this only relates to the solar system, and KEPLER'S LAWS only apply in the contents of the solar system. The three laws governing the orbital motions of the planets, discovered by J. Kepler is as follows: The first law states that the orbit of a planet is an ellipse with the Sun at one focus of the ellipse. The second law states that the radius vector joining planet to Sun sweeps out equal areas in equal times which as it says refers to time and not the circle. The third law states that the square of the orbital period of each planet in years is proportional to the cube of the semi major axis of the planet's orbit. The first law gives the shape of the planet's orbit; the second describes how the planet must continuously vary its speed as it follows its orbit, moving fastest at perihelion and slowest at aphelion. The third law gives the relationship between the planets' average distances from the Sun and their periods of revolution. Instead of placing the true value to Kepler's laws, I. Newton placed his own interpretation to Kepler's laws, and in doing this he wilfully destroyed the principle working of the Creation. Through Newton's tunnel vision, he applied his own miss interpretations to the correct presumptions of Kepler. Newton reduced the implication that Kepler's findings hold. This is clear when putting Kepler into a wider context and that only can become clear after a development of some four hundred years in time after Kepler. Kepler is put in chains when one looks at the interpretation that is connected to the simplicity in which Newtonians regard Kepler, when they go about enslaving Kepler's broad vision in terms of the most simplistic mathematics as Newtonians work with and those simple mathematics does not find true regard in any of Kepler's work. It is obvious by using Newton's variation of what Newton wished Kepler's work would provide Newton with, when Newton was introducing his interpretation to the law of gravitation. Newtonians use mathematics as a tool whereas Kepler uses mathematics as a language of the mind to express and convey the brilliance that stretches beyond naming and calculating. He then went about and changed it to three laws of motion. I. Newton generalized Kepler's first law, verified the second law, and showed that the third law should be amended to the form; $4 \pi^2 a^3 / T^2 = G (m + m_p)$. In this, the value of T and a are the period of revolution and semi major axis of the orbit of a planet of mass m_p about the Sun of mass m, and G is the gravitational constant. The major aim of this book is to correct these misgivings of Newton. I shall return to the statement about $4 \pi^2 a^3 / T^2 = G (m + m_p)$

What Kepler observed and formulated was more of a dimensional coming together by the cosmos in nature. He saw more of the principles guiding the cosmos than what one will find in the cosmos. It was practical mathematical forms of symbols finding value by form instead of true figures, which Newton wished to see. Kepler saw the first and second dimension forming the third dimension by allowing the third dimension space to flow and room to duplicate under the control of singularity. In the argument Kepler made he had hidden so much more facts into one formula than what I think even he realised. Well, it is much more than what the Accepted Policy Protectors Of Science ever came to realise. He officially formulated space-time, he officially coined not the name but the origins of the Universe being the Big Bang and he was the first to put the speed of light in relation to cosmic development…and all of that with his rather simple formula. He said the space a^3 not the circle (a) or the circumference a^2 but in the circle a^3… where such a circle represents a factor in the third dimension.

The formula he compiled was not rather but very specific about the area being a third dimension area and to prove it beyond doubt he placed it in the relevancy of the formula in a ratio of presenting the third dimension in space. He said a^3 is equal to $T^2\ k$. Newton and Newtonians came afterwards and played with mathematical toys as to challenge their mental capabilities. Newton introduced a $4\Pi^2$ to indicate the presumed circle on the one hand and on the other hand he brought this lot equal to $\{G\ (m + m_p)\}$ which he then presumed to be the general Universal gravity constant (G) and the sum total of the two structure mass. Newton saw a ring circling around a centre having $4\Pi^2$ to indicate such a ring outside a centre and he positioned $\{G\ (m + m_p)\}$ where the two mass factors combine the gravity effort in the general grand gravity constant in space. I have had so much resistance in the past from all Academics but that is not what I see what Kepler saw. I shall trace this back to the centre of creation.

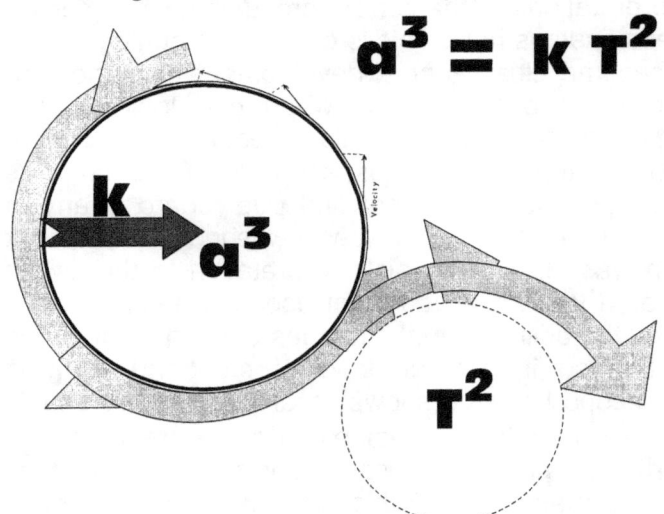

The one position in space will place the electron in time where the electron then will be below…behind and above the atom sub particles with which the electron shares frequency. The position k indicates implicate the electron T^2 establishes. in the environment a^3

In their eagerness to calculate they calculated a formula to measure the circumference a^2 of a circle being $\Pi\ r^2$. I have seen an Astro physics examination where they use $4\Pi r^3 / 3$ as the formula to calculate the Sun and other stars volumetric space! They formulated the measuring procedure of the circle being in the third dimension that will show how big the volumetric space is of a sphere at a^3 being measured with the procedure being $4\Pi r^3 / 3$. This too was a fanciful devise allowing mathematicians to be much superior to the rest of the commoners and to dictate to the lowlife how and what they should think when they think and if they indeed can think of anything to think of. Then some Mathematician and an Englishman of Substance came onto the idea of gravity. Being a mathematician the Englishman placed the Universe at the feet of mathematicians. He saw circles where Kepler saw three dimensions. He saw three dimensions where Kepler saw nothing. He knew time had to be somewhere as something and then covered it by denouncing the circle as nothing.

What then is it that Kepler saw as he formulated $a^3 = T^2 k$. At the normal flow of time it takes the electron a certain time to spin around the atom. The atom uses space a^3 and the atom is a certain length **k** that forces the distance the electron has to travel in one cycle period T^2. The atom a^3 connects the electrons travel **k** to gravity T^2. The relevance **k** produces to support a^3 is to point T^2 to two positions the electron will be in the duration of one specific time. The electron travel will be cyclic and periodic in relation to the space the atom holds. The space stands related to the gravity with which the Earth confines the space of the atom to the space and speed with which the liquid heat confines the atom space.

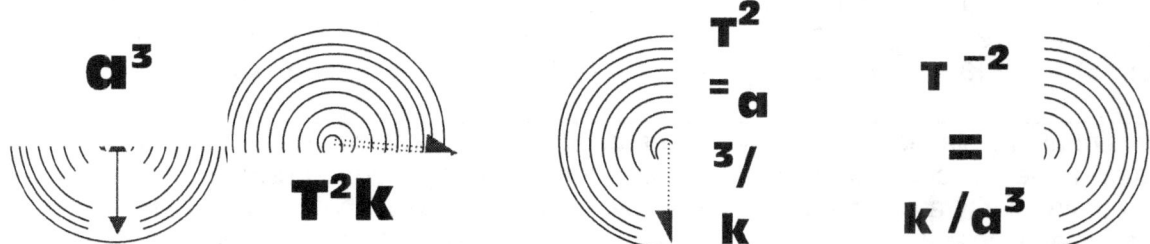

The Universe divides between space that was and will be where time is and time where it just released space is where it is accepting space. The one side is while the flanks are in motion of releasing or accepting the position space has.

The time frozen on paper in a single t is effective in remembering the viewer of an event but that is not the event in the present any longer. That was how the event occurred during the time from where the camera shutter opened T_1 to where the camera shutter closed T^2 and the time frame T^2 was then during the open period of the camera shutter. But afterward it represented **t** when looking at the picture and the looking of the picture became an event during a specific T^2 that went from where one is taking the first look to where one is looking away from the paper carrying the first dimensional image of an event gone by and that is at that stage a representation of **t** in another milieu of $a^3 = T^2 k$.

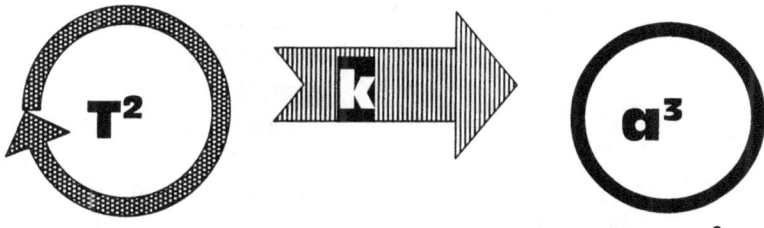

The **t** in the single is when mathematically presented as only **t** indicating a mathematical single flat dimensional view of time and is then correctly applied because it represents a reminder of a four dimensional event $a^3 = T^2 k$ that went single dimensional because the moment in the fourth dimension was then frozen in a single dimension on paper while the fourth dimension $a^3 = T^2 k$ soldiered on and time will always be representing T^2 as Kepler stated in the square allocated to space having a cube $a^3 = T^2 k$ at a time even before gravity got a name.

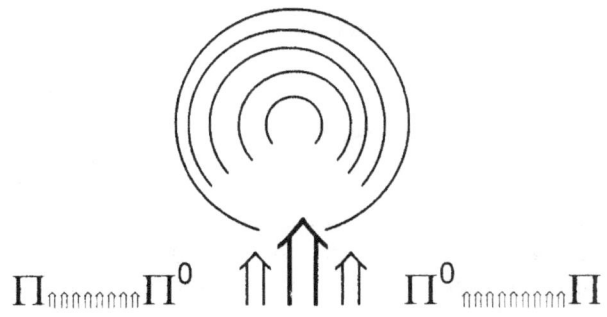

Planet	Period T years	T^2	Distance k	Space a^3	Ratio
Mercury	0.241	0.058	0.39	0.059	0.983
Venus	0.615	0.378	0.728	0.381	0.992
Earth	1.000	1.000	1.000	1.000	1.000
Mars	1.881	3.54	1.524	3.54	1.000
Jupiter	11.86	140.66	5.20	140.6	1.000
Saturn	29.46	867.9	9.54	868.25	0.999
Uranus	84.008	7069	19.19	7067	1.000
Neptune	164.8	27159	30.07	27189	0.999
Pluto	248.4	61703	39.46	61443	1.004

At the first glance Kepler's formula seems to be numbers and positions applying between the sun and specific but different planets in the solar system.

With singularity placed in infinity within the centre of every rotating object every atom and its relation to its surroundings including other atoms form space-time diverting from the point holding singularity as far as rotation goes because every object holds three relative positions in as far as where it was, where it is and where it will be in relation to singularity providing time. I elaborate on this else where.

Newton said a sphere is $a^3 = 4/3 \, \Pi \, r^3$

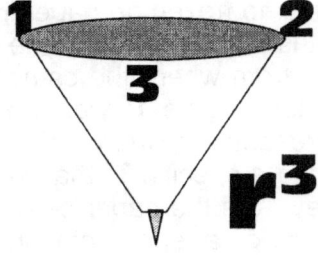

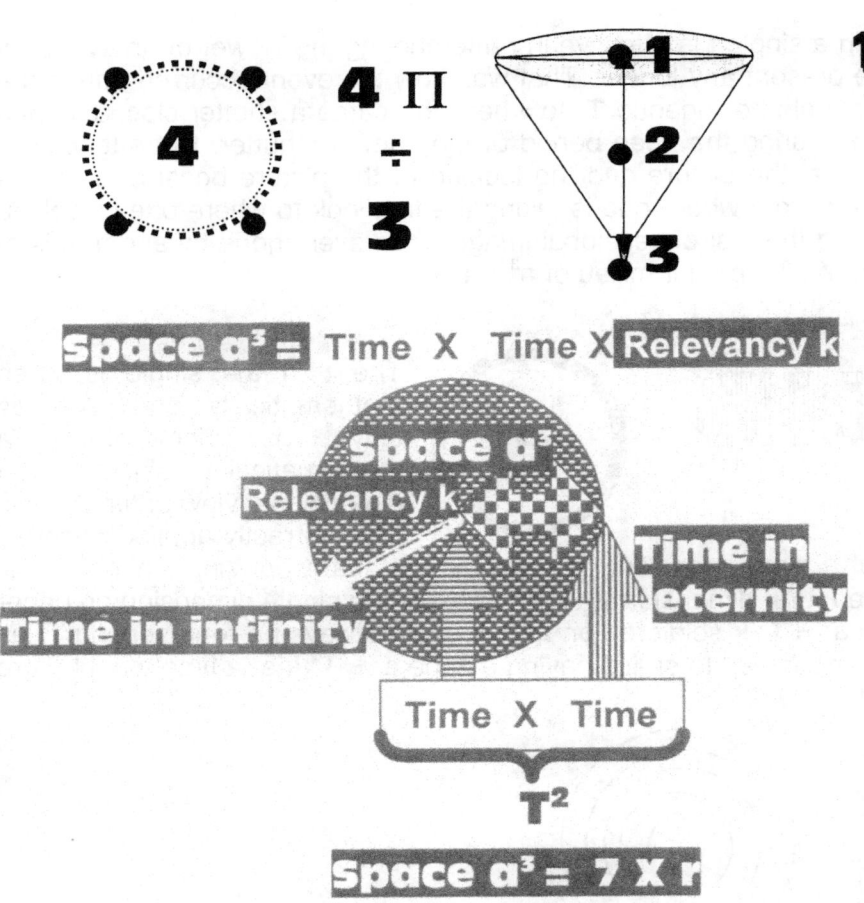

There is a Universe of difference between that which Newton saw applying as physics where mass becomes a factor when relevancy is removed. Newton can only apply when the independent relevancy of **k = 1** because the mass factor issues the motion to that of the Earth. Then when the motion the object has is directly the same as the Earth has, the object has mass and all Newton's predictions ring true as the mass removes the distance factor and **k=1** because $a^3 = T^2$ = one and the same as the Earth has. In that event the spin and the drive becomes the same and becomes not zero as Newton said but one as Kepler stated. Then all that which Newton said does become true but it is not equal to what Kepler said when Kepler said $a^3 = T^2 k$.

In $4 \pi^2 a^3 / T^2 = G (m + m_p)$

$$a^3 = T^2 k$$

$a^3 / k = T^2$ but at the same margin is

$$k / a^3 = 1 / T^2$$

$k = a^3 / T^2$ singularity

$$a^3 / T^2 = G (m + m_p) / 4 \pi^2$$

and $a^3 / T^2 = k$

then $k = G (m + m_p) / 4 \pi^2$

But I showed that $k = a^3 / T^2$ and Newton's claim is that $a^3 / T^2 = G (m + m_p) / 4 \pi^2$

The only definite place one will locate zero is in between the starting point of the lines going in opposing directions in the position the lines hold before there was the least of directions applied, but that is only because there is no such a position, not because any line is coming from there. The two lines are still one holding the opportunity of parting as an option but have not yet parted and therefore are on the very precise same spot. The line coming from there is already there because it already has the choice of going in any and all opposing directions and when it starts running it will place filled space in that location because the space was already filled with a line starting and not with a line not there at all. When reversing a line we might find a better idea of what is in place and where it is in place.

In the action of the inseparable drawing closer and moving closer gravity finds the dual value of linear and circular gravity. There is no separation of the two factors acting as one but both have different applications and values in the unit. This is the result of singularity having three parts acting as one but giving three distinctions in application. Gravity is as much part of dismissing space as it is about making contact with space in time.

But since the connection comes about as a circle, the connecting points will relate to Π as the value. Due to the spinning nature of such a point with all surrounding the point will be alternating direction favouring change every second and in that the value of such a point can only be Π because of its constant changing. Using r would specifically oppose another r from every angle because the use of r will bring about a static relation to the previous and following instant and therefore it will cancel the constant spin flow. The new direction pointing to a new location in relation to the previous point will oppose the previous point it had in relation to direction considering the centre point.

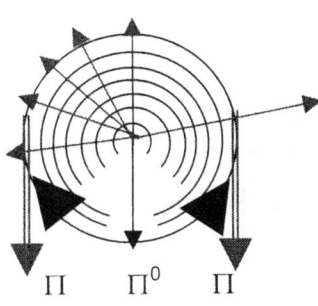

Π Π^0 Π

Pinpoint positioning of singularity Π^0 with Π positioning space to either side forming the border set by singularity
The new direction pointing to a new location in relation to the previous point will oppose the previous point it had in relation to direction considering the centre point.

The motion of a liquid confirms a centre and the confirming of the centre provides a flow of space-time in either direction, which produces the gravity. Without establishing and activating such a centre, the centre is not active. It is there but it is inactive in being present.

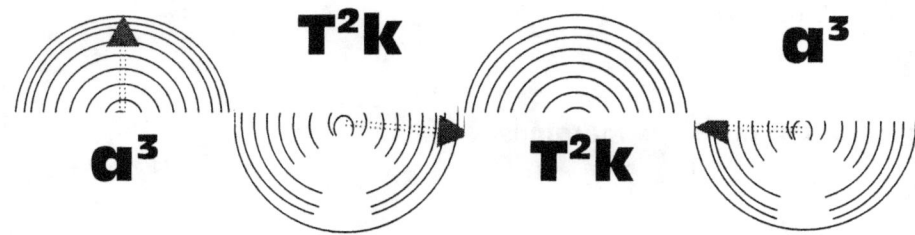

If the one side is a^3 then the other side is T^2k because the one side opposes the other side by providing control a^3 to the motion T^2k on the other side.

By taking k into a negative the space will reduce the time because the space cannot sustain the demand of space growth.

$$k^0 = a^3 / T^2k$$
$$1/ k^0 = T^2 k / a^3$$
$$a^3 / k = T^2$$

$$\Pi^0 = \Pi^3 / \Pi^2 \Pi$$
$$1/ \Pi^0 = \Pi^2 \Pi / \Pi^3$$
$$\Pi^3 / \Pi = \Pi^2$$

In all my other work, I make exclusively use of the value of singularity Π since it makes a lot more sense, but when I use the value of singularity, which is Π then no one seems to have a remote idea to which I am referring.

$k^0 = a^3 / T^2k$ forms

$k^0 / k = T^2 / a^3$ that becomes

$k^{0-1} / a^3 = a^3 / T^2 / a^3$

$k / a^3 = 1 / T^2$

The replacing of the symbols Kepler used with the value of singularity the mathematic equation comes into practise.

$\Pi^0 = \Pi^3 / \Pi^2 \Pi$

$\Pi^0 / \Pi = \Pi^2 / \Pi^3$

$\Pi / \Pi^3 = 1 / \Pi^2$

However, the equation looks far more sensibility when using the value of singularity

$k^0 = a^3 / T^2k$ forms

$1/ k^0 = T^2 k / a^3$

$1/ (k^0 k) = T^2 k / (a^3 k)$

$1/ k = T^2 / a^3$

Expressing the equation by using the value singularity has instead of the symbols Kepler designated to the formula he introduced it makes far better sense expressed mathematically

$\Pi^0 = \Pi^3 / \Pi^2 \Pi$

$1/ \Pi^0 = \Pi^2 \Pi / \Pi^3$

$1/(\Pi^0 \Pi) = \Pi^2 \Pi/(\Pi^3 \Pi)$

$1/ \Pi = \Pi^2 / \Pi^3$

The same motion is contradicting the motion it was or will be and because of that the contracting part is going to be expanding while the expanding part is going to contract. The one will accept while the other will release but the cycle can never be broken. Still, in the end it is the same motion and the motion never interrupts the flow of time.

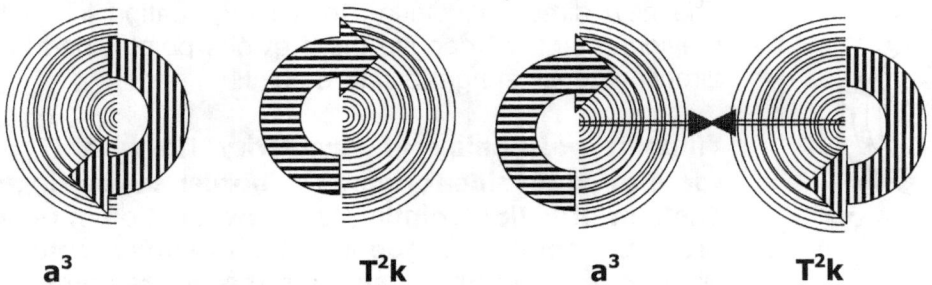

a^3 T^2k a^3 T^2k

However, keeping as one ($\Pi^0 = 1$) we keep the Universe in the first dimension. This point, which I now am referring to, is the point where Π is the fully appreciated value while the diameter D still remains a dimensional factor of one. This is the dawn of the second dimension where space was there but space was sparsely shared because the motion at that point was so excessive it allowed space no place to be but to find a factor in time. It was when $\Pi^0 = 1$ to become Π for the very first time. Please also keep in mind that my referring to this event in the past tense also places the event in the present since the event has no where to be but to also be part of the present.

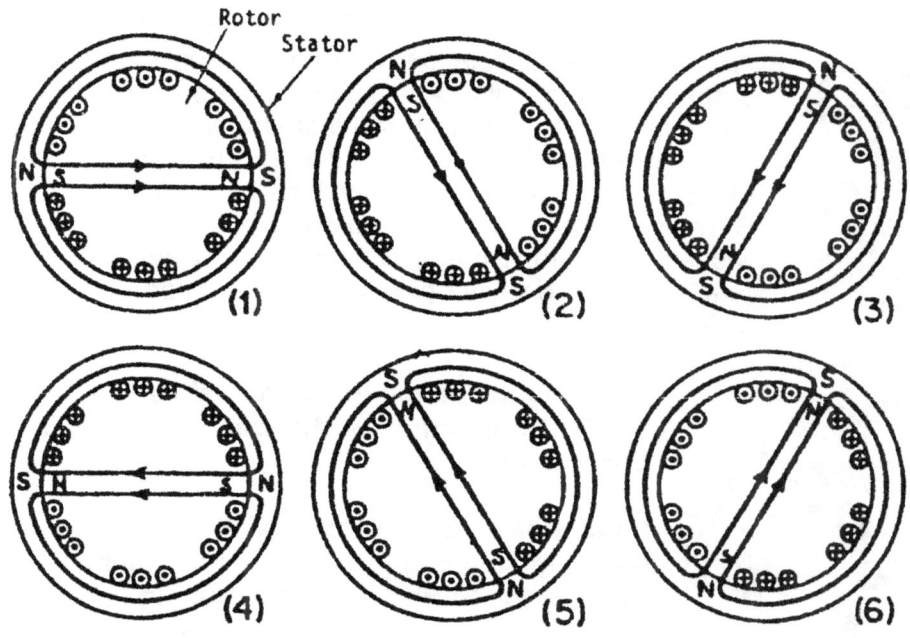

Generating electricity is using the very same cosmic principle as charging gravity because electricity is just intensifying gravity to a point we will find the intensity of gravity as it is within the Earth somewhere near the core. The moment of the cosmic birth during the arrival of the Big Bang time was at a point where the electricity or gravity was eternal in voltage, amps and resistance and because it was eternal intensity it brought material from infinity. Electricity is gravity and gravity is charged by the same formula as we charge electricity. It is the rotating of Iron through a

field induced with copper where copper reduces space within that field placing space as an infinite value in relation to time and this conducts the displacement of space in time.

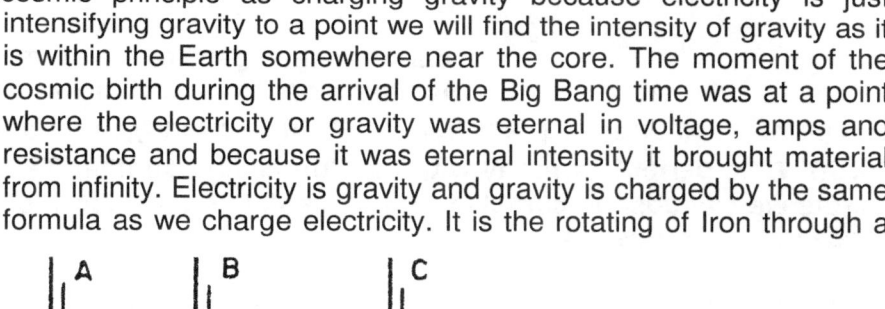

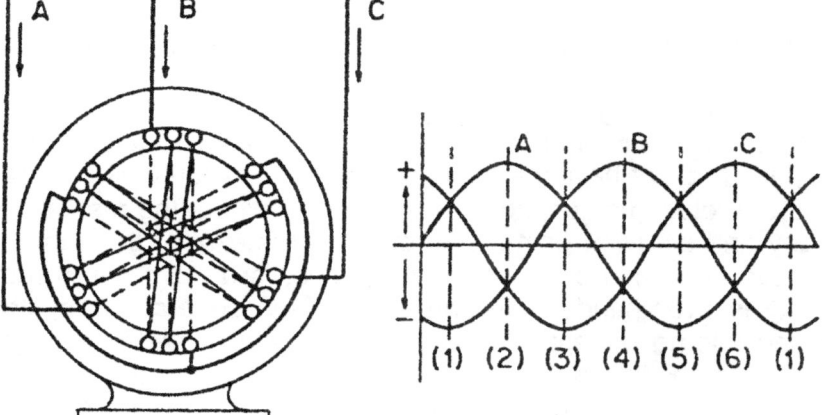

In the sketch above, it is exactly the principle applying when electricity is generated. The motion being accelerated to C^2, reduces the space to C^3 which is the photon and the relevancy becomes C. This pushes the space-time within the flow of electricity to what it was during the GUT or the period referred to as the grand unified theory. It was a time when space was the electron, motion was the neutron and dimensional implication was the reserved domain of the proton. In the context of dimensions one finds coming from the centre Π^0, an established eternal flanking of

Π to six positions since Π^0 forms the centre to the six sides and all six sides not having a diameter yet must apply Π to indicate specific value. In the very centre, which I am referring to, rotation must end or start depending from what vantage point the relevance is placed.

The very centre form an eternal divide that will not allow what is on the one side to present

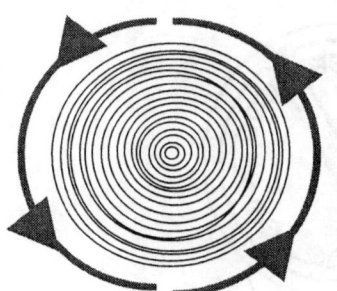

an influence on the other side. It divides spin. It divides direction of spin. It divides all rotation from the outside that one may detect and such divide is there because at one point spin will run to the left coming from the right and just immediately next to that point must run a direction from left to right. It cuts without contributing or participating in movement. It divides without any favour.

By expressing a wish to accomplish time travel, such a person wishes to accomplish that material must collide with itself a mean feat if ever there was one. Such a person wishes to have one side collide with the other side as he stops and reverses time while the rest of time is motoring on.

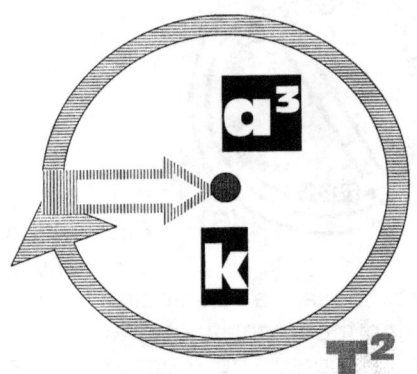

a³ forms the space the atom claims while travelling in the Earth spinning all the way and travelling with the Earth around the Sun

k positions the electrons travel in the space relating to the space the atom holds while travelling with the Earth in the Earth around the Sun in relation with a specific position k will indicate in relation to the Sun .

T² is the time it takes the electron to be relevant to the position the Earth places on T² while the Earth captures the space of the atom by providing the space for the atom to be within while the Earth travels from one point T₁ to another point holding T₂ in frequency to the atoms T² relating to the Earths T² in perfect harmony with the Sun having another T² relevant to all the other factors we call cosmic particles. Big or small it is only about cosmic particles holding space in time in relevancy.

Looking at the effect of gravity it shows the precise quality of no distinctive point, as gravity never seems to end at a point but flows all over affecting all that holds a position in its sphere of influence. The gravity coming from China meets the gravity coming from America at no particular spot but intermingles without distinction. That is singularity not having a dimension of space and not having a dimension of time, or a radius connecting the rotating distance to Π. Every rotating object holds a centre from where the rest of the rotating direction will differ at any and all given points. Not one point is exactly the same, but in the very middle, the centre no one can draw, measure or see, is a point not in motion. The first condition for gravity is even-handedness through out the sphere holding the applying gravity and the second is to have most or the strongest gravity located where the space is least. That gravity then has a position in the very centre of the sphere and from that centre the gravity produces all the edges or borders that the sphere consists of. In the case of the sphere, this factor makes the sphere much more dominating than any other form does. From the centre point controlling all sides is gravity and with gravity applying control, the sphere has seven sides to the square in any other possible form having at least six sides.

The cube can come in whatever form there may be but the sphere adheres to precise measure and behind this principle is all that forms the Universe. The cube has six sides connected loosely and can change form just by changing the relevancy between one side (or more) in relation to the distance brought about by the other sides. The sphere being a complex circle, stands related where the sides has to apply precise measure in equality. This becomes a law because in the precise middle one will find the strongest gravity as that gravity holds the object in form and true to form. If there is even gravity spread in all directions, the form must be a sphere and the sphere insists on seven points relating to sides or borders.

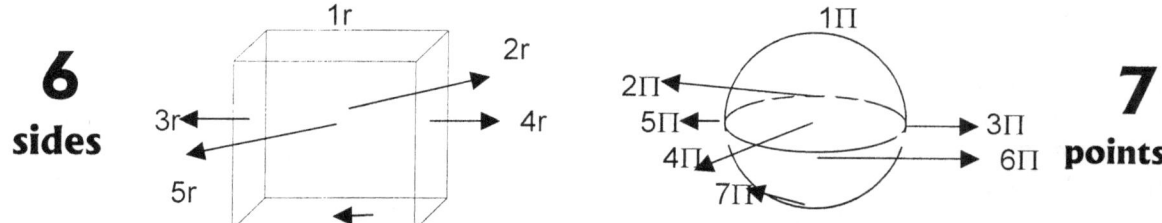

6 sides

7 points

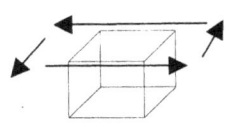

In the sphere there is no radius but only the extending of Π from the centre Π in six opposing directions relating to one another by the square but remaining Π because of the unity the matter holds in relating to space. It is not possible to draw a precise line that would form a precise ring and not

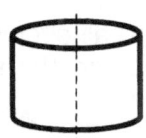

cut some atoms in parts. Because there will always be an atom disallowing the precise positioning of the circle the circle continues on a solid basis holding Π as a positional reference and not r. In every sphere there then are the seven Π relating in precise dimensional and positional equality forming equilibrium to the centre Π as well as to one another by 90^0 and 180^0 implicating the dimensional positioning. Therefore the sphere holds $_7$ $^{\Pi}$ and the cube holds $6 \times r^2$. Where space comes into contact with the sphere the cube loses one of the six dimensions it has to the more dominating seven dimension of the sphere whereby the seven dimension in equilibrium will dominate the six dimensions loosely connected by r bringing about that the cube then has 5 sides to the seven of the cube. Because the space surrounding the sphere takes on the shape of the sphere and not the other way round where the sphere resolves in accepting the form of the cube, one may presume the form of the sphere is the most dominant of the two choices.

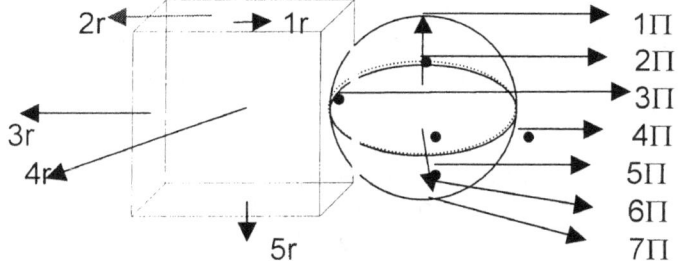

5 sides in the cube vs. **7** sides in the sphere

The sphere is a multitude of innumerable circles that forms one unit with all the innumerable circles that compile a sphere all put together. The circle is a constellation of Π where every Π flow from one into another and such flowing varies the number arrangement of r. In order to measure the surface of a sphere, the radius carries the torch by going square. It is the radius, which is just another line that comes from the outside and runs all the way inside to bring the value of the line into the circle.

In a circle, there is a radius that initiates the circle. The calculation of such a circle is $\Pi \times r^2$.

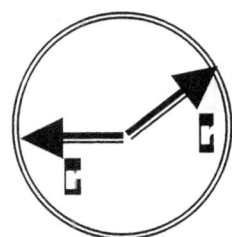

Only the circle has a point on the outer edge of the form that is at a precise distance from the centre and every point in the form measures a precise and equal distance from the outer rim. In a circle, there is a radius that initiates the circle. The calculation of such a circle is $\Pi \times r^2$.

The radius r runs from the circle outwards, from a circle centre point towards Π, the value of the circle. In the centre of the circle, there is a point where the radius starts. It runs outwards from that point in all directions towards the circle Π. Technically, there then has to be a point where r is infinite and not zero, an absolute infinite. However, the circle therefore remains Π. The circle does not disappear; it remains there for all to see. It is only the radius that almost disappears into the infinite, but it does never become zero! $\dfrac{\Pi r^2}{r^2} = \Pi$. If one removes the radius from the circle, the circle remains, only holding the value of Π. By removing the value of r, Π becomes singularity with no place to be. Singularity is the place where there is no space to be in place. However, Π remains because once r receives the slightest of space Π will find space. Then the circle will grow to Πr^2 and r would determine the space. Without space, there is no r but there is a circle with the value of Π. Singularity is in every single rotating object, be it the proton or the Universe. This situation is part of any and all circles and is therefore part of any and all spheres. The line will end at a point where the line starts going in the opposing direction and that value is indicated by the use of $r^0 = 1^0 = 1^1$

To that end the shortest possible line (hypothetically) must be so short it must have **an initial and ultimate point sharing the same spot.** Any theoretical line being the shortest possible line cannot have the line holding the initial starting point at point zero and advance from there. If it used zero as a start, the zero part would not count, because the line will only start at a point past zero where the line then will start. Zero ultimately means not existing and then that point, as a start does not exist. At one point the reducing attempt of the line would start making the use of mathematics seem silly. The reducing would seem tedious and leading nowhere. But as sturdy as mathematicians can be, they would carry on (or so I am made to believe…). Then, when the man doing the calculations gets carried off in a straight jacket, while the man is making funny noises, when he is totally cracked mentally, the calculations can still go on and on and on and…and that is where sanity prevails and someone says "drop the affair". It is at that point I would have loved to see Einstein carry on counting stars in so many galactica in his attempt to determine the critical density joke. That is not how we locate infinity. That is where man's brain gets blistered but infinity is still far off. As any one can see, the Universe is far beyond some insignificant and senseless formulae invented to impress Academics while others are kept busy and free from boredom, but in the real Universe the attempt is not worth the thought it takes to disregard the attempt.

When the line **has a beginning and an end at the very same spot** and it wishes to extend the position as to further the possibility it has, which direction should it favour? Extending the line in any one direction will favour one direction without any clear reason not extending in other directions. The only mathematically sensible option about extending will be in all directions equally in order to give a meaningful non-bias flow of mathematical equilibrium. That is where one would have to go look for the beginning of the Universe. The Universe is about lines connecting but where does that which does the connecting end. The Universe used form to this point in development, but then at some point the line came and established the presence it still has. The first form was moving from $\Pi^0 \Rightarrow \Pi$. Again I wish to press the issue, at that stage form was in use, and not mathematics. The Universe was just simply too big to measure. If radius did apply, one could use r and r^2 but since only Π was in use there was no radius to r^2, be used. There is forever one circle leading to the next circle, which is followed by the following circles. Where the light does not reflect the image that is there, we will still find a concentration of circles leading another and another and another. The end is eventually endless.

The spot forms a full circle, but the line running through the circle is forever present because that is the future radius of the circle that will one day develop the circle, which is equal to the present diameter. The fact of the presence of such a possible line in such a possible circle dividing the possible circle into two parts makes the centre line equal to the half circle. The line forms the half circle but not only that the line presents the half circle as much as the line is the half circle. The line then is 180^0 and the half circle is 180^0 because in singularity the two factors are the same. The same value is of course $\Pi^0 = 1$.

Everything at the time that was outside singularity and was in form at that time was equal. Think how big they were. They filled larger parts of the Universe than our brains can cover by thought. They formed the holding tanks that still hold us in the massive Universe. They were at the time too small to have size, but since then they grew into structures that are too big to have size. Those dots still are bigger than mathematics can apply because there still is no measure quantifiable mathematics can reach. Just because they compare with what we seem to preserve as small in cosmic relation they are too big and too large for us to comprehend. Even if they were immeasurably many, they filled an immeasurable Universe in the same way they still fill the immeasurable Universe and we are so small we and our surroundings are measurable and quantifiable. They were so enormous there were no relevancies applying to compensate for distinguishing. Distinguishing only followed later when size started to matter. That meant the Coanda effect was in place without the Roche limit or the Titius Bode law. It was the start of the relevancy principle from which the atom later came and which is the result of the Kepler expression.

It eventually gets so small we humans can fit into it… remember, we are seeing the reverse of the truth. It was never so big that it contained nothing because all that and we came afterwards being smaller than the dot that filled the dot.

With no line possible there had to be another dot that formed since the Universe has many dots that formed lines. But let us not to get confused and lost in the range of possible diversions but let us stick to two dots. One dot was next to the dot next to the dot, but as I said, we stick to one dot next to the second dot Π is the first step with which gravity began. That leaves us with a huge problem in as much as when r = 0 then $r^0 = 0$ and 0 dividing any value will leave 0 as the answer. If the particles were inseparable at the start, it must bring about that gravity would not be forming since the distance will not permit any dividing. By allowing the distance separating the particles to be zero, the particles melt into a unit. Again this is Mathematics and not my incoherency as some Academics dismissed my work. Let me run through the argument one more time, because I have been insulted by Academics in the past telling me I am bending mathematic rules with my applying double values to try and produce some argument. The two particles formed by an inseparable unit separated by a sharing of a spot. We know that at least two spots formed because there are many more than just two that remained to become part of the visual Universe. Let us name the spots because that is what humans do best if they do not know what to do with what they have to do. Let us call the on dot and the other one spot next to dot we then call dot two. Between dot and dot two there were nothing because dot and dot two were inseparable. By they're being inseparable we would naturally be inclined to think that the separation value should be nothing or at least zero. But putting zero in that place is a mathematical excluding procedure leaving future mathematics excluded. With m multiplying m_2 and then dividing ÷ r with zero (r=0) such a procedure will leave the lot at zero and with that nothing is going nowhere. That means although we think the space between the two parts are nothing the non-existing space has to be at least one to be a future factor.

At first $\Pi^0 = \Pi$. Then after a while $\Pi^0 = \Pi^3 / \Pi^2\Pi$ and gravity comes about forming space-time by motion of form. Being the sphere that formed the 7 holding relevance in the form of the sphere took shape. But the Universe is a layer in dimension forming the next layer in dimension forming the following layer in dimension. The Universe was $\Pi = \Pi^3 / \Pi^2$, which is taken from Kepler's formula he received from the cosmos as $k = a^3 / T^2$. Where there is a sphere involved there is a natural tendency to grow by developing the sphere.

Every part of the argument is sound but was never yet used. I repeat once more if my argument reflects on inconsistencies those inconsistencies are not about my work. In order to disprove my argument replace Mass one and Mass two with any number possible, then divide such a number with the square being zero. If there was no space, then the value of the particles had to be one. If there was no space between the particles, the particles then had to form a unit. But if there is a mathematical possibility of reducing a line to the single dimension then there had to be a factor representing r as a factor of one. Take $(M_1 \times m_2) / r^2$ and substitute any of the factors with zero and the result coming about has to be zero. The factors in the equation have to have any and all the elements at a value of at least one. Only if r was a factor of one can gravity bring about any mathematical equation developing from this argument. That means the mass on both sides must have a factor of one being a limit, which does not allow such further reduction of r and any further reducing of r beyond the limit, will not be tolerated. Only if $r = 1$ then r^2 can be 1 and mass can be apart. Like it or not but believing in the Big Bang must also bring about the accepting that the cosmos moved apart somewhat. The fact that r brought increase in the space separating the mass produces a problem that was solved already. About a century and a half ago Roche found just such a limit. Once again I were confronted by zero becoming growth. There is a huge hole that needs filling when bringing into a relation any forming of an alliance between a cosmos coming from nothing and filling with nothing and a cosmos growing spontaneously through balance shifting prominence. Mathematically the fact of applying nothing as a vale applying in the cosmos is not a strong and convincing argument. The minute one brings in zero as a multiplying factor forming a definite value working into the calculations of the cosmos, growth disappears. If growth was not a factor, the zero factors could be involved with some form of maintaining stability and where then further growth will accept the responsibility of zero.

Any point will be opposing itself within the **rotating of 180°** where it **then changes every aspect** of its **previous flowing** characteristics it had or **will once again have** in 360⁰ from there. While in rotation from the view point of a bystander, it all may seem static and never changing but to the object in spin every next instant in time will be diverting from every aspect it had every second passing, and the direction it held in relation to the direction it held the previous mille, mille second will totally be incompatible with the direction it holds the very next mille, mille second of rotation. This is why we can use degrees measuring the circle by (6^2) (forming the square relating to matter through singularity) $\times 10$ (square if space) = 360⁰ however it is always in motion. That proves no point can be static or constant, though it may seem that way to outsiders. Although matter is matter, matter can also be anti-matter and moreover form its own anti-matter at the same time. This degeneration of structure is very likely to occur with overheating. Revaluing Π to Π^2 will bring about a new contact point where Π meets **r** forming another relation in Π^2. **Time is** the **changes in relation** where Π **contacts a different r** not withstanding the many r points there may form because **every r constitutes a different value** to the universe through other ratios and relevancies brought about **by heat and light. Time is the duration it takes** Π **to rotate between any two given points of r** and therefore must always amount to **a square (T^2)** moving from point to point through the **cube of space (a^3)** in that **duration of time (k)**. With that it proves **Kepler's a^3 (space) $= T^2$ k (time in the instant of motion)** but motion must continue through a specific value in space where the space-time is maintaining relevant equilibriums throughout singularity connecting.

In the circle using $r^2\Pi$ the r has to have distinctive qualities placing it as a factor apart from Π. Where the growth shows no separate distinction but a continuous flow from the precise centre to the precise edge the flow would become in relation with Π depicting the circle and

Π replacing r as reference to any point on the circle. By using r distinction in the circle is possible but by using Π there is no distinction possible.

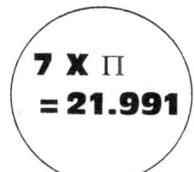

The sphere is 7 X Π = but from the other view the sphere is singularity relating to ten. That starts the Titius Bode principal which in relation to the Roche limit at the limit of Π use $\Pi^2 / 4$ to form gravity Π^2. However this is a little more involved, as it seems because the one sprouted because of the

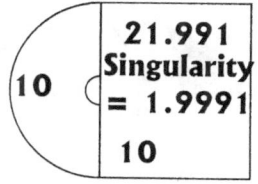

consequence of the other bringing the one into the Universe as a relative that will sprout to bring about gravity.

However if the attachment was $7 + 10 + \Pi^2 / 2 = 21.93 / 7 = \Pi$ is the circle that serves as an attachment meant being a sphere and holding as well as sharing singularity. This is what Newton saw, but this is not gravity in the cosmic sense.

7 A sphere being formed by the six ends crossing as it incorporates the centre singularity

+ 10 another sphere with an identifiable motion keeping the independent singularity apart but within the relevancy of the unit formed.
+ $\Pi^2 / 2$ Singularity by gravity shared by two in one unit.
= 21.93 / 7 still holding the unit to form where the overall containing form will be a cosmic sphere.

The sphere holds many dimensions relating to seven but also by the square of space which is 5+5 = 10. This brings about gravity generated by means of the Titius Bode law and is principle proof of this statement because it indicates an infinite number of numerical positions influenced by quarterly divided sectors around a point holding singularity.

Newton said the rotation delivers no work and therefore the effort of the rotation results in a zero. Firstly it bring us back to the zero idea where with all the reasoning in the world and all the leniency I allow I cannot find zero as a value being part of mathematics. Let's move back to the circle to try and find the zero Newton saw in the rotation.

Π X r^2 = CIRCLE

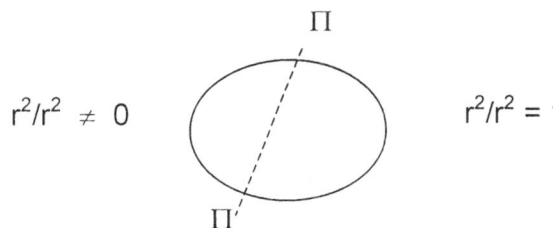

$r^2/r^2 \neq 0$ $r^2/r^2 = 1$

means a factor of 1, not zero.

If you remove r it then is $\Pi \times r^2 / r^2$ = CIRCLE.

You cannot then say $r^2/r^2 = 0$ and therefore Π x 0 = 0. That is nonsense. $\Pi r^2/r^2$ will always be Π x 1, and that is the eternal circle. When looking at any rotating object, there has to be a point of no rotation and no rotation means "no rotation", not no existence. No rotation

Not only does atomic individual singularity maintain self- preservation, but in doing that it also sustains a governing singularity holding structural composition and form within a cluster of matter for example a star. As there is between stars, so there are in the same manner, a mutual or bonding singularity between atoms in stars, which we see as fusion.

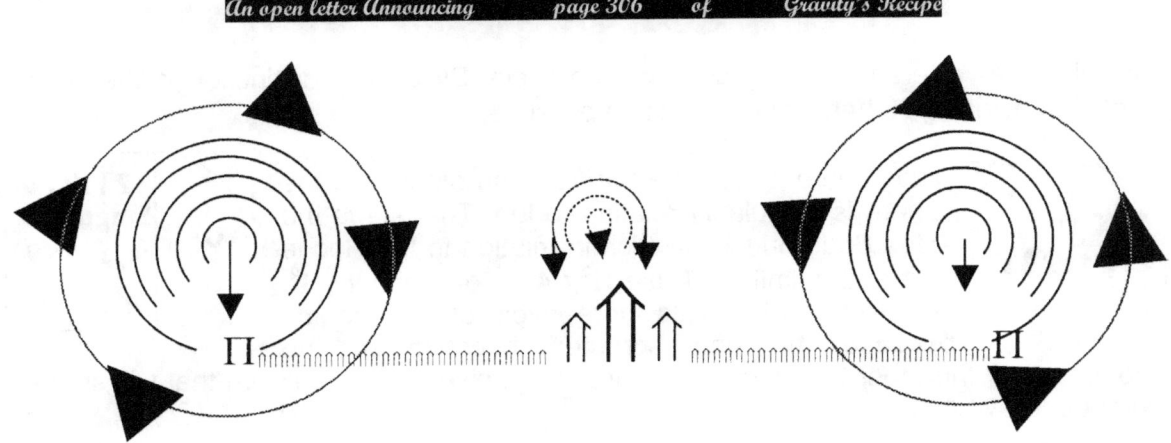

Any object in rotation will have a middle point, a very specific centre point that does not spin. That point once again hypothetical but none the less must be standing still because every line running from that point in opposing directions are also in opposing directional spin to each other

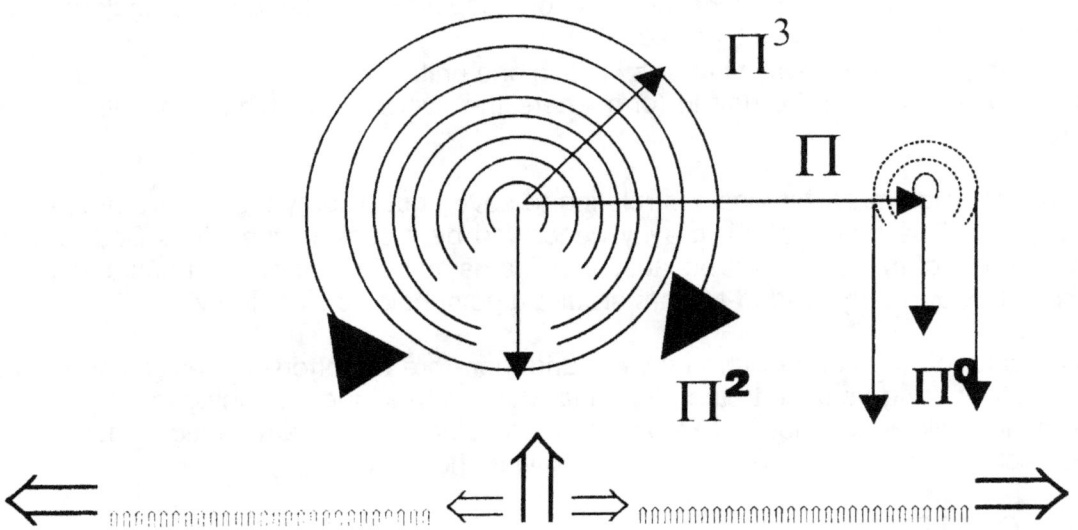

As the stop starts to spin, the motion establishes the centre line, which activates singularity, which activates space-time that activates gravity at a specific relevancy. Where we locate singularity there is not nothing because gravity cannot come from nothing because only nothing comes from nothing.

After all it is gravity that keeps the top as it is spinning in an upright position while it is spinning because it is gravity that stabilises the cosmos. Moreover, what is actually in progress from the top spinning is the Coanda principle activating gravity and that happens in accordance with Kepler's formula.

This means that in the cube at the point of contact between the cube and the sphere the cube experiences such a contact point as if the "bottom falls out" of the cube and without a "bottom" to support objects, they fall to the sphere as objects does fall to the Earth. Remember that a body "floats" in space, but at one specific point it starts to "fall" to the Earth. That is gravity and it is a dimension change much more than any force. I shall explain this last remark later on. That too is the Lagrangian system with five cosmic structures holding relevancy to the centre structure where the centre structure stands in for seven positions diverting from singularity and the orbiting structures standing in for five positions in space.

From such a point every other point will be opposing any other point not pointing in the direction to which the first point is pointing, whereby it extends the direction it holds. No matter what the point is or where the point leads, such a point holding a specific direction will be unique in the direction it is rotating because at that or any other specific point wherever, it will be directing not in the direction it spins but in the direction flowing from the centre point outwards.

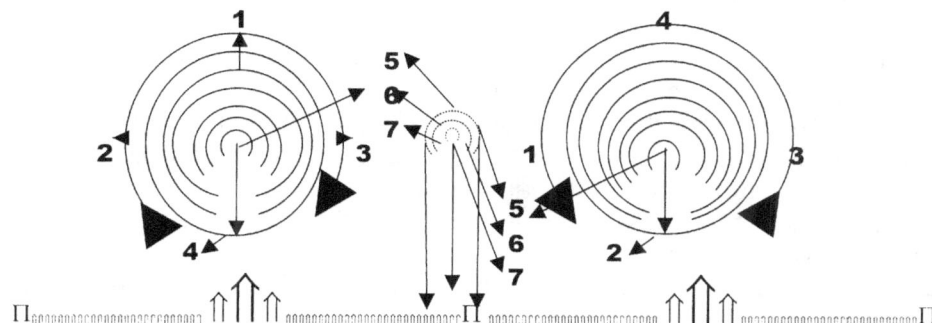

Any point will be it opposing itself within the rotating of 180° changing every aspect of its previous flowing characteristics it previously had or will once again have in 180° from there. While in rotation from the point of an outside observer all may seem static and never changing but to the object in spin every next second will be a diverting from every aspect it was in every second passing, and the direction it held in relation to the direction it held the previous mille, mille second will totally be incompatible with the direction it holds the very next mille, mille second of rotation. That proves no point can be static or constant, all though it may seem that way to outsiders.

In the very centre of the sphere the form of the sphere dictates that the shape will relinquish space as the line runs from the outside towards the very centre. With this natural state of affairs the sphere is naturally inclined to dismiss all space that it can form in the form as the sphere holds space inside and the form will finally be without dimension. The line shrinking by reducing actually takes place in every sphere as the diameter reduces to the centre. In the centre where the radius line goes single, the form relinquishes the three dimensional form it has inside. Being without dimension in the very centre means that at a point in the extreme centre of all spheres there is a point that holds singularity because this point with no space has a mathematical position although it is invisible since there are no sides to such a point to give that point any dimensions. The shape of the sphere is calculated by using the formula $4\Pi (r^3) / 3$. By reducing r to a point where r is r^0, singularity steps in because only the form remains as Π. Going even further, we find that there then comes a point where Π goes singular Π^0. At that point absolute singularity is present but so is absolute gravity present at that point. When holding the strength of the shape of the sphere in mind as well as taking into account that all cosmic objects of importance are in the form of planets or stars and they are all in the form of a sphere, we therefore may contemplate that it is where gravity originates. We now only have to find the reason why gravity will hold a base in a space less ness as Einstein predicted. It is clear to be seen that gravity is in the centre of the sphere controlling from the centre everything that is outside the space less centre. We can reason with confidence that gravity is the strongest where space is the least. We can further reason that it is gravity that is holding the sphere in true form and since the sphere allows gravity the best working opportunity, gravity can form the sphere in as strong a shape and form as the sphere seems to have. From every point on the surface of the sphere is where that point connects with the other side of the surface of the sphere by a line that runs through the space less ness of such a centre of the sphere. Such a line also connects by an angle of 180° as well as 90° to six other lines running from top to bottom, right to left, and back to front, where all join and cross in the centre of the sphere. There are therefore six lines crossing and connecting by a centre from any given point on the surface of the sphere. Such points connect in total six surface points on each side of the sphere while they all support one another through the space less centre. In that absolute space less ness in the centre holding singularity we find gravity supporting and controlling all space within the sphere as well as space connected to the sphere. That is where gravity controls and guides the space, which falls in the parameters as well as under the influence of the form of the sphere. In the gravity centre space goes singular meaning space becomes space less or flat.

Also it is true that the entire form that is the sphere is controlled from a centre within the sphere. That centre holds the sphere in form and shape. Therefore the strong form is dictated from that space where there is no space and no form left. The natural inclining is in the form of the sphere. It is part of the roundness that the overall shape of the sphere represents and this structural strength is carrying down to the very centre. Because the circle is forever reducing, that reducing which is inherently part of the form of the sphere becomes a tool in distorting space in the sphere and is eventually removing all forms of space from within the centre of the sphere. The very centre ends up as having no space because of the reducing that continuous down to become the space less inner centre. The all roundness is the ingredient that forms the backbone of the absolute strength that the sphere has and that is the component that the sphere is so famous for. The form the sphere has allows the sphere to have a control that is coming from the centre deep inside the sphere where the space vanishes and being without space seems to keep the entire structure rigged. The strength of the sphere comes from the centre of the sphere, which is inherent of the shape. That is why the sphere has such strength in form and the fact that all connecting sides refer to a centre brings credence to the strength that the shape has. How does it work in its most basic analyses?

It is from the layout that the sphere uses as a natural form that we are able to locate singularity. In the case of the sphere the material naturally reduces by measure of the radius becoming smaller to a point where the radius is r^0. At that point the line that will form the radius has gone single dimensional r^0 and that is equal to 1^0, which is singularity.

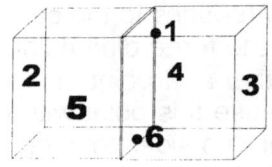

There is one more point in the sphere in the centre forming an addition in the sphere. That point holds gravity secure.

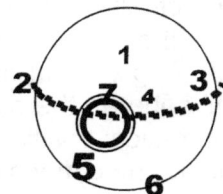

The cube has sides and the sides form a rather weak and flat surface that connects four corners. The flat surface produces a rather indifferent contact point with no special features on the surface. The corners connect to other sets of corners and those corners form a weak structure without any direct support coming from the other five sides. Without material to fill the body of the cube the cube has no direct connection between any of the sides other than corners connecting at the edges of the sides.

Taking the vantage from the point the sphere is holding from the centre out into space there are ten points connecting to the centre. In that are the dimensions of singularity connecting to space where five connect to space in the second dimension of singularity, and five connect in the third dimension of singularity. On the other hand, the cube does show a very different characteristic, which involves only six sides (at least) connected.

In the very centre of the sphere the form dictates that the shape will relinquish all grounds in space that it can hold and the form will finally be without dimension. Being without dimension means that at a point in the extreme centre of all spheres there is a point that holds singularity because this point with no space has a mathematical position although it is invisible since there are no sides to such a point to give that point any dimensions. When holding the strength of the shape of the sphere in mind as well as taking into account that all cosmic objects of importance are in the form of planets or stars and they are all using the form of a sphere, we therefore may contemplate that it is where gravity originates. We now only have to find the reason why gravity will hold a base in a space less ness as Einstein predicted. It is clear to be seen that gravity is in the centre of the sphere controlling from the centre everything that is outside the space less centre. We can reason with confidence that gravity is the strongest where space is the least. We can further reason that it is gravity that is holding the sphere in true form and since the sphere allows gravity the best working opportunity, gravity can form the sphere in as strong a shape and form as the sphere seems to have. From every point on the surface of the sphere is where that point connects with the other side of the surface of the sphere. All other possible points connect by a line that runs through the space less ness of such a centre of the sphere. Such a line also connects by an angle of 180^0 as well as 90^0 to six other lines running from top to bottom, right to left, and back to front, where all join and cross in the centre of the sphere. There are therefore always no less than six lines crossing and connecting by a centre from any given point on the surface of the sphere. Such points connect in total six surface points on each side of the sphere while they all support one another through the space less centre. In that absolute space less ness in the centre holding singularity we find gravity supporting and controlling all space within the sphere as well as space connected to the sphere. That is where gravity controls and guides the space, which falls in the parameters as well as under the influence of the form of the sphere. In the gravity centre space goes singular meaning space becomes space less or flat. That is where Einstein's Universe goes flat because that is where gravity is at its strongest. However my bringing up this statement brings me directly to the point where I get very confrontational about how the brilliant mathematicians treat those they suspect are less inclined to think.

By examining the form of the sphere, we find that there are 6 points on the surface of the sphere holding the form at a specific and equal distance from the centre. Lines run from the centre into space at $90°$ and $180°$ angles of each other from six opposing sides. There then are six lines at $90°$ and $180°$ connecting to the centre from six points on the outside edge of the sphere. As a result of the basic shape that a sphere has, there is a spot in the extreme inner centre of the sphere where the lines in $90°$ relevance cross each other and others connect by $180°$. There is also at that point a spot where all space relinquishes a position and only singularity 1^0 as form remains. At such a point we find the measure of the sphere being Πr^0 with $r^0 = 1^0$. That is where the line that represents the radius as a line disappears, as it becomes singularity r^0. After more reducing continue we get to such a point where we find only Π^0 left. At that extreme point is where space in all form disappears, as the circle providing the sphere the form the sphere has, removes all possible form by going into singularity $\Pi^0 = 1^0$.

Then in that area all form of any possible space disappears leaving only the dimensions of singularity 1^0. I cannot delve deeper into the argument. However, from such a point there run lines that connect to space on the outside where six points on the outside points connect to the space less point in the inside. In this book I take this argument much further but for now I leave the argument at that. Those lines carry the structural strength the sphere has. Contact with one point has support of six other points across the whole structure where the other six support every one of the six by singularity and the support runs through the entire sphere including the middle. Where there is no space, there must be singularity 1^0 just because the space filled with material removes zero and only material filled space is present. That means material fills the lot although in singularity 1^0. If zero was a factor where all space finally halted in zero as the value, then zero would be able to remove the space from the centre and such removing would continue to remove the space until all space was removed. It will finally abolish all space in the sphere and it would remove the sphere. Zero removes all possibilities of anything coming about. Since the sphere is there, a zero factor in the centre cannot be present. Only infinity can be a factor from where space may grow because infinity can extend and grow into and up to eternity.

The implication of this is that following the line down to the centre of the sphere we located the centre of the Universe. That is where gravity is. There is a lot more to that but be patient, we are getting there. In every centre we find a point, which is in truth not there but is the mainstay of all that is within the sphere. The mathematical value of such a point is $\Pi^0 r^0 = 1^0$ and 1^0 is singularity. That is the point where the Universe started and that is where the Universe will finally end. That is the Universe without space-time. That is $k^0 = a^3 / T^2 k$ which proves the Universe is without doubt a sphere...and we just located the centre of the Universe!

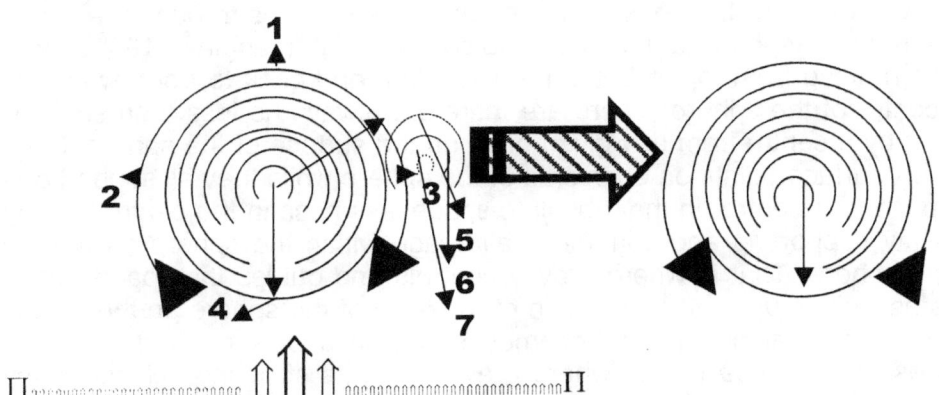

As one can see with the spinning top delivering the Coanda principle, every point overheating can spawn space-time by centralising singularity.

One can see from the top that singularity is established wherever spin occur. The motion generates a position of seven in relation to ten and singularity manifests as 1.9991 as is explained elsewhere. That means any point formed by the sphere spinning can and does start a centre in which no motion holds no space and of which motion surrounds such a point by forming space. Although everything at the time was in the form as a multiple circle, which results in a sphere, the sphere was not the only form present. This too has to do with singularity interpretations. We see a cube, as we know the cube but at first when form came about the cube were not yet a form.

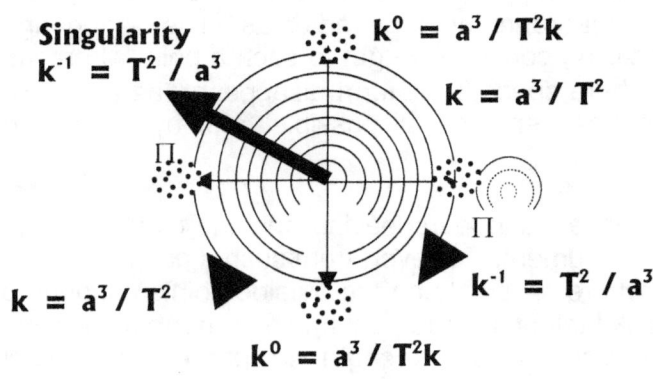

While the one sphere forms on this spot where the dominating sphere secures an edge the dot may be reserved as an edge marker to the dominating sphere. To the forming sphere in progress of emerging heat gathers at that point because the rotation is a result of duplicating and duplicating is the tendency of naturally growing in space-time $k = a^3 / T^2$. In order to find duplicating coming about there has to be heat in order to duplicate what will form heat. The duplicating process is a process of one factor going softer or less solid and therefore more dynamic than the other. To have singularity is to have gravity but to have gravity there has to be a point of motion and a point of sturdiness. The point of sturdiness may be in the centre of singularity, but then the solid must be motion. However even today it still applies: what moves, forms liquid in the presence of a solid and at that point singularity presents the solid therefore what we might think of as solid was the liquid because it moved around the solid. Where the one factor is duplicating the other factor is compressing $k^{-1} = T^2 / a^3$

The points duplicating is four moving around a centre by the square of gravity. The motion is the source of heating because the movement brought about by the heat. The heat growth therefore provides the action because the action is what energises the points to provide the motion. The motion is purely space-time duplicating and the duplicating is feeding heat to the centre from the four points overheating, thus the points shows expanding.

But also the duplication leads to the spawning of one point of singularity that provides the installing of the next centre for the next sphere.

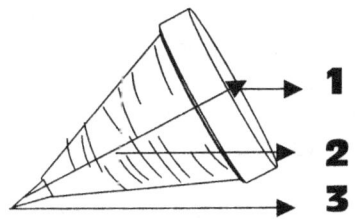

Because of the principal in which the Coanda works the motion will centralise a new sphere and by appointing six positions around the centre, three points will not move while four will move about the three points forming the centre line. The result is that the four points by duplication will reserve the point moving as the next point in singularity because of $k = a^3 / T^2$ singularity will be a natural result of the motion. Then that point will secure a position $k^{-1} = T^2 / a^3$ which will secure six points about such a centre. The centre will bring about four points spinning around three points holding a line in singularity. The line in singularity will stand in relevance to the contacting factor $k^{-1} = T^2 / a^3$ and the duplicating by expanding points will be four and serve the relevancy by contributing $k = a^3 / T^2$ as space-time only in form. From this the rest of the Universe bursts into the next phase of Creation.

The gravity is in relation to the spin, which is in relation to the four points spinning which are $\Pi^2 / 2$ and that is the Roche limit. It is the dividing of singularity sharing space-time just as we on Earth share singularity by division between the Earth and us others that are not part of the Earth. The total that forms from the point that spawns is seven plus five plus pi square in division of four totalling twenty one that stands related to the first seven and once again another sphere is formed.

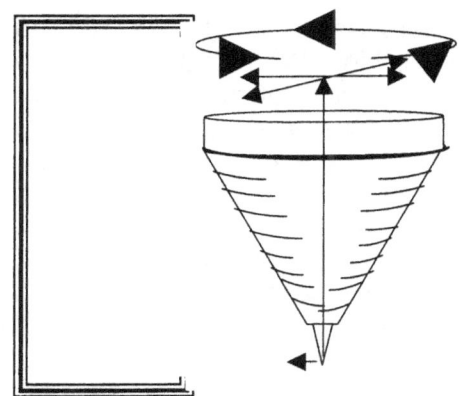

However this is an eternal relevancy that can never break. Any object in rotation will have a middle point, a very specific centre point that does not spin. That point once again hypothetical but none the less must be standing still because every line running from that point in opposing directions are also in an opposing directional spin to each other. Although the points had the same characteristics only seconds before, they oppose the characteristics it had just before and just after the very second in which they are and to which they relate by similar points also in rotation. Due to the spinning nature of such a point with all surrounding the point every varying second, the value of such a point can only be Π because of its constant changing.

The sphere has seven points. The cube without truly being a cube but is just in consideration of having a cube in form holds five points to singularity. In the centre runs singularity to the value of Π^0, which means that which surrounds Π^0, holds a position of Π^2.

The spinning sphere activates the seven points, which places gravity in relation to a centre. Outside the centre there are five sides by dimension. The sphere has seven points of which four is spinning. The four spinning stands related to the gravity of spin,

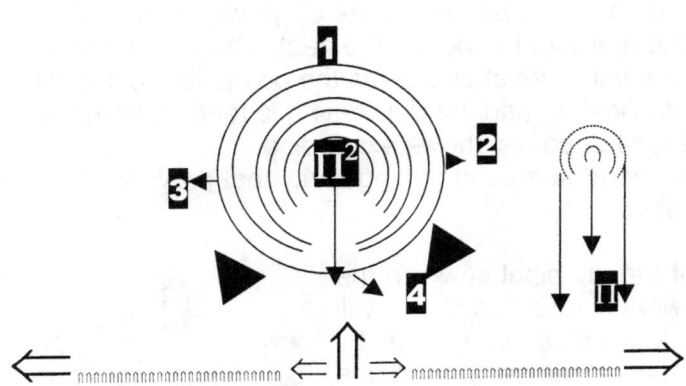

The extending of the motion also produces an ending point where the limit will either attach or dissolve the material next to the motion. Such a point liquefies what ever is in the location. That is the principle behind the Roche limit but also not only that; it is also the principle behind the Coanda effect. All that is closer than what the Roche limiting point at $\Pi^2/4$ will allow, becomes dissolved as liquid and removed into the unit forming the sphere.

Using r would specifically oppose another r from every angle. From such a point every other point will be opposing any other point not pointing in the direction in which the first point is pointing, whereby it extends the direction it holds. No matter what the point is or where the point leads, such a point holding a specific direction will be unique in the direction it is rotating because at that or any other specific point wherever, it will be directing not in the direction it spins but in the direction flowing from the centre point outwards. Any point will be it opposing itself within the rotating of $180°$ changing every aspect of its previous flowing characteristics it previously had or will once again have in $180°$ from there. While in rotation from the point of an outside observer all may seem static and never changing but to the object in spin every next second will be a diverting from every aspect it was in every second passing, and the direction it held in relation to the direction it held the previous mille, mille second will totally be incompatible with the direction it holds the very next mille, mille second

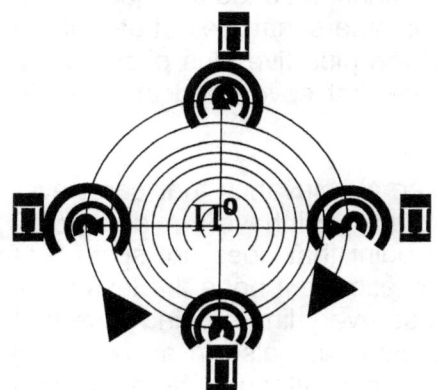

of rotation. That proves no point can be static or constant, all though it may seem that way to outsiders. Although matter is matter, matter can also be anti-matter at the same time.

At this stage time was still eternity being interrupted by infinity. To say the Universe is or was 13.5×10^9 years old is shear Newtonian thinking. Was it 13.5×10^9 years and how many days in the year of our Lord and what about all the years that passed since this date was revised? Time was flowing according to interruptions in eternity changing from what was to what is to what will be. Time is a norm that comes as things in the Universe change about things that are places around and scattered throughout the Universe. We may presume time at this point somewhere became a factor since the sphere sprouted from points on sphere edges and differentiation in development came in place. Considering the role that the Roche limit played, one can see how points in singularity grew from contraction and secured ever-stronger centres by divulging the points within the realm of singularity control. When a point in form developed at a position that was close, than the original Π^0 to Π, the singularity in control took control.

With every one of the four points taking position on the side of Π^0 running to the allocated position of singularity extending, that forms the value of Π at a measure of $\Pi/2$ each brought about the Roche value of $\Pi^2/4$ in relation to the developing centre. One has to remember that the star of today takes on the characteristics of the form of that era.

Consider what happens to a star that developed closer than the Roche limit of Π to $\Pi^2 / 4$ would allow, it is easy to see how the singularity centre grew by concentrating the heat, the points in singularity brought about

$$(\Pi^2 + \Pi^2)$$
$$(\Pi^2\Pi) = 7$$
positions holding singularity

$(\Pi^2 + \Pi^2)$ $(\Pi^2\Pi)$ 3. =1836, which afterwards the atom was about and the Big Bang proceeded

I again wish to repeat the centre in the sphere and the centre of the sphere because in this is where the realisation comes from how the Cosmos started. If there was gravity at the very first instant, then there was a sphere at the very first instant because gravity can only be in the sphere because of what the sphere represents. If that which extends from singularity does not meet the form of singularity by measure of $\Pi \Rightarrow \Pi^2$ in precise duplication, it will tend to destruct and that we call imbalanced spin.

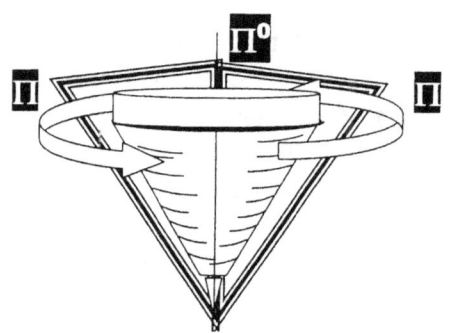

The truth about gravity in the cosmos and that gravity will contract the cosmos one day is the fact that the sphere has singularity as a natural substance. The entire form that is the sphere is controlled from a centre within the sphere. That centre holds the sphere in form and shape. Therefore the strong form is dictated from that space that is not space, as we know space to be because the space eventually becomes lesser at the point where the many lines are crossing at the very centre. The crossings of the lines become so small in space that there where the actual crossing occurs, there is no space and no form left. The natural inclining is in the form of the sphere. It is part of the roundness that the overall shape of the sphere represents and this structural strength is carrying down to the very centre. Because the circle is forever reducing that reducing which is inherently part of the form of the sphere becomes a tool in distorting of space in the sphere and is eventually removing all forms of space from within the centre of the sphere. The very centre ends up as having no space because of the reducing that continuous down to become the space less inner centre. The all roundness is the ingredient that forms the backbone of the absolute strength that the sphere has and that is the component that the sphere is so famous for. The form the sphere has allows the sphere to have a control that is coming from the centre deep inside the sphere where the space vanishes and being without space seems to keep the entire structure rigged. It is the controlling dynamic of such a shared centre that allows the sphere as a shape to show the strength that the shape has and as a form to be as tough as it is. How does it work in its most basic analyses?

If the cosmos had any other shape that might be available, contraction in the end as the final conclusion would then not be possible. Only in the form the sphere offers as a sphere and more so in the circles that are all joining to form a sphere, is singularity present. Singularity comes as part of the construction we find in the sphere. By singularity forming the base of the sphere the Coanda effect will forever be present and with the Coanda effect gravity applies.

As is evident from the experiment of the Coanda effect there will always be an attraction by singularity controlling the space-time from the centre of the sphere. Such control has the name of gravity, but it is no force. It is a combining of liquid (10) and solid (7) in the presence of three in time that will produce such control by motion. But why would the Coanda effect establish gravity by rotation and moreover by motion?

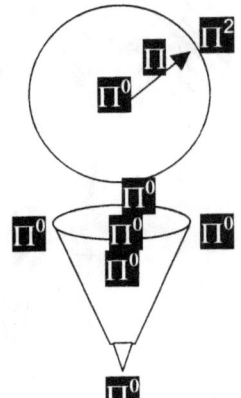

It is because one Universe arises with opposing qualities where the one attract as the other repulse. It is in the motion that the action of gravity is vested.

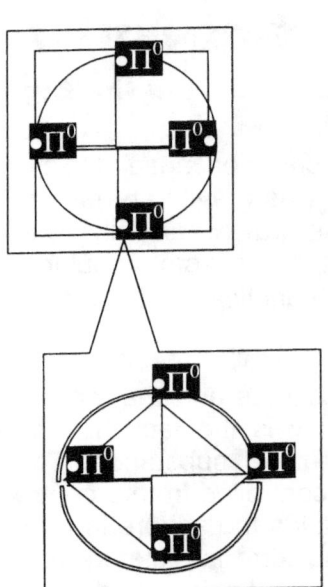

The spinning of Π^0 around the centre Π^0 establishes Π and Π is what produces the singularity. Singularity is always present in the form the sphere presents. However, in the sphere singularity presents no influence except in the case where the sphere starts to spin. In the spinning, a line comes about which caries the term of an axis. The axis has never been acknowledged as the most vital piece of any particle in the Universe mainly because the role that the axis plays has been neglected and subdued by the incorrect attribute Newton placed on the motion and the axis.

The moving of Π^0 to Π involved relegation and not motion as we consider motion. It was Π^0 getting a side and that is all. There was no true side but only a form that came into place. Singularity (A) received singularity (**A**) and no more of anything but the shift to comply with having a relevancy forming in relation to singularity. The dots had no sides, had no length or diameter. There was not measurable space or measurable time involved. The time could have been a micro, micro second as much a trillion millennium because time had no relevance. It was eternity interrupted by infinity, as it still is the case, however the line that eternity followed was no line because there was no space to hold the line.

The line was momentarily interrupted by infinity, however with no one there, there was no one to notice. The lines were not lines but relations to sides being formed.

Inherent to the form the sphere offers, there is a specific location of singularity where the radius first goes single $r^0 = 1$ and then form goes into the realms of singularity $\Pi^0 r^0$. The cube also may have such a point but having such a point does not connect directly to six points located on the edges of the cube or any other form the is.

In relation to such a centre where $\Pi^0 r^0$ forms singularity there are always four cubes related to such a centre where the centre is part of seven points in total representing the sphere.

Every cube has lost one side to a point of the sphere where the sphere takes control of form and removes one side of the cube. In relation to the time factor that is inherently part of singularity by the extending of singularity, there are five sides connecting to four points standing related to singularity by the Π^0 factor and that gives 5 X 4 = 20. That is always directly in relation to seven points singularity offers.

This should lead any person to investigate a centre that forms because evidently, there is a centre but that centre comes about by motion of rotating around a fixed point serving all points in motion. That leads us to the centre of everything in rotation because everything in the Universe is in rotation. As Kepler said the Universe is centred by $k^0 = a^3 / k\,T^2$ and we have to find $k^0 = 1$

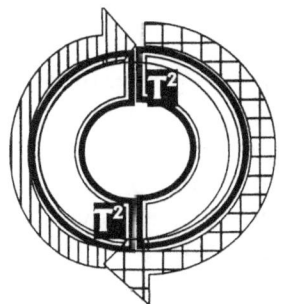

At this point Newton's second law come into affect. Motion by means of the Coanda effect introduces space as motion introduces time. For the first time ever time was interrupted when motion provided time the space to interrupt. From motion by the way of the Coanda principle gravity came about as a centre formed a point where motion surrounded space. By motion space-time was established in relation to singularity

If the universe did start from one single point and time matter and space flowed from that point, then that point must have a relative connecting base because such a point holding singularity must be eternal as space, matter and time link eternal. There therefore must be one point linking the entire universe when regarding the fact of singularity. Then according to the theory off relativity there has to be one exact point holding time in relevance notwithstanding the fact that time departs from that position and relates differently to all space-time away from such a point.

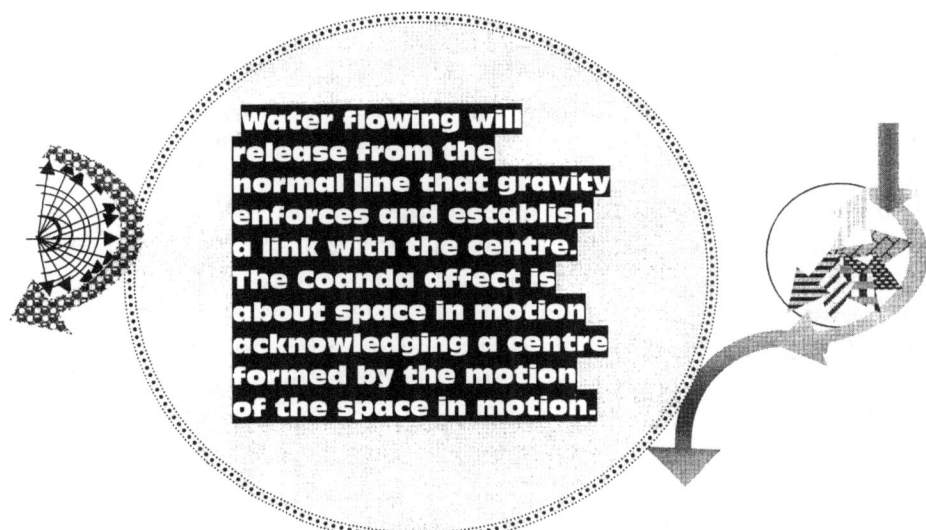

Water flowing will release from the normal line that gravity enforces and establish a link with the centre. The Coanda affect is about space in motion acknowledging a centre formed by the motion of the space in motion.

Every person, with whom I have discussed facts about creation, recollects images in the trend depicted in a presentation as one may find in an array of total chaos. That would be the most unlikely way Creation came in place. The recalling of pictures representing images about creation must have form, but to mathematics it had no form. From this thought the very opposite arise where Creation came from nothing but such an idea is mathematically simply not possible.

The thought of nothing is just what it is, a thought of nothing and although it is in the human mind common nature to present nothing as a value in the recalling of something, nothing is a presentation of the figment in the human mind. There can be no number such as nothing and that was (possibly) Newton's biggest error. Nothing represents non-existing and that is just what nothing is, it is non-existing.

In order to prove my point, I wish to ask the reader to define the shortest line there can theoretically be. If he should answer anything but that the shortest line will be at a point where the beginning and the end of the line falls on the very same spot, he will be wrong. The shortest line that can ever be anywhere must have a point representing the start that is sharing the exact same point as the point that represents the finish and both allocations are holding the exact same spot. The line will be humanly impossible to create but we humans are capable of very little.

When the line has a beginning and an end at the very same spot and it wishes to extend the position as to further the possibility it has, which direction should it favour. Humans in the west would naturally think of extending from left to right while in the east humans may want to go from right to left. Some persons will tend to go up or down, but all of the options are about human preference and not mathematical conclusions. Extending the line in any one direction will favour one direction without a conclusion about not extending in other directions. Such a conclusion has no sound mathematical foundation. The only option about extending will be in all directions equally in order to give a meaningful non-bias flow of mathematical equilibrium.

The shortest line in the realm of possibilities must have a start and finish holding one spot and such a line will also be a dot or a circle. Not favouring one direction puts all directions at equilibrium, meaning that any form what ever may be can develop from such a spot with the end and the start being the same. This reasoning prompted me to look for singularity in such a spot because if the prime spot from which all came was a spot, then the spot must hold the shortest line but more prominent it will hold the smallest form including the smallest circle.

One possibility that the shortest spot can never have is having a starting point on the zero mark. If the mark of zero holds the start it must also hold the end because the end and the beginning has the same position. If the position of zero then is the beginning, the end will also be zero leaving the line without an end as well as without a beginning.

While very line is circling bringing about time in space to the value of Π repeating Π to form Π^2 at the same time Π is extending in one specific centre to the value of Π^0 and only the spin value keeps Π not becoming r The spin keeps the immovability from becoming Π and maintaining Π^2 by performing duplication, but with any slightest reduction in spin reducing Π^2 to $\Pi^{2/4}$, Π will start extending and as one can see from the behaviour shown in the Roche limit, the heat will be concentrated at the centre and the singularity in the centre will grow four times in concentration. Only at points exceeding Π in diameter was time as Π^2 able to retain form and also grow. From that, space slowly developed because at Π could Π^2 bring about a

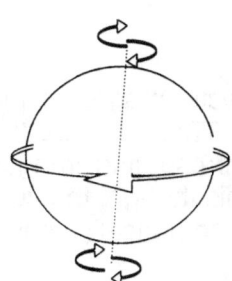

form which provided motion. In the centre there developed Π^2 and Π^2 kept all form at a safe distance of Π to bring about the needed solid immovable centre with the form $\Pi^2\Pi$ about the double Π could $\Pi^2 + \Pi^2$. That secured the makings of the atom by applying the Coanda principle of enticing gravity in the centre of motion, which then provides space-time by measure of $(\Pi^2 + \Pi^2)(\Pi^2\Pi)$. This totalled seven in dots and with three of those seven circling singularity at 1.9991 the atom came about. **I again wish to repeat the centre in the sphere and the centre of the sphere because in this is where the realisation comes from how the Cosmos started**. If there was gravity at the very first instant then there was a sphere at the very first instant because gravity can only be in the sphere because of what the sphere represents.

On the inside, there are the seven markers of which singularity is the focus point in the centre of the centre. The markers are representing one aspect of space, which for argument's sake let us call it cold. Then there are three more markers on either side being part of the space but not captured in the space. It is space in motion by the influence of the motion of the Earth.

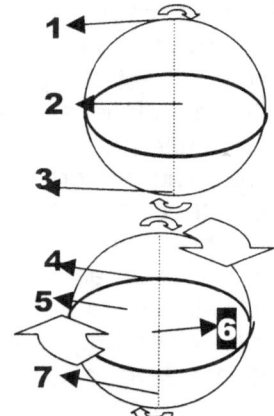

By not having motion, the lines also have no space as the space extends to form space forms space and the line includes serving the three points to the outside. Where there is no motion, there is no space and where there is little motion, there is little space. The only space the line may relate to can be a point that is on the border of the sphere that is crossing singularity and connecting the two edges on either side of the sphere that is forming the sphere. That means the line from one point holding singularity to another point holding singularity that line will cross the centre line which gives the line in singularity valid space-time to control. Singularity does not have the ability of motion, therefore singularity does not hold space. Singularity is also eternally indifferent to motion and motion can excite singularity but singularity cannot be shifted by motion.

Three points form the line.

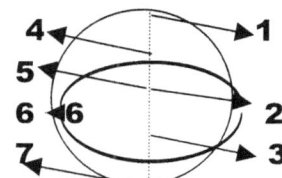

Every time motion takes place, the centre line holding positions 1, 2 and 3 stand still. There is little generating going on and reducing to a point where there is no generating going on. On the outer edge of the rim however there are four points that do shift. The points shift from one location to the next location by generating space.

All this is happening while the crossing is all concerning singularity moving from one sector of singularity to the other sector of singularity which is (Π/2) X (Π/2).

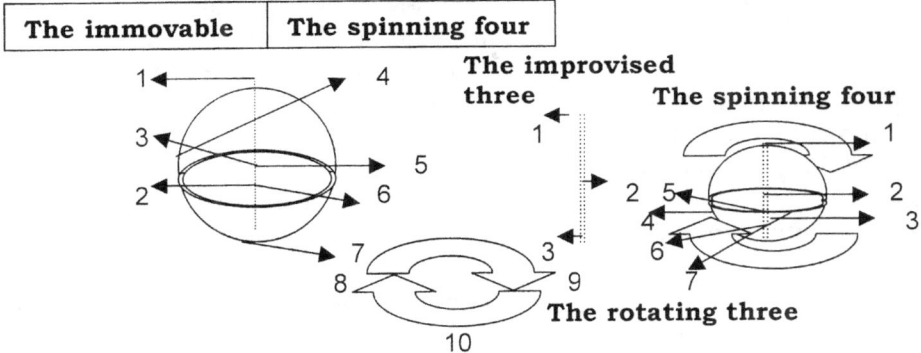

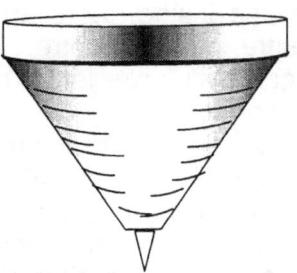

The relevancy forms part of the duplicating and dismissing displacement of space-time we call gravity. In that, we are looking at relevancies and no precise specifics. However, the Universe was built block by block in this manner. As it was but is no longer only form that applies in the Universe but concrete measurements also come into play. Therefore, even the relevancies may apply in different relations as they switch over to compensate for other factors alternating as they are coming into prominence. The lesser developing sphere orbits the dominating sphere and between them, there are definitive relevancies. The centre circle singularity line of three is unaffected by spin which I shall call the immovable three. However, the immovable three holds such a stout position as far as the centre sphere is concerned. In relation to the orbiting circle the centre line is part of the building and destructing process that manifests as duplication as the centre singularity maintain domination and control over the orbiting structure. In addition, it has a major part in the motion building of the sphere in orbit and moreover building by generating the singularity line that generates the lesser and the orbiting sphere. In that relation the centre sphere reflects the centre line to serve the orbiting sphere by supplying the reference needed to establish motion in the orbiting sphere as one Unit. In that there is an indisputable reference of seven orbiting the centre and the centre providing three as a reflection of the seven which in all accounts for ten relating to the four which also is spinning as time and in total forms the seven taken in relation from the orbiting ten.

Points seen from the side form a line that never moves and is a line in singularity holding three points

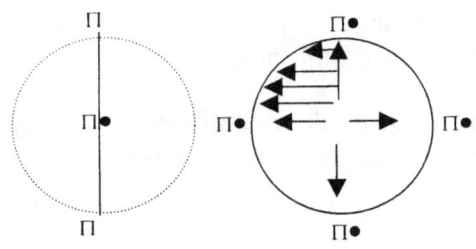

The seven can never totally separate from the ten, but by singularity being the same, but being on the other side, it is withdrawing space-time altogether. See it as seven (let us think of that as the cold basis of space) spinning or turning in the ten (which then will represent the hot part in the cold basis) and the ten is part of the seven but the seven is not part of the ten. The third factor is the axis around which hot as well as cold will turn. Therefore, when reading the next page, please envisage a cold base turning in a hot and cold space. The purpose of this is not to define whether the argument is correct or not but it is to help the reader gain understanding of the principles of the process involved. But motion also converts space to relate to space by changing relevancies through motion. Matter is in relation (part of) to the total dimension of space but is not the total dimension of space.

Space-time is allocated in progress from the line singularity offers. That space-time consists

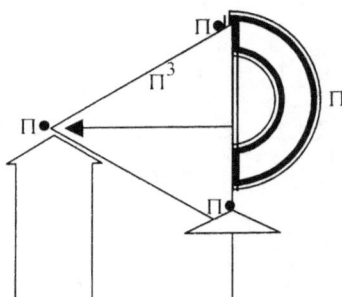

of heat and can therefore store more heat, which stands in contrast to the cold that singularity presents. By providing, more space-time at the equator there is more space-time to allocate to heat being there.

In this one can clearly see that it is the motion that sets the top free and independent from the gravity of the Earth. By motion the top generates individual gravity that allows the top an individual gravity and that motion frees the top from the gravity by which the Earth restrains the top. The motion gives the top independence that the top immediately loses when the motion subsides. This fact is the utmost important issue of all physics in the Universe.

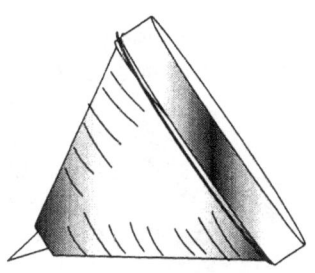

The top is one of (perhaps) the easiest and most common examples one can find to demonstrate the cosmic generating of gravity. By spinning a "force" which is no "force" keeps the top erect and it is only motion that accomplishes the act of gravity.

The top lying still holds the same singularity principle that the sphere holds because if the shape the top has. The roundness protects singularity at a seventh position deep inside. However, the top is a dot Π or even going down to a spot Π^0 and is only by the form the material has which puts singularity in place.

Singularity 1° expanding 1¹ by duplicating time

When the top is in a retired position laying on the Earth the top lost all gravity while it receives maximum mass. The mass is the Earth suspending the top's personal gravity and enlisting the top into the gravity the Earth has. The top can have mass but then the top has the Earth to supply gravity while being retired to what the Earth conforms the top to be.

According to Newton it takes no effort $\dfrac{dJ}{dt} = 0$ to get the top from where the top was motionless to where the top is spinning. I say this on the work that Newton suggested comes about from the effort it takes the top to circle.

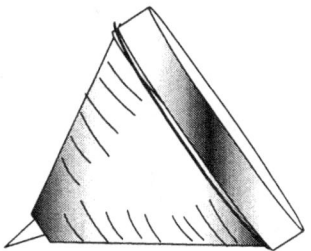
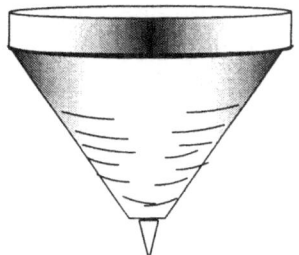

From a position where the top is lying down being collapsed, the top generates a position by spinning and the motion puts the top erect. It takes motion and not nothing to establish such an independent and secure position. It puts the top in the centre in a newly established and independent Universe as the top then finds courage to fight the gravity of the Earth up to the last "breath" that is fought.

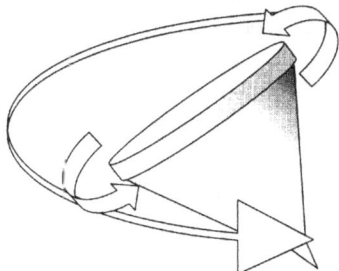
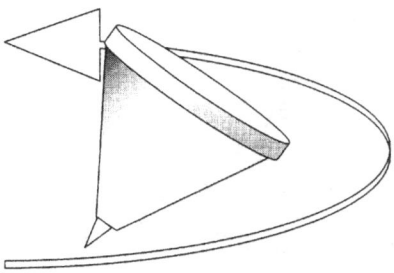

In the motion a line comes to life running in the centre of the top. This line is not just another line but can focus the top to spin upright and erect. The line was not there when the top was on its side. By motion the line can concentrate an effort that will unleash such dependence to the top that the top will come into a position where the top has the tenacity to take on the Earth gravity in a struggle for life and supremacy. The

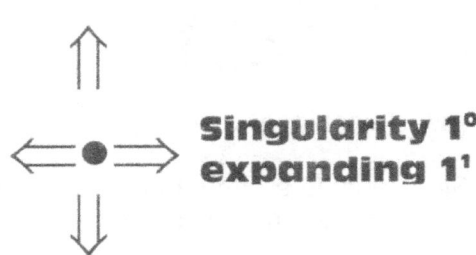

Singularity 1° expanding 1¹

motion is a result of heat parting singularity.

From whichever angle one looks at the top, the top seems possessed and I can even be slightly forgiving towards Newton for calling it a force, because

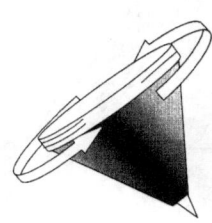

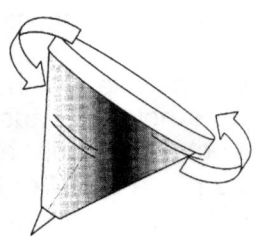

although not a force, the stance the top takes when spinning upright leaves on with an impression of forcefulness being part of the situation. One must not see a force but one should see the manipulating qualities of life extending to the top and by life's ability to manipulate space-time and control motion in space-time with space-time, the throwing of the top is as little a cosmic event as the apple Newton saw falling from the tree. In every event, the top as well as the apple, the drive was life controlling events and as far as there is proof, there is no possibility of such an event taking place anywhere in the Universe by something as small as the top or the apple. In the case of stars such development does start the star on a course of independence and it sets the star development on the road of becoming independent from the galactica. The drive generates gravity but the driving that allows such rotation is inspired by the accumulating spin of the entirety of all the atoms in motion within the young developing star. The heat blanket in which the star cradles has a lot of influence but it is the atoms becoming a driving factor that inspires the motion and such inspiration allows the top the initiative to form independent gravity as a star. In the case of the top, the spin is completely cosmic unnatural. If you start to imagine about life in the Universe, you may just as well start believing in ghosts, fairies and all other fantasy creatures. Science must decide whether they wish to speculate about the life's abundance and being a dime a case, found all over and everywhere you look throughout the Universe, but in such an event distance their fantasies from science and reality, or stick to science in reality and believe only in facts as science presents facts.

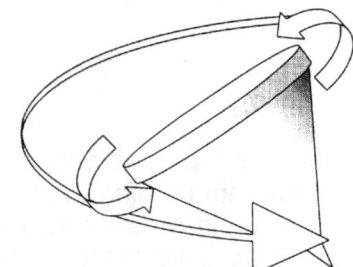

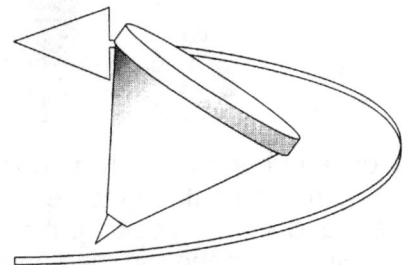

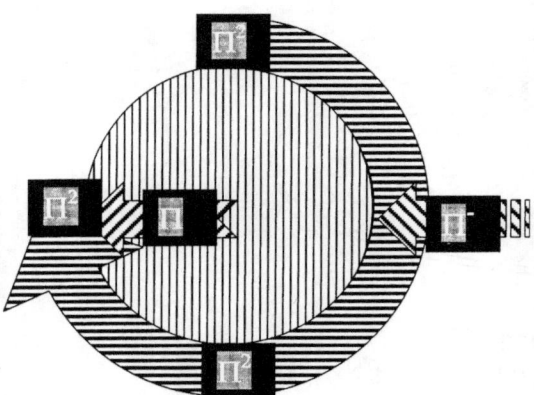

Let's consider what are facts with the top spinning as the top does. This is no fantasy or life coming from some imaginary source but it is a cosmic reality in which life found a way to manipulate.

The effort it takes the top to spin gives the top a distinction of extreme significance. The top is promoted by the motion initiative to that of a star in motion because it charges singularity into existence where singularity then controls space-time. There is no difference between the top spinning and the Coanda effect and in both cases singularity is generated by motion that is charging a centre to activate a Universe.

Singularity is a mathematical point hidden in every sphere. Singularity is very much inactive in every sphere but to keep a centre of structural bonding in every sphere. Something happens to the top in having a singularity just like any other sphere has to a point that takes charge with all the cosmic dynamics the sphere may show.

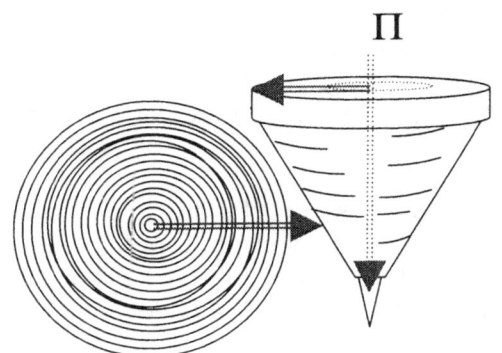

The top is charged with an energy, which not only takes charge of the top as well as the body of the top, but also the immediate space surrounding the body of the top. When a person with skill manages to put a high degree of spin into the motion of the top the top then spins in the surrounding space so vigorously the top stars to whistle vigorously

The linear remains linear because the linear redirects its intentional direction because of the rotational change that the linear motion always ends up doing. The line forms an eventual circle because the linear line must constantly entertain the centre.

Our gravitational falling to the Earth is a result of a circle going straight and forcing us straight down to an everlasting directional alternating circle, we have as we spin with the Earth as we spin around the Sun. As we fall straight down, we change direction while we are falling straight down because that point we are heading to what we are falling to, is changing too. From the centre of the axis, everything seems neutral. The axis does not spin at all, because the axis brings about spinning motion changing eternally. That is in nature and not man-made motion. As the top comes to motion the top finds the characteristics that the top shows very indicative of the characteristics that all moving objects in the cosmos show. The top spins in a straight line that bends by a 7^0 inclination.

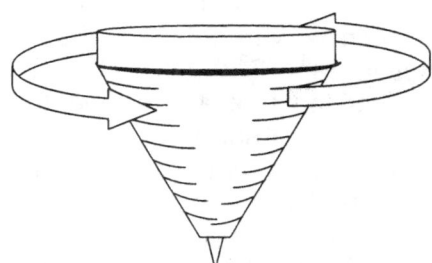

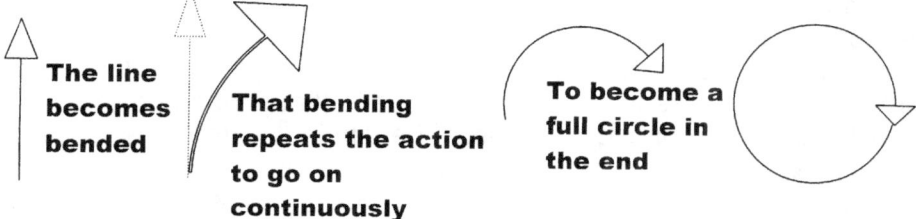

The line becomes bended   **That bending repeats the action to go on continuously**   **To become a full circle in the end**

Apparently an idea concerning the subject of gravity Einstein came to was about him falling off a multi story building and the gravity mass that he then would experience. Remember, this was long before flying and parachuting with free fall acrobats showing on TV how a man and a car with a man in the car can descend five or six kilometres while falling side by side. This happened while Einstein was still being a patent clerk in his younger days. Apparently Einstein was looking out a window of the multi story patent office, when Einstein suddenly realised that had he, Einstein fallen out of the window from the roof to the ground of the patent office where he was working at the time, then he (Einstein) would feel as if he was weightless during the time of his fall. By falling with him, those articles would feel equally weightless should they accompany his fall down as being part of the falling process in his imagination. As the objects were travelling alongside Einstein down the building to the ground, the lot would travel at the same speed from the top to the bottom of the building. Then I went one step further by supposing the Einstein group's falling was real and no imaginary thoughts were set in the fall, then what was the imaginary factor then? Let's pretend Einstein did fall with his pen, his chair and his desk and Einstein was not imagining his fall. Einstein as a human being can imagine but his falling companions can't. If Einstein was imagining his weightlessness, it might be psychological, but in the case of the other travelling companions it was not possible to imagine anything. There is an immense difference in size between the falling companions and that notwithstanding they travelled the same speed while descending. If they travelled the same speed as Galileo proved and they all hit the Earth the same time, which then indicated that their weight and mass, that which gravity used to drive and what propelled them downwards and that which was causing the drawing of what the mass was instigating to allow the motion of fall to commence, was equal. Kepler found space a^3 being equal to the motion thereof T^2 in relevancy to a centre point k. Kepler found space had to move.

When reading this that evening so many years ago, I came to realise that Einstein could only feel weightless if it was true that he (Einstein) was weightless. He could not feel as if as if was part of his imagination because he was truly falling, and in truly falling the falling was then without his imagination doing the pretending. Einstein had to feel his weightlessness as a cosmic fact in the true sense because if he was truly falling, then the part, which was the falling experience, was what he was experiencing in reality by three dimensions with one dimension in time. If Einstein was experiencing weightless ness, it would be because he was weightless while falling, then Einstein would not imagine the weightless ness because Einstein was truly falling, thus carrying out his cosmic state he was in. His body being in motion ($a^3 = T^2k$) was at that moment truly weightless while experiencing unrestricted gravitational motion. Einstein, the pen, and the chair had the same weight since they were all weighing the same in falling. If there were any mass differences there had to be speed differentiation for the force of the one would generate more motion than the force of the other onto the different mass components but since there is not mass discrepancies amongst the falling while falling, the lot is having the same state of weightless ness. They adopt the same speed in the fall. After all it supposedly is the mass that is doing the pulling and more mass does more pulling…except if the mass is not doing the pulling in the first place. All four items including Einstein, would be equally weightless during the falling…that was what Galileo found, because objects of different size and different mass travel at an equal pace (distance over time or space moving divided by time flowing while the object changes position in relation to the Earth ($a^3 = T^2k$)) while descending. The bigger objects do not fall quicker than a smaller object and that can only be attributed to one fact; it can only be true if the four weighed the same while falling and no one weighed anything while falling. That means the gravity applies while time flows in relation to the space that is applying the motion, which is what gravity is $k = a^3/ T^2$ according to Kepler. The single line falling is represented by the factor k being the relevance of space a^3 that is relocating its cosmic position, while all that is happening in relation to the motion of the Earth T^2, which is in relation to the Earth spinning around the Sun and that rotation gives us our time T^2. While in motion the four different objects weigh the same since they travel at equal speed downwards. By standing still, the objects have mass differences and when they are in motion they weigh the same. When the motion becomes frustrated by being blocked by another space that is also filled with material and that is holding the spot to where the motion is directed, they then have different weight. The pushing resulted from the bodies striving to remain independent. The two objects are in a fight to claim the position each desires, and that is to fill the centre of the Universe. Being ($a^3 = T^2k$) is being in the centre of the Universe because the centre of the Universe is $k^0 = a^3/T^2k$. Then one may conclude that gravity is motion of space and mass is the restricting of the motion of space. Having mass does not bring about gravity but it does restrict gravity's motion, which is what brings about the mass and weight. Gravity produces mass but mass does not produce gravity or in fact the mass that produces weight but mass is not responsible for the intended motion. The intent on moving while being blocked by another object is frustrating the motion of gravity in both cases and the higher the frustration on motion is, the more mass there is coming the way of the bigger object who then has the greater desire to move. There is a reason why it has the desire to move and why space is equal to the moving thereof in ratio with and also in time. By the moving of the space in relevance to the centre of the Universe (which at that point might be the Earth or it might be the Sun, or any and all atoms forming either or both objects, or the total effort of all such atoms combined) is what keeps the Universe in the state we find it. The motion is the duplicating of space by contracting time in an effort to maintain the position of both space and time. Gravity is the maintaining of any and all objects' independence in time while when receiving mass, such mass destroys all independence of material by forsaking individual independence for the accepting of the time of a much larger object. Mass is the restraining of motion and gravity is material moving about by committing gravity. Mass only comes into the application thereof when two objects filled with space moves into a position where both want to claim the very position in space the other occupies.

It is the motion and the independence they show to hold onto their individuality as independent cosmic structures, that prevent them of sharing the space, which in turn prevent further motion that causes mass. Gravity is in essence where mass is present, still in a tendency to commit motion, but is then in the frustration of motion and gravity at such a point is the commitment to move once the blocking of space is relinquished. Because the one objects that has more "mass" would put in a more assertive effort to move in relation to a smaller object and the effort to move will constitute to a greater resisting effort by the blocking objects in a fight not to relinquish its position on the space both object claim that the tendency to move and the tendency to block the movement will bring the effect of greater or smaller mass being present during the effort and in line of resisting the effort. However, while any space is in motion, the gravity of motion is equal to all and puts everything on an equal basis. Therefore there are no big and small and the big Sun does not pull the small Earth closer. Mass is the result of the scenario when the motion that supports individuality is prevented and by that a differentiation in motion effort comes about which then changes the picture.

Do not be fooled by the seemingly innocent explanation that space is the motion thereof which is what gravity produces, because of all things the cosmos creates, motion of space through time is the utmost complex manoeuvre and without bringing a restraining of mathematics into science, it is so complex there is no viable explaining in physics about how the cosmos produces the act of motion of space in time. In order to get every atom to spin, as every atom follows the lead of the atom in front, heat from time is contracted as the device that excites such spinning. This exciting gives direction to the atoms to follow the atom just in front and lead the atom just behind, while giving coherency to the structure. By following the one in front and being followed by the one behind the lot of atoms are holding as an individual unit, times the units there are going around in the entire Universe. The measure of this complexity is beyond what the human mind can absorb. While the atom in front is vacating space to fill, the space of the atom in front is vacating at that instant, the atom behind is filling the space that the atom in front has vacated in order to vacate and relinquish the previous position in favour of the following position to honour the direction gravity is insisting upon. Times that with every atom there is in the Universe and one may grasp the significance of the calculation. Removing material from space by filling material into a position of new space sounds simple because the complexity has never been realised. I am in the hope that in this matter I will be able to will reveal what the factors are in understanding the commitment of material to move through time. This was all a result of understanding the dynamics of Einstein's arguing about gravity and mass. Then I kept this information in mind and it helped me to further realise gravity is motion differentiation between objects.

It is the independent motion providing a different speed while sharing a common centre of attracting that allows a discrepancy to establish mass under specific conditions applying between the two in relevancy. While falling the gravity applies as moving of space that is putting time in relation to the distance travelled. That means there is a speed relevancy between particles in motion and synchronised motion, which would bring about equal orbit around a shared centre. That is the result of gravity functioning. While the object falls, the motion confirms gravity. When motion ends, mass sets in and becomes the constraining of the object, preventing further motion. The motion is still there but now it is reduced to a tendency to move thus establishing the object mass as the limiting of further motion. Preventing the motion by implementing mass is the resting of objects against each other by resisting the motion to continue, which then is where the mass takes the place of the motion. Where a confronting of objects restricts gravity, the action then implements an introducing of the mass as a substituting factor to motion that then replaces motion as substitute to the motion that would be and the mass is providing the tendency of gravity being the motion of space.

However mass then restricts motion and becomes motion in a tendency to apply motion. While falling, gravity applies and motion neutralizes size, mass or weight. Mass counters motion being when the Earth restrains further motion of the falling object and the moving object is stopped from further movement where mass is then preventing or hindering gravity. This is the result of objects claiming an individual and personal claim to space occupied in a dual or in fighting for their individuality and independence of each other while wanting to be in the centre of the Universe. While falling or moving, there is no opposition to the body being independent. When the motion seizes, the falling object remains individual and still tends to move while Earth individuality resists further movement of the falling body's movement.

Further movement is disallowed as other material fills space that the falling body wants to lay claim to. The only manner to remain independent by the falling object will be to relinquish to motion in the securing of mass as a substitute to motion where it then finally comes to rest. Mass then sets in not causing the motion but substituting the motion and from that motion restriction becomes resistance that becomes mass. While falling, the object is experiencing gravity because the object is in gravity but when on the soil the object experiences mass which is the restricting of gravity or motion by other space filled with material.

Looking at the top spinning and not spinning, brings a question to mind: why would the motion of the top beat the gravity of the top and that of the Earth hands down when the top is spinning. Surely the mass is in effect while the top is spinning just as much as the mass is in effect when it is not spinning and yet when it is spinning the pulling subsides to give way to

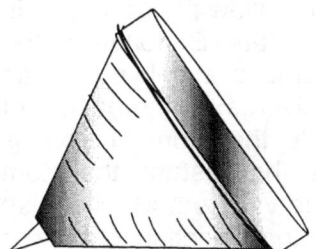 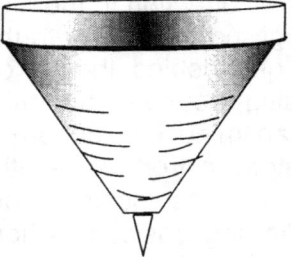 free the top from the mass restriction and charges top with much excitement. The excitement is so much it seems to relieve the top of the pulling there is between the top and the Earth. That even strengthened my suspicions more about the fact that gravity is motion. By spinning, the top finds additional motion on the side of the top that brings the side of the top in a more favourite position than was the case before the spinning commenced. The top finds additional gravity in the spinning and the gravity in addition has to bring reconsideration to the position the balance in gravity sets in margins. The top secures a better margin or stretches the parameters of location because the top inherits the Motion of the Earth and in addition to the gravity motion that the Earth provides the two secures more gravity motion. With more gravity motion in addition to the Earth's gravity motion, the top has to have more gravity motion than that which the Earth has. That will allow the top the spin while facing such an enormous disadvantage there is on the side of the top in mass difference when considering the size disproportions. Let's face it, if it was about mass pulling the top in relation to the top pulling the Earth, then the top had no chance ever to move by the very slimmest of chances there may ever be.

However, the fact that the top finds additional motion is the result of the motion coming with the aid of what motion life can add to that of what the top inherits as the gravity motion, which is coming from the Earth. The Earth represents cosmic motion, and the cosmic motion is the part, which the Earth already contributes as motion, in addition to the motion already contributes the motion. The extending of gravity or motion to supply the top with independent motion, is adding to what mass destroys by collecting the object and retaining the object independence.

Considering the mass the Earth provides then it will not require that much a bigger effort to get the top going. Singularity charges motion by instigating motion without ever moving. Coming from a spot to a dot and then producing a line running from dot to dot though a spot signifies the birth of the cosmos. A line holding time by the square and by the square of ninety degrees announces another birth in the Universe, which also is one more birth of one more Universe. That which was not there suddenly is there by not being there. That which was undetectable suddenly is detectable by being undetectable. It is not my forte to write riddles but in this case there is a Universe within a Universe, which is not in the Universe and does control the Universe from a point no one may ever locate inside the Universe. The Universe is built up by innumerable dots and each dot is charged with being the Universe while being in a representative position since it is not in the Universe.

Space-time is a four dimensional position of the Universe where the position of an object is specified by three coordinates in space and one position in time. According to the theory of special relativity there is no absolute time, which can be measured independently from the observer. Events that are simultaneous as seen from one observer can occur at different times when seen from a different place. Time must therefore be measured in a relative manner as are positions in three-dimensional Euclidean space, and this is achieved through the concept of space-time. The trajectory of an object in space-time is called world line.

General relativity relates to curvature of space-time to the positions and motions of particles of matter. In view of the definition of space-time I wish to elaborate on my view of singularity and my deriving of space-time from the likeliness that singularity may produce space-time. In the past singularity was mentioned in the manner one would speak of a ghost hiding in a haunted Black Hole. Let's put singularity in the clear. Singularity is within every sphere due to the natural shape or form the sphere is committed to.

Locating and finding the presence of singularity

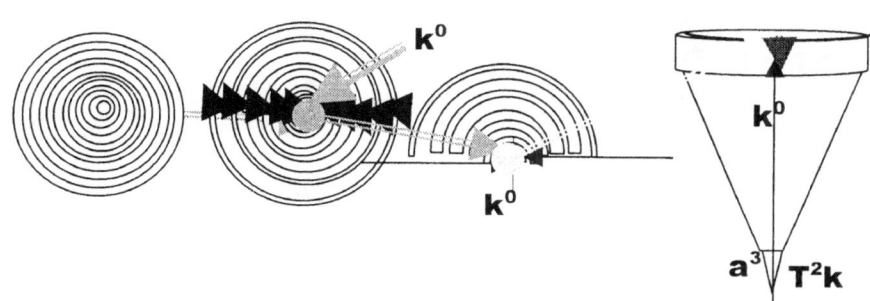

While the toy top is spinning one will find singularity by moving the rotating line or radius progressively to the middle by reducing the length the line has from the edge to the middle. At one point all further reducing must end but the ending cannot include zero or nothing because the rest of the line still attach the rest of the top.

What is in the Universe is spinning. In the precise middle of all objects in rotation is a precise centre dividing the object in sectors that will start the spinning initiation from that centre point.

$k^0 = a^3 / T^2\,k$ states that whatever is, is also spinning in order to be present. **Thus, the spinning object** will have a middle point, **a very specific** centre point that does not spin **and only holds Π as a specific value because no radius can apply. But also the one value such a line** cannot have is zero **because the line** is there and holds contact **to the rest of the material bringing about that** zero does not start any **line and therefore the** value of the line must be infinite, **just as described in accordance and by** the definition of singularity.

As I am introducing a very new idea, I wish to explain in better detail what I try to convey. That point albeit hypothetical, is also as much a reality none the less and is placed where that point **must be standing still** because every line **running from that point in opposing directions** is also **in opposing directional spin the other or opposing side.**

As the rotating direction moves inwards, the rings holding Π will become smaller and smaller. The reducing of the radius r will eventually end where the spin direction ends at $Π^0$. However that point where the directional spin ends is the point where the actual spin takes place. The spinning is on the precise location the point is not spinning.

The definition of space-time is as follows:

According to Einstein, singularity is a mathematical reality within the Black Hole but much more so in every sphere. Einstein may be the first to name it and Galileo (unwittingly) may have been the first to define it, as Kepler was the first to formulate singularity, but in mathematical terms, singularity is the most basic principle. At this point I wish to establish a fact that seems lost in all other grandeurs of cosmology. When tracing the radius down into the sphere, the radius starts where all lines start and a straight line cannot begin at zero or nil, it can only start at infinity. Such a statement will hardly seem appropriate but the relevancy of this fact has no limits. If gravity is motion then motion starts with a line. Let us follow the line as motion abides by the rules of the line.

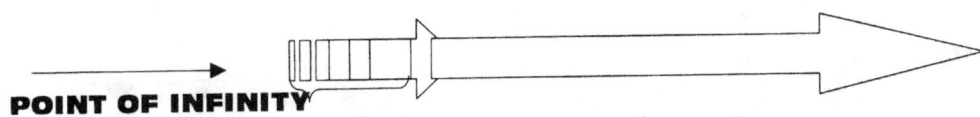

POINT OF INFINITY

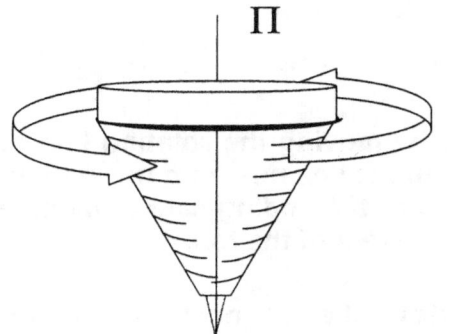

Π

If the line started at zero there was no line to start because zero multiplied by whatever, results in zero as the answer. That must also be the cosmic starting point. Einstein introduced such a point and named that point singularity. When looking at the cosmos from whichever angle, all indications lead to the fact that the whole cosmos is in motion in its entirety. It is forever spinning and it is going to as much as it is coming from. Everything is on the move and always encircling something of greater importance. A top can spin but the

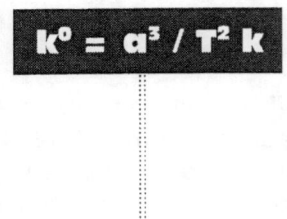

$$k^0 = a^3 / T^2\,k$$

parameters of its spin are limiting the motion it can apply. By not spinning the top is still spinning as the Earth is doing the spinning on its behalf.

When the top starts by spinning too fast, it is clear that the top is in a fight with something that is restricting its spin. This we can see by the re-aligning and the swaying the top manoeuvres with to try and circumvent the restriction. As the top spins, something starts to tarnish and erode the spin. When the top starts to spin too slowly, the top tries the same manoeuvres but in that case it then seems as if the top is in a struggle to keep the spin alive. These manoeuvres that the top displays, triggers questions in need of answers. Why would the top stand upright when spinning? It can only be that the spin activated singularity into manifesting the gravity of motion.

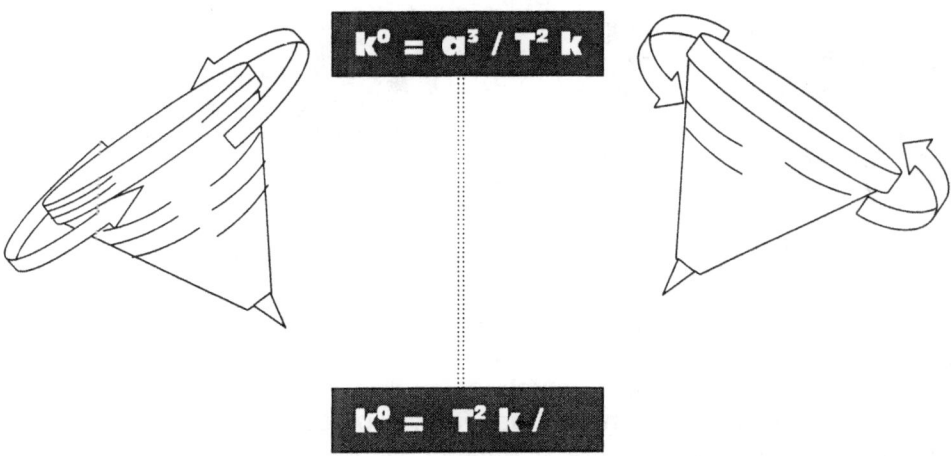

The spinning of the top is all the evidence one needs to come to a conclusion that the motion establishes a drive and the drive establishes independence of the surrounding space the top finds a control over. The top divides the Universe by establishing a generated Universe independent of the Universe we think of as the Universe.

The centre is never there because the centre is eternally there. The centre is drawn into action by motion but the centre does not change in principle. That which activates the centre changes by directional contradicting of its nature.

The motion establishes singularity, which implicates the Coanda effect as much as the

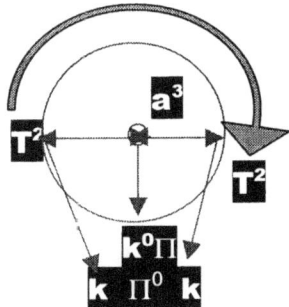

motion establishes the Coanda effect. The spin realises a space limit while the space limit attaches the motion in the form of time onto the space of the material, which then allows time inside connecting to time outside the space filled with material. Judging from the behaviour of the top while the top is spinning, it seems very obvious that not only is there a singularity running inside the centre of the top, but another point holding singularity forms next to the top. It is this singularity point that seems to carry the restraining the top wishes to find release from.

In dimensional terms, which I explain later on, the value of **2k** relates to T^2. That relation extends to the next value where T^2 relates to **k,** which relates to T^2. The first space in the circle will then be T^2 **k**. From the centre being in infinity, one can realise by applying mental power, the single dimension factor not seen but present all the same. Extending that into the 3D comes six **k** and any one of the six will further extend to form a seventh point as T^2 All this is a multiplying of $k^0 = a^3 / (T^2 k) = 7$

The Coanda principle indicates that the gravity described in the previous page is generated by motion of liquid in relation to a solid anywhere motion can produce gravity. There is no mention of mass because mass is a derogative of the gravity which the motion creates. A centre is formed where the surrounding space-time forming the one group is relating a position from the "centre point". That forms one inclusive relevancy between points within the gravity field. The gravity field is holding "back" and "front" running through "the centre" where the other line is relating from "side" to " side" running through the "centre point". The fact of the line in the centre is that "it is there", but we cannot see it. Try as you may, no one will be able to calculate the very position that forms the lines, but as they change all particle characteristics, the lines are a reality as the spin of the matter is real. Being too small to hold atoms, the space holding such a

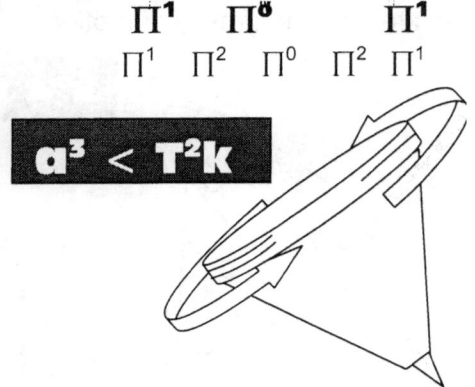

$$\Pi^1 \quad \Pi^0 \quad \Pi^1$$
$$\Pi^1 \quad \Pi^2 \quad \Pi^0 \quad \Pi^2 \quad \Pi^1$$

$$a^3 < T^2k$$

centre line is no space at all and with that knowledge we may presume then therefore, what ever the line constitutes of must become part of singularity, where singularity is a spot in the centre with two lines crossing the spot at an angle of 90^0. That is the basis of singularity, and since all the positions still relates to a centre of a circle, forming a part of a spinning circle, Π must form the basic value. The second major reality that one has to recognise is that the only way singularity was broken was by motion. The only way motion can come about and break space less ness is by establishing heat which establishes expansion and the Universe became a possibility and later a reality by expansion. The heat swell into space and the space swelling is the motion that produces the gravity we find visible in the Coanda principle. The space at first was presumably filled with material because the expanding could only be material. The Coanda principle alters time and establishes with such alterations to space-time a new Universe with borders and all. By introducing motion, it sets a new time standard by which the space created will apply a newly generated gravity.

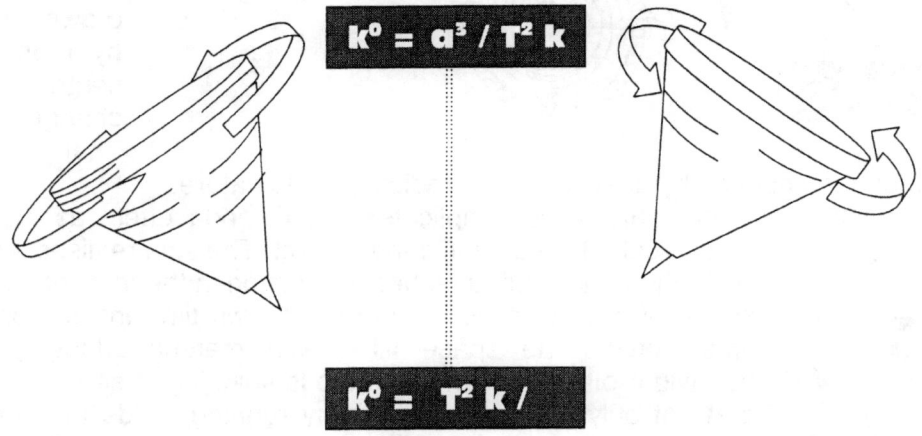

$$k^0 = a^3 / T^2 k$$

$$k^0 = T^2 k /$$

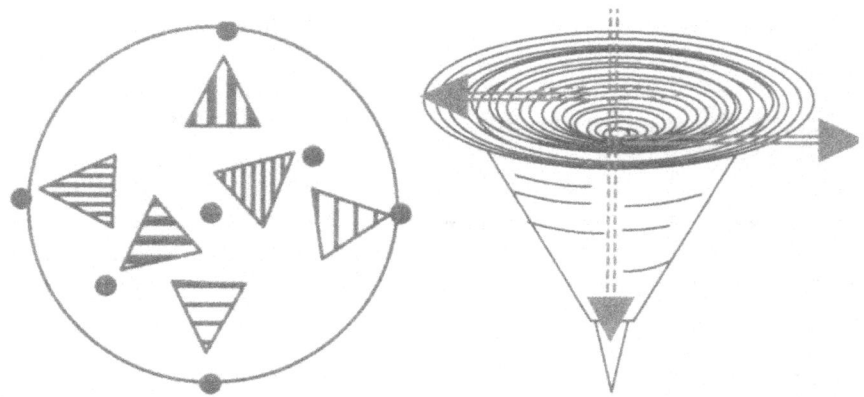

The motion introduces space between singularity in infinity and singularity in eternity when space splits time. In this movement an entire new Universe comes about and space –time again establish patrician in time with no start and time with no end.

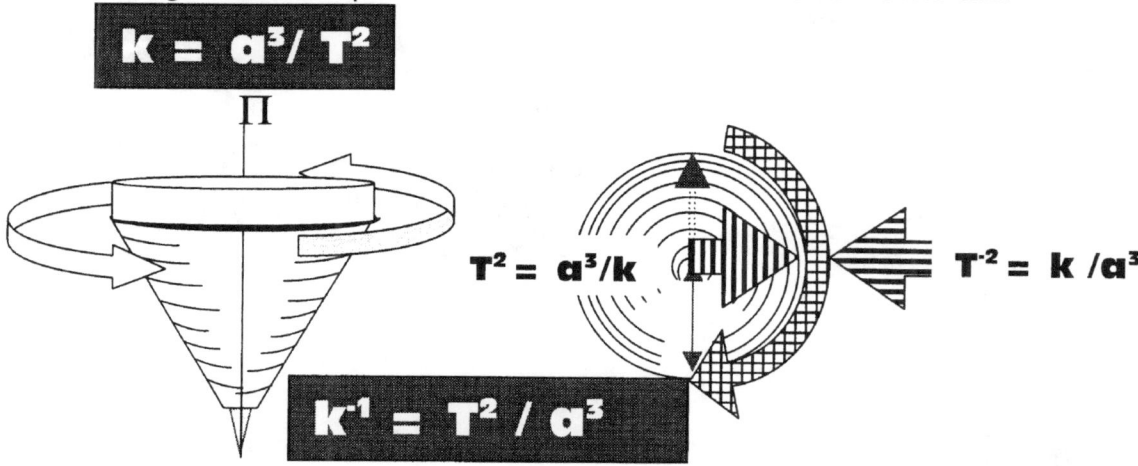

$$k = a^3 / T^2$$

$$\Pi$$

$$T^2 = a^3/k$$

$$T^{-2} = k /a^3$$

$$k^{-1} = T^2 / a^3$$

The motion activates singularity but also establishes singularity by creating limits and borders. Those limits and borders serve as the gravity, applying a differentiation the spin direction brings about, by normal flow of rotational spin. This gravity establishing factor we call the Coanda principle and the division also divide the liquid relation as a factor from the solid as a factor. We find that the centre secures a point of control and the borders form an expanding of the limits. When the spin exceeds the limits, the expanding of the borders tries to find a way of release or relieve from the controlling centre. When the limits of rotation can no longer sustain a motion, it is the borders that become unable to balance the control and the centre control diminishes the spin. The liquid reduces as it wishes to contract while the space claims as it expands.

We have to be clear about what we think of when we think of the Universe. Most people think of a picture recalling the black night sky when thinking of the Universe and that thought is most incorrect. Einstein was most correct when he declared the Universe was going flat where gravity is at its utmost, but the concern we should have is not with the mathematics being valid or not but with the vision about the Universe being what we think of and where we place the Universe. The Universe is in the centre of what is spinning and the biggest single particle that is spinning in total independence of the rest of what forms a total Universe is the atom. The atom spins and by the motion the atom evokes the Universe forming what must be the group effort of all the atoms then spin by the motion the atom renders the rest of the larger Universe. The Universe is the part that allows the rest of what the Universe establishes to spin. What spin you may ask. Kepler said it without saying it: $k^0 = a^3 / T^2 k$ and not even Einstein with his super human mathematical skills could say it better or more accurately.

The motion established by singularity results in the implicating of the Coanda effect as much as the motion establishes the Coanda effect. The spin realises the space limit while the space limit attaches the motion to the space in the time within the time. With the top spinning, the Coanda effect steps in and does justice to Kepler's formula. Time is always a displacement of space in relation to the implication of singularity, and comes about between two points in space relating to the centre of singularity as positioned by **k**, either to the value of **k** or to k^0.

With the top spinning as it establishes the Coanda principle, it brings justice to Kepler's formula.

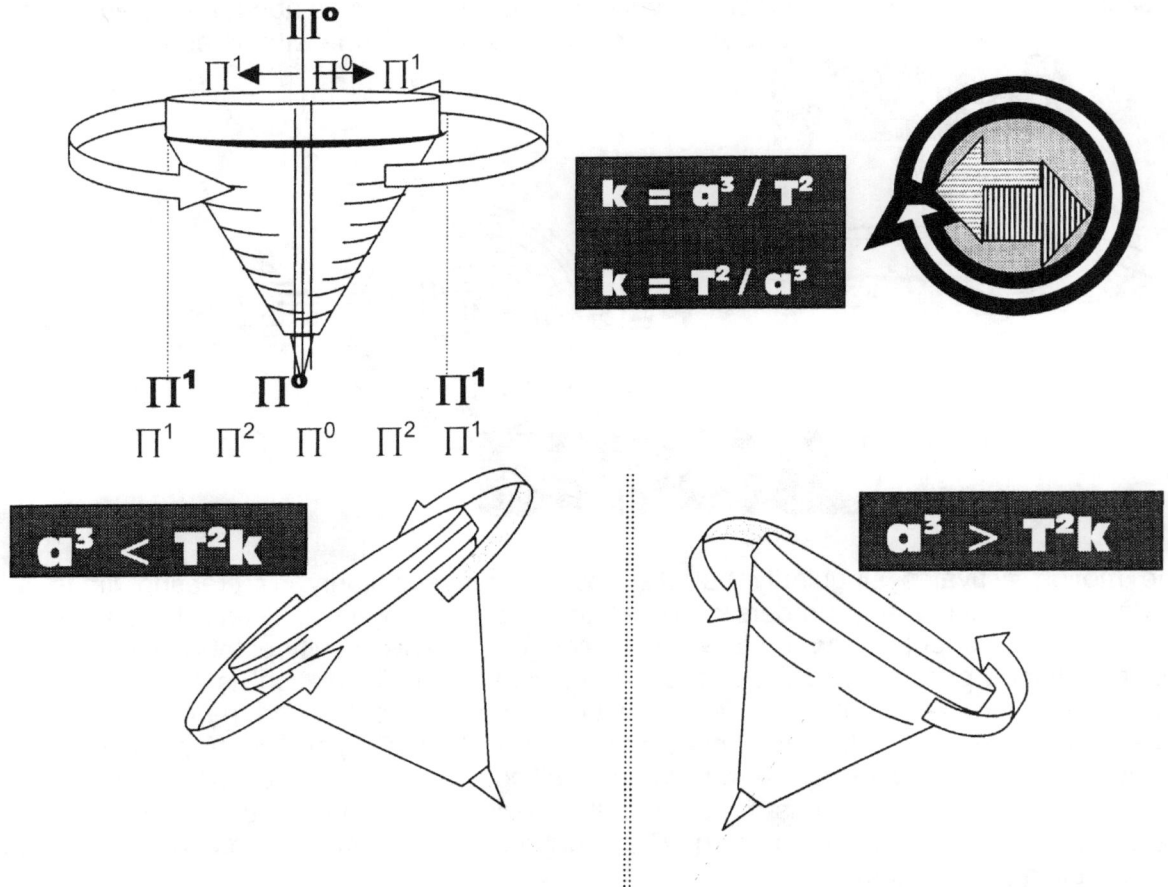

$$k = a^3 / T^2$$

$$k = T^2 / a^3$$

$$a^3 < T^2 k$$

$$a^3 > T^2 k$$

Time is always a displacement of space in relation to the implication of singularity, and comes about between two points where time forms time by being divided by space. Time in the centre is generated by parting from time in motion where that establishes space as **k** departs from k^0

We not only have to start thinking exclusively space-time, but we have to stop not thinking space as a substance standing apart from time because it is the same thing. The one is not merely complimenting the other it is providing the other with substance of being something. Without space there is no time and without time there is no space. It is having a left side when there can only be a left side when there is a right side to match the left side, and without the right side the left side disappears.

We cannot accept space without accepting dimensions and by accepting dimensions we accept space because time is the duplicating of sides to form the space that time is matching. It is space doubling or it is time halving but it is the same thing. When space falls away inside a Black hole this comes about as time goes eternal inside the Black hole. The Neutrons leaving the Neutron star provides the Neutron star with time to dismiss the neutrons as it finds the time to provide the neutron with space to leave. It is not only modern and cool to think space-time (as I one day heard a lector tell his class) but it is of utmost understanding the concepts to think space-time because there is no other way of thinking than to think of space as time and time as space.

In the centre we have a space that is not there because the space has no space. On the outside we have a space that never ends because such space goes without limits. In between we have time parted by space, which is heart or light and that light or heat forms a time delay. What we are in is a time delay between time with no end and time with no start.

Singularity by Time

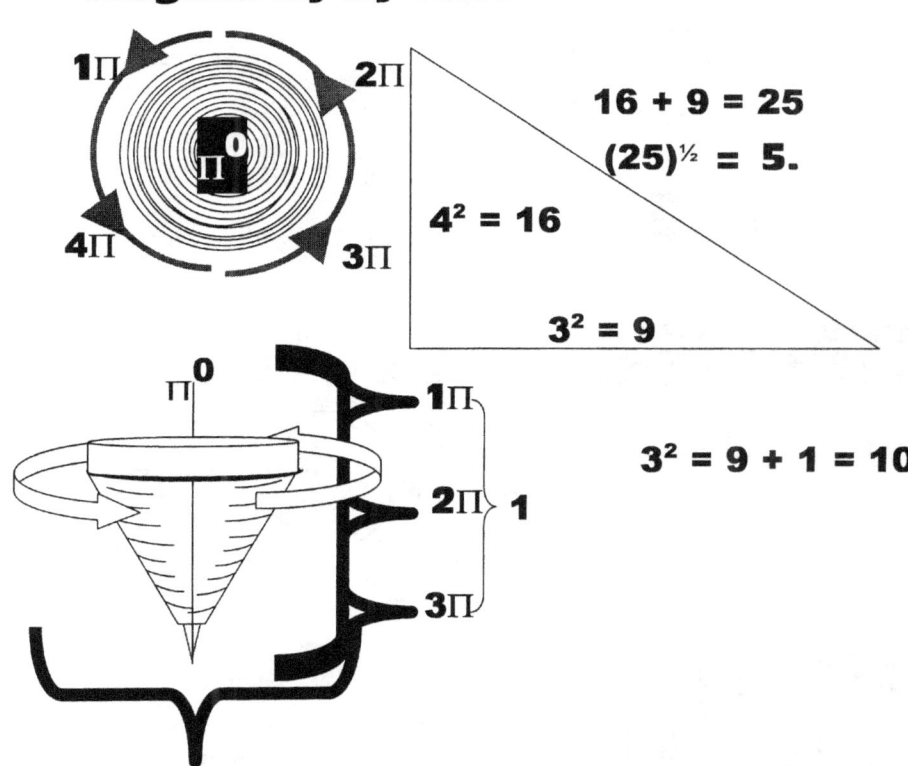

$$16 + 9 = 25$$
$$(25)^{1/2} = 5.$$
$$4^2 = 16$$
$$3^2 = 9$$

$$3^2 = 9 + 1 = 10$$

At the point where space began, time began because time and gravity is the same thing. It is motion that is creating space. That is why the Universe is that which it is today because to the Universe everything that is in the past is taken through the present into the future. Expanding is more gravity by being antigravity. It is antigravity applying heat and will always expand. It is securing heat to cool the rest through applying gravity or space conservation. But material grows as space grows and as **k** extends so will T^2 and a^3 extend. When the dimension, walls reduce space reduces but so does time increase because the gravity providing the density claiming, the space increases. Space compressed is heat denser and in that is time because time is motion of heat in space. By accepting that, there is some conductor (not the ether of old) between two cosmic structures can one accept that there are certain invisible undetectable influences on the edges outside the surrounding of material.

One then can see how space conforms as it converts to liquid heat. In the same effort one can see how material confirms heat from space to material. By reducing and confining the heat drawn to the centre, the space becomes more concentrated as the heat levels begins to increase. Take any bicycle pump and compress the plunger and the result will be that the heat created by such action will burn the finger you use to cover the valve hole. The heat comes about from concentrating the space, which holds the air. But the air does not concentrate because one does not bring in more air than there was before. The relevancy changes as the space reduce to change the space back to heat. This action is the very opposite of an explosion. But reducing pace, the action brings about that the space turns to heat. This is most crucial in accepting because this is the precondition about the understanding as much as accepting a new concept, which I try to introduce, and at the same time I try to produce a concerned effort in dismissing myths from cosmology.

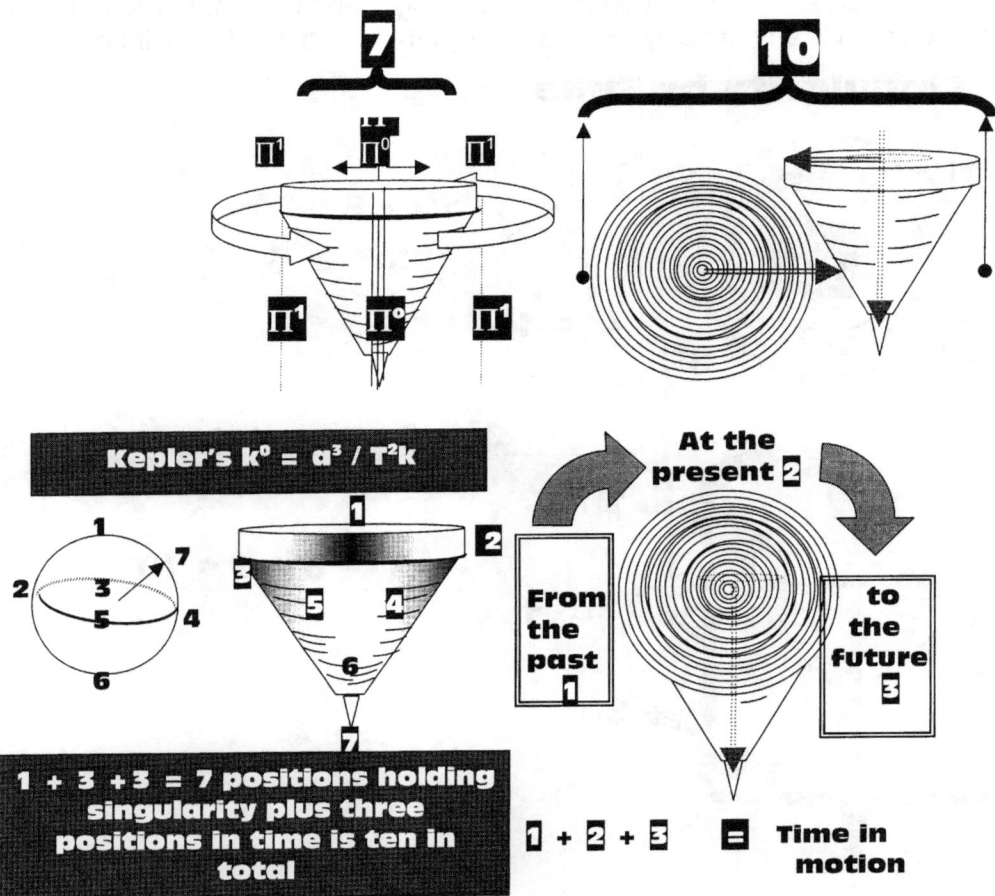

Gravity is about turning space into hotter denser space as it reduces space and that is the reason why there is no visible or measurable stronger gravity in the centre of objects. The centre does not indicate more gravity because the gravity that is the measure of the accumulation of heat that the gravity produces in that space is $a^3 = kT^2$. The heat increases as the gravity becomes more intense because the more intense heat is the more the gravity increases. By the reducing of the space coming down towards the smaller area, such coming down leads to space reducing, which brings about heat increases. The stronger gravity personifies in the denser heat produced by the reduced space. As space increases, heat dissipates and that we find is what happens in an explosion. The bigger the heat release, the more space becomes available as winds (shock waves to use the name hiding the truth) blowing across fields. The winds come about as space multiplies through the release of heat creating new space that was not there before the time. In the explosion will the heat decrease be the decrease of gravity that was before the explosion bounding the heat into condensed space and the explosion is the release or antigravity reducing the heat as it increases the space. In my search I stumbled on two accepted but not intergraded laws and when I found and located singularity the two laws became very much plausible and factual.

Take a circle and reduce such a circle constantly to where it no longer can reduce. Reduce it to a point where only form remains part of the circle because the radius has gone beyond human measure and becomes so small it is not noticeable with what ever tools man may use, then what remains is pi since pi does not indicate size but indicate form, and form is all that then will remain. I believe one can begin to see where my suspicions are heading because the flaw comes about in the manner mathematics are practised for thousands of years. Space is nothing because that means space is a standard fit all issued out before the time. Space cannot increase and winds are ghost blowing their breath. Winds are as much antigravity returning reduced space back to increased space. Before coming to the mathematics, I would first like to bring your attention to the practical side. I am promoting a theory in which I am able to prove there is as much contraction going on in the cosmic universe as there is expansion and the contraction is as much part of the expansion. The universe rides on a balance and we have to locate such a balance. To prove my theory, I firstly had to locate the centre of the universe. Even admitting to such a notion sounds like madness, but please give me a chance to explain in more detail. I realised that my effort to locate the point holding singularity enabled me to backtrack the exploding universe to its origins. By applying some basic effort I have located the position from where all movement came and the direction it took moving forward in time…and yes, even time as such. Gravity is the dimensional changing of space holding r as reference in the cube as to the sphere holding Π as the reference. In order to generate spin producing time in matter occupying space, therefore creating dimensional change, Π has to be a factor indicating the possibility of spin because implementing Π the circle sides will follow one another without establishing separation. The answer must be in finding Π, and thereby locating singularity. If singularity is in affect the original point of the cosmic birth, the reducing path we should follow will indicate the whereabouts such a point must be.

In the normal applied mathematics there are two standard formulas used to calculate a circle. The one use an r to indicate the radius and the other use a D to indicate the diameter, which is double the radius and therefore needs to be divided by a four to eliminate the Newtonian inverse square law amounting to the difference there will be between the two. The one using the radius is Πr^2 and the other formula using the diameter is $\Pi D^2 / 4$. However one looks at the mathematical expressions and Kepler's formulating of space-time, there is an exceptional difference between the two scientific uses. When investigating Kepler's formula one do find it appreciably differs from the normal Mathematical equation like $a^2 = r^2\Pi$ and $a^3 = 4/3\ \Pi r^3$.

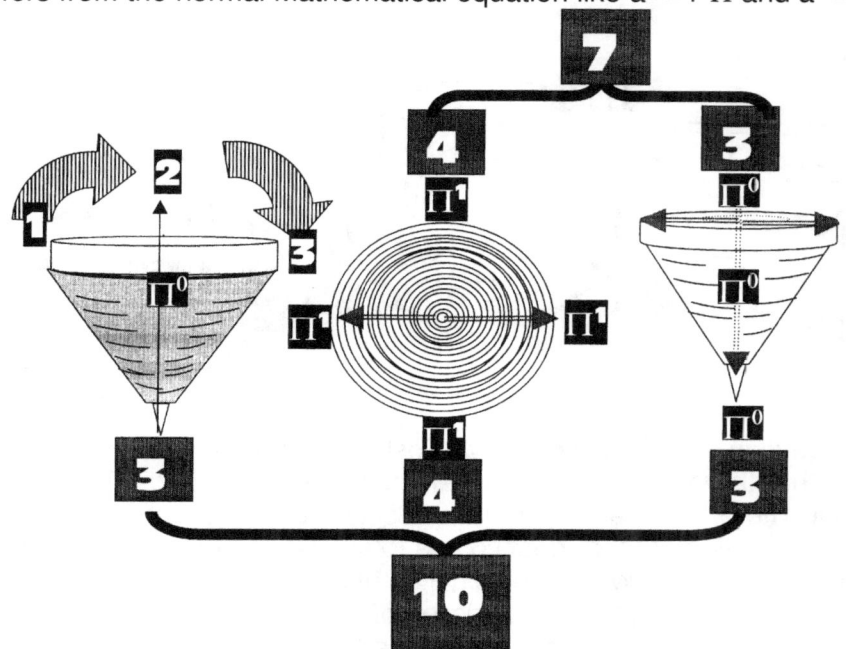

In the normally used mathematical expressions, such equations tend to concentrate on the volumetric aspect. In the case of Kepler's expression it is something else that wants to surface. It is another idea that is coming to mind. In Kepler's formula a^3 stands to symbolise the third dimension and such a third dimension becomes equal to two other dimensions grouping and sharing value to equal a^3 efforts. It is not the circle of the rotation because with such a normal circle the radius is in the square and Π evaluates form. Here there is no mention of a factor Π, which one would suspect to be somewhere applying since the circle is Π and Π is the circle and the two are inseparable. But not in Kepler's a^3, where there is no mention of Π at all.

The fact that there is a radius of some sorts used to indicate a position cannot hold the square as it normally does in the case of the normal equations. In the mathematical equation the factor indicating the position of the circle edge has the square value being called the radius or in some cases the radius doubles and which then is the diameter, and the circle indicator is Π. But in this event the formula value will bring about a square value to the answer one receives. It will bring a value to the surface of the circle. In Kepler's formula it specifically does not. I am not the one that brought Newton into disrepute. Before me the cosmos did. The comets with they're not colliding did, and so did Roche and Lagrangian principles. Hubble was another one and it becomes apparent that every one that made a study about matters in the cosmos was in some disagreement about Newton. By Newton's effort to improvise on behalf of Kepler, Newton made a statement that Kepler never made. In all honesty nature reacted strongly against the claims Newton made on behalf of Kepler and not about Kepler's work but about Newton's modifying of Kepler's work. In short: how can a comet sail past the Sun time after time without colliding and still apply a contraction in the manner which Newton suggested by the one claiming a freezing grip on the other? This strongly contradicts $F = G \, (M.m) / r^2$ How can five structures as the LAGRANGIAN POINT form around a centre structure while the centre structure keeps the five in position at equilibrium? By rejecting Newton's improvising, this strongly contradicts $F = G \, (M.m) / r^2$.

The normal perception is that any circle growing spontaneous would grow by the radius, which is r. In mathematics that may be true but it is not true in nature. In nature that cannot be the case because r is an indication of a straight line. By growing with the aid of a straight line from the centre to circle, the influence that that would have on the circle would result in many circles following one another and not a continuous growth.

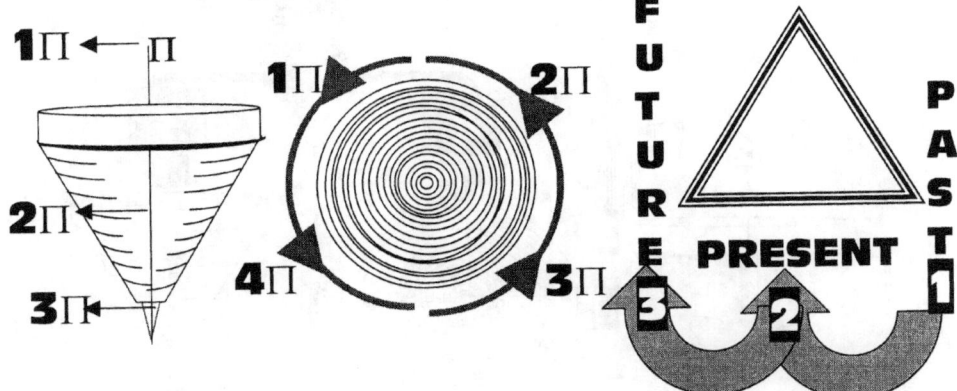

If we wish to believe in the Big Bang and we wish to accept the factor of singularity, then we have to accept that there was a period where there was no space. We can backtrack the space to a point where the space is no longer space on the precondition that the space is coming about from motion. If more heat comes to such a centre the centre will produce more motion. The motion will produce more duplication of space and the duplication of space is the gravity we experience as a contracting direction of motion where as heating is the expanding of space through motion. But it had to have started with a space less motionless dimensionless Universe wrapped in singularity. The differentiation coming from motion is a dimensional barrier that changes many aspects in cosmology. The dimensions came about as the Universe came about and each had its individual introduction period. Space and time parted at $(\Pi^3)^2 = 961$, material formed identities at $\Pi x \Pi^2 x \Pi^3 / 5 = 192$ and $\Pi^2 x \Pi^2 x \Pi^2 / 5 = 192$ where space either had material or had heat without material and space separated from heat and matter at space holding $10/7\pi^2/2(\pi^2 + \pi^2) = 139$ material $7(\pi^2 + \pi^2) = 138$ and space having liquid within $7/10\ \pi^2/2(\pi^2 + \pi^2) = 136$ This is suggesting that these are meaning this was the first time liquid became part of the cosmos while all were still part of the same unit as the Roche principle would suggest $(\pi^2/2)$ as well as the $(7/10)$ and the $(10/7)$.

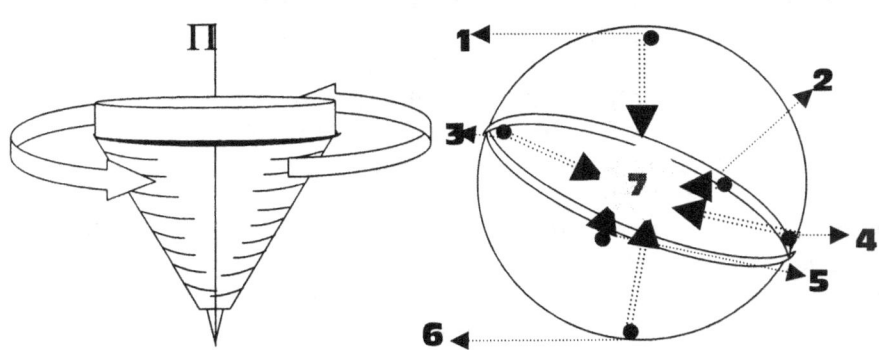

Material seems to get glued to the earth by some force, which holds the name of gravity. Moving such a gravitated particle needs some drag by motion. The secret of lessening the effort in applying motion to the object in need of shifting is reducing the drag that is not drag at all, but motion not in motion and hiding behind the name of mass. Let us find this drag in nature and work from there to find a better natural understanding of being in a solid state on ground or a liquid flowing down into the ground or a gas floating above the ground. It is accepted by science that water can rub together and form static electricity. Can you believe respectable men say this shamelessly! The scientist that thought this one up had some or other big problem with his hair and thought that rubbing his hair with a plastic comb will be the same as water rubbing against water. I cannot believe that science can indorse such shit and shit it is! How can water form static electricity because lightning is the product of gas going into liquid by motion by turbulence of heat. Let's remove gravity and find mass.

It is well documented that the volumetric content also known as space, increases where the heat levels increase. This is because heat turns to space and space becomes more when heated. By increasing heat levels, the space also increases and becomes volumetrically more than what the space was. The heat forms the space, which increases the space that the matter is occupying. The space is more than what it was before the heating. By heating material there is an introducing of space because the space needed after heating becomes more than what the space would have required before the heating started. By heating, the material with a sudden burst will bring about so much space available it brings along the destruction of matter in the position and form it holds. This destruction we know as an explosion. The advancing of the newly formed space reconstructs the position layout the matter holds. With the knowledge and countless demonstrations brought about by war and other destruction, the opposite must also apply. With the reduction of space in forming heat, the reconstruction of matter therefore also must come about from such a manner. Where matter removes heat to reconstruct its element worth and element position in value, the reconstruction is the direct opposition to the deconstruction of matter by explosion, therefore the relevancy changes and with the relevancy changing the result, therefore must become reversal to the explosion.

In outer space an object floats. On the moon nothing solid will float but there is no liquid air either. Those not familiar with this statement must think what the difference is of space in outer space and space in the atmosphere and why objects entering the atmosphere suddenly acquire the ability to heat up and burn out. The reason why outer space is a gas is because the density applying between space and material is very little in comparison to what we are use to in the atmosphere. Outer space is not colder but hotter, much hotter. In space all objects are very loosely connected and move quite freely about. When this occur it reminds one of a gas because in a liquid there is much more density in the matter relation and when solid the matter is as close as can be found. Therefore the conditions in outer space form a gas and a gas is the hottest of the three conditions there are available to substance material. Comparing the likeness with anything we can compare to with our vision of what is on earth, we must move to something we all consider to be a natural in all three forms, one being solid ice (very cold), two being liquid water (less cold) and three being gaseous steam (very hot). Conditions in outer space come down to steam because there is much space between the particles bringing about more space in the density than there is material. By introducing heat to water, water changes from being a solid we call ice where there is much more material in the ratio between space not filled and material filling space whereas with through liquid such as we call water is more space unfilled than there is in solids but much less there is than in gas and gas we call steam.

By introducing heat to water we change water from a solid (cold) to a liquid (less cold and more hot) and with the introducing of much more heat we get the heat to become a gas such as it is in outer space. By introducing heat to water we get clouds forming. By introducing heat to air we get clouds moving, moving excessively, where the movement in fact displays a density increase. Because there is a density increase the wind can uproot large trees. To suggest that something as light as say oxygen and nitrogen can blow down a tree with the quantities present in such a density as one find in space proves how little science are able to think! With more increase of density in the wind we find spiral motion in lateral movement. With wind circling it has terrific density because in such a form it not only uproots trees but also takes on houses and much of what man can build. The wind in access, blowing extensively, produces the same qualities by producing destruction damage than water flooding in a river can match. When the density increases by adding motion, much of the increase goes along with vapour, a form of air that is thick with water, (I distinguish the terminology because why not only use one word, either steam or vapour. After all it is the same thing!) In clouds we find lots of vapour but we find little water. The difference between water and vapour is that vapour has more unoccupied space and less space filled with water material in ratio. The thick density is there, but the air is so thick the vapour and the air combines to form a gaseous liquid we can see as a cloud. Remove the heat in the cloud (which the cloud needs to be being vapour) the water returns as a liquid and produces a form of heat we named lightning. The vapour liquefies to electricity as liquid heat and falls as a solid in the form of water we named rain. The liquid which was the water did not vanish but became liquid air which we call lightning. The question about density increase always comes from material being more prominent and more abundant in such a space. With the increase of density it always accompanies the increase of heat and the discharging of heat.

If one would think that it is vapour in the wind that increased to such an extent that the wind can uproot trees, then why does hail with such a lot of solid water not uproot the trees. There is a world of difference between windstorms and hailstorms because of the abundance of electricity or more bluntly phrased heat in spinning motion. Windstorms having the ability to uproot the trees have a very sticky substance between the molecules and the more sticky evidence there is, the bigger the ability to cause damage. The air substance shows a bigger resistance to part or create space than the tree shows to remain secured by its roots. The substance can be sticky to the point where it breaks branches that will require an effort of many hundreds of Newton meter to break.

Even spraying the tree with water which man created artificially by pressurising the water would hardly break the branches and less hardly uproot the tree. If it is done, the water flow will be enormous, but many times I have seen trees uprooted without one drop of water visible. The only logic remaining of such a density increase must be the heat. In this there are changing relevancy dynamics, which I then introduced as equal to the substance found in atoms.

 When one takes Kepler's equation into consideration the whole process of motion starts to make sense. It is motion that keeps the top erect and that was accepted from the time of Newton. However, by close scrutiny as well as considering Kepler's equation, where he underlined the fact that space move through time by the statement $a^3 = T^2k$, it emphatically reads that the space a^3 is equal $=$ to the motion T^2 of the space in relation to a very specific centre k. He acclimated the role of space that is moving through both ends of time and being pertinent, it then forms the space in between time.

But this relation works both ways and not only from one side. It has nothing to do with pulling because if the two were pulling the top had no chance of any motion. The top uses the motion of the Earth to the advantage of the top and then on top of the Earth's motion, the top applies individual initiative and claims even more independence than the structural independence it had before. If it were merely gravity of mass and nothing more, then the gravity disappearance the Earth produces would be such an imbalance that the top would never stand a chance of committing individual motion. However should my view be correct about motion and mass being the frustration of motion hindering the motion of gravity, then yes, the top by motion free the distorting of the Earth mass, which is a frustration to the top gravity motion and that would enable the top to spin even with such slight energy applied. Merely taking into account that the top has to overcome the considerable mass the Earth brings the Mass theory into dispute because how can the top with such slender energy find freedom from the enormity of the Earth mass.

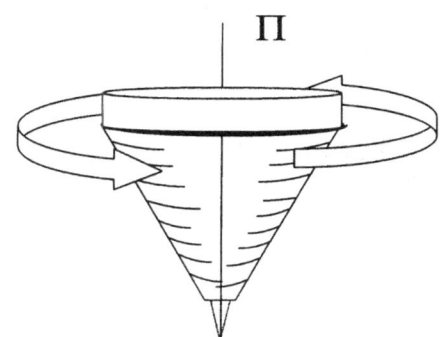

When the line **has a beginning and an end at the very same spot** and it wishes to extend the position as to further the possibility it has, which direction should it favour? Extending the line in any one direction will favour one direction without any clear reason not extending in other directions. The only mathematically sensible option about extending will be in all directions equally in order to give a meaningful non-bias flow of mathematical equilibrium.

Every round object has a point establishing a very centre, a middle dividing one side from the other. That division determines the space from one side away from the other side. At one point there must be a point that does not fall on either side of the divide. Such a point will still be a circle, because from that side the circle divides into two sectors.

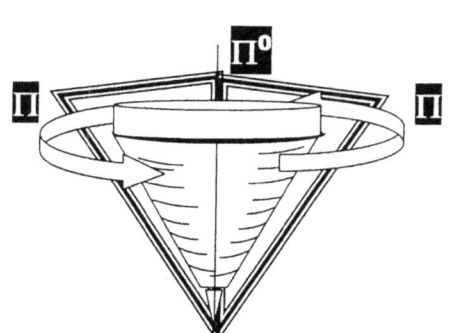

In every object there is a centre but the centre we find in the sphere is the only centre that can take complete charge of gravity because it is the only centre that is controlling every point on the edge of the circle at any point the circle can offer.

In all units there is a singularity seeking independence in relation to singularity elected seeking dominance. From one side and any singularity there will be present in the unit one factor of singularity carrying the value of 1 but also there will be one divided by space square singularity absent because only one singularity can apply to the unit in dimension. Therefore, one tenth of the space is absent where singularity is one and holding .9991 valid as part of the other side of the Universe it is attached to but not connected to. With the unit being connected to motion the motion will stand related to seven from the one side as well as the full ten from the other side on both sides of the Universe. That relates as ten in space-time on both sides of the Universe (10 + 10 = 20) plus one factor of singularity present and one factor in the tenth not present (1+.9+.09+.009+.0009+.00009-.0001=1.99991)

Many dots spawned from the spot and such spawning was in flurries but equal.

Then the four cosmic pillars set laws and progress started happening as the cosmos applied and stuck to conditions set under these principles.

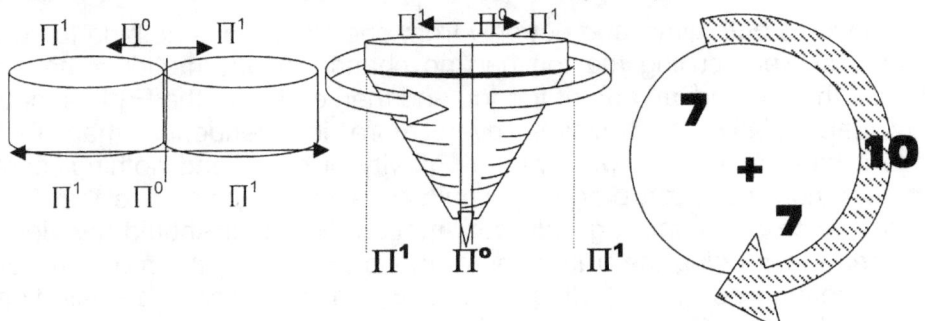

Some singularity formed a dominating role. As they were in dominating, some particles formed a subordinate role but space-time formed since there had to be motion and motion provide form, which can be space. But I wish to place a distinction between form and space because by calling what was in progress space, the mediate human connection would be to adapt space by dimension to what was in place. If there were space it was Π, which is form. There were no mathematical equations because for mathematics to be in place there has to be space. There was no space of that particular type yet invented. Please think clearly as this is very important. Is that not precisely the commitment we find in gravity, where gravity is flowing from singularity outwards but never favouring any side? This reasoning prompted me to look for singularity in such a spot because if the prime spot, from which all came, was a spot holding all, then the spot must hold the shortest line but more prominent, it will hold the smallest form including the smallest circle or for that matter the smallest sphere. With gravity always being in the centre of a sphere where the space is least available in the entire structure (there is not even space left to fill) one finds a flow of gravity from that centre spot outwards in all possible direction even-handedly. The fact that the original gravity will begin as a circle or will be a circle is the direction it will take when being the first spot created. All progress will be evenly in all direction because no direction will stand out or be in favour above any other direction at first.

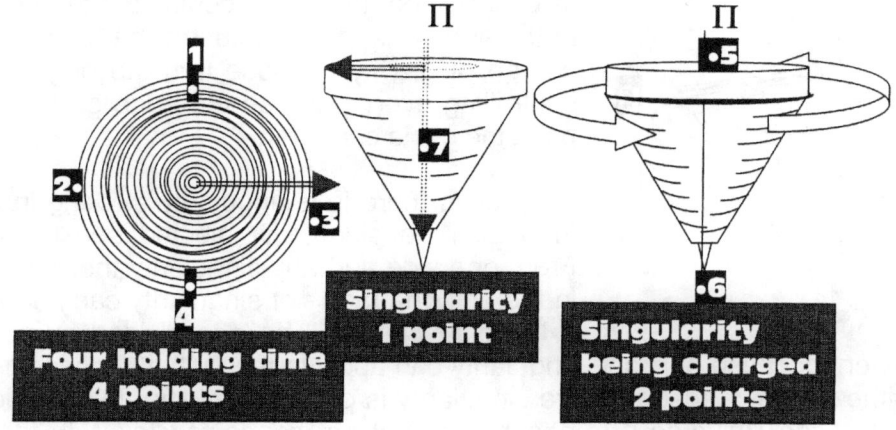

Four holding time 4 points

Singularity 1 point

Singularity being charged 2 points

Again I have to stress: where one divides into zero and where zero divides into one is where the beginning of all there are, is. That is where one would have to go look for the beginning of the Universe. The Universe is a about lines connecting but where does than connecting end. The Universe used form to this point in development, but then at some point the line came and established the presence it still has. The first form was moving from $\Pi^0 \Rightarrow \Pi$. Again I wish to press the issue, at that stage form was in use and not mathematics. The Universe was just simply too big to measure. If radius did apply, one could use r and r^2 but since only Π was in use there was no radius used.

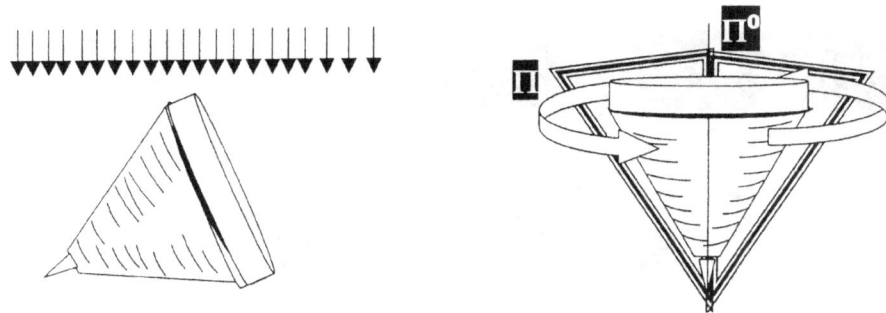

There is one Universe of difference between the top lying down showing no independent motion and the top spinning erect. A top on its side that is not spinning exists as part of another Universe. The top's entire Universe collapsed when the motion seized to activate singularity in infinity. While being in a motionless state the top submits to the singularity lines running towards the centre of the Earth. There are gravity lines that to us are invisible. Nevertheless, as they are invisible, still they are there running through from the top (or the sky) at 180^0 to the Earth centre, placing the lines at 90^0 angle in relation to the rotational direction of the top. The curve ball is but one example proving the influence that these lines exert. They are just as valid in outer space where we find satellites have to spin and establish singularity in order to allow meaningful control from the Earth. These lines suppress the structure of the top to confirm in mass the gravity the Earth applies. The top succumbs to the flow of the lines running to the centre of the Earth and it is the responsibility of these lines to eventually burry (grave, gravity) the top as part of the structure of the Earth. That is the purpose of the mass because by applying mass to the top the Earth will eventually have the top relinquishing its structural independence to the Earth and totally submit all individuality to become part of the Earth. But when motion is added to the structure, the top holds another dimension of independence that are accomplished and the top receives almost the same cosmic independence as the Earth has. It is the task of the top to uphold the motion and maintain independent singularity while the Earth singularity will fight to submit the chances of independence by restraining the top even to a point where the top is liquid.

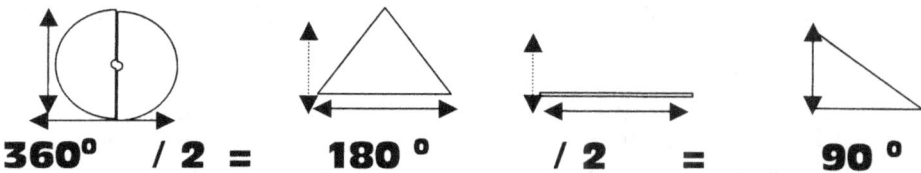

360⁰   / 2 =   180 ⁰   / 2   =   90 ⁰

The circle is a square holding a round shape, as the straight line is a square holding one side to infinity. Calculating a circle involves two aspects where the one is either the radius or the diameter that is double the radius. The other is the factor Π. Because gravity works both ways and not singularly in one direction as the Newtonian myth would have us believe, there is the interaction in the neutron position between the total of material in relation to time formed in space as space and time formed in space in relation to the total of material. In contrast to Newtonian view about a spontaneous effort there are in the cosmos of joining and sharing. That is untrue because in fact, quite the contrary is true. There is a natural tendency to remain independent t and away from each other and where the tendency of staying apart is bridged, there is a tendency to destroy and conquer, to control and delete the lesser by an onslaught of the more superior. There is no mass fighting to join mass and to become one in all. That part is fiction as much as the part about a force is fiction. There is a struggle for superiority and there is a fight for freedom from dominance. The whole idea about masses joining and uniting runs very much against the basic fabric of cosmology and in particular the Big Bang theory, The sound barrier principle, the Coanda affect of motion bringing about space-time control and so many more. There is no contraction by mass, as Newtonian myth would have us believe.

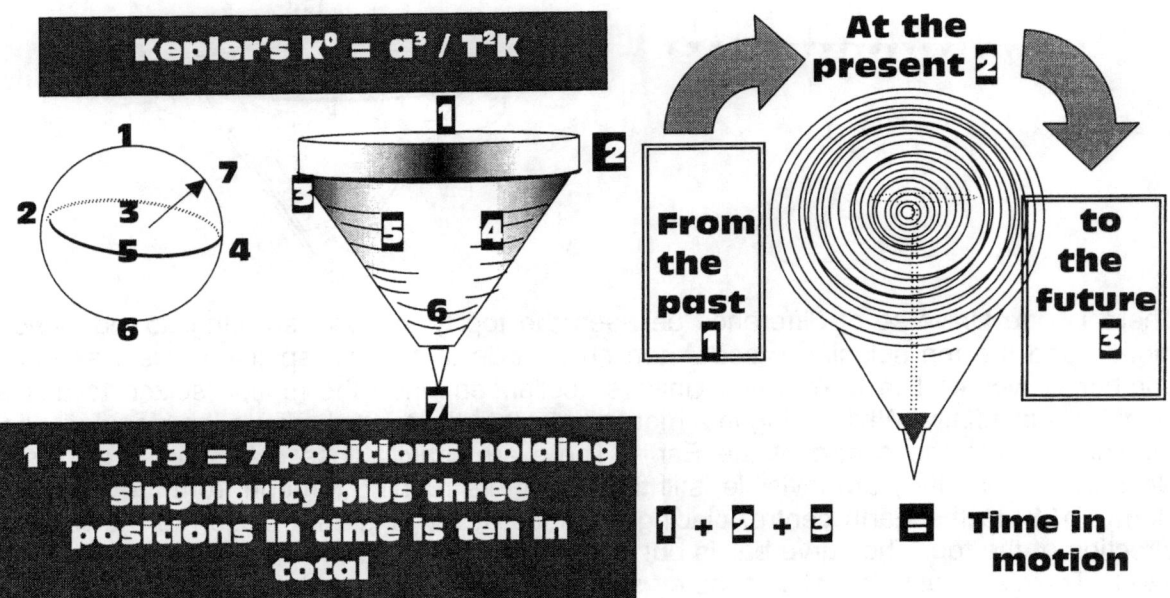

If the joining that Newton claimed were with merit, we would by now not have known a moon orbiting apart and on its own coarse around the Earth around the Sun around the Milky Way. While there are those attachments they are only attachments and not obsessions of joining and uniting. The moon holds a separate identity, which it refuses to relinquish. This refusal we call a lunar cycle. The moon is on a running spree ever since it's Independence Day. The moon is taking a route that would progressively carry the moon further way from the Earth as the Earth is rerouting its orbit further way from the Sun. The question to ask is what makes the Sun, the Sun and the Earth, the Earth and the moon the moon. It is $a^3 = T^2 k$ or better put it is $\Pi^3 = \Pi^2 \Pi$. It is the space-time as collected by singularity using the Coanda affect. The Coanda effect shows that it is through motion of space that gravity comes about. It is the way Kepler introduced gravity and the way Kepler predicted gravity before Newton named such a concept, that gravity applies. It is established by spin of space in time.

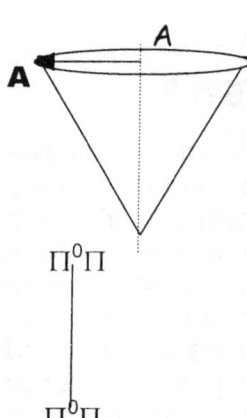

The spin splits infinity from eternity by placing space in between time. In the centre is infinity where the motion of eternity parts infinity in time by space. In the centre is that which has no beginning because the spin does not apply. It has a value of infinity and infinity at the time was combined with eternity. This very same process applies as electricity is generated. The moving of Π^0 to Π activate a line that was not there before. It was Π^0 moving to Π that evoked a line but the fact of the matter is that the line is still not there.

The line shows a presence in *Singularity* (A) establishing (**A**) and that shift announced the rise of another Universe. The movement did not do nothing as Newton would indicate, but the motion evoked the birth of an entire Universe surrounding singularity. The motion brought the top into space just as Kepler announced with his formula $a^3 = T^2k$ where it says space is produced in equal measure of the motion of the space…and the top is the undeniable proof of Kepler's statement. The top by motion brings space into the Universe and without motion the space is denounced as a Universe by the Earth motion. Should the motion of the Earth end all space accountable to the Earth will seize. It is once more proof that time cannot stand still as Newton, Einstein and Mainstream science would declare because if time stands still, all fall back into and to singularity. Singularity is one being $k^0 = a^3 / T^2k$ and if T^0 then all the other factors would follow the same path.

The top has space but there is a space in which the top spins that covers the time part of space. That is the time part Einstein identified (1) as coming from (2) being at and (3) going to and the position holding singularity is represented by the entire body that holds all the space of the spinning top. As soon as the spinning of the top commences, the time aspect releases space in which the top spins from the space holding the time of the Earth and the rest of the Earth within that time. It is this space in time that becomes so hot when the aircraft is speeding because the motion takes the time back to what the time was when time was nearer to the Big Bang. By receiving space, singularity received a value from where it was in eternity Π^0 to just one point outside eternity as Π. But the motion of seven relating to ten brought about gravity as Π^2

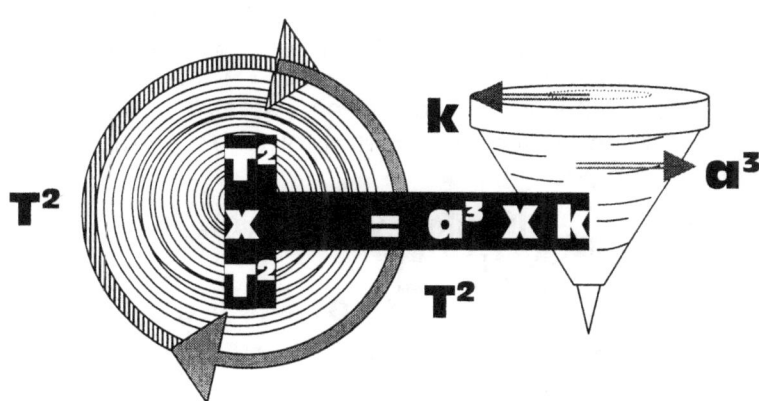

With everything in a cube or a circle or a potential of the two, brings about the implication of eternity in a form of singularity or the point of creation. Removing the radius of a circle does not remove the circle, because the circle is there, securing the ring. If the line (or imaginary line if you wish) holding the value of $\Pi^0 = 1$, there has to be a point where the circle is no longer in infinity but claims existence outside the imaginary. At that point the radius may be slightly more than infinity, but to all calculating purposes, it still remains as infinity. The spin was going on for eternity because the spin does not apply, it has a value of zero and zero is another expression for eternity. The full square of the motion $T^2 X\ T^2 =$ is equal to the full space $a^3\ X\ k$ created by the motion and that relevance became the atom.

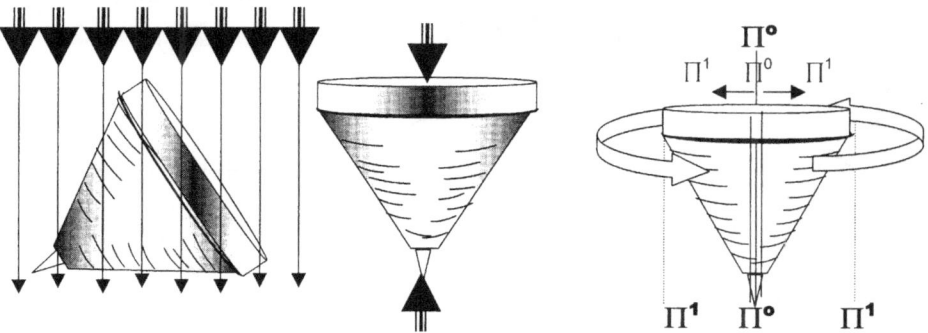

Singularity forms lines running towards the centre of the Earth.
Having edges where Π^0 duplicates to present the edges, singularity lost the value of Π^0 to the value of Π^1 with the same value singularity had being Π^1 to the one side and Π^1 to the other side, Π^0 must be the point splitting singularity into two parts of eternity, the eternal

value of the first dimension outside eternity. It was the square of Π^1 being Π^{1+1}. That was the first dimension outside singularity Π^0 where singularity has a value of Π^1 in the form of $\Pi^{1+1=2}$. The first claim to space had a value of Π^2. This applied to both sides of the claim to space outside singularity, and the double proton became the dominant factor on matter.

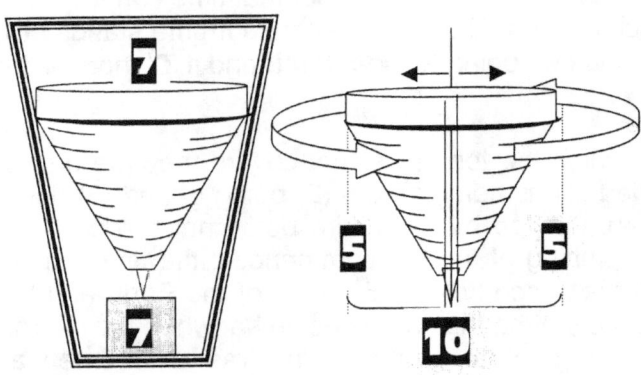

The seven is part of infinity parting eternity. Infinity is a point that one finds inside any and all solid spheres and the point is where all lines cross at a 180^0 as well as 90^0. The points form, as the rotating motion establishes a charged line that take control of space –time. Then, where singularity at seven points ends and the eighth point holding singularity begins another sector comes into action. This holds points in time eternity at eight nine and ten. The points are physical but needs to be generated while it never actually is there.

With their use of such logic makes science appear foolish. Since the time of Newton, the arguments made by those in the time of Newton, tarnished from being brilliant to clever to fair to poor and a hundred years ago it reached the point of being stupid. That is what Kepler's formula is all about. That is what Kepler indicated with his formula $a^3 = T^2 k$. The space of an object (a^3) is equal to the time (T^2), which it is in, in every given instant (k). If the space becomes smaller, the time duration becomes longer every instant of time's progress. The motion follows the graph in relation to motion and time.

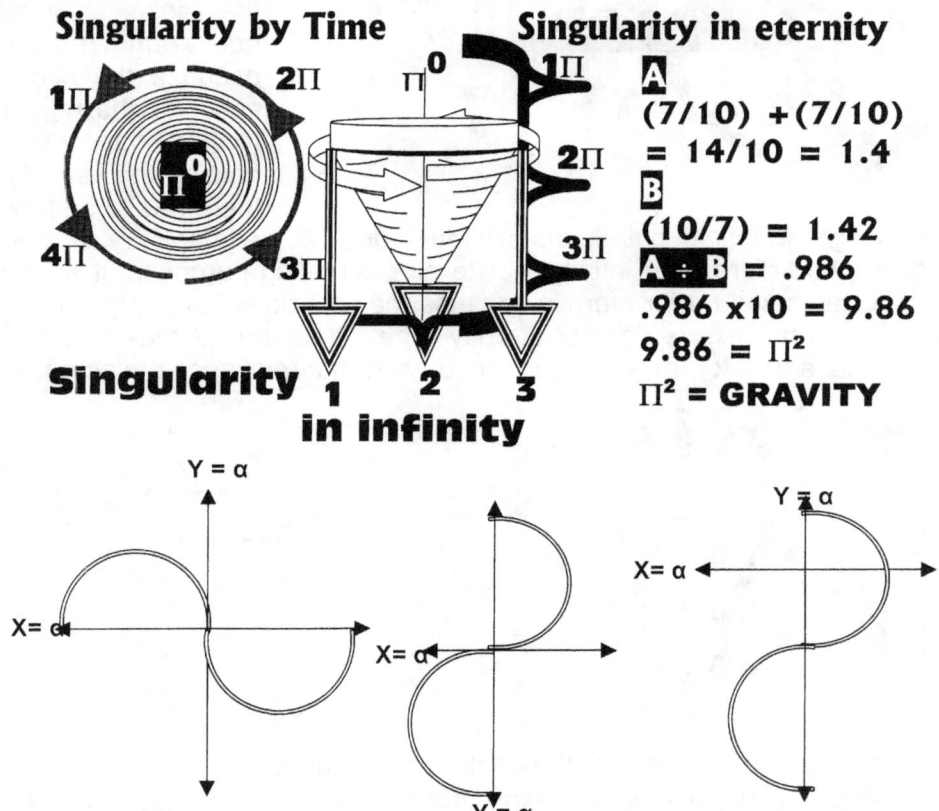

From the graph one can establish the link in the circle's rotation around a conforming unit being singularity. Saying that, one therefore has to admit that the smallest spot has to hold space because the most insignificant dot can transmit light and being able to accomplish that, that one must accept it to carry a value of something. If that spot had the value of nothing, it means that spot was not there to begin with. If the graph connected by zero, would mean there is no connection at all. With no connection the graph would be a mathematical tool with no value or use in any way.

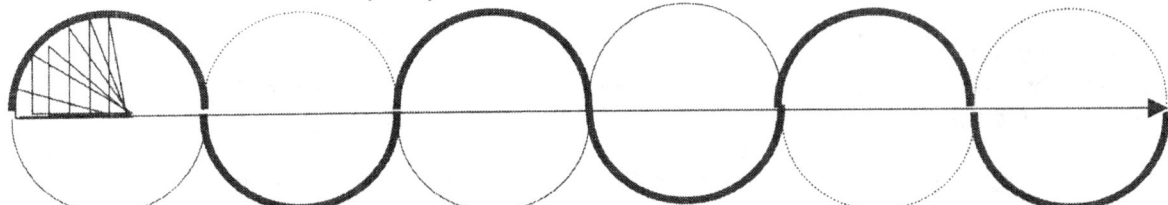

Holding space-time, one should return to the original formula indicating space-time in as much as $a^3 = T^2 k$ where $a = R$ and $T = T$. Being time, it has to alternate positions and that can therefore only apply to **k** where **k** will indicate a relation to the space-time in question or the relevancy to singularity being $k^0 = 1$. By receiving k on top of the already $k^0 = 1$ that is in place, the top becomes an atom by erecting the line of singularity from $k^0 = 1$ to $k^0 = a^3 / T^2 k$

It started with a dot, because that is the only form, size and dimension mathematical logic will allow our brain to accept. From the one dot had to come a second dot and a third dot. The dynamics of such a dot is smaller than we can understand because such a dot is in negative relation to what we see Π to be, and delving deeper, we will eventually find the smallest fragment where space started, in the spot where time is still eternal as much as we can accept eternity to be.

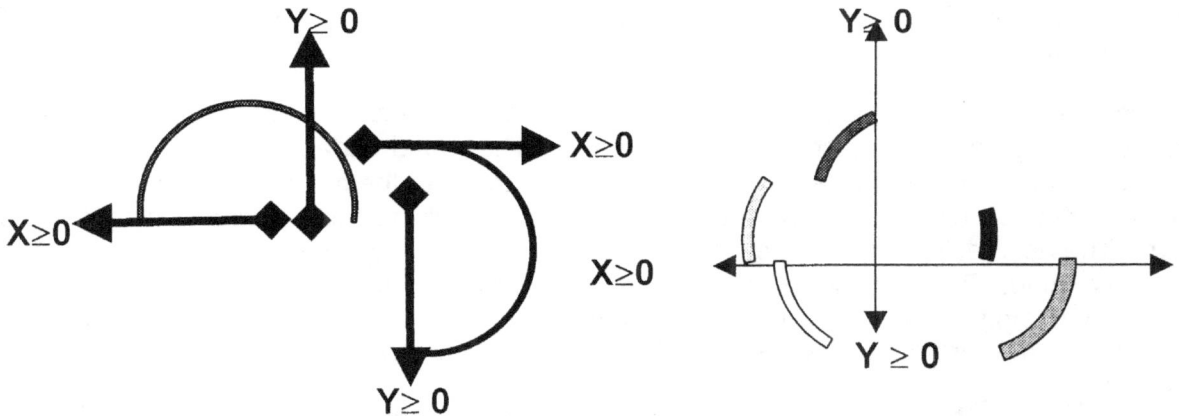

The graph with no connation between points because zero or nothing connects the points will render the use there of in mathematical terms quite obsolete.

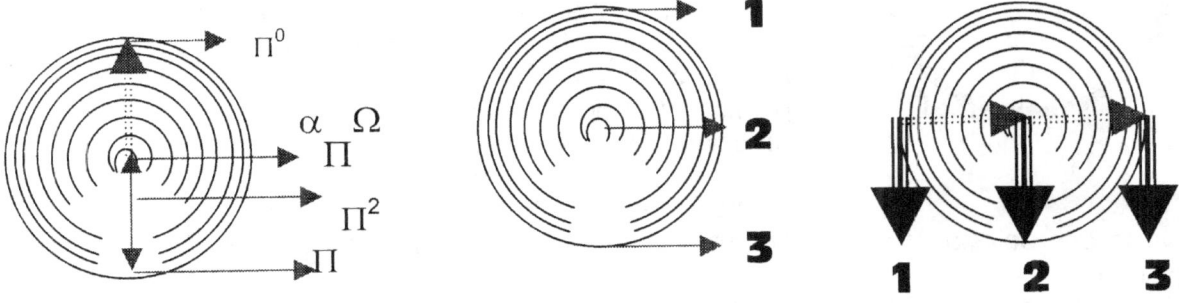

The reason why we should first locate the spot is because we can only work from that point forward. By working forward we have to work backwards to locate where we are heading.

The cosmos started at a point and where such a point is, we will find the Universe. Every one knows where the Universe is, because we can see where the Universe is, but if we can see where the Universe is, then we should find the centre of the Universe in that spot. Einstein theoretically positioned the point of beginning at a place he indicated where singularity should be.

By the duplication, it therefore insists on a relevancy, because without relevancy there can be no motion and no motion means no space. The strongest proof there is about this is the manner in which the Coanda principle applies the reproducing of space taking shape from a round object and involving motion to produce such duplication.

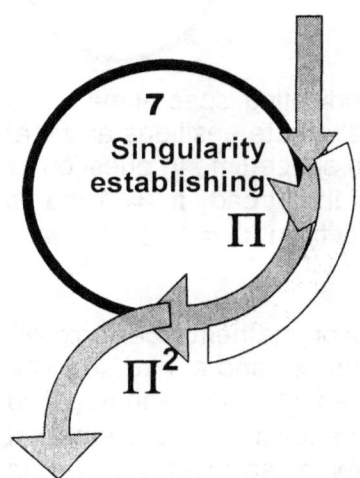

The relation forming the duplication of singularity is duplication but applies as a dimensional forming of Π and placing 7 in relation to 10 forming Π^2. The liquid applying motion forms the 10 disciplines. No motion leaves no Coanda as well as no gravity because gravity is motion that duplicates singularity

The Coanda principle which in fact should be seen as a law because it is that strong, is the principle of gravity duplicating with the motion that provides such duplication a relation of the particles having the seven factor and such a factor of seven produces through motion another three dimensions. This total that material fill while in motion is ten and when ten crosses the line of singularity to duplicate the seven on the other side of the Universe the crossing cuts singularity in two as much as it puts singularity in the square. But it involves the motion of concentrating space to be or hold fluids around solids that may or may not move. In this must be a solid, a round basis Π, fluids concentrated in space and motion applying to one or all of the factors.

Conditions that prescribes the enactment of the Coanda effect is that the one surface has to duplicate singularity by establishing Π as a form. The round surface Π will bring about the shape of singularity Π that becomes enticed by the action of the motion of the liquid or of the solid or the motion of both around Π, which then establishes and confirms singularity by form. The next factor is the presence of liquid. Air or atmosphere is liquid and water is liquid. Heat is liquid. The third factor being just as important is the motion establishing Π^2 by duplicating singularity as singularity becomes relevant through the applied motion that produces gravity from the singularity spot that provides the form.

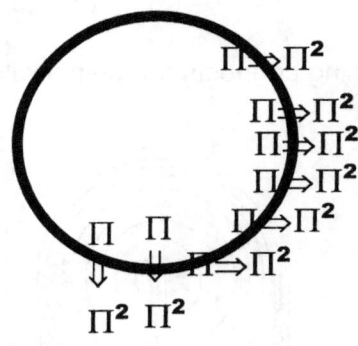

That too forms the answer about the question concerning how the Titius Bode implicates gravity in cosmology. The seven sides are linked by rotation. Nothing changes because there is a steady linking to the inside centre of the sphere. But it is to the outside that this rotation brings about dimensional complications. There are five T_1 points moving to five T_2 making contact with five moving points. The moving non fixed points is the point before reducing by five to the point after reducing by five that bring along the ten points in stead of the

five to one point as it is the case with the Lagrangian system.

The heat brings about expanding singularity from a one sided affair to filling a volumetric Universe. But all of it is a relevancy where ten positions will sacrifice individuality and compromise singularity in order to secure two positions in singularity. The spin that comes about from such expanding and the duplication has the Coanda principle as result, where in the same motion of the

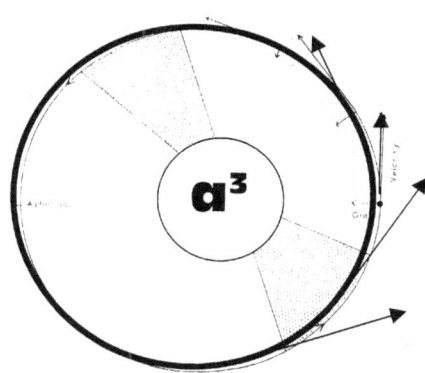

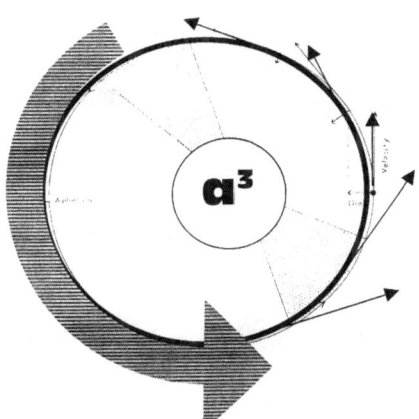

same unit forms opposition in spin direction, and in that is what forms both factors in gravity. It is this liquid relating to solid, or motion relating to motionless ness that forms the Coanda effect where two forms (solid and liquid) bond as a unit. The solid in relation to the liquid substantiate the contradicting nature there is in circular motion. The liquid will ale\ways substitute the solid because the liquid always contradicts the solid.

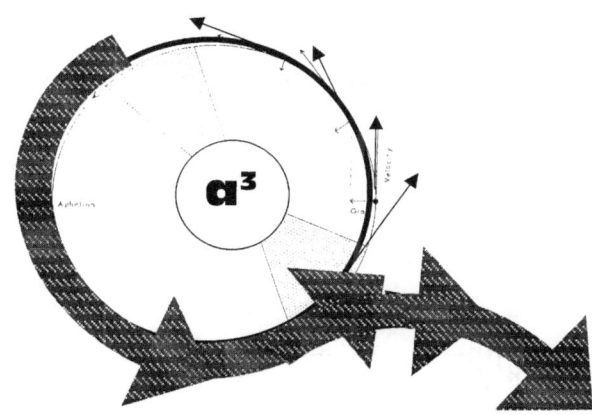

The orbit of any cosmic object around a controlling centre object is a fight between two relevancies born from the conflict in rotation. In every rotation there is one part that moves in the opposing direction of the other part notwithstanding that it is the same object or that it is changing relevancies or that the one will very shortly be in the role the other side has at the present motion. It is the one going to the future while the other is in the present holding singularity and the third is coming from the past. One can not literally give value or connect a measure to either but giving an non-relating example just for the sake of making conversation it is similar to the one coming from the past being k^{-1} $=T^2/a^3$ and the other is at the same time on the other side of the Universe at that time $k = a^3 / T^2$ and the present confirms singularity at $k^0 = kT^2/a^3$. There is always an opposing to the other side of the present form. While the one is striving to advance by confirming new space, the other is conforming space by retracting and the third holds the balance of circling around. In that we find the Coanda principle.

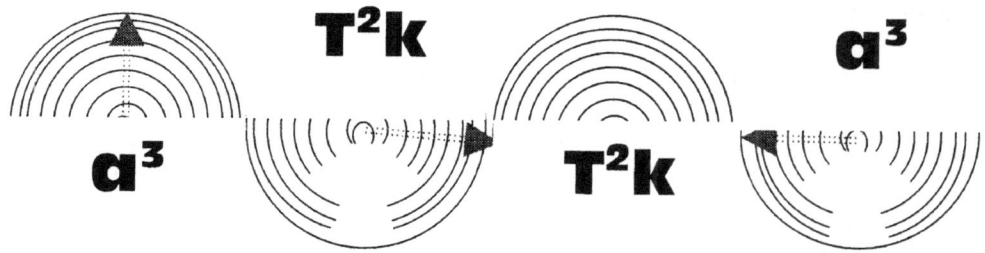

There are always the conflicting sides, which is a built in characteristic of rotation. The one side forms the space while the next is in motion of going away or coming towards at the same time. In that we find the Coanda effect producing gravity by applying opposing directions in the same unit as a result of the spin contracting as well as expanding simultaneously but on different sides of the divide.

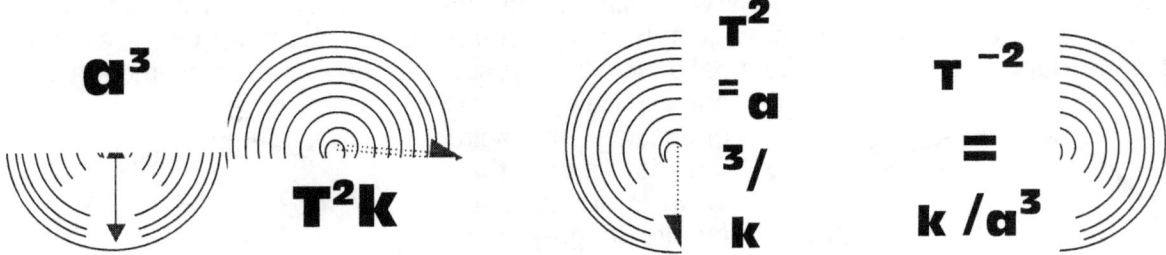

The lot is more evidently moving further apart

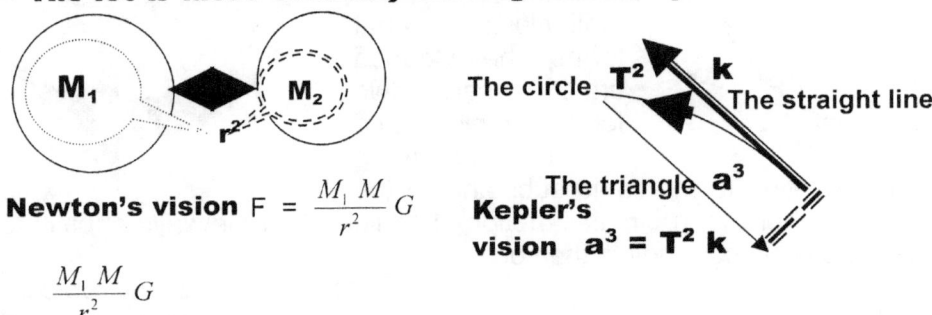

Newton's vision $F = \dfrac{M_1 M}{r^2} G$

Kepler's vision $a^3 = T^2 k$

The circle T^2 k   The straight line

The triangle a^3

$\dfrac{M_1 M}{r^2} G$

We have this tendency of a dual in rotating action that is the principle that is bringing about parts of the same unit orbiting other part of the same unit and still being in conflict with the other part of the same purpose and that is to rotate a centre.

If we look at the rotation we find from a human aspect that we put much claim to the circle. The circle holds importance but however important the circle may be, the circle forms part of eternity, which is the perfect part of creation.

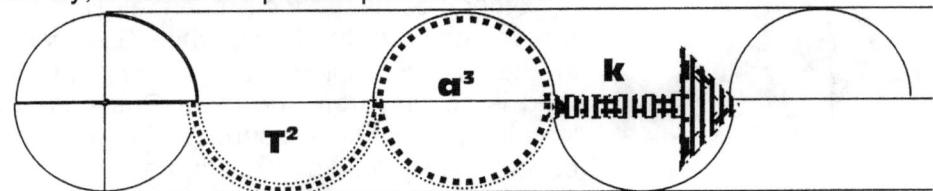

From the graph one can establish the link in the circle's rotation around a conforming unit being singularity.

Newton said that the rotation brings no influence and that he accomplished by allowing time to stand still. What he said was quite true because if time stood still all the Universe will tumble down into singularity or into the centre where from our perspective we find that **dJ / dt = 0.** This however will be a totally destructed cosmos

When $k = k^0 = 1$ then at the time also $a^3 = T^2$ and $a^3 = 1$ leaving $T^2 = 1$

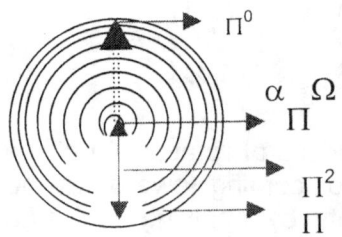

When Π^0 expands to Π we think of the rotation coming from the expanding. That is very much true but there is another aspect every one is missing. While the circle comes about time also shifts the centre to another location. In the duplication time brings the motion forward and although "forward" would not be a direction that is possible to point at, so is singularity not a point visible. Time in infinity does bring about the expanding of the sphere in the infinitive number of circles but the motion in time in eternity repositions such location to a new relevancy where the entirety of all the Universe rematches to find altogether new relevancies all over.

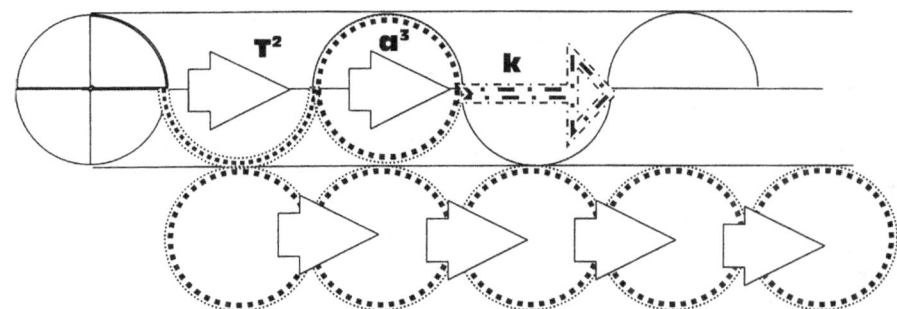

Matter or material is the concentration of heat that has gone dense because of time delay. I do not wish to elaborate that at this moment but there is a book out in which I explain this statement in much detail. It is the time delay of the shift in time from the past running through the present into the future that produces the material density we call the Universe in matter space and time. The duplication of the material is so extensive that the entirety of the Universe has to be demolished and again re-established to allow motion to take place. The one atom pulls the other atom while the one atom pushes the other atom to take its place the very next instant. Singularity cannot move but is totally rigid. Look at the centre of the top activated and one can see how immovable singularity truly is.

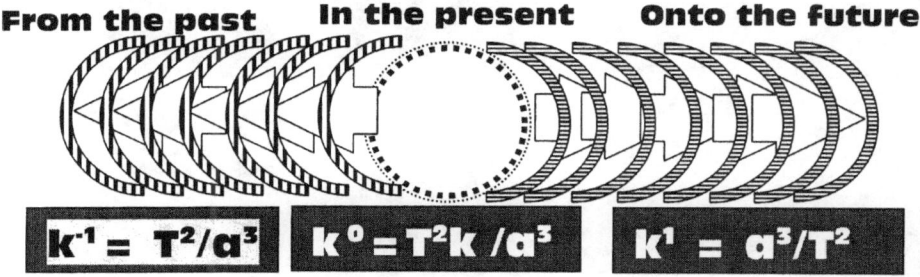

The entirety of time that we think of as the Universe is singularity holding every possibility that it can become when generated by heat. The circle remains the circle but the circle as a unit relocates while the circle complete the circle motion. By the way, that black stuff we see at night is not black stuff but it is light so bright our eyes kill off the brightness to allow us to see by daylight. That black stuff we see at night is heat and heat is time in motion.

The material complete a circle by rotation and thereby confirm the heat that is sealed in the unit as the unit and that forms the unit we call material. That however is half the story where

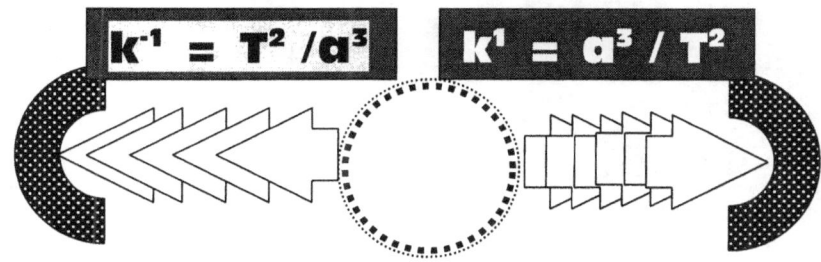

the other half is the directional redirecting of all that has spun into new allocated locations where each fin a relevancy that only apply at that specific moment. The motion redirecting movement plays as much a critical role as does the circle confirming the rotation and sealing the heat into the confined space that confirms material.

Kepler said the solar system and therefore the cosmos is space a^3 being equal to the time thereof T^2k. When one translates the formula it explains the Big Bang precisely. It says that the relevancy $k = a^3 / T^2$ and $k^{-1} = T^2 / a3$. With this formula one come to understand the flow of the cosmos much better. When using the formula in this back to front and opposing front to back we get to realise how it is that a cosmos can contract while it is expanding. It is a translation of one definition seen from opposing perspectives. As much as the one is filling with heat the other part is emptying with space and as much as the one part of filling with solid the other part is emptying with liquid. In all the balance remains the same no matter form which perspective one approaches the solution. What is contained remains contained but flows from one side of the container to another side of the container. By the flow of time space is emptying of liquid concentrate as much as it is filling with space concentrate while the other is filling with liquid contained and cooled in material that then forms a solid as it freezes in the motion that exceeds the sped of light. That means the atom inside spins faster than the speed of light. The product in space is time because it takes more time for the planet to either grow towards or to grow apart from one another. That which is between the two cosmic objects is time because space is time. The atom filling with liquid concentrate is also demanding more space and the demand on space is proof of the growth in time. When everything was new the lot was without space. The more time develops the more space there is. That means space is equal to time.

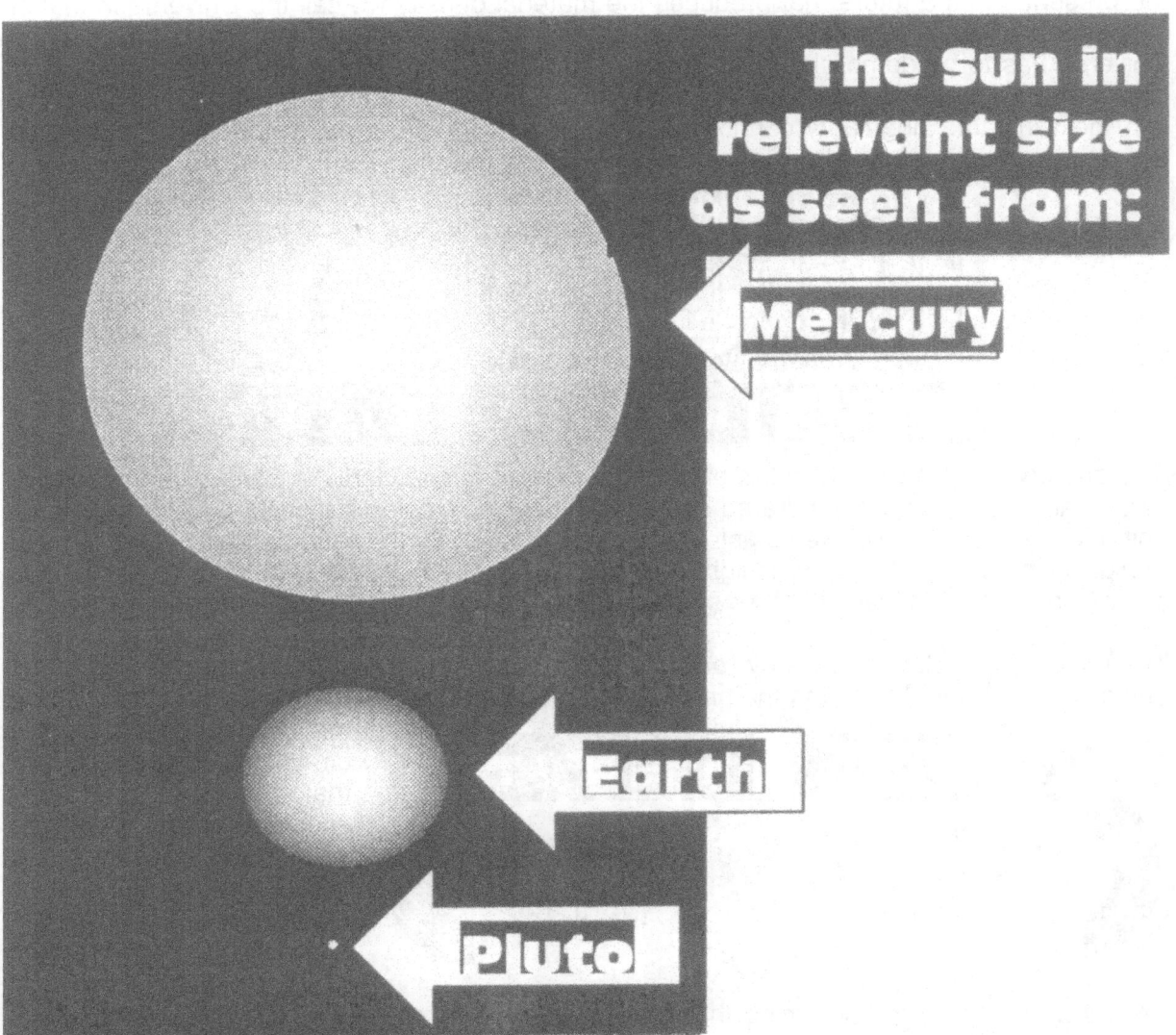

The Sun in relevant size as seen from:

Mercury

Earth

Pluto

As the material duplicates and the duplication forms motion, the material drags what it left behind as much as it pushes what it is catching in the future. The atom is following the other atom but to fill the time slot in the future the following atom has to relinquish the position it holds in the next one's future. In order to move into the next position in the future pretender the atom has to relinquish the claim it has on the present location. The only easy way to do that is to drag the next tenant into the vacant to be slot the atom at that instant fills. To move, it has to push the atom in front in to a new location while dragging the one behind to fill what it wishes to vacate. With that action we find that the relevancy **k** goes minus and goes plus which allows space to grow as time decline or to allow space to shrink as time becomes more prevalent.

Apparently, according to informed sources we will find the Sun viewed from each planet indicated by name, as the photo would suggest. From Mercury the Sun seems large therefore the Sun is close and from Pluto the Sun seems small which we think of as further. That way of thinking is very indicative of Newtonian thinking with a very explicit Earthly connotation. It is far from cosmology. When an object such as one of the photos suggests shift further away by getting smaller, it is not the distance we should consider, because the distance has no meaning in cosmology. It is the time it would take to reach the object that has to be considered. The "further away" the object seems, the longer it would take to reach the object and the "closer" the object seems, the less time it would take to reach the object. There is more time between the Sun and Pluto than the time being between the Sun and Mercury.

In the case of Mercury, time is more and therefore it divides space into a smaller factor **100k** $= a^3 / 100\ T^2$ where as the space the Sun has, seems to be more space at Mercury because $k = a^3 / T^2$ or in relevance to Pluto the Sun must be a **100** times bigger with the time being a hundred times less of a factor $k = 100a^3 / T^2$. That mathematically with the aid of Kepler proves that the Black stuff being the Biblical light that was mentioned at the explanation of creation is time and not space. Material fills space but the filled space of material moves through time we incorrectly think of as space.

When the one is pulling, the other is in opposing mode not only by spin but also by having a singularity to protect.

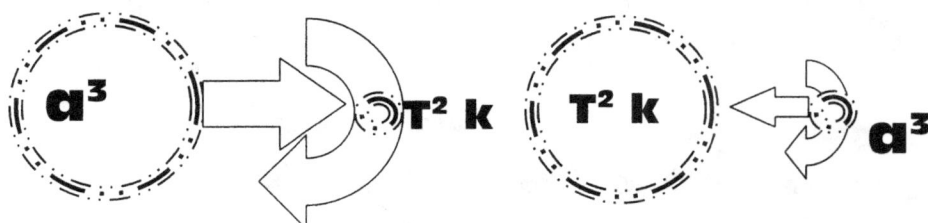

In the cosmos there is no big or small but only singularity generating space-time, Looking at the Sun and any planet, we see two points holding singularity where in both cases singularity is equal at $k^0 = a^3 / T^2\ k = 1$. On the one side of the divide the one factor holding singularity takes prominence and then crossing the divide the other point holding singularity holds dominance. Crossing the divide of singularity puts one of the two in dominance as far as controlling time and controlling space. The one holding lesser space would affectively control lesser time and the one

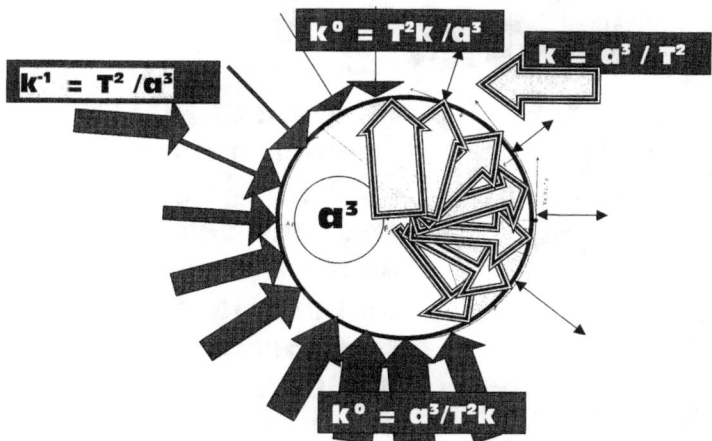

controlling more space holds more time in control.

As I mentioned a while ago, material is the delay of time bringing about heat where the motion contains the heat in confirming the location of the preserved heat. The material fill time with space by relocating singularity while singularity generates material filling time in a specific allocated location. Through the motion there will be some area where the expanding $k = a^3 / T^2$ is more frivolous and where the material is pushing the next out of position to fill the next position. Then there is the other where the delay becomes more because the delay is caused by matter not generating space filling fast enough so that the follow on filling can come about becoming motion or heat relocation of time.

In this changing of the relevancies, we find that at one point the Sun gives the planet a nudge onwards and the next part where the planet goes past the centre singularity the Sun drags the planet onwards. The one part of the circle that the planet forms in orbit is smaller than the other part that is bigger but small and big is no issue. It has more time concentrated by motion effort or less time concentrated by motion effort. In that way the Coanda effect forms gravity and locates the time intervals as structures orbit a centre of concentration.

Before time moved, there was a spot that repeated by being perfect.
Then the imperfect entered creation and the spot moved to a dot.
►By returning to the spot the dot advanced to form a new dot.

opportunity to

The motion of parting from the perfect to move into a position held by being imperfect gave time in infinity the part from time in eternity.

The returning to the previous position was always there and is part of the eternal perfect. Moving in one direction brought about a future that institutionalised time by implementing heat. It was the departing from the position the perfect held, that is time but that time is a delay caused by time going imperfect. As time becomes more imperfect so would the relevancy of repeating the perfect in ratio of the imperfect, grow and space filled with material will compact from that to become more, denser and dominating the entirety of the Universe.

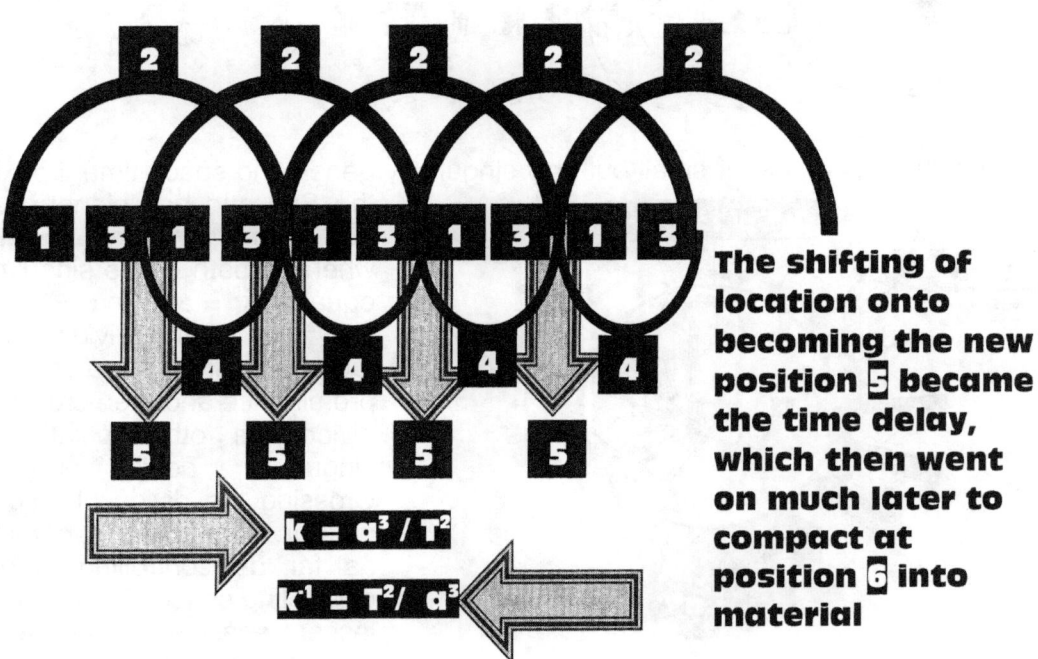

$$k = a^3 / T^2$$

$$k^{-1} = T^2 / a^3$$

The shifting of location onto becoming the new position 5 became the time delay, which then went on much later to compact at position 6 into material

Even high and low tides have nothing to do with "the pull of gravity". If it did have anything to do with the pull of gravity, then there was no reason for cyclic change since the mass of both the Earth and the moon remain at a constant. The changing of tides is the rotational cyclic contradicting nature that motion has and the Moon crosses the divide twice daily where the cycle then begins to oppose what it was before. It remains just the crossing of singularity where singularity indicates the divide the motion establishes.

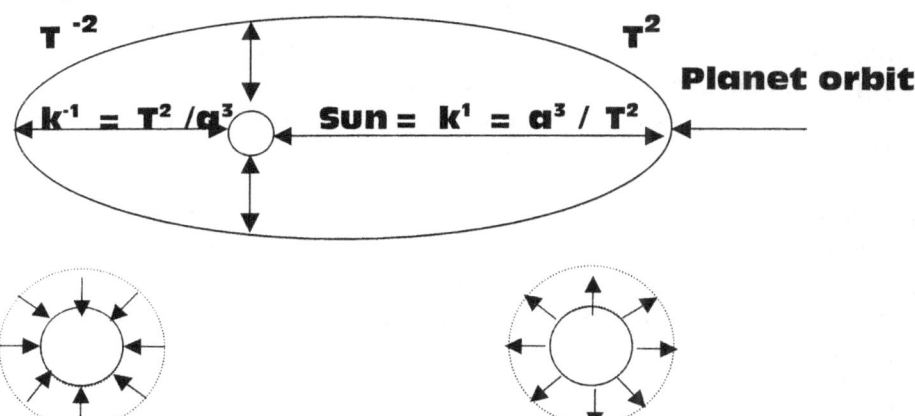

Planet orbit

T^{-2} T^2

$k^{-1} = T^2/a^3$ **Sun** $= k^1 = a^3/T^2$

Gravity is about reducing space

Expanding is all about heating. Heating takes up more space and gravity reduces space.

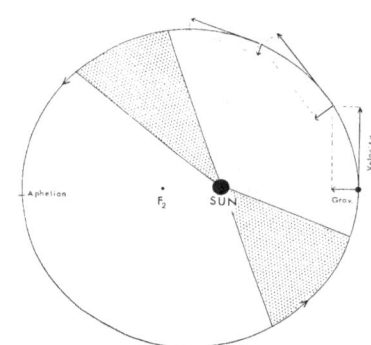

The same pattern is still very much visible in the way structures follow the centre of contraction. We still have the relevancy shifting from one to the other with not one point holding singularity absolutely domineering, We still have the one trying to expand while the other is trying to preserve and the relevancies do not go all out the way of the controlling structure. We find in this manner that the straight line goes bended by 7^0, and the curve follows the guidelines of singularity by measure of Π.

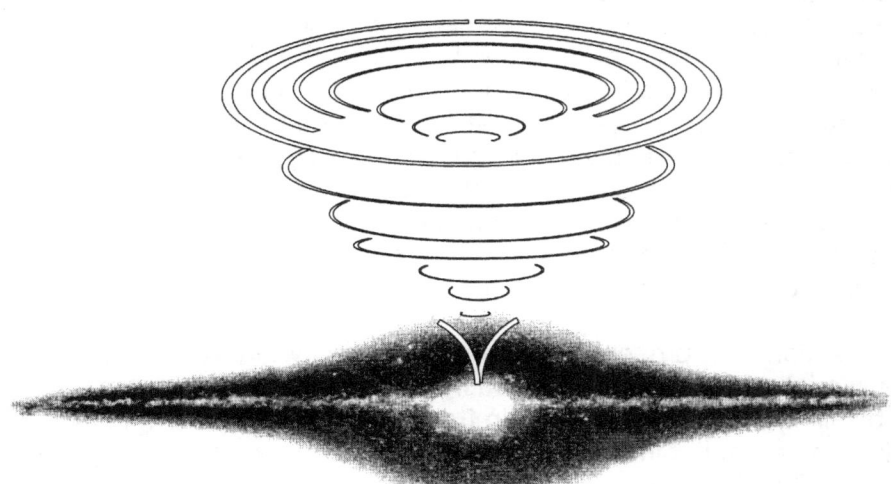

Even in galactica where they are the enormous size they are, they still generate by measure of atoms spinning in future stars that is in a cocoon blanket of heat or liquid time. Still notwithstanding size, the atoms form a unit where the rotation of such a unit generates a singularity governing in the centre with such intensity it forms the spiral we associate with Black Holes.

Let us investigate and try to find a way by using logic how a star applies gravity. Therefore it is not the number of dots that is important. It is not the size of the number of dots occupying the position or the size of the space the dots occupy that is prominent. It is the relation in the dismissing of space and the duplicating of space that becomes important. The less space there is, the more the favour will be to reduce the space because of the advantage the dots have in securing space-time that will prevent overheating. On the other hand the more space secured will also prevent overheating and therefore those will opt to duplicate space in order to find space to secure and prevent overheating.

Since the Earth has no singularity demand that is much better developed than the universe sustains, we find on Earth a relevancy of Π to $(\Pi^2+\Pi^2)(\Pi^2\Pi)3$ is adequate. But in bigger units the space-time displacing relating to space duplication presents much more demands on atomic structures occupying space within the star containing through set boundaries. In the presumed to be bigger stars there is much space filled with atoms occupying much space. In the stars more massive but holding lesser space the atoms must also hold lesser space but they also hold more protons by number in the lesser space.

I would suggest we think of stars in the following terms. Stars that generate and transmit a lot of light are weak on gravity because we can see from the progress they show that their development started recently. They command a lot of space-time but the demand they have to keep their cooling acceptable is very low. In that they can generate a lot of light but with the demand on cooling low and the gravity in the centre not very developed, those stars cast a lot of light back into outer space. It is just because of the size the stars holds that enable us to tell with a certain degree of certainty, that the stars are still young and have a weak developed governing singularity. The stars will have very prominent hydrogen and helium layers, with the inner core not very prominent. The control of the star is still very much in the individual atoms and in that the motion the atoms have to produce in order to maintain their individual singularity will only come about through motion. The atom has to make contact with as much space-time through motion as possible since it has a very poor ability in contracting space –time in support of the cooling system.

The entire motion and the entire contraction of every atom culminates as one effort and this produces a single combining effort which is then displaced to the centre of the sphere of the star where singularity is normally nurtured as a result of the shape of the sphere. The star is not a star. The star is a combination of atoms where every atom serves a purpose to give the star a specific character. Every layer of the star serves a time era and represents the star as one developing period. As soon as ythe period has gone by, the star sheds the layer in a flair phase and set about forming another layer development era. All the atoms in all the layers through out the star causes the displacement in rotation as well as contraction and both time factors provide the star with specific qualities. Stars are what all the various the atoms within the star produce as gravity that is spinning the star.

The contracting action is at present the only part of gravity that Newtonian science credit as gravity. There is a lot more to gravity than such simplicity. Every atom in a star is pushing the atom in front by filling the space the atom in front vacated. Every atom in front of every atom behind is pulling the atom behind as the atom behind is urged to fill the space that the atom in front vacated. That is motion, which is the most complex issue one can find in the Universe. Since every atom is driven by singularity and no singularity is able to move, it bring about that every singularity must remove and rebuild the space every atom fills or vacates as the atom moves along.

There is a building of an entire Universe going on in every split second and this split second is so fast we cannot name it. By naming it, there will be so many time units gone by, by the time we said the name, the Universe might not even be recognisable. We might call it energy but I hate to call it energy because energy is a lot like Holy water. It can come from anywhere and you can use it for everything and in the end it does not even become something durable because its use eventually comes to nothing.

The atom restricts dismissing of space by the containing structure to the atoms relevancy being Π^0 in singularity bringing on Π relating to $(\Pi^2+\Pi^2)(\Pi^2\Pi)3$.
As the layers swap the alignment between duplicating and dismissing the atomic relevancy adapts to comply

Since the star performs as an accumulated atom where innumerable atoms inside the confinement of the star combine to select one centre spot forming singularity that represents the star, I have chosen the to use the same symbols that I found in atoms to describe the relations in space –time to singularity within the space-time of the star. I refer to a star as a cosmic atom in other books.

Early stars still in the envelope of heat within the centre of the Galactica have only space duplication and growth through the cover of such enormous heat. These class stars are not visible but are shrouded in a blanket of heat covered by light. The atoms forming the stars are small and under developed. They remain cool because they contrast with the heat surrounding the star where the star material supports the cool space and does not form part of the liquid heat forming the outer limit. I would like to draw your attention once again to the fact that the Sun at one stage was a cool $18 \times 10^6{}^0$ on the inside and a freezing cold at 6500 ^{0} on the outside while all the time outer space was a blistering 10^{34}. This was considered the coldest place in the Universe because the Sun was still part of the deep frozen space inside the blanket of heat. Look at any galactica and see in the centre there are stars surrounded by a blanket of heat with stars conversed by heat sitting like a duck frozen in this pond of liquid heat.

With the cosmos the size it is and space so large compared to our smallness, we have no chance in finding the centre of the Universe. The Universe started where singularity is and singularity, is the sure indicator of the Universe. With all spinning objects holding singularity we then have located singularity in as much as finding the centre of the Universe. The Universe started with a dot forming. That answer arrives from taking mathematics back to a point of being the smallest possible position, far smaller than we may be able to calculate form. The ten dimensions I named the atomic relevancy is also showing the double value of singularity as singularity extends into as well as beyond space. The atomic relevancy is $(\Pi^2+\Pi^2)(\Pi^2 \times \Pi \times 3) = 1836$ that is the mass relation between the electron (3) and the proton. Proton = $(\Pi^2+\Pi^2)$ Neutron =$\Pi^2\Pi$. The atomic relevancy holds the dynamics of singularity control. In the ratio and dimensions we find in the atom, all space-time derives from the atom, whatever the atom is. Our instincts, our logic and our calculating process all indicate that the sphere holds a centre point from where six evenly positioned point's position matter to be. Using The formula $F=G\ (M_1.m_2)/\ r^2$ it indicates to a force pulling objects closer, where each force is coming from each centre point the body in question has. The contraction must commit the two bodies towards a point in each case being spot on in the middle, not withstanding what direction the force is applying, the body will draw to the centre. If the Universe spins around a centre point holding singularity, and singularity confirms the centre of the Universe, then every particle holds the centre of the Universe making the number of universal centres immeasurably many, and every atom and sub atom particle presented outside the atom in smaller bits, are all not pieces of the Universe but they are a Universe surrounded by many Universes. If every atomic particle, no matter how small, is holding the centre of the Universe, then the gravity is coming about from that point because that is where the gravity applying in the Universe is applying contraction. If the Universe did start from one single point and time, matter and space flowed from that point, then that point must have a relative connecting base because such a point holding singularity must be eternal as space, matter and time link eternal. There therefore must be one point linking the entire Universe when regarding the fact of singularity. Then according to the theory of relativity there has to be one exact point holding time in relevance notwithstanding the fact that time departs from that position and relates differently to all space-time away from such a point.

In the final analysis it is the atom that controls the Universe because it is the atom that is the Universe.

It then is the atom in the most centre part where space and time meets singularity, that Einstein found a Universe collapsing to a single dimension, and every atom at a point post of the proton where gravity initiates in according with the proton dimensional colas of $(\Pi^2+\Pi^2)(\Pi^2 \times \Pi \times 3) = 1836$

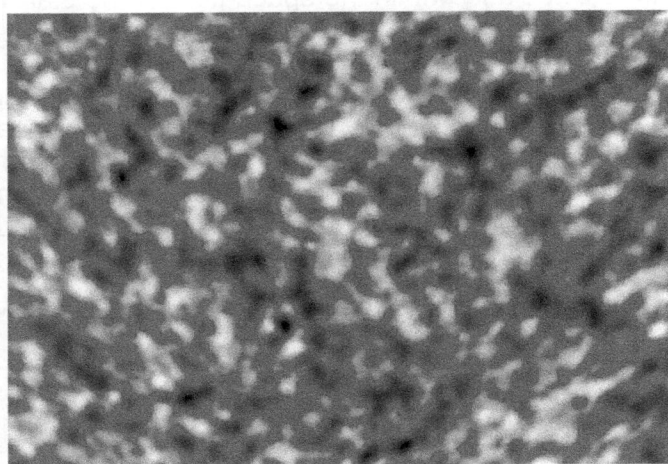

Every person with whom I have discussed the facts concerning creation recollects images in the trend depicted in a presentation as one may find to the above. That would be the most unlikely way Creation came in place. The recalling of pictures representing images about creation must have form, but to mathematics it had no form. From this thought the very opposite arises where Creation came from nothing but such an idea is mathematically simply not possible. The thought of nothing is just what it is, a thought of nothing and although it is in the nature of the human mind, to present nothing as a value in the recalling of something, nothing is a presentation of the figment in the human mind. There can be no number such as nothing and that was (possibly) Newton's biggest error. Nothing represents non-existing and that is just what nothing is, it is non-existing. In order to prove my point, I wish to ask the reader to define the shortest line there can theoretically be. If he should answer anything but that the shortest line will be at a point where the beginning and is at the same spot, he will be wrong. The shortest line that can ever be anywhere must have a start and finish holding the exact same spot. The line will be humanly impossible to create but we humans are capable of very little.

When the line has a beginning and an end at the very same spot and it wishes to extend the position as to further the possibility it has, which direction should it favour? Humans in the west would naturally think of extending from left to right while in the east humans may want to go from right to left.

Some persons will tend to go up or down, but all of the options are about human preference and not mathematical conclusions. Extending the line in any one direction will favour one direction without a conclusion about not extending in other directions. Such a conclusion has no sound mathematical foundation. The only option about extending will be in all directions equally in order to give a meaningful non-bias flow of mathematical equilibrium.

The shortest line in the realm of possibilities must have a start and finish holding one spot and such a line will also be a dot or a circle. Not favouring one direction puts all directions at equilibrium meaning that any form what ever may be can develop from such a spot with the end and the start being the same. This reasoning prompted me to look for singularity in such a spot because if the prime spot from which all came was a spot, then the spot must hold the shortest line but more prominent, it will hold the smallest form including the smallest circle. One possibility that the shortest spot can never have is having a starting point on the zero mark. If the mark of zero holds the start, it must also hold the end because the end and the beginning have the same position. If the position of zero then is the beginning, the end will also be zero leaving the line without an end as well as without a beginning. The conclusion from this is that no line can start at zero because that will be a mathematical impossibility. A line or spot starting at zero would therefore be shorter than the shortest line possible. A line growing or extending from zero can never leave zero because of the influence of being zero disqualifies any possibility of growth. If the line then had to grow in all directions at the same pace, the line must therefore be a circle. The value of the circle is Π, and that is where creation started.

Singularity by Motion

Singularity by Time

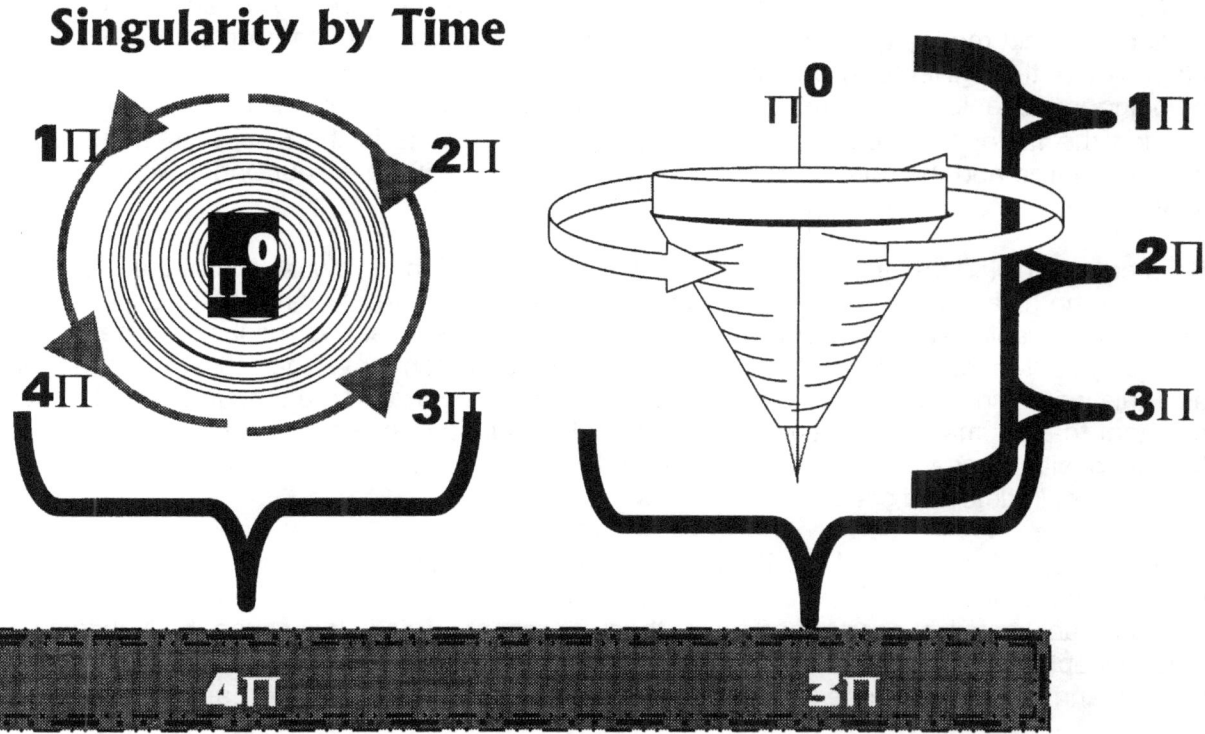

In the centre that holds the line the line, is a generated notion that is so thin the line is generated by motion and still the line is not part of the cosmos, while it is supporting the entire cosmos by controlling the entire cosmos. While it is not there, there is no denying that it is there and the control it has over all of the entire cosmos goes beyond question. It is establishing all the dimensions by seven supporting ten.

Every object that spins also generates such a line through the spin. The purpose of the spin places coherency into the Universe to generate the control by command and every atom is the seven. The atom places the seven in relation to the ten and all other atoms in direct linking of the seven points form the ten, which the atom holds as motion or liquid. The line is the diversion of the four by three establishing a parting between infinity and eternity.

If the alignment is in ninety degrees to each other then Pythagoras has to apply strictly. Should my argument be sound, which it is, we have to be able to use Pythagoras to determine the value of time in space. When the material rotates or moves, the filling of the material is in perspective to the time. However, material can only be in one location in one split time. Since material has to cross over to the other side of the Universe in order to duplicate, which is how material moves, then the material, can be only on one side of the Universe.

Every time matter is generated and moves, it is singularity that is complying with it activating another point in singularity being charged with the motion. The Universe started from allocating singularity charged by heat into positions where such positions contributed to space- time. Every time the spot overheated the spot expanded into four dots and by expanding the spot cooled. In cooling the spot retained heat by which it spawned the dots allocated as time. In overheating objects expand and by expanding objects cool. That is gravity.

Gravity is the expanding in relation with the cooling which means it is duplicating material in relation to a generated centre that is contracting the motion by cooling. Every inclination of motion is in fact motion and every movement be it contraction or expansion is moving to the other side of the Universe by bridging singularity because singularity is immovable. Therefore by being immovable, motion has to cross the division singularity applies and by crossing the division the factor that comes in place is $\Pi^2/4$, which results in the Roche limit. But such motion is three and the square of three in addition to the square of four brings about time in space.

Science acknowledges growth as the Hubble constant and then refuses to put the growth in line with the solar system. The growth they reluctantly admit to, they refuse to connect that growth to the solar system in any way. They take a Universal year as a solar year being that of one cycle it takes the Earth to rotate the Sun in the present day. Then they reflect on this as if this was going on since time began, because by doing that, there then is a nice crooked constant that fits mathematicians. Push this double standard applied back to before the Sun took its position and there was not Earth to indicate the year. How small was the year circle at that point in time and space?

Take this right down to the:" Big Bang" where "the whole Universe were the size of a man's fist" (To use their words), how far did the circle goes to indicate a year then? The year was immeasurably smaller, shorter and faster than at present. This is logic even the Newtonians must accept. There is no space outside insanity to apply time to the past at the value it is at present and far worse, to use something so extremely insignificant as the Earth to measure it by.

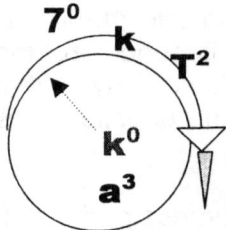

Again I feel that the use of this type of constant just to fit mathematicians to corrupt the truth they in science are using such logic to rubbish the truth. It is as if the one crooked posture corrupts all in general.   Since Newton, the Hubble constant, space-time, the realisation of singularity, the Big Bang and atomic cosmic control (which the GUT theory describes), all came into science as new truths. That is what Kepler's formula indicates in the simplest of terms and is what $a^3 = T^2$ is all about. That is what Kepler indicated with his formula $a^3 = T^2 k$.

The space of an object (a^3) is equal to the time (T^2), which it is in, in every given instant (k). If the space becomes smaller, the time duration becomes longer every instant of time's progress.

Only by creating a total independent heat centralised in a point holding singularity and feeding k^0 with an independent heat supply can an area a^3 establish a k that will release the independent a^3 from the secure larger k^0. The overall condition is that the escaping a^3 must establish a route following k^0 as k^0 places a diverting 7^0 where that 7^0 then forms part of the object creating heat to secure a release from the established a^3. Providing the heat will bring about a release placing a new object into outer space.

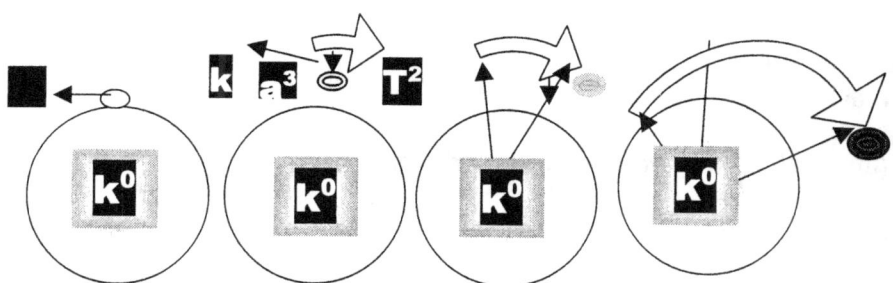

In the accumulating of heat, the object finds motion. The motion brings along structural independence and such independence puts distance between k^0 and a^3. The heat increase will accompany a larger T^2. By increasing heat, the distance between the objects will grow.

Time has been three since eternity started and time will remain three until eternity ends. When the heat came, about eternity spawned the one in infinity that generated a line and then time became the three positions of past present and future all depending on the one line in infinity. In the triangle that time established in conjunction with the law of Pythagoras the three of time goes square that forms nine and when the one marker of infinity is added, time by the square in space becomes ten.

The line however, forms three and in conjunction with the four positions in space –time (three in eternity and one in infinity that is there eternally) there are four positions relating to the three in the line and from that lying between eternity and infinity is the four eternal position plus the three generated positions which forms the seven in space-time.

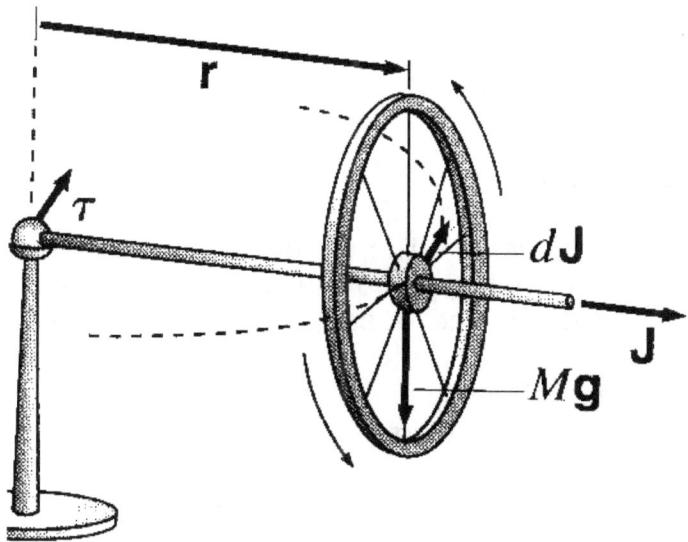

However because of dimensional duplication the square of time is ten and five will be on the one side of the Universe and five will be on the other side of the Universe. That then is why the Lagrangian system holds five positions in relation to singularity.

When the four in time spins off one more in infinity as time moves on, a fifth spot becomes valid that erects a line by heating and that fifth spot then reverts to the first spot that again parts eternity from infinity. When it has spawned a fifth position that position also goes square and forms by the law of Pythagoras the

Lagrangian fifth position.

This puts a huge question mark on the correctness of Newtonian presumptions that currently fondle the idea that rotation has no influence on the cosmos and all gravity goes down to mass where mass has all the influence and control.

All spinning matter has the point where the spin is still there but the radius is too small to measure by any means. That point is standing still in relation to the rest of the spin. In relation to that logic I do not accept Newtonian science holding the radius of s spinning object unaccountable in the spin, whether the spin is applying or not.

Applying Newton's second law F=ma
One arrive at the formula
$GMm / r^2 = m (\omega^2 r)$

By replacing $(\omega^2 r)$ with $2\Pi / T$ we obtain Kepler's third law

This law predicts that $T^2 = a^3$

What this statement implies is that r does not exist. When anything has a value of zero it is for all purposes non-existent. Only when an object is following a straight line can the radius be non-existent because the radius alters value through time development.

Taking the argument back to Kepler's law,

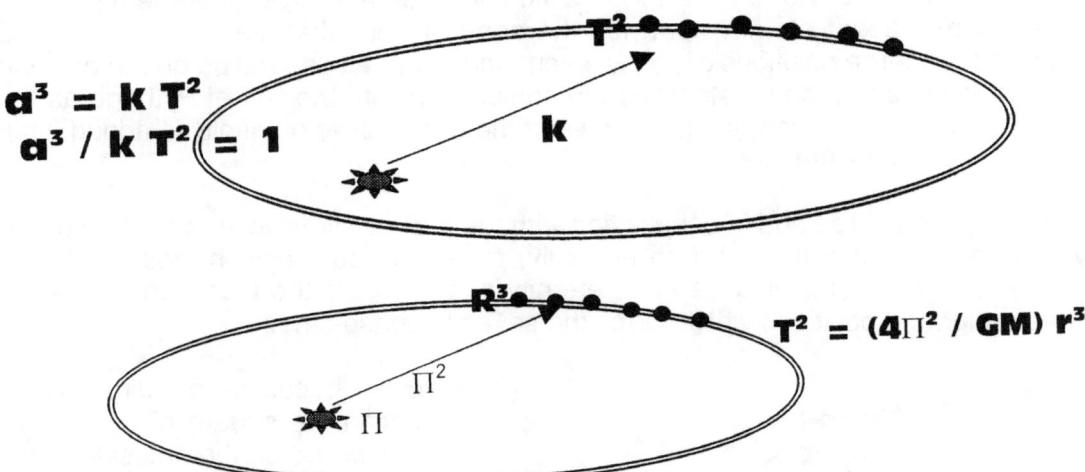

$$a^3 = k\ T^2$$
$$a^3 / k\ T^2 = 1$$
$$k$$
$$T^2$$
$$R^3$$
$$\Pi^2$$
$$\Pi$$
$$T^2 = (4\Pi^2 / GM)\ r^3$$

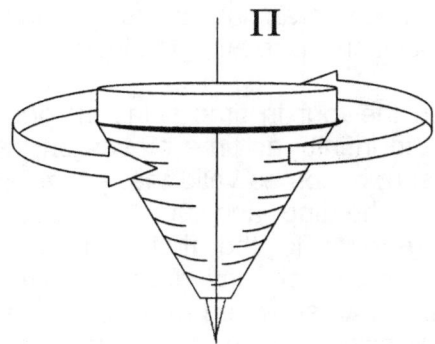

The spinning or not spinning is not part of the issue because at the point of absolute singularity the object never spins. Therefore spinning or not spinning does not apply to the point of singularity because singularity never spins in any event. In the whole structure with a pivotal centre as the control to the motion of the space the fact of Π is a natural outflow and any adding of Π is totally incorrect. According to Newton the result of spin is zero, however the top will tell a much different story.

On the surface, at first glance the top is an ordinary piece of dead wood that is machined into a sloping shape. The top is normally fitted with a sharp needlepoint at the bottom and the sharper the point is the better will the spin balance be. It is obvious that the spinning of the top inspired the entire Universe into a reality that is not there while it is in control of the entirety we find as real as life itself

When translating Kepler's mathematical expression into a verbally spoken form of communication such as English we can see what Kepler said also reads as $k = a^3/T^2$ where k is one point from a centre point that is space a^3 relating to time T^2. From a centre comes space-time

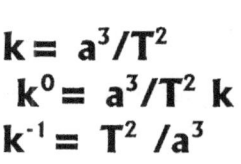

$$k = a^3/T^2$$
$$k^0 = a^3/T^2 \, k$$
$$k^{-1} = T^2/a^3$$

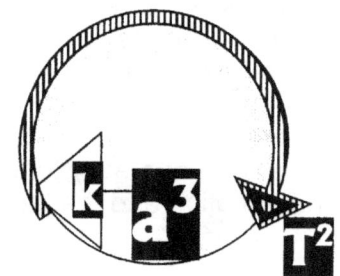

Kepler said $a^3 = T^2k$ but that could also be $k = a^3/T^2$

Others like Newton and Einstein came much later and coined the phrases but Kepler formulated the concepts. They named Kepler's innovations. That is very clear but only on the condition that Kepler is read correctly and Newton is gossip about what Kepler is saying is ignored. What Kepler said in mathematics all the brilliant Mathematicians through so many centuries were unable to read although the coded language was written in mathematics!

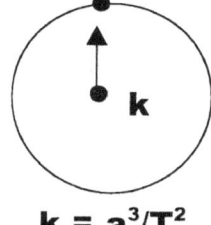
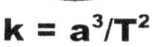

$$k = a^3/T^2$$

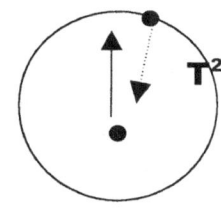

$$T^2 = a^3/k$$

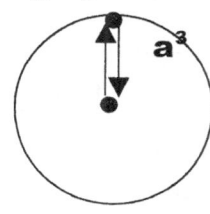

$$a^3 = T^2 \, k$$

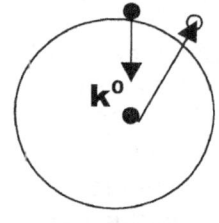

$$k^0 = a^3/T^2 \, k$$

But as one can see I also realised gravity is relations of motion applying in two factors. There is no separation of two of the factors acting as one but both have different application and values in the unit. It was what gravity was because this action prevented expanding. This is the result of singularity having three parts acting as one but giving three distinctions in application.

Gravity is as much part of dismissing space as it is about making contact with space in time. Since the connection comes about as a circle, the connecting points will relate to Π as the value. Due to the spinning nature of such a point with all surrounding the point will be alternating direction favouring change every second and in that the value to such a point can only be Π because of its constant changing. Using r would specifically oppose another r from every angle because the use of r will bring about a static relation to the previous and following instant and therefore it will cancel the constant spin flow. By reducing the line to its maximum possibility, one end with Π being the minimum but that Π is actually Π^0 which can also be k^0 or a^0 or T^0, which all indicate positions in singularity. Only when forming a value past singularity does independent identification come about. When the atom formed, that atom applied a relevancy of ten positions where seven positions are included in the atom spinning and three positions are partly the exterior of the atom spinning but all the positions relate to singularity but as space flight taught us such relevancies can change when an object is within the space boundaries of a larger structure or roaming free in outer space.

Within the boundaries of the atmosphere where the sphere border touches the space borders, the space borders hold six positions and the sphere hold seven points. But at the precise place where the points make contact with the sides one side falls away in favour of the point it connects to leaving five sides relating to seven and where one of the six sides takes control in removing one of the cubical sides by replacing that side with a sphere point position. The object then becomes directly controlled by singularity positioned in the centre of the sphere. The object seems then to fall from space and enter the atmosphere becoming a shooting star. What the Coanda effect proves above anything else is that gravity in control of space-time comes about from a centre and such a centre can be created by motion applying to a liquid in relation to a solid. That means there is undisputedly a flow of space-time towards a centre and the centre has to diminish the space-time reaching such a centre to create the flow and therefore the control from such a centre. That's the one pivot of gravity.

Since the Coanda effect shows gravity is control of space-time by motion flowing towards a centre that also proves as it explains the one part of gravity that reduces space by increasing time towards a centre that is established by motion and the lack of space establishes a lack of motion in that centre.

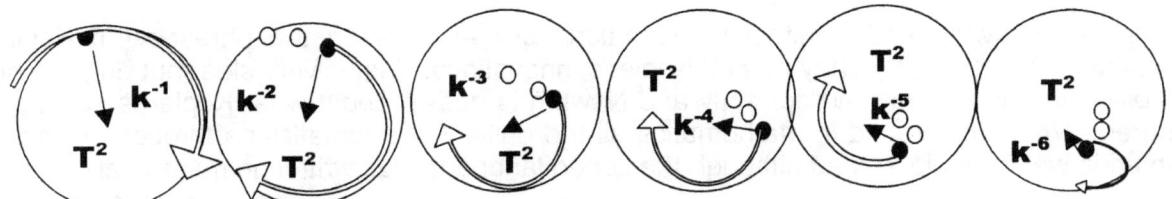

Because the smaller object holds much less space, the duplication is in a lesser relation than the main object and because the time factor enforces the duplication period to match therefore something in the applying ratio has to give in to allow the major relevancies to remain in place. Since a^3 has to rematch to apply to the conditions set by the larger object, a new relevancy comes about where the new a^3 will bring along a reducing T^2 with the diminished k that the Earth enforces. Since the space that motion reproduces is smaller in relation to the Earth, but the earth enforces the same time value, the relevancy of the time value will deplete by reducing k, but not in a straight line because all factor changes will then only be carried by one factor. I this way the diminishing space produced, helps the cyclic time factor to decrease with the distance that grows smaller.

When the object is released from the atmosphere of the dominating space, this very same gravity $k^0 = k\, T^2/\, a^3$ ratio will still be enforced since it is not the law of the Earth prevailing but it is the law of the Universe applying. Outside the atmospheric borders the Earth no longer has the means to remove one of the cube sides that forms the lesser object space and where the cube reinforces position by keeping the rotating object in position floating above the Earth.

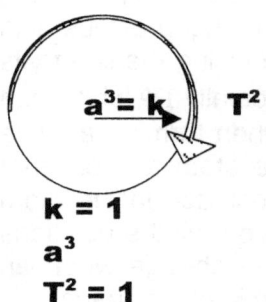

The space became too small to allow the time it takes to enter because the distance k decreased faster than the space a^3 could compromise with the time T^2 changing from what is present in outer space comparing that to the time in to atmospheric space. With this information being in hand for a period of four hundred years, one should think that the wise could derive a conclusion. Where the information forms the basis of modern cosmology since the information formulated gravity and not merely produced a name for gravity as our English friend did, it is amazing that such accidents can happen and it is more amazing that no one in Mainstream physics has the slightest idea why this is taking place!

Our most impressive astronautic engineers are assembling a machine that will scramble the ratio Kepler introduced to a level in outer space where the ratio will be more than what the ratio in the Sun is. Surprisingly they are not in the least surprised that not one object in outer space is using an excessive velocity.

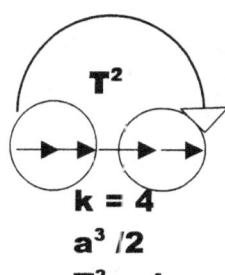

T^2

$k = 4$

$a^3 / 2$

$T^2 = 1$

In realistic physics it means double the space will fill in half the time. We know that that is not possible because it can only bring about half the space in double the time or twice the distance in half the time. Space time and distance is a mesh where the lot integrate because Kepler said so. Kepler said the space forming space is the same space forming the distance of the space and that is the same space taking the time to fill the space. If the ratio changes then changes come about the entire ratio. In order to bring about such acceleration much more heat has to be released to gas in order to find such a drive that will sustain such a high velocity. The drive can only be the result of massive quantities of heat being stored around the singularity the atoms generate.

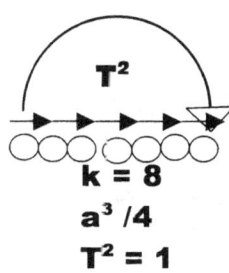

T^2

$k = 8$

$a^3 / 4$

$T^2 = 1$

The way the cosmic has designed the fight to relieve overheating is by motion. By duplicating the overheating space through motion the duplicating reduces the heat by half because the heat is spread over half the area that is distributed in double the space. By increasing the relevancy **k,** such increase reduces the area or space by quantifying the number of positions the space holds per time unit in the time or in the atmosphere or outer space.

k = 4 and **a³ /2 if T²** remains the same but that will not happen and that we know from past experiences. If that happens, we have the challenger 2004 disaster repeating once more.

Increasing space-time displacement by six will decrease space by six and the distance the space progresses from a centre by twelve. The heat factor of the craft will rise by twelve times as the space decreases by six times.

Increasing space-time displacement by twelve will decrease space by twelve and the distance the space progresses from a centre by twenty-four. The heat factor of the craft will rise by twenty four times as the space decreases by twelve times.

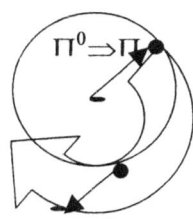

T^2

$k = 16$

$a^3 / 8$

$T^2 = 1$

$\Pi^0 \Rightarrow \Pi$

$\Pi \Rightarrow \Pi^2$

one is looking at honest or there the centre and

Motion of anything in any form is about duplicating the existing into following on images of the same thing. That is connecting space to last a certain period in relation to a specific point holding singularity before the next singularity is enticed or charged to maintain the space-time in motion. Every time (and in this case the referring to time proves to be most accurate) is having another singularity building and breaking down the space it represents for that duration of time. The time duration leaves singularity selected in charge of producing the roving space the extent in which it can duplicate the space it has to duplicate. By reducing the period the particular singularity may lay claim to the space, will inadvertently produce smaller space it is able to reproduce in the shorter period of time.

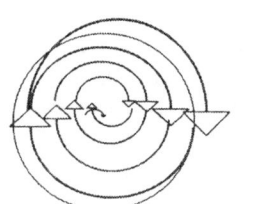

There are two ways of looking at this issue. The it from the centre that is keeping the rotating object is the rotating object forming space in relation to placing the centre in the centre. It will always be one taking prominence to the other and where Kepler introduced the formula it is indicating motion producing gravity which is gravity that is keeping form outside the sphere. Gravity is motion but the motion we see is much different from the gravity we experience while we know it has to be the same with only relevancies changing.

No matter how one looks at the Kepler formula, it signals the same principle. It shows how motion erects the Universe by mathematical equations. It puts singularity, as one in relation to six and that is the Universe decoded.

By rotating around a centre that is standing still, such a centre forms a divide that separates the unified unit. **Any point will be opposing itself** within the **rotating of 180º** where it **then changes every aspect** of its **previous flowing** characteristics it had or will once **again have in 360º** from there. While in rotation from the viewpoint of a bystander it all may seem static and never changing. However to the object in spin every next instant in time will be diverting from every aspect it had every second passing, and the direction it held in relation to the direction it held the previous mille-, mille-second as it will totally be incompatible with the direction it holds the very next mille, mille second of rotation. This is why we can use degrees measuring the circle by (6^2) (forming the square relating to matter through singularity) X 10 (square if space) = 360^0 however it is always in motion.

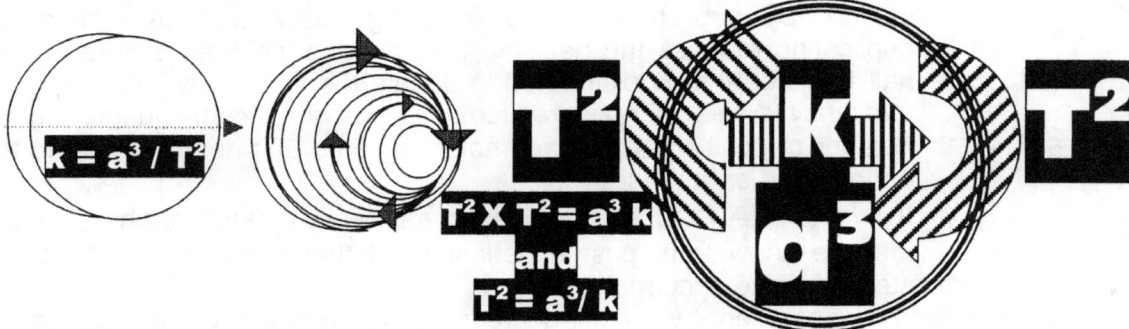

The square of motion T^2 X T^2 forms the square of space a^3 X k . Space a^3 is reducing by the motion of space with the implementing of T^2 having k as the constant. It comes about as the earth spins around the Earth axis. I call this positive space-time displacement

$$k = k^{3-2} = k^1$$
$$a^3 = a^{2+1} = a^3$$
$$T^2 = T^{3-1=2}$$

$$a^3 = T^2 k$$
$$a^3 = T^2 k^1$$
$$a^3 = T^{2+1} (k^1)$$
$$a^3 = a^{2+1} = a^3$$

is the same as

$$T^2 = a^3 / k$$
$$T^2 = a^3 / k^1$$
$$T^2 = a^{3-1} = T^2$$
$$T^2 = T^{3-1=2}$$

It is all the same

$$k = a^3 / T^2$$
$$k = a^{3-2} (T^2)$$
$$k = a^{3-2} = k^1$$
$$k = k^{3-2} = k^1$$

is the same as

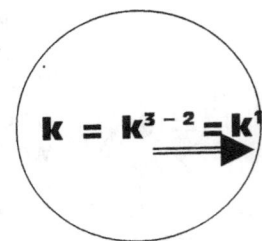

$$k = k^{3-2} = k^1$$

If space were zero or nothing as Mainstream science so affectively teaches us, then Kepler's principle formula would need the changes Newton brought about. It is true and stands tested like no other research ever coming either before or after Brae and Kepler's work. By reducing the line to infinity and raising the line again back in the direction of space, the line would erupt as a natural sphere having Π as the natural basic value. That is the value Kepler interpreted. However not realising what he saw he chose to use different symbols.

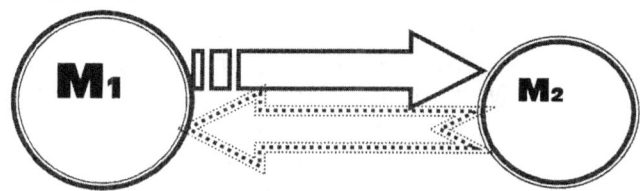

$k = k^{3-2} = k^1$ is in direct relation to $a^3 = a^{2+1}$ is in direct relation to $a^3 = T^2 = T^{3-1=2}$. With this information staring mainstream science in the face and scream pleading at them to recognise the information, they turn around and ask why can man not fly off to other galactica at the speed of light.

When the astronaut is departing from space on Earth or filling Earth space it will take the departing astronaut k^2 time to reach k^1 and fill out k^3. At present and in this moment our most impressive astronautic engineers will devise an engine that would cut k^1 by say half. This achievement will come as they increase the power output say for argument sake to double what it is at present. There was no friction of particles destroying the frame of the craft because there are not enough particles in space to do it.

However Newton recognised just the opposite and even allowed a freezing of motion and therefore time.

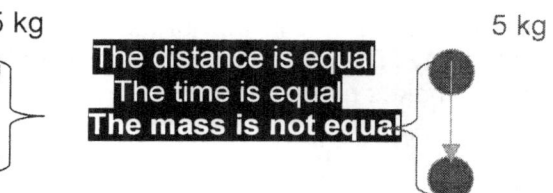

How does one reconcile the behaviour of the top with the foundation of science?

Mass has no influence on gravity in spite of all Newton's unproven claims. The fact that mass is inversely related to the radius as Newton's first formula proved, is the proof that mass is not gravity but something after the fact of gravity. $F = \dfrac{r^2}{M_1 M_2}$ The only true way that mass influence the radius by

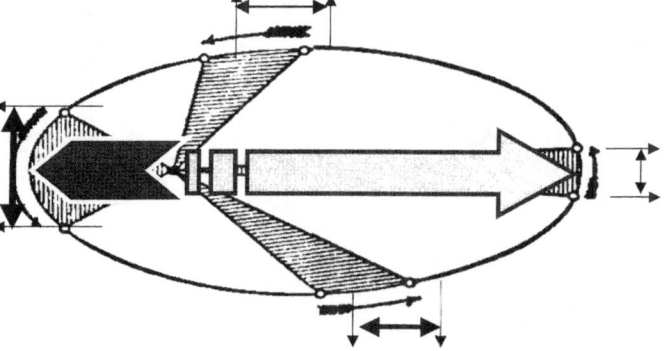

diminishing the length is when the mass of both is placed inversely in relation to the radius.

When viewing the findings of Galileo, one finds that objects falling have no mass. To calculate the speed of the object one would

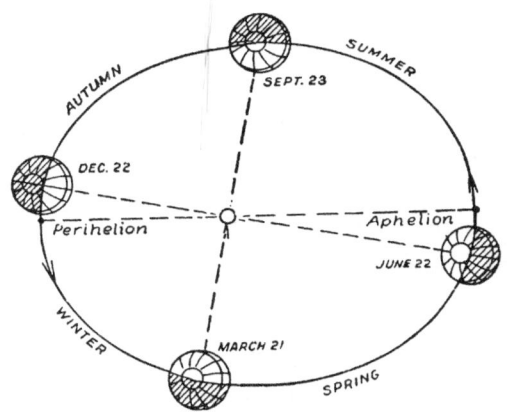

require the driving force and since mass must be part of such driving force it has to accelerate the object. At this fact Newtonians gave me so many answers that holds a variation, it can cover from north to south, where all are senseless and meaningless and not relating to the issue of proving mass as a factor.

Fact remains if I fall and my mass has any factor in my falling then me being heavier must have a profound affect on the speed of the falling. When I fall I have motion and my having motion eliminated my mass because by moving unrestricted I have no mass.

I only have mass when the motion gravity give is restricted by some influence that restrains the gravity in motion. All objects rotating around the Sun has motion, which is the duplication

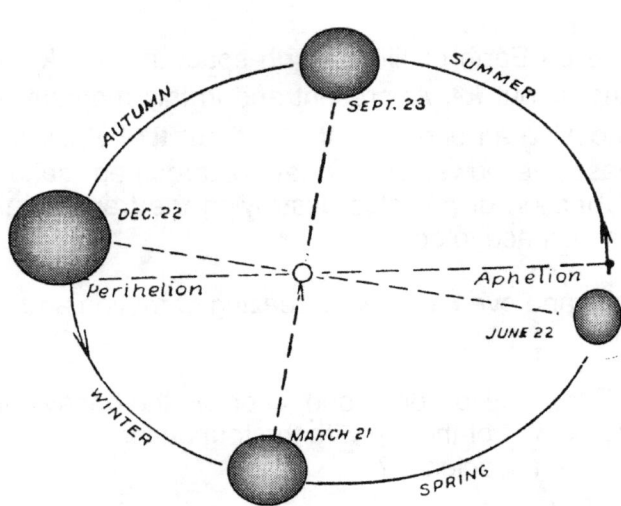

of the space in ratio with the containing of the time. You Newtonians out there, try to be realistic for once in your life in your thinking of cosmic physics without being brainwashed by your education. If the planet mass had the influence of producing the gravity that held the planet in orbit in relation to the centre of the Sun then the planets had to orbit by using the perfect circle.

By having a variation radius between planets and the Sun centre, it has to mean that either the planet mass show strong variation during the orbit of the year or the Sun shows variation that affects different planets at different times or both must show mass differentiation where the planets

become bigger sometimes and other times reduce in size. Since we know that is not the case and we know the orbits do have an eccentric anomaly by the measure of $E - e \sin E = m$ it is the M that I dispute. Another aspect of contention about the fact is that if mass did play a part in the orbit, it is the largest of the lot that should be closer and the smallest being further away. They are as scrambled as coffee with milk and sugar, which again shows mass and size makes no distinction.

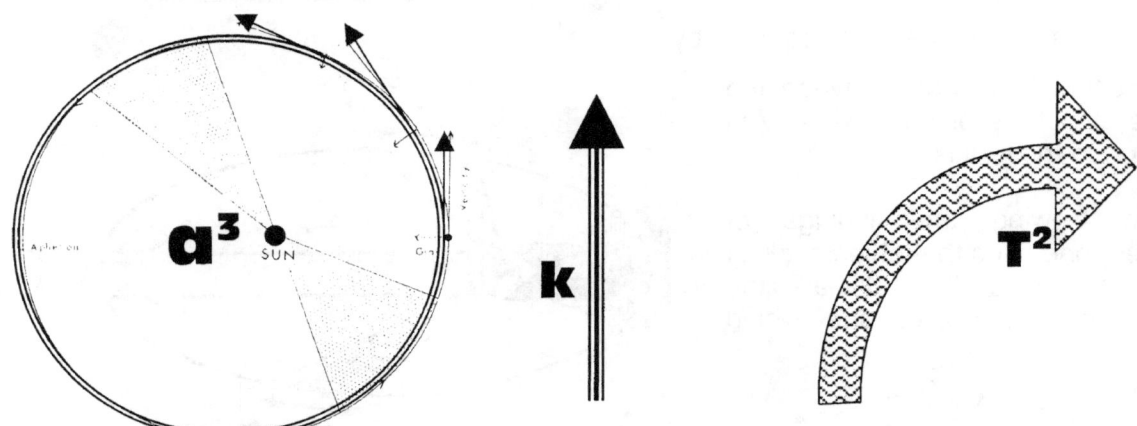

All spinning matter has the point where the spin is still there but the radius is too small to measure by any means. That point in the very and precise centre of all rotating objects is standing still in relation to the rest of the body that is spinning around such a centre. In relation to that logic I do not accept Newtonian science holding the radius of a spinning object unaccountable in the spin, whether the spin is applying or not.

To be realistic there is no comparing a wheel spinning on Earth to the planets spinning around the Sun. The wheel spinning on Earth can only do so if aided by life whereas life cannot make a planet spin or stand still. To have a wheel spinning on Earth one requires to intervention of life and life is a most alien aspect in the cosmos. The wheel spinning can only be associated with mass and mass can only be associated with life. Where life is not a factor, such as on planets that do not support life as a factor, mass will represent the state of something that is integrated with the larger and hosting object and is therefore cosmically dead. The wheel can never spin by independent initiative without life supporting such a spin. In that sense it is illogic to compare $a^3 = T^2 k$ with Newton's second law F=ma

One arrives at the formula
Applying Newton's second law F=ma

One arrive at the formula
GMm / r^2 = m ($\omega^2 r$)

By replacing ($\omega^2 r$) with 2Π / T we obtain Kepler's third law

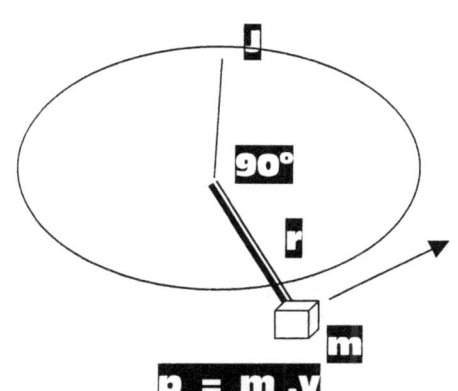

This law predicts that $T^2 = a^3$

The mass (m) multiplying the speed (v) forms a new value J **AND THEREFORE** j **CONTINUOUS TO IMPLY** J = I ω

= r X p **where** p = (v = r x ω)

J = r.m.v = m.r^2 .ω = I. ω **and becomes interpreted as J = I ω**

This establishes that r = dJ / dt

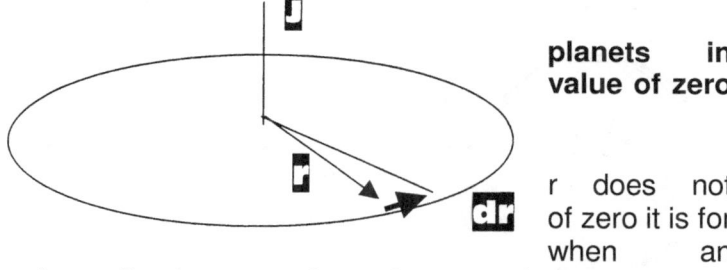

r = dJ / dt In the case of orbit around the Sun r forms a because dJ / dt = 0.

planets in value of zero

What this statement implies is that exist. When anything has a value all purposes non-existent. Only object is following a straight line can the radius be non-existent because the radius alters value through time development.

r does not of zero it is for when an

GMm / r^2 = m ($\omega^2 r$)

This can only be a reality if life provide and actively participate in the support the energy supply that will realise the spinning motion a wheel would have. To work with mass in physics is very earthly bound and that is precisely what life is. But as mass is a very Earthly aspect of physics, we must never spare any intensity in the effort we have to keep mass and the likeliness of life from our minds when considering the cosmos and all aspects about the cosmos.

By replacing ($\omega^2 r$) with 2Π / T we obtain Kepler's third law and that is trash because then the third dimension becomes equal to the second dimension and all goes to hell as this formula then would suggest. This law predicts that $T^2 = a^3$

Newton had the revelation of all the above mentioned as an apple fell from a tree apparently very close to him. He was admired as an instant genius and the one the world was waiting for to be born. I do not, for one second, deny or dispute the revelation. What I do encourage is to place the event into its correct context. It was merely, and simply an apple that fell from its branch to its roots. The apple did not pretend to be a meteorite that fell from the heavens. If it were a meteorite, I am sure, with the man's genius, science would be somewhat different at this stage. However, as a young man, being very impressionable, as all young men are, and with the attention this brought about in the world of science, the matter overshadowed the fact.

I am not disputing Newton; I am disputing the relevance of Newton's scientific breakthrough. It was not two objects of cosmic proportions, colliding in a show of the spectacular. It was, after all, only an apple falling from a tree and not that big an event. With this miracle he revealed, Newton found he was competent to improve on the work of Kepler and what Newton saw about what Kepler found was to Newton's mind the proof of total mathematical incompetence. He (Newton) saw a circle and without Π there can be no circle. Further more, since he was the founder of the invert four square principal, the principle also had to be included the make the picture a smart Newtonian picture and with that remove Kepler as such.

$\dfrac{dJ}{dt} = 0$ Newton, and science, made one enormous blunder, from this stance.

They took the radius of a wheel not to have any influence on the wheel. In doing that, they removed the very fact that keeps that which makes the wheel a wheel and serves as the universal attachment that keeps the cosmos together. After all, if you believe in the Big Bang, it is that part that is expanding and it is that which stands in place as the Universal development in progress.

They put two objects in an attaching relevancy and then announced no relevancy. Doing

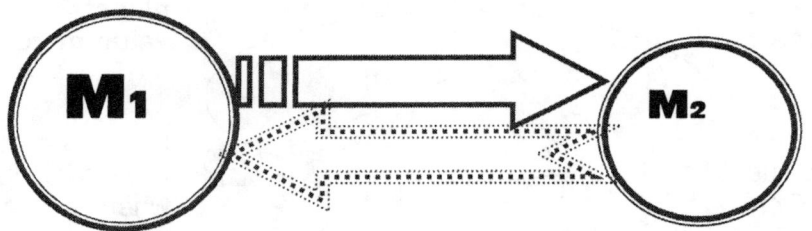

that is breaking the most fundamental mathematical principle.

$\dfrac{dJ}{0} = dt$ or $\dfrac{0}{dt} = dJ$ This disputes mathematics.

DJ / dt can have any number from eternity to infinity, excluding only one possibility; it cannot be 0. By placing the one in division of the other, you bring in relevance. You cannot then say there is no relevance. By doing such, you proclaim that one of the factors is non-existent. In both cases, one of the factors then does not exist. Such a claim is incoherent, because you proclaim that a circle has no radius, or a radius has no circle. When calculating a circle, you multiply either the square of the radius by Π, or the quarter of the diameter at a square by Π.

$\dfrac{dJ}{dt} = 0$ constitutes a circle and is also therefore $\Pi \times r^2 = $ CIRCLE

If you remove r it then is $\Pi \times r^2 / r^2 = $ CIRCLE.

You cannot then say $r^2/r^2 = 0$ and therefore $\Pi \times 0 = 0$. That is nonsense. $\Pi r^2/r^2$ will always be $\Pi \times 1$, and that is where Kepler placed singularity. By hiding this fact, Newton went and threw the baby out with the bath water. There is little standing further from the truth than this statement and reality disproves Newton completely. In the motion every wheel has to have a pivot around which the wheel turns. That is called the axis.

Newton's claim of mass pulling is totally incorrect when compared with reality
Instead time can never stand still as time delivers space in ratio to singularity.

 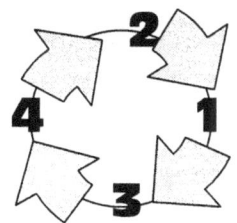 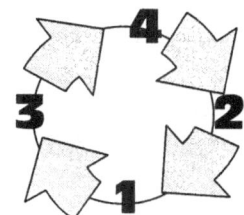

Notwithstanding all the protesting and objection Newtonians may have about the correctness of Newton, the concept is completely fraud in principle.
Built into the nature of rotation is the conflict there is between the two opposing sides of the

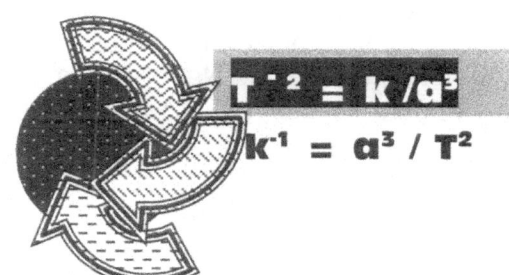

same rotating unit. By crossing the divide the fundamentals in nature changes as every aspect of what was valid change completely to the opposite. On the one side there is a contraction in spin. The thrust draws the spin into the centre by direction of the spin that favours such contraction. This has nothing to do with mass.
Then by crossing the divide where singularity

changes the direction of motion every aspect concerning the ration changes as it actually alternates. That, what previously by rotation

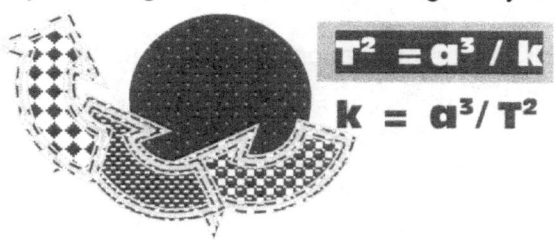

came down then goes up in the opposing direction. However that is not surprising because every slightest motion involves just such a change in direction and it is the process of interacting changes that manifests in charging motion into singularity.

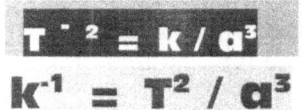

In the way planets rotate around the Sun this characteristics are also present and that forms the criteria for cosmic order or gravity. The same characteristics we find in the rotational spin around

singularity as well as a governing centre. On one side there is a directional preference to favour the one side and on the other side of the divide this favouring will swap ends.

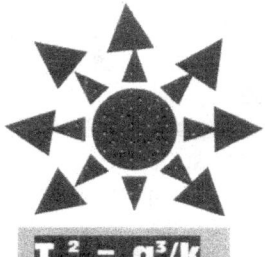

Since singularity is all the same and equal it is also true that that which singularity generates are not equal but depends on the motion that provides the space-time which charges singularity to define space-time. In that the Coanda effect has the role it plays.

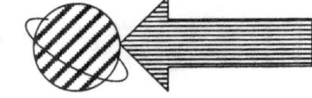

The motion of the relevant liquid establishes the rotation as the motion bonds the liquid to the solid while the solid uses the liquid in motion to extend the space confined by the motion thereof. Since it is two aspects in one unit, the defining of the dividing comes about as the two parts perform each its role on either side of the divide. On the one end the expanding party takes privilege position and on the other side of the divide the concentrating partner takes a privilege stance. From where we stand we see a small and a large, but from the singularity it is one side contributing more motion than the other side by performing duplication or contracting. However, there is no big or small. It is all based on the contribution in motion that maintains singularity.

In every cycle of every orbit we find four in time forming different allocations in positions and the varying depends on the relation the allocated position has with the centre.

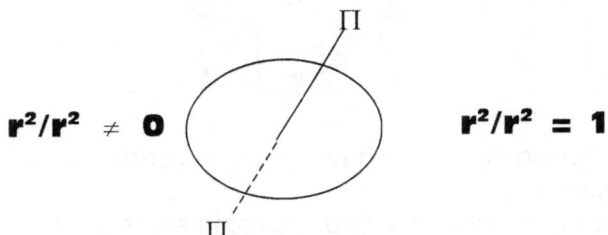

$r^2/r^2 \neq 0$ $r^2/r^2 = 1$

There are relevancies that form location preference to allocated positions just like seasons do. That is a product of time, which holds four positions in one cycle.

By denouncing Kepler and his formula, one must be prepared then to denounce all motion in that manner, and Newton more than most should have realised that. When looking at any rotating object, there has to be a point of no rotation and no rotation means "no rotation", not no existence. No rotation means a factor of 1, not zero. That then is singularity. The eternal Π, the Π that may not have significance but still it is Π of value.

The relativity remains one, eternally one, but it cannot be zero. Therefore, dJ/dt cannot be zero.

dJ/dt can be eternal or infinitive or at the worst it can be dJ/dt =1 but dJ/dt $\neq$ 0

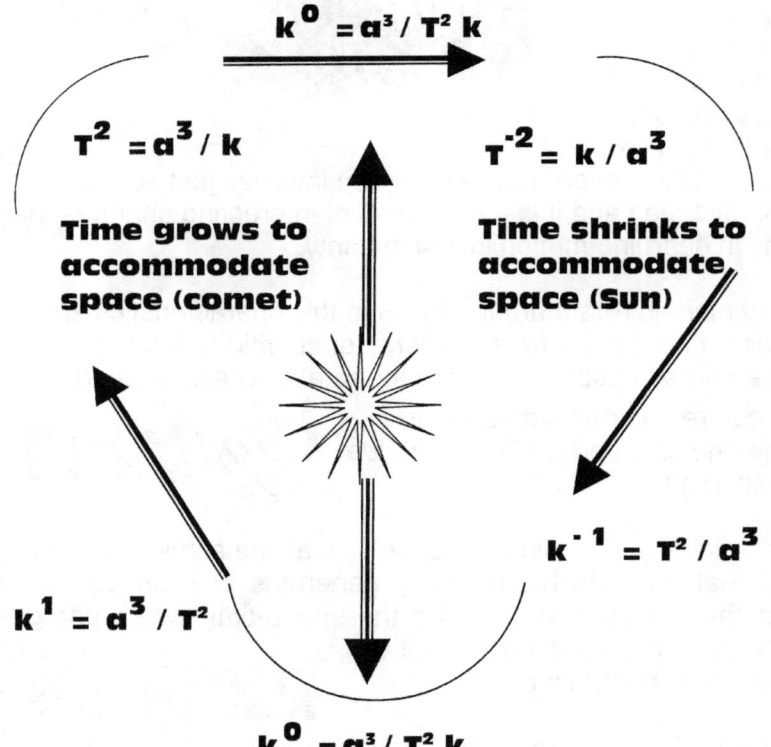

$k^0 = a^3 / T^2 k$

$T^2 = a^3 / k$ $T^{-2} = k / a^3$

Time grows to accommodate space (comet)   **Time shrinks to accommodate space (Sun)**

$k^{-1} = T^2 / a^3$

$k^1 = a^3 / T^2$

$k^0 = a^3 / T^2 k$

Looking at Kepler's statement it seems like a mathematical blunder made by an incompetent not understanding the most basic principles one might have in mathematics. It reads that the space in the third dimension is equal to the calculated motion of the space in the second as well as the first dimensions.

It is in this concept that Newton completely failed to realise the total extent of what Kepler's mathematics comprised. It is here that Newtonian mathematics failed to realise what true intelligence referred to. It is where mathematics was used as a tool of intelligence and not a tool of a basic need.

In every aspect we find the circling of the comet, or the circling of the electron or the relation of 1^0 to 1^1.. There are points where motion moves past the Pythagoras point of 90^0 ands 180^0

and then the motion falls on the other side of the Universe where everything on that side of the Universe strictly opposes that which is on this side of the Universe. It is gravity acknowledging singularity and singularity being in control of the Universe which that specific point representing singularity control

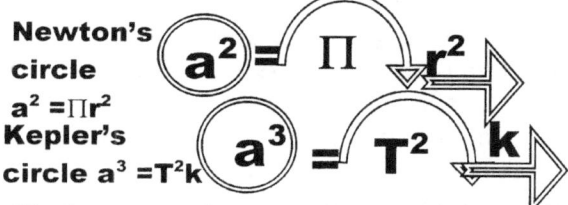

Newton's circle
$a^2 = \Pi r^2$
Kepler's circle $a^3 = T^2 k$

Kepler saw a circle because space is motion provided by singularity from a centre.

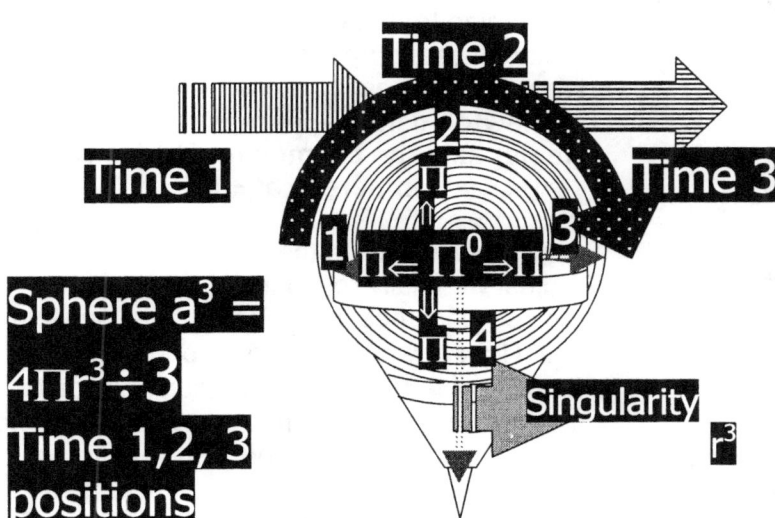

Sphere $a^3 = 4\Pi r^3 \div 3$
Time 1,2, 3 positions

A child will know better than that because it is hardly the manner one might use to calculate the surface of the space let alone the cube of the space. Far better is the use of the correct formula to calculate the cube in space as the one used to measure the volumetric displacement accurately as follows: $a^3 = 4\Pi r^3/3$. Using $a^3 = T^2 k$ is mathematically a fools argument, and yet even when someone perceive it to be wrong, was it accurately changed by Newton?

Newton said a sphere is $a^3 = 4/3 \, \Pi \, r^3$

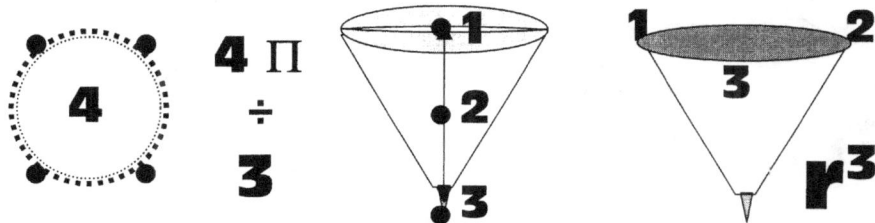

Following the mathematical volumetric formula one can see what is required to generate and duplicate every time the motion allocates space a new position. The centre line holding singularity ($\Pi^0 = r^3$) forms and then charges space by the cube in relation to the outer edges where singularity in Π meets time and where the Coanda principal puts the edge on space. The space in time redistributes the volumetric charged space in relation to three sectors time hold space in. That is the volumetric formula.

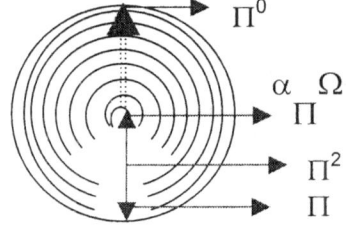

When the top starts spinning, the spin generates the time difference that singles out the space independence to bring about the space spinning. This is not what Kepler's formula show. Kepler's formula shows the space turning by rotation as well as lateral motion and by duplicating and contracting the space generates new space as the space carries material through time.

The space **a³** duplicate the position it had, duplicates the position it has and will duplicate the position what it will have **T²k** in relation the where it was, where it is and where it will be the very next instant. It does not mathematically reflect on a volumetric space to enable mathematicians to play a game and prove to their compatriots as well their own personal vanity how skilfully they can play a game with numbers and rules. It indicates a cosmic principle on which the four cosmic pillars rests. Dare I say one has to be more than a mathematician to appreciate this difference? This I say because I am of the opinion that if just one mathematician in four hundred years tried to find out why the formula used by their skills to calculate a volumetric displacement do function in the purpose they use it, that

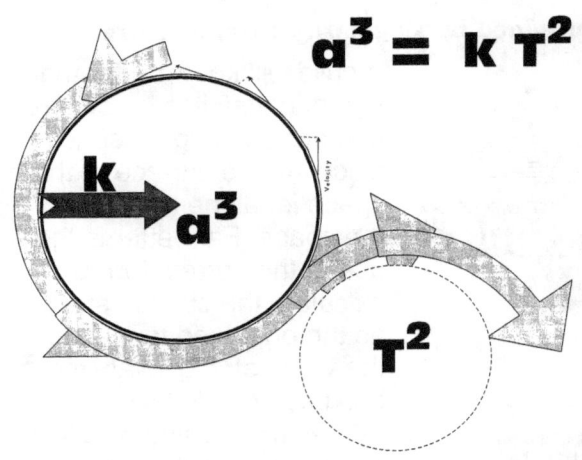

$$a^3 = k T^2$$

mathematician must then have had the ability to see the difference there is between the Newtonian formula depicting the mathematical purpose and the mathematical language suggesting a cosmic principle.

Newton said a sphere is $a^3 = 4/3\ \Pi\ r^3$ Kepler said the cosmos told him a cosmic sphere is $a^3 = k\ T^2$.

Going down the line that will reduce the radius, we finally reach the spot where the radius r becomes just a factor $r^0 = 1$, leaving only Π as a valid measurable part of the formula $a^3 = 4/3\ \Pi\ r^3$, which then only leaves the form Π as a measure of any value and no measure to legal space as we know space has. Yet, that is not the end because we find in mathematics one more possibility carrying singularity as a mathematical value and that is $\Pi^0 = 1$. This then will bring the ultimate position of singularity to the space concept in the mathematical formula of calculating the cube by $a^3 = 4/3\ \Pi\ r^3$, and that is where singularity defines the two prone but single value of 1^0 and 1^1.

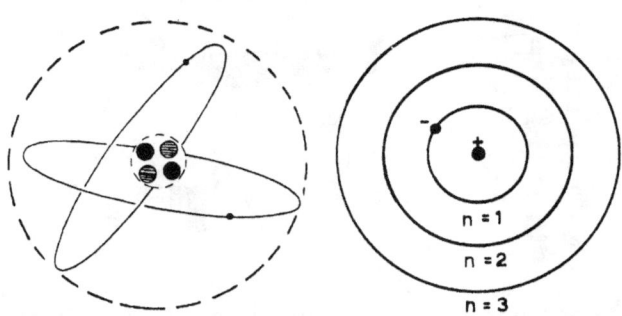

I have tried for so many years to accommodate Newton or parts of Newton into how I see cosmology but with no success. Newtonians leave everything half explained and from that try to formulate perceptions. What is matter? What is in the finest essence that which form material? When in search of an explanation to define material I find that there is an ongoing report on the particles forming material, which is not ongoing but is lumps called atoms. Then I suppose you may dissect that into as far as we can see with an electron microscope but that still says nothing on what atoms are. What is it that confines the circling electron to the atom nucleus? What is the electron putting into a circle that then becomes a confined unit? What fills the space in the cube of material?

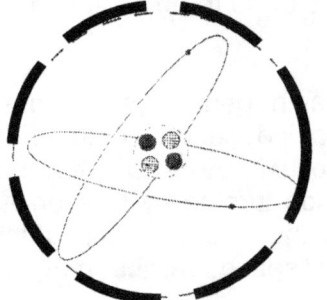

It is well accounted that there is this spectroscopy and the energy bands shifts as the energy levels rise or reduce. It is accounted and it is better well documented that the energy requirements allow the bands to rise or deplete. The more the energy is in the atom the wider circle the electron has to travel to accommodate the more there is in the atom. In the atom we find a neutron with no mass and an electron with mass. Then down at the bottom there are two protons that expand and that reduce the space they hold. That is so scientific accurate and correct and the only way I have to be more accurate is to remember every useless name some incompetent Muppet gave that enormous discovery that that incompetent Muppet found to be his remarkable creation. The more illustrious the meaningless name sounds the more important it will sound. The most important name will hide the biggest discovery of the lot...that the discoverer has no idea what he discovered and therefore his discovery in the long run has no meaning to science what so ever. Therefore, to hide the truth he has to invent a more useless name than the previous useless named other particle has. Never is there a reference to why the electron is spinning in the first place. What is in the purpose of the spin? Why would the electron spin and why would the electron enclose what is in the spin it protects.

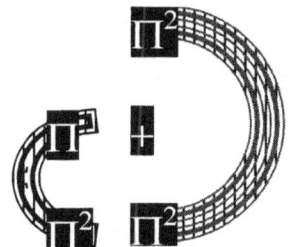

What is on the outside and what is on the inside of the electron that is spinning in a wider or smaller orbit. What is the enclosed material that the electron is protecting? What is behind the electron and what is in front of the electron and what is the electron?

That again brings me back to my first question: what is material?

We know that the atom has two positive particles in relation to what seems to be a neutral particle that is between the positive and the negative particles. The positive is a double 2/3 and the neutral is as much negative (2X 1/3) as it is positive 2/3 and

at that point there the explanations stops. To continue the discussion we than have to establish what a positive is and what a negative is to find a neutral. What will bring about something being negative because being negative is only a position in relation to being positive. What is being positive then? What is positive material and what is negative material because then the neutral has no clear definition in existing.

All the negatives will eat up all the positives and that will leave nothing as neutral. Most senseless is the part that I am the one they frown upon, the one they reject and the one that is uneducated.

In the atom I found a relevancy where the proton has double motion ($T^2 + T^2$) and the neutron has the linear motion of $a^3 \times k$. The total positions are equal to every possible position the sphere may offer in relation to a centre singularity

Present 2

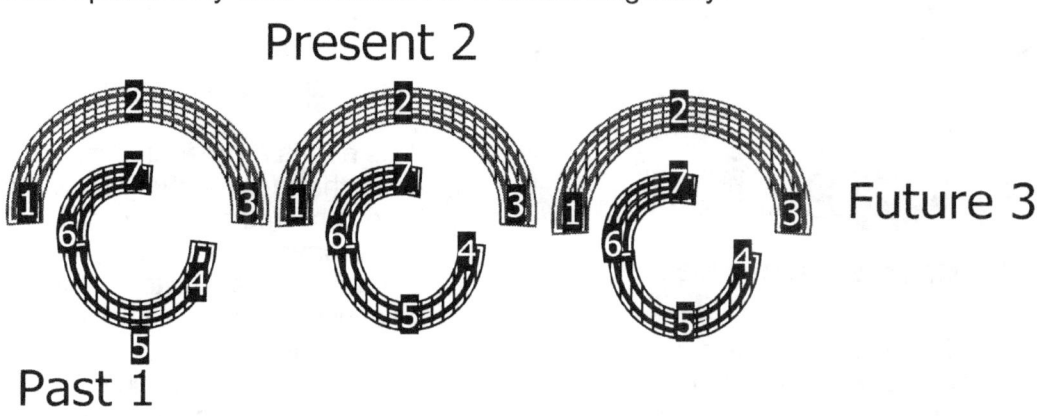

Future 3

Past 1

Then when this seven positions are put in relation to time, we find that the seven multiplied with the three positions time has the total including singularity expanding and singularity contracting is (7X3 = 21+0.9991 = 21.9991 / 7) = Π. That concludes that the Universe has gravity at all levels because the atom is a sphere as much as the atom is the Universe.

If we look at nature there is no position we can give time to rest at zero because time is the constant flow of all material in ratio to each other. In that the notion of t=0 is incredibly outdated. The fact that there is time is also the fact that there is repositioning of material in relation to each other on an ongoing constant basis. When I see a planet such as the Earth in rotation around its axis (the centre line), then we find the rotation is a flow of atoms repositioning in relation to one another. I stop at the atom because I have to stop somewhere and the atom forms the conclusion of the Universe but our conclusion of the atom forms our conclusion of the Universe and not necessarily the atomic conclusion that the Universe comes to define as the final conclusion. In other words there are mostly and many other forms of atoms in the Universe outside our spectrum.

Every time an atom is moving, such moving consists of a series of duplicating and relocating the position it has to what it had in relation to what it will have. It is the atom holding reference to positions as time flows and without such relevancies, there can be no atom. This

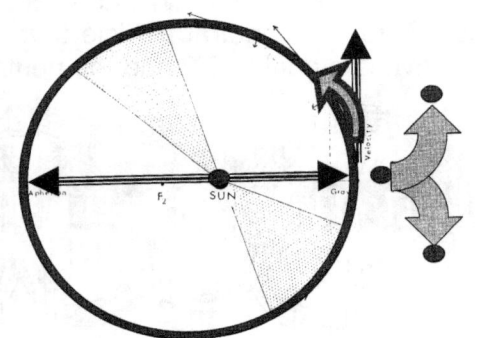

flow through time in time is where every individual atom is constantly proceeding through a line in time, where the atom is replacing its entire being there by relocating the location and the position the atom had, has and will have. It is shifting through time by taking up the place the atom in front has, had the previous instant, while constantly and cyclically having a relation to the centre where singularity controls such movement of the atom. That centre is incapable of moving therefore every time motion is established such motion is the generating of the entire field that is between the two locations of eternity versus infinity.

The atom is not only moving forward it also is moving to the side in accordance with the centre. It is following the leader but the leader is following a line that finds a relation in line with the position the atom in front had and the atom at present then fills. The filling of the position is in accordance with the centre and the centre is the point that claims the dominance of the moving. While from the point the atom has, such moving is straight in line with the allocation the atom in front has. The relation however is a centred one where there is a point that never moves and cannot move.

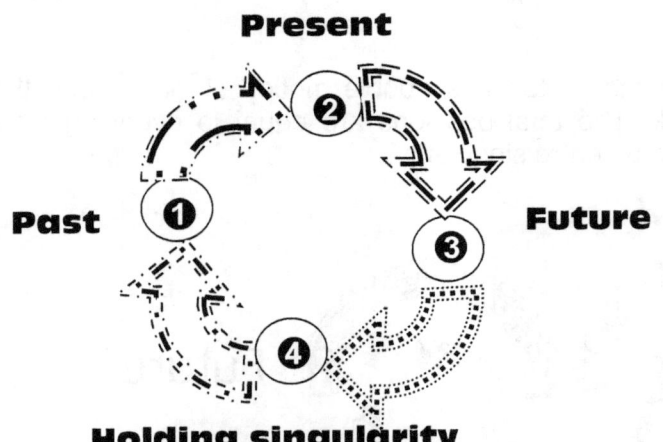

This part is represented by one side of time that forms the eternal side in time. The rotation is never concluded but always repeats the previous into the future. It is Newton's invert square law where there is four positions always following each other and this position was present when time was eternal. Please take note that time could never stand still and therefore Newton's assumption of t=0 is much misleading. The time found on location in which it rotates and that position holds all four points on one exact spot but that does not remove the location by giving the location a value of zero. However, in that there is another aspect where the infinite secures a position and still maintains the point having all sides sharing one point. That point is the very inside and to detach infinity from eternity, that point has to be moving where that point holding infinity also has to relocate while in truth it cannot relocate.

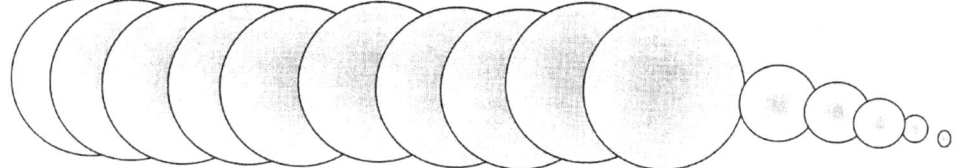

Taking this motion to another level is the repositioning of that which cannot move into a new position to where it has moved. This is part of the rotation but also it is the moving of time in a lateral direction. While the lot is rotating, the lot is shifting into a new position going from where it came to where it is to where it will be next to where it was every time it is going

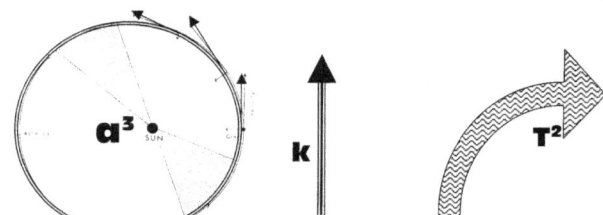

forward. There are seven points moving in the circle while there is this seven points coming from and going to while it is in the present position. That makes it the seven it is plus three, which is the time aspect ten.

That action is what brings on motion and involves a line as well as an incomplete circle. The space a^3 duplicates what was to the present going to the future in a semi circle T^2 as well as a straight- line k. The motion of the space a^3 involves the rotation T^2 as well as the straight -line k. In order to be in space, the space has to be on the move. Time cannot stand

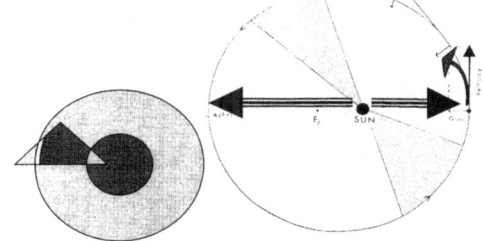

still for then there can be no space because the space is the directional movement in a lateral as well as a circular direction simultaneously. **There is no possibility of time standing still at t = 0. If there is t = 0 there is only 0 and since there never can be 0 there is no possibility of time standing still.** If you find time standing still, chuck away your watch because the watch and not time gave up the ghost. In order to be in space, space has to duplicate by presenting what was in the past, take it through the present and fling that which filled the past and is filling the present into the future. There are both actions serving space where space is going sideways to go forward as it is going forward in a sideways direction. That is motion whether it fits Newton or suits Newtonian perception it is of no consequences because the atom is moving along as the atom is spinning.

The Newtonian idea that the one direction eliminates the other direction by the sharing of a mutually established centre has no base in reality. Kepler clearly brought to science the fact that space forming includes motion of the space but as important is the fact that the motion cannot exclude any one of the factors in the circular or the linear and that is the product of both factors that form the flow of space through time.

By supposing that the one factor removes the other factor from the Universe is quite frankly forming an effort to absolutely destroy what Kepler said! Then we get back to the question that I asked earlier as being what material is? What is filling the atom? What is inside the proton, the neutron and the electron? If getting more energy is getting the atom fuller by expanding then the inside must be energy. What is energy because the word or term energy has become an escape goat used whenever Newtonians run out of answers? Einstein was of the opinion that when matter goes past the speed of light it would become pure energy. So what is matter before it goes past the speed of light? Is it then less energy or does it become something other that energy?

That brings about the next question: what is mass. Why would that, which is inside the atom have mass and that, which is outside the atom have a different mass. Why would the proton being so much smaller than the electron be so much more massive? Newton came up with this idea of mass and everyone accepted mass because no one could disprove it and since Newton never was set the task to prove what mass is, it was left at that. Why would material have mass?

We all know that all objects have mass when the object is on Earth but when the object is in outer space what would give the object mass? If mass was conducting the motion of planets, then surely they would be arranged according to mass. The biggest must be to the most inside and the smallest to the further outside. If it was mass that was generating the motion then the most massive must be flying while the smallest must be crawling. That is not the case at all. The distribution of planets relies neither on mass nor speed of motion or any distribution in size at all.

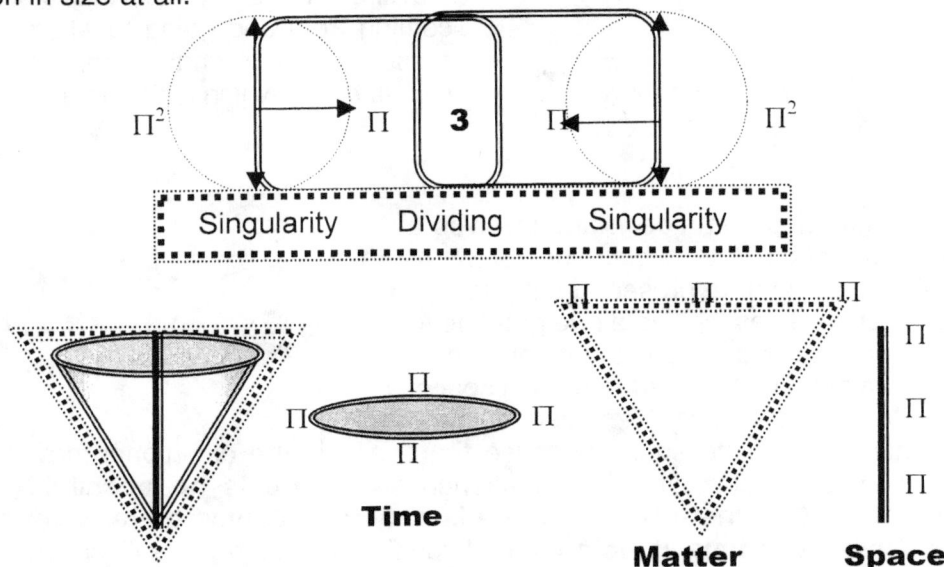

There is no correlation that would even suggest that mass plays a part in the orbits of planets or any cosmic structures. The dynamics we contribute to the realisation of mass should be far better defined if there is to be any clarity on the matter. According to Kepler, space needs motion to produce space $a^3 = k\ T^2$. Let us see how true that is.

How does that which fills the inside of a particle move from one point to another point during time? The rear must back the front by pushing the front forward while the front is pulling the rear by vacating the front. If the front does not move the rear will smash the font out of line and in some cases destroy the front from behind. Think of a car crash where the front becomes blocked by other objects while the rear is coming to fill the front from behind.
That which fills the front has to vacate the location it holds in an agreeable direction within a suitable time while that which is filling the relocated position of the front requires the filling from behind as to find the required flow so that the motion can go about spontaneously. If that which is coming from behind is retarding in motion as to fill the vacating position in front the unity in the movement will tear into parts. In order to accomplish motion, the motion has to accomplish the sphere.

The motion not only fills the vacating space by rotating in a "follow my leader" process but also establish from one point in infinity a line of points running along a line that is never there. The line establishes four rotating positions parting infinity from eternity. In that I do not find grounds to blame mass for anything. I do find that the matter behind the matter in front is only retarded allocated filling of a position in time. The following is in a retarded stance to that which it flows and that makes that which follows behind that which is in front being ahead.

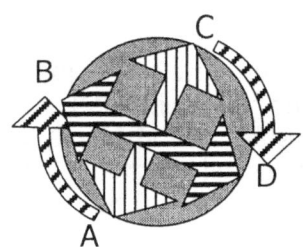

The one is always behind the other which is in front and that is material because being behind that which is in front gives both that which is in front as well as that which is behind an independence from one another. The proceeding to follow and the following that proceeds placed both in separate compartments during time and that puts the one in motion while the other is filling the motion. I still do not find mass. What I do find is the relevance that Galileo found. All objects notwithstanding size or mass fills the position the one in front left vacated to be filled by the one coming behind.

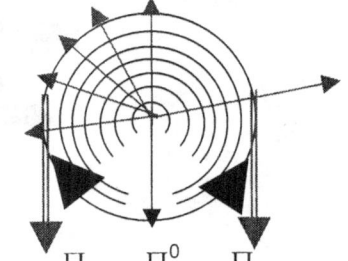

There is more to motion than just that because the rotating plays follow my leader while the lateral plays follow my leader and with that there still are two identifiable substances in relation to the space and the motion that time represents. The lot is duplicating and by duplicating every aspect is relocating what it represents in terms of space –time to a new location.

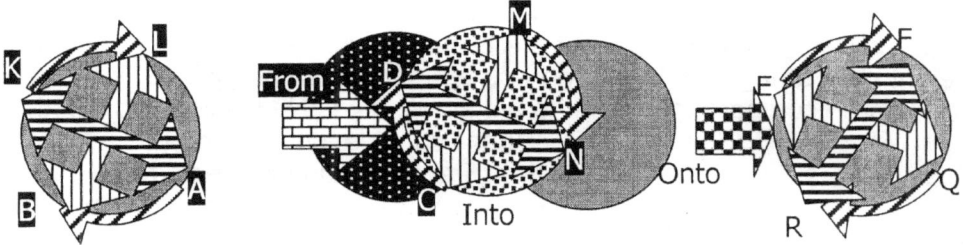

We have the rotation there is where the one point serves as a guide to the previous point and follows the next point into the future while honouring singularity. That is one part of the story. There is more and that is time in the lateral or eternity running without ever stopping.

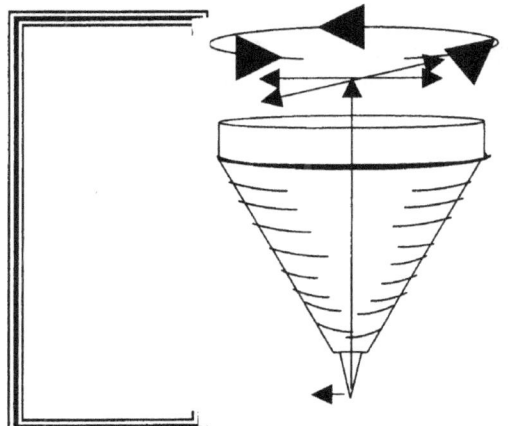

However that which cannot move does move not only by repositioning the lateral but also by relocating the rotated into a new relevancy.

By shifting point A to point B, that shifting also involves moving point K to where point L was before. However, that centre is not movable and on both flanks the lot shifts. That means what ever is relocating a position is breaking down all there is and shifts the lot to where it is going to be by generating what there was in the previous location onto where it will be in the next location.

Then this lot in rotation also has to relocate that which rotates along a single file.

The point in singularity cannot move and by it being unable to move it has to break down all relevancies there was and shift all that to new positions in relation to an entire new Universe.

In that there is no mass. What there is we can describe as controlled duplication where the duplication is represented by the expanding while the control is represents by the contracting. It is just as Kepler said it will be when it is　= k T² and on the one side there is **k** = **a³ / T²** as singularity generates space-time in the space generated by time finding a new location during the time such generating takes while **k⁻¹ =T² /a³** the motion produce a new space in the allocated position it has to be at that particular time. Still in all the dissecting I find no evidence of mass doing anything and even less generating gravity.

It is a pity Newtonians never get more specific as to where one may find mass and how mass goes about producing gravity. Being as explicit as I am by going into time between eternity and infinity I still find no mass. I do find eternity forming the three positions on the side that has no end and those three positions do correlate to the most accurate detail in relation to that which has no start. In between that which has no end and that which has no start space-time is generated through the motion that activates space-time.

Newton made the error Newtonians still do after so many years. He took the fact of life as a cosmic reality. He took the motion that life can achieve as standard cosmic occurrences. Newtonians go much further than Newton did by claiming life comes at a dime a gross throughout the Universe while there is no evidence of that. To swing a weight on a string from the hand has as much cosmology in it as trying to pump a tire with air and then compare that result to a star…and yes Newtonians do just that!

When considering the motion applying then first see that being tested fits freely in cosmic reality. A rock cannot roll up a hill and a brick cannot be chasing his partner while a cloud can't cycle a bicycle.

That which Newton supposedly discovered being gravity is that which Newton denounced as zero. When Newton discarded the motion part by putting the value at zero he threw away

$$10/7 \, (\Pi^2(\Pi^2+\Pi^2)) = 136.37$$

Contracting in liquid

Expanding into in liquid

$$7(\Pi^2+\Pi^2) = 138.1$$

The Solid

$$7/10 \; X \; (\Pi^2 /2 \; X \; (\Pi^2+\Pi^2) = 139.$$

that produces both sides of the motion.

which gravity on border of

Where the motion of time forming the liquid interacts with space forming the solid we find the two parts motion offer on either side of the divide. The motion is the same but crossing the divide that motion then falls into the other side of the Universe where all changes to become the opposite of what was. It is vital to realise that at this stage of cosmic development the neutron was the liquid because this was pre-Big Bang or then the introduction of light in dark or bright visible form.

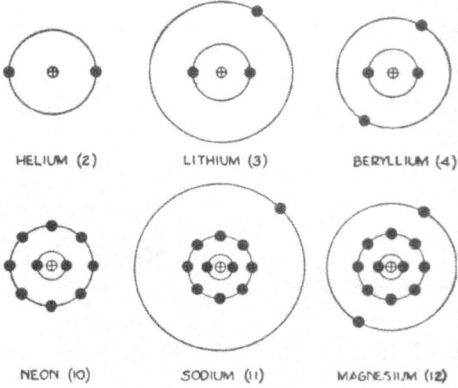

HELIUM (2)　　LITHIUM (3)　　BERYLLIUM (4)

NEON (10)　　SODIUM (11)　　MAGNESIUM (12)

Because of this we find that by nature and not by mass there are elements that favour duplicating more than contracting notwithstanding the number of protons or the resistance the number of protons may show to the blocking of motion. Then there are other elements that favour contracting much more than the duplicating side and again mass plays no part. Mass has a place in Earthbound Newtonian physics where one may substitute the correlating motion with the restricting of the motion because there gravity has the tendency to move whereas mass is the restricting of the motion, turning the motion into a tendency to move.

That however is not related to cosmology because as long as there is unrestricted motion there is no mass and then all the planets are equal as they circle about the centre of the Sun. Mass is the frustration particles experience when unable to contribute to free gravity - motion.

Getting back to the fictitious mass and the question about what material is. What is material and why is material what it is?

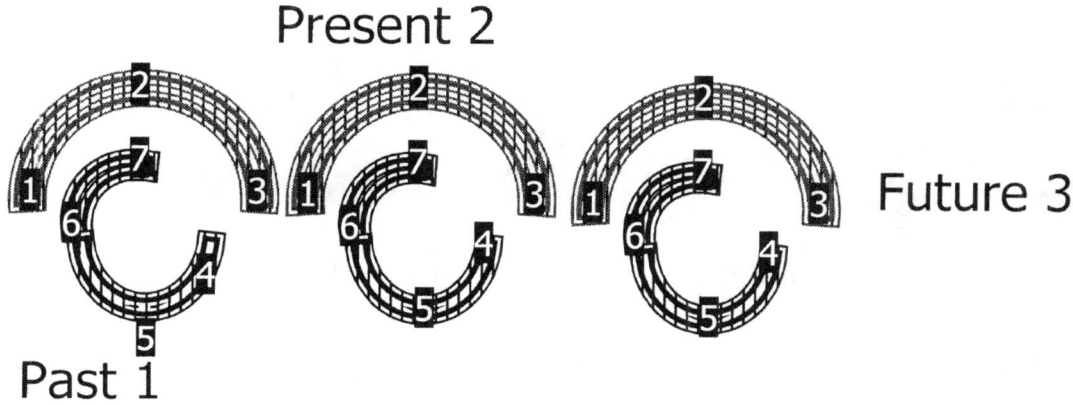

The flow holds seven points that forms the atom and the atom was able to retard time so dutifully and preserved heat so faithfully an entire Universe with immeasurable Universe coming from an array of immeasurable possible Universes can now serve as multitude stages of universe developing eras.

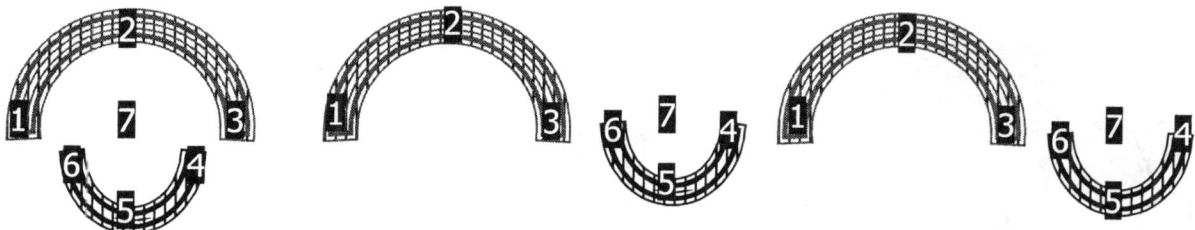

Yet in all that, it never deflected one measure from the original retarding by rotating principle it had in the culmination of starts we preserve to put a Universe in.

Expanding

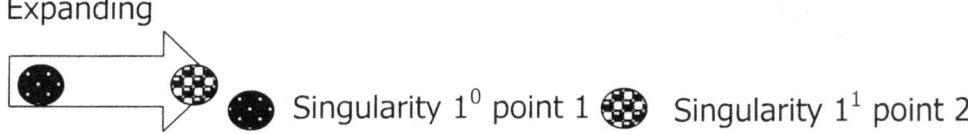

Singularity 1^0 point 1 ⊕ Singularity 1^1 point 2

There was a point in eternity that held infinity and combined singularity. Both are still in our presence and both are wall established. We still gauge the one in the centre of all spinning material and where there is no point, there is a line that is not and in the line the line holds seven points all holding the precise same position. Then light or heat came about and since light or heat expands because it takes more space than what was taken before, there is more of the same that was before.

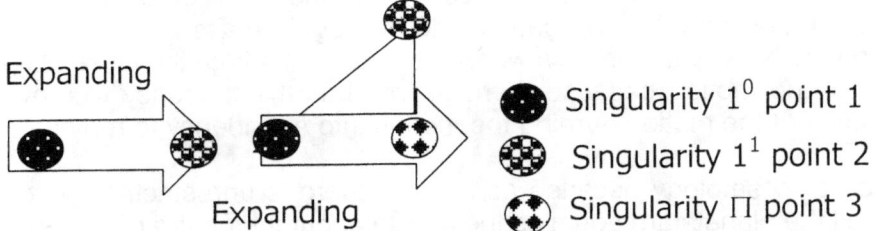

Expanding

Expanding

Singularity 1^0 point 1

Singularity 1^1 point 2

Singularity Π point 3

Since the expanding brought what was not in between what is not, eternity parted from infinity. But time is motion and since the one became parted from the next the next placed the previous on the other side of the Universe while the following did the expanding. At such a point the law of Pythagoras started to develop the cosmos by putting in the triangle in relation to sides.

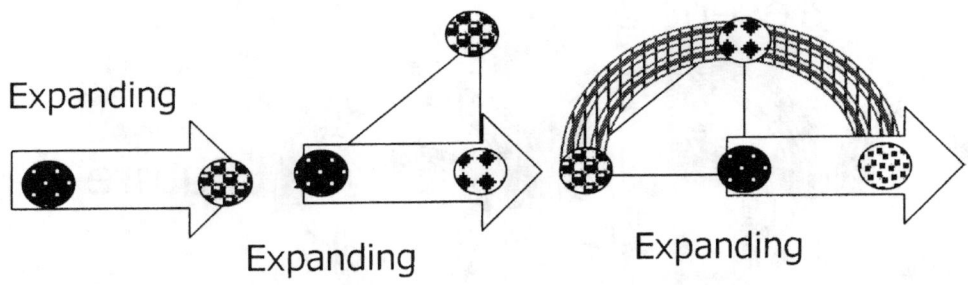

Expanding

Expanding

Expanding

Singularity 1^0 point 1   Singularity 1^1 point 2

Singularity Π point 3   Singularity Π point 4

The flow of time continued as new positions established, a point once established did not vanish because once anything is part of the cosmos it stays part of the cosmos, as there is no other place to go but remain in the cosmos. The points continued as time moved on.

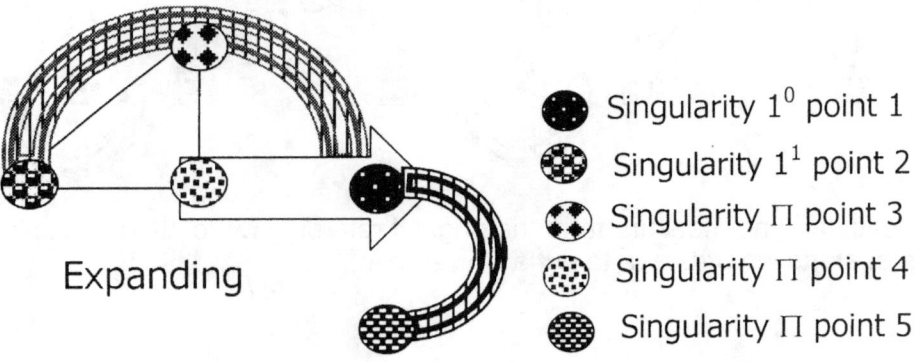

Expanding

Singularity 1^0 point 1

Singularity 1^1 point 2

Singularity Π point 3

Singularity Π point 4

Singularity Π point 5

Then after the four points expanded, eternity produced an additional point at a location where infinity parted company with eternity. At such a point, point five came in place and this was where the cosmos became different from what was previously applying.

The cooling set in where the expanding brought more but also the expanding distributed more of what expanded and since that which expanded was covered by more, the more made the expanding less and by cooling the expanding retracted. It was not the progress that retracted but the direction the progress had that retracted.

At this point eternity parted from infinity. Infinity is there for all to witness. It is not a hypotheses but a reality. Eternity is there where everyone is seeing it as long as man has a mind. That too is not new and that too is a reality and not a hypothesis. It is a reality within every human and is as concrete as the blood running through the observing person's veins. It is as real as the Universe itself because it is the Universe itself. Any one arguing this reasoning has no mind to understand the smallest concept any human can form. Then came the rest.

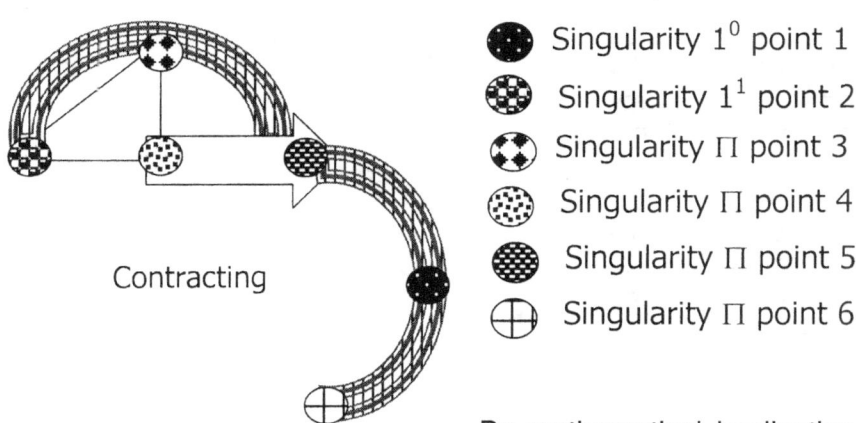

Contracting

- Singularity 1^0 point 1
- Singularity 1^1 point 2
- Singularity Π point 3
- Singularity Π point 4
- Singularity Π point 5
- Singularity Π point 6

By mathematical implication and the influence of the law of Pythagoras material found a limit at point six.

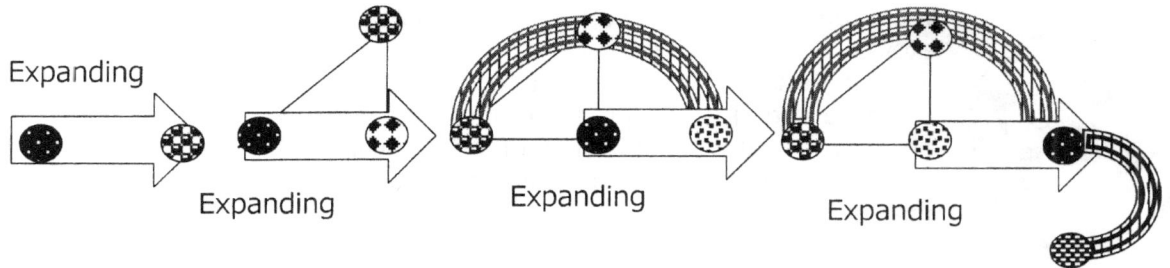

Expanding

Expanding

Expanding

Expanding

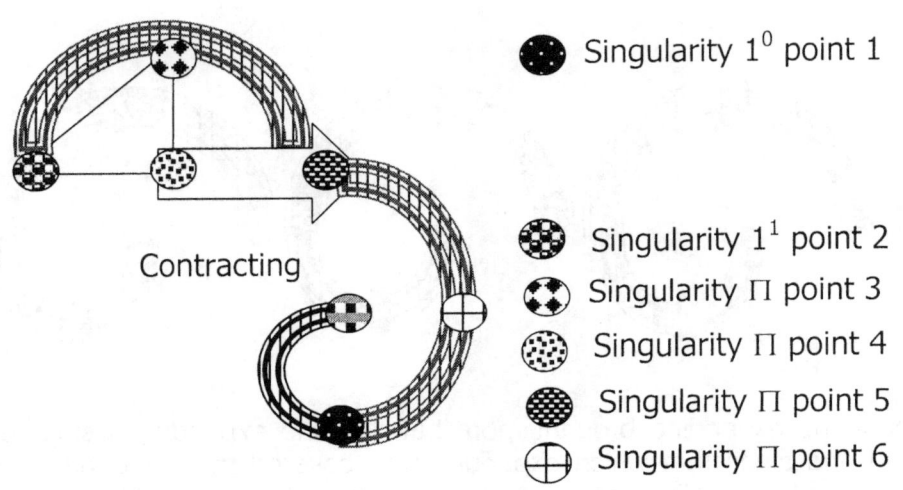

Contracting

● Singularity 1^0 point 1

◉ Singularity 1^1 point 2

◉ Singularity Π point 3

◉ Singularity Π point 4

◉ Singularity Π point 5

⊕ Singularity Π point 6

◉ Singularity Π point 7

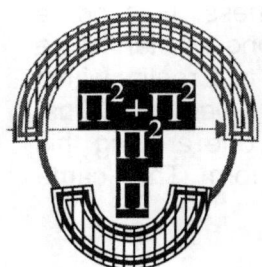

That which was inside the seven connecting points serving singularity is contained heat by spin. That which is outside the seven points is expandable heat that expanded without control. That inside was controlled by motion that those outside provided. Any one with doubt that it is controlled heat look at photo's of Nagasaki and Hiroshima and the Bikini island later and see what a tiny bit looks like when the control is released. It is heat that is retarded that is forming time-controlled by spin, (we call it an atom) and the time is heat dragged on by the lagging behind of heat in a different era of time.

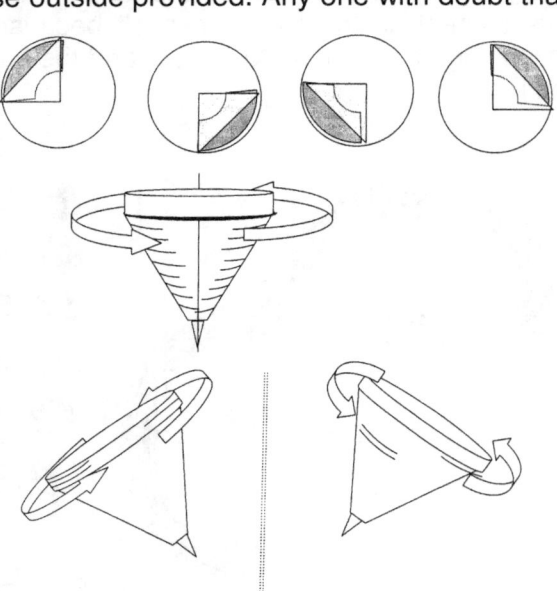

That is material and every seven points holding singularity confirms the backlog of heat dragging behind time. Still, I see no evidence of mass and if mass was not then, then mass cannot be now except for in the head of Newton and in the imagination of Newtonians suffering from mental programming. Material is heat that is responding according to a time delay where the spin reduced the time and in that, controlled the expansion as it expanded the reducing. That is the essence of any atom.

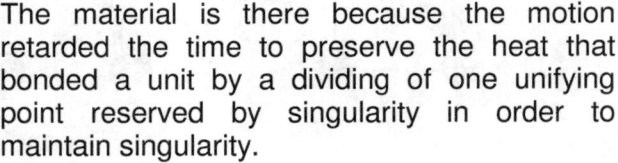

The material is there because the motion retarded the time to preserve the heat that bonded a unit by a dividing of one unifying point reserved by singularity in order to maintain singularity.

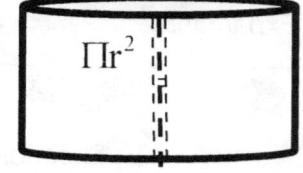

What is it the Newtonians fail to see? If an electron is orbiting around an atom, the inside of the atom must be a circle. If the atom was not a circle, it then had to be a cube. The electron cannot rotate around a cube; therefore, the inside of the atom is a circle.

The radius r runs from the circle outwards, from a circle centre point towards Π, the value of the circle. In the centre of the circle, there is a point where the radius starts. It runs outwards from that point in all directions towards the circle Π. Technically, there then has to be a point where r is infinite and not zero, an absolute infinite. However, the circle therefore remains Π. The circle does not disappear; it remains there for all to see. It is only the radius that almost disappears into the infinite, but it does never become zero!

In a circle, there is a radius that initiates the circle. The calculation of such a circle is Π X r^2.

$$\frac{\Pi r^2}{r^2} = \Pi$$

If one removes the radius from the circle, the circle remains, only holding the value of Π. By removing the value of r, Π becomes singularity with no place to be. Singularity is the place where there is no space to be in place. However, Π remains because once r receives the slightest of space Π will find space. Then the circle will grow to Πr^2 and r would determine the space. Without space, there is no r but there is a circle with the value of Π. Singularity is in every single rotating object, be it the proton or the combining effort of all particles in the Universe. That is what light and the photon is. It is concentrated heat and that is the manner how Sun (or any other generator of electricity) connects heat to singularity where the heat receives either temporary connection to singularity or a small piece of individual singularity.

All spinning matter has the point where the spin is still there but the radius is to small to measure by any means. That point is standing still in relation to the rest of the spin. In relation to that logic I do not accept Newtonian science holding the radius of spinning object unaccountable in the spin, whether the spin is applying or not.

What this statement implies is that r does not exist. When anything has a value of zero it is for all purposes non-existent. Only when an object is following a straight line can the radius be non-existent because the radius alters value through time development.

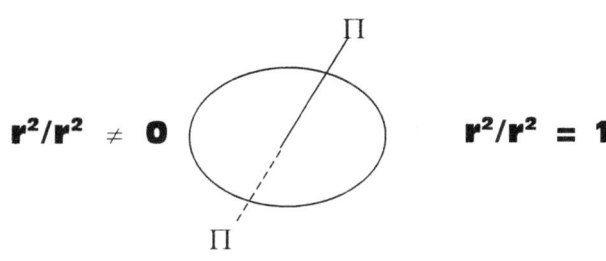

$r^2/r^2 \neq 0$ $r^2/r^2 = 1$

The spinning or not spinning is not part of the issue because at the point of absolute singularity the object never spins. Therefore spinning or not spinning does not apply to the point of singularity because singularity never spins in any event.

Π X r^2 = CIRCLE

If you remove r it then is Π x r^2 / r^2 = CIRCLE.
You cannot then say r^2/r^2 = 0 and therefore Π x 0 = 0. That is nonsense. $\Pi r^2/r^2$ will always be Π x 1, and that is the eternal circle.

That then is singularity. The eternal Π, the Π that may not have significance but still it is a Π of value. The relativity remains one, eternally one, but it cannot be zero. Therefore, dJ/dt cannot be zero.

dJ/dt can become eternal or infinitive or at the worst it can become one
dJ/dt = 1

When explaining this to any child, they can immediately see that. Explain this to any Newtonian High Priest and he may have you removed forcefully from campus. I cannot find one Newtonian, of any significance being large or small to accept that. By not having a wheel rotate, the wheel becomes the factor of one, and the rotation becomes zero. The wheel does not disappear. In the cosmos, everything is rotating because nothing ever stands still. Therefore the mean equilibrium, the common factor there is to share, has to be one, eternity, the eternal Π, because all rotating objects have Π in singularity, and sharing singularity, gives every object in space a relation with all other objects in space. After trying for many years to bring them the candle, I concluded that Newtonians are incapable of realizing that mathematical principle as reality.

If Newton had said that $dJ / dt = 1$ then that is exactly what Kepler said when he said that in the centre of space–time singularity is allocated a position of control $k^0 = a^3 / T^2 k$. Kepler also said the motion brings about the filling of singularity $k^0 = 1^0 = 1$ when he said that the space is filled by the matter in the motion through the time period. Motion establishes space in time and cannot be zero because THAT is what gravity is. It is the motion of space-time and that can't be zero. If gravity were equal to zero, the entire Universe would stop existing.

The centre and that which connects the centre is of such importance that every Universe holding singularity at $k^0 = 1^0$ pivots around it. With every one of the four points taking form to the value of Π at a measure of $\Pi / 2$ each brought about the Roche value of $\Pi^2 / 4$ in relation to the developing centre. One has to remember that the star of today takes on the characteristics of the form of that era. The barrier there is relates to this precise limit being the end of the Universe at $\Pi^2 / 4$. If there was no bearing of the centre that was implicating on the why would that be a factor in the formula we use to calculate the size and why would it have any role to play.

In the same manner the ring cannot be removed, because the spokes will then still imply where the ring must be. The only way to cheat yourself out of the situation is to remove the wheel and spokes altogether, and you are left with what you say there is: NOTHING. But that does not apply in cosmology. The object rotates the centre structure and therefore there has to be a radius holding the circling orbit in relation to the centre structure.

Removing the radius from a circle does not remove the circle, because the circle is there, securing the ring. If the Universe started from a point of singularity, then there was initial spin at a pace where the spin did not apply and that spin included the entire Universe, still in non-existence. That is singularity. That is the only singularity there can be. The spin was going on for eternity because the spin does not apply, it has a value of infinity and infinity was running along the line of eternity.

By receiving the command, singularity received a value outside eternity as Π^0 received edges. Granted the fact that the edges were so small there still was no r to present a circle.

Having edges where Π^0 duplicates to present the edges, singularity lost the value of Π^0 to the value of Π^1 with the same value singularity had being Π^1 to the one side and Π^1 to the other side, the cosmos received the eternal value of the first dimension outside eternity. It was the square of Π^1 being Π^{1+1}.

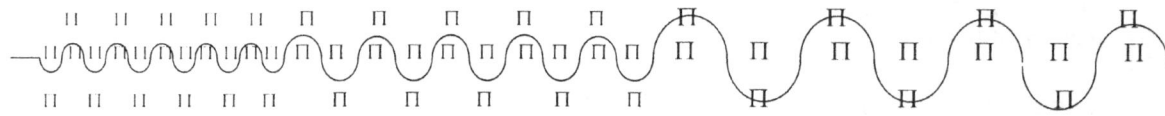

That was the first dimension outside singularity Π^0 where singularity has a value of Π^1 in the form of $\Pi^{1+1=2}$. The first claim to space had a value of Π^2. This applied to both sides of the claim to space outside singularity, and the double proton became the dominant factor on matter.

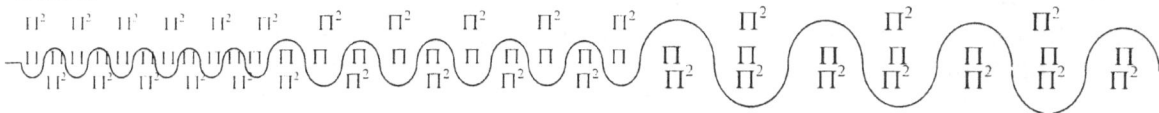

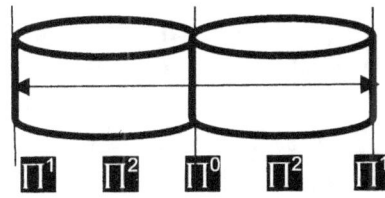

That, which formed the Universe was the growth of time expanding while overheating. It is the expanding by the measure of Π^2 that has much significance as where the expanding went Π by Π. The expanding was limited to the value of Π^3. As singularity burst out into matter forming space as much as occupying space inside singularity, the protons started flying around, spinning around singularity, as each individual proton occupies matter in space. For every space there has to be spin and every spin is defined by a relevancy. It truly

makes me feel bitter thinking about the many times I tried to explain the facts to the Brainy Bunch with no luck. You know you are correct, but that person holding the establishment secured for Newton just pushes your argument aside, because he has the authority to investigate and lacks the interest to initiate change.

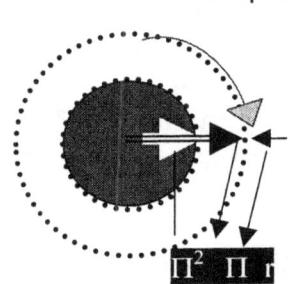

Individual singularity and governing singularity and group singularity enhances the gravity every time singularity finds an accumulation. With looking at Kepler's in a mathematical sense it is clear that from singularity comes space by three, duplicating space in time by three.

$k^0 = a^3 / (T^2k).$

In the action of the inseparable drawing closer and moving closer, gravity finds the dual value of linear and circular gravity. There is no separation of the two factors acting as one but both have different applications and value in the unit. This is the result of singularity having three parts acting as one but giving three distinctions in application.

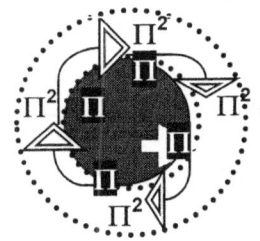

How many dots was there is a question no person can answer because everything was un-dividable solid and yet it did group together to form every atom located in the 3D.

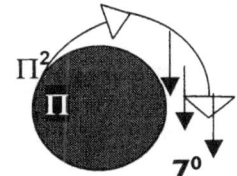

In the circle T^2k which consists of the atmosphere, the space surrounding the rotating object will also extend by **k** as the concentration of the spinning motion draws or drags on past T^2 extending the influence of T^2 by the value of **k**. Very clear evidence about this one can see in the Coanda effect. This extending of T^2 to accommodate **k** we refer to as the atmosphere, but physics apply to this extending in the normal fashion. The soil of the structure represents the solid proton being $\Pi^2+\Pi^2$. From the spinning motion T^2 does not stop at the end of the solid structure but the influence of **k** extends and this then becomes the atmosphere. The influence of T^2 stops at the end of the solid structure but the influence of **k** extending plays a most dominant role in the cosmos, although not yet recognised and that factor is most crucial to a better understanding of the implications of laws governing the cosmos.

With the circle being T^2 k the T^2 will reflect the circle in the square with **k** forming the extending of T^2. This is an extending of the six **k** forming in alliance with the centre **k**. This produces that any extension of 6 forming material, one further extending goes into space and relates to a seventh dimension. The extending of **k** will not end immediately but will carry to the surrounding space the circle influences through rotation. The influence immediately above the circle will have the biggest influence and reduce gradually as the value of **k** reduces in the leverage that the space has on **k** and a gradual but definite change from Π to r will affect the extending of **k** progressively more. The decline of **k** will follow the same contour of the circle at 7^0. Every one of the dimensions indicates an individual significance as I shall show later and the increase into space runs by 7^0.

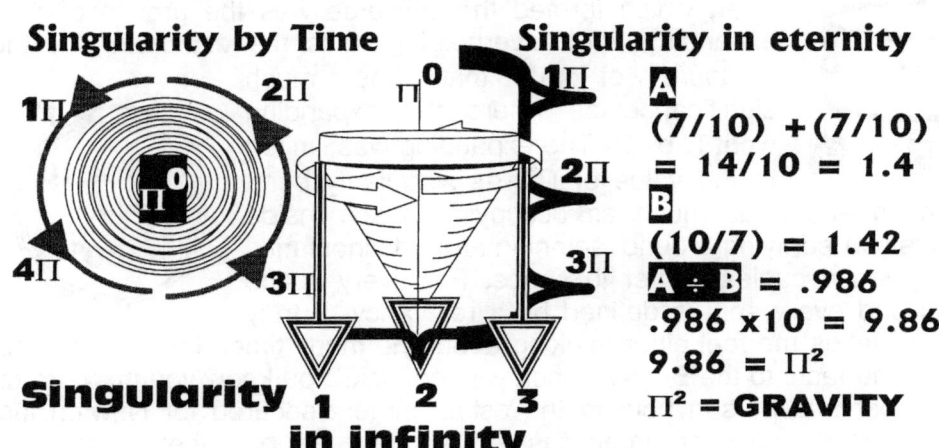

Singularity by Time

1Π 2Π Π^0

Π^0

4Π 3Π

Singularity in infinity

1   2   3

Singularity in eternity

1Π

A

(7/10) + (7/10)

= 14/10 = 1.4

B

(10/7) = 1.42

A ÷ B = .986

.986 x10 = 9.86

9.86 = Π^2

Π^2 = GRAVITY

2Π

3Π

With the dimensional change from space in the cube to space in the sphere a relation of 5 to 7 comes about depicting gravity. The principle of 5 sides in space relating to 7 in the sphere holding matter, forms the basis of the Titius Bode and the Lagrangian principles.

Any point will be opposing itself within the **rotating of 180°** where it **then changes every aspect** of its **previous flowing** characteristics it had or **will once again have** in **360°** from there. While in rotation from the view point of a bystander it all may seem static and never changing but to the object in spin every next instant in time will be diverting from every aspect it had every second passing, and the direction it held in relation to the direction it held the previous mille, mille second will totally be incompatible with the direction it holds the very next mille, mille second of rotation.

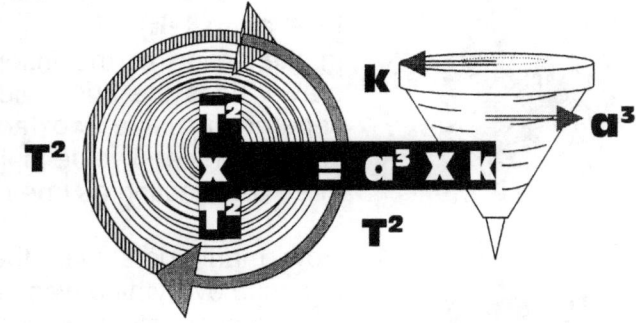

$$T^2 \times T^2 = a^3 \times k$$

This is why we can use degrees measuring the circle by (6^2) (forming the square relating to matter through singularity) X 10 (square if space) = 360^0 however it is always in motion. That proves that no point can be static or constant, though it may seem that way to outsiders. Although matter is matter, matter can also be anti-matter and moreover form its own anti-matter at the same time. This degeneration of structure is very likely to occur with overheating. Revaluing Π to Π^2 will bring about a new contact point where Π meets **r** forming another relation in Π^2. **Time is** the **changes in relation** where Π **contacts a different r** not withstanding the many r points there may form because **every r constitutes a different value** to the universe through other ratios and relevancies brought about **by heat and light. Time is the duration it takes Π to rotate between any two given points of r** and therefore must always amount to **a square (T^2)** moving from point to point through the **cube of space (a^3)** in that **duration of time (k)**. With that it proves **Kepler's a^3 (space) $=T^2 k$ (time in the instant of motion)** but motion must continue through a specific value in space where the space-time is maintaining relevant equilibriums throughout singularity connecting.

Newton, and science, made one enormous blunder, from this stance. They took the radius of a wheel not to have any influence on the wheel. In doing that, they removed the very fact that keeps the universal attachment together. They still insist that rotation results in nothing.

The state that is the time component called outer space, is coming about from the fact that outer space is the Titius Bode law because the Titius Bode law is evidence of how the Universe was compacted by motion. The time zone called outer space is the motion called gravity and is the neutron factor in the Universe.

$$\frac{dJ}{0} = dt \text{ or } \frac{0}{dt} = dJ \qquad\qquad \frac{dJ}{dt} = 0$$

$\Pi \times r^2$ = CIRCLE If you remove r it then is $\Pi \times r^2 / r^2$ = CIRCLE.

You cannot then say $r^2/r^2 = 0$ and therefore $\Pi \times 0 = 0$. That is nonsense. $\Pi r^2/r^2$ will always be $\Pi \times 1$, and that is the eternal circle. When looking at any rotating object, there has to be a point in the infinite middle where the one side rotates in one way and the other rotates in the other direction opposing the opposing direction. That point in infinity is the point of no rotation and no rotation means "no rotation", not no existence. No rotation means a factor of 1, not zero.

That then is singularity. The eternal Π, the Π that may not have significance but still it is a Π of value. The relativity remains one, eternally one, but it cannot be zero. Therefore, dJ/dt cannot be zero.

       dJ/dt can become eternal or infinitive or at the worst it can become one
       dJ/dt = 1

When explaining this to any child, they can immediately see that. Explain this to any Newtonian High Priest and he may have you removed forcefully from campus. I cannot find one Newtonian, of any significance being large or small to accept that.

The comet rotates the Sun, and the Sun by itself has a point of singularity where Π remains without r. The comet, holding the orbit, also has a point of singularity, but since there is space separating the two objects, they cannot share a mean point of singularity, the very point of existing. Since singularity means just that, being single, there cannot be two. The comet and the Sun have a mean point of singularity but the space they occupy divides their common singularity. That is why they orbit in an oval path, a path where the one structure holds on to more space from its point of singularity towards the space it claims.

Since they do not claim equal space, BY THE DENSITY they hold, the space will not be in proportion. Singularity is a mathematical reality. Einstein may be the first to name it and Galileo (unwittingly) may have been the first to define it as Kepler was the first to formulate singularity, but in mathematical terms singularity is the most basic principle. It is singularity that attaches the top to the orbiting comet. At this point I wish to establish a fact that seems lost in all other grandeurs of cosmology. A straight line cannot begin at zero or nil, it can only start at infinity. Such a statement will hardly seem appropriate but the relevancy of this fact has no limits.

POINT OF INFINITY

If the line started at zero there was no line to start because zero multiplied by whatever results in zero as the answer. That must also be the cosmic starting point. Einstein introduced such a point and named that point singularity.

The Universe does not change because there is not one single item that is in the Universe that can change. What the top evokes is what was established at Moment-Alfa when time was interrupted by space for the very first time. The spin or expanding that was introduced, still present, as it was in the beginning. The very same principles are still the spin or expanding that was, as it was when it was for the very first instant of time when space parted time in infinity and eternity. The line of time is eternal and was interrupted by space with infinity bringing an end to eternity in time. However, in infinity the line was interrupted extensively but briefly while the line continued intensely small but eternally long.

Einstein introduced matter time and space and I can see where Einstein was heading with the three concepts forming one Universe but Einstein got his wires slightly crossed, because for one, what Einstein saw as space is time and what Einstein saw as matter is time in general as it is a connecting that singularity has with the flow of time throughout the universe. In that there are two strands of time that formed with one massive time delay that compacted the heat in the time delay and that time in delay compacted in nice units, which now forms the part that are the matter Einstein referred to. In that there is overall just time relating to time in time. However the only existing Universe is the Universe, which is not part of the Universe. The rest is creation by generating motion and it is not created as such in measure of time flow so very long ago and back in the most distant part. It is created by motion of time delay through time in time. By moving from one point to another point the flow of time goes square while the points duplicate. The duplication is a product of overheating in one specific spot where that spot exaggerates the space by expanding. However in duplicating there is cooling and cooling is reducing heat. Heat is what there is becoming more of the same thing and therefore cooling is what there is taking away some of that.

The motion brought about the square of the value but in the square there initially was only singularity.

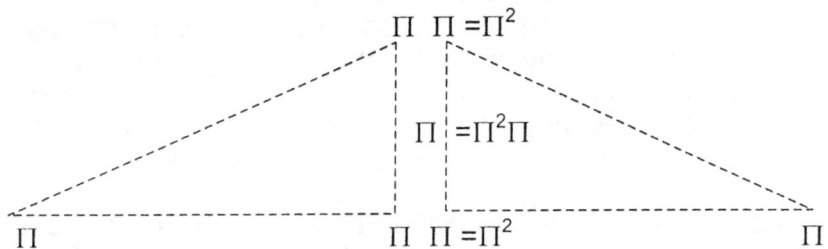

As we already determined on a previous occasion we now accept that expanding comes about from heat and only heat brings about expanding. By heating whatever one may find in the universe expand.

By expanding the space becomes more and the space in doubling cut the heat by half. It is an immaculate and genius way of controlling heat.

As it is still the case, the contraction results in duplication and by distributing the overheating space over a bigger area, the cooling contracts half (in the beginning but at present the proportionate) the heat back to secure the original singularity while the other position secures a new point that activates singularity. The distribution results in the contraction. When 1^1 expanded from 1^0 there was a linear motion established. The motion took what had no start

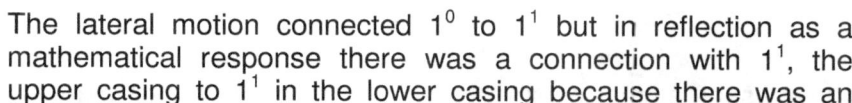

away from what had no end. In that the expanding had a direction as lateral that connected to the cross of the lateral that connoted what remained of the eternal to the lateral.

The lateral motion connected 1^0 to 1^1 but in reflection as a mathematical response there was a connection with 1^1, the upper casing to 1^1 in the lower casing because there was an immediate contraction to the expansion.

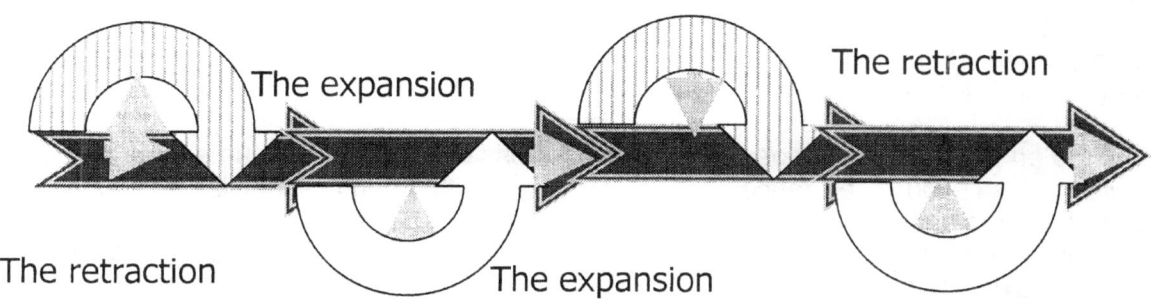

The flow of time from the past through the present onto the future had three dimensions resulting from the oblong time in the perfect eternity strayed from the perfect that it always had been before, by going imperfect. This brought to the future where half confirmed the past, half conformed the future and believe it or not but half converted the future.

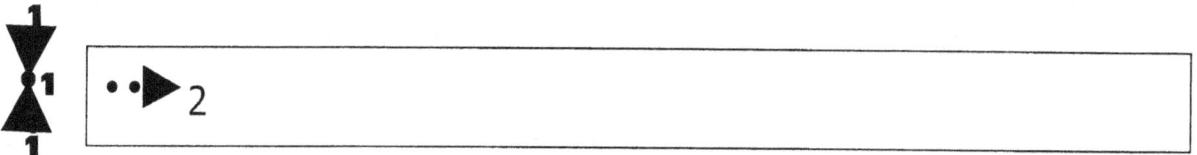

Therefore with one the Universe was in eternity and the value of the one was confining everything in singularity. By duplicating two relevant points forming three positions, singularity came into being a form in the universe. The number arriving in the Universe was two, but I am somewhat reluctant to say that what ever formed at this stage, was already part of the Universe. That is still miles off.

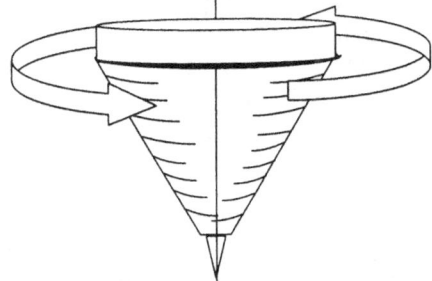

$T^2 = a^3 / k$ and $T^2 = k / a^3$. In this period of development the time associated with eternity much more prevalent than it did with a break on continuity in infinity.

What happened back then is precisely what we are able to gauge from the behaviour we find the top shows because the Universe doesn't ever change.
This brings us back to the spinning top I presented at the beginning.

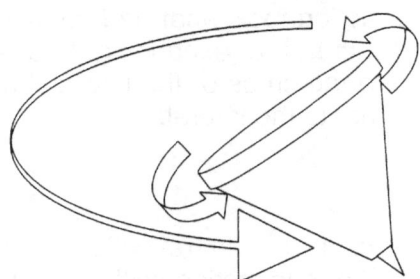
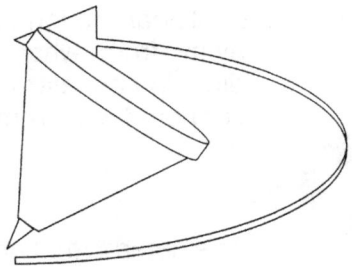

I have asked as many persons as I do not care to remember why the top spinning will remain spinning around one point while turning. The answer I receive from the most educated to the schoolboy is always about momentum. That is a very simple answer and to say the least a little too simplistic by further analysis. Why would the spinning top go off centre when spinning higher than a specific velocity and lowering the velocity it would stabilize and run square to the Earth only after that it will go oblong and then fall?

I could go on about different positions bringing across different momentums of thrust but I do not wish to insult your intelligence because I am aware that you are familiar with all the laws. When the top is spinning it is spinning about its own axis and when it is not spinning it still remains spinning about the Earth's axis, therefore when it is spinning it is also spinning about the Earth's axis.

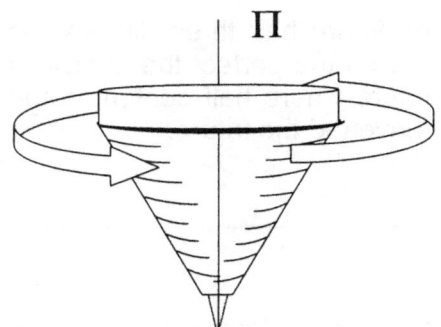

Therefore the limitations applying can only result as an influence coming from the Earth's axis. The second question now comes screaming across and that is in what manner could the Earths axis ever affect a spinning top since the spin and the spinning top is a gross mismatch to what ever standard the Earth may introduce. It is clear that spinning objects do influence each other in contrast to Newtonian opinion.

Every round object has a point establishing a very centre, a middle dividing one side from the other. That division determines the space from one side away from the other side. At one point there must be a point that does not fall on either side of the divide. Such a point will still be a circle, because from that side the circle divides into two sectors.

In every spinning object there is a point of infinity, a point that does not turn because it holds the dividing spin. From that point running in all directions the spin is opposing the other side. All spinning activity starts at that point diverting outwards and from that point the spin is either clockwise or anti clockwise in all directions. As I pointed out, no line can start at zero

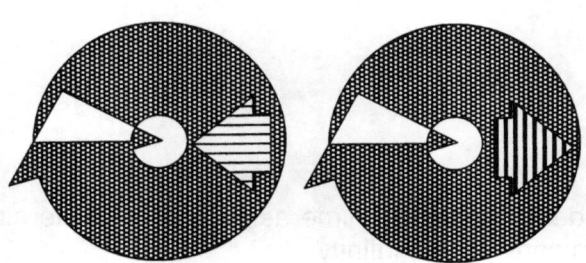

because then there is no line and no rotating point can start at zero because then there is no rotation.

I have indicated that motion creates space $T^2 \times T^2 = a^3 \times k$ and space finds limits of space in motion $k = a^3 / T^2$ as well as $k^{-1} = T^2 / a^3$ where the motion confirms the space while the space conforms the motion. The rotation of the motion in both relevancies completed the space that formed by the motion in linear as well as rotary.

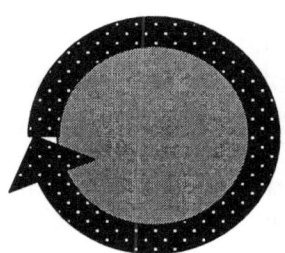

Saying that 1^0 moved to 1^1 sounds senseless but in that motion is so much potential locked in that no other movement ever came close to such giant step. It happened when the first relevancy came about and everything that ever could be in the Universe and would be was locked in one spot becoming one dot. To our mathematical genius of numbers and measure, the movement had no meaning because it still remained what it was before in **k = 1.** To us with small minds that reality left little cancelation with **k** being either 1^0 or 1^1, which remains at **1,** but the moving of the space gave the space a reason to be and the space gave something that could be moved a distance from 1^0 all the way **to** 1^1. The emerging Universe in its entirety was in prominence.

By reducing the one line the other line can never reach zero because then there is no such a line to begin with. That makes a straight line also inevitably always a potential square and that makes the straight line half the value of the square being 180^0. At a later point I shall continue with this argument, but for the mean while I wish to come back to the circle. This same principal applies to the cube and that means everything there is and ever will be is either a square being part of a cube or a circle. With the straight line forming half the value of a square $360^0 / 2 = 180^0$ in as much as being one line and reserving one line in infinity to eternity. The straight line is just half the value of a square. In that manner the triangle is also half a square and therefore holds the same dimensional value as the straight line being also 180^0

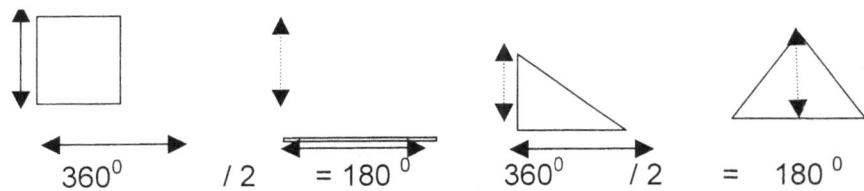

360^0 $/ 2$ $= 180^0$ 360^0 $/ 2$ $= 180^0$

The circle is a square holding a round shape, as the straight line is a square holding one side to infinity. Calculating a circle involves two aspects where the one is either the radius or the diameter that is double the radius. The other is the factor Π

$\Pi \times D^2 / 4 = $ circle and $\Pi \times r^2 = $ circle

The point of singularity cannot be in space at large because space is not there and secondly what ever is there spins too slowly to have a connection with singularity directly.

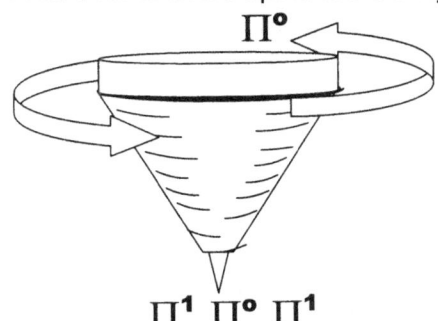

Π^0

Π^1 Π^0 Π^1

With everything in a cube or a circle or a potential of the two, brings about the implication of eternity in a form of singularity or the point of creation. Removing the radius of a circle does not remove the circle, because the circle is there, securing the ring. If the line (or imaginary line if you wish) holding the value of $\Pi^0 = 1$ there has to be a point where the circle is no longer in infinity but claims existing outside the imaginary. At that point the radius may be slightly more than infinity, but to all calculating purposes it still remains as infinity. The spin was going on for eternity because the spin does not apply, it has a value of infinity and infinity is another expression for eternity.

Having edges where Π^0 duplicates to present the edges singularity lost the value of Π^0 to the value of Π^1 with the same value singularity was being Π^1 to the one side and Π^1 to the other side, Π^0 must be the point splitting singularity into two parts of eternity, the eternal value of the first dimension outside eternity. It was the square of Π^1 being Π^{1+1}. That was the first dimension outside singularity Π^0 where singularity has a value of Π^1 in the form of $\Pi^{1+1=2}$. The first claim to space had a value of Π^2. This applied to both sides of the claim to space outside singularity, and the double proton became the dominant factor on matter.

Right at the start before space and time became developed the motion produced space in the principle of the Coanda effect. By receiving space, singularity received a value outside eternity as Π^0 received edges. Granted the fact that the edges were so small there still was no r to present a circle. The manner that the top uses to evoke singularity, which enables the top to maintain independent motion could be, traced right back to the very first line that came about as singularity initiated spin

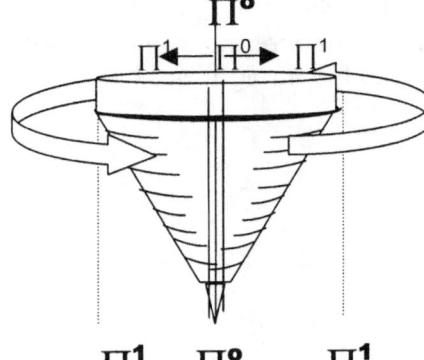

In the beginning there was no space in which to move so therefore the only way to move straight was to move in a circle. The movement **k** producing the line was the same as the motion **T²** that produced the circle and the space **a³** achieved was the compliment that two factors combined and that formed the space developed.

Taken from the point of rotation the two sides are in opposition to each other in every aspect that they may contain and with all that they hold. The motion is the extending of singularity and singularity reacts on the motion by establishing a proton value.

With Π^0 little more than a figment of the imagination there is actually to values of Π^1 facing each other in a relation combining Π^1 to hold the value of $\Pi^{1+1=2} = \Pi^2$ and with two sides being the very same but opposing each other there will therefore also be Π^2 to every side that holds Π^1.

At last I can come to the one part that I disagree with Newtonians, and what I regard as Newton's second biggest infamous or famous blunder. Science, made one enormous blunder, from this stance. They took the radius of a wheel not to have any influence on the wheel. In doing that, they removed the very fact that keeps the universal attachment together. Singularity controls the Universe by establishing a Universe but that is done in a specific manner.

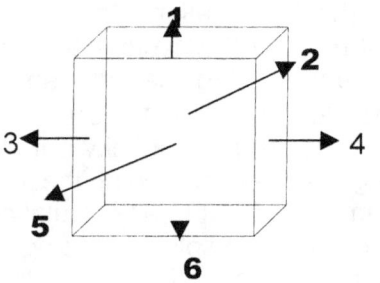

6

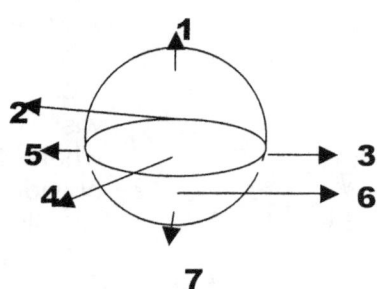

7

In the very centre of the sphere the form of the sphere dictates that the shape will relinquish space as the line runs from the outside towards the very centre. With this natural state of affairs the sphere is naturally inclined to dismiss all space that it can form in the form as the sphere holds space inside and the form will finally be without dimension. All the reducing by shrinking that I attribute to finding the centre by reducing the radius as a line, actually takes place in every sphere as the diameter reduces to the centre. In the centre where the radius line goes single the form relinquish the three dimensional form it has inside. Being without dimension in the very centre means that at a point in the extreme centre of all spheres there is a point that holds singularity because this point with no space has a mathematical position although it is invisible since there are no sides to such a point to give that point any dimensions. The shape of the sphere is calculated by using the formula $4\Pi (r^3) / 3$.

By reducing r to a point where r is r^0 singularity steps in because only the form remains as Π. Going even further we find that there then comes a point where Π goes singular Π^0. At that point absolute singularity is present but so is absolute gravity present at that point. When holding the strength of the shape of the sphere in mind as well as taking into account that all cosmic objects of importance is in the form of planets or stars and they are all in the form of a sphere, we therefore may contemplate that it is where gravity originates. We now only have to find the reason why gravity will hold a base in a space less ness as Einstein predicted. It is clear to be seen that gravity is in the centre of the sphere controlling from the centre everything that is outside the space less centre. We can reason with confidence that gravity is the strongest where space is the least. We can further reason that it is gravity that is holding the sphere in true form and since the sphere allows gravity the best working opportunity, gravity can form the sphere in as strong a shape and form as the sphere seems to have.

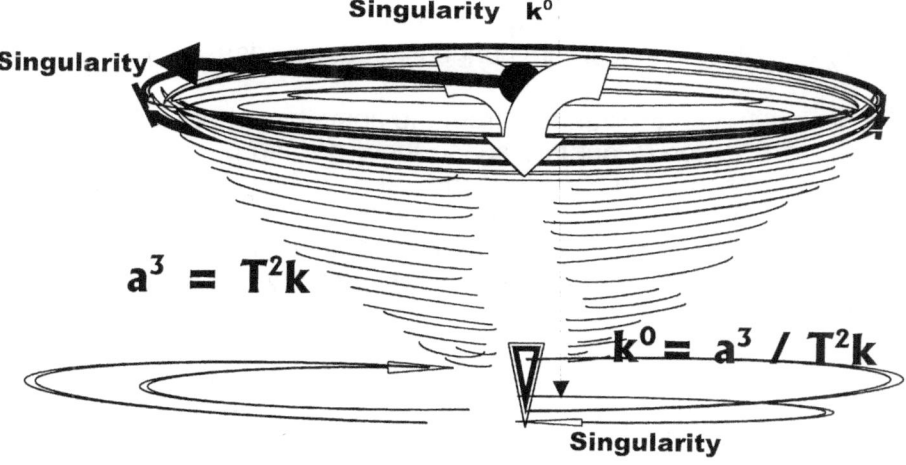

Singularity k^0

Singularity

$$a^3 = T^2 k$$

$$k^0 = a^3 / T^2 k$$

Singularity

From every point on the surface of the sphere is where that point connects with the other side of the surface of the sphere by a line that runs through the space less ness of such a centre of the sphere. Such a line also connects by an angle of 180^0 as well as 90^0 to six other lines running from top to bottom, right to left, and back to front, where all join and cross in the centre of the sphere. There are therefore six lines crossing and connecting by a centre from any given point on the surface of the sphere. Such points connect in total six surface points on each side of the sphere while they all support one another through the space less centre. In that absolute space less ness in the centre holding singularity we find gravity supporting and controlling all space within the sphere as well as space connected to the sphere. That is where gravity controls and guides the space, which falls in the parameters as well as under the influence of the form of the sphere. In the gravity centre space goes singular meaning space becomes space less or flat.

It is from the layout that the sphere uses as natural form that we are able to locate singularity. In the case of the sphere the material naturally reduces by measure of the radius becoming smaller to a point where the radius is r^0. At that point the line that will form the radius has gone single dimensional r^0 and that is equal to 1^0, which is singularity. Also it is true that the entire form that is the sphere is controlled from a centre within the sphere. That centre holds the sphere in form and shape. Therefore the strong form is dictated from that space fewer centres where there is no space and no form left. The natural inclining is in the form of the sphere. It is part of the roundness that the overall shape of the sphere represents and this structural strength is carried down to the very centre.

Because the circle is forever reducing, that reducing which is inherently part of the form of the sphere becomes a tool in the distorting of space in the sphere and is eventually removing all forms of space from within the centre of the sphere. The very centre ends up as having no space because of the reducing that continuous down to become the space less inner centre. The all roundness is the ingredient that forms the backbone of the absolute strength that the sphere has and that is the component that the sphere is so famous for. The form the sphere has allows the sphere to have a control that is coming from the centre deep inside the sphere where the space vanishes and being without space seems to keep the entire structure rigged. From the centre the sphere shape shows strength that the shape as tough as it is. How does it work in its most basic analyses?

The cube has sides and the sides form a rather weak and flat surface that connects four corners. The flat surface produces a rather indifferent contact point with no special features on the surface. The corners connect to other sets of corners and those corners form a weak structure without any direct support coming from the other five sides. Without material to fill the body of the cube the cube has no direct connecting between any of the sides other than corners connecting at the edges of the sides.

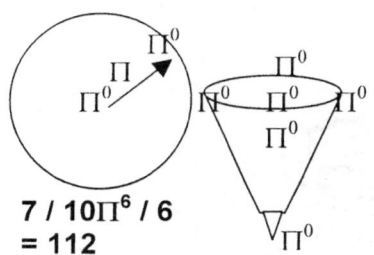

$$7 / 10\Pi^6 / 6 = 112$$

Taking the vantage from the point the sphere is holding from the centre out into space there are ten points connecting to the centre. In that are the dimensions of singularity connecting to space where five connects to space in the second dimension of singularity, and five connects in the third dimension of singularity. On the other hand, the cube does show a very different characteristic, which involves only six sides (at least) connected.

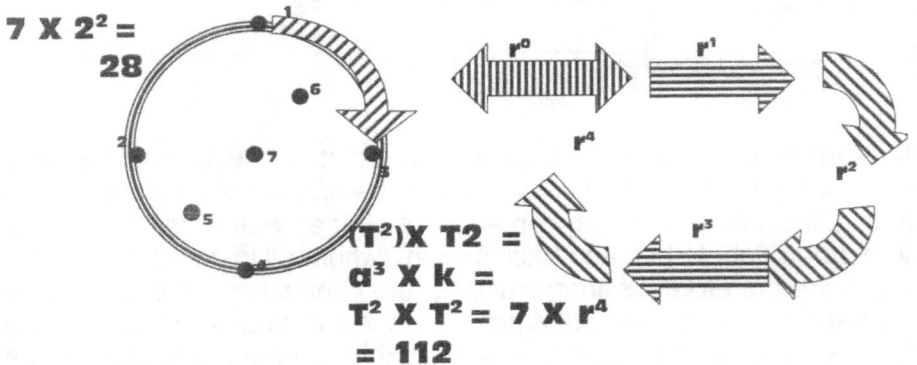

$$7 \times 2^2 = 28$$

$$(T^2) \times T2 =$$
$$a^3 \times k =$$
$$T^2 \times T^2 = 7 \times r^4$$
$$= 112$$

The spinning of Π^0 around the centre Π^0 establishes Π and Π is what produces the form gravity has. Still it is the relation or relevancy there is between the centre Π^0 and the spinning Π^0 that gives status to the form that Π represents. In out Universe we are accustomed to and are familiar with the rules. We want to place seven points holding singularity to the centre holding singularity in a relation of $7/10 \; \Pi^6 / 6 = 112$. In that Universe everything less that a duplication ability to the value of 112 protons fit but only atoms to a maximum of 112 protons fit. That puts us not in a three dimensional Universe but a four dimensional Universe that can accommodate the movement of three dimension in a fourth dimension of time.

If gravity is motion T^2 the process may sound simple but it is a deception since it is very complicated. By motion T^2 space comes about forming a^3. But the motion T^2 will mean a crossing to the other side of the universe since singularity divide the Universe into sectors. Material produces the dismissing or the concentration of space by applying the motion.

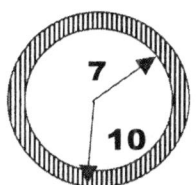

Surrounding all elements is a layer we call the atmosphere and even Pluto and the moon must have the atmosphere because they have gravity. In this the relevancy of ten to seven forms this layer and it results in forming a circle because of the combining of the motion duplicates the singularity factor Π forming from that gravity as Π^2

THE PROCESS PARTED USING THE ROCHE PRINCIPLE

By establishing motion and creating motion, singularity quadruples to 4Π in rotation. But since the rotation is motion duplicating the space established as four times the value of singularity, the motion divides the space coming about by halving such space by the dimension, which is putting a square root over the quadrupling of space. In this comes about the direction gravity takes the universe. The expansion is always double the square root but the square root is neutralising the expansion and that brings about that the neutralising of the expansion creates a contraction that seems dominant to us but it is not. The contraction is doubling the expansion by halving the effort of the expansion.

Singularity is a mathematical reality. Einstein may be the first to name it and Galileo (unwittingly) may have been the first to define it as Kepler was the first to formulate singularity, but in mathematical terms singularity is the most basic principle.

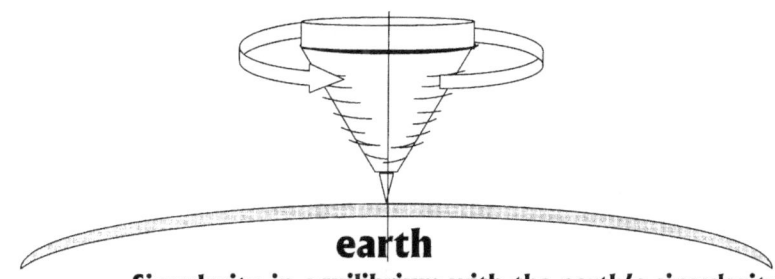

earth

Singularity in equilibrium with the earth's singularity

Singularity applies gravity by charging motion and the motion (not the pulling) of space-time is gravity. It is motion that moves the gravity that moves the Earth and it is also motion that moves the top that forms gravity. Gravity is not the pulling of but the motion or tendency to move to the centre of the Universe, which is the next domineering point, holding singularity in a control or a governing mode. The motion or tendency to move is that which forms gravity and mass is the occupying of space and therefore restricting the motion of gravity.

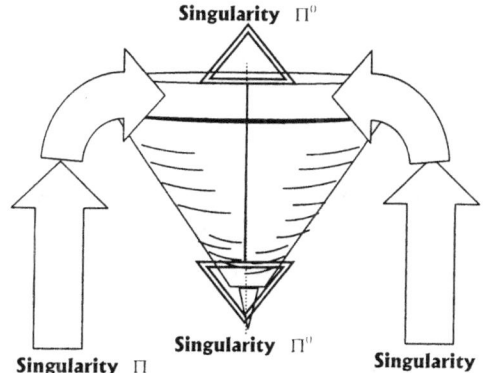

Singularity Π^0

Singularity Π^0

Singularity Π

Singularity

The greatest minds in the entire world are missing the smallest line in the entire Universe.

By rotational motion, the top creates a line confirming singularity running down the line and by generating the line the line charges gravity. The gravity is what drives the top as the top and as long as the top spins. There is an influence generated by the spin of the top that keeps the top upright while the top is spinning.

The line is generated but the line is far from magic. The line is where the centre of the Universe is which the Universe is then that what the top filled by particles from the line to the edge of the sphere. The particles in motion generate motion by electing a centre from the centre of every particle in the spinning top. Such an elected centre becomes the centre of the Universe as far as the top relates to a Universe because all the atoms in motion elect the centre of the Universe.

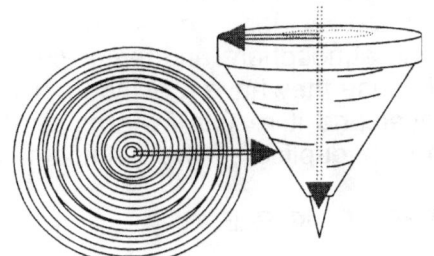

In this, it is clear why the Titius Bode ([10 + 10 + 1 + .991] / 7) and the Lagrangian 5 \\ 1 systems part their ways when applying the different processes they hold. With all the differentiating, the observer must also consider the dual message that light uses in travelling through the vastness of universal space. The thought of nothing is just what it is, a thought of nothing and although it is in the human mind common nature to present nothing as a value in the recalling of something, nothing is a presentation of the figment in the human mind. There can be no number such as nothing and that was (possibly) Newton's biggest error. Nothing represents non-existing and that is just what nothing is, it is non-existing.

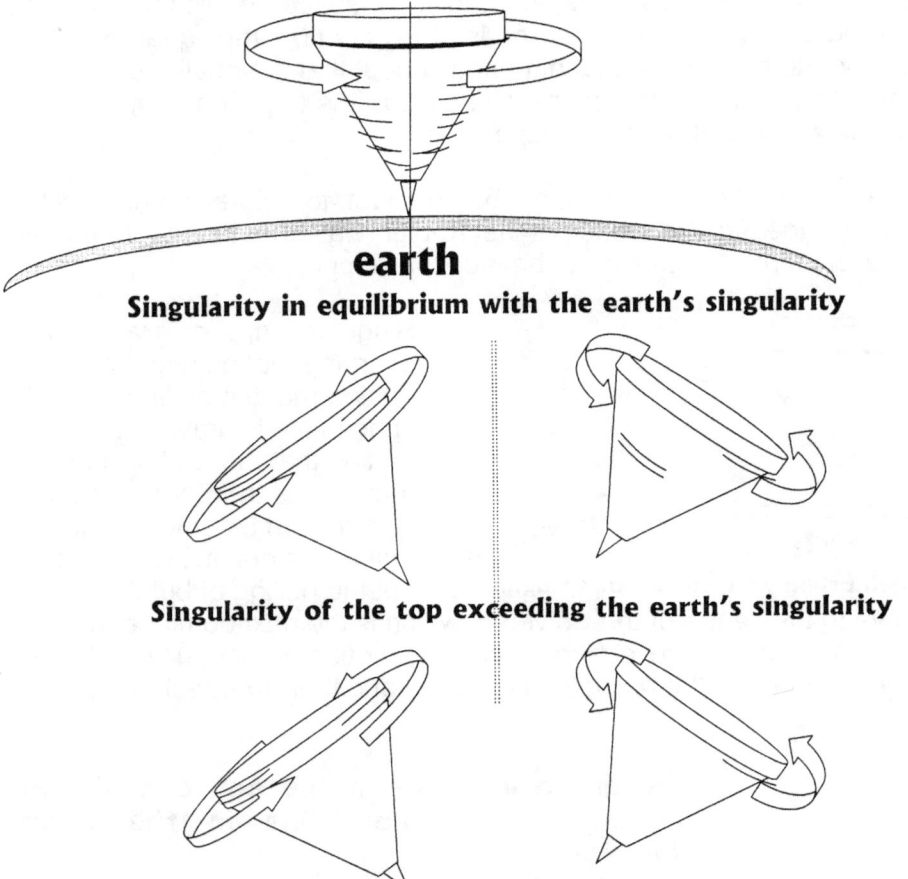

earth

Singularity in equilibrium with the earth's singularity

Singularity of the top exceeding the earth's singularity

The earth's singularity dominating and exceeding the singularity top

The centre may or may not spin and the fact that it does or does not spin is all the same because that centre part never spins in any case. Therefore, the boundaries set by the spinning motion does not depend on the spinning motion of the object but has to stand related to another body bringing about a larger spin influence.

Granted the fact that the influence the Earth has on the top may be that of gravity but if that is the case then surely the Sun also has influence on the Earth and other rotating objects through gravity. It needs more investigation because it may bring about evidence we are not aware of.

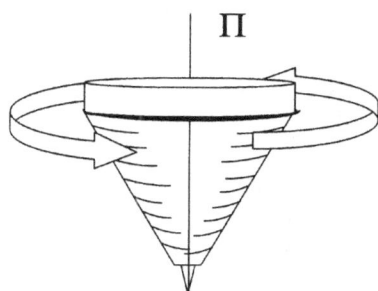

This observation places a much bigger question mark on the statement of Newton where he proclaims no influence on two rotating cosmic structures.

By rotation such rotation is a duplication of what singularity retracts. The rotation involves the three factors in time coming from the past through the present and onto the future, which holds three positions excluding the one position allocated to singularity. The four I named the eternal motion because singularity being without motion was contracting as well as expanding at the same time it was not moving. That placed singularity in eternal motion without ever moving being singularity.

Understanding all the following is connected intimately and conditionally to accepting the fact that all individual particles in the universe use motion and therefore spin.

Every quarter provide a distinct value that indicates the progress of the flow of time from the one point Π to the next point Π.

Any changers occurring in Π will lead to a an unequal triangle providing two different values to r and will alternate the link between r and Π^2 bringing about a difference in form (Π) and time (Π^2). When singularity forming the lines of the triangle is not in equilibrium, the triangle will destroy the matching of the half circle.

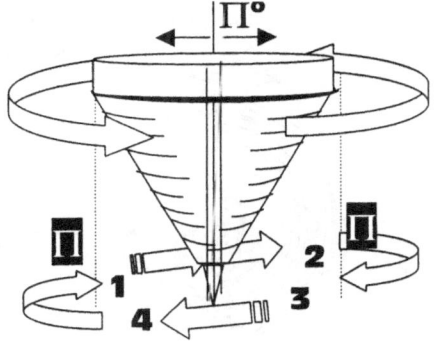

In considering the spinning motion in the fraction of time in the detailed instant every aspect of rotation will turn in every instant of change in time. Although the points had the same characteristics only one instant before, they oppose the characteristics it had just before and just after the very instant in which they are and to which they relate by similar points also in rotation. The fact of the graph proves my point in quarterly opposing dimensions and values. As the rotating direction moves inwards, the rings will become smaller and smaller. Move the rotating line progressively to the middle by reducing the length the line have from the edge to the middle. At one point all further reducing ends.

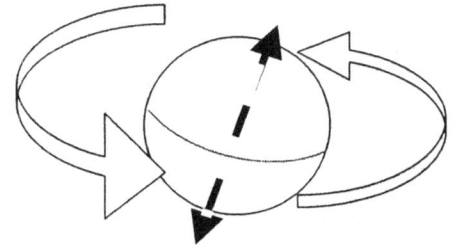

The drawing is the circular Π^2
The movement is the linear r
The change over of dimensions is

r meets Π Π^2

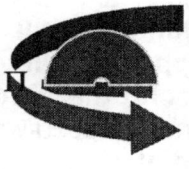

In considering the spinning motion in the fraction of time in the detailed instant every aspect of rotation will turn in every instant of change in time. Although the points had the same characteristics only seconds before, they oppose the characteristics it had just before and just after the very second in which they are and to which they relate by similar points also in rotation. The fact of the graph proves my point in quarterly opposing dimensions and values. Due to the spinning nature of such a point with all surrounding, the point will be alternating direction favouring change every second and in that the value of such a point can only be Π because of its constant changing. Using r would specifically oppose another r from every angle because the use of r will bring about a static relation to the previous and following instant and therefore it will cancel the constant spin flow. There must come a point where the ring is infinitely small, where it can reduce no more, where it reached its ultra limit, but at that point it cannot be zero, because the point is there notwithstanding that it is at a location beyond our Universe. But the spinning object **will have a middle point**, a very specific **centre point that does not spin** and only holds Π as a specific value. One value such a line **cannot have is zero** because **zero does not start any** line and therefore the **value of the**

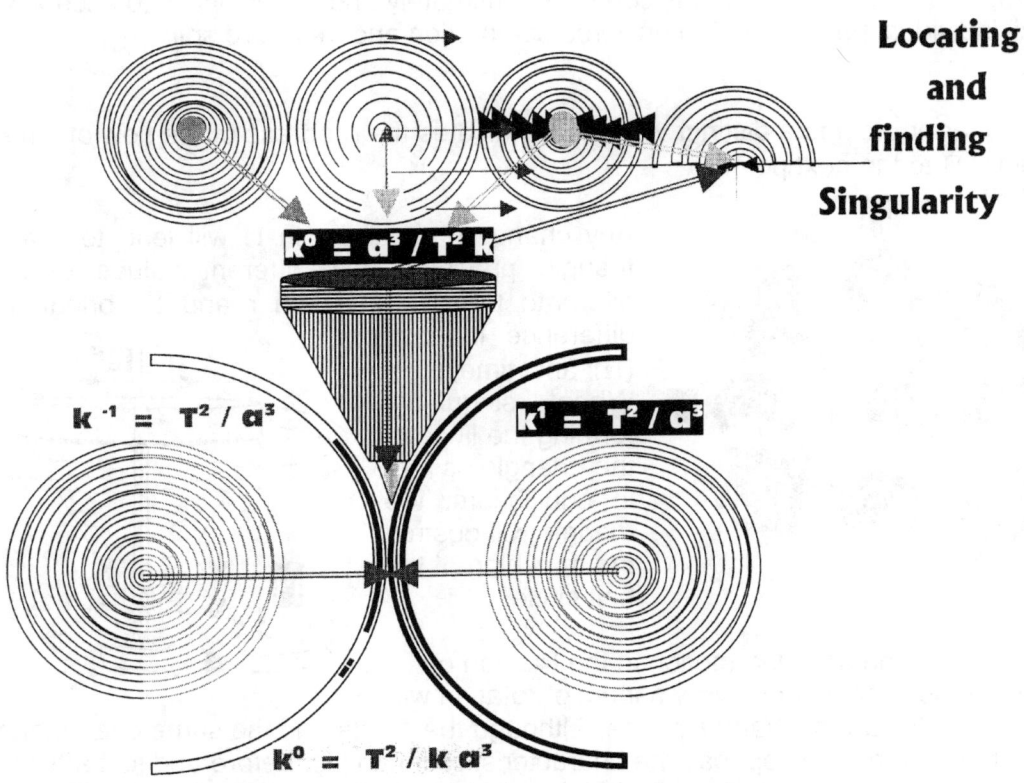

Locating and finding Singularity

$$k^0 = a^3 / T^2 k$$

$$k^{-1} = T^2 / a^3$$

$$k^1 = T^2 / a^3$$

$$k^0 = T^2 / k\, a^3$$

line must be infinite, just as described in **accordance** and by **the definition of singularity.**

In the **precise middle** of all **objects in rotation** is a precise centre dividing the object in sectors that will **start the spinning initiation** from that centre point. **That point** albeit hypothetical. It is also as much a reality none the less and is where that point **must be standing still** because every line **running from that point** in **opposing directions** are also **in opposing directional spin to each other. That point** is completely hypothetical, is also as much a reality none the less and is placed where that point **must be standing still** because every line **running from that point** in **opposing directions** are also **in opposing directional spin to the other or opposing side.**

From somewhere outside the Universe a line rises while remaining in a position allocated to space being outside the Universe. The line is activated by the rotation that sets the top in

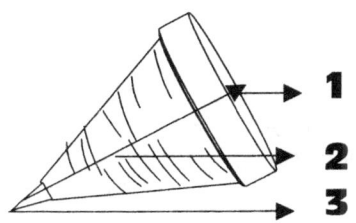

coherence with time and the motion grants the top individual status by establishing independence. The motion holding a dual action while being a unit lays down the ground rules for the Coanda effect. The four points turning around a fifth centre also charges a line to place the space-time the top holds to an erect status of independence. The line is at the centre and it is motion by rotation that activates the line. The erecting of the top underlines the status the top receives as an independent Universe, which is maintaining an individual singularity. Even when the motion no longer finds the ability to charge the line in singularity and the top stumbles before it falls, there is still a desperate fight for keeping the motion and with that the independence active. The fight is not a fight for balance but a fight for survival.

Space parting eternity and infinity activates the line and Newton was of the opinion that rotation brings about no work. The line runs from the top of the top down to the bottom of the top without ever being present in the space of the top. The top form the space being fully independent and it is most critical to consider that line that the motion activates also activates the third dimension of time in space. Space in time and time within space separates by finding separate identities in the cosmos and this gives singularity independent identity. It parts eternity in time from infinity in time. A universe is born through the rotation of the top in spin. By parting infinity from eternity, the space in between fills with material that allows the top the position the material in the top has while the top is spinning. When the motion falters in sustaining the singularity it requires to remain independent, the point holding singularity goes cold and looses the acquired independence it had.

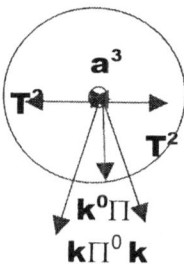

That point albeit hypothetical, is also as much a reality none the less and is placed where that point must be standing still because every line running from that point in opposing directions are also in opposing directional spin to the other or opposing side. Moving the rotating line progressively to the middle by reducing the length the line has from the edge to the middle. At one point all further reducing ends. In considering the spinning motion in the fraction of time in the detailed instant every aspect of rotation will turn in every instant of change in time. Although the points had the same characteristics only one instant before, they oppose the characteristics it had just before and just after the very instant in which they are and to which they relate by similar points also in rotation. The fact of the graph proves my point in quarterly opposing dimensions and values.

This only applies in relation to time because time is the square or then if you wish time is the flat to space being the cube. Time in the square draws space in the cube flat and that is the why the Universe holds the sphere in place.

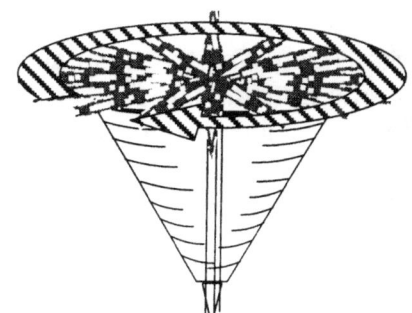

Understanding all the following is connected intimately and all conditionally to the fact of accepting that all individual particles in the Universe use motion and therefore spin.

In dimensional terms, which I explain later on the value of **2k** relates to **T²**. That relation extends to the next value where **T²** relates to **k**, which relates to **T²**. The first space in the circle will then be **T² k**. From the centre being in infinity, one can realise by applying mental power the single dimension factor not seen but present all the same. Extending that into the 3D comes six **k** and any one of the six will further extend to form a seventh point as **T² All this is a multiplying of $k^0 = a^3 / (T^2 k) = 7$**

From this line of reasoning I dismissed the theory of the presence of a force being gravity but rather consider it as a dimensional changing contributed by the spin of the Earth and the spin comes from singularity located in the centre of the Earth. It is all about dimensional changing that influences space as a factor of ten to reduce to Π^2 on a continual basis from a point forming new dimensions through billions of such points.

In conditions found in the Universe spin can only come about from heat that is concentrated around singularity. Dropping water on a red hot metal will lead to feverish motion as the singularity in the water absorbs the heat and expands the space by accelerated duplication. The singularity finds the time differentiation there is between eternity and infinity excelled and space excelled brings about motion in duplication amplified.

That what we see in the top results from the interaction of life that in principle is some mistake the Creator allowed to happen in a very small region and as far as man is concerned, taking into account all the proof man has to his disposal the phenomena of life is no where else in the Universe. By having T^2 overheat space will reduce in ratio that brings about linear motion k being shorter per time unit but much more frequent per time unit accelerating because $k = a^3 / T^2$. However in accordance to Newton's law on motion there will be a reaction $k^{-1}=T^2/a^3$ and with a reduced space the time in ratio will increase allowing for the heat rising and the amplifying of motion.

More spin increases both lines that forces gravity by the increase of T^2 that extends the influence of k, k^{-1} in the formula as factors because it reduces the moment of a^3. The extending of the liquid heat will increase the motion and increases the contracting gravity $k^{-1} = T^2 / a^3$ and the reaction to that is that the space reduces by a larger time contraction $k = a^3/$

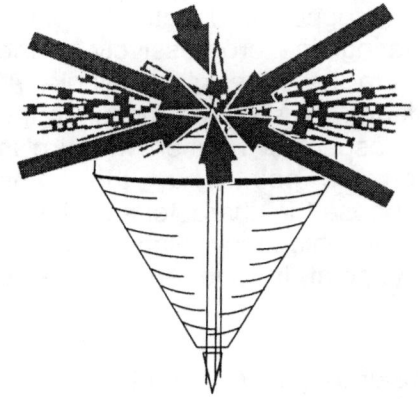

T^2. The space then has to duplicate more vigorously in order to cope with the rise in the time aspect.

Therefore the space wants to exceed the boundary time slaps on space as a result of the Coanda principle by applying more contracting motion because of the rise in the liquid heat levels in time while space has to extend because the rise in heat produces a need for countering the overheating. The duplication of space by an increase of the number of **k** in the time unit T^2 will allow as far as $k^0 = a^3 /T^2k$ will permit. In this there is a living up of standards in space and in motion. However not one of the mentioned normal aspects will apply to the top since it is the manipulating abilities of life that charges the top into action.

In the circle using $r^2\Pi$ the r has to have distinctive qualities placing it as a factor apart from Π. Where the growth shows no separate distinction but a continuous flow from the precise centre to the precise edge the flow would become in relation with Π depicting the circle and Π replacing r as reference to any point on the circle. By using r as a distinction in the circle, division is possible but by using Π there is no distinction possible making it a solid flow.

Any object being in outer space floats and such floating is seemingly random with no specific detectable interfering favouring a movement in a particular direction. Such a devise is depending on influences not in our scope of detection. But then the object comes closer to the Earth and reaches one specific point where the six dimensions that influences the object suddenly changes. At one point, one of the six dimensions falls away as it disappears and the object quite latterly falls to the Earth. The support of one side disappeared and the centre point of the sphere took over the control. At that point the object is under the influence of one centre point in the sphere and we all also know that in such a centre point one will always find the strangest or the controlling gravity.

Space-time is a four dimensional position of the Universe where the position of an object is specified by three coordinates in space and one position in time. This evidence we find as matter grows into the dimension we now share with billions of stars in the cosmos.

With the dimensional change from space in the cube to space in the sphere a relation of 5 to 7 comes about depicting gravity on one side of the divided Universe. The principle of 5 sides in space relating to 7 in the sphere holding matter forms the basis of the Titius Bode and the Lagrangian principles.

The TITIUS BODE Principle Outside the sphere

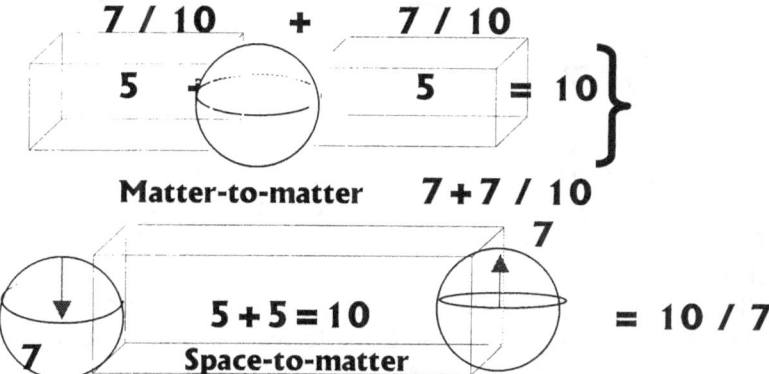

The Titius Bode law is an extending dynamic deriving from the law of the gravity dimensional factor where the space factor in a square of ten relates to a matter factor in the square by half (half since nothing can be in two places in the Universe simultaneously) of the matter factor of $_\Pi{}^{7+7}$ or the square of space (10) relates to the matter factor of 7. From such a point every other point will be opposing any other point not pointing in the direction to which the first point is pointing, whereby it extends the direction it holds. No matter what the point is or where the point leads, such a point holding a specific direction will be unique in the direction it is rotating because at that or any other specific point wherever, it will be directing not in the direction it spins but in the direction flowing from the centre point outwards.

When the foundations were laid in place with singularity expanding even before it was growing The Roche limit became one condition. But while that was taking place another principle came about which is as secured in the foundation of the third dimension as the sides supporting the third dimension. Sides came about through the dimensions that are framing the dimensions, as we know them.

There was the dot. The dot had no borders therefore there was no separation and still we know there were more than one in a group of one. The evidence of this is very present in the cosmos at present and one can find such evidence all around us. The dimensions personify the Titius Bode principle and understanding the relevancy between the dimensions will also mean the understanding of the interlinking values of the Titius Bode law.

Everything is space-time by confirming space in establishing time

When the Universe was in the beginning with the entire cosmos still in a single dimension there were no limits as we know limits to form in the Universe we use and no borders indicating limits because after all it is the single dimension where there is only one dimension holding so much diversity. The dots referred to in this case have no space but were as close as singularity is when singularity has no sides but only shapes and the lot were the same, the very same one with a time delay parting them. The borders were part of development because we can witness the legacy of such borders in the present day holding the 3D in place. There will forever be smaller particles that combine to produce larger units. The forming of particles start at infinity and there no human can reach except with his understanding, and his mind power.

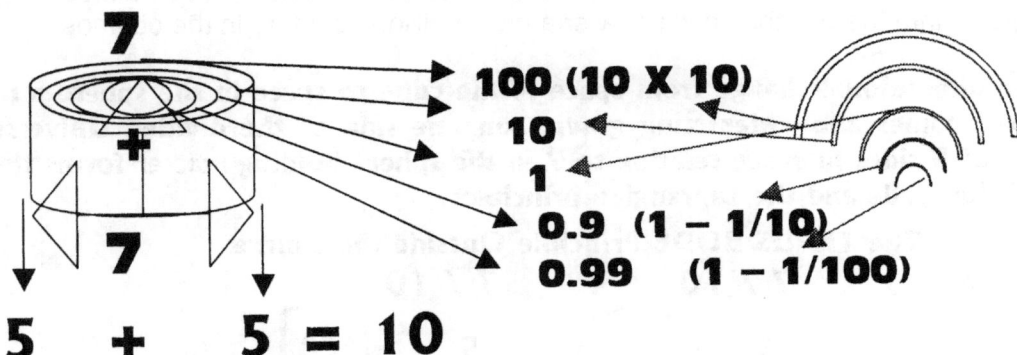

The normal flow will allow singularity extending to 10Π but when singularity blocks another sphere in singularity the two will form a joint value and by this joining the larger will dominate the space as well as the time of the lesser taking control of the surface and the atmosphere. Through this the Roche lobe comes about with all its other dynamics I describe. The principle is the same, which we know as the conducting of lightning and Jupiter uses it extensively to implement this action. In the Roche limit the straight line forms part (1) and the half circle is part (2) and the triangle forms part (3) to singularity (4) Holding 5 points outside singularity. Every aspect connecting to the universe changes everything it holds totally and becomes the anti-matter to which it was matter 180° previously.

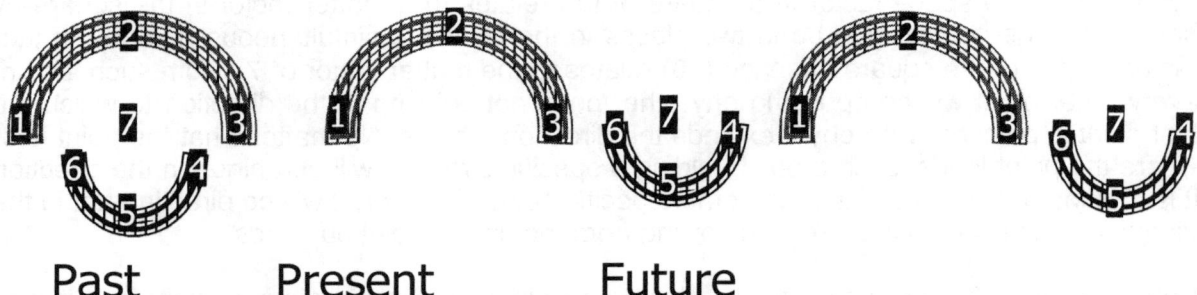

It starts where the first seven points serving singularity meets three points holding time. At present we named the proton combining with the neutron and served by the electron as the atom and from where we gauge the Universe to us the Universe is the entire atom.

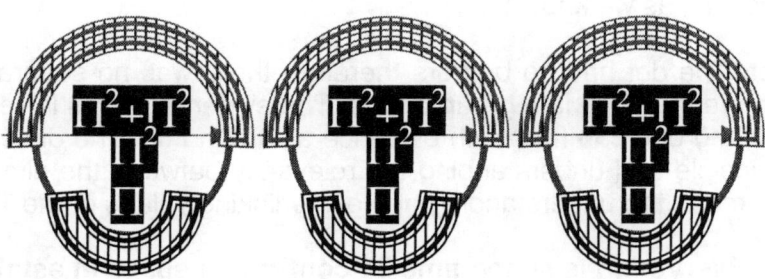

The atom holds seven points $(\Pi^2+\Pi^2)(\Pi^2\Pi)3=1836$ as the Universe but that Universe is seven points in Π being 3 points serving dimensional time to form the Titius Bode law, and a law it surely is! The gravity extending from the Titius Bode law forms the entirety of the building of the Universe by constructing the Universe in the using of the atoms to form the Universe in the entirety thereof.

Gravity produces mass but mass is only the result of gravity. Mass do not produce gravity and the manner in which science uses mass can only apply when using the calculations in terms of the Earth. However applying it to stars as science indicates by their formulae used in their calculating of gravity on structure beyond the solar system is very inaccurate. Heat stored in motion produces gravity. Any one not in agreement convinces you by comparing the neutron star with the massive red giant. To calculate a Black hole they go and throw C^2 next to the dividing radius and throw the square onto the C that presents the speed of light. Then they sit back and feel smart in the way they manage to cheat once more to prove their incorrect views correct because after all who will ever fly down a Black hole and return to support or deny their calculations.

The Gravity of the Black hole is a speed because the entirety of gravity is speed or better said it is motion. Then the speed that light has is gravity. The gravity of the light can be gravity as much as it at that very same time can be antigravity. What the hell has C^2 got to do with a Black hole because you can pop what ever nuclear device far away from a Black hole and it would be at the most and at the worst very much insignificant. The light will not even escape from the gravity of the Black hole. When this became apparent that the radius of a star reduces as the star develops through progress, the whole logic of cosmology became apparent. Yet, it passed Newtonians by as if it was a mystery. Newtonians were so enrolled with mass and how mass allowed them to play mathematical games; the truth of cosmology simply went pass them unnoticed. It is some time ago that someone was supposed to say: hey, there is a dead rat I smell. For my saying so I am the clown in the courtyard, the one with the two dead brains cells and have no more to use as spare.

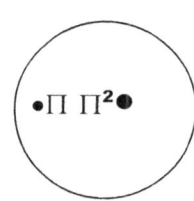 Space generates the mass where the space has to reduce the size by becoming more intense and concentrates space-time to the time of 1836 time more when entering the point of the proton being on the verge of singularity. In single dimension seen from one aspect, with single dimension contacting the edges forming the sphere it will still keep the seven positions because the sphere remains a unified structure though apart because of singularity. In the core of the sphere the proton connects in alliances as Π^2 +Π^2 with the solidity of the neutron holding Π^2 as a second forming a π value. That brings the atom unit of π to a number of seven.

By reducing the space-time the lesser singularity is claiming singularity independence by offering reduced space, which will result in promoted time with heat being the net result. That heat is filling the space, which should then be entered by the independent space in motion if the motion of duplicating is brought closer in a relation to what the matching tempo requires.

It is clear that the density of material in motion is $\Pi^2 + \Pi^2$ but since that is k, which extended we know that that extending cannot sustain the initial speed. Since the speed is reduced the space in motion will value less. Taking these atomic relevancies into account, we can detect what relevancies brought about the atomic Universe of $(\pi^2 + \pi^2)\ (\pi^2\pi)3 = 1836$.

The first substance that formed from singularity was solid and if that were the case the contra substance would then be a fluid with less motion filling more space taking shorter time duration in duplicating. The fluid substance that then formed was one less than the proton in motion which makes it slightly more in mass since it duplicates more with less space that then has to form $\Pi^2\Pi$, which has one Π less from $(\pi^2+\pi^2)$ which is resolved becoming a fluid like substance relevant to the first solid substance which is the proton. The loss of the one Π then became the factor claiming more space that is holding less substance. In this fluid state the neutron has more duplicating of the substance than is required of the proton. That what we find in space we also must find in the atom because the cosmos is not keen on inventing but is passionate on duplicating. This fact will also apply to space-time in many forms. That means investigation must prove the same results and what we find in the atom then also has to present in the cosmos at large.

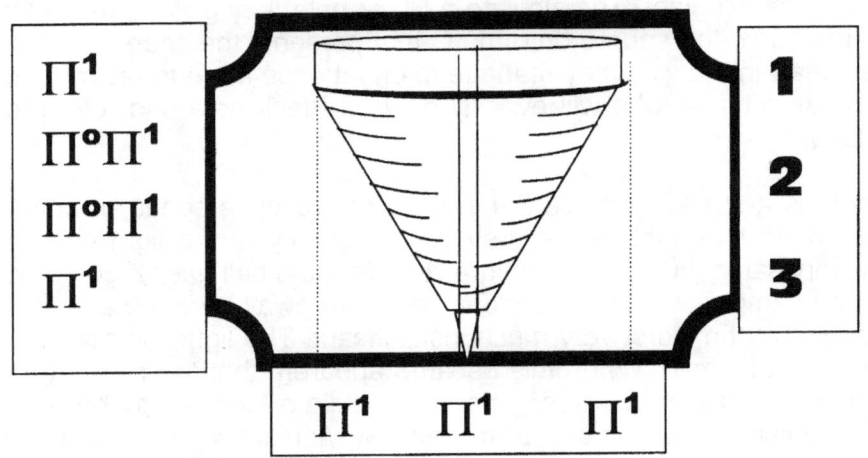

The overall picture resulted in a ring and all rings hold Π to secure the form. The only form that existed then was Π and therefore even today the borders use Π to indicate positions. But in the single dimension such definitions were far from clear and the only distinctions came from securing singularity in preserving the position of singularity to apply gravity and thereby absorb all anti -gravity. But anti gravity could not control expansion by counter acting contraction through gravity so the overheating continued forming non-existing borders in some thing infinitely solid just as Einstein predicted because this took place before light came about and therefore before the speed of light became part of the cosmos. The cosmos formed a partnership with one side overheating forming antigravity by expanding into space through the applying of the overheating and the other side formed gravity or

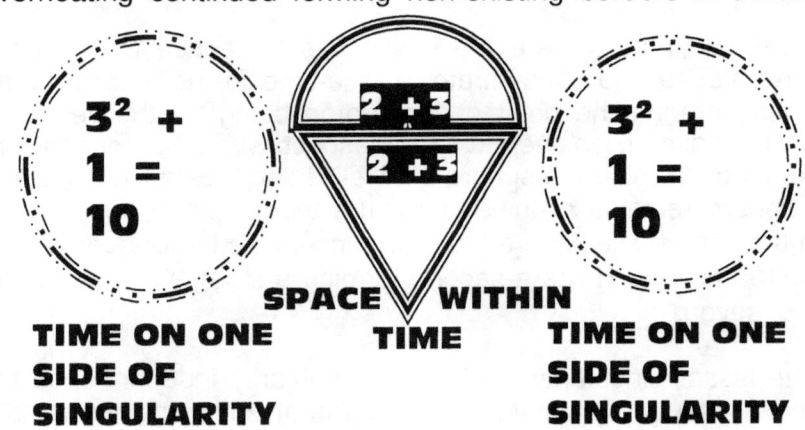

contracting of space.

If **k** is the middle being $k^0 = 1$ then $a^3 = k^0 = 1 = T^2$. When time is in a shift freezing then $a^3 / T^2 = k^0 = 1$. In order not to overstep my limits by changing valid formulas I changed Kepler's formula to $R^3 = T^2 = 1$.

In the relevancy where space divide eternity from infinity the three holding Π in relation to singularity holding Π^0 there are three points forming a square in relation to 90^0 which is implicating the law of Pythagoras while on the one side of the Universe the duplication is forming the same result and three points goes square. The result is that on the one side the square of space is ten and on the other side the square of space is also ten.

Newton said a sphere is $a^3 = 4/3 \, \Pi \, r^3$

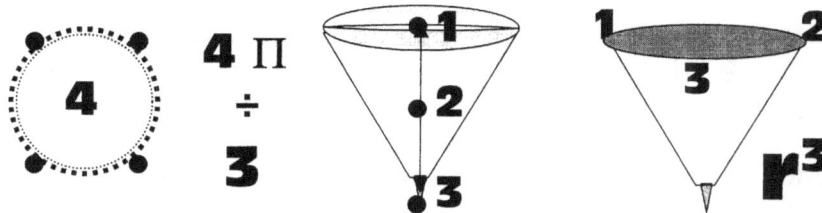

Keeping these factors in mind it is clear that Π^2 are the choice of gravity and not r^2.

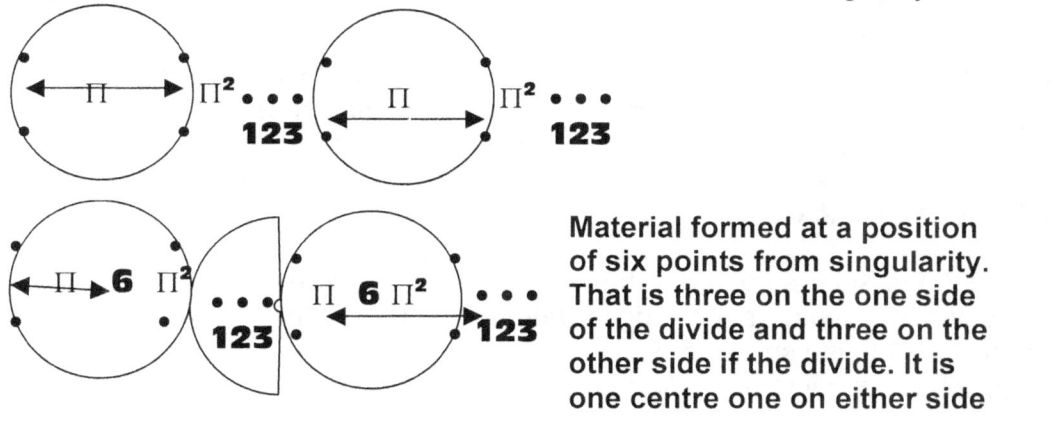

Material formed at a position of six points from singularity. That is three on the one side of the divide and three on the other side if the divide. It is one centre one on either side

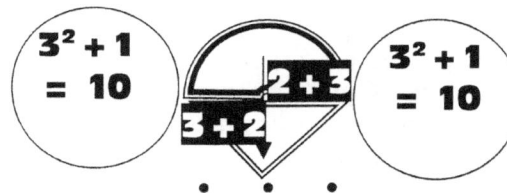

In relevancy from one another material held five inclusive positions of two in time including the three positions as material. That made being in one quarter of time five in all. That makes the Lagrangian system dominant.

$F=G \; (M_1 \; x$  $m_2)/r^2$
Is this truly the   answer…?

In the investigation of light and gravity and objects and gravity, the mathematical rule of the invert square law must apply without question. But according to the observation of Roche that is not the case. From what one gathers through the Roche limit implicating two orbiting structures the opposite is applying. One must accept that although **k** proves as an indicator it is also much more when complying the thin influences brought about by singularity in the values carried on by singularity.

As a^3 increases, so does T^2 as well as **k** increases and with that the influence of gravity per space unit increases with the concentration demise of a^3. But why would that be and what are we missing? Light shows there is an influence out there in outer space, that redirects light's route through space when passing large gravity fields. It is about the relevancy of **k** influencing the a^3 to allow the T^2 of light to divert in route because of influences established by **k** on a^3 and slowing down or increasing the line diverting. In this measure one may also find the Roche limit applying, but to truly understand how the Roche limit comes in place and how the Roche limit works, one has to replace Kepler's factors with singularity and singularity extending being Π^3 $\Pi^2\Pi$ and **3.**

In the Roche limit the space factor provides space to a solid structure and therefore the value of r is replaced by the value of Π bringing about a square in half of Π. The cube holding 5 to either side removes allowing the extending of Π to indicate position to space.

Where Π extends to lock onto the next sphere's extending indicator, Π has to connect to Π forming the square of space and translating that to the half of Π being $(\Pi/2)^2$.

5/2

Five sides divided by two spheres.

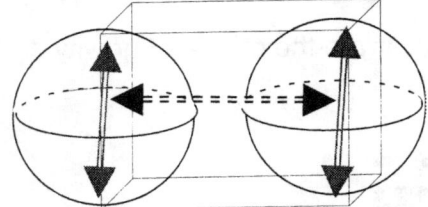

According to normal mathematics the half of space should have been 5/2, but at the time this divide took place,

space was all in motion and motion was Π in motion Π^2 crossing the divide (/ 2) forming $= (\Pi^2/2 X\Pi^2/2) = 2.467)$

The space between the spheres divide in half, but because of the extending of Π and not applying r as ordinary mathematics will suggest where Π replaces r the singularity extending from Π^0 will be half of Π in the square of $\Pi = (\Pi/2)^2 = $ **2.4674.** In this lies the dynamics why planets have a positional (be it rather a dimensional) relation of 7/10

The Titius Bode law must not be seen as some obscure event that took place just before and / or after the Big Bang or when the solar system formed it fell into place. The Titius Bode law applies when the top is spinning, when an atom is spinning, when a motorcar wheel runs on the tar, when a jet engine fires up. It takes Place whenever the Coanda principle comes into effect and the Coanda principle is wherever there is motion in relation to singularity in a centre forming a centre.

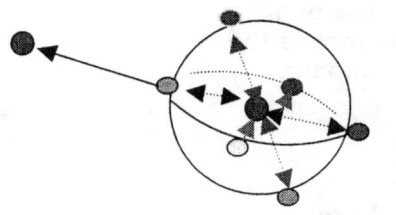

This is why we can use degrees measuring the circle by (6^2) (forming the square relating to matter through singularity) X 10 (square if space) $= 360^0$ however it is always in motion. That proves no point can be static or constant, though it may seem that way to outsiders. Although matter is matter, matter can also be anti-matter and moreover form its own anti-matter at the same time. This degeneration of structure is very likely to occur with overheating.

Time is the changes in relation where Π contacts a different r not withstanding the many r points there may form because every r constitutes a different value to the Universe through other ratios and relevancies brought about by heat and light. Time is the duration it takes Π to rotate between any two given points of r and therefore must always amount to a square (T^2) moving from point to point through the cube of space (a^3) in that duration of time (**k**). With that it proves Kepler's a^3 (space) $=T^2 k$ (time in the instant of motion) but motion must continue through a specific value in space where the space-time is maintaining relevant equilibriums throughout singularity connecting.

Using the concept that gravity applies Π as the circle factor Π as well as Π^2 replacing r^2 the replacing by Π brings two values as Π and Π^2. That I found is the case with gravity and will be apparent when explaining the sound barrier as well as the Four Cosmic Pillars. In order to create a distinction I remained using r as the indicator of the cube or non-circle that has vacant space and by vacant space I refer to non-solid structures. In the solid structure I use Π as a value for reasons that will become apparent in due time.

Gravity does not apply mathematical equations to the letter as we would like, but rather uses Kepler's thinking by enlisting an average gravity applying through out because it never favours any particular direction above the other five directions it possibly can expand into and is equal every where. In gravity one find the extending of Π implementing Π^2 on average as a unit and not the radius r as a specific.

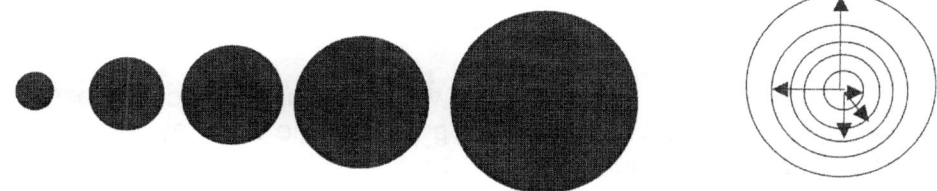

Looking at the affect of gravity, it shows the precise quality of no distinctive point as gravity never seems to end at a point but flows all over affecting all that holds a position in its sphere of influence. The gravity coming from China meets the gravity coming from America at no particular spot but intermingles without distinction. This takes mathematics back to another fact beyond normal explaining.

We take a line running between two points as being 180^0 and the rest of the explaining is saved in the accepting part of mathematics. Any one of the two points the line starts or end at is a point in infinity. The start and the end depend on the viewer putting the relevance to favour the side of choice. That puts the point of end or beginning in the spectrum of choice and not fact. Any direction is as equal as all other directions.

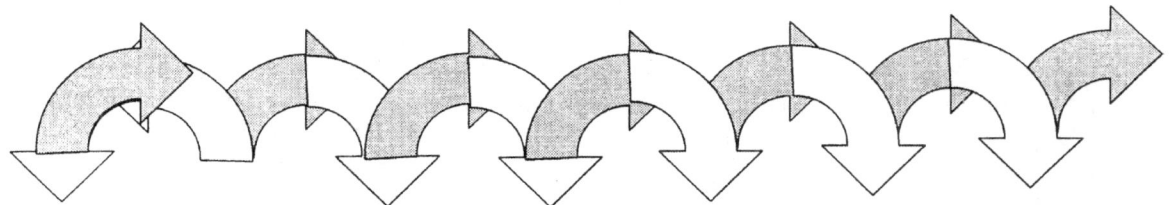

Following the flow of any line such a line is an extension of the previous dot in infinity to the next dot in infinity without any ability to skip or bypass any of the other dots in the connecting line. Any direction change including the remaining of travelling in the same direction is in relation to a line travelling, all being the very same. Change does not affect the line.

A straight line, triangle and half a circle will always have equality in dimensional capacity providing equilibrium being 180^0 because each one shares a common denominator in singularity to the value of Π. As the straight line averts a zero it holds another straight line in place to set about such an averting where the two lines will always carry a relevancy in relation to progress (the triangle) and a common denominator in the start from singularity.

This concept we apply as the graph or the vector. By going back to a line, any lines and all lines, the line is a connection of dots in infinity, running from one specific to another specific and avoiding zero. At every point in infinity it dips into infinity coming out on the other side by choice of direction and the direction is unforced and change presents any angle including the straight line, which incidentally is just another angle.

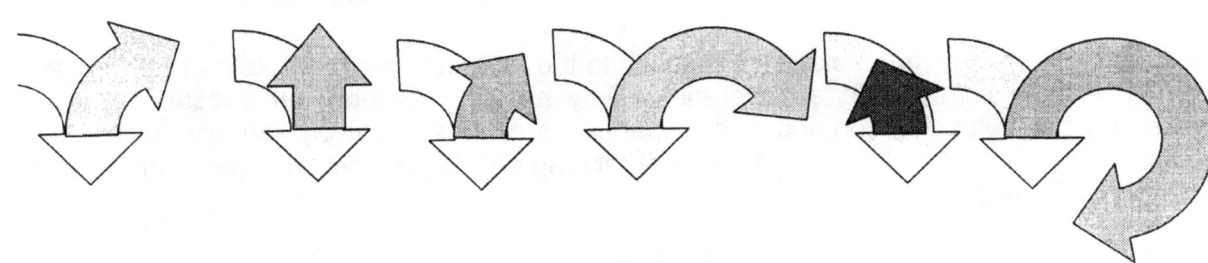

When connecting to the dot representing infinity, the flow can be in any and all possible directions, including in the same direction. We all live in a graph, as the universe with all in it is nothing less than a three-dimensional graph flowing according to time. That means in the case of Pythagoras the mere fact that the line shows changes in direction does not implicate or affect the line as a tool of mathematics. Whether the line changes into a half circle meeting at the other end again or meeting in a triangle in forming a half square by joining the point where it began, the result still indicates a line flowing between points. Motion became an integrated part of space because motion is what establishes space. If motion redirects space and such redirection is not complying with a balance, there will be even further delay in time producing motion, which will bring about more time distortion and heat.

Since the very first space from point to point was Π and motion produced a value of Π^2 while the four points indicated time, it is presumable that from that the Roche factor of $\Pi^2/4$ came into place.

The line dips into infinity every time it passes infinity when it cuts through infinity. The line going into infinity comes natural as the line progresses because all lines are infinite dots linking one point to another point. That means that coming out of infinity might slightly change the angle but that directs the route to the future and not the form because the form still remains equal whether the form is a triangle, half circle or straight line.

The form remains a factor that confused every one in the past. When replacing the value we

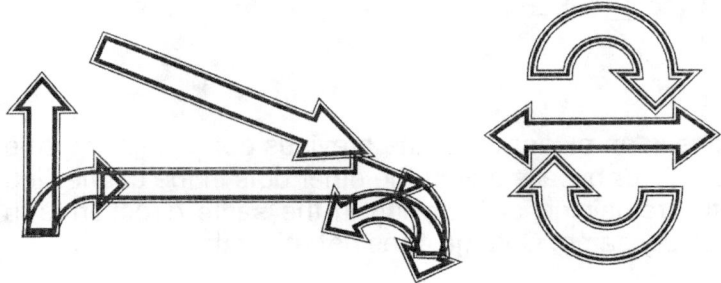

normally attach to a circle being r with Π, the law of

Pythagoras becomes quite meaningful and mathematical.

In that way a circle is a straight line following a loop as it comes out of singularity at a different angle and a triangle is a straight line that dipped into singularity but at three stages changed the angle with which the line then left to follow different directions at specific points. From the point singularity observes it still remains a straight line because there is no direction alternation in the first dimension and in that dimension it still remains a straight line in which we on the outside may experience as three forms but is in fact one single line. Only when the direction changes completely in reverse the line doubles in value but comes from multiplication for instance 2Π become Π^2. But the Lagrangian system proves much more than dimensional interlinking, it proves Pythagoras in principle.

LAGRANGE (-TOURNIER), JOSEPH LOUIS DE (1736-1813)

French mathematician, born in Italy. In celestial mechanics, he studied perturbations and stability in the Solar System. He examined the three-body problem for the Earth, Moon and Sun (1764) and the motion of Jupiter's satellites (1766). In 1772, he found the particular solutions to the problem that give rise to the equilibrium positions called Lagrangian points. Lagrange also studied the Moon's liberation.

The entire concept of motion rests on the centre forming time and having one point outside time to be delayed or behind time. The delay parts motion in eternity from motionless infinity, which results in forming the Universe. Since the satellites are located as electrons the motion gathered from that falls in as a time delay. All motion is about time trying to cross that space to form a unity with infinity. That is the essence what keeps the top straight when spinning. The spin puts the outside of the top in another time zone than that the inside of the top is in and the four inside has to align with the fifth one on the outside where the fifth one is one in three positions allocated to the flow of time. By having time parted from time there is a flow coming from the fifth to the centre.

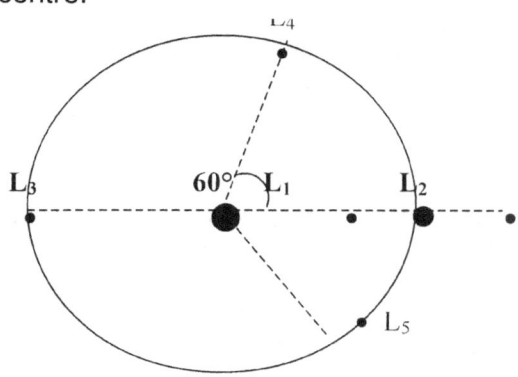

LAGRANGIAN POINT:
*The Lagrangian points
are five equilibrium points
in the orbit of one body
around another, such
as a planet around the Sun*

LAGRANGIAN POINT

One of five points at which small bodies can remain the orbital plane of two massive bodies; also known as liberation points. Three of the points lie on the line joining the two massive bodies: L_1 lies between them, while L_2 and L_3 have the two bodies between them. These three points are unstable, slight displacements of a body from then resulting in its rapid departure. The fourth

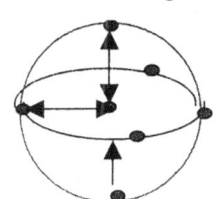

and fifth points (L_4 and L_5) each form an equilateral triangle with the two massive bodies, 60° ahead of and behind the smaller body in its orbit around the larger one. A well-known example of bodies flying at the L_4 and L_5 Lagrangian points are the Trojan asteroids in Jupiter's orbit. Among Saturn's satellites, Telesto and Calypso lie at the L_4 and L_5 Lagrangian points in the orbit of the much larger Tethys. In similar fashion, tiny Helene precedes Saturn's satellite Dione, keeping 60° ahead of Dione. The Lagrangian points are named after the French mathematician J.L. de Lagrange, who first calculated their existence.

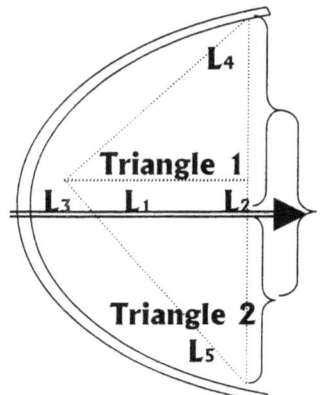

A STRAIGHT LINE, TRIANGLE AND HALF A CIRCLE WILL ALWAYS HAVE EQUALITY IN DIMENSIONAL CAPACITY PROVIDING EQUILBRIUM BEING 180⁰ BECAUSE EACH ONE SHARES A COMMON DINOMINATOR IN SINGULARITY TO THE VALUE OF Π. As the straight line averts a zero going down infinity it holds another straight line in place to set about such an averting where the two lines will always carry a relevancy in relation to progress (the triangle) and a common denominator in the start from singularity. This concept we apply as the graph or the vector.

The Lagrangian System implicating the five positions extending from singularity
Each triangle claiming a side of the universe

The half Circle = 180⁰ combining as a Sphere when comprising

Singularity dividing the cosmos

1 Half circle   = 180⁰   L₃ L₄ L₅

2 Triangle 1   = 180⁰   L₃ L₄ L₅

3 Triangle 2   = 180⁰   L₃ L₄ L₅

4 Straight Line   = 180⁰

Singularity in the matching of the value of the straight line forming the half circle and combining as the triangle and all are equal 180⁰

The normal flow will allow singularity extending to 10Π but when singularity blocks another sphere in singularity the two will form a joint value and by this joining the larger will dominate the space as well as the time of the lesser taking control of the surface and the atmosphere. Through this the Roche lobe comes about with all its other dynamics I describe further on in

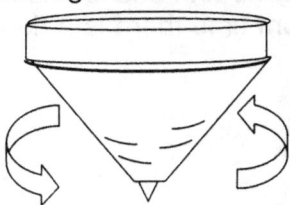

the theses. The principle is the same, which we know as the conducting of lightning and Jupiter uses it extensively to implement this action.

In the sphere there is never only one direction implicated in movement. Movement are always in relation to the centre position because as a line goes up it also goes in or out. When a line goes north or south, it also comes towards the centre or going away from the centre.

The Coanda affect is proof of the functioning of gravity inside the atom. It proves that motion (T^2) of the neutron establishes a centre in line where the compliment of material forming the atom will secure a controlling singularity that is governing the entire atom. That forms the centre of the Universe. Singularity then finds a position at the distance of (**k**) and such motion claims the space (a^3), which is the atom by construction from a centre within that motion (T^2). The motion (T^2) creates a centre at the line of (**k**) and a centre of the space (a^3) the motion

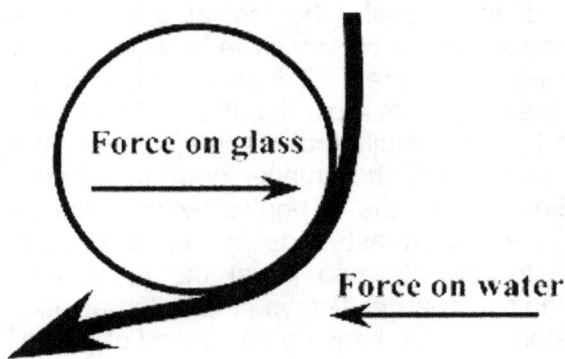

Force on glass

Force on water

(T^2) establishes a gravity field all along the lines and at the distance of (**k**) in the space (a^3) that the motion (T^2) created.

When singularity by the straight line increases, the singularity by the triangle, it will also bolster giving equal potency in singularity by the half circle. As the singularity of the major component revives the lesser singularity to equality, the **triangle in singularity** will match the performance and so would the half circle respond in precise ratio setting equilibrium in order. The major partner's singularity in the straight line excites the minor partner's singularity in the straight line affecting all other aspects holding singularity in both objects to match equilibriums in all aspects of singularity. That is the Roche lobe.

From this the lesser partner will fill by the extent of the larger partner and as soon as equilibrium sets in the growth will duplex to matching on both accounts, normally to the fatality of the lesser partner, as the lesser partner will be capitulating under the strain of the dual. In that way the inner planets came in place as I explain in part 7 ***of Matter's Space In Time The Theses***.

The Titius Bode configuration in accordance to orbiting formation holds a slightly different explanation to the explanation that applies to cosmic structures surrounded by space. It is moreover the individual singularity in maintaining the major singularity, which sustains the governing singularity providing equilibrium in space-time. Not only does atomic individual singularity maintain self-preservation, but in doing that it also sustains a governing singularity holding structural composition and forms within a cluster of matter for example a star. As there is between stars, so there is in the same manner a mutual or bonding singularity between atoms in stars, which we see as fusion. From this one may freely deduct that gravity is not forcing material closer but is destroying space whereby it converts the space to a density the senior partner has in the atmosphere of the senior partner. Where does all the information given thus far take us you might ask? For one, it can help to explain something Newton science can never understand. To start with, we have to realise that the Coanda principle is the manifesting of Kepler's gravity and we have to accept Newton's version of gravity is a load of rubbish.

Years ago it dawned on me why we all would be so egocentric. This was a problem that was eluding every thinker ever thinking. I admit as a thinker I am quite average but still we are all thinkers, which puts us apart is what we think about and in that I am then equal to the attempt of any other average person with the right also to think. There is something that makes every person in his or her eyes having the opinion that that person is the greatest there ever was. Let's call it a Jesus syndrome.

Either the person frequents with Jesus or the person has a special prayer linking such a person directly with Jesus or the person may recognise Jesus or Jesus has come in person to meet with that person in particular and others just simply become Jesus. We all know what I am talking about. What is it that gives every person on Earth the idea that that person is superior to all other persons except those we regard as being more advanced than us? Why would every man walking on Earth think his sperm is just what every woman on Earth would give her front teeth for? Why would every man that walks this Earth do so with the idea that every woman is just waiting on him to impregnate her and that his her sole purpose in life…to wait on him to fertilize her? Why would we be so God damn ghastly superior in the way we see our status we have? Why would every person see him or her with the superior capabilities of reinventing life? Some would not eat meat. Others would bullshit through their teeth about health implications and the misery of death just to get the world to stop smoking. If we are that scared about death, then we better ban the wheel first before any other thing because the wheel in whatever form is killing a hell of a lot more people than smoking can ever achieve. It is the thought that a person can impersonate God and that would allow and enable such an individual to change the course of man forever in all time to come… Some would go to war for any reason because only leaders that killed millions are worthy of the remembrance by Historians.

The more any leader killed off his fellow beings the greater role his memory has in the history of man. Others would not war for any reason even in the face of being threatened by death. Some would drop a Uranium bomb on others with the pretext that they did it to save lives. Others would drag a whole world into a war for the benefit of monetary gain, because lets face it, in the back ground behind the drawn curtains there are those bankers and industrialists that makes enormous profits from other fools fighting "for justice". Something is making every person feel horribly special. Something allows every person to know that that individual is in the centre of the Universe right where God should be. There is a very good reason we all feel that way because we are not wrong to feel that way, and we are in the centre, the very centre of the Universe. Step outside into the night sky and the reason is in front of you. Every sparkle of light coming from where ever is coming to your honour. All the light that was released from any and all points in the Universe is coming to the place you stand. That makes you the most important person ever born because you are the **centre of the Universe**.

When any person is standing on any place anywhere, while viewing the Universe, that person is filling the **centre of the Universe**. Let's get more personal. When you, the person that are reading this, are standing at night and are looking at the Universe you are seeing the

Universe from the position that one only can have if that person is filling the specific spot in the **centre of the Universe**. All the light, every single beam that ever left any destiny at any time acknowledges this fact. You are the most important person in the Universe because you are holding the most important position in the Universe. All the light that come across and travelled all of the vacant space from any and all possible positions in space runs directly towards your position using a straight line towards you where you are filling the **centre of the Universe**. Not excluding the effort of one photon, all light is heading to meet you where you are in that centre spot and not one photon will pass you by. Not one photon dare miss you because if they do they miss the effort that all light has to accomplish and that is to locate you as the person filling the **centre of the Universe**. This may sound controversial but there is more truth in this idea than there is in most other concepts. Wherever you are, you will find that all the light coming from the Sun comes directly to you or later it comes indirectly. The light might first hit other objects but then the light streams from the object straight towards where you stand in the **centre of the Universe**. The Sun sends all the light it has at its disposal to you because the Sun acknowledges that you are filling the **centre of the Universe**. Even light that went astray to objects come back to you so that you may recognise the other objects. Even coming from other planets the light has to return to you and acknowledge your position where you are in the **centre of the Universe**. All the light that ever left the Sun has to return to you because you can see it returning by seeing objects and the planets that reflect the light as it sends the light back to you. It filled the planet, but then it seemed to have realised the mistake it made and by correcting the mistake it made the light returned to the position you fill in the **centre of the Universe**. The light corrected its error and travelled all the way just to find you filling the **centre of the Universe**, right where you are. By you're standing anywhere, you fill the **centre of the Universe**, and the entire Universe admits to that because all the light comes to meet you there. If you shift from the North Pole to the South Pole you will shift the **centre of the Universe** because all the light travelling throughout the Universe will find you where you then moved the **centre of the Universe**.

The light left its destination billion years ago as it travelled through space at the speed of light anxious to acknowledge you're being in the very **centre of the Universe**. No photon will be able to pass you by where you are in the **centre of the Universe** because all light is heading your way from their starting positions. No wonder every person born has the idea they were born to fill **centre of the Universe**, which we do fill. The Universe is spinning around you or I, who is filling a centre where all motion is connected. That is the Coanda effect on the uttermost grandest scale imaginable; nevertheless it is only a manifestation of the Coanda effect. It implicates gravity as wide as can be…

Some things mathematics is able to explain but other explaining goes beyond mathematics. Try to explain mathematically the colour of the sky being blue on a clear sunny day and changing to black when nighttime falls. Do the explaining in mathematics to a blind person that had no vision since birth in such perfect mathematical detail that would allow the person afterwards be able to explain the difference between blue and black to other blind persons by using only mathematics. Some aspects of the Universe go beyond mathematics and some even go beyond words. It is our task to find space, to find time and moreover it is our optimal task to find the Universe.

We have to see what is solid, what is liquid and what causes gravity. Please keep this part in mind because in a short while I am returning to this to show how this becomes a cosmic reality. Gravity **is to move or apply the intension to move** space a^3 **at the** distance or relevancy of **k** while T^2 is the time it is going to take to **apply gravity** or move the space filled with material space a^3 at the distance of **k** in the time period of T^2. That confirms Kepler's attribution to gravity where according to Kepler space a^3 is equal to the movement T^2 (time it takes to move) at the distance **k** from the centre specific.

The Sun   Mercury   Venus   Earth   Mars   Jupiter

$\dfrac{7}{10}$ $\dfrac{7}{10}$ $\dfrac{7}{10}$ $\dfrac{7}{10}$ $\dfrac{7}{10}$ $\dfrac{7}{10}$ $\dfrac{7}{10}$ $\dfrac{7}{10}$

$\dfrac{10}{7}$ $\dfrac{10}{7}$ $\dfrac{10}{7}$ $\dfrac{10}{7}$

Then I took Human nature and science and combined the two, which gave me the vision on the findings Kepler received from the Cosmos. It puts all aspects of gravity in the Universe in new dimensions. But the visions formed the beginning because the visions unleashed many new questions. If gravity is motion, what causes motion? What stops motion? That answer is in the Black Hole. In truth the explaining of the Black Hole is as complicated as the Universe may represent and as simple as the cosmos truly is. If a star is about fusing atoms and with such fusing of atoms is thereby growing, what happen when all the atoms fused into one all collective atom in one already all—atom-accumulated star?

What is the gravity if the star has melted all atoms it had into one all-inclusive atom and this all-inclusive atom is providing all the gravity that the star had when the star still had massive volumetric space? If all that space that once filled an entire giant star fused into one specific space less centre holding singularity 1^0 then the enormous gravity is applying to the centre of such a non existing space-less atom and that entire enormous force has been secured in the space, which is less than the space, which one atom holds. In that case the atom would then show a force that would pull the surrounding Universe flat. The purpose of fusion is to reduce space and magnify space less ness inside the sphere. Where does the gravity of the star end when all the atoms in the star became one giant atom by fusing all atoms into one nucleus? Gravity is smallest where space is least. Where space of an entire massive star is left in the size of one atom the gravity coming from that will pull the Universe flat at that point. Newtonians have the opinion that it is energy that keeps the planets in rotation and the system is equal to the rotation one will find on Earth. There is one slight problem and that is that all the mass used in the calculation is not worth a penny in practise. In nature all the planets orbit in an equal ratio while in their opinion the mass is the key factor, which implicates all aspects of the energy requirements in the planet orbit.

They say that E = - (GMm) ÷ 2r and the gravitational constant (G) is one factor of three where the product of the three factors holding the Mass of the Sun multiplied by the mass of the Earth (or what ever planet apply at the time) giving the M (Mass) X the m (mass) X the G (Gravitational) constant and this is in division of the radius (r) from the Earth (or what ever planet apply at the time) added (2) from both ends. There is a problem looming on the horizon...

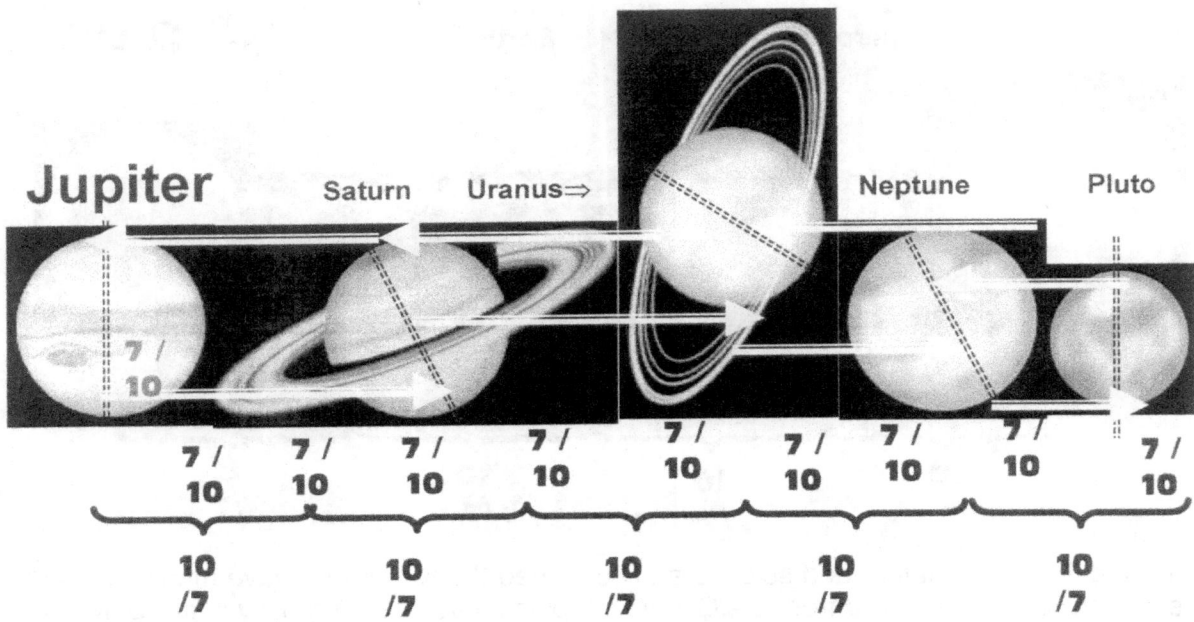

Notwithstanding mass differentiation and mass discrepancies of the large planets in relation to the small solid planets, all the planets are in a similar ratio in space and time around the Sun. That means big or small, they travel alike. You can say what ever you like about Newtonians but stupid they are not. They know how to think and think they can…for instance try and beat this:

Mercury	Venus	Earth	Mars	Jupiter	Saturn	Uranus	Neptune	Pluto
0.055	0.86	1.0	0.11	318	95	14.5	17.2	0.002

Notwithstanding the enormous mass discrepancies we see illustrated in the table, all the planets orbit equal in ratio. That means we can ignore the fact that Jupiter is 318 times more massive that is than the Earth because they use the same time to space ratio. One might think that if the one mass (the smaller mass) in the case of the Earth stands to be used in the formula $E = - (GMm) \div 2r$, in comparison to the case where Jupiter is 318 times more, or in the case where Pluto is 0.002 times that of the Earth, the mass will bring changes. As I said, one thing you may not call the mathematicians is that they are stupid. They did notice that all the planets orbit equally and at the same ratio. That did not stop them from implicating mass, no they just went on to blame the gravitational constant being guilty of eliminating the mass discrepancies.

If it were true that it is the gravitational constant that is eliminating the supposed effect of mass on the potential gravity of a star then it would be that the formula would read as follows:
$F = (M \times m) \div (G \times r^2)$ where $(G \times r^2) = (M \times m)$ because that will mathematically show that the Gravitational constant (if there were anything of that nature applying) cancels the effect the mass factors has on the orbiting structures. That would mean that the gravitational constant eliminates the mass factor on both ends of both the radii and not as it is at present where the gravitational constant incorporates the mass as the mass on both ends incorporates one another in order to compliment gravitational constant to calculate the required planet orbit. As I said, they are not stupid, they will use any bullshit to wiggle them out of a loop. They do with that problem just what they do with me as a problem they pretend it never was a problem and ignore the problem.

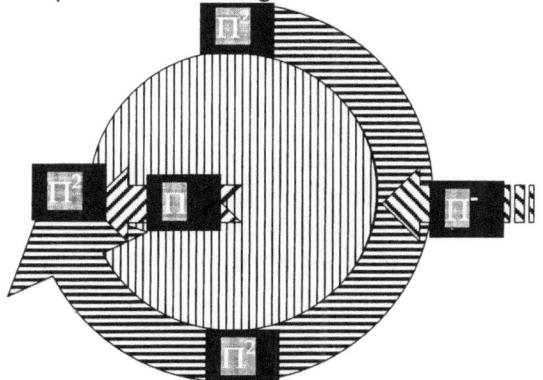

In another part of the book I went into the criminality of falsifying evidence in order to colour a picture to the likings of the person acting criminal or to falsify in order to bring about purposely an incorrect situation. In this part I wish to elaborate on the incorrectness of this approach and the magnifying of the intended incorrectness. It is acceptable that there was no one in the past that saw the Titius Bode law for what it is but in the same manner if there is deliberate protectionism of the corrupt and a deliberate effort to cover falsifying evidence and statements, then it will be a natural tendency to over acclimate the process where further investigation is required.

In the Titius Bode law one finds the distribution of planets in response to the allocation of singularity respectively. By having a distribution of twice time seven divided by ten in relation to ten divided by seven that is the location or position that serves the outside planet. Where the one is twice the other we find that the distribution is coherent with one marker and one planet. The location of the other planets has no role in the position the outside planet has since to the outside planet only its immediate inside planet is a seven. Any planets closer than the immediate inside planet is to the Sun or farther away from the Sun than what that outside planet is, has no function in the allocation of the planet distribution. To every planet that planet is forming the outside and all other planets except the one to its immediate inside is of no significance to the planet forming it's most outside border. This result in the way the distribution uses (7+7/10) in relation to 10/7.

That brings us to another Newtonian problem in the way that they handle concepts that do not match Newtonian mass and all the misconceptions accompanying such inconsistencies. Newtonians ignore the obvious and denounce facts while those facts do not salute their preconceived, misjudged ignorance. They deal with it in precisely the same manner, as one would expect when judging how they deal with my work; they ignore it and declare it never existed in the first place and any one mentioning it must first prove that it ever existed by proving that it never was a coincidence to start with.

One can clearly see how the singularity of the atoms form the building form used to increase the space –time growth. The seven that material holds are in double relation to the ten that time holds. By valuing the atom as $(\Pi^2+\Pi^2)(\Pi^2\Pi)3=1836$ we find that the seven reflect as the material component and the seven on both sides of the Universe is in regard to the five it is in contact with. But on the other hand the five doubles to ten on every side of the Universe since no one can determine precisely where the five begins to form seven and the five will always be a square to the seven it is in contact with.

It is so obvious that mass plays no part in the orbit of planets. I just cannot believe any reason or excuse put forward why the world's most intelligent that will hide the truth about mass not playing any part! Yet where the Titius Bode is so overwhelming in evidence of being the process used to form the allocated orbits of the planets, there is such a strong and deliberate attempt to by pass the issue. The blatant misleading reasoning about why the mass will be illuminated by the gravitational constant without having that reflected in the formula used is shocking but even much more shocking is never having one person investigate (in earnest) the Titius Bode law.

The square however dates back from a time when the square still was just a doubling to bring a duplication of one to the other side of the Universe. For every seven in singularity holds relating to material (7/10+7/10 = 1.4) the time doubled by remaining the same ratio (10 / 7 = 1.42) That allocates one line in singularity in space holding time to twice the ratio of time holding space while the ratio remains the same. That means the radii (if one could call it that) in distance doubled (.7 + .7) by allocating one time unit in relevance (10/7).

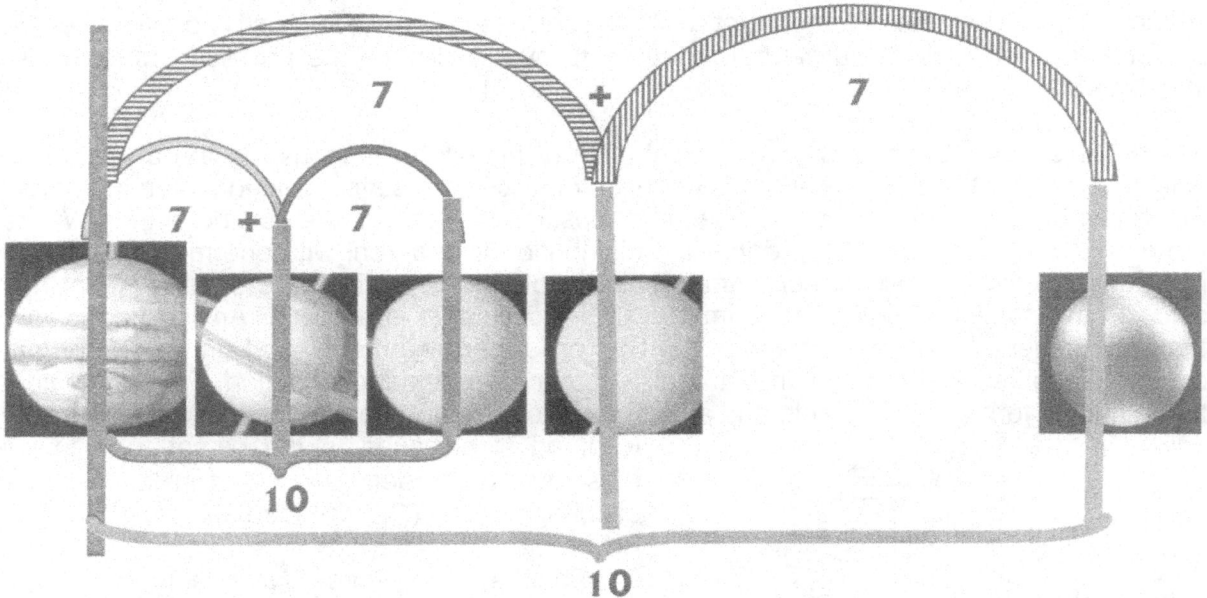

The make up of the Titius Bode law is the material to time configuration where a sphere expands into time and contracts from time. This grants the material part the value of seven points forming part of the ten points of time.

Bode's Law:

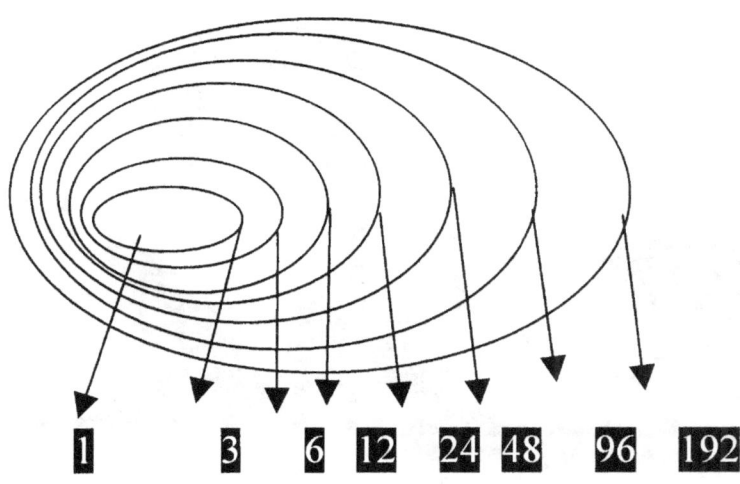

1　　3　6　12　24　48　96　192

Bode's Law

Planet	Mercury	Venus	Earth	Mars	Ceres	Jupiter	Saturn	Uranus
Bode's law distance	4	7	10	16	28	52	100	196
Actual distance	3.9	7.2	10	15.2	28	52	95	192

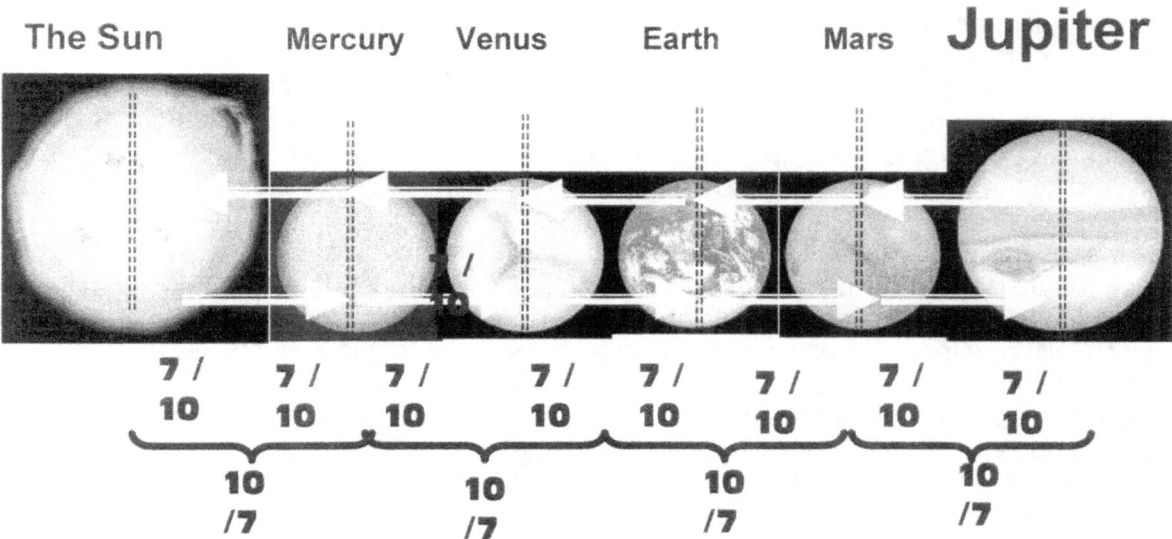

There is no one of the seven directions we move in because time takes the seven directions and move infinitive number of sevens in three time positions from the future to the past. The direction we see is the Universe coming towards us and disappearing into the infinity we have. We are moving from eternity in the direction of infinity. Infinity is guarded by material, which then forms seven in relation to material. The seven is in motion although from the point singularity holds it is time that is ever changing

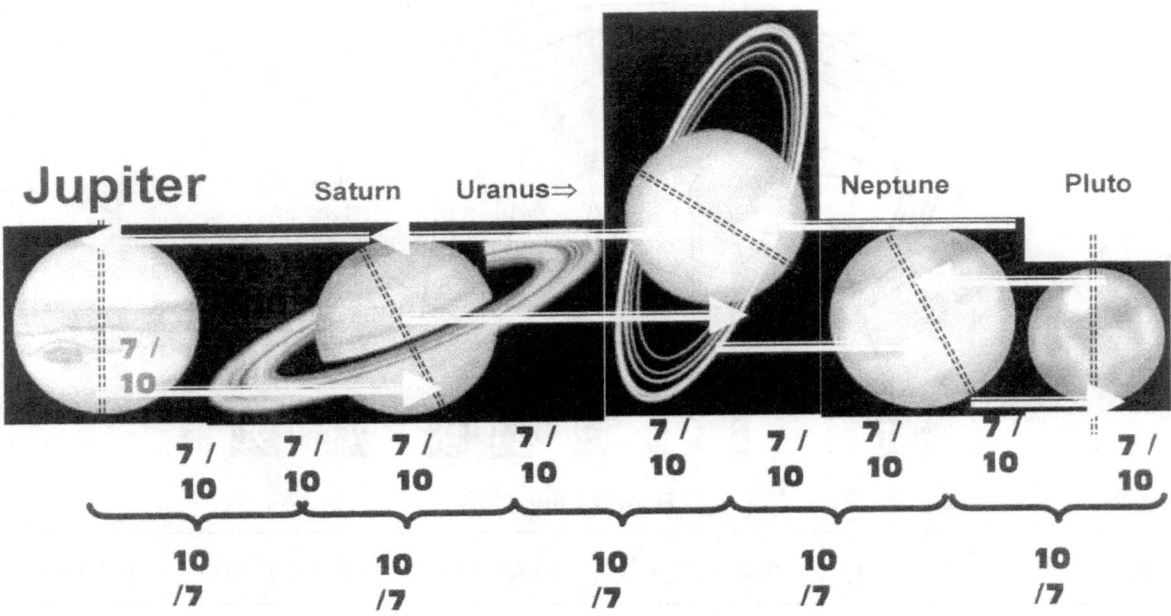

That easy part explains the frequency Titius and Bode mathematically could interpret. The outer space region is the neutron. The neutron provides gravity by producing motion. Motion is $(1.4 / 1.42) \times 10 = \Pi^2$ and that makes outer space the compliment of motion Π^2 going to Π. That is why the location (Π) is in double the time (Π^2).

Another bone of contention I fail to see is how does Newtonians compromise logic in order to justify Newton in terms of Galileo. Yet I have been, to put it very frankly insulted on more than one occasion because I fail to see how Galileo says mass plays no part in the falling and Newton formulate that the whole affair is mass orientated. $F = G (Mm) / r^2$. On one campus in particular there was one professor that truly got nasty about this and he insulted me in a way I cannot forget. However that same professor failed to show me how Newton's mass brought any object faster to the ground since $(GMm / r^2) = mv^2/r$ which suggests that the square of the velocity multiplied by the mass is the same as the gravitational constant multiplied by the product of both the masses and then divided by the square of the radius.

That means the mass m has to multiply X with the velocity in the square (v^2), which then will reduce (demolish) the distance (r) there is between the Earth and the object on a continuous basis until the distance is reduces. That's rubbish. How do they console this statement with that of Galileo where Galileo said all objects fall equally to the ground! Galileo said that notwithstanding mass discrepancies all objects will hit the ground at the same moment when dropped the same distance and at the same moment. Newton insists on mass while Galileo insists on equality of mass during the fall. The biggest bogus part of the lot is that I have not come across one Newtonian that was able to see this distinction. It is as if they all have an inborn blind spot.

Galileo said that the atmosphere is a neutron that is providing unrestricted mass in the time period that the earth set. Galileo unwittingly suggested 7 / 10 and that is what gravity is. I found the sound barrier as $7(3\Pi^2) = 207.2616$km per hour. That is applying to what ever is falling whether whatever is falling or intending to fall at that moment. That is the neutron state of a body in the atmosphere.

A while back I indicated how man's senses evolved around his view that man (every one alive) is in the centre of the universe. Everyone and I can see how all light coming from wherever is heading directly towards one self. By standing outside and gazing into the dark eternity that never ends I see from eternity light flows towards me and that places me in the centre of the Universe. That is a cosmic reality.

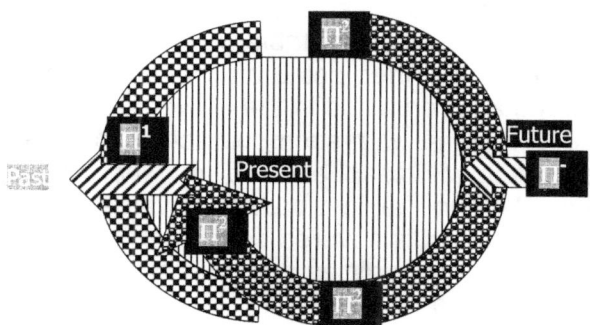

The atom holds seven points $(\Pi^2+\Pi^2)(\Pi^2\Pi)3=1836$ as the Universe but that Universe is seven points in Π being 3 points serving dimensional time to form the Titius Bode law, and a law it surely is! The gravity extending from the Titius Bode law forms the entirety of the building of the Universe by constructing the Universe in the using of the atoms to form the Universe in the entirety thereof. That puts the atom in charge of the Titius Bode law since the atom forms the Universe.

The three points we find time to be moving in is a direction unlike what we in the past thought about as a direction. There are seven basic directions being front and back, north and south, top and bottom and in or out. To our view that is the only way anything can move. It is either one of the lot or a compliment of two forming one of the lot.

By seeing light travelling towards me, I am seeing time travelling. I am the direction that time flows. I can see where light was. I can see where light will be. I cannot see where light is going because that is within me and my singularity presents the future. Any one in disagreement should just go outside and see the light coming towards you. See how the light meets from all over the Universe precisely where you are.

The flow of time must never be confused with any part of the seven dimensions in space-time. The flow of time is away from the structure towards the structure then into the structure, through the structure where time disappears into singularity by measure of infinity. The flow of time is the motion that concerns the part science at present thinks of as outer space but which in essence forms time in eternity. Because Mainstreams science has the name incorrect they also have all attachments they connect with time in eternity incorrect, for instance that the Universe has an edge and they give the cosmos a place where the Universe ends. That part on the outside of the atom never ends because that part is continuous to the point where time ends and time cannot end in the part holding the seven dimensions of time in space. Therefore time ends within the atom or as I have life my time ends in me. Therefore I have my position where I am in the centre of the Universe. Time

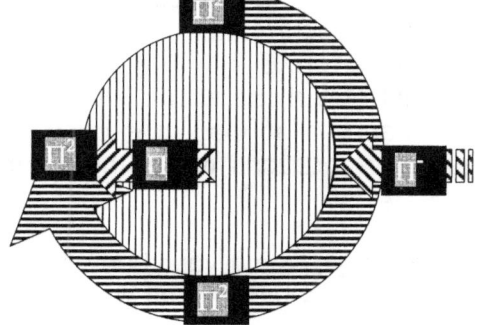

comes from the outside as far and as wide as things can go but time ultimately ends within me.

Time is on the outside of the atoms and the atom is a means to express matter. All material that encircles infinity is a black hole in progress of becoming larger. The material is covering and defining infinity from eternity and is parting that which has no start from that which has no end by giving that with no start an end and that with no end a start. Time is the inside of the atom where time is excluded from eternity by giving time specifics in motion and a defined value confirming space while the space is conforming time. Then time is taken to infinity where infinity absolves time into a unity once more. Time is what is between infinity and eternity and while eternity is parted from infinity time in eternity as a unit is also standing between eternity and infinity. Eternity is part of the part that is standing between eternity and infinity and therefore I am able to see eternity as a reality.

Have you ever thought about whereto the light is going after it was coming toward me and has hit my particles? Sure you may dismiss it as that my particles block the further flow of light but again that answer would be so Newtonian it is even predictable. After the photon has hit my body, the question then is whereto is the photon then going after it is upon me. Why would it disappear from the Universe? If it hit me and remained in the Universe it would ricochet from me like it does with glass mirrors and material with such characteristics. The only logic is that the light must be going to the past. It will disappear into me where I am another Black Hole in the Universe. But from where I stand the light coming towards me, is representing my future while I am in the present and my past is covered by the part I hold as infinity covered by material. As my future is in infinity and my particles are covering infinity or my future, then that means that I am taking my past down my infinity into the future. I am taking the space and light into my present down to my future, which is where I will take my life one day when I leave the present holding material inside the Universe. I am representing the end of my time and I am holding the end of the Universe as far as the Universe relates to me and I relate to the Universe. There is no going down any spiral Black hole to escape from my past by finding a short cut to my future. Should I try to go down towards singularity I would introduce eternity to infinity and extend the instant into becoming the eternity, which it truly is! The space being between the parts of time forming eternity and infinity represents that part that represents me within the time I am part of the Universe. I am the end of the Universe because as far as I can see light is bringing the past in the Universe towards me as my future. That is why the Universe is shrinking into the oblivious while it is expanding limitlessly beyond boundaries.

That is why everything into my future is shrinking into the oblivious as time engulfs material into the future. You were in eternity because the light is coming from eternity towards you. You are where you are because I can see where you are plus the time it takes the light to

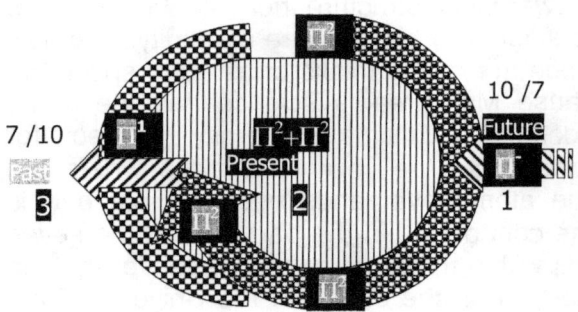

come from you to me added to where you are in time. The light is going to disappear into where you are but that is not true. You are dragging the light that reached you into infinity because light tries to escape time by going infinitive. Infinity, which is that which has no start, is in you and you with your eternal life are generating time that parts infinity and eternity. That is not religion because that is raw physics. I have my doubt that any Newtonian will understand this concept since they can't even see that mass has no application on objects in orbit in outer space. If they

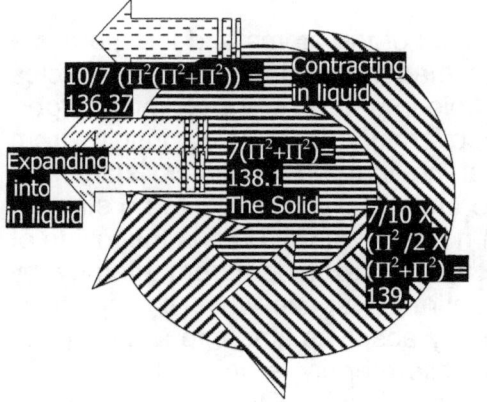

are incapable of seeing the obvious how the hell will they be able to see what only those with intellect can see. That is why they can see no God. It is because they see mass applying in locations where there can be no mass applying.

Where you are and where you hold your body is the closest Black Hole to you because at that point time converts to space and space disappears into the gateway of singularity. Time ends where you stand but that only applies to your Universe and while your Universe is in contact with the rest of the Universe your Universe is solo and alone in time.

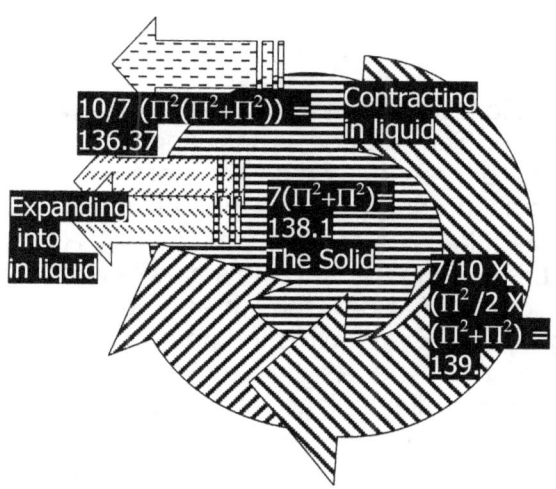

Time is taking the seven that was into the seven that is through singularity (.0999991) onto the seven that is going to be and that (3X7 = 21 + .99991) / 7 of material to which I relate I can be sure the Universe having time forms a sphere. By forming a sphere it gives meaning to the growth we see as the Hubble constant indicates without Newtonians trying to rape any common decency out of it by their equation of 13.5×10^9 years. How could or can any one be that crude? The Earth alone is one million times older than that because what they use to measure time is the readjusting of the atom in relation to the factor the space represents. That is how the star inside accumulates the liquid by freezing the star. However, I put more on this in another book where that belongs.

One thing we must not forget is that outer space is what material that is orbiting through outer space is allowing outer space to be. The Universe is the proton. The Universe is 7 / 10 in relation to 10 / 7.

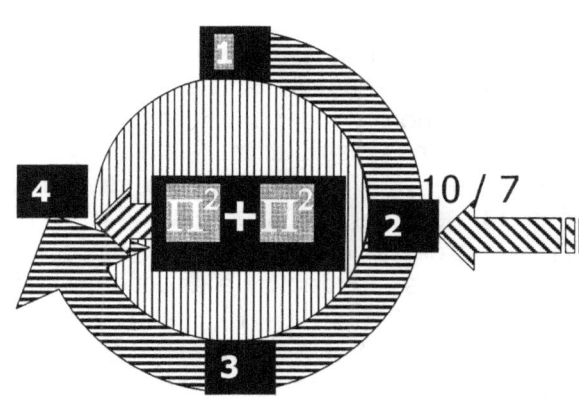

The Universe was what we now have from the first instance but in our perception that which was then does not apply to what we see in the Universe. We have an individual Universe from the one that will apply one day when one hydrogen atom will be a full star at an era of $7/10\ \Pi/ 2 = 1.09955$. According to my opinion and that is my opinion, what we see as the Universe first applied when liquid and material stood apart from singularity. It was when liquid transformed into substance and formed space in order to allow time a means to combine again. That was when the neutron as we know the neutron first found a measured value in the Universe. Before that it was a factor but motion in time was at that point only a definition in our standards we now apply. It was when $10 / 7\ \Pi^2(\Pi2+\Pi^2) = 136$ formed the one wall of the then applying Universe while $7(\Pi2+\Pi^2) = 138$ formed the solid and the material was $7 / 10\ (\Pi^2/2)(\Pi2+\Pi^2) = 139$.

Today in our Universe we have the wall of time at $10/ 7(4(\Pi2+\Pi^2)) = 112.8$.

That is from where liquid flows to singularity. That from where gravity is generated by the iron core of the star. The core must have a relevant displacement of $7/ \ 10(4(\Pi2+\Pi^2)) = 55.267$ in proton displacement to have gravity establish the concentration of heat. That puts the Universe within the borders of the Titius Bode law at 10 / 7 and 7 / 10 in relation to the proton $(\Pi2+\Pi^2)$ forming time (4).

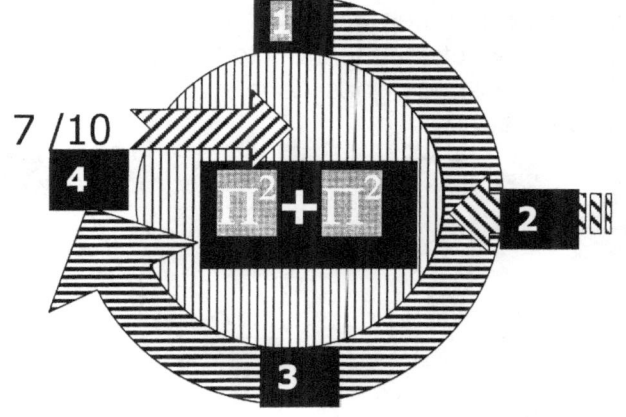

That is where liquid ends and material begin. That is where contraction of gravity begins within every structure that in our era has the ability to generate gravity. It therefore has to have an Iron core.

At the point where the neutron disengages from the atom we find our Universe catch up with time as time then takes control of space once more. The neutron is the lagging of time between 7 / 10 and 10/7. When gravity intensifies to a degree that the neutron removes from the space-time the atom offers and becomes absent as a factor that influence the displacement from the atom, we find the electron link directly with the proton at $3(\Pi^2+\Pi^2) =$

59.217. At this point in the star, the star is developing past the scope of what the three dimensional Universe can offer. By rejecting the neutron, the star is collapsing into time as it removes space by ejecting volumes of liquid light. As one can see, the neutron removes all influence from the atom and when that happens we have a neutron star, which is no longer valid in our Universe. Outer space is not mass implying the gravitational constant. It is not mass that is producing the product by multiplying mass. Outer space is the Titius Bode law. It is gravity or motion or the neutron or movement. It is what the Titius bode law says it is. It is seven where four relates to three. It is where the building blocks of the atom leave their layers in the forming of time.

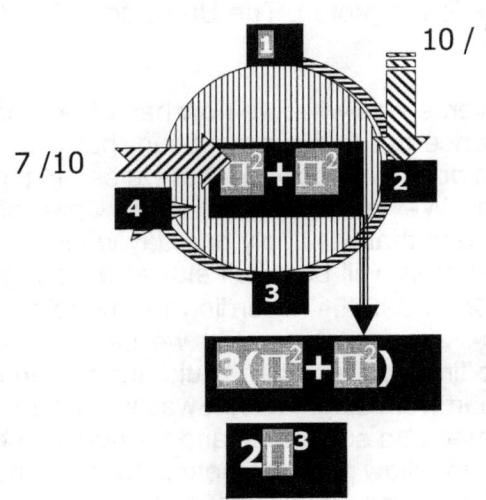

Our Universe is the flow of space-time in the form of retarded heat coming from the region 10 / 7 $(4((\Pi^2+\Pi^2)$ where gravity is generated by a revolving planet similar to electricity being generated by a spinning armature. There is no difference between electricity and gravity except that electricity is more concentrated in generating and is of most intense distribution. The gravity flow is directed by the iron core within the structure, which has to have a displacement of is 7/10 $(4((\Pi^2+\Pi^2))$ in order to establish a flow direction. Just as electricity is charged by a directional flow between Iron 7/10 $(4((\Pi^2+\Pi^2))$ and copper $(\Pi(\Pi^2+\Pi^2))$ the flow is between 10 / 7 and 7 / 10. By collapsing the space-time within the core of the planet, there comes room available and with that reducing of space-time it starts a need to fill the collapsed space-time where that flow in space-time or heat contracting is gravity. This can only be when the atom freezes into a position where there is no required motion available to host the neutron. As soon as the proton $(\Pi^2+\Pi^2)$ links with the electron 3 by forming $3(\Pi^2+\Pi^2)$ the star is going outside our Universe and then becomes a proton star. By further cooling the atom will then directly link singularity outside the atom to the proton $\Pi(\Pi^2+\Pi^2)$ and in that the space catches up with the time. The space goes double $2\Pi^3$ and eliminates the requirement for motion. Singularity feeds itself.

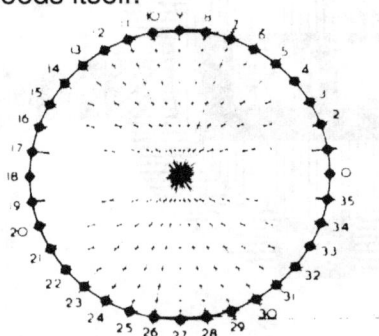

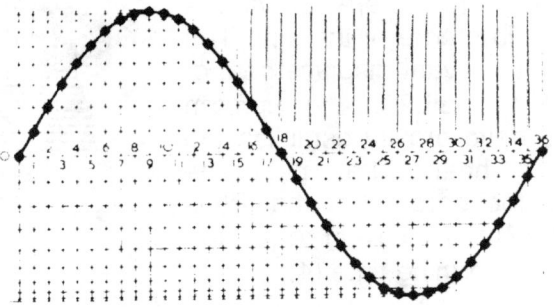

That is why gravity and electricity is not only managed by the same principles, but gravity and electricity is the same principle. Electricity is the conducting of electricity by using one strand of singularity and gravity is employing the entire core for the conducting of gravity.

In the star we have the same flow of space-time, which is the result of the same displacement, and in that we can see that charging electricity is the same as generating gravity except that the scale such displacement produces the flow is conducted much different. The motion of 7/10 (4 $(\Pi^2+\Pi^2)$ (which represents the space-time displacement of iron) in relation to $\Pi(\Pi^2+\Pi^2)$ (which represents the space-time displacement of copper) allows the neutron stage of space-time flow $3(\Pi^2+\Pi^2)$ to come about. That displacement we call electricity and when it is done on a larger scale we think of it as gravity, however in the end it is the same.

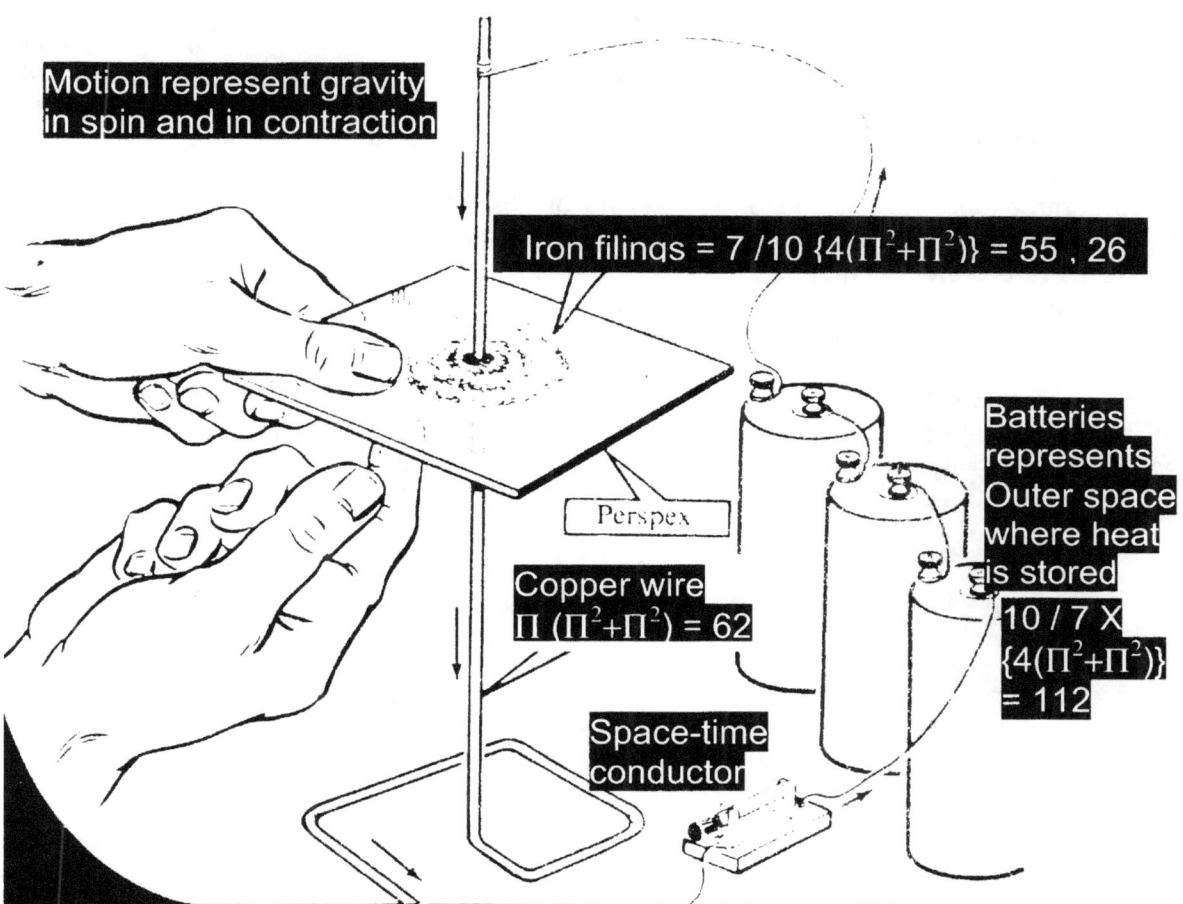

Motion represent gravity in spin and in contraction

Iron filings = $7/10 \{4(\Pi^2+\Pi^2)\}$ = 55 , 26

Batteries represents Outer space where heat is stored

$10/7 \times \{4(\Pi^2+\Pi^2)\}$ = 112

Perspex

Copper wire $\Pi (\Pi^2+\Pi^2)$ = 62

Space-time conductor

By using the element that is closest to the displacing required to manage the flow of space-time from the extended through conducting to dissolving space-time we find stars has to have an iron core, and it has to have cobalt $3(\Pi^2+\Pi^2)=59$ in order to remove the neutron displacement from the star. Then to bring about the final collapse of space into time the displacement of copper is employed as $\Pi (\Pi^2+\Pi^2)=62$.

The manner in which stars charge gravity and the manner we use to charge electricity is the very same thing except the scale is in the case of human use considerably less. There is a conducting flow of heat through the Titius Bode law or neutron passage to the inner core.

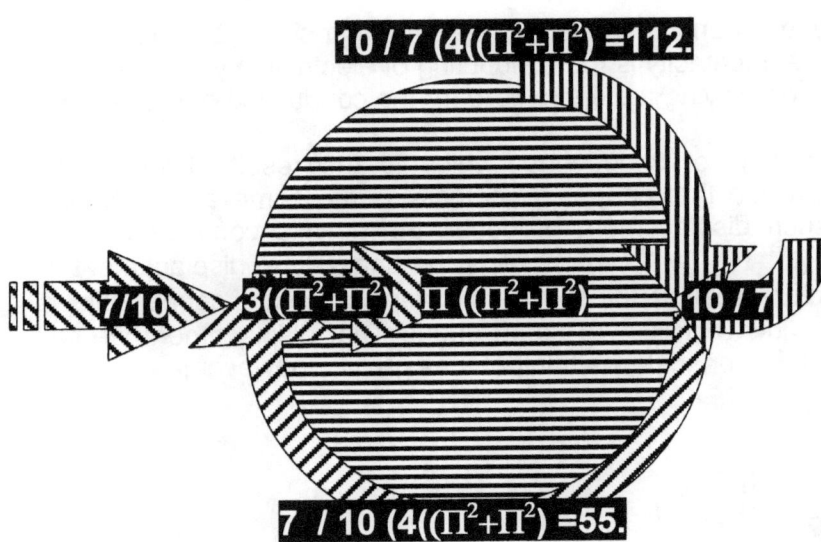

$$10 / 7 \, (4((\Pi^2+\Pi^2) = 112.$$

$$7/10 \quad 3((\Pi^2+\Pi^2) \quad \Pi \, ((\Pi^2+\Pi^2) \quad 10 / 7$$

$$7 / 10 \, (4((\Pi^2+\Pi^2) = 55.$$

As the star moves through outer space in is in contact with outer space in more than one way. By moving through outer space the star is disturbing outer space. Outer space is pushing against the star by measure of the star moving through outer space. The maximum displacement in duplication and contraction that outer space can accommodate is the total sum of the atom in relation to singularity which is $((\Pi^2+\Pi^2)+((\Pi^2 + \Pi)+3) = 35.75 \times \Pi = 112$ and with that in relation to singularity it forms the atomic displacement limit of 112 protons to one cluster, with the motion that is the maximum expanding there can be when duplicating. However motion stands in relation also to contraction. The contraction is freezing of heat into a state of liquid coming from a gas. At 112 the state of singularity is expanded at a maximum and cannot cope with more heat than 112 protons will manage to control in one atom cluster. But relative to that must be a cold where such a cold will not hold space under a specific level of freezing. Beyond a specific limit the cold of space freezes time into singularity. The flow of electricity is not the transporting of some unattached electrons lazing around and then put to labour. That is Newtonian thinking. The shifting of electricity is involving motion, which stretches the neutron that then is running space-time all the way from (10/7) to (7 / 10). That also is gravity and electricity and gravity is the very same thing. It is the condensing of heat through motion.

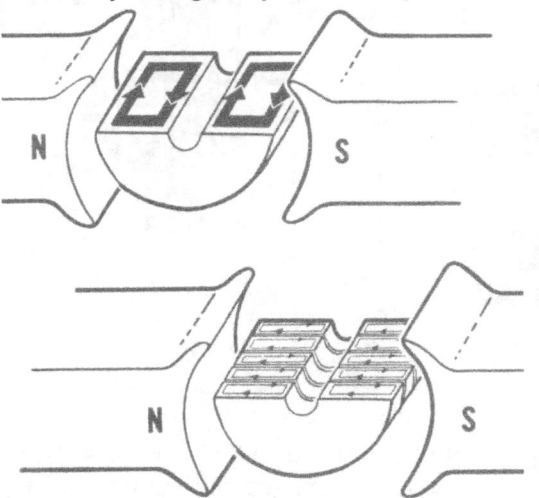

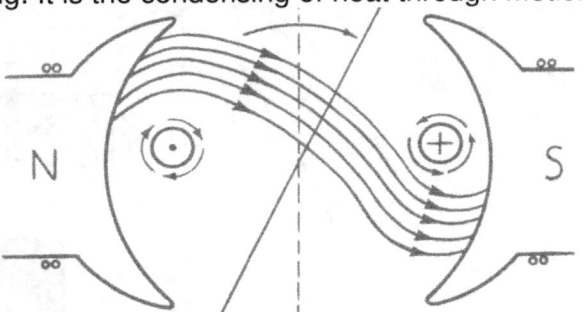

By imitating gravity humans can charge electricity. Humans turn (creating the curvature of space-time) a round iron (imitating to create and establish the provoking of singularity by employing motion of space a^3 through time T^2k) in the presence of copper and call it the generating of electricity. They created nothing but they only found the way to imitate the manner that stars employ as stars generate gravity. The process works on the principle of taking space and allowing the space a passage through time in relation to time to reunite time. It is the very same process and yet, by employing Newtonian fraud instead of trying to discover the honest truth in the cosmos they could never come to realise the connection that is so obvious. It is therefore no wonder most of the brilliant Newtonian brigade is atheist or atheist sympathisers. With that vision they show they can't see light from darkness and even that I prove later on in this book.

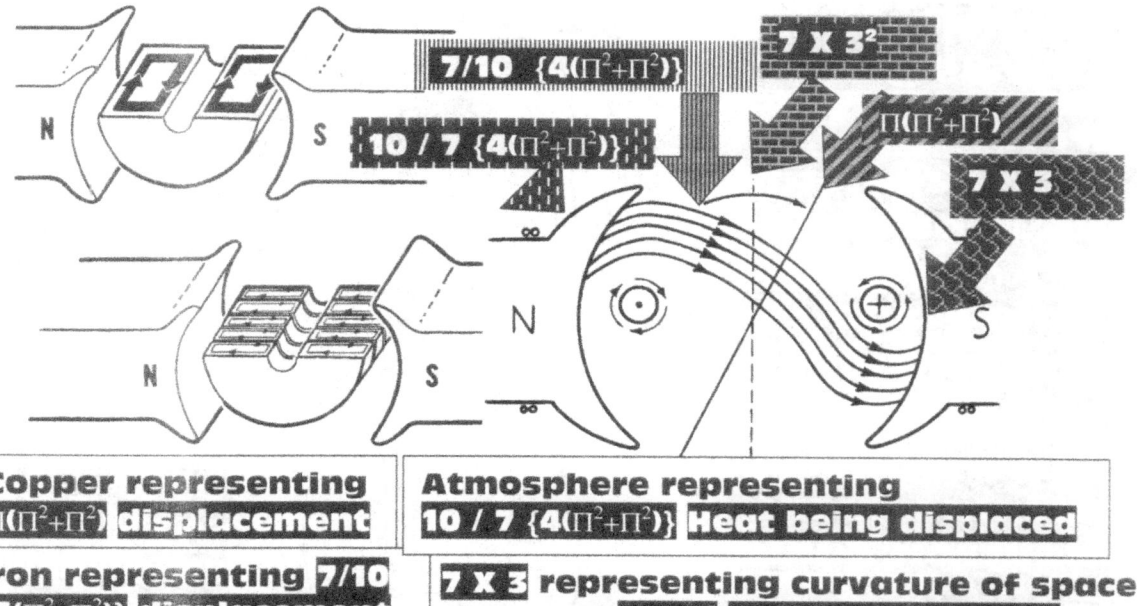

Copper representing $\Pi(\Pi^2+\Pi^2)$ **displacement**

Atmosphere representing $10/7\{4(\Pi^2+\Pi^2)\}$ **Heat being displaced**

Iron representing $7/10$ $\{4(\Pi^2+\Pi^2)\}$ **displacement**

7 X 3 representing curvature of space-time and **7 X 3²** **end of space-time**

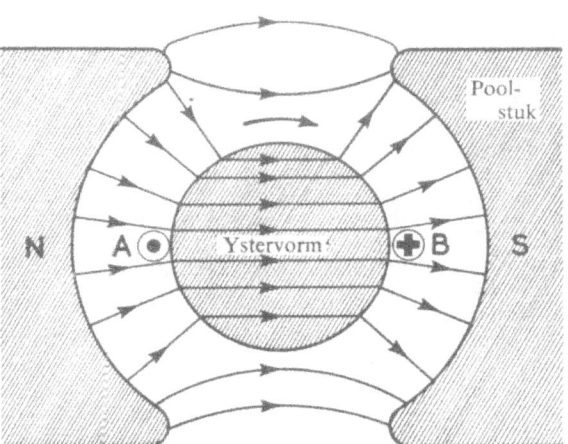

Electricity is the gravitational process to take what is so hot it can expand no more and freeze such heat into a state so cold that that can contract no more. It is taking the most expanded and the hottest in the Universe and by motion through singularity to double singularity. It freezes what has all the space into becomes that which is what has no space. By rotating the material seven in relation to time (3) through the bridge of singularity (.99991) we establish space-time, the curvature of time as well as singularity expanded Π. **We take 7 by motion of time (3) in relation with singularity (.9991) and this then forms the curvature of space-time, which is Π. The motion then converts Π to Π^2 and that is gravity. Gravity is taking time to reunite with time forming time in the square. That is the purpose and the role that the Titius Bode law has.**

In electricity Humans take a few atoms and spin the atoms in relation to other atoms and by doing that we generate electricity.

The cosmos takes a few atoms and spin the atoms in relation to other atoms and by doing that the cosmos generates gravity.

It is not the copper, but it is the displacement value the copper atom provides. It is not the iron, but it is the displacement value the iron atom provides in its service that bring about electricity or gravity or strong forces or weak forces or all the other names man might think up to hide his ability in not understanding what he claims he does understands.

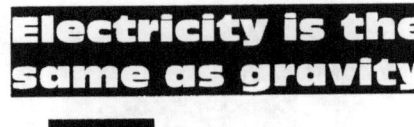

Electricity is the same as gravity

Iron displacement Neutron value = 7 / 10 Time value = 4 Proton value = $\Pi^2+\Pi^2$

Outer space displacement Neutron value = 10 / 7 Time value = 4 Proton value = $\Pi^2+\Pi^2$

Copper displacement Neutron value = Π Time value = 1 Proton value = $\Pi^2+\Pi^2$

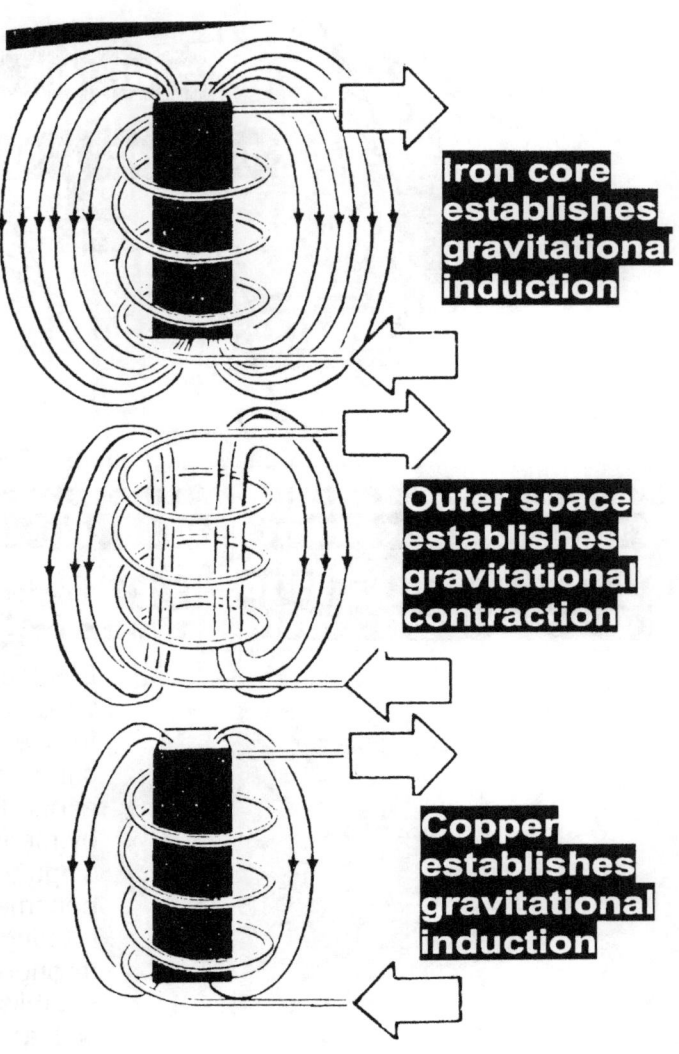

Iron core establishes gravitational induction

Outer space establishes gravitational contraction

Copper establishes gravitational induction

That which is between unbridling expansion and total collapse of space is the Neutron. The neutron stretches from 10/7 to as small as 7/10 from where it can reduce little more before abandoning the atom altogether. The Universe is gravity and gravity is the neutron where the neutron can have no mass because the neutron personifies motion of space-time.

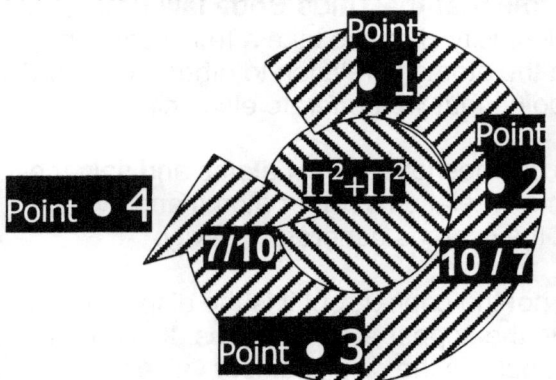

The motion is contraction but the contraction is not directly reducing. It is more a filling of vacant heat and displacing the heat in order to align the occupied heat with unoccupied heat. The expanding on the other hand is not expanding by going bigger but repositioning in order to duplicate. Such duplicating is exaggerating of space by instantaneous reducing of space. In the reducing of space the virtual contact grows substantially more. In relevancy the star moves while outer space is motionless but because outer space is motionless the star is putting the friction of motion onto the account of outer space. The point of contact between outer space and the state atmosphere produces heat as outer space is, it is reduced as it is accelerated. That spinning produces the light, which we see as photons. It is cooling the gas of outer space by reducing outer space to the point where outer space holds friction and the particles spinning in friction that comes across as photons. That is at a displacement level of $3^3 + 3\Pi^2 = 56.6$ the photon is the product of intense cold coming into contact with intense heat. By motion the atom is removing heat from inside the electron orbit to outside the electron orbit. The motion the atom is subjected to become more intense with every time there is a duplication of the space-time.

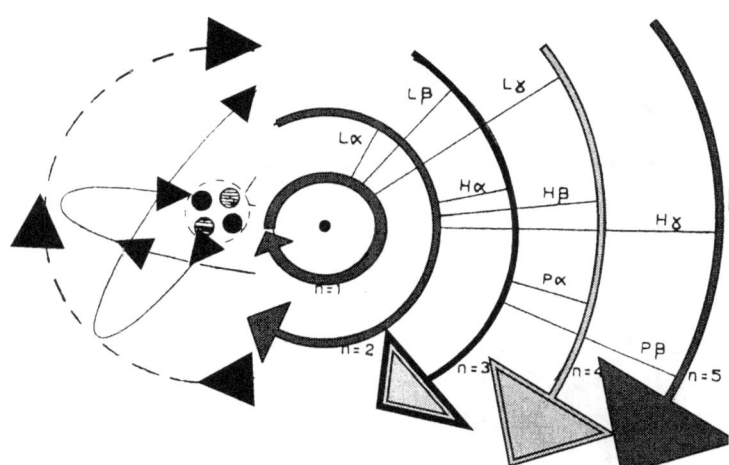

In this duplication presented as cooling, the star puts the atoms of difference under the same state of affairs. The one duplicates more then the other does because the one consists of more protons in the cluster as the other does. The one has a more frequent cycle of time repeat that the other has. In this array of possibilities confusion sets in, in terms of duplicating while the other is contracting and fusion takes place where in motion the one incorporates another atom because the cycle period does not match. As I have indicated how one serves as a liquid to another particle serving as a solid the same process apply within the star. The contraction is presented by atoms with a larger indication of consuming space that the other atom can. This is what Newton got confused as mass. Because the one has a different setting in relation to heat one would stand prone to duplicate more and the other would rather contract. We see this as one being a gas while the other element is a solid. It is the measure in which the element favours duplication above contraction or the other way around.

The fact that neither element of different standings show identical patterns in behaviour the one will try and incorporate the lesser developed particle by duplicating at the same pace while also duplicating considerably more at the time of duplicating. It takes considerable more duplicating time to contract an element holding say fifty proton displacement duplication than it takes an element with a duplication displacement of say 1. By duplicating the one element holding one proton in the nearness of the element duplicating 55 protons it can quite easily become a question of finding the element of one proton to be liquid and by the nature of the Coanda effect the space also extends to incorporate the additional liquid supplying the motion. The Coanda effect works on the basis that the extravagance of the liquid providing the motion suppresses the space when the motion is absorbent in nature at that point. There is a hot spot in all the cold and the hotspot causes sudden motion acceleration, as heat will do. The surge in heat has nowhere to go although the surge in heat at that spot insists on having more space. We know a liquid that heat takes up more space because it expands. If the heat at that point has no where to expand, but the heat is there altogether and the same, then the only expanding must be to incorporate the one proton element into the element holding fifty fife protons and that takes the total up to 56. But in this one must see the liquid surging in space as the heat level rises but at that very point the space will reduce because the atomic relevancy will freeze the atom into less space. If the liquid becomes more heated and surges for more space but there is no more space to supply in a star that is predominant and overall engulfed with liquid the reaction of the solid will be to become colder. Becoming colder is also freezing in the face of the liquid heating and with the liquid heating the liquid will have much more motion.

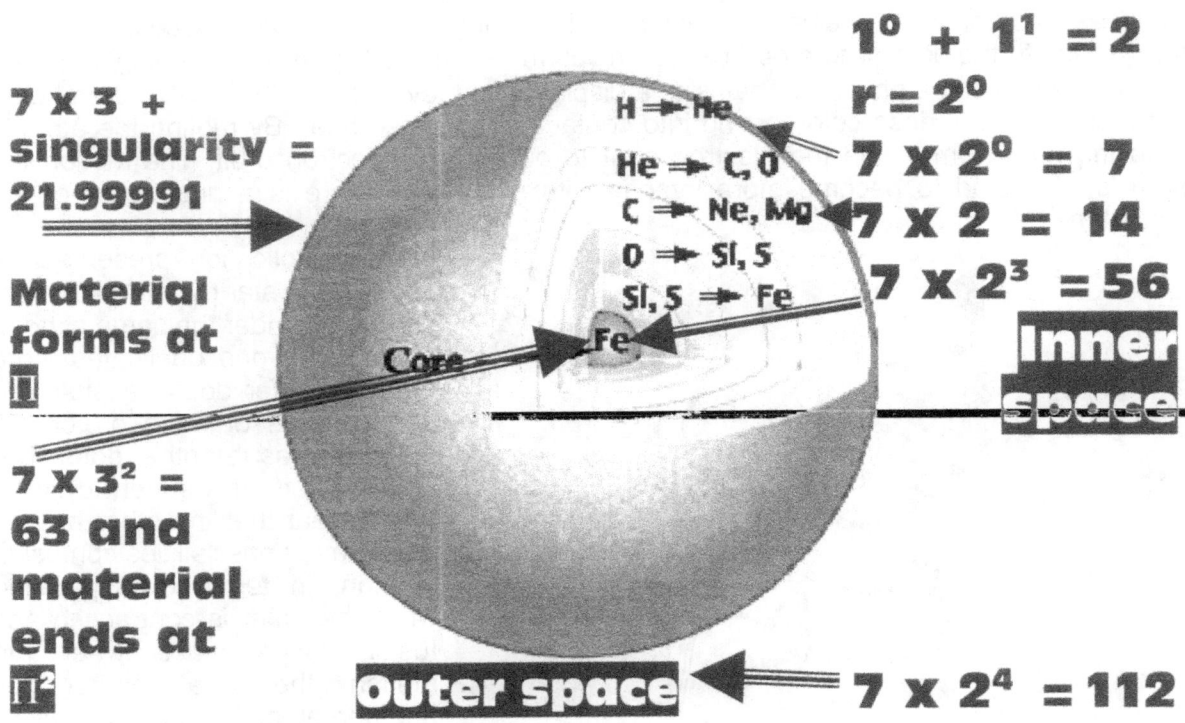

$$1^0 + 1^1 = 2$$

$$r = 2^0$$

$$7 \times 2^0 = 7$$

$$7 \times 2 = 14$$

$$7 \times 2^3 = 56$$

Inner space

7 x 3 + singularity = 21.99991

Material forms at Π

$7 \times 3^2 = 63$ and material ends at Π^2

H ⟶ He

He ⟶ C, O

C ⟶ Ne, Mg

O ⟶ Si, S

Si, S ⟶ Fe

Core ⟶ Fe

outer space $\quad 7 \times 2^4 = 112$

The material sector starts at seven (that is representing material) in relation to 3 (representing the flow of time from the past through the present and into the future) moving about the present (or singularity) that is forming Π. On the inside we find time coming together or forming a square of time. This translates to seven (material sector starts at seven that is representing material) in relation to 3^2 (representing the flow of time from the past through the present and into the future) in the square because of motion of time towards time and at the point where such moving about meets the present (or singularity) that is forming Π^2

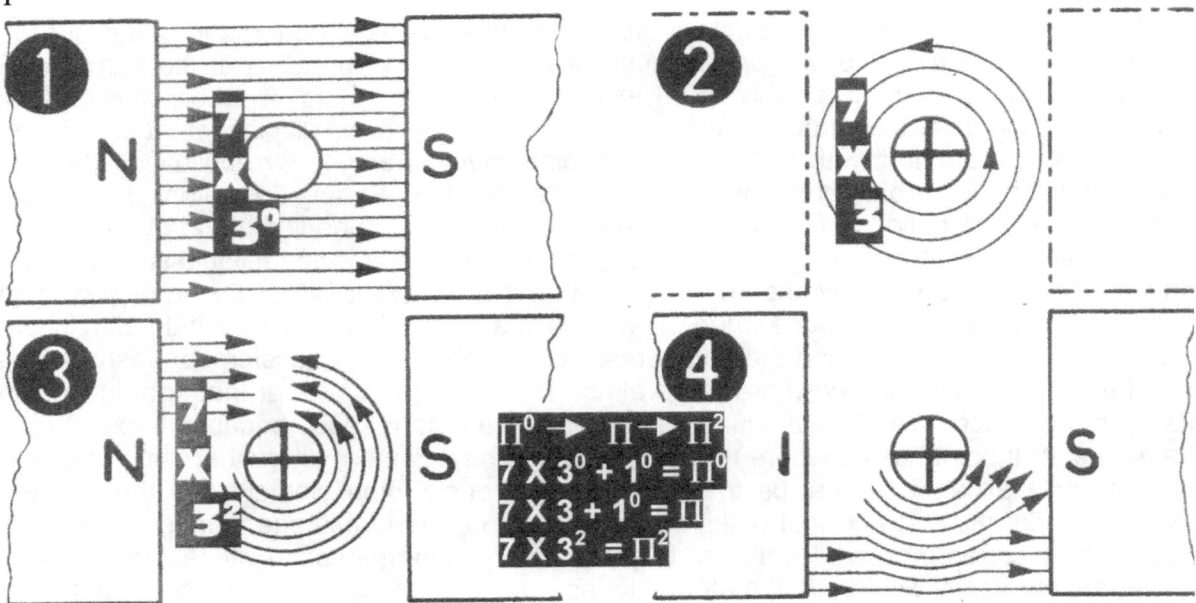

1   N 7×3^0 S

2 7×3

3   N 7×3^2 S

4 $\Pi^0 \rightarrow \Pi \rightarrow \Pi^2$
$7 \times 3^0 + 1^0 = \Pi^0$
$7 \times 3 + 1^0 = \Pi$
$7 \times 3^2 = \Pi^2$
  S

Time is going from Π^0 onto form Π by the creating of the curvature of space-time and ends as time in the square Π^2 that forms the final gravity. What is electricity is the flow of space in time through time. There is 7×3^0 that we regard as air or as space. That motion then converts to singularity by Π in duplicating 7×3^2. The movement then takes time going by 3

(7X3 + singularity) to 3 square which then also forms the uniting or the square of time (7 X 3^2 = 63).

The more motion comes with supplying more gravity where more gravity is pushing time longer in the face of space reducing. The situation is running at that point back in the direction of the Big bang and when the motion bridges the Big Bang era, fusion comes about between the two particles. In the end the particles joining space was the result of motion differences and mass is the result of motion difference. At one point the motion differences extended to a point running into eternity leaving the space at that point in infinity and infinity joins eternity at that spot where the element grows by one more proton. The mass of motion discrepancy did not create the enormous space - time deficiency but the moment the mass became eternal and infinitive the element joined space while enduring eternal time. That is the use nature has for the principle Newton named mass. It is a motion discrepancy whereby atoms would then comply to combine space, should the motion applying validate such a drastic step.

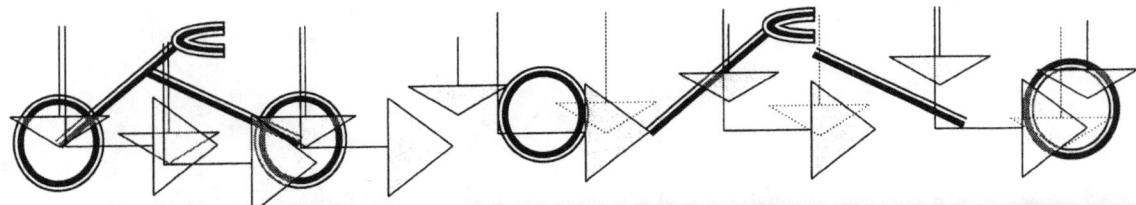

Motion distributes space and therefore decreases heat. By spreading the space over a larger area the heat in the space is reduced because the density of the heat allocated to the space reduces. Motion decreases the heat and therefore the Sun is the coldest place in the solar system while outer space is the hottest part of the Solar system

Once more I have to return to the sound barrier to explain the point I wish to bring across. Where the sound barrier becomes evident, there is a particle at one point displacing space-time at a rate of $\Pi^2/2$. In reality without the assistance of life to intervene it would then be the entire atmosphere that was moving at that pace having the particle maintaining such a motion in relation to all the liquid. In the case of the aircraft there is expanding without the earth compromising. But is the entire liquid atmosphere reaches such a point it could only come as a result of the earth in relevance duplicating at such relevance. The motion refers to the solid, as the liquid is motionless. Therefore the duplicating of the earth crossed the $\Pi^2/2$. By reaching $\Pi^2/2(\Pi^2+\Pi^2) = 97.4$ and that takes all material into the cosmic state of liquid. That would make the earth and the aircraft both having a state of being liquid where both can join. But also the earth would duplicate by reducing to the tune of $\Pi^2/2(\Pi^2+\Pi^2) = 97.4$ in order to reach such a state. In cosmic terms the earth is an atom and the aircraft is a lesser-developed atom but both adhere to the same atomic law, as would hydrogen and an Iron atom do. It is about solids and liquids and motion.

Then a point is reached where there is no more room to allow motion within the Universe. The atom in the star reached the cosmos limitation on the motion permitted of $7/10(4(\Pi^2+\Pi^2))$ and has surpassed it. The atom has cooled to the point it could no longer sustain the neutron at a point of $3(\Pi^2+\Pi^2) = 59.217$ and the neutron at 7/10 is no longer part of the atomic space. It is where cobalt is and for that reason we find cobalt radioactive. In that region the neutron is no longer any part of the atom. Then more growth increases the atom to a displacement value of $\Pi(\Pi^2+\Pi^2) = 62.0$ which is a point where all space totally collapses. That is the end of space-time in an atomic environment because this is where the atomic space-time relevancy reaches $2\Pi^3 = 62$. At that point in the Universe the Universe at that point became a Black hole and as development evolves the dynamics of the Black Hole will eventually consume the entire star.

That is not the road that serves the Earth situation.

Let us first forget the accuracy of the bicycle moving in cosmic terms and concentrate on deliberate motion. It is said that if a butterfly flaps its wings in China, a hurricane will hit New Orleans. That is not true because we have to see what does exist and what does not exist. When a bicycle is motionless the bicycle is not part of the cosmos. It has atoms but the atoms forming the unit do not charge a governing singularity where that governing singularity responds to the motion of the Earth singularity and by doing so establishes the unit into independence. The unit is a unit within the Earth unit. The unit is in motion with the Earth as part of the Earth. There is no additional motion confirming the bicycle as a force that is promoting itself by promoting the Earth motion in addition to what the Earth does to promote motion on behalf of itself as well as the motionless bicycle. The atoms spinning would form a unit as far as confirming the bicycle independence in the unit. That does not make the unit independent. It does confirm the unit as a group of atoms forming a unity in form. Some might even see it confirming

$$F = \frac{r^2}{M_1 M_2}$$ which is what Newton saw at first. That is absolutely correct when some of the sharp edges of incorrectness are removed. There is only gravity applying between the Earth and the bicycle and the gravity confirms the restriction of the radius parting the objects to the very limit. That is because the only independence the bicycle has is in form and without cosmic motion.

Then motion enters the scenario on the part of the bicycle. When the bicycle was motionless the gravity the Earth developed restricted the bicycle to one line placing the bicycle in a direct line with the centre of the Earth. That is a value of $7(3\Pi^2)\Pi^0$. That is the motion the Earth bestows on the motionless bicycle. The line running through the bicycle to the Earth centre has a value of Π^0 while the Earth reserves the motion on behalf of the earth and also of the bicycle at a premium of $7(3\Pi^2)$. Should some cosmic miracle wonder come about such as life is. The bicycle might just find the opportunity to achieve cosmic independence from the Earth such as the moon has. The bicycle then holds its cosmic independence at a value of anything between $7(3\Pi^2)\Pi^0$ and $7(3\Pi^2)5\Pi^0$. The moon however holds its cosmic value at $4(7\Pi^0)$ days. And in return the earth holds the moon at a reference of $\Pi^0 / 2$ days. The one day of the moon is 28 days made up of one Moon day = $4(7\Pi^0)$ days in the life of the earth while the Moon is $\Pi^0 / 2 = \frac{1}{2}$ days in the life of the Earth. There is a definite division of cosmic liquids that is time between the earth and the Moon and the moon holds a stronger identity of independence in relation to the Earth than the Π^0 that the earth offer the bicycle.

The liquid space surrounding the Earth confirms the bicycle as part of space which time in motion draws onto the space. If the bicycle has independent motion the bicycle sides with the liquid time by moving with time as it holds the motion in the relation. If the bicycle has no independent motion the bicycle sides with the Earth putting all relative motion in the basket of time. One must keep in mind that although it is the earth having the motion the Earth projects the motion onto the liquid time since the Earth holds a steady point on the surface of the

Earth, which is steady in relation to the centre singularity. The bicycle can be part of space **k** = **a³** / **T²** that is confirmed by time or the bicycle can be part of time **k⁻¹** = **T²** / **a³**, which forms an extension of space. When I fall down a cliff I fall at a steady pace. It is the same pace that I would have when I fall down a waterfall holding a cup in my hand the water in the cup will not stay behind. The water in the cup will not spill. The cup will not fill with water. If I had to fill the cup with water I will have to supply motion in access to the motion with which we fall. The water and I will have the same pace therefore we will fall at the same gravity. My density will not leave me superior. The water mass will not have the water fall more forceful or less forceful. The motion considering all objects is not discriminating on any basic grounds. In such an event I will be submitted to $7(3\Pi^2)\Pi^0$. That is gravity and gravity is motion. Forget about Newton's mass controversy because blaming it on mass is instituted fraud.

Having independent motion requires more than gravity. One may even be able to apply

$$F = G\frac{M \times m}{r^2}$$ under conditions about where what fits. This is not a cosmic principle. It

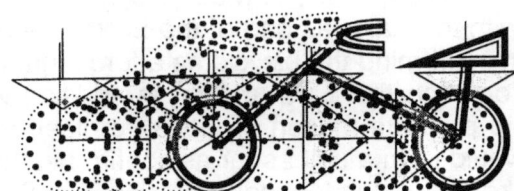

applies only to the Earth under conditions serving the earth and applying to all objects that submit to the Earth. The motion can be seen where the one M holds the motion the earth provides while the other m indicates the independent object's independence while the G then will be the additional motion and the radius by square is the balance there is between being liquid and being solid.

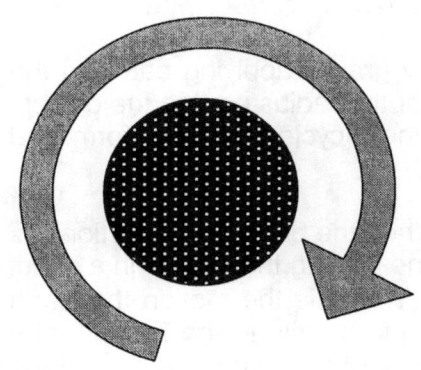

In relation to the Sun the Earth is in motion and the Earth is a part of the liquid that outer space provides. To the Sun the Earth is a factor that flows with outer space and although the Earth shows the ability to counter the flow that outer space has, the Sun still regard the Earth as equal liquid to outer space. Being a solid or a liquid has no bearing on the state of the matter but it all depends on the flow in relation to a securing solid.

Understanding what the Universe is becomes the important key about realizing the dynamics of cosmology. The bicycle being without motion is a part of the Earth because of the Coanda and Kepler principle where the bicycle without motion sides with space **a³** and when achieving motion the bicycle sides with liquid by moving **T² k**. This dynamic is the key in understanding what measures apply where in the cosmos. When not moving, the atoms that are moving while they are in the unit the bicycle forms as a unit and as a unit the unit hold relevance relating to the general or governing singularity in the centre of the earth. It taps in to sustain the governing singularity by providing motion that forms part of the earth singularity. The unit uses the atoms to the advantage of the Earth motion and supplies the Earth with relevance in order to sustain as well as promote the earth moving. The bicycle is the Earth because it is space of the Earth within the space boundaries of the earth. It is **a³** and stands relative to **T²k**.

Then, when for some reason the bicycle finds the ability to move, the bicycle splits infinity from eternity. A universe is born. That which has no end parts from that which has no start, leaving space-time generated. It is motion that puts the bicycle in three positions relating to time. That splits eternity and infinity just like it splits eternity and infinity when moment-Alfa came about. There is just more heat in the backlog and less in direct relevance.

The rolling with time sets a differentiation between eternity and infinity and the measure of the time delay forms matter in time. By providing motion, the bicycle no longer only keeps the

earth singularity generated but also it supplies a potential singularity by establishing the individual generating of singularity which sets out maintaining the individual singularity that is apart from the earth singularity while still being within the Earth singularity. The singularity it now generates is no longer Π^0 but forms an independent singularity by as much as $5\Pi^0$. It shifts the line of currently to at the most five

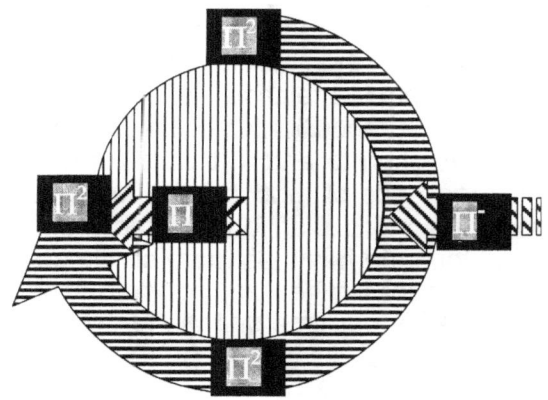

positions in delay of currently. Any shifting further brings about serious conflict.

The Universe we have (not the earth filled with life that we have but in the era we landed) we find the Universe going from 10 / $7(\Pi^2+\Pi^2)$ towards $7/10(\Pi^2+\Pi^2)$ ending at $3(\Pi^2+\Pi^2)$ while eventually all space-time will form $(2\Pi^3)$ in the star limit. The proton disappears when the proton goes to singularity at $\Pi(\Pi^2+\Pi^2)$ which then becomes double space $(2\Pi^3)$ where space being double catches with time being single and loses it's lagging behind time quality. That is what they call a Black Hole or what I named a proton star. When the proton goes singularity then $\Pi(\Pi^2+\Pi^2) = (2\Pi^3) = 62.01255$

All objects are classed by heat either being in motion through duplicating (overheating and expanding) or being in motion through heat contracting (heat being reduced through motion removing space), but most of all is that all material is about motion forming the space-time and classifying the space-time. This is most pivotal in understanding cosmology notwithstanding Newtonian views.

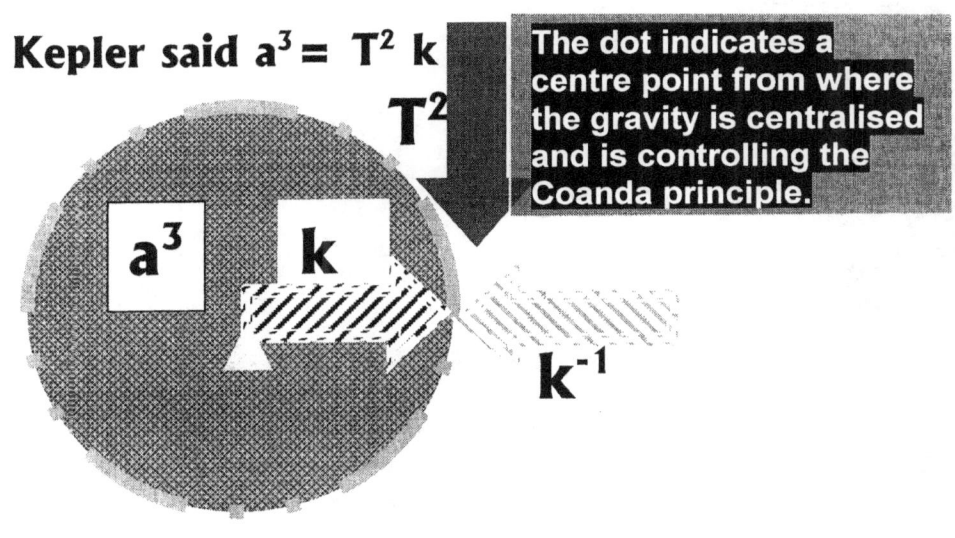

Kepler said $a^3 = T^2 k$

The dot indicates a centre point from where the gravity is centralised and is controlling the Coanda principle.

The motion of the liquid which the neutron is, proves to be the time (T^2) aspect because as it increases, it claims the space (a^3) in that a distance k of time (T^2) that the running has increased. The faster the motion is, the stronger is the gravity that the motion generates in the space it claims by the gravity it generates.

The Coanda effect is proof of gravity coming about through space forming motion. In the case where water diverts the normal directional flow the space that translates to the motion is deflecting singularity with the flowing water charging the motion. In the centre of the object having the round form, singularity is duplicated and by transferring Π to form Π^2 and the motion of the water creates a line of gravity that pushes the flowing water to follow the direction that the newly gravity applies to the water. This again proves Kepler's statement of **$k = a^3/ T^2$** that specifically states that space (in this case the object transferring singularity to a new position within the round object) and with the motion of the water redirects the gravity flow of the water to new space in new time. Only Kepler can explain the phenomenon but only when Kepler stands alone, correctly interpreted and divorced from Newton's opinion about Kepler's statements. There is a flow of time created by motion and defined by direction that produce expanding as well as contracting where expanding is contracting while it is on the other side of the Universe.

All objects are either cold and reduce space-time in relation to others being hot and expanding space-time. Being cold puts the object in the role of conserving space-time in contraction and that puts the object in a position of being a solid. Then in relation to the first conserving factor there is the overheating factor, which brings into the relation the expanding, or moving away from the singularity. The duplicating requires a repositioning of the aligning of **k** from a certain position to a more forward position in relation to and in that **k** will also have to extend a value when moving from k_1 to k_2. The essence of motion is to duplicate material that is in a process of overheating. By producing more than one of the same material unit the heat is distributed over a wider area and thus the heat has more space per time unit but less space in a time frame. That is motion. By applying heat to provide motion the Universe sees that as overheating and the longer k_1 to k_2 is per time unit the lesser will T^2 be because a^3 is spread over a larger area.

This letter you are reading is my effort by which I hope to interest you in reading my Introducing letter an **Open Letter Announcing Gravity's Recipe** The book on offer has the title of **an Open Letter Announcing Gravity's Recipe** and is the actual letter I sent to various establishments.

What brings about the expanding?

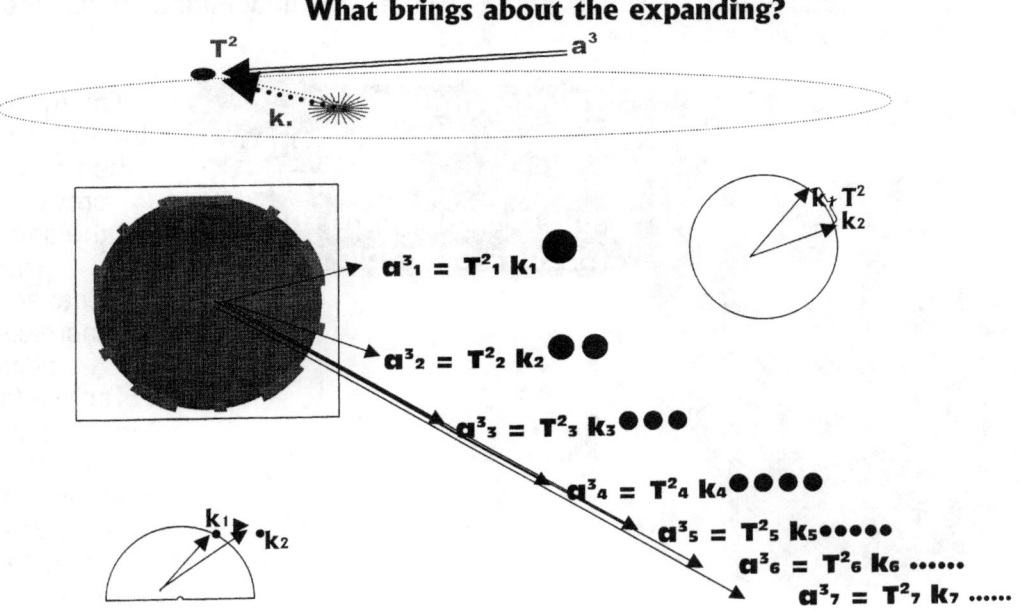

$$a^3{}_1 = T^2{}_1\, k_1$$
$$a^3{}_2 = T^2{}_2\, k_2$$
$$a^3{}_3 = T^2{}_3\, k_3$$
$$a^3{}_4 = T^2{}_4\, k_4$$
$$a^3{}_5 = T^2{}_5\, k_5$$
$$a^3{}_6 = T^2{}_6\, k_6$$
$$a^3{}_7 = T^2{}_7\, k_7$$

Kepler was the very first person to mathematically introduce space a^3, the centre **k** holding infinity and time T^2 that is representing time in eternity. Not only did he introduce space-time a^3 / T^2 but he also placed space a^3 and time T^2 in a relevancy long before Einstein did and placed gravity in space-time a^3 / T^2 even before Newton named gravity. Kepler was the person who placed gravity as the ingredient in the Universe that determines space a^3 and time T^2 and much more. Kepler was the first one that saw that gravity comprises of two factors being **k** or linear gravity and circular gravity or T^2 as gravity keeps space in form while all is staying together.

Since gravity also influences the space outside the sphere, the space we call outer space has seven plus three points bringing about ten positions of gravity influencing space.

The influence inside the sphere also captures the space outside the sphere.

This means that in the cube at the point of contact between the cube and the sphere the cube experiences such a contact point as if the "bottom falls out" of the cube and without a "bottom" to support objects they fall to the sphere as objects does fall to the Earth. Remember that a body "floats" in space, but at one specific point it starts to "fall" to the Earth. That is gravity and it is a dimension change much more than any force. I shall explain this last remark later on. That too is the Lagrangian system with five cosmic structures holding relevancy to the centre structure where the centre structure stands in for seven positions diverting from the centre and the orbiting structures standing in for five positions in space.

Gravity has all to do with dimensional changing and reforming of forms to re-affirm alliances supporting the centre. It is the reforming of space converting space to more concentrated heat.

The Universe is in the three dimensions using twelve dimensions that is visible to us and indefinite number of stages in size differences ranging from the immeasurable small to the immeasurable large where mathematics becomes a short fall to the next and the previous dimension.

Up to now every one in science is normally acting as if gravity is a commonly explained factor, which every one knows every aspect about all the principles that are involved in gravity down to the smallest detail. In truth, no one in science anywhere remotely knows what brings gravity about and I used Kepler to unravel this mystery called gravity. But no one in science will admit this fact about Kepler being the one who formulised gravity decades before Newton came and gave gravity the name.

Newton did not underwrite or define gravity and even today the most informed in Science at best can only assert their suspicion on a rumour presumed about what causes gravity to perform as the part interlinking the cosmos but no one can go any further by explaining the concept. Newton started this realising of gravity but it had and still has no more substantial proof than a rumour has and Newton admitted to it being a concept he could not explain. In Newton's ignoring to test Kepler's findings, Newton missed the opportunity to find what gravity is. Since Newton every person in science also ignored Kepler and every one is guilty of missing the opportunity Kepler maid available.

By my efforts of studying the implications that results from Kepler's finding I can now un-emphatically declare I know what gravity is. Gravity is the entire following locked into one compiling unit: Gravity is not being some magic force found between particles grabbing onto everything. I mathematically explained the following phenomena:

  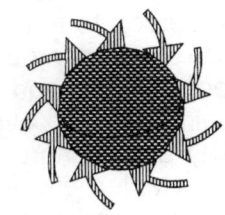

Gravity is singularity as a factor forming space-time

Gravity is finding space-time

Gravity is proving space-time and aligning space-time with gravity

Gravity is the working principals behind all cosmic occurrences that pre dates the Big Bang period.

Gravity is the Roche limit.

Gravity is the Lagrangian system

Gravity is the Titius Bode law

Gravity is the Coanda effect

Gravity is the sound barrier

By being able to pin point prove what Gravity is that enabled me to unravel the other entire phenomena that form gravity. Each of the phenomenon I mention above has one part or role in what is forming the totality that which we know as gravity.

Should you think this is rather a wild presumption I challenge you to spend a little more time and please think about what I say when you read about what I say in the next few pages. The first thing you should admit in private is what study did you personally so far made about the work of Kepler?

Still, to this day nobody in science at present will denounce the principle of gravity as vaguely researched. Gravity has never been explained as a principle. Even when one is considering what the importance of gravity is, gravity never yet has been understood. It is by now very clear that little if nothing of all objects is pulling closer in outer space. Comets are missing the Sun on a regular basis and no planet has come much closer toward the centre of the Sun. Still everyone in science acts in a manner as if Newton's gravity ideas are the best detailed proven fact and only occasionally does someone quietly admit that even Newton admitted not knowing what gravity is. No one ever comes to the front and boldly state that gravity is just a rumour spread by scientists pretending to know all there is to know and knows little to nothing about what there is to know. Newton admitted that much when he introduced the name (not the concept). Mistakenly Newton corrupted the concept he named as gravity.

I prove gravity is strongest where space is least (not where the Universe goes flat as Einstein promoted).

My theory I propose is one seeing in context Newton's law of motion. It is that which forms where every action has a reaction and gravity has a relation that is based on this very law of Newton. Newton's law says that for every action there is an equal but opposing reaction to the action and we find this where outer space forms one part and inner space (the iron and copper core) forms the next part of the dual relation. Gravity is not a one- way traffic but is a relation of relevancies that are applying equally and without the relevancy applying between the action and the reaction in a balance, there is no gravity.

Going according to what Newton introduced, Newton's concept will by now have the moon much closer to the Earth than it was in the time of Kepler's studies, yet we know the moon is moving away instead of coming closer. By the same measure Kepler suggested that the space a^3 is content with the motion kT^2 as long as the motion T^2k is equal to what the space a^3 will allow. Kepler suggested motion of space remains in equilibrium as long as motion of space a^3 duplicated space a^3 by motion thereof T^2k. That is much more true than objects rushing towards one another by the pulling power of mass. Newton agreed that he could only declare gravity as a vague concept. This fact was at that time drowned by the man's stature and was relieved from the manner of requiring the proof that later in Academic science became an absolute necessity. The proof one would demand now a day was never given to put Newton's rumour beyond doubt. When Newton announced a force he also admitted the force could be anything. No one ever came after Newton and proved the fact better. That still underlines the fact that the force to this day can be anything. Not once could one person in the past or present provide substantiating proof on gravity as a reality by defining the very principles thereof.

That includes every one since Newton as well as including Einstein and even Hawking. Scientists can declare gravity was a factor at 10^{-43} seconds after the Big Bang but what brought gravity about or why gravity became or still remained, as a presence is still tightly concealed information which all are speculating on. Even in the best and most informed circles and amongst the most educated there is no one that knows what gravity is because they all ignored Kepler and for them to ignore Kepler, the price they pay is not finding the principles bringing about gravity. Using Kepler even makes the method to follow and understand Einstein's discoveries shockingly simple. Gravity is the motion of space relating to time in movement.

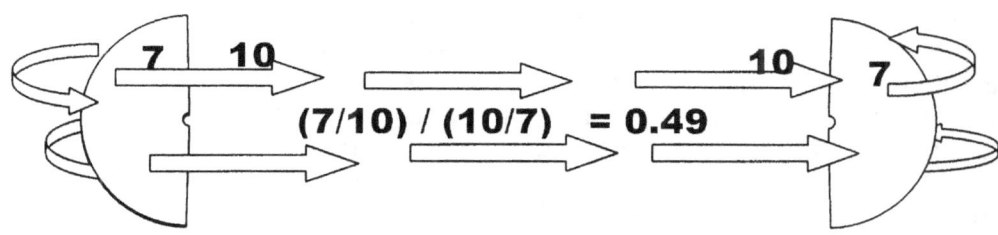

SPACE DIVIDED INTO TIME

(7/10) / (10/7) = 0.49
.7 / 1.4285 = 0.49

Taking also from both orbiting influences

THE PROCESS PARTED USING THE ROCHE PRINCIPLE

10 / 7	$(\Pi/2)^2$ The Roche influence on Titius Bode
7/10	$2.04 \times (\Pi/2)^2 = \quad 5.033$
$(\Pi/2)^2$	$2.04 \times (\Pi/2)^2 = \quad \underline{5.033}$
10 / 7	$5.033 + 5.033 = \quad \overline{10.066}$ from both objects

Crossing the singularity divide and activating the Roche principal $(\Pi^2/4)$

(10 / 7) \ (7/ 10) = 2.04
1.4285 / 0.7 = 2.04. Taking from both orbiting influences

SPACE MULTIPLIED WITH TIME

$$7/10 / 7/10 = 1 \text{ and } \quad 10 / 7 \text{ X } 7/10 =1$$

From dissecting the formula I prove that:

In the relevancy there are opposing motion where each participant provides a motion which is contradicting the other relevant but opposing motion. I base this conclusion on what Kepler introduced when he introduced the fact that space is half the motion and the motion is the other half of space. Therefore space cannot be as if space is not moving because the movement is space repeating what was before in the instant and onto the future. Space is one part and the repeat of space from the past onto the future taking space through the present is the part forming space by repetition in motion. Space cannot be if space is not moving to provide space a past into the future bringing the one part of space that in which it is and the other part either what it was or what it is going to be.

hat part is time in the formula space-time. The one part of space is what there is in the present but the second part, which forms the time aspect, is what there was or what there will be. Time depends on space moving while space depends on time, a position coming from or going to. In an attempt to explain my view I am prepared this one time to grossly simplify my view and relinquish accuracy in the process. There is motion expanding the confined space just like a rabbit tries to flee a dog chasing the rabbit. The rabbit is trying its earnest to escape from the dog. The dog again is trying to contain the rabbit by reducing the space that forms between the fleeing rabbit and the chasing dog. But the space between the two is what is of a major concern and not as much the running of the dog. Focussing on the running will lead to missing the process that really applies. From the view the onlooker has, it may even seem as if the rabbit is pulling the dog by an invisible string in an area where both participants are remaining in an area that is repeating the continuing chase as the dog holds the rabbit in chase and the rabbit holds the dog at a distance. The rabbit has itself set a task where it wishes to never see the likeliness of the dog ever again while the dog sees dinner.

The rabbit would love to leave the dog at a distance where the rabbit finds freedom from the dog as the dog's dinner opportunity. While the space between the two is merely a common fact, it unifies their differences. Both in relevancies have to appreciate their differences by the space in the unit that is keeping them apart. The space is the factor that has to resolve the issue being the differences in motion but cannot because different relevancies sustain equilibriums. I prove that as much as there is Newton's pulling there is Kepler's running around and the running around is equilibrium of the other factor providing the running away part.

At the angle science is looking at the issue, science either dismisses or cannot explain the characteristics or principals, which is there none the less. The explaining of the phenomenon is quite impossible when using the pulling rope magical attachment idea in the manner science tries to explain gravity. Therefore instead of dismissing the rope they dismiss all other factors present by gravity unleashing free motion but they would not release their idea about the rope. Gravity is motion between two particles that brings about mass. In the book I explain this in much detail but frankly there is not enough room to explain this in this letter.

Mainstream science knows about the fact that gravity has never been defined, the Bode principal that is there in all the planets and even the fragmented planet, the Roche limit, the Coanda affect, the Lagrangian system and the sound barrier, but cannot explain any of the phenomena all though the presence of these phenomena is without dispute. It is the explanations about what causes the phenomena that is part of the dispute but in science the manner in which they defend Newton, science would rather discount the obvious phenomena than question the legality of Newton's cosmic views. I only dispute Newton as far as his cosmic principles are inclined. The phenomena being there or not becomes disputed. Science fails to give acceptable explaining of such occurrences we see in the phenomena and therefore disputes the validity of the phenomena and this failing to explain the presence becomes disputing the presence thereof. I on the other hand found a way where these explaining of the phenomena took me past the Big Bang era and introduced me to the start of all starts. Science cannot get past one specific date because they do not accept or understand the phenomena, which I prove started the Universe.

In such a light Scientists must somehow realise they are barking up the wrong tree with the information they have to use to do some explaining. They cannot refuse the phenomena and not realise they must have the cat by the tail as far as cosmology goes. Please remember that with this I am referring to cosmology and not general physics. There is an Earth versus a Universe with huge difference between the two concepts but Newtonians fail to see that because Newtonians cannot appreciate the differences thus they're not able to understand cosmic gravity; they go about blurring the understanding of gravity. If there are that many phenomena (it represents all there is in cosmology) to explain and such little ability to explain (science fails to explain even one) by using the information Mainstream science is using to explain the cosmos, then someone somewhere has to realise there is something drastically wrong in the way they present the knowledge they claim to have. One cannot be serious about science but defend your view by dismissing the validity of all unknown indicating factors presented as such. There then is some gross incorrectness in the way one reasons. The Roche limit is there and no denouncing thereof can remove it from the cosmos. They may refer to evidence received from the Hubble telescope as "the star is blowing bubbles" for the lack of explaining or comprehension of what is occurring but occur it does. One cannot say it is some unknown gesture presented on occasions because by not explaining the pictures present certain foolishness. It leads to tragedies in aviation, which they are incapable to understand or explain. For fifty years they lost many pilots but still has no idea what brings the sound barrier about, or find the link gravity holds in the process we call the sound barrier. Instead they try to interpret some effect established almost two centuries ago with steam trains that is travelling at the same speed that horses run. No further investigation with the science in hand brought them closer to new facts! That they should rather see as a sign telling them they are going about incorrectly because by ignoring the cosmos one produce a fantasy and not science. Nevertheless, Newtonians are admirably known to ignore anything that may compromise Newton and since Newtonian thinking does not tell them the facts in hand, it is as good as no one told them anything. They ignore obvious facts deliberately just because Newton did not say so. When I first came upon the amount and the totality of the unknown quantities in cosmology as well as the complacency those involved have about such unknown factors being discarded, it stirred a sense of disbelief and I decided to respond.

All principles I use in the theory I introduce with the publishing of this book. All principles I apply are part of nature. I base my theory on heat stabilizing through space using motion to produce cooling. That is gravity.

I believe some of Creation remained as some particles formed by applying gravity in motion and the lack of motion in others became the lack of gravity, which inspired overheating which then formed plasma. Plasma is the result of heat where light is the epitome of heat. How light became plasma is rather obvious, which again I believe (within reason) I do prove. I believe heat is the destructed form of material and this information the atomic thermo explosions give us.

Analysing Kepler's formula without Newton interrupting Kepler's work helped me realise science has been running on an error for the past three hundred and fifty years. Please let me explain: Tycho Brahe and later Kepler made a study of outer space as never repeated afterwards. From this Kepler concluded that $a^3 = T^2 k$. We all know that a^3 is space and with the space indicated as being in the third dimension and the third dimension is unmistakably a cube that forms volume, which by definition is presenting space. We also know from the way calculations come about by using the formula of Kepler that T^2 is the duration of a specific period of time relating to a specific centre. On the one hand we have space a^3 and on the other hand in direct relation to the space Kepler introduced motion coming from a centre that forms time $T^2 k$. Kepler gave us space-time a^3 / T^2 centuries before Einstein gave the concept a name but no one ever took any notice. In the formula is space a^3. In the formula the space a^3 has direct relation to time T^2 If k is a^3 / T^2 it means that from the centre holding the gravity is space-time. Space is a^3 and the motion of space a^3 we accept as time $T^2 k$ and such accepting is part of our understanding for the past three hundred and fifty years. Kepler gave us gravity before Newton named it as a force. Kepler gave us space-time long before Einstein named the notion. With Newton's meddling he missed Kepler introducing gravity as $k=a^3/T^2$ space / time.

I believe that I achieved an all time breakthrough success because I can now explain what gravity is. Remember that not even Newton could explain what gravity is or where it comes from, but Kepler did that without any person ever noticing. Scientists over the years paid the price of ignorance about gravity by their unwillingness to investigate the father of gravity, which coincidently is not Newton but Kepler. From Kepler's formula, without the interference of Newton who changed the formula on Kepler's behalf, I prove the Titius Bode principal also known just as the Bode principle. I prove that the Bode principle forms gravity when using the Roche limit. Moreover the Titius Bode principle is gravity because the Titius Bode law composes the motion in liquid, which the neutron is. These phenomena were never explained or understood by Mainstream Science although they appear more than regularly in the cosmos. In the same breath I might add that Kepler also was never investigated. My achievements came from my effort where I separated Kepler's work from the opinion that Newton formed and that he (Newton) gave his compromised views about Kepler's work to the world. For instance from Kepler's work I can explain the operation of the Black Hole, which not even Prof. Stephen Hawking understands. That is because Hawking ignores Kepler. In my opinion my explanation of gravity makes much more sense than the accepted force of Dark Age proportions…and the best part is you do not have to be a genius to realise or understand it.

Even a simple person such as myself can see it clearly! From my view a force is just motion applying and that is what Kepler said gravity is. Kepler said $a^3 = T^2 k$. I dissected **k** as a factor in the Coanda effect and found that the Coanda effect is proof of my view about gravity and the ability of establishing gravity by centralising space, which forms singularity that produces the Coanda effect. The Coanda effect is the establishing of individual space a^3 by applying motion $T^2 k$. Where the Coanda effect is producing gravity and such producing is stronger in a small space than the gravity produced by the Earth in that spot, I use that principle to show that there was some manner in which the reducing of **k** brought about a stronger T^2 just as Kepler said. This was a crucial part during the Big Bang and therefore had to play a part during the period of the Big Bang. Einstein came to this conclusion but failed to refer his view back to Kepler.

As presumptuous as it may be on my part of trying to disprove Mainstream Physics, such a presumption does not change the truth about Mainstream science being incorrect about gravity. After all they admit they do not know what gravity is. I am not disproving anything because they agree they do not know, which paves the way for my showing what gravity is.

By admitting not knowing what gravity is they then also admit there is a chance that they can be incorrect about gravity but unfortunately mainstream physics do not see it that way (yet). The question in hand is finding what role gravity played when the Creation came about for the first time. I had to find a method that would allow me to explain why gravity played a role.

My ambition is proving the Universe not coming from nothing and therefore outer space cannot hold nothing. By taking Kepler's $k = a^3 / T^2$ and using k as a line I show through using the line as an example that the cosmic Universe holds everything and all concepts. However the only thing it does not hold is also the only aspect not present in the Universe at all. That is the value of nothing or zero in as much as carrying the definition of the absolute absence of any value. This means the Universe is filled to the point it is overflowing and we even gave the concept a name: we call it the Hubble constant. By overflowing nothing cannot be and is not there. With a room overflowing it is indicative of the fact that the room can be anything but be empty. With the line that light uses to flow the lines eliminate any such a possibility of nothing being present. Mathematics is a means of communication about matters concerning the cosmos. As an intercultural language spanning across race and ethnicity or as a principle as such, mathematics cannot have zero because mathematics indicating lines, which is about not applying the numerical number or value of nothing. Everything came about from singularity and Einstein proved that. From singularity nothing ever had the chance to enter space. I challenge any person that disagrees with this statement to show mathematically where nothing as a factor ever entered the mathematics of the Universe. If there is any one there believing there is nothing in outer space I challenge that person to prove mathematically where nothing is a factor in the cosmos. Your attempt may either be before or after reading my work but my challenge will stand since mathematically nothing cannot be part of mathematics. Multiply whatever with zero or nothing and such multiplying results in nothing where nothing is then, can establish no multiplication. Kepler gave us the relation between cosmic objects as $k = a^3 / T^2$. From the formula k forms a connecting straight line filling the first dimension and not the single dimension because k in the single dimension is not zero. It is unproven how k can backtrack to become $k = 0$. I deliberately press this point and make it an issue because that removes all the theory of mainstream science from any logical base they have in support of their views that space is, holds and comprises of nothing.

We stand on the outside 150×10^6 km from the spectacle and from such distance we judge the Sun. We don't even judge the Sun from what we can see but we judge the Sun from what we feel. We feel heat coming from the Sun and from that we argue that the Sun is hot. We see the Sun has heat rising from the surface as a liquid soup. That puts the hydrogen layer as the outer layer in a liquid. Hydrogen freezes on Earth at a temperature, which is the coldest amongst all other elements. Yes, the Sun is 6500 ^{0} and that is on the outside. To a human that is hot but a human has no mind judging the Sun. If the Sun squirts pure heat turned to liquid from the surface and the heat falls back into the surface the Sun is a lot colder than the Earth is. The earth requires an enormous effort to cool hydrogen down to a liquid state. We must mind the way we think of the hydrogen in liquid. The hydrogen remains a solid. The element is untouched by temperature differences. It is the heat environment surrounding the hydrogen that changes it from a gas to a liquid to a solid. One removes or one amplifies the heat in which the hydrogen is and that turns to liquid or solid or gas. The hydrogen is untouched in the elements worth.

I do not elaborate or explain the broader aspects or form an overall view. I found if I do that before a solid understanding of the basic concept is established, no concept becomes established. The most basic to explain is that the line cannot start at zero because then there can be no line to follow zero. The cosmos has lines forming cubes and lines forming circles, which in applying 3D manifests as spheres. Between the circles and the cubes run lines, so the key to understanding the Universe is the following of a line. The Big Bang was a time when the Universe was incredibly small making the running lines small. Understanding the Universe is taking the line connecting particles through space back to its limits where such limits were during the Big Bang. But the reducing cannot go to zero because zero removes the line all together. By reducing the line to where the line will not reduce any further, we will find at that point that all points land on the same spot. The spots all share one position because that is the only position there is to hold in the form singularity presents.

That is singularity being one to all but it is not zero. Finding form in that point shared by all will give a value of singularity. Extend that value received to a Universal centre and bring that value to align with Kepler's $a^3 = kT^2$ and understanding the Universe by finding the centre of the Universe makes the Universe simple as can be. The Universe becomes sensible making the entire different yet unexplained phenomenon as easy as children schoolwork. There are suddenly no more mysteries in the Universe. It is only possible when we see gravity not as a grabbing force, but instead seeing gravity reducing the space between particles. Gravity is not being some magic force found between particles grabbing onto everything.

The following is the mathematical proof that through the atom, time illuminates space by applying motion. Following the mathematical proof I explain how that is achieved.

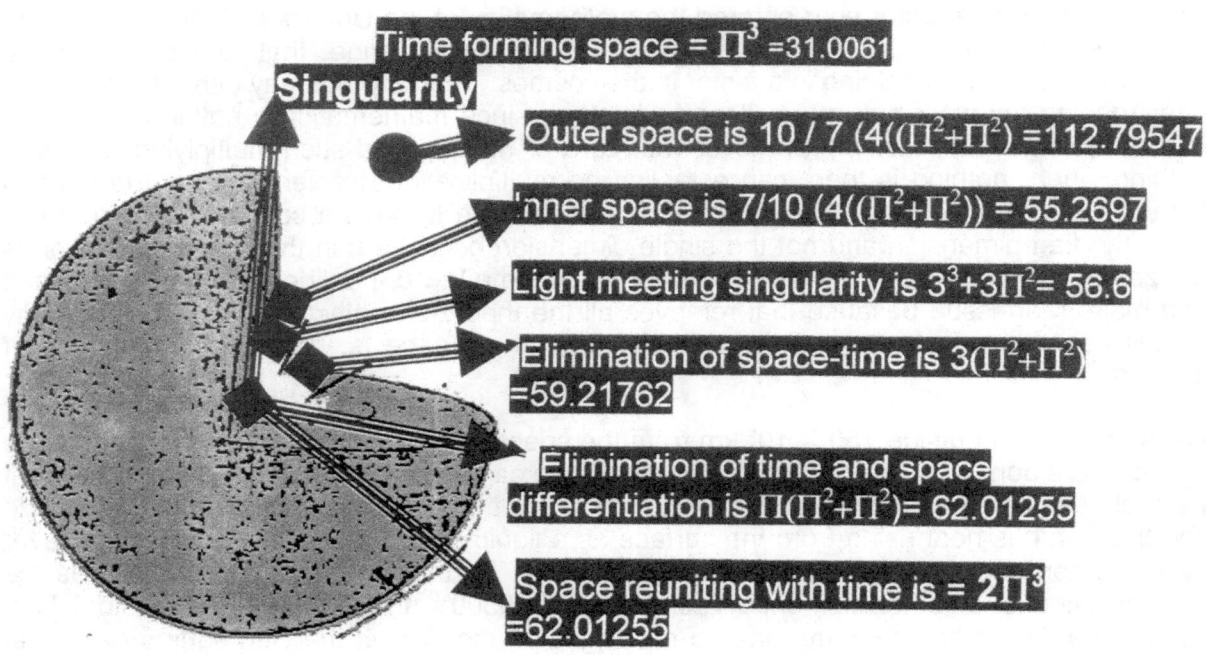

Time forming space = Π^3 = 31.0061

Singularity

Outer space is 10 / 7 $(4((\Pi^2+\Pi^2)$ = 112.79547

Inner space is 7/10 $(4((\Pi^2+\Pi^2))$ = 55.2697

Light meeting singularity is $3^3+3\Pi^2$ = 56.6

Elimination of space-time is $3(\Pi^2+\Pi^2)$ = 59.21762

Elimination of time and space differentiation is $\Pi(\Pi^2+\Pi^2)$ = 62.01255

Space reuniting with time is = $2\Pi^3$ = 62.01255

Yet we see the heat flow amongst the hydrogen as a liquid. Nevertheless we remain adamant that the liquid is a gas and the hydrogen is in a gas and the Sun is a gas bowl filled with hydrogen because to our mind hydrogen must be a gas. After all, our element table classifies hydrogen as a gas and that is the way we think of hydrogen. We do not consider hydrogen to be in a liquid state when we see the heat is flowing just like a liquid and shows all indications that it is a liquid. No, the Sun is hot because the Sun feels hot!

In the Universe there are no hot or cold but a state of differentiation produced by time. The Universe parted by parting heat from cold when eternity parted from infinity, when Π^0 singularity parted from Π singularity, when 1^0 parted from 1^1. There is no hot or cold but there is a relevancy where one factor cools and another factor overheats. By retaining and contracting space-time, the Sun is the coldest space in the solar system and outer pace is the hottest there can be.

From since the time that man discovered intelligence (if he ever did) man has been with the presumption that the Sun is the hottest centre in the solar system. Later on in the present time, it came to someone's attention that the Sun also holds the solar system in gravity. The Earth by its standard and dominating its sphere of which it can control with influence is the hottest centre in the space of its domain and it holds the moon centred to the Earth. The gas planets are the hottest centres in relation with the most heat and they all hold their satellites captured by a hot centre. All space structures hold in every centre there is that is confirming their independence at that point of securing independence the centralizing of the most heat it is able to concentrate and from that centre holds all material captured or controlled in the domain of what that forms the independence of the structure. I can go on and on but heat in the centre couples gravity to space-time, just as if Kepler said before he was spoken for on his behalf and without his permission or him agreeing to it.

$a^3 = (T^2 k) = a^{3+2+1=6}$ with the sphere presuming the position of singularity as part of $k^0 = 1$ = **singularity**. Einstein proved that at the point where space reduces and such reducing reaches a point where space as a factor in the third dimension disappears into the single dimension (space going flat) gravity is overwhelming. Einstein interpreted this, as the complete Universe going flat but while it may be true that the Universe is going flat, that can only be within singularity since singularity represents the Universe as flat as it can get.

The centre of any sphere has to be at the very point where space completely falls away. It is at the point where all the points of line centres meet by crossing the centre of their individual connection coming in to contact as a group. In that way one may assume that the lines connecting the controlling points on the other end are crossing on a centre point that all that is participating in the constructing of the sphere are democratically electing such a centre. Please note this conclusion very well because this forms the heart of the Coanda principle. That point where the lines cross will put that position, which in itself is centralising all space in the sphere, at that point where all the lines connected forms a centre crossing point. Such a point will become very distinct and controlling where that point forms in the single dimension and singularity is the single dimension. Kepler also solves another riddle by resolving that, which truly got Newtonians stuck. This, to which I now refer, is what is referred to when they refer to the Hubble constant.

The growth we see in the Universe is an adding of space in every cycle completed by every cycle, which all the protons complete. The adding is the smallest addition that can come about in the shortest period of repeating by cycle rotation there can ever be. This growth of space-time next to singularity confirms the growth of singularity as singularity recalls the space it uses to grow in the time it grows. The margin of growth will be by the extension of **k** in the formula $k = a^3 / T^2$. Every cycle completed in the relation to space by the initial value of **k. $k = a^3 / T^2$** leaves ultimately a^1 extending as space or as Kepler chose to indicate it as k^1. That too has to be compensated by the duration of time reducing the time aspect by the margin that the space expands. This confirms what is evident in the Hubble Constant. The further one looks at time, the more time seems to race because time has the invert properties we give to space.

There is a position that is in motion that is forming the very edge of the outside. To be in motion the position must be in relation to a point from a centre. From the centre, there must be a specific allocated space ending at the object in motion and starting from a centre that has no dimensions. The object in motion determines the one limit and the centre with no sides and no space, which is standing still in singularity, determines the other limit. By that we can see there is only one way of looking at what we can observe and that is from the outside in.

The atom must represent that which is the utmost coldest we can think of. Moreover is the proton even much colder. When in a nuclear heat release where the atom splits and when that cold escapes, it turns to heat forming space in such violence that no one can understand what the process involves. When the spin of the atom allows the cold of the atom to release the heat, it had frozen to space the atoms hold but when this heat releases from the containing form of the atom it brings about much more heat than the Human mind can cope with. One may not look at the material and judge the surroundings. The fact that hydrogen remains a gas and so does helium in outer space must serve as enough proof that outer space is hot, regardless of our interpretation of the temperature gauge telling us what we wish to hear. One must look at outer space and judge outer space from the findings only considering outer space. If helium remains a gas, it is hot. The removing of heat makes the centre of the Earth cold although we see it as being terribly hot. The only reason why it can seem to be hot is because it is cold and in such a cold environment, the heat can gather and space can collect heat because the particles find the surroundings extremely cold.

The cold in the Earth centre causes the concentration of heat by space reducing, as all cold surfaces tend to do. If it was hot, the space within the Earth would expand and the space within the Earth where we think so much heat is concentrated does not expand therefore it must be cold. To gather and accumulate the space in a liquid means it became much colder being a liquid. Finding the surroundings terribly cold will allow the heat to gather and not expand but when the surroundings are hot, it will not tolerate more concentration of heat and thus will expand to rid the balance of excess heat within space. Look at the Sun and see how the Sun turned the hydrogen to a freezing cold liquid at 6500 K. Hydrogen is in a fluid state within the Sun and is colder than the hydrogen that is in a gas form in outer space.

The Sun is the coldest place in the solar system. That is when the protons oversupply the removing of space to produce the cold that is so apparent. By the reducing of space, it can concentrate heat to a fluid state by producing the opposing cold that finally freezes the heat to a solid state. The expanding of space is a way of duplicating space without reducing space and by duplicating in the form of expanding it becomes just the opposite to duplicating by motion therefore reducing space by halving space in time. That is what gravity does. By motion, space duplicates and by space halving, it removes heat in space as well as by dismissing space. In all the applying of gravity, space dies. The density of the protons brings about space dense enough to harbour the heat in such quantities and visa versa applies in outer space.

The fact that hydrogen remains a gas and so does helium in outer space must serve as enough proof that outer space is hot, regardless of our interpretation of the temperature gauge telling us what we wish to hear. One must look at outer space and judge outer space from the findings only considered in the terms which outer space insists upon. If helium remains a gas, it is hot. The removing of heat from the space that contained the heat makes the centre of the Earth cold. In our universe we see it as being terribly hot because the heat then forms a separate substance but remains a form of material (8) but that is because we see the heat and not the space derived from the separating of the heat.   The only reason why the space can seem to be hot is because the space is cold and in such a cold environment the heat can gather in a much concentrated state and space can collect heat because the particles hold concentrated heat in the space separating the particles.

By removing such high concentration of heat from the space that used to be expanded heat, the space then must contradict the heat by being extremely cold. We look at the heat in the space, which by that time is another form of material and find the surrounding heat in the space hot while the space is extremely cold. The cold in the Earth centre causes the concentration of heat by space reducing, as all cold surfaces tend to do. The proton contributes to that reducing of space. If it was hot the space within the Earth would expand and explode but the space within the Earth where we think so much heat is concentrated is so much it does not expand, therefore it must be cold. To gather and accumulate the space in a liquid means it became much colder when the space parted from what then is being a liquid. Finding the surroundings terribly cold will allow the heat to gather and not expand but when the surroundings are hot, it will not tolerate more concentration of heat and thus it will expand to rid the balance of excess heat within space. The concentration or release of space with heat or space from heat is a direct contribution of the singularity in control of the space-time. The regard of the singularity stipulates the conducing of heat in space or the release of heat to form space by means of bisecting the occupied space.

Look at the Sun and see how the Sun turned the hydrogen it holds captured in its atmosphere to a freezing cold liquid at 6500 K. Hydrogen is in a fluid state within the Sun and yet it is still colder than the hydrogen we find in outer space that is in a gas form in outer space. The Sun is without any doubt the coldest place in the solar system. That is when the protons oversupply the removing of space to produce the cold that is so apparent in the heat levels that the atom cannot absorb in normal growth and therefore cannot find accommodation in the walls of the atom. By the reducing of space, it can concentrate heat to a fluid state. By producing the opposing cold that finally freezes the heat to a solid state, we find that is what matter is. The expanding of space is a way of duplicating space without reducing space and by duplicating in the form of expanding it becomes just the opposite to duplicating by motion, therefore reducing space by halving space in time. That is what gravity does. By motion space duplicates and by space duplicating, the material must be dividing or bisecting - halving it removes heat in space as well as by dismissing space and in that heat is concentrated heat. The density of the protons brings about space dense enough to harbour the heat in such quantities and visa versa applies in outer space.

The particles claim more space when heated to preserve the cold. The claim to more space produces more space and reduces more heat. Such expanding brings about cooling. When particles heat or cool motion applies in some form. Motion started at a point when the Universe was extremely hot and there was no space. By introducing motion, space formed and the lack thereof produced friction that became heat that became space. It is natural, it is simple, and above all, it makes believable sense.

The application of gravity is that which condenses space by bringing about heat with the compressing of space. We apply the progress we have as a species in the way we go about with our skills to unveil ways we can tap into the energy that nature provides. Internal and external combustion engines all rely on this application for harvesting motion by driving power. Compress space even today with a piston in a cylinder and then pump the compressed air into a container and such confining of space will increase the heat by the piston effort to reduce the space brought about in the container.

The heat coming about inside the cylinder has no relevance to particles colliding because all compressor cylinders cool down with time moving and not necessarily with the loss or release of particles. It is not only the discharging of air that will reduce the temperatures inside the container. The time flowing bringing motion about where the motion is not about particles escaping but heat escaping in the replacing of the heat density (not the density of the particles forming the material content within the container) but the space that compressed to heat will also bring about that the heat displaces through the container wall to the outside. This is bringing about equilibrium where heat will always flow from more dense areas to the lesser dense areas. This has no influence on the status of the particles on the inside of the cylinder but only concerns the density levels of the particles inside versus outside. After the pumping of air increased the heat in the cylinder which even can go to dangerous levels, will reduce back to room temperature when further pumping ceases and that stops further air movement into the cylinder and such surging of pumping air is what brings about heat stabilizing.

Mainstream physics ignored the clear connection completely, notwithstanding it being so very obvious. There is this far in their recognising of principles in natural physics not one single reference made to prove their appreciation of this matter. They are bent on particle colliding. When particles collide, such collision forms an atomic thermo release and that action we call an exploding atomic bomb. What principle this argument about particles colliding ignores, is that all atoms use negative charged electrons forming the atomic limit on the outside forming a definite border to the boundaries of all atoms and in both electrons from different atoms are being negatively charged.

In being negatively charged, it means both will come out and totally reject the other. The closer they come, the more violent the rejecting will be and such rejecting is the production of heat that will turn to space. The electrons repel other negative charged sub atomic structures, which the electrons are that form the outer borders of all atoms. With all electrons highly negatively charged (being as negatively charged as any possibility will allow to match the utter extreme) such electrons could not touch.

It is about time scientists start looking with their minds and not their eyes at the Universe and see what is truly out there to see. All the difference we find is seated in the human mind. We humans set differences because we look at the cosmos by placing humans and the life we find on Earth in a pivotal centre in the cosmos instead of placing singularity in the centre and life where it belongs; only found on Earth. Einstein proved mathematically that in the presence of a strong gravity such a strong gravity slows time down.

Surprisingly with that evidence being around this long, nobody in science since Einstein's discovery took those statements and made any further progress from that. Science still sticks to the opinion that time did not change, not even slightly, since the beginning of the time it held the same pace ever since the start of the Big Bang notwithstanding the implications this concept carries. Before the Earth took one year to circle around the Sun and even before the Sun was there a year was still the same duration of one year. How odd... don't you think ... that the only aspect in the entire Universe that is beyond change is the aspect of time? With the entire Universe including all the gravity now present and not excluding one Black Hole or dust speck pressed in such an area that was possibly the size of a lepton even then the gravity extending from that circumstances must have been beyond what words can ever describe.

When everything was that small when the Big Bang took charge, the gravity at the time was beyond light, because even today in the Black Hole the gravity is beyond the speed of light. If the gravity was that high and Einstein already proved that strong gravity slows time down, then there is one logical conclusion and that is that time was in fact at the time of the Big Bang standing still. Mathematically it is incorrect to allow gravity to compress the Universe into a spot smaller that an atom and exclude any other factors and relevancies to change.

As usual Newtonians has the relevancies mixed up. It is not the Sun that is cooking outer space to cinders but it is outer space that is boiling the Sun to steam. The light surging from the Sun, as if it is steam we see that is coming from a kettle is what mainstream science promotes as light coming from way inside the centre of the Sun. This is not factual because of characteristics hot objects have. If the Sun were hot, as they'll have us believe it is on the inside, the Sun would have exploded, just as stars do when they overheat. Supernova are stars that overheated and if a star can overheat when everything is going wrong then the star must under heated when everything is going right. If the star expands into destruction when it overheats, it must be cold when it is contained and kept in a workable form. The star must be cold when normal to be able to overheat and explode when it is destructing. Therefore a star is a particle that is frozen into liquid and in a process of would be one day frozen into the oblivious.

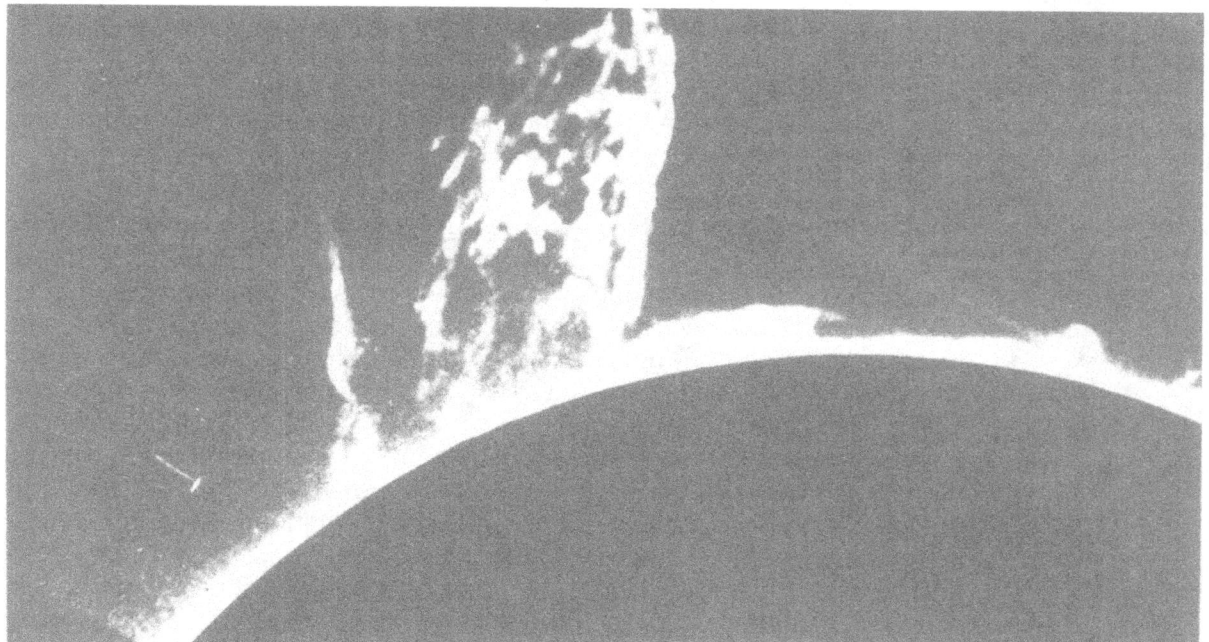

The fact that the prominence squirts out liquid in vast amounts is because there is a lot of space reducing going on in the Sun, so there is less space inside the Sun. The fact that the prominence expands to outer space means that the prominence was expanding into a hotter area and does so as a liquid. However that fact that the prominence fell back into the Sun can only result from the prominence not being hot enough to return to a gas state and then through such density discrepancies had to return to a less hot area. That is science, not magic. By using Newton, one cannot even begin to explain any one of or the combined efforts of the above cosmic phenomena that are all over the cosmos and form all the laws in the cosmos. Newtonian definition cannot even recognise any of the principles but only Newtonian science are taught to students. No student can have the fortune to disagree with Newton and remain a student. If the student will dare to disagree with Newton it is the end of such a student's academic career. By setting this firm condition, Newtonian science becomes institutionalised mind conditioning of the concepts of thought forming in physics. With my saying this I have not made one academic friend but neither have any one proved me wrong. Students are taught to accept Newton and to ignore Kepler and any student doing it the other way around will fail all examinations and other testing at Universities. Students accept Newton or they accept a ticket taking them home. Newton is an institution force fed to each following generation but saying that reserves only resentment towards me amongst Academics. According to Newtonian science space is simply nothing with no qualities but gravity separating space and space does not mingle, as one would expect if space was nothing because space does form borders.

Disasters of unprecedented magnitude arise from such borders. The Challenger disaster of February 2003 is pertinent testimony to those borders that were powerful enough to break the aircraft into pieces while the explanation contributed by Mainstream science is evidence of a shocking lack of understanding about what took place as cosmic laws were breached. I do not pretend to be of superior understanding and do not place myself on any pedestal. On the contrary the information is so simple and so easy to understand that the lack of any

Academic understanding frustrates me almost witless. But academic taught culture demands all persons to miss the evidence, which is so clearly visible because academics demand researchers looking in other directions because students are forced to accept Newton's vision about Kepler's work. By the time they reach researchers status, they too have tunnel vision that can only acknowledge Newton and ignore Kepler. Our not understanding laws, provide a platform for future disasters occurring because it will lead to us ignoring more of applying principles that leads to space tragedies of magnitudes we have not thought of as yet. By not understanding the sound barrier, tragedies have and will again come about and will increase as misconceptions become more present in the future because the demand on space travel increases.

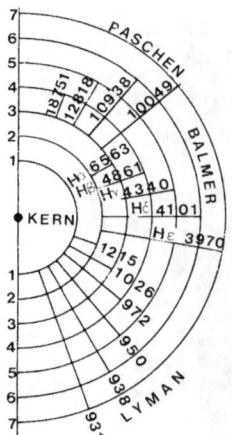

The book **an Open Letter Announcing Gravity's Recipe** ISBN 0-9584410-9-X is about that process adapted by the Big Bang, never ended and it is still bringing over, that which is in unoccupied space to material being in occupied space. Occupied space holds matter and unoccupied space is empty of solid materials. There is contraction, which we know by the name we gave as gravity. Then there is expansion, which we gave many names being the Big Bang and the Hubble Constant or better known as simply exploding or forming plasma with all the terminology accompanying that simple idea. This I show, is antigravity. Apply heat and space and a balloon lift where such lifting is antigravity. There is a balance in the Universe where gravity contracts and reforms space and heat expands becoming space and produces space.

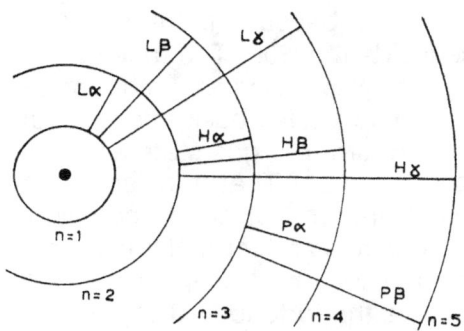

This puts my theory in line with reality. The only way anything can get bigger is when heat is added to what there already is. The Bomb at Hiroshima and Nagasaki showed how intense the heat in atoms are and how well packed the heat is which is contained in the atom. Bringing more heat to the atom brings about the atom having more of the same but in a higher measure. Only heat can make material expand and with the Universe unable to expand because the Universe is what ever can be, it must be the material inside the Universe that is expanding. In the same measure we find where space reduces, heat is removed from that space. Gravity is about contracting and that is true. In the manner that Kepler put it material that is filling space (a^3) is equal and the same as the motion ($T^2 k$) of the Material moving. That means to have time then time must move and the only way time can move is to move the space in the time.

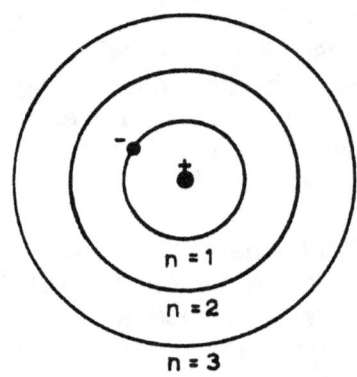

However to move is either to come closer and using Kepler we can see how Kepler would reduce space in $k^{-1} = T^2 / a^3$. That means the expanding subsides to a point where contraction is and contracting is about cooling or reducing the Heat that was expanding the space. Then there is expanding of material which is as Kepler would put it $k = a^3 / T^2$. That would be when the space becomes less and the only valid way for space to become less is when what there is becomes less than what there was. That amounts to cooling where the heat is being removed from the space.

Whenever the electron jumps from a higher into the lowest (innermost) orbit, the atom gives out radiation at a wavelength corresponding to a spectral line of the Lyman series. The jumping down into the second lowest level contributes to the Blamer series. The greater the jump is, the greater is the emitting of radiation to the limit of the series, which is reached when an electron enters from the outside of the atom. Outward jumps involve the absorbing of heat and that is inclining to provide space to accommodate the increase heat levels because of the increase or rise in the absorption

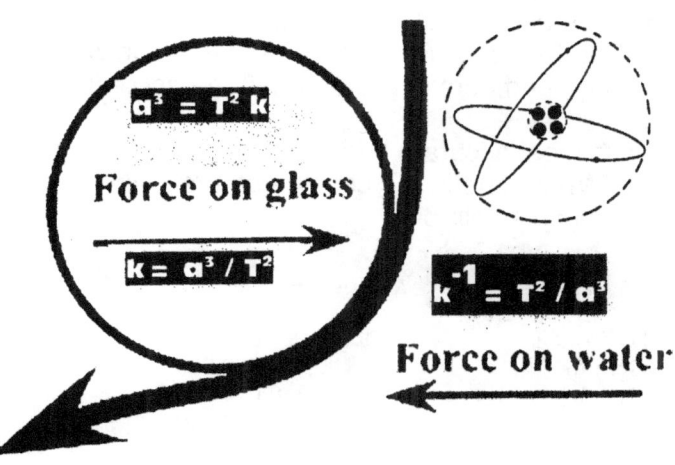

lines. When the heat level in the atom rises, the electron jumps to a higher band and when the heat reduces it moves down one band. The heat coming about in the surrounding area of the atom produces more space because the atom increases the space by applying the electron in a higher orbit ring. The moving of the electron is coupled to the giving out of radiation at a wavelength corresponding to the spectral line of the Lyman series.

When the heat level rises or lowers, the space within the atom decline or increases.

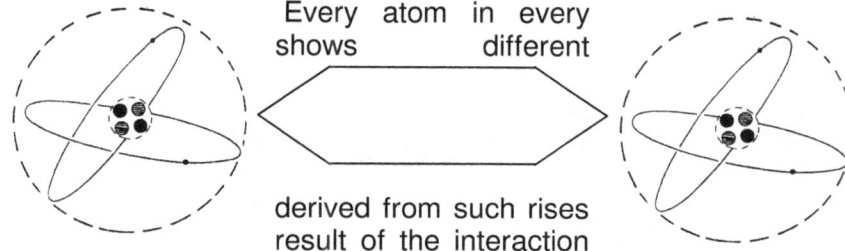

Every atom in every shows different

derived from such rises result of the interaction

element association correspondence to the heat it is in association with. The corresponding of the atom and the reaction of space is a direct there is in the gravity contracting and the gravity in expanding depending on which of the actions of the Coanda principle is in dominance at the time. The rise in heat is a rise in the liquid part that extends the contracting by giving rise to the adding of motion. The heat is liquid because the heat is motion and the atom inside becomes the solid since the motion is conserved by the spin of the electron.

Today it is the rise of levels that is in focus but this same principle had to be in use when atoms were formed. If it was true about mass pulling mass by reducing distance, the big bang was not possible and individual elements were not possible. With the cosmos down to the size of a neutron and mass confined to that space within the neutron that would be the recipe for the biggest crunch there could ever be. It the Big Bang was brought about mass confining mass by reducing the radius that implied the space within, then what would ever be more applicable than at that moment to bring in all the forces the hell can unleash and destroy what ever was not yet in the Universe? The entire idea of mass pulling mass to reduce space is a prehistoric thought and explicitly incompetent in explaining science. The following is a far more suitable explanation and is as true as Kepler is.

The most sensible way atoms formed is by the method of the Coanda principle. Those ones that were first were most of all very dense atoms, were the first to come in place when time was eternal and dominant and space infinite and one notch off singularity.

We still find the liquid time having a vital role in the space the atom uses. At the time when T^2 was almost eternal, space was infinite because T^2 will not permit the space a^3 much room to be. But the opposite is also true that if time was steady then time being so long could pack in large numbers of protons with the accompanying neutrons at the time into the time unit in space that formed.

This is most accurate but this is only concerning a unit in rotation rotating in conflict of its own spin. When time in the cosmos at large views time in progress, we find that the development is not quite so simple.

Element	Relative number of atoms
Hydrogen	1,000,000,000,000
Helium	90,000,000,000
Carbon	350,000,000
Nitrogen	85,000,000
Oxygen	590,000,000
Sodium	1,500,000
Magnesium	30,000,000
Aluminium	2,500,000
Silicon	35,000,000
Phosphorus	270,000
Sulphur	16,000,000
Potassium	110,000
Calcium	2,100,000
Chromium	300,000
Iron	3,200,000
Nickel	120,000

BERYLLIUM (4)

MAGNESIUM (12)

SODIUM (11)

NEON (10)

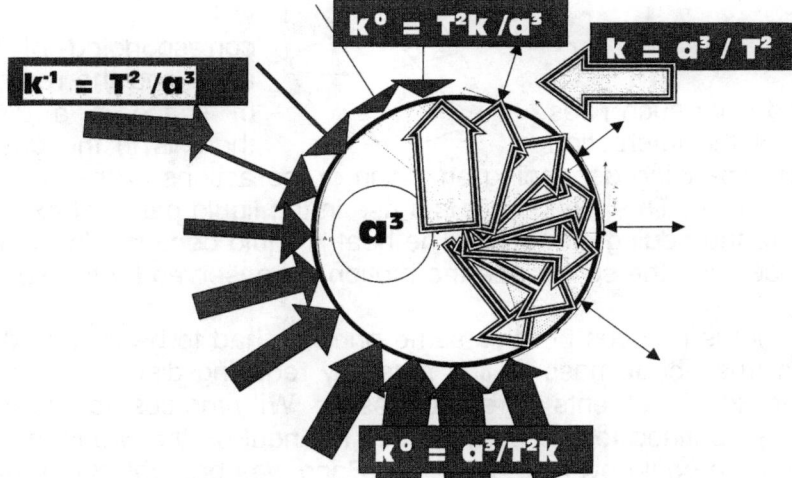

$$k^0 = T^2k/a^3$$

$$k = a^3/T^2$$

$$k^{-1} = T^2/a^3$$

$$a^3$$

$$k^0 = a^3/T^2k$$

There is no trace of Newtonian mass pulling mass. It is the intensity of time forming the liquid that reduces space forming the solid to accomplish gravity. When time was the in the epitome of eternity and was moving away from infinity, time in eternity could place an immeasurable number of protons into one proton cocoon. It could fit because the intensity had space next to infinity and the motion coming as a result of the intensity was so enormous no calculation by any human can relate to what that was. The symbol of that is still with us in the form of proton stars or better known as Black Holes. As time reduced the intensity space gained ground and less protons got confined to a specific space. It is very clear that at one point time had so little motion left it came to the same value as what the electron has and the Big Bang came into place. The era of forming atoms came to past and the era of forming stars by dimension of atoms came into place. Today a star is a group of atoms forming one atom where at the moment of the Big Bang the space holding a neutron formed one Universe. The proof that the liquid motion of time started the solid Universe at the end of the forming of the Big Bang is proven by the Grand Unified theory.

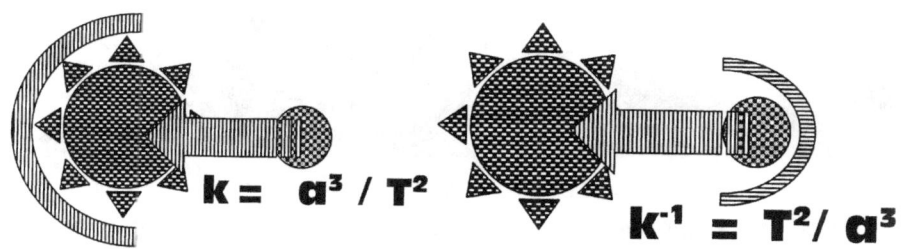

$$k = a^3 / T^2$$

$$k^{-1} = T^2 / a^3$$

The rebound is always less than the progress and the reason for that is the flow of time. But that also forms the reason why there is a Big bang and why there is a Hubble constant in the midst of all the contracting that is shaping the Universe.

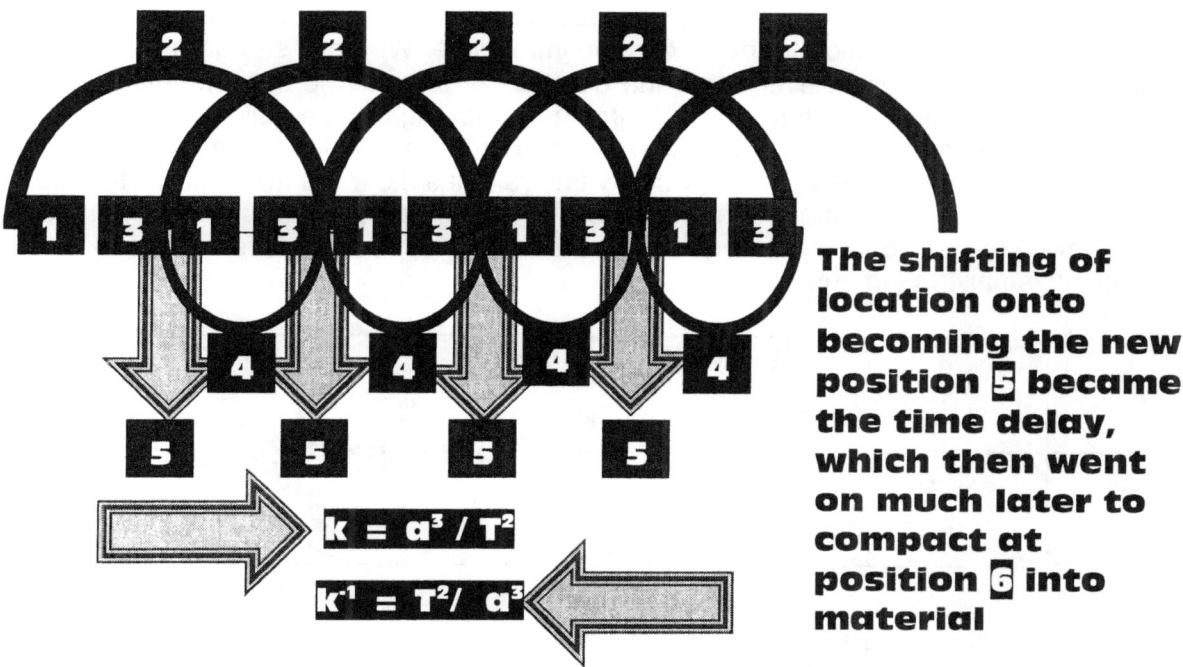

The shifting of location onto becoming the new position 5 became the time delay, which then went on much later to compact at position 6 into material

$$k = a^3 / T^2$$

$$k^{-1} = T^2 / a^3$$

Newton presented laws on motion without realising how complex such motion is when considering what Kepler's formula proposes. We see a train standing still but while this training is standing still the Universe allotted motion to the stationary train to the velocity of $7(3\Pi^2)\Pi^0$.

This motion is the relation that the Earth allows the train. While having gravity the motion the Earth permits the train to duplicate its structure, is what enables the train to represent its entire structure in the flow of time in relation to what the Earth represents, because with mass the train spins as a unit that is part of the unit we see as Earth. Being one with the Earth due to it having mass, all of its entire construction is in relation to the Sun, and the entirety is representing its entire construction plus all the time lapse in between the Earth and the Sun. This then also includes the Sun having a relation with the Milky Way and so on where this relevancies carry on and I can go on with this comparison until time catches the present duration. The space that forms the unit we recognise as the train is in motion with the Earth as part of the Earth and that part is what Newton recognized as mass. Every one of the smallest aspects of material refurnishes singularity with a new aspect of time in relation to what it was and where it is going. The more the train moves the more train there is and the more train has to confirm with time in duplicating space.

If the train is standing still there are a number of trains following one another from the past through the present to the future. The relation is $7(3\Pi^2)\Pi^0$ and if we covert that space –time to what we humans measure by it will be kilometres (space) per hour (time). It is the number of space that repeats its relation with singularity that furnishes time with space in the position (instant) in the duration of the instant. That means while the train is motionless and getting rusted from not being in action it is moving about at a rate, which in it is duplicating and affirming the previous position to the next in the immediate. We don't see a train but we see many trains a^3 being rebuilt every time the duration T^2 is confirmed by the instant k. We don't see one train but a constant flow of trains that is flowing at an equal pace to our rate of flowing. If the train moves, it just means there is more trains in relation to time than there was before in relation to time and the more velocity the train has the more frequent will the repeat of the train be in relation to the repeat of the rest of the Universe. The faster the train moves the more train there is during the duration in relation to the number of instants breaking the duration into smaller segments.

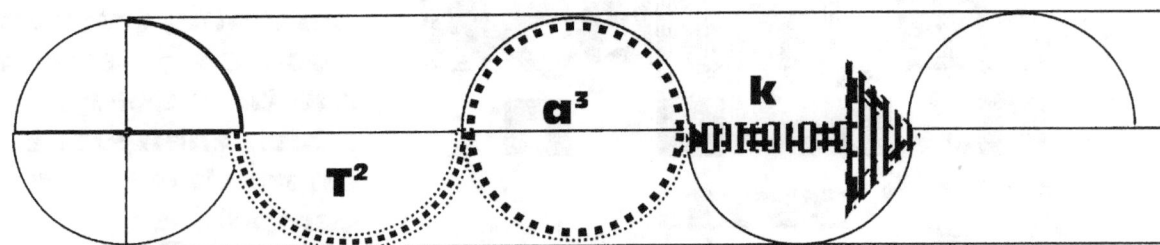

.

Time moves by the measure of T^2 but time in progress is by the measure of k. Therefore space in progress of time is $a^3 = T^2k$ where progress is T^2 but in relation to gravity we find that $a^{3\,-1}k = T^2$. In the relevancy we find the action and reaction of space-time flow is $a^3 = T^2k$ and that translates to being $T^2 = a^3 / k$ on the one side and $T^2 = k / a^3$. In the times we now live in we can and do produce an optical illusion of $T^{-2} = k / a^3$, but that is implementing the use of a telescope. In the true time we find as a cosmic reality the fact of $T^{-2} = k / a^3$ is rather a mathematical statement and no more than that. In reality we have $T^2 = a^3 / k$ on the one side as time expands and on the other side we find $k^{-1} = T^2 / a^3$. This we know is true because while it is possible by using an optical illusion the reality is that time can never reverse. In truth the reality about the opposing actions is that we find normal growth and that which Hubble first saw, is the process of expanding space-time by the margin of $T^2 = a^3 / k$ while on the rebound we find the opposing while contracting space-time is $k^{-1} = T^2 / a^3$.

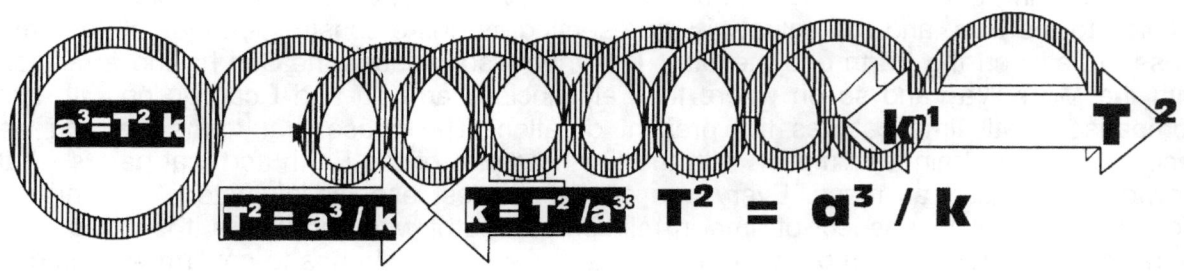

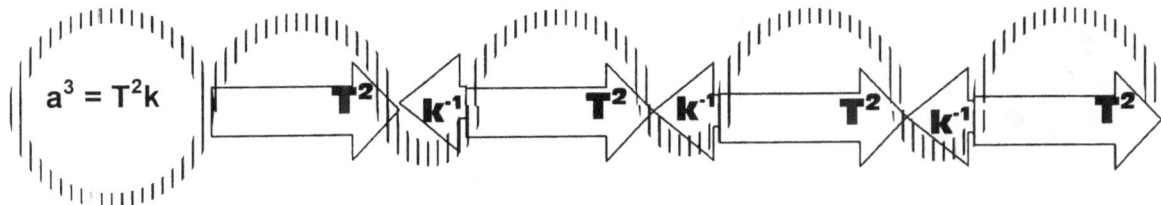

$$a^3 = T^2 k$$

$$k^{-1} = T^2 / a^3$$

$$T^2 = a^3 / k$$

$$k^{-1} = T^2 / a^3$$

The duplication is presenting time in eternity a new motion where the duration of the spin T^2 is a new position in which it affirms infinity **k.** The duplicating in relation to the confirming presents the growth of time in relation to space in relation to material. The duplicating represents a new relation to singularity in infinity where singularity in infinity is completely immovable.

Every proton confirms what every neutron duplicates and that stands in relation to what every electron conforms. Every proton although in one cluster with many other is a time process that takes matter through time back to what singularity affirms as time in the immediate.

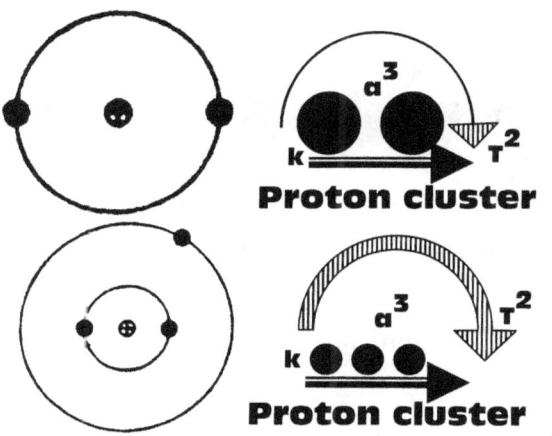

$$a^3 = T^2 k$$

$$T^2 \qquad k^{-1} \qquad T^2 \qquad k^{-1} \qquad T^2 \qquad k^{-1} \qquad T^2$$

In our ability not to see the imperfect while experiencing the imperfect and at the same time see the perfect in eternity and not being part of the eternity being perfect it is little wonder the lot of us are so mixed up in what we see and cannot understand. Eternity lasts forever and that is why no changing is visible but we can only experience infinity because eternity is an ongoing repeat of the same without changes. Infinity on the other hand is what interrupts eternity and therefore what we see in eternity is what infinity is interrupting. The proof of this is in the top where motion distinguishes infinity in the centre from eternity surrounding the time position of the top designating the motion from the past through the present onto the future. Those are there for all to see and the fact of that being there goes beyond dispute.

Proton cluster

Proton cluster

That is the manner how stars move time back to the point of having eternity sharing infinity infinitely. The one can absorb the other just by reducing the relevancy and increasing the flow of time. As the relative flow increases the relation is space subsides and the one becomes infinitely in relation to the other that then provides an eternal time. The match form and the element gain one more proton.

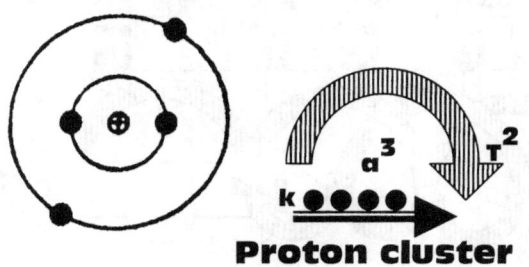

Proton cluster

The imperfect part is the part we find in infinity and with life holding part in eternity we cannot see infinity. We experience infinity while we see eternity. We see what is always there because we are unable to see what is changing. Look at your own fingernail growing or your hair growing or even a wound healing and you will see the nail, the hair, the wound but never the addition by growth. You might find what I say at this point not to be physics but it is more physics than anything currently used as cosmic physics are part of true physics.

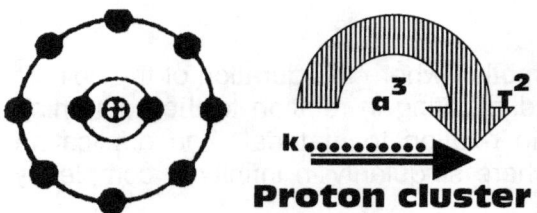

Proton cluster

When creation established space-time, eternity was interrupted by infinity as much as infinity interfered with eternity. It is not the same thing although we as humans tend to regard the matter as such. The one is having a look at it from the one side and the other is looking at what happened from another perspective.

Time moved on and space came about from the imperfect moving of time as well as the

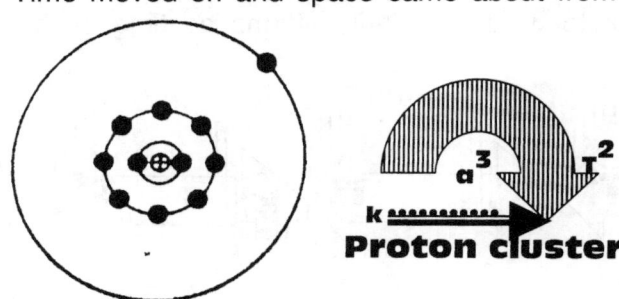

Proton cluster

perfect moving of time. The roundness and the perfect shape is part of eternity and that in eternity is what we see with life being part of eternity (not the human body which is a cosmic result of the imperfect) but life seeing it self in the position where it is occupying the centre of the Universe by studying light and night.

That is what Darwin missed with his species being from one ancient origin. Yes that is true but the one did not develop into the other. Time did produce changes but the donkey has no family ties with the horse. If it had, the mule would have been able to multiply and be fruitful and the mule is a lot of things, but that it is not. Things go along in eternity while infinity interrupts and then one-day

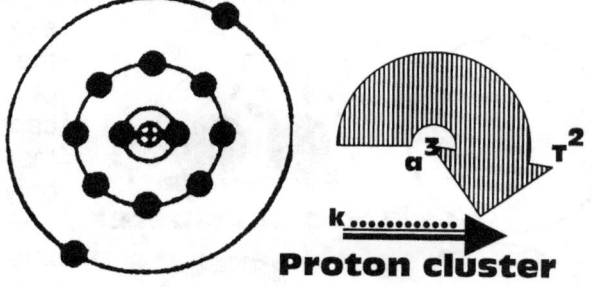

Proton cluster

infinity brings a change no one noticed before. The same building blocks are used but one day a new corner stone is laid and a new specie arrives that has no family ties with the previous lot.

In this manner elements came about. But the placing of protons within the atom formed elements whereas the atom was there the first instant heat parted from cold. I go into that part in the **Cosmic Birth...Dismissing Nothing** and since the issue is rather complex in explaining I would prefer to leave that explaining to the book mentioned.

Nonmetals

IA	IIA	IIIB	IVB	VB	VIB	VIIB	VIIIB			IB	IIB	IIIA	IVA	VA	VIA	VIIA	VIIIA
1 H																	2 He
3 Li	4 Be				Transition metals							5 B	6 C	7 N	8 O	9 F	10 Ne
11 Na	12 Mg											13 Al	14 Si	15 P	16 S	17 Cl	18 Ar
19 K	20 Ca	21 Sc	22 Ti	23 V	24 Cr	25 Mn	26 Fe	27 Co	28 Ni	29 Cu	30 Zn	31 Ga	32 Ge	33 As	34 Se	35 Br	36 Kr
37 Rb	38 Sr	39 Y	40 Zr	41 Nb	42 Mo	43 Tc	44 Ru	45 Rh	46 Pd	47 Ag	48 Cd	49 In	50 Sn	51 Sb	52 Te	53 I	54 Xe
55 Cs	56 Ba	57-71 *	72 Hf	73 Ta	74 W	75 Re	76 Os	77 Ir	78 Pt	79 Au	80 Hg	81 Tl	82 Pb	83 Bi	84 Po	85 At	86 Rn
87 Fr	88 Ra	89-103 †	104 Rf	105 Ha	106 Sg	107 Ns	108 Hs	109 Mt	110	111	112						

$$T^2 = a^3 / k$$

$$a^3 = T^2 k$$

$$k^{-1} = T^2 / a^3$$

Once again I have to draw the attention to what is out there in the cosmos serving as evidence. The proof we still find in the manner in which Galactica and all other orbiting objects develop. There are $T^2 = a^3 / k$ that is in favour of the promotion of one point holding singularity in the relation and to that there is another and opposing point holding singularity in prominence which is in relation to the expanding contributor that holds a relevance of $k^{-1} = T^2/a^3$ in ratio to the conserver.

The relevancy was there from moment-Alfa and brought relevance from 1^0 to 1^1. We can see that there were seven in ratio of ten and we can see how the seven produced the gravity of motion relating to the ten in time. We can see when the dominance started creeping to the other side and when $k = a^3 / T^2$ got the better of $k^{-1} = T^2/a^3$ because at a point where the sum total related to the singularity the **proton ($\Pi^2+\Pi^2$)** + the **neutron ($\Pi^2+\Pi$)** + **the electron** 3 = 35.89 × **singularity Π =112.75 outer space**.
Past such a point the expanding factor began to gain lost ground and the expanding got predominant as the containing factor started to store and preserve more than contain.

The major issue in hand is to recognise that when 1^1 overheated it parted from 1^0 because 1^0 was to cold to harbour 1^1 and by measure of 1^1 being hot, the same measure places cold on 1^0. The one cannot heat without the other establishing a border for the cold. In the Coanda principle the liquid establishes itself onto the solid by gravity as much as the solid allows the extending to lock on. The liquid is not locking onto an unattached solid. The solid gains as much as the liquid gains but that what the solid gain is, is not the same in likeness to that which the liquid gains. The solid will as much prove to be colder as what the liquid proves to be hotter.

One should see the electron as the indicator where the liquid attaches and where the solid ends because where the liquid ends is where the solid starts. The fact of the matter is not entirely that simple because the electron does view the neutron as a stabilizing solid while there is nothing in the cosmos more liquid than just the neutron is. The entire idea is in judging what is solid at the time in ratio to what is liquid at the time.

We see our bodies as solid while the truth is that it is life that keeps the body liquid. When life detaches from the body, the body loses its motion ability and then the body goes rigid. Then the body becomes a solid. However, while life is in the body the body is used as a mobile object in relation to the Earth being solid.

When going colder, the atom raises its solid level to lose some of its liquid level and the other way around. But the fact is that it is the atom going hotter that reduces the atom going hotter. Hot and cold is not by measure of temperature but it is by measure of space moving through time. If there is more space moving more rapid the atom becomes colder and the opposite is also very true.

As soon as motion commences, k increases because although the relevancy from the Earth perspective remains the same the relevancy from the aircraft changes drastically. From the view the aircraft holds, the k that the Earth has becomes the T^2 that the aircraft has and the T^2 that the Earth holds becomes the k that the aircraft uses.

The sound barrier is just another manner in the way the Universe brings about gravity. The aircraft has to produce excessive heat and by more heat delivers the space between the Earth and the aircraft increases. The motion becomes more and that stretches the space connecting the Earth and the aircraft applying heat to produce extended motion where the extended motion leads to extending of the space the aircraft covers in the same duration of time.

The heat (formed by the release of motion in the engines) allows more motion to apply than that which the Earth generates which puts the aircraft in a higher atomic bracket where the aircraft holds more space that the Earth normally would grant the aircraft. The aircraft has more gravity (granted that it is using the motion of the Earth which is the gravity of the Earth) than it would have being stationary. With an additional source of heat, the aircraft can add to the earth gravity and the Earth gravity is unrestricted motion. It has no bearing on mass whatsoever. Gravity is about motion and mass is the restricting of such motion.

It is not the motion we must be after but what causes the motion in the first place (other than being Newton's pet force). We must find what produces motion and from that then we must think further than what we can recollect from thousands of years of culture that got us this far but is getting us no further. We must see why that which moves or tends to move as we do on Earth in our gravity. It is nothing to do with mass pulling everything about but it is a flow of space-time.

When one applies heat to an object, it expands. That is primary school science. This states that more heat applied, leads to more space acquired by the heated object. In sharp contrast to this is the growth in space when heat levels rise but freezing brings about the opposite result. When I freeze an object, that object reduces its occupied space as it shrinks. Removing heat reduces space. That comes directly as nature responds to heat and I can prove that easily.

By expanding it accumulates space to increase the improving of the size of the material. The accumulating of heat is for the sake of securing singularity, which accumulates the heat in the material whereas the freezing tarnishes the overheating symptoms by the removal of material in unoccupied space using external matter and setting motion to the material until it contracts into a form which we see as visible heat. The heat is in the form of dissolved singularity that became material as material used it as growth. That is why by freezing, it will diminish the space as to accumulate the heat absorbing into the material to maintain the equilibrium needed in space.

The atom is the optimum proof of the statement. The atom is the absorber of heat as well as the release valve of heat. The atom regulates heat in relation to space acquired as well as space required. The atom is as much about heat as controlling heat and when the atom expands space, it accumulates and stores heat. When it cools it reduces and absorbs space. The cosmos is the atom and the atom is what the cosmos use to regulate heat.

Taking this equation of nature to outer space, we seem to confuse the natural law. With outer space as expanded as anything can get, we regard outer space as incredibly cold. As heat sets in, the normal flow will bring about expanding of heat into the form we think of as space that limits the heat overheating. Outer space is the very edge of expanding, of space where heat cannot expand into space any more. Outer space is the limit, the epitome of expanding where heat meets space at the edge of all limits once more. Therefore being the representation of the very limit of expanding outer space has to be the hottest place there is. By applying heat to a kettle holding water, the adding of heat manifests as steam and steam is hot water that traded heat as it reviewed space. By allowing the receiving of the heat to continue the container will let loose steam in order to match the contributing of space.

The manner in which heat expresses itself when confronted by overheating is to provide additional space through expanding of space. Outer space is outer space because outer space has expanded all it can, it is still expanding to the speed of Hubble's $1/H_0$ which inevitably does not only affect far-off places where we cannot be, but effects us on a daily basis. As outer space is stretched to its limit, its limit will continue to stretch but while it is stretching it has to have more than it had before, in that outer space holds the limit of heat expanding possibilities. Singularity has been expanding since way back when but that means singularity is still releasing heat as space-time that turns out as space in the universal time of outer space. In outer space heat cannot expand any further because it is growing with time to become more and the becoming more is part of the Big Bang, therefore except for the continual growth that benefits all singularity throughout on a continuous bases concerning all outer space that is known as the Big Bang expanding, the heat cannot turn into more space than what outer space can permit.

Every element is in relation to the heat level it uses in forming the gravity it has. One can see how the forming of the numbers of elements available in the Universe stands related to the density of the elements total numbers. More pertinent to note is that the effect of gravity is not in the mass of the element but shows a much stronger relation with the density and the density is the relation the element has with the heat that marks a boiling point or a freezing point The density factor shows what we use to classify the element in relation to being a liquid, a gas or a solid. This factor is much more prudent than the mass factor and that I show later on as the book develops.

If singularity expands when heated and there is a limit to the point it can heat, and where that point forms the maximum expanding possible, then it has been reached in the area we think of as outer space. Outer space has expanded through the unleashing of heat, where overheating is turning liquid heat into space. Any explosion is a vivid reminder of this fact and the unleashing of space is so real it destroys the space holding solids by rearranging the construction of the solids. With that in mind we can declare with great confidence that outer space is the hottest place there is. Whatever expanding there possibly is, was done to secure the cooling and all cooling that can be introduced to bring about further cooling was performed in the area we think of as outer space. Forget schoolboy culture and the temperature scales and other Newtonian scientific defects I call "**_Xepted mistakes_**". Think of reality and throw out culture teachings. Use the mind and not the thinking power of the past. Any place that can expand no more is the hottest place there is just because of the shear implication that it can cool no further, is as hot as it gets anywhere. If that is the case then it is safe to say that galactica is freezing cold notwithstanding our concepts of heat and space and heat in space given to us by our collective culture and not by our ability to reason.

The galactica is little frozen islands in a vast see of heat. That is the reason we can see the galactica because the galactica is space concentrated by a frozen space. The galactica is slowly heating and therefore it is expanding into outer space. Outer space on the other hand has expanded to the maximum that it can yet we think it is cold when it is the extreme there is in heat that introduced the maximum expanding. What I now am saying might be deemed by the most purist as the contradiction of the century and that much I do realise. At the inner core of a star all space shrinks into the oblivious but we consider the inner core area of a star to be the hottest spot in the solar system. That just cannot be because when material shrinks it becomes cold and by shrinking into the oblivious it has to freeze into a fusing element as newly formed units. Again that is the contradiction of the century. Why will that be? The space inside the star shrunk to the minimum there can be and that tells us the space has to be cold because of the shrinking that took the space to a position where no space can shrink anymore.

That shrinking of no more space can only be inside the inner star and in that region is where we locate the strongest gravity. With outer space as expanded as nature may allow the space that grew, could only grow in conditions of heat because heat produces expanding and expanding is the result of heat coming about. Space shrinks because it is cold: that we know and taking this law to the star centre it means regardless of our interpretation of hot and cold, that area in the star centre is as cold as it can get notwithstanding what our nature may tell us. Then obviously the same must apply to outer space for precisely the same reasons because it is so hot there it can expand no more.

At this I have to redeem myself from being human. Only in the eyes of humans are there hot and cold, but as a reality in the cosmos we will find this nowhere. We look at the hotness of space and the coldness of space but it is the relevancy to the solidity that forms the actual heat and cold limits. It is so hot no expansion can produce more space in outer space, as the outer space seems to be the epitome of what can be cold while it is truly hot and quite the opposite reveals as the true scenario inside the star in the centre of a star structure. That means the number of protons in motion has a lot to do with the cold and hot scenarios because where the protons are most dense the cold is in extreme…well in most cases. Only in the absence of space can so much heat gather in excess and the opposite is true about outer space where the least denseness found brings about the space in heat found in outer space. Our human selecting of hot and of cold and what is hot and what is not prevents us the clear vision we would have when truly understanding the applying temperature. Temperature comes about from spin and the smaller the spin density is, the colder the space becomes because the more duplication produces the most cold. We think of outer space as 0^0 Kelvin but in fact it is as hot as no other place can be in the Universe. The coldest is where material is freezing solid as material does when frozen solid and the hottest is when by boiling the material is going into a gas with liquid being the intermediate position where heat acquires the space to perform as a flexible substance.

When we look at particles in outer space, we see the particles being frozen. It is because there is such a severe contrast between the particles and the environment surrounding the particles and not the particles that is so frozen. The particles are in a gas state because the particles do not form a part that is part of the space unit. Hydrogen clouds of hundred of light years in diameter are a common sight in outer space. The heat we find filling space is not part of the space but like the particles the heat is a separate issue. That heat filling the space is another form of material that could conduce by diverting from space or marry the union of space by becoming more space. If it were that cold which we think it is, it would not have expanded into such a massive cloud but would have contracted forming a cube of frozen hydrogen. But as we can see, the cloud expanded the gas as far as the gas can expand.

That expanding is indicative of heat and has extremely little to do with gravity or is it just a matter what we think of as gravity? If you are of the opinion that those hydrogen clouds will contract one day into forming a star, well then think again as there is just no such a chance that that will ever happen because that is not the manner in which gravity functions. Because outer space is completely overheating, the condition it has in support of the particles makes the particles appear to be in a state of freezing but the particles is counteracting the heat limit it meets. However, the particles do not contract, as the heat is immense. The space in outer space has absorbed all the heat by means of expanding and will appreciate still further as it will never depreciate. That is not because outer space is freezing the particles but it is because in contrast to the heat of outer space the particles seems to be frozen.

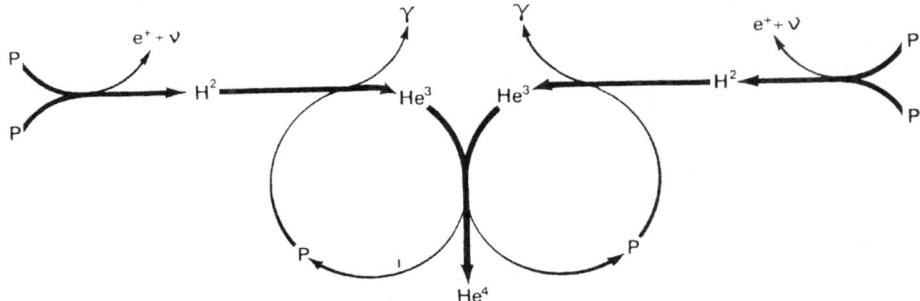

The atom must be the utmost coldest because the proton is even much colder than when the electron can freeze. In fact the proton is 1836 times colder than that which the electron is able to freeze. We find that when cold escapes, it turns to heat and the heat relieves by forming space, however it seems that no one can understand that. Motion brings about cooling. When the spin of the atom allows the cold of the atom to release the heat it had, which it had frozen, the heat returns to space. This is what the atom shows in the electron bands or rings the atom holds. This must not be confused with uncontrolled release of heat. When the motion of the electron is interrupted, such motion reducing results into the utmost expanding there possibly can be. When this heat is released from the containing form of the atom, it brings about much more heat than the human mind can cope with because no human mind can comprehend the total devastation a nuclear release of space may bring forth. In this I am not referring to the normal way material relates to heat. That is a totally different matter altogether.

One may not look at the material and judge the surroundings. The fact that hydrogen remains a gas and so does helium in outer space, must serve as enough proof that outer space is hot, regardless of our interpretation of the temperature gauge telling us what we wish to hear. In the vent of outer space truly being the coldest we have, then hydrogen and helium should be frozen crystals clotted in balls of material. One must look at outer space and judge outer space from the findings only by considering outer space without the prejudgement of teachings about ideas when persons were still held in prison for being suspected werewolves. If helium remains a gas it is hot. However we can witness hydrogen being a liquid in the Sun. The squirting that we see, coming into prominence from the Sun in flowing streams of ejected liquid, is liquid heat. It is heat that became so dense, the heat formed a liquid by the extreme cold it is subjected to within the Sun. The heat is a pure liquid and not a gas. It is the heat that keeps the hydrogen and helium as a liquid substance within the Sun. The liquid is pure heat that is frozen as a form of material in the hydrogen layers and holding the hydrogen in form in the hydrogen layer.

We might think it is hot in the centre of the Earth but that type of thinking is as Newtonian as can be, when thinking of big stars as mighty gravity pools. The removing of heat into a liquid makes the material in the centre of the Earth cold although we see it as being terribly hot. The only reason why it can seem to be hot is because it is cold and in such a cold environment the heat can gather and space can collect heat because the particles find the surroundings extremely cold. Then again we confuse heat and time altogether and completely but more about that later on…

The cold in the Earth centre causes the concentration of heat by space reducing, as all cold surfaces tend to do. When material reduces space, it parts the material from the heat within and places that heat within the electron bands on the outside of the electron bands. By removing the heat, the atom contracts and by contracting the atom reduces space. That heat forming space has to go somewhere. If it was hot the space within the Earth would expand, and the space within the Earth where we think so much heat is concentrated does not expand therefore it must be cold. To gather and accumulate the space in a liquid means it became much colder being a liquid. Finding the surroundings terribly cold will allow the heat to gather and not expand but when the surroundings are hot it will not tolerate more concentration of heat and thus will expand, to rid the balance of excess heat within space. That is the terms in which to think in when thinking in terms of cosmology.

Look at the Sun and see how the Sun turned the hydrogen to a freezing cold liquid at 6500 K. Hydrogen is in a fluid state within the Sun and is colder than the hydrogen that is in a gas form in outer space. The Sun is the coldest place in the solar system. The gravity of the star is vested in every atom within the star and the source of motion of the star is the total motion accumulated as a result of every atom moving in the star. The atoms combine to supply motion by gravity but because the individual motion or gravity is much higher than the compliment of the star, there is an over supply of not only motion but also contraction of heat. That is when the protons oversupply the removing of space to produce the cold that is so apparent. The atoms form a cold that is much stronger than the overall coldness that the star produces. The excess liquid has to go somewhere, since the atom cannot absolve the heat totally. The heat remains as a liquid where the atoms float in. By the reducing of space it concentrates heat to a fluid state by producing the opposing cold. Through much development of the star, the star finally abandons the liquid state to employ a state of solidity where it then starts to discard the atoms in the star. By discarding the atoms we find that the star finally freezes the heat to a solid state. As much as the star contracts, is the space expanding and the one is as much a compliment as it is a result of the other. The expanding of space is a way of duplicating space without reducing space and by duplicating in the form of expanding, it becomes just the opposite to duplicating by motion therefore reducing space by halving space in time. That is what gravity does. By motion, space duplicates and by space halving it removes heat in space as well as by dismissing space. In all the applying of gravity, space bites the dust. The density of the protons brings about space dense enough to harbour the heat in such quantities and visa versa applies in outer space. However, it is not purely the density of the protons that produce such cold but the exquisite motion forming a rapid duplication of material and such duplication brings the contraction by removing space. Removing space is also removing heat that is separating material.

We have to accept that the coldest place in the solar system is in the very centre of the Sun because there the most number of protons sharing the least amount of space produce the coldest area that can be, therefore the hottest density of heat within the cold environment. Later I will show why the star is so extremely cold it freezes material together and outer space is over boiling with heat expanding into more space. We have to see what forms space and why space can be the absolute basic container through which gravity can relay the influence it carries.

We have to realise that whatever forms space, has to be that same ingredient which also is the basic component that forms the lot of everything in the entire Universe. It is that, which becomes more making everything seems more and it is also by removing that which reduces every aspect of the Universe. That which becomes more is what the Universe is built with and it is that which the Universe uses to form its entirety. When particles heat up, the particles expand the space that the particles holds to the limit which the rising heat demands in relation to the heat rising. The particles claim more space when heated to preserve the cold that the material is protecting. The claim to more space produces more space but that in turn reduces more heat exaggeration. Such expanding brings about cooling. When particles heat or cool motion applies in some form. Regarding this fact, we can claim that motion started at a point when the Universe was extremely hot and there was no space. However, I have indicated that hot and cold are only factors with little specific or formal value in the Universe. By introducing motion, space formed and the lack thereof produced friction that became heat that became space. That must be the way the Universe then started.

The application of gravity is the same as the condensing of space and bringing about heat by the compressing of space we apply in the way we go about tapping into the energy that nature provide. Internal and external combustion engines all rely on this application for harvesting motion by driving power. The heat coming about inside the cylinder when being compressed has no relevance to particles colliding because all compressor cylinders cool down or become colder when that cold escape through the walls of the cylinder. As soon as the pumping stops, the heat releases from the inside space. There is an immediate stopping of the increase of heat as soon as the pumping stops. The material inside the container forms a secondary form of material that comes about since the space reduces and the forming of space is in a turnabout. The compressing of the space inside brings about a rise in the heat levels within the container but apparently that no one in Newtonian circles can understand. By compressing, the spin of the atom increases and the motion of the material remove additional heat from the ranks of the inside of the atom. Thus, when the spin of the atom increases, it allows the cold within the atom to release the heat the atom holds into uncontrolled space. This releasing of heat and unifying the released heat once again with space, increases the levels of heat in the atmosphere of the containing cylinder. What that heat does, is the heat that the material absorbed as material within the atom was captured as frozen heat, because of the motion of spin to space that the atom holds remains in a frozen state under the guard of the spinning electron. But when this heat is released from the containing form that is the atom in being the biggest cosmic heat container, the heat changes to a frozen state through the motion within the atom. Forming a frozen substance by producing motion that is faster than the speed of light, the heat is frozen by the spin of the electron. The spin of the electron brings motion and such motion reduces the heat to a frozen state which is the frozen state of heat we named material. Therefore one may not judge the element by looking at the material and the state of form on the grounds of its surrounding, which is then measured in a specific temperature. Parting atoms is heat that surrounds its electron to the outer side of the containing spin.

Again we must look at the state of material in outer space and realize that the fact that hydrogen remains a gas and so does helium in outer space must serve as enough proof that outer space is hot, regardless of our interpretation of the temperature gauge telling us what we wish to hear. One must look at outer space and judge outer space from the findings only considering in the terms which outer space insists upon. If helium remains a gas it is hot. The removing of heat from the space that contained the heat makes the centre of the Earth cold. In our Universe we see it as being terribly hot because the heat then forms a separate substance but remains a form of material but that is because we see the heat and not the space derived from the separating of the heat.

The only reason why the space can seem to be hot is because the space is cold in such a cold environment that rejects the heat within the atom. There the heat then must gather in a more concentrated state and space can collect heat because the particles hold concentrated heat in the space separating the particles. By removing such high concentration of heat from the space that used to be expanded heat, the space then must contradict the heat by being extremely cold. We look at the heat in the space, which by being in a liquid state should be by our standards considered as another form of material and find the surrounding heat in the space hot while the atomic material in space is extremely cold. The cold in the Earth centre causes the concentration of heat by space reducing, as all cold surfaces tend to do. But the numbers of protons contribute to the reducing of space and the removing of heat captured by the material. If it was hot, the space within the Earth would expand and explode, but the space within the Earth where we think so much heat is concentrated, is so much it does not expand therefore it must be cold within the solid parts. It is the motion of so many protons in such a little space that allows the heat to be contained as a liquid and the extravagant motion by the many protons in such a reduced area forms the ability to contain the heat as a liquid substance without allowing the expanding of the heat into gas or space. To gather and accumulate the space in a liquid means it became much colder when the space parted from what then is being a liquid. Finding the surroundings terribly cold will allow the heat to gather and not expand but when the surroundings is hot, it will not tolerate more concentration of heat and thus it will expand to rid the balance of excess heat within space. The concentration or release of space with heat or space from heat is a direct contribution of the motion controlled by the space-time. The regard of the space-time providing the motion, which provides the cooling of the space, stipulates the conducting of heat in space or the release of heat to form space by means of seizing the occupied space.

Look at the Sun and see how the Sun turned the hydrogen it holds and which is captured in its atmosphere to a freezing cold liquid at 6500 K. Hydrogen is in a fluid state within the Sun and yet it is still colder than the hydrogen we find in outer space that is in a gas form in outer space. That must be because of the enormous motion of the particles within the confinement of the Sun. The Sun is without any doubt the coldest place in the solar system and that is because of the ferocious motion within the Sun. I repeat that, that is when the protons oversupply the removing of space to produce the cold that is so apparent in the heat levels that do not join outer space. By the reducing of space it can concentrate heat to a fluid state. By producing the opposing cold that finally freezes the heat to a solid state we find that it is what matter is. The expanding of space is a way of duplicating space without reducing space and by duplicating in the form of expanding it becomes just the opposite to duplicating by motion therefore reducing space by halving space in time. That is what gravity does. By motion space duplicates and by space duplicating the material must be by dividing or halving. Halving the material, which is heat, is at the same time doubling the space, which is bringing about cooling. By doubling the space as the duplicating of material removes half the heat from a single space and distributes that same quantity of heat over a double amount of space, it removes heat in space as well as by dismissing space and in that concentrating heat. Again it is apparent that in all the applying of gravity it is space that bites the dust. The density of the protons brings about space dense enough to harbour the heat in such quantities and visa versa applies in outer space.

We have to accept that the coldest place in the solar system is in the very centre of the Sun because there the most number of protons sharing the least amount of space producing the coldest area that such intense motion can allow therefore the excessive motion brings about the hottest density of heat within the cold environment. It is the duty of scientists to look far beyond the ordinary and find why the inner star will be so cold and as to why outer space will be so hot while being seemingly so utterly cold or hot in humanly applied standards. It is the duty of the professionals to find matters as they are and not as they would seem to look from a human vantage point. Later I will show in much better detail why the star is so extremely cold and outer space is over boiling with heat expanding into more space. We have to see what forms space, and why space can be the absolute basic container through which gravity can relay the influence that it carries. We must come to realise that whatever it takes to form space it has to contain something that is the same ingredient, which also is the basic component that forms the lot of everything else in the entire Universe. When particles heat up the particles expand the space the particles hold to limit the heat rising.

The particles claim more space when heated to preserve the cold. The claim to more space produces more space and reduces more heat. Such expanding brings about cooling. When particles heat or cool, motion applies in some form. Motion started at a point when the Universe was extremely hot and there was no space. By introducing motion, space formed and the lack thereof produced friction that became heat that became space. It is natural and it is simple and above all it makes believable sense.

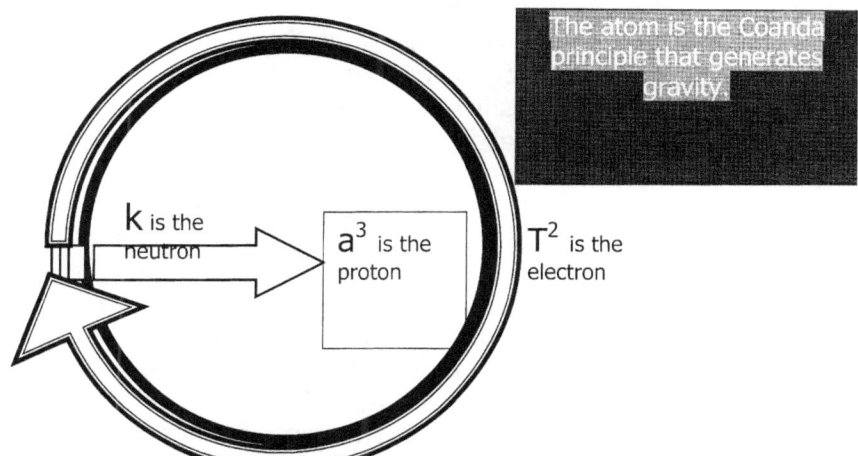

The application of gravity is that which condenses space by bringing about heat with the compressing of space. Compress space even today with a piston cylinder wall in an engine cylinder and then from that action pump the compressed air into a container and such confining of space will increase the heat by the piston effort to reduce the space brought about in the container. The heat coming about inside the cylinder has no relevance to particles colliding because all compressor cylinders cool down with time moving and not necessarily with the loss or release of particles. It is not only the discharging of air that will reduce the temperatures inside the container but the time flowing bringing motion about where the motion is not about particles escaping but heat escaping in the replacing of the heat density (not the density of the particles forming the material content within the container) but the space that compressed to heat will also bring about that the heat displaces through the container wall to the outside. After the pumping of air increased the heat in the cylinder, which even can go to dangerous levels, the heat will reduce back to room temperature when further pumping seizes and that stops further air movement into the cylinder and such surging of pumping air is what brings about heat stabilizing.

Mainstream physics ignored the clear connection completely, notwithstanding it being so very obvious. There is this far in their recognising of principles in natural physics not one single reference made to prove their appreciation of this matter. They are bent on particle colliding notwithstanding the nonsense such an idea promotes. Atoms cannot touch simply because electrons are all negatively charged and will therefore repel one another long before there is any possibility of touching coming about. However, in the case when particles do collide such a collision forms an atomic thermo release and that action we call an exploding atomic bomb. What principle this argument about particles colliding ignores, is that all atoms use negative charged electrons forming the atomic limit on the outside, forming a definite border to the boundaries of all atoms in both electrons from different atoms are being negatively charged. In being negatively charged, it means both will come out and one totally rejects the other as much repel the other or cast the other away. The closer they come the more violent the rejecting will be and such rejecting is the production of heat that will turn to space. However that rejecting will increase the motion and the increased motion will reduce the space occupied. The electrons repel other negatively charged sub atomic structures, which the electrons are that form the outer borders of all atoms. With all electrons highly negatively charged (being as negatively charged as any possibility will allow to match the utter extreme) such electrons couldn't touch. When the pumping of the air container commences, the balance at first favours the forming of heat from the space coming in and being reduced in the containing size they are squeezed into, that is reducing the space from what it was on the outside. The space distribution inside then changes considerably and reduces a great deal compared to conditions outside the cylinder wall and with the decrease of the space distribution inside compared to conditions outside that space then becomes reduced and charges with excess heat on the inside.

The electrons will disallow any direct contact between atoms. No force can be big enough to enforce such touching. It is because of that contact rejection electrons bring about that science has to use an overload of neutral neutrons putting them in the atom nucleus to fake a complying of charges that will eventually lead to atom touching each other but that is through enticing a neutral stance which is enticing a positive overload for a short while. When the touching of electrons do take place, the event is called a thermo nuclear reaction where heat is released in unmatchable quantities and the atoms in reaction dissolves into a liquid heat. The increase of heat by the distribution of particles in the space that is forming the connecting space still keeps particles separate. The heat rising is a separate issue that has nothing to do with contained particles colliding because why does it stop when pumping is seized. This ratio of heat reduction is time connected as much as it is motion dependent. Motion reduces space by expansion as much as time contributes to space distribution by allowing the flow of heat. When the pumping stops the heat immediately starts the reducing thereof. Most important is the realising that every atom constitutes of two parts. In fact the entire Universe constitutes of the two parts, which I go about mentioning in this entire book. On the inside of the atom there is a circle formed by a rotating electron that contains the outer wall of the atom forming the sphere and holds material in contact with the protons. On the outside there is heat surrounding the inner material part within the sphere and distance the inner material from the space between it and the next atom. The electron forms the division between heat uncontained and heat contained. This is why the Roche factor is so very important. There can be friction between particles in reduced space under controlled circumstances where such particles are grouped together in a unit and as a unit elects a group singularity forming the centre of the chosen form of the unit.

The Universe separated heat from material by covering the exterior of material with heat that forms space. Some material became softer by uncontrolled overheating while others remained more solid by containing form through controlling the overheating. On the outside of all elements there are a layer that is the heat the element uses in relation to place relevancies between such an element and the rest of the cosmos. In the case where many atoms form a unit such as an aircraft coming in from outer space the space surrounding the craft becomes liquid heat as the space becomes more intense within the atoms combining as the structure in concentrated space that forms heat. In an aircraft coming in from outer space at altitudes that high, there can be no particle in friction and even more so way up there in the atmosphere at the altitude where the cosmos meets the atmosphere just because the particles up there are so sparsely distributed in that part of the atmosphere. Above and beyond this lies the fact that all the so called air particles are very volatile and excitable by nature and they are known to turn the slightest heat into rapid motion, thus establishing a scene where the particles that supposedly are in contact with the aircraft sheeting will move away from the hot incoming aircraft. The gasses will become more gasses when the heat levels surge. If then there can be no friction, then it is because the particles are highly volatile and exceptionally sensitive to heat. Airborne particles are prone to motion just because it is the airborne element's nature to change heat into motion and the motion comes about from their sensitivity to duplicate. No particle in the air being part of the space we call air, which is in free floating in that air, can produce friction because of the volatile nature that those elements have. The craft's coming into the atmosphere produces a point where $a^3 = T^2k$ changes to $k^{-1} = T^2 / a^3$ (the explanation is forthcoming a little later on) The distance separating the incoming object from the Earth centre reduces rapidly, therefore the object starts to cescend towards the centre of the Earth. We must also acknowledge the fact that there is one specific point of specific entry where this will occur more than before.

That point will rapidly increase the time factor where the incoming object crossed such a very visible border. By the reducing of distance k, space a^3 will have to compromise in the relation of all the factors forming the equation, since T^2 will very suddenly grow more acute. What happens is that the applying gravity reduces the space a^3 and the compromising factor comes about since the time factor T^2 moves back to a time where outer space was as dense back then as the density we now have within the atmosphere that then became the Earth atmosphere. It is outer space that remained denser than what the outer space currently is. I am now referring to a process that I introduce as this book unfolds which is by nature completely different to what is accepted by mainstream science (as you might have noticed in this short space of reading). That area, which I now refer to as being outer space, was back then the same density as that which the Earth now supports. As the Earth moved apart from outer space while establishing individuality, outer space in the meantime expanded. Outer space became less dense while the motion that the material that forms the Earth structure provided, came about. The parting of the Earth from outer space came at a point just befcre the Earth established an atmosphere. The atmosphere we now enjoy, was outer space when the Earth had no gravity to place an atmosphere that stood apart from outer space. The atmosphere we now have is that which grew through gravity and by the measure of the Earth gravity, the atmosphere remained as dense as before while outer space became separated from the atmosphere. While the gravity of the Earth contained the space surrounding the Earth in a much denser packed envelope, the area not under the direct influence of the Earth governing gravity became more spacious.

The Earth contained its atmosphere and it relatively grew much denser as the solar system developed into what it is today and outer space reduced its density. It is a matter of the kettle not being able to call the pot black. As the atmosphere released from what we think of as outer space that releasing from outer space made the atmosphere much denser in the space just above the Earth, which is using a reducing time factor. It is there that the applying gravity makes the Earth atmosphere more compact. That established the T^2 factor to be that more condensed when one compare in ratio the density with outer space. The density that was there at the time when the separation came about in outer space when such parting between the limit of the atmosphere and the limit of outer space separated and such separation allowed outer space as a separate object to move away. This parting brought a barrier that is in place between the Earth and the outer space and any object coming from outer space into the Earth's atmosphere will have to negotiate its entry by passing through that division. The incoming object then would have to reduce the measure of the space the craft holds as the containing singularity sets new standards applying to the incoming object with which the craft then needs to confirm its form and its status within the contained space of the Earth.

 The reducing will then suddenly no longer use space as the compatible factor but the focus will shift to the time factor that dictates to the space what the space can be. Such reducing comes from the switch there is in space – time where it was in outer space performing as being $k = a^3 / T^2$ to what it has to be within the Earth atmosphere $k^{-1} = T^2/a^3$. There are two ways of looking at the event when the atmosphere grew apart from the outer space. One can think that outer space expanded by the implication of the Hubble constant or that gravity withdrew the atmospheric space of the Earth at the time that the parting of space came about. But however you look at it, there was a time when both outer space and the Earth's atmosphere shared equal density as we find it still applies on the moon and on Pluto. Then the Earth became dynamic and now they do not share any density at all. Things were overall more compact back then than at the present time and that included all things in the Universe. The space component is reducing the time component by compacting space to alter the space – time ratio.

This is portrayed by Kepler's formula $a^3 = T^2 k$ It shows space as the density of space decreasing. The Earth still compacts space by reducing the volumetric confinement of space $T^{-2} = k / a^3$. This we call the atmosphere, as the atmosphere becomes denser towards the soil of the Earth. There is a change in the time component. Most evident of this is when studying the pendulum. Just as we can see in the pendulum swinging, we can see that the swing reduces. Such reduction is because the space diminishes by reducing through compacting every time the arm rocks from side to side. With this there is proof that in the developing atmospheric space of the Earth the ratios change from outer space. This is proved by the pendulum arm that Galileo's experiment used to show that the swinging pendulum indicates $k^{-1} = T^2/a^3$. Further more, it proves that Galileo was correct after all and unnoticed by science; Kepler helped Galileo prove Galileo's point. In this the net outcome establish Kepler as being correct and the Newtonian argument of friction brought on by gasses falls apart. At that altitude where such friction supposedly should take place, the material in friction is not even present in the atmosphere.

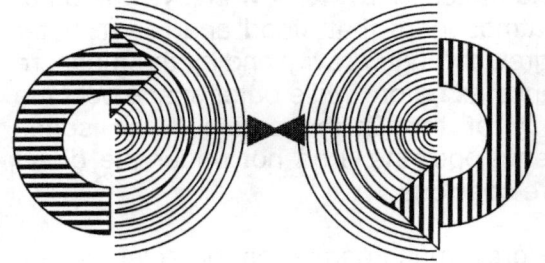

Nevertheless science will stubbornly cling to the old theory with persistency that would warm any warring Field Commander's heart. In retrospect the following information is established in the past few pages: Every element stands in different regard to the heat surrounding the material, which makes us consider the material to be either a gas or a liquid or a solid. The material in every element there is as such is all three forms and not one of the forms in particular. It is the way under which the circumstances is presented that the element allows the heat to gather and accumulate as the surrounding heat occupying the surrounding space. Every particle is unique in the way it regards the heat to material ratio and how much heat it uses to form either the gas liquid or solid state. The fact of being a gas or liquid or solid is so much more complex but in time we will get to that explaining. If space a^3 declines then so will motion in relevance have to compensate by reducing k and limiting T^2 because space a^3 must always be equal to motion $T^2 k$.

Space shifts as heat releases space and converts the Universe in one direction bringing about expanding into more space but less dense space. Remember how the heat came down from 10^{34} to 0 K at present? The density of heat in space surely diminished considerably since then to now. Gravity on the other hand is exchanging heat through the concentration by removing space bringing about space loss with increased density of particles and therefore heat concentration. In the centre of all spheres, which all stars are, it is hot. In fact the heat in the centre of the star is the product of the space it concentrates to form heat and in that we can read the gravity the star can produce. The ability to secure heat by reducing space becomes the measure of the star. Momentum is the second form of gravity symbolised by Kepler, as k. The Big Bang is the result of heat expanding into the forming of space. Gravity, on the other hand is about concentrating space back to heat, and take recouped heat through to material, acting out a balance of expanding while contracting.

This way gravity is applying the onset of the Big Crunch by destroying space while space is converting heat to material occupying space. The Big Crunch is coming about because the Universe is expanding where the two processes are one principle.

The relevancy there is between the aircraft and the Earth is precisely the relevancy we find between the proton and the electron in the atom. When heat released provides more space between the aircraft and the Earth, the distance between the aircraft gravity relevancy and that which the Earth allocated to the aircraft by only providing Earth gravity without the adding of heat by the aircraft allows the aircraft to respond exactly as the electron does in the case of the Atom. The aircraft falls into the role of the electron, the atmosphere takes up the role the neutron has and the Earth retains the aircraft therefore the Earth has the role of the proton. When the aircraft has more heat than that which the Earth provides through the atmosphere the neutron position has to expand in order to facilitate the new dimensions, which the additional heat that drives the aircraft provides. The ratio that the Earth initially holds becomes stretched as the aircraft suddenly finds more heat and therefore more motion that becomes more space between the allocated position and the position the aircraft claims by individual motion in addition to the motion the Earth has provided. It is all about heat released that generates motion and motion provides space differentiation.

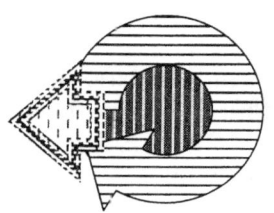

Throughout the entire cosmos is leaning on the four pillars which is the phenomena and the four culminates in one which accommodates all the others and it is the Coanda principle that establishes space which provides the gravity which allocates the motion a position within the space that forms. This very principle of electron / proton is in gravity. Gravity I shall prove is motion and not mass inspired. In fact mass being a factor corrupts gravity by restricting motion. Gravity is anti mass and mass is anti gravity because the neutron is all motion with no mass. The gravity of motion is heat driven because it is heat that drives gravity. When an atom is in outer space it is surrounded by an atmosphere of 0 K. That puts a limit on the atom as far as structural differentiation goes.

$$T^{-2} = k / a^3$$

$$k = a^3 / T^2$$

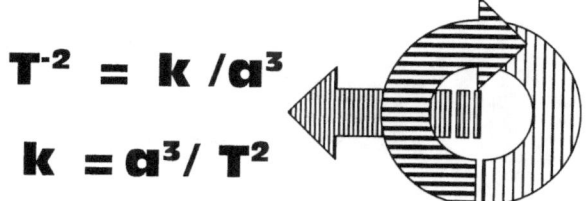

By looking at the construction of the Coanda effect we find that the space takes the liquid as an extending of the space that increases the domain the space claims. The space is always the solid acting factor that holds

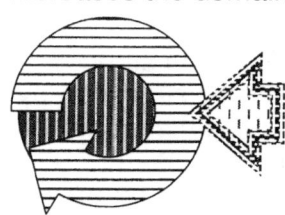

the space a^3 in relation k to the liquid T^2 and the Coanda effect is the personifying of Kepler's formula stating that the space holds a direct value to the motion connected to the space $a^3 = k T^2$. When putting Kepler's formula into the correct connotation, the Coanda principle is the materialising of Kepler's formula. The motion of the liquid limits the space by adding the motion to the claimed space.

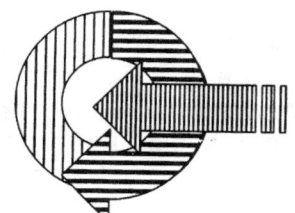

Take this formula into context from the liquid's point of view and we find that there is quite another and opposing connotation to the same formula. From the liquid's perspective, we find that the liquid T^2 attaches k to the space a^3 by adding one more layer to the unit $k^{-1} = T^2 / a^3$ and the motion T^2 is an

$$T^2 = a^3 / k$$

$$k^{-1} = T^2 / a^3$$

addition k to the space $= a^3$ by measure of $T^2 = a^3 / k$. By removing the extending that the motion T^2 of the liquid offers the space, this reduces the space by the value of k.

$T^2 = k / a^3$. This means the liquid extends the boundary of the space while the space includes the liquid as the motion attaches to the space.

The inside as well as the outside must be zero Kelvin because outer space has no other scale that being zero Kelvin. When the atom is on the Earth the relevancy goes that the atom is 40^0C, which is 313 K.

If the outside of the atom is 40^0 hot, then the inside of the atom must be 40^0 cold. The heat on the outside must generate a condition on the inside, which opposes the condition on the outside. The inside is in relevancy or in division of the outside because there is a mass differentiation of 1836 times.

It is true that when concerning the Earth and outer space, standing in comparison with a true star such as the Sun, one can see how little there is to choose between the Earth atmosphere and outer space. It then becomes evident how much time a star, such as the Sun took to develop in contrast to the measly five hundred million years Newtonians add to the sun in development. When comparing what changes took place when the Sun departed from outer space and by comparing the time it took to develop the Sun, one can see that the Sun is a multitude of billions of years better developed than the planets. Look at what is occurring in the atmosphere of a star, and then think how long it will take the Earth to generate the gravity to sustain an atmosphere as intense as the star and start laughing at the Newtonian view that the Sun is five billion years old and the Earth is four point five billion years old. That is rubbish…that is Newtonian calculations…need I say more? The Earth is as close to outer space as any development would allow and calling the Newtonian calendar hogwash is about the term that common civilized decency will allow.

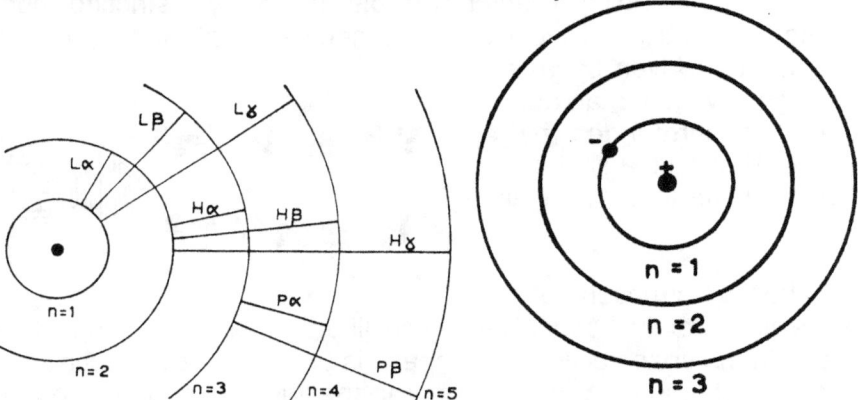

When an atom finds a location in a minor star such as the Sun, we are filled with surprise. It seems to us that Sun is very hot and with the Sun that hot the atom has little validity to stay intact. The atom should explode being in such a hot environment, and yet it is there and it is very much undeterred. Any atom we would heat to a temperature of 6500^0 C as the Sun's temperature is will destruct with an enormous bang. Well it is good and well to say gravity keeps it from destroying but when saying that, we should use that as a clue and not as an answer. It puts what is in the Sun in another class of structure and confinement. If the temperature on the outside of the atom rises to 6500^0 C, then the temperature on the inside should respond to what applies on the outside. It is quite true that when temperatures rise, the electron jumps a band. The electron moves apart from the proton as the circle widens. It is not the amount that the circle widens that should be of any interest to us but the total response. On Earth the electron ring would enlarge but at the same time the proton should equally respond by reducing. Place the atom in the circumstances we find in the Sun where the atom heats to 6500^0 C. On earth the atom would explode but in the Sun the atom remains well formed and very intact.

The atom does not explode because the atom does not get bigger and extends to outside its proportions. In such an event where the heat rose enormously and the atom remained as it is in outer space, it would mean that the atom therefore must have gotten smaller because the enormous atmosphere kept the atom in tact. Yet with such temperature rising there has to be a change to the form the atom has and that means the atom shrunk in size. The proton became smaller when the temperature rose because the atom had to respond in some way to the rising of the temperature. Putting all this down to gravity is tiresomely attributed to laziness on the part of the human thinking capability because it proves how far we will go to restrain our ability to think. If gravity controls size by heat contribution then gravity has more to do with temperature than it has to do with what mass contributes.

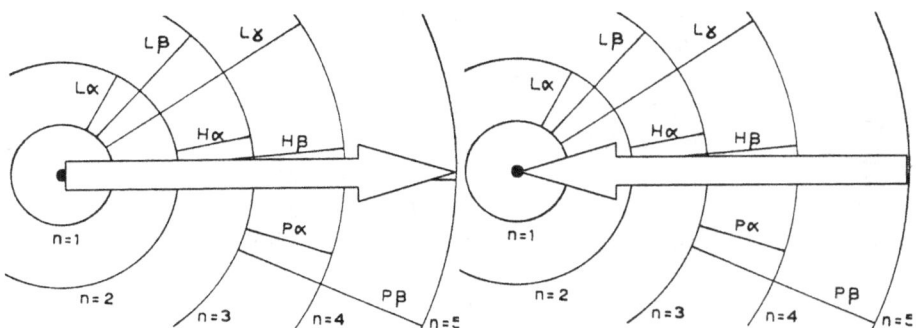

What we see as heat is relevancies because as the relevancy within the Sun changes the atom adapts to the changes. The atmosphere of the Sun becomes denser, which we see as being hotter and the containing becomes stronger. The atom has to reinvent it by adapting to the changes or different surroundings. In this manner the motion that the star provides, which is so much more than the motion we find in outer space, that the hot / cold dynamic changes all together.

When an object is in outer space that objects encounters a specific relation with what we presume is space. This comes about by motion and through material volumetric size. The space the object encounter, by moving through outer space puts a value of a ratio between the space it moved through and the space moving through which Kepler introduced as $a^3 = T^2/k$. That means there is a contact ratio between space containing and space contained by.

When the atom is in outer space, the atom is surrounded by a temperature of zero Kelvin and that is because as far as human thinking goes, zero Kelvin is what humans would presume to be the coldest any temperature can get. Being zero Kelvin on the outside and with zero Kelvin being the coldest temperature there can, it would make the atom also zero Kelvin on the inside since there can be nothing colder than that. That would mean the entire atom is then zero Kelvin.

However applying motion reduces temperature and there is much motion going on inside the atom. That means the fact that zero Kelvin produces the coldest there can be makes a little nonsense of such a statement. When the atom is 40^0 C the outside of the atom must affect the inside of the atom because from the fact of what the Balmer and the Lyman series would represent, proves that the outside temperature of the atom does influence the inside temperature of the atom. The normal summer's day temperature on my farm normally in the shade is 40^0 C. I suppose they measure the temperature in the shade because at that temperature it would be a little loony to venture outside of the shaded area. We consider that the atom must be 40⁰ C because that is what the daily temperature is outside the atom. We feel and experience the 40⁰ and we presume that all around is suffering from the heat of 40⁰ C.

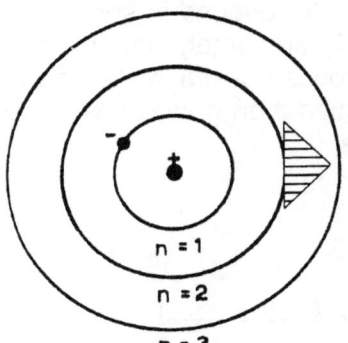

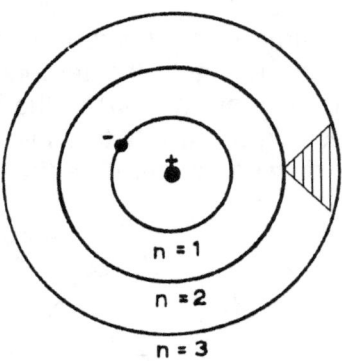

We know that the action brings about a reaction and the actions leads to a response. If the atom heats on the outside by measure that it finds a need to reposition the electron by one band, then also the inside got smaller in relation to the growth by one band. We associate such repositioning with the heat on the outside to amplify or reduce. However, the adding of heat brings on a faster flow of liquid, which results in higher motion and it is in the motion that we find the answer to the cosmic principle applying. In the cosmos there is no hot or cold. There is higher or less motion.

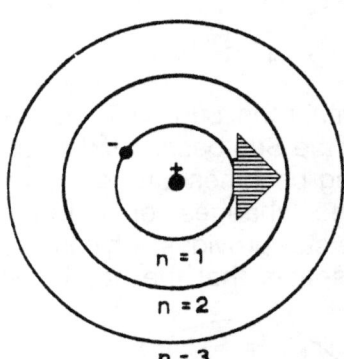

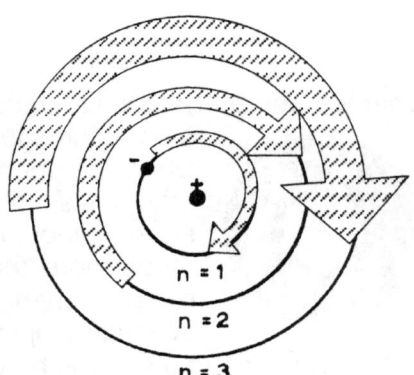

The relocating of the electron into a new position where the electron jumps a band is done by implication of the Coanda effect. From the Coanda effect we know that the liquid attaches to the solid using the formula $T^2 = a^3 / k$ where as space identifies new boundaries by identifying the allocated boundary set by the liquid as $k^{-1} = T^2 / a^3$ where the space then forms the limit at $k = a^3 / T^2$. Every time the motion of the liquid intensifies the motion will attach to the solid by applying a new relation, which alters the relation of the solid by extending the space the solid has differently.

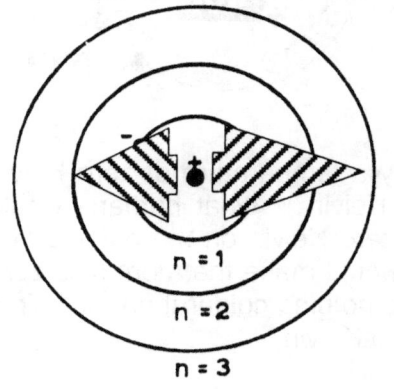

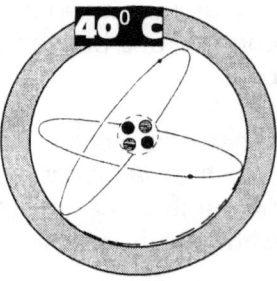

However, we must not lock our focus on the heat but we must refocus on the motion that intensifies or weakens. It is the motion that produces the new electron allocation and the motion produces a heat that establishes a cold. The focus is on the motion because the motion brings on accelerated duplication and accelerated duplication produces cooling that brings on a relevant cold within the atom.

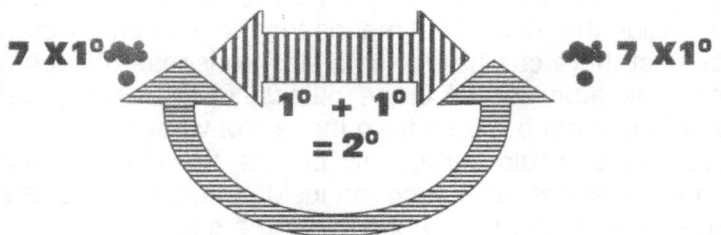

Heat and temperature is just a shift in relevancy that places singularity in infinity apart from singularity in eternity. The space in between that which has no start and that which has no end is a position allocated to time.

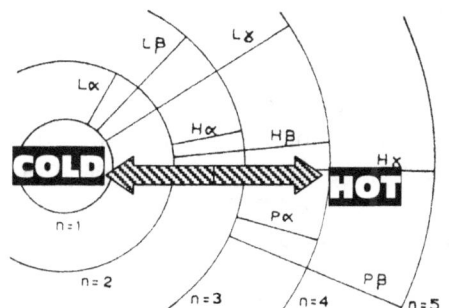

If the temperature on the outside of the atom changes from zero Kelvin to 400 C it is not the temperature that changes but the atom is responding to higher motion. With the atom in outer space the atom is subject to lesser motion since the atom is only in a distinct and personal orbital motion in relation to the Sun. That is why the atom can be subjected to zero Kelvin. When the atom is within the boundaries of the Earth and circling around the Sun in a location set by the singularity of the Earth, the motion is distinctly more than what it would be if the atom were located in outer space.

The outside of the atom calls for a direct response to condition inside the atom since the outside can change very little if the inside does not respond in an opposing manner to what the outside produce. In such a relevancy there are always three factors performing as gravity and in that is the Coanda effect in charge of committing the standards by applying the gravity or the motion in relation to the solid.

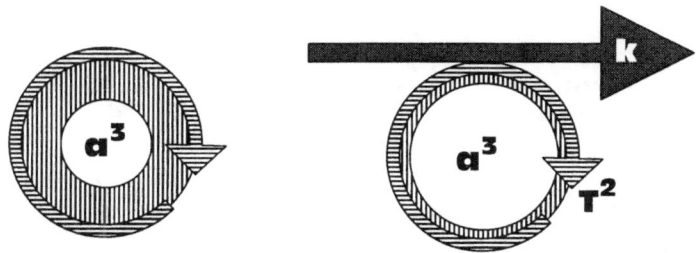

The material revolving through the space holding the material and allowing the material the privilege of motion is in the amount of material per time frame that makes contact with the space which serves time and that it encounters as the space duplicates its position it holds coming from the past through the present into the future

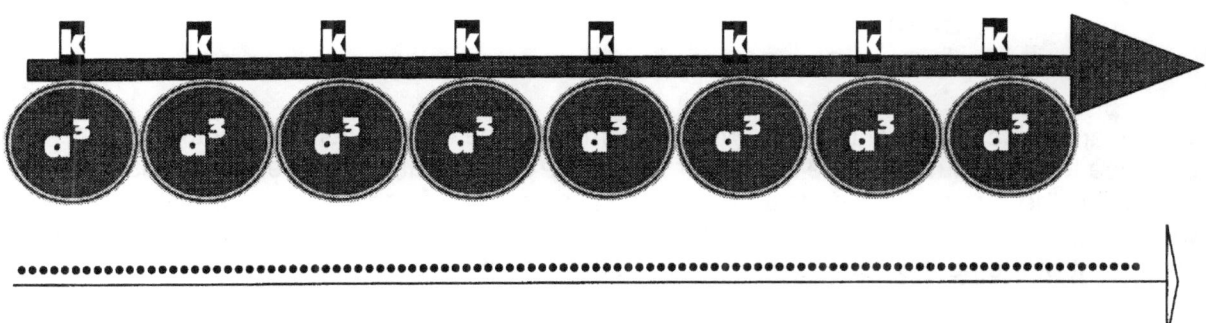

The movement reduces the size the material occupies by duplicating such vat amounts that the duplicating freezes the material into the oblivious.

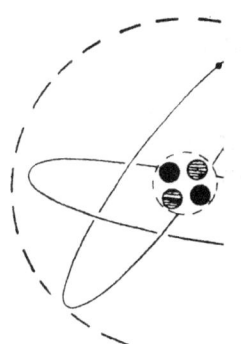

There is a definitive relevancy between the electron and the proton and that factor is what fills the neutron. The neutron is unrestricted gravity or liquid motion whereas the proton as well as the electron is very much restriction of motion of space-time flow, hence the mass. It is proposed that when the atom becomes hotter the electron jumps a band but that statement is not altogether the truth. The proton shrinks as much as the electron jumps a band just as much as the neutron fills the vacant space.

By jumping a band the space within the electron becomes more and the neutron fills that relevancy therefore the neutron becomes more. But if the neutron becomes more the neutron is there to bridge the gap between the electron and the proton and that will have it that the proton needs to respond just as much by becoming colder in the presence of the electron facing more heat. The heat is not the factor but the motion contributed by the heat is what brings about the larger jump in spin.

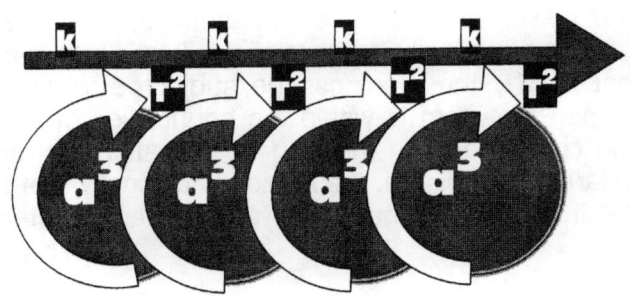

The neutron facing off the electron as well as the proton will respond on both sides of that which it influences because the response is that of bringing over more motion from the electron to the proton. One cannot gauge the electron's behaviour without extending such behaviour to the reaction that the proton would have since the neutron fills the gap and also provides the response on both sides and the changes is what the neutron contributes by suffering the greater discrepancy in changes. However, in the ratio or relevancy there will never be any change. The changes come in the form of an amplifying of the motion, which is a relation the space has with time.

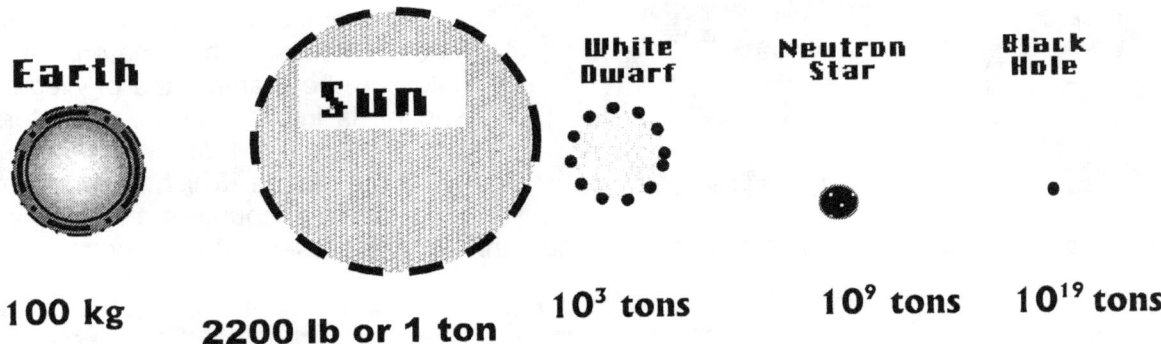

When an object is in a location with little motion the duplication presents a lot of heat because the distribution of the heat over the space in duplication has very little possibilities of spreading the overall heat over a wide area. The motion of something as small as the earth will confine the atoms into a relative hot area since the space in duplication does not reduce the extent of the heat by distributing the heat over much space.

In a structure with the size of the Sun, the motion of space is enormous by the sure quantity of space in need of duplication. Shifting that volume of space needs duplication that is millions if not billions of times more extensive than that which the Earth may produce. By duplicating such a vast area in a period, reduces the individual atom to a fraction of what the situation on Earth would allow. The more the spin of the liquid is in relation to the solid state of space, it reduces the space and it extends the material by such a quantifiable measure, that it is many billion times more than that of smaller stars. It is not the space that holds the matter but it is the spin in relation to what the matter holds that puts the relevancy of hot and cold within the star. The more cold there is because of the more liquid heat bringing about motion, the colder would the atomic material be and the higher the relative contracting gravity that the star produces. This we see in the admitting of Mainstream science confessing that the reducing of space produces an increase in mass and because mass is the frustration of material unable to move, it admits to the fact that mass in volumetric size has no influence on gravity.

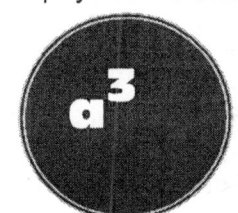

The physics we encounter on Earth allow us to use a common and a constant, a fit all and an all-purpose because we find us captured by the Earth singularity. The Earth provides the space we may claim as well as the time in which such material duplication will take place. The earth does not provide conditions found in outer space and neither are the conditions found in the Sun remotely compatible with the conditions the Earth prescribes. On Earth we find conditions little different from the conditions applying in outer space.

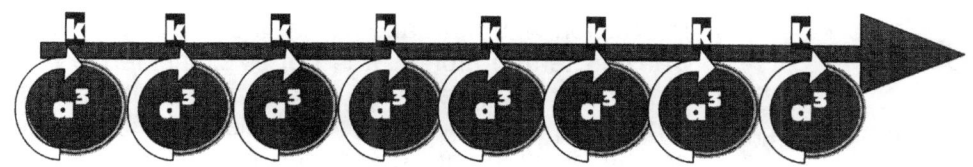

There is a certain ratio of heat to space that allows the material the motion to reduce heat to the extent that will grant the material a certain volumetric size in space-time. The material is hot because it is holding large quantities of heat in the structure of the atom. By moving slowly, more space is duplicated in less instantaneous time, which is allowing less heat being distributed in a longer eternal duration with less interrupting of the duration that is covering a larger area with a smaller interval to duplicate that which will reproduce the material. By the motion the conditions give a specific ratio of time that allow space to duplicate to that specific required ratio. Since outer space is as hot as time can be, there is no more expanding of singularity possible in outer space.

We call this dynamic speed or velocity, which is just another name for a motion in ratio with time. There is a volume of filled space (material defined by time in cubic meters) which is moving through time by using time (seconds flowing in relation to the meters moving) and that ratio produces the size by which the object is measured, that is in relation to the time the object allows the ratio to be in contact with the time the object moves in distance. The space stands related to the time it takes the space to flow over the material and this is also known as wind resistance in some cases. We also know by blowing over a body the "air" cools the body. That means the more "air" that the body is in contact with, the colder the body will get. To this argument there is a lot more and later in this book I return to the matter. However it is a ratio that is coming about.

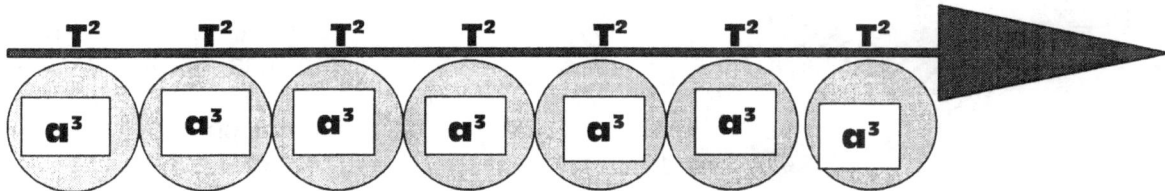

If the motion of the material is more it is more in contact with "air" which is not "air" or even "space" but it is time, the time (or space or air) effects more of the material since the same volume of material moves through more time. The time is a constant and therefore the material cannot increase the time but the ratio can produce more material (in contact with time) than moving at a slower speed. The material reduces in relation to the time it moves through and therefore the material shrinks allowing heat to flow from the material to the time aspect. The same, when compressing air into a container used for storing compressed air.

That means the relevance between "space" which the material moves through or is in contact with, reduces the size of the material in ratio to the space it encounters. But we know that this effect the heat balance more than the size because the material moves through more time therefore more heat is transferred from the material to the space surrounding the material. It is for this purpose that we blow or radiators with fans. By the excessive motion of the massive Sun the material reduces allowing the material to become so cold that the space outside the atom, becomes 6500^0 C in a normal day on the Sun.

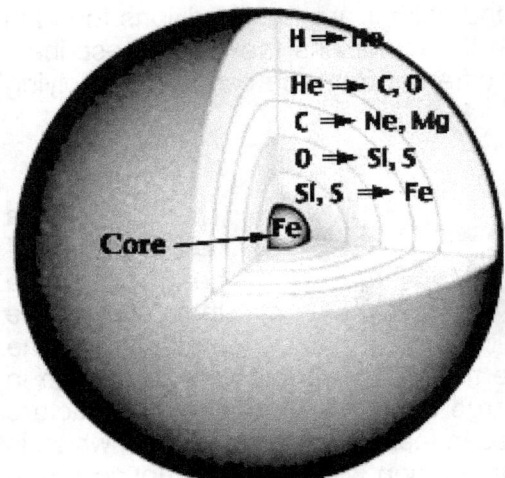

With the enormous container that the Sun is, there is nowhere remotely the duplicating by motion going on anywhere else in the solar system. The outside layers are formed by elements known for their volatility, which is another term for duplication. Hydrogen and helium has a very high ratio of interaction with heat and in that they have very high freezing temperatures. It is not by coincidence that the most mobility is on the outside and as the layers reduce space towards the inside we find the containing elements preserving space on the very inside.

The Sun is as enormous as it is because it freezes material in ratio as the motion shrinks the material to a fraction the size it holds on the lesser solar structures. By the massive duplicating of space-time it contains heat in vast quantities because it freezes hydrogen at 6500^0 C to a liquid. Deep inside the Sun it gets so cold that the restriction in motion freezes hydrogen to other elements and this process is called fusion.

It is a case of the motion cooling the material and the cooling is shrinking the material while it

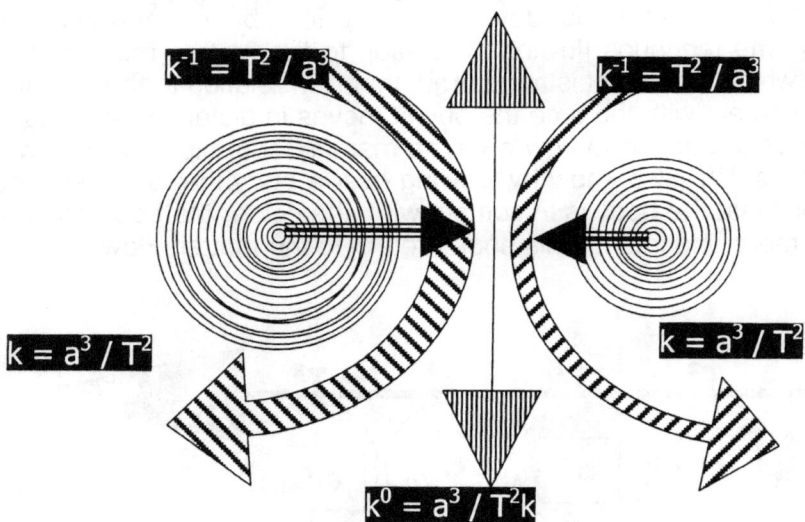

is excelling the heat in response to the material cooling. That way gravity is all about motion and heat that contracts material as motion cools material in relation to the outside of the atom that has to rise because of the lowering of the coldness and size of the material. Gravity has very little to do with mass and has so much to do with motion and it is gravity in motion cooling down material that shrinks material to accommodate more dense heat on the outside of material which becomes prevalent within stars.

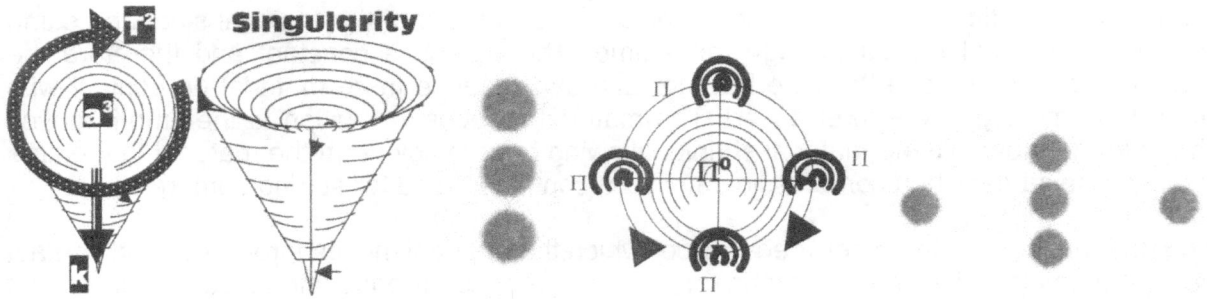

If the Sun is 18 X 10^6 on the inside of the Sun one have to take into account that the inside of the atom therefore would presumably be zero Kelvin. If that is not the case then the exercise is fruitless because the 18 X 10^6 will be meaningless. If there is a limit to the side being hot, then the side being hot must have another side being cold in order to give "the being hot side" any validity. In that case we also have to mention that when the hot side is zero, the inside of the atom that is zero Kelvin from the other end must be minus 18 X 10^6 when gauged on the reaction side because for all actions there has to be equal reactions. Everything is in a relevancy and only nothing can be unattached. There is always a relevancy coming about as one part in the relevancy does thee expanding factor and the other part is producing the containing factor. As the rotation commences around a centre point, there are changing relevancies as the one cosmic object orbits the other cosmic object and there are forever cosmic objects orbiting another cosmic object. Even in the case of the Black Hole eternity is orbiting infinity as eternity melts back into infinity. That is the reason why so much "mass" would fit into so little "space" by such a lot of material, and even moreover is the fact that the smaller the star gets the more material can fit into so much less space while the temperatures get so much higher. Gravity is motion that reduces heat which brings on cold which reduces material size which compacts space into more dense heat that multiply the gravity or motion of the relation there is to "space" or one actually should call it time and space which we actually call matter.

Once upon a time a very long time ago everything we see sprang from one point. Since then all points are a precise duplication of such a point where one forms the expanding factor and the other forms the contracting factor. Both factors are equal since both are the same but one appreciate space to the benefit of the unit by expanding while the other part is conserving the space by contraction also to the benefit of the unit. Since the unit is equal in the value notwithstanding that the unit is served by different factors, the factors hold the equal value in all aspects of the rotational gravity. Each one of the factors is covering one side of the Universe they form since both have the goal to preserve and maintain singularity.

So often Newtonians talk about pressure within stars and heat pressuring to commit to fusion. A Star that is under pressure is a star that is destructed. For that they have a fancy name. They call it a new star. Can you believe it that after the star has come to pass and blew up like a cherry cracker on New Years Eve, they call that star new? In the star there is supposedly pressure and with the enormous pressure the star pushes elements into fusion …and best of all is that they walk around with the doctorates in physics! Let's have a close upon the process that applies when the air or pneumatic compressor is pumped with air.

Light is the only factor that responds directly to time by joining time as light is in the space sector $a^3 = 3^3 = 27$ and the motion part is $\Pi^2 3 = 29.6$ giving a total displacement relevancy of 56 .6 within the star. Yet even in this case where light is overheated, singularity there still is a cyclic flow of time T^2 through the four quarters, which is in the orbiting contexts of cosmic structures forming seasons. It is where singularity changes the dynamics in the relation k has with T^2 and a^3. The locations of positions and the allocations of positions of material in motion places the dynamics within motion in different concepts in the ratio they have to each other.

In the cosmos every aspect there is indicates an atom's behaviour pattern. Even the behaviour witnessed when objects move, shows expanding relating to contracting and the one forms the electron or expanding factor while the other forms the proton or contracting factor and the two factors are joined by a liquid that holds pure gravitational and unrestricted motion. The part that connects the two factors form a neutron dynamic that is free flowing within the unit.

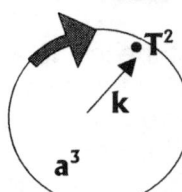

$a^3 = [T^2 = 7(3\Pi^2)] [k = \Pi^0]$.
The Earth holds a specific size in relation to the motion of the individual object being the Aircraft $T^2 = 7(3\Pi^2)$. Since the craft is stationary the distance between the craft and the object is $k = \Pi^0$

$a^3 = [T^2 = 7(3\Pi^2)] [k = \Pi^0]$.

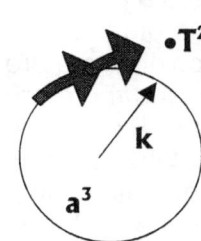

As the motion of the aircraft accelerates the Earth still holds a specific size in relation to the motion of the individual object being the Aircraft $T^2 = 7(3\Pi^2)$. However the connecting flexible link being the atmosphere, which plays the role of the neutron, has to extend in order compromise to being $k = \Pi^0$

$a^3 = [T^2 = 7(3\Pi^2)] [k = 1 - 5\Pi^0]$.
The Earth holds a specific size in relation to the motion of the individual object being the Aircraft $T^2 = 7(3\Pi^2)$. Since the craft is now in motion the first beacon to arrive at in relation to singularity Π^2 extends the distance between the craft and the object to $k = 2\Pi^0$ then $k = 3\Pi^0$ and so on.

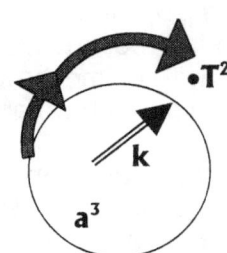

$a^3 = [T^2 = 7(3\Pi^2)] [k = \text{exceeding } 5\Pi^0]$.
As the motion of the aircraft further increases the Earth still holds a specific size in relation to the motion of the individual object being the aircraft $T^2 = 7(3\Pi^2)$. However the connecting flexible link being the atmosphere now has to extend to being the most furthest that the neutron can possibly extend and such extending can compromise up to being $k = 5\Pi^0$
$a^3 = [T^2 = 7(3\Pi^2)] [k = 2\Pi^0]$.

When the motion of the aircraft further increases, the Earth expands beyond what the limits will allow being the Roche factor of $\Pi^2 / 2$. This puts a cap on the specific size in relation to the motion of the individual object being the aircraft $T^2 = 7(3\Pi^2)$. In this the neutron of the aircraft parts in a dimensional time value from the neutron the earth holds and the connecting flexible link being the atmosphere then cannot extend beyond what the neutron can possibly extend and such extending breaks down at $k = \Pi^2 / 2$.
The entire issue is about the atom of whatever proportions

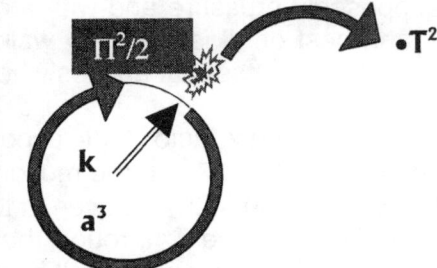

containing heat in relation to a specific centre.

When pumping a cylinder with air, the air inside heats up. The scientific explanation is that the molecules bumping each other to the extent that friction must occur because if not how does the heat come into place to cause this? It is hydrogen, oxygen, nitrogen and a bit of helium that is pumped. It is not copper vapour tinted with iron and tungsten. The so-called gasses which is extremely volatile and very much a gas, finds the air inside the cylinder so cramped they collide.

This is rubbish, as the next day the container is cold. What made the molecules calm down because through all evidence, the air is still there and the container wall is cold. Pumping the container will increase the heat levels inside the container. The inside gets hotter but we are taught at school level that heat will flow from hot to colder areas. There is another way of thinking about this issue, which might seem more accurate in the final analyses. When any object is heated it expands and

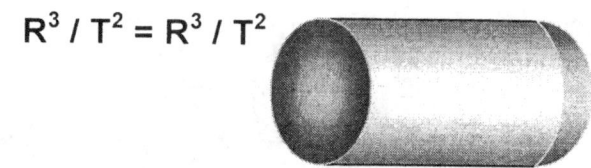

$$R^3 / T^2 = R^3 / T^2$$

when the object is cooled it shrinks or that is what those carrying the flame of knowledge tell us. When we pump air into the compressor the air gets more inside the cylinder. The compressor gets hot while the air gets more. The air gets more while the size of the container remains the same. Seen in another way we can think of the air remaining even while the compressor is getting smaller. The compressor is containing more while the air level is at a constant.

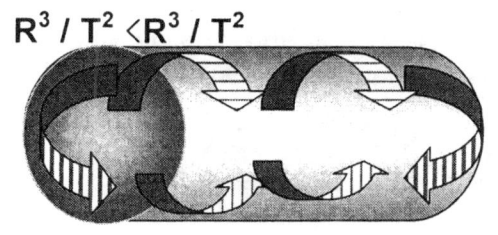

$$R^3 / T^2 < R^3 / T^2$$

The heat on the outside of the cylinder at first has the same value as the heat on the inside of the cylinder. Then by pumping the air into the cylinder, the molecules enter the cylinder but as they enter, they also take with the heat (unoccupied-space time) they contain and hold as their envelope. Inside the container the relation to heat gets more because the volume of the container remains the same except that the container walls get hotter as more air enter the cylinder that holds the air in place. Because the walls get hot we may assume the walls try to stretch because by heating the walls should expand as it gets hot inside and therefore it will force the walls to expand. By this token it is clear

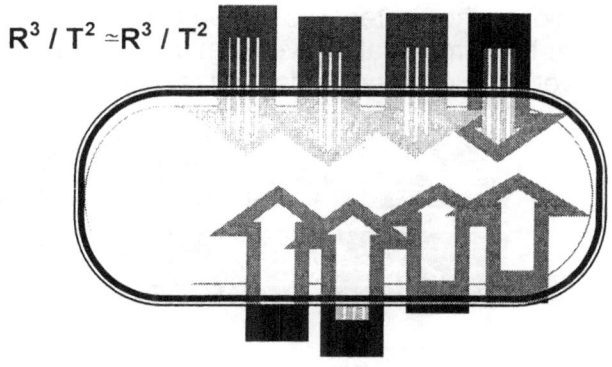

$$R^3 / T^2 \simeq R^3 / T^2$$

that as much as the container is filling the container, at the same time is also shrinking. Because it is also the size of the compressor that shrinks as much as it is the content growing more, the space outside the cylinder has to accommodate the increase in flow of heat coming through the compressor walls because the overall practise of science is that nature rules by equilibrium.

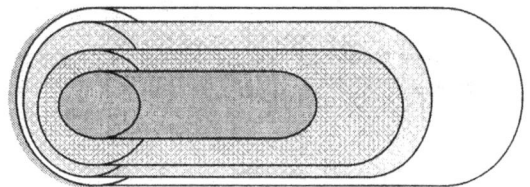

As the air becomes more the walls of the cylinder will reduce by the same token. There is more air connecting with the cylinder wall and therefore there is less cylinder wall with which the air can connect. The flow of air inside the container encounters less of the cylinder wall and more space and in that the truth is about cosmology. The air that came in brought with it the same volume as heat as what it had related to per volumetric ratio as was applying in the atmospheric space when the air was outside.

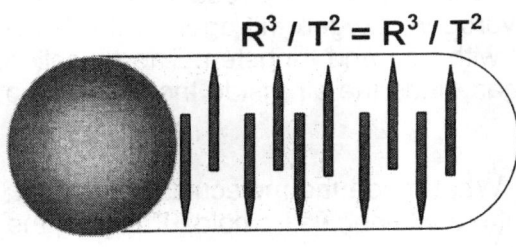

$$R^3 / T^2 = R^3 / T^2$$

The volume of air expanded but when anything expands it gets hotter. There is no evidence of anything expanding without increasing heat and even in the spectroscopy we have evidence of just that. It seems the bouncing has the increase in heat except that when gasses increase in volumetric capacity the gasses become volatile. If that was the case then there was more heat within them container that the air brought in and since the container got smaller the space held more heat per measure of atoms than was the case when the air was outside the container.

With the air increasing, by the very same ratio will the size of the cylinder keep reducing. That is why the compressor walls get hot. It cannot stand the reducing and ties to grow and expand, as it should while the air remains the same volume. From the container side there is no growth in the sir volume but there is a decline in the wall size of the container and that is why the container tries to expand the shrinking walls by allowing heat to try and expand the heating walls. The molecules are then more to the inside than the outside, the heat containing them, is also more on the inside than the outside (bigger ratio inside than outside).

What we find taking place in the wall of the container that is shrinking is the same that is taking place when concerning the position of the space not filled by material. The space that is holding the material is becoming less while the space within the material is becoming more. If the material per volumetric molecule is taking up more space then the space holding the volumetric molecule per unit is getting less.

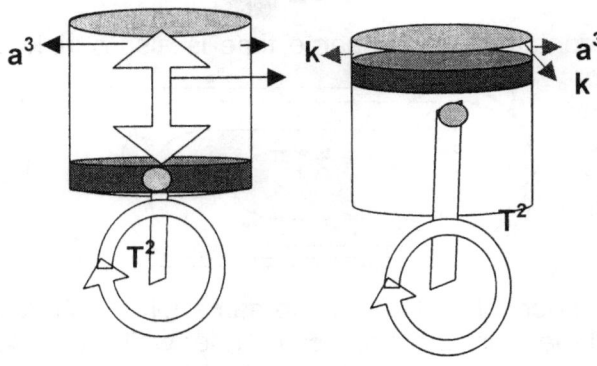

The more the molecules are the less the space must be that the molecules claim and the more the molecules has to reduce volumetric space to compensate for becoming more. Then the same applies in relation to the space parting the molecules as the space in ratio also have to reduce in order to accommodate more space claimed by the molecules as well as space pumped in that was accompanying the increasing number of molecules entering.

However we find the same process in the internal combustion engine with the only exception in that the process is put in reverse. Notwithstanding the different application the end result remains the same. In the case of the Diesel oil engine we find spontaneous combustion occurring when the process becomes at its peak of compression. However there is not a pumping of air but normal airflow into the container. At a point an intake valve ends all further airflow. Then the piston moves up and the piston reduces the space.

This time it is the space of the container that becomes less by motion reducing and the volumetric reduction of the space. In this the heat level rises to a point where the air gets so

hot it makes oil combustible. The volumetric space reduced and in that the particles became more. With the increase in the number of particles per space available the space available between the particles also became more in ratio. What becomes very clear is that the reducing of space brings about an increase in temperature.

The very opposite is also true and we use that principle for cooling in everyday life. By blowing with a fan over a surface reduces the temperature of that surface. By making the space available more, the space between the particles also becomes more. Then the space being more reduces the heat level surrounding the object, which is cooled. There is air blowing over a surface normally not moving and therefore by blowing over a surface that is not moving one gives the surface not moving the opportunity to be in a position that it can move in. In that way we enlarge the surface that is not moving by duplicating the surface not moving as the surface finds a location where it enjoys a larger ratio with the space it does not occupy in the same period of time.

Past going onto present and becoming the future

We have two persons standing still. The one is a thinker and the other is an accomplished and distinguished but sincere Newtonian scientist and which one of the two is which, that is for you to decide… The problem we investigate is how does both come from the past move through the current and leave for the future. The defining characteristic about time is that as time moves on, the position of every object changes in relation to the future and past positions.

We find that in any given area there is a ratio of Matter filling space and

In this ratio is built in another ratio of Matter holding time.

Matter determines THE RELEVANCY OF Matter to space during time.

Space holds heat in A RELATION OF SPACE-TIME unoccupied- occupied-, densified and singularity. Motion or moving by time or otherwise is the most complex issue there ever can be. Time relocates the structure by breaking down the entire structure as to relocate the entire structure and re assemble the entire structure to the previous specifications and by perfect duplication.

The position of the following instant neutralizes the previous position as it takes the place of the previous position.

In order to understand this concept it will be best to return and see how space and time started. The location where it all started is still present in the entire Universe in use today.

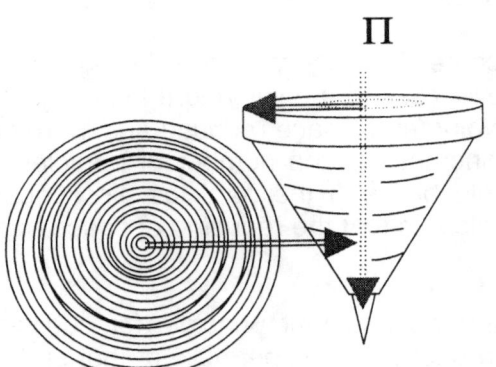

Fortunately we do not have to move back that far but investigate how the ordinary top is enabled by motion of rotation to stand erect. In the centre is a point that was there before time began. Time evoked the point back then as time still evokes the point in the present.

The point is so small it holds all points in one position. All four points are there and are rotating but by rotating from point 1 to point the point number 2, as it

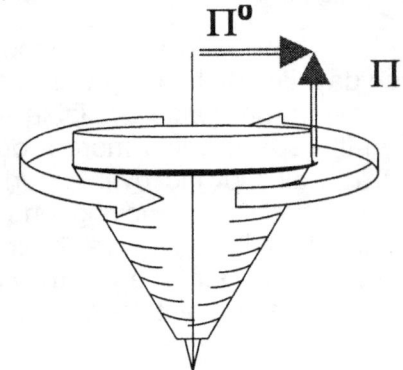

leaves 1 it lands on three because from there it moves to four which is also 3. All the points rotating are on the very same point. The point was eternally rotating and the rotation was there but the points became undefined and blurred because they were allocated to the same position. In such a simple concept as motion there are so many relevancies that has to establish new relevancies before relocation by motion can take place.

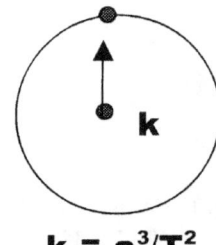
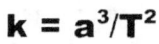

$$k = a^3/T^2$$

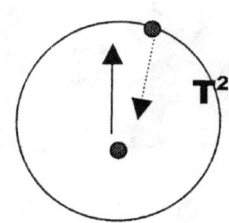

$$T^2 = a^3/k$$

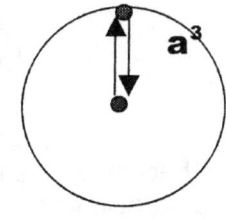

$$a^3 = T^2 k$$

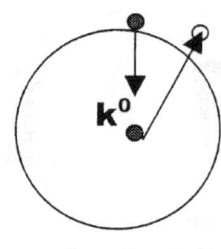

$$k^0 = a^3/T^2 k$$

$k = a^3 / T^2$ This proves the fact that in the moving of space-time brings a new identifiable location for space to centre although the space in the centre is immovable.

$T^2 = a^3 / k$ This proves the fact that the motion will establish such a centre

$a^3 = T^2 k$ This proves the fact that the space provides the motion to continue into the future while the space fills the one side of the Universe holding a position in eternity.

$k^0 = a^3 / T^2 k$ Singularity establishes and relocates space-time successfully by completing the motion.

Simple wasn't it? Let's run through the process once more and find the simple matter of motion in time.

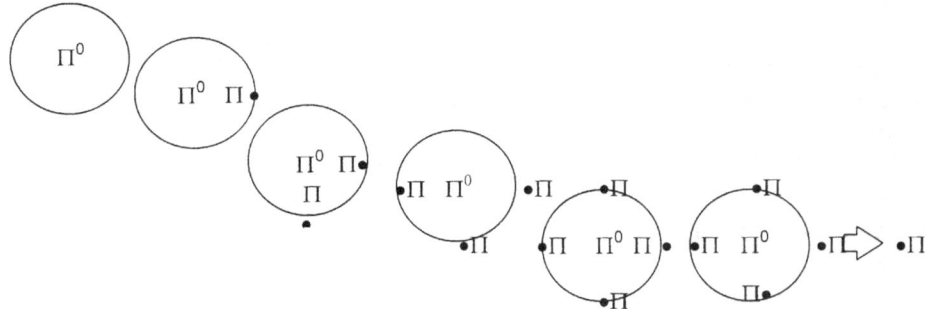

Singularity shifts from Π^0 to Π, which is a spot forming a dot. In our simplistic Newtonian way of thinking a spontaneous sphere formed as the spot expanded into a dot. Beware, all is not that simple because we have then the small matter of $k = a^3 / T^2$ to deal with. Every spot has to find a position in accordance with rotation as well as a position with relocation.

The same dot that was 1 became 2 because it was relocated and not reinvented. Then the dot was allocated a position in position three by reinventing 3 as well as relocating 2. This became $T^2 = a^3 / k$. At the very same instant $T^2 = a^3 / k$ did not disappear because what once is in the Universe is always in the Universe. As the motion took the dot to 4 then 4 became the new 1 because motion took singularity from Π^0 to Π where Π^0 was placed into a new allocated position by establishing a point as point five and relocated 4 as 1. Simple is it not. Try do that to every point that has a possibility of holding 1, 2, 3, and 4 in one position where all share the same position.

All this is true because $k^0 = a^3 / T^2 k$ singularity positions place in time by circling the straight line and repositioning the allocated spot.

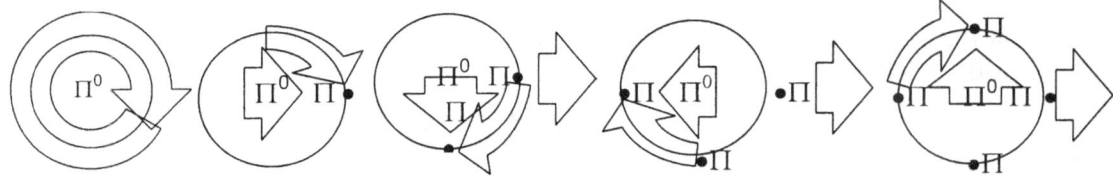

This is the prominence we find in the Lagrangian system using 5 points in the system where singularity forms four plus one. The motion in time in eternity is in a direction, which we might call progressive but also there is another relocation of the dot taking k from k_1 to k_2 that will form T^2. In all of this it is vital to see that there is the rotation as well as the linear motion and that forms the allocation of space where material is the time delay caused by locating the position of distribution and not being able to remove the previous allocated positions quick enough. Material is the time delay of heat distributed.

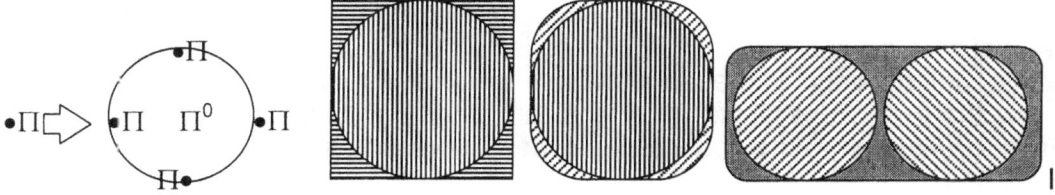

In the cosmos we have space filled with material filling space not filled with material while both are filled with heat. By moving, the material fills faster with heat than that which the cosmos can relate to by the filling of heat. This will reduce the space we think of as being unfilled with material, which is heat. The ratio guarding the specifics during such action will

surpass the limits of a specific ratio and that ratio we call time. By blowing air over the hot surface that is not moving we are relating that area with a larger unfilled space and therefore without moving the filled space becomes larger, in relation to the increase in unfilled space. With the filled space becoming larger the heat within the filled space becomes distributed through a larger area because the relevancy of the filled space has increased the filled space in size by matching the filled space to a greater ratio in unfilled space. That means by moving the air we are increasing the size of the material and by increasing the size of the material the material has to duplicate more often and by duplicating more often the material is shrinking in size.

With this information fresh in mind let's return to our compressor cylinder filled with air.

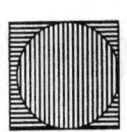 The cylinder had an initial size to begin with. While the air was expanding through the pumping, the inside of the cylinder became smaller as a result of the pumping of air into the cylinder. The Newtonians say the molecules are colliding and bumping and that friction brings about the heat. Then why is the cylinder cooling with time because the particles doing the bumping is still doing the bumping if they were doing the bumping in the first place. Yet, the heat does subside and no air has to leave the cylinder to get the heat to subside.

As the space reduces, the air gets more and as the space reduces the material becomes smaller and with the material becoming smaller the material has to dump heat from the inside of the container to the outside. Therefore, not only does the air not fill with space, it compresses that which is pumped inside and which becomes heated, but also the heat inside the atom has to disperse of some of the heat to decline and dispense of some filling because it has to reduce the initial size it had.

The process just described relies on pumping, on pressure, on retaining by an outside wall, by confining through deliberate replacing of material, which is confined into a cylinder that offers more confining.

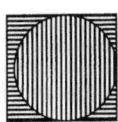    Most important is that the entire process relies on life and if not for life intervening in cosmic conduct, then none of this would be possible. So how does this comply with conditions in side a star? Well it does not comply even by a stretch of the imagination and only a Newtonian that is prepared to forsake logic in favour of madness and forces of unknown origins can see any connection between the star and the cylinder having pressure.

Looking at the Sun as a cosmic object, I do not see any retaining cylinder wall and therefore there is no material seeking to find a way out. There is no pump putting material against the flow of nature into the container. There is no forceful relocation of material from one side to another side and there is no escaping from what are unnatural circumstances. All these factors contribute to what makes me not accept the view of pressure inside a star.

There is no bursting of what is inside to what wishes to be outside. There is no evidence of retaining what is inside. It is a round structure and therefore it holds what is inside in accordance to singularity applying. There is no possibility of life intervening in any way or life interfering with the process. The scope of cosmic affairs just is limitlessly beyond what life has as possibilities.

Yes we do see what is inside trying to spill to the outside but it is far more evident that the spilling out is the forceful behaviour and the retaining is what comes naturally. There is no deliberate escaping from the pressures within but when released that which was inside flows back as a natural reaction and defies the whole idea of unnatural pressures building up inside of the retainer. There is no comparing the cylinder of pneumatic principles to the star that holds liquid and not gasses inside that star. There is no evidence of any gas although the flow of photons emitting light rays is by some imagination some part of a gas.

Inside the star the movement of all atoms combine in motion that establishes a centre governing as a principal all conditions applying in the star. The rotation of the atoms forms a synchronised motion that establishes the line, which parts infinity from eternity. The motion confines the material to the star but it is because the material elects a principal to confine the conditions to a status which all material inside the star agree on and benefits by the

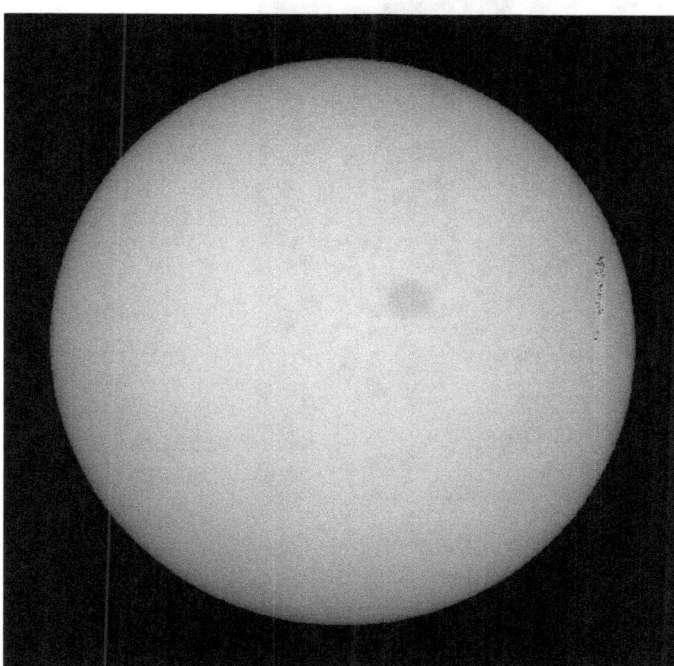

conditions of space-time we find in the star. As the conditions serve the star, gravity will come about and gravity sets freezing conditions within the star.

What happens inside the cylinder is the part that is compatible with what is applying in a star when we exclude the pumping, the pressure and the container idea. Lets go back to the fan blowing wind over an area in need of cooling.

When heating, the object increases its initial size from normal to becoming larger. The heat increases the size the object has to a larger ratio than that which applied before. To cool the object an increase in the ratio is needed on the airside to keep equilibrium and to bring in cooling even bigger ratio is required. By increasing the air we are decreasing the material and by increasing the heat we reduce the material by progressively anticipating more duplication as the ratio of space to material is increased by more space duplicating more material. The heat has to increase the size of the object within in order to match the ratio set by time on the outside. At this point I think it worthwhile to remind the reader that during the Big Bang the lot outside seemed hot and today the lot seems a lot cooler. I say that it seems that way because it is not truthful. The heat has to increase the size of the object in relation to the match it has to find in the space it is within. With the heat coming into the object the relation that the object has with the heat or air outside makes the object that many times bigger, because the ratio in the heat balance is disturbed. If we blow air over the object we increase the size of the object by allowing the surface of the object to make contact with much more air in the same period of time, which will bring the size of the object back to the normal ratio it was before,  because in relation and considering the contact with air, the object expanded by the motion that increased the amount of air being in contact with the object. In the normal flow of time the object has a heat to space relation set by the time the dictates. Then we go and increase the heat of the object and in that event we actually increase the size the body has in relation to the heat in the air. By blowing air over the body we increase the air and therefore we increase the size of the body during the same period of time. There is now a dispensation of many times more air where the body is carrying more heat and making more contact with the surface whereby it is contacting heat or air which brings the equilibrium back to normal what ever normal then is. There was a body size ratio and by applying heat the balance shifted to the reducing of the body size in relation to the heat. The body then had to expand in heat because the body was too small to incorporate that larger heat. Then by blowing the air over the body it increased the size of the body and heating the body decreases the size of the body in comparison with the air it comes in contact with. The body is either expanding or the body is reducing and the balance in heat places the body in relation to either gravity cooling by contraction or by expanding by overheating. The very same principle applies in the sound barrier.

Earth in relation

The gravity motion of the Earth is $7 (3\Pi^2)$ which is the distance of space in relation to the time it takes to displace that space and any motion above that is an extension of the atmosphere where the atmosphere accepts the role it has as being the neutron of the Earth. The extending can go from Π^0 to $5\Pi^0$, which will then be the moving object extending its neutron part while still attaching to the Earth atmosphere. Above that Lagrangian limit of 5 the Roche limit is sharing neutron status sets in at $\Pi^2/2$ and the attachment there is between the atmosphere linking the aircraft and the Earth is severed.

In physics there are always two relevancies at work, which has nothing to do with mass but is solely directed on singularity achieving motion.

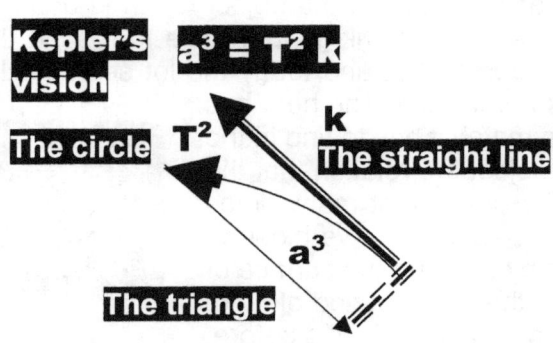

Kepler's vision

$a^3 = T^2 k$

The circle T^2

k The straight line

a^3

The triangle

The current notion of mass pulling mass has no comparing with reality because as I explain mass has only a counter effect of gravity. In the Universe there are a flow of space-time and a relevancy bringing about expanding as well as contracting gravity. This has no bearing on mass and is a control mechanism we find in the cosmos. It is a product of the cosmos we named the Coanda effect. We have to see that there are two parts in gravity where the one is expanding while the other is contracting and centre to this is the control we call gravity.

Many moons ago long before I dreamed of becoming an author of any of the books including this letter I started my search on the basis of a certain remark that Einstein once made on a realisation or a conclusion that Einstein came to in his younger days while still being a clerk at the patent office. Apparently the idea Einstein came to was concerning the subject of gravity. Apparently Einstein was looking out a window of the multi story patent office, when Einstein suddenly realised that had he, Einstein fall out of the window from the roof to the ground of the patent office where he was working at the time, then he (Einstein) would feel as if he was weightless during the time of his fall. Not only that but also so would all the articles in his office that surrounded him at the time being his office chair, his desk and a pen. By falling with him those articles would feel equally weightless should they accompany his fall down as being part of the falling process in his imagination. As the objects were travelling alongside Einstein down the building to the ground the lot would travel at the same speed from the top to the bottom of the building. That is what Galileo concluded about five hundred years ago. Then I went one step further by supposing the Einstein group's falling was real and no imaginary thoughts were set in the fall, then what was the imaginary factor then? Let's pretend Einstein did fall with his pen, his chair and his desk and Einstein was not imagining his fall. Einstein as a human being can imagine but his falling companions can't. Then during a true fall Einstein may have had an imagination that could tell him about his feeling and in particular about the condition of his weightlessness, but the pen, the chair and the desk had no such imagination and they were travelling at the same speed as he did downwards and therefore had the same weightlessness as he

▲ **The pulling away of the smaller space. a^3**

▶ **The double counter-acting referee. T^2**

▼ **The pulling towards within the larger space k**

(Einstein) had while they all were being in a downwards fall. If Einstein was imagining his weightlessness, it might be psychological, but in the case of the other travelling companions it was not possible to imagine anything. The falling companions had no such a luxury as having an imagination, however they too had to be weightless as they travelled next to Einstein all the way. There is an immense difference in size between the falling companions and that notwithstanding they travelled the same speed while descending. If they travelled the same speed as Galileo proved and they all hit the Earth the same time, which then indicated that their weight and mass, that which gravity used to drive and what propelled them downwards and that which was causing the drawing of what the mass was instigating to allow the motion of fall to commence, was equal. Size changed nothing to the equality there was in speed. Einstein should only have thought a little further than he did at the time because that would have made him realise what gravity exactly was and what Kepler found

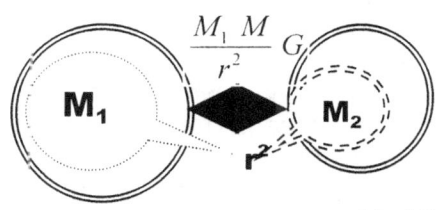

Newton's vision $F = \dfrac{M_1 M}{r^2} G$

gravity to be. Kepler found space a^3 being equal to the motion thereof T^2 in relevancy to a centre point k. Kepler found space had to move.

Realising this part made me doubt the correctness of comic science. In the cosmos all things are moving therefore all things are falling never to get there. Galileo said that all things falling fall equal. If there were Newton's mass discrepancy when falling this would not be possible because part of the driving force of such a fall is the mass. Newton even put the product of the mass in relation to the destruction of the radius. The mass forming the driving force then has to allow for heavier objects falling faster by measure of mass discrepancy. But that would sideline Galileo and Galileo could not be sidelined. It is either mass driven with Galileo being incorrect or it is as Galileo said that all things fall equal rendering mass equal during everything falling.

When reading this that evening so many years ago, I came to realise that Einstein could only feel weightless if it was true that he (Einstein) was weightless. He could not feel as if when the as if was part of his imagination because he was truly falling, and in truly falling the falling was then without his imagination doing the pretending. Einstein had to feel his weightlessness as a cosmic fact in the true sense because if he was truly falling, then the part, which was the falling experience, was what he was experiencing in reality by three dimensions with one dimension in time.

Then he (Einstein) was feeling weightless through falling and that feeling came as a result of what was happening to him as a cosmic interpretation of reality. He was not pretending to fall whereby he then would feel as if…he was really falling and with that there is no "as ifs". What he then would have experienced came by means of what he was experiencing in reality because of his cosmic state in relation to his relevancy with gravity. If Einstein was experiencing weightless ness, it would be because he was weightless while falling, then Einstein would not imagine the weightless ness because Einstein was truly falling, thus carrying out his cosmic state he was in. His body being in motion ($a^3 = T^2k$) was at that moment truly weightless while experiencing unrestricted gravitational motion. Einstein, the pen, and the chair had the same weight since they were all weighing the same in falling.

If there were any mass differences there had to be speed differentiation for the force of the one would generate more motion than the force of the other onto the different mass components but since there is no mass discrepancy amongst the falling, while falling the lot is having the same state of weightless ness and they adopt the same speed in the fall. After all it supposedly is the mass that is doing the pulling and more mass does more pulling…except if the mass is not doing the pulling in the first place.

With more force applying to different masses there had to be more speed involved and an increase in mass in some participants has to generate more force. All four items including Einstein, would be equally weightless during the falling…that was what Galileo found because objects of different size and different mass travel at an equal pace (distance over time or space moving divided by time flowing while the object changes position in relation to the Earth ($a^3 = T^2k$)) while descending. The bigger objects do not fall quicker than a smaller object and that can only be attributed to one fact; it can only be true if the four weighed the same while falling and no one weighed anything while falling. That means the gravity applied while time flow in relation to the space that was applying the motion, which was what gravity is $k = a^3/T^2$ according to Kepler. The single line falling is represented by the factor **k** being the relevance of space a^3 that was relocating its cosmic position while all that was happening in relation to the motion of the Earth T^2, which was in relation to the Earth spinning around the Sun and that rotation gives us our time T^2.

While in motion the four different objects weighed the same since they travelled at equal speed downwards. However, when they stopped moving and came to a standstill, they then weighed different, which then indicated a difference in mass factors amongst them. By standing still the objects had mass differences and when they were in motion they weighed the same. When the motion became frustrated by being blocked by another space that was also filled with material and that was holding the spot to where the motion was directed, they then had different weight. The two objects standing against each other had different levels of frustration with the larger party being more frustrated in the inability to move. The pushing resulted from the bodies striving to remain independent. It is the independence of the two bodies and the desire the bodies have to remain independent and not to share space that bring about the mass or weight. The two objects were in a fight to claim the position each desired, and that was to fill the centre of the Universe. Being ($a^3 = T^2k$) was being in the centre of the Universe because the centre of the Universe was $k^0 = a^3/T^2k$. $a^3 = T^2k$ $k^0 = a^3/T^2k$

From this one can deduct that gravity is motion or the intent to commit motion and mass is when the motion of gravity is frustrated by some solid structure blocking or preventing the continuing of the motion. Then one may conclude that gravity is motion of space and mass is the restricting of the motion of space. Having mass does not bring about gravity but it does restrict gravity's motion, which is what brings about the mass and weight. Gravity produces mass but mass does not produce gravity or in fact mass produce weight but mass is not responsible for the intended motion. Gravity on the other hand is the intention that the body has to move the very instant the blocking is removed.

The intent on moving while being blocked by another object is frustrating the motion of gravity in both cases and the higher the frustration on motion is the more mass there is coming the way of the bigger object who then has the greater desire to move. The reason why it has the desire to move and why space is equal to the moving in time of the space in relevance to the centre of the Universe (which at that point might be the Earth or be the Sun) is what I am trying to explain. Mass is the restraining of motion and gravity is material moving about by committing gravity. Mass only comes into the application thereof when two objects filled with space moves into a position where both want to claim the very position in space the other occupy.

It is the motion and the independence they show to hold onto their individuality as independent cosmic structures that prevent them the sharing of space which in turn prevent further motion that causes mass. Gravity is in essence where mass is present, still in a tendency to commit motion but is then in the frustration of motion and gravity at such a point is the commitment to move once the blocking of space is relinquished. Because the one object that has more "mass" would put in a more assertive effort to move in relation to a smaller object and the effort to move will constitute to a greater resisting effort by the blocking object in a fight not to relinquish its position on the space both objects claim that the tendency to move and the tendency to block the movement will bring the effect of greater or smaller mass being present during the effort and in line of resisting the effort. However, while any space is in motion, the gravity of motion is equal to all and puts everything on an equal basis. Therefore there is no big and small and the big Sun does not pull the small Earth closer.

The big Sun allows the small Earth to glide past in a circle year after year without interfering because the two do not claim the space each other has. Mass only becomes a factor, when the motion of an object in duplication, is prevented from being in individual duplication and in that a differentiation in motion effort becomes part of the picture. Mass can only be attributed by the extending efforts that life can produce. Anything not with life and not aided by life, or serves as an extension in motion to what life produces, cannot have mass.

Do not be fooled by the seemingly innocent explanation that space is the motion thereof which is what gravity produces because of all things the cosmos creates, motion of space through time is the utmost complex manoeuvre and without bringing a restraining of mathematics into science, it is so complex there is no viable explaining in physics about how the cosmos produce the act of motion of space in time. To get every atom to spin as every atom follow the lead of the atom in front and gives direction to follow to the atom just behind while giving coherency to the structure the lot of atoms are holding as an individual unit times the units there are going around in the entire Universe is beyond what the human mind can absorb. While the atom in front is vacating space to fill the space of the atom in front that is vacating at that instant, the atom behind is filling the space that the atom in front has vacated in order to vacate and relinquish the previous position in favour of the following position to honour the direction gravity is insisting upon. Times that with every atom there is in the Universe and one may grasp the significance of the calculation. The coordinating of moving one atom from one point to a next point requires the skills that the human mind may never conquer.

We may see the moving of an object through space being as simple as merely accepting it as a given fact, as science has done in the past, or we may reason about the complexity as civil person's should do, and come to realise that the complexity of motion of matter is beyond the scope of human understanding. Removing material from space by filling material into a position of new space sounds simple because the complexity has never been realised. This was all a result of understanding the dynamics of Einstein's arguing about gravity and mass. Then with this information I further realised gravity is motion differentiation between objects. It is the independent motion providing a different speed while sharing a common centre of attracting that allows a discrepancy to establish mass under specific conditions applying between the two in relevancy. While falling the gravity applies as moving of space that is putting time in relation to the distance travelled. That means there is a speed relevancy between particles in motion and synchronised motion would bring about equal orbit around a shared centre.

That is the result of gravity functioning. While the object falls the motion confirms gravity. When motion ends mass sets in and becomes the constraining of the object preventing further motion. The motion is still there but now it is reduced to a tendency to move thus establishing the object mass as the limiting of further motion. Preventing the motion by implementing mass is the resting of objects against each other by resisting the motion to continue, which then is where the mass takes the place of the motion. Where a confronting of objects restricts gravity the action then implements an introducing of the mass as a substituting factor for motion that then replaces motion as substitute for the motion that would be and the mass is providing the tendency of gravity being the motion of space. However mass then restricts motion and becomes motion in a tendency to apply motion. While falling gravity applies and motion neutralizes size, mass or weight.

Mass counters motion being when the Earth restrains further motion of the falling object and the moving object is stopped from further movement where mass is then preventing or hindering gravity. This is the result of objects claiming an individual and personal claim to space occupied in a dual or in fighting for their individuality and independence of each other while wanting to be in the **centre of the Universe**. While falling or moving there is no opposition to the body being independent. When the motion seizes the falling object remains individual and still tends to move while the Earth individuality resists further movement of the falling body's movement. Further movement is disallowed as other material fill space that the falling body wants to laid claim to. The only manner to remain independent by the falling object will be to relinquish to motion in the securing of mass as a substitute to motion where it then finally comes to rest. Mass then sets in not causing the motion but substituting the motion and from that motion restriction becomes resistance that becomes mass. While falling the object is experiencing gravity because the object is in gravity but when on the soil the object experiences mass which is the restricting of gravity or motion by other space filled with material. It is a fight of objects to secure and retain the position they have of being in the **centre of the Universe**.

Moreover, I then came to another conclusion of equal importance. When any person is standing on any place anywhere, while viewing the Universe, that person is filling the **centre of the Universe**. Let's get more personal. When you, the person that is reading this, are standing at night and are looking at the Universe you are seeing the Universe from the position that one only can have if that person is filling the specific spot in the **centre of the Universe**. All the light, every single beam that ever left any destiny at any time acknowledges this fact. You are the most important person in the Universe because you are holding the most important position in the Universe. All the light that come across and travelled all of the vacant space from any and all possible positions in space runs directly towards your position using a straight line towards you where you are filling the **centre of the Universe**. Not excluding the effort of one photon, all light is heading to meet you where you are in that centre spot and not one photon will pass you by. Not one photon dare miss you because if they do they miss the effort that all light has to accomplish and that is to locate you as the person filling the **centre of the Universe**.

Should you decide to shift your position to any other place in the Universe, you will shift the **centre of the Universe** to that location as well. If you install a camera on Mars, the light is obliged to acknowledge your relocating the **centre of the Universe** at your will to reposition you're being that **centre of the Universe**. All the light that ever left its destination crossing the vast spaces of the Universe, excluding no particular light, travelled all the way just to find you filling the **centre of the Universe**, right where you are. By you're standing anywhere, you fill the **centre of the Universe**, and the entire Universe admits to that because all the light comes to meet you there. If you shift from the North Pole to the South Pole you will shift the **centre of the Universe** because all the light travelling throughout the Universe will find you where you then moved the **centre of the Universe**.

The light left its destination billion years ago as it travelled through space at the speed of light anxious to acknowledge you're being in the very **centre of the Universe**. No photon will be able to pass you by where you are in the **centre of the Universe** because all light is heading your way from their starting positions. No wonder every person born has the idea they were born to fill **centre of the Universe**, which we do fill. The Universe is spinning around you or I, which is filling a centre where all motion is connected. That is the Coanda effect on the uttermost grandest scale imaginable; nevertheless it is only a manifestation of the Coanda effect. It implicates gravity as wide as can be… Some things mathematics is able to explain but other explaining goes beyond mathematics. Try to explain mathematically the colour of the sky being blue on a clear Sunny day and changing to black when nighttime falls. Do the explaining in mathematics to a blind person that had no vision since birth in such perfect mathematical detail that would allow the person afterwards be able to explain the difference between blue and black to other blind persons by using only mathematics. Some aspects of the Universe go beyond mathematics and some even go beyond words.

It is our task to find space, to find time and moreover it is our optimal task to find the Universe. We have to see what is solid, what is liquid and what causes gravity. It is therefore very important to see what is a solid and what is a liquid. Again we must put culture in the background and value the cosmos by using cosmic standards. Everything that moves, do so in relation to another object that has a direct relevancy connection to the moving object. The moving of the one must serve to show that while it shows motion, it is relevant to another body that holds a relevance of being stationary.

By moving, the moving object will serve as the liquid partner and is therefore a liquid notwithstanding that it may or may not contain material. By moving, it is a liquid nevertheless. Everything that is relatively stationary is a solid in relation to that which is moving and by motion, serves the purpose of the liquid that moves about. The one that serves as the solid is the solid that anchors the liquid by gravity. The liquid adds by motion while the solid extends by not moving. Gravity **is to move or apply the intension to move** space a^3 **at the** distance or relevancy of **k** while T^2 is the time it is going to take to **apply gravity** or move the space filled with material space a^3 at the distance of **k** in the time period of T^2. That confirms Kepler's attribution to gravity where according to Kepler space a^3 is equal to the movement T^2 (time it takes to move) at the distance **k** from the centre specific.

Do not frown on this being in the **centre of the Universe** or regard it too lightly because from that I can prove life being eternal and life being part of the other side of the Universe. That is not part of this letter since that I do prove in another book with another title under the article heading **Man – in- Motion**.

I then subsequently reviewed my vision I received from the vision Einstein received and applied such a vision on the findings Kepler received from the Cosmos. It puts all aspects of gravity in the Universe in new dimensions. But the visions formed the beginning because the visions unleashed many new questions. If gravity is motion, what causes motion? What stops motion? That answer is in the Black Hole. In truth the explaining of the Black Hole is as complicated as the Universe may represent and as simple as the cosmos truly is. If a star is about fusing atoms and with such fusing of atoms is thereby growing, what happens when all the atoms fused into one all collective atom in one already all—atom-accumulated star?

What is the gravity if the star has melted all atoms it had into one all-inclusive atom and this all-inclusive atom is providing all the gravity that the star had when the star still had massive volumetric space? If all that space that once filled an entire giant star fused into one specific space less centre holding singularity 1^0 then the enormous gravity is applying to the centre of such a non existing space-less atom and that entire enormous force has been secured in the space less than that which one atom holds. In that case the atom would then show a force that would pull the surrounding Universe flat. The purpose of fusion is to reduce space and magnify space less ness inside the sphere. Where does the gravity of the star end when all the atoms in the star became one giant atom by fusing all atoms into one nucleus? Gravity is smallest where space is least. Where space of an entire massive star is left in the size of one atom the gravity coming from that will pull the Universe flat at that point. However fusing means freezing together because only by reducing the heat can the removing space be accomplished and by reducing space to the point of freezing material permanently together is getting material frozen permanently. That means the Sun and all stars are as cold as they can get and not hot!

The motion is a product of heat and the motion produces a cold that sets in as the motion comes about. When the object moves, it moves because the heat becomes excessive, but by moving it is doubling the area it holds by halving the area as it divides the space between where it goes and from where it came. As soon as the motion halves the space used to move the halving of the space halves the heat which produces the cold which brings about the containing or the reducing of the space. Then with the motion completed the contraction retains heat where the heat increases to bring more heat rising that leads to more expanding coming about and the cycle once more repeats it self.

I am not getting into that argument now, but because of the size the Sun has and the size moving through such distance the Sun in its very centre is the coldest place in the solar system and outer space is so hot it is over boiling. That is why outer space is expanding. It is because the heat rises as much as the stars reduce the heat by containing the heat as material inside the atoms. Nevertheless, gravity is motion and motion comes from overheating whereby the motion then produce the cooling that contains the overheating by accumulating the contained heat inside the atom. To do that the Coanda principal is employed and the Coanda principle sets the Titius Bode law in operation and the Titius Bode law produces gravity. By spinning liquid heat around solid space a relation between 10 / 7 and 7 / 10 produces a relevancy that contracts heat.

It is about. Gravity is all in Kepler's $a^3 = T^2k$ and $k^0 = a^3/T^2k$ where then relevancy $k = a^3/T^2$ and in response to keep equilibrium applying $k^{-1} = T^2/a^3$

On the inside, there are the seven markers of which singularity is the focus point in the centre of the centre. The markers are representing one aspect of space, which for argument's sake let us call it cold. Then there are three more markers on either side being part of the space but not captured in the space. It is space in motion by the influence of the motion of the Earth.

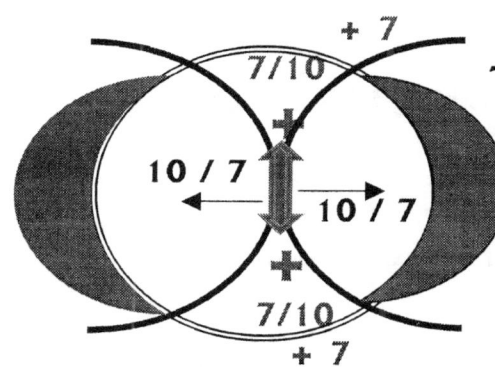

TITIUS BODE LAYER CONNECTING

Inside the cosmic sphere

$7/10 + 7/10 = 1.4$

Singularity in the square of matter

$10 / 7 = 1.42$

Singularity in the square of space

$1.4 / 1.42 \times 10 = \Pi^2$

MATTER HOLDING THE SECOND PROTON COUPLING THAT TO THE NEUTRON TO COMPLETE THE NEUTRON. Due to the influence of the matter dimension on the space dimension, the curvature of space-time comes into affect by dominating outer space. The Titius Bode Principle is equal to gravity @ $= \Pi^2 = 9.8696$. Proving that the Titius Bode Principle is a product flowing Directly from the growth of singularity forming space-time The Titius Bode principle directly valuating TIME to SPACE $= \Pi^2 = 9.8696 =$ MATTER HOLDING THE SECOND PROTON COUPLING.

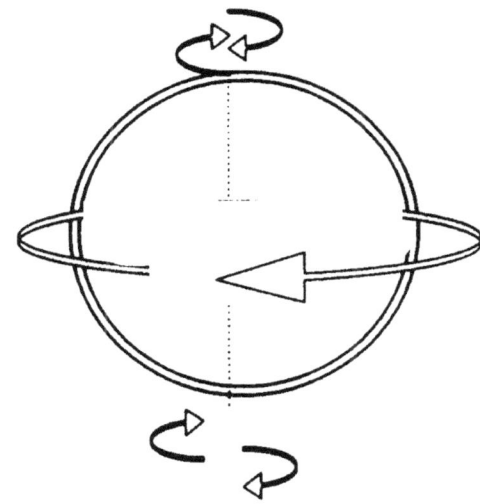

This proves the reality of the Titius Bode law, which too I have to add, the Newtonians put down to a coincidence. This proves that the Titius Bode law is part of the chain that brings about gravity. Most of all, this disqualifies mass as having any importance or prominence what so ever in the producing of or the conducting of gravity. This proves me correct where I was adamant from the start that mass has no influence on gravity.

This proves me correct when I say that the Titius Bode law does not only represent gravity, but it is gravity. This proves me correct when I say that the Titius Bode law is the neutron at 7 / 10 and also at 10 / 7 and that the neutron represents the Universe. This proves that gravity is the motion where space interacts with time to give singularity the significant control it has notwithstanding the part it has as not being part of the Universe and yet being responsible for all action in motion taking place in the entire Universe. The Titius Bode law is that which is between eternity in time and infinity in time.

This is what keeps the top erect while spinning and it keeps the Earth in gravity as much as it built the solar system to a mould that built the entire Universe. It is the way the Universe develops through motion interacting between time and space. The fact that motion brings about gravity in line with singularity must be proof to all Newtonians that their perception on mass has no grounds. Even where those Newtonians are unable to show what brings about gravity even after so many centuries of investigative research while trying and getting no results does this simple arithmetic prove more than all the multitude calculations on their part that prove zero about the manner they promote gravity as being a pulling force.

Matter in relation (part of) to the total dimension of space.

(10 / 7) \ (7/ 10) = 2.04

1.4285 / 0.7 = 2.04 Taking from both orbiting influences
SPACE DIVIDED INTO TIME

(7/10) / (10/7) = 0.49
.7 / 1.4285 = 0.49 Taking from both orbiting influences
SPACE MULTIPLIED WITH TIME

7/10 / 7/10 = 1 and   10 / 7 X 7/10 =1 Therefore not influencing change
THE PROCESS PARTED USING THE ROCHE PRINCIPLE

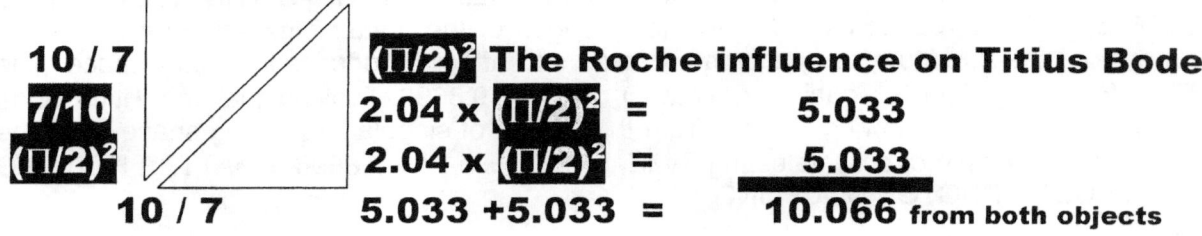

10 / 7
7/10   **$(\Pi/2)^2$ The Roche influence on Titius Bode**
$(\Pi/2)^2$ **2.04 x $(\Pi/2)^2$ =   5.033**
       2.04 x $(\Pi/2)^2$ =   5.033
    10 / 7   **5.033 +5.033 =   10.066** from both objects

SPACE DIVIDE INTO TIME

         7/10
              7/10 / 10 / 7= 0.49
10 / 7   **0.49**

    10 / 7   **10 / 7**
    7/10 =.49   **7/10 = .49**

     .49   +   .49   = .98
     .98 X 10.066 = 9.8 =Π^2
       TIME SPACE = Π^2 = 9.8696

TIME SPACE =Π^2=9.8696= Space and time in a dimensional implication.

In this maintaining of cross referencing of singularity located in individual atoms providing spin to the governing singularity that maintains structural form in solids, many factors of singularity all form a close knit network that is inseparable as one unit. By the same margin it also is strictly individual to a point of destructing. From the inner or governing singularity outward all is concerned as space-heat. The relevancy in the material sector always includes the governing singularity and the very next one to the inside. All the others do not form any part of such a relevancy to the object forming the relevancy. On the time issue it is only the relevancy forming a connection with the one in question and the governing singularity. All other objects in the line are merely space-time with no value to the object that holds the relation. The fifth object will link in the material sector to the fourth and then directly to the governing singularity skipping or excluding all others from one to there. In the case of say three, it will connect to two and skip one while all points holding singularity to the outside is of no consequence to the rotating object.

SINGULARITY BY DIVIDING SPACE INTO MATTER AND MATTER INTO SPACE, AND ALL OF THIS ACCORDING TO THE TITIUS BODE LAW OF 10 / 7 AND 7 / 10 IN CONJUNCTION WITH THE ROCHE PRINCIPLE OF $(\Pi/2)^2$.

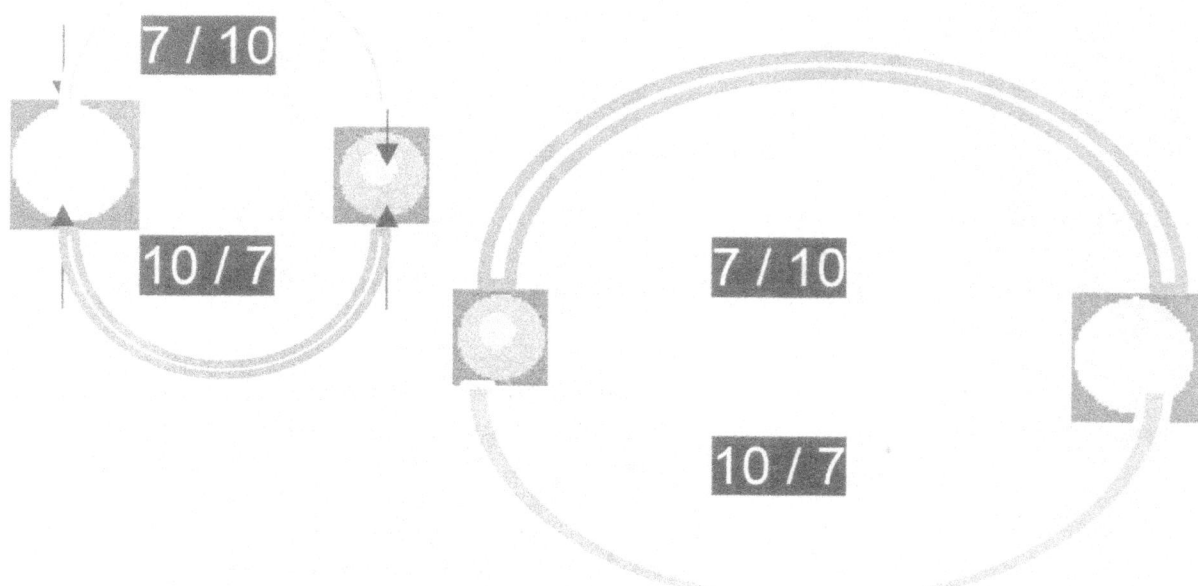

Time started at zero, eternity, whatever you wish to say, as long as you say time did not move at all. Then the command came and time overheated for the first Π^2 in time. That brought space into play.

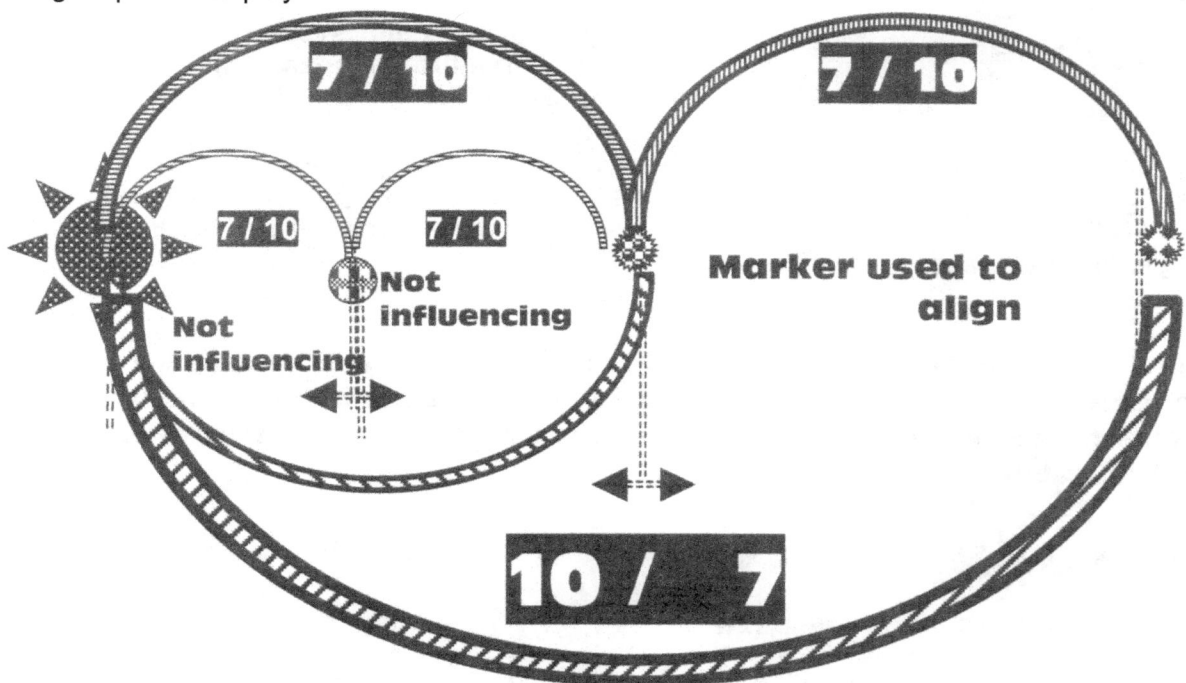

From the dividing singularity only one reference holds a matter value forming the position next to the governing singularity and therefore 7+7 becomes a factor and not all the dividing singularity between the point of reference and the governing singularity. That way the star to the outside takes a position doubling the distance every time. In balance everything in space to the outside of the governing singularity is space be it space or matter that makes no difference, therefore that is 10.

The spherical positioning layout forming the Titius Bode Principle

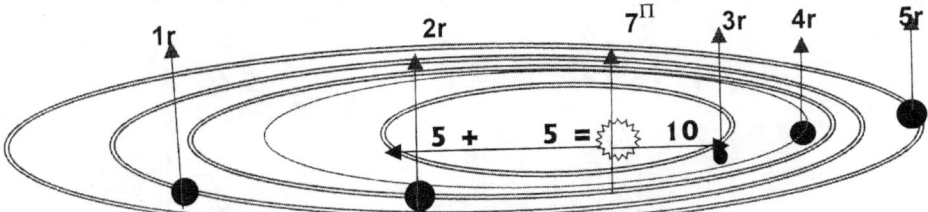

From the matter-to-matter relation in the Titius Bode configuration there are 7 / 10 + 7 / 10 = .7 + .7 = 1.4

From the space-to-matter relation in the Titius Bode configuration there is 10 / 7 = 1.42

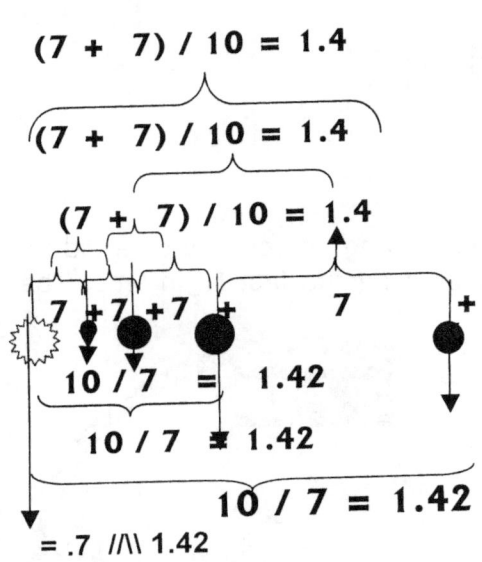

$(7 + 7) / 10 = 1.4$

$(7 + 7) / 10 = 1.4$

$(7 + 7) / 10 = 1.4$

7 + 7 + 7 +   7   +

$10 / 7 = 1.42$

$10 / 7 = 1.42$

$10 / 7 = 1.42$

= .7 /Λ\ 1.42

degenerating space.

The 5 + 5 = 10 is a position of dimensions as space loses value to singularity. The 7 that matter diverts in points from singularity may seem as coincidental but is valid. Still in accordance to our perception valuing the number in degrees, it seems coincidental but if it is coincidental, it is nevertheless a figure of diverting proven as accountable in all other calculations and plays a most dynamic role.

The Lagrangian 5 point system results as much from the Curvature of space-time as does the form the Black Hole holds. The Galactica is the opposing equivalent of the Black Hole and has identical but opposing similarities being the five points positioned to singularity. The galactica is generating space and the Black hole is

= 1.4 /Λ\ 1.42. Because the space-to-matter is in the square at 10 placing the matter-to-matter at a square of .7 + .7 = 1.4 the space-to-matter forces the matter-to-matter to double the distance by number as structures are placed father from the mainΠ^0 maintaining singularity.

1   3   6   12   24   48

Reasons why this does not fully apply to the solar system I give in book # 7.

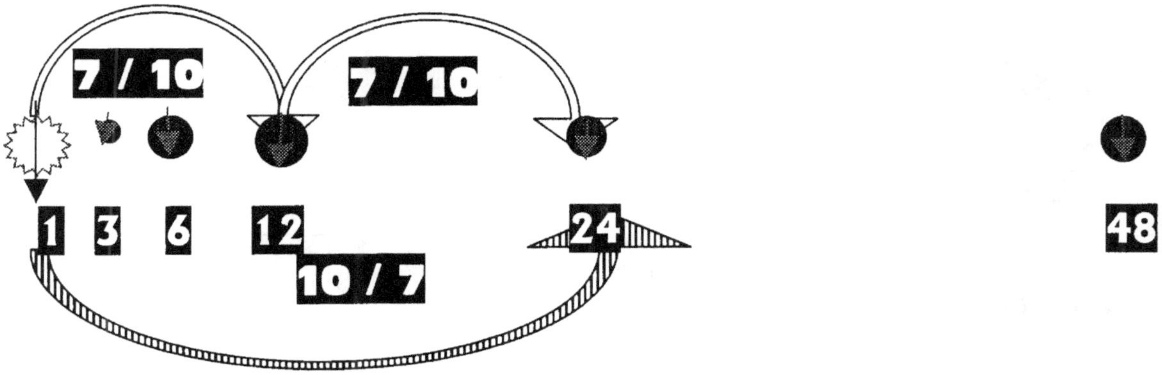

7 / 10   **7 / 10**

1 **3** **6** **12**   **24**   **48**

10 / 7

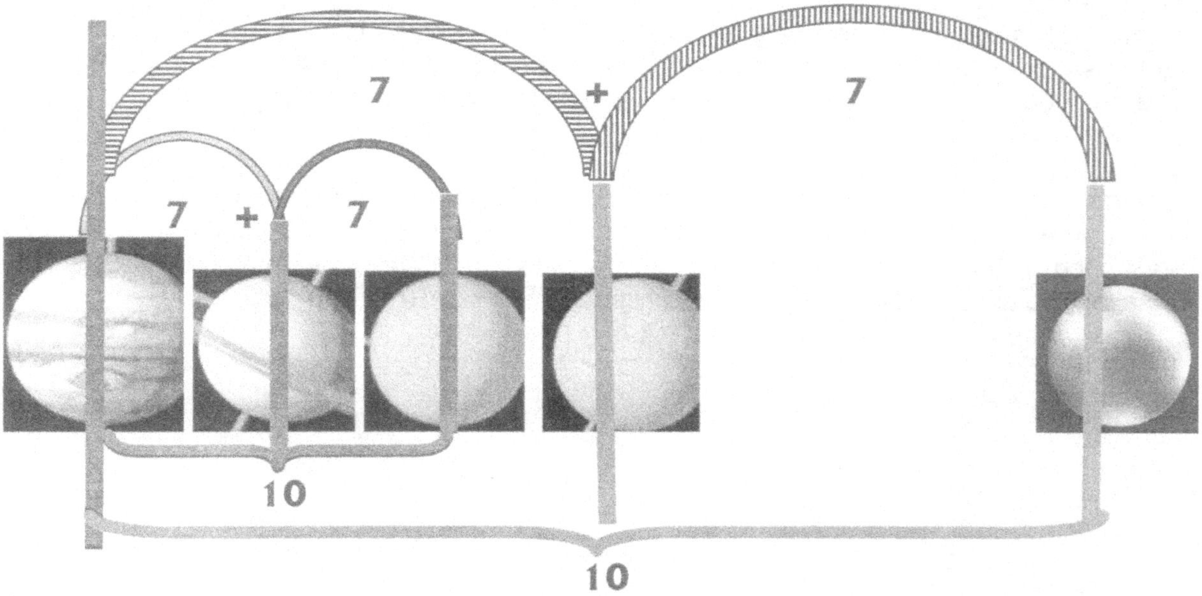

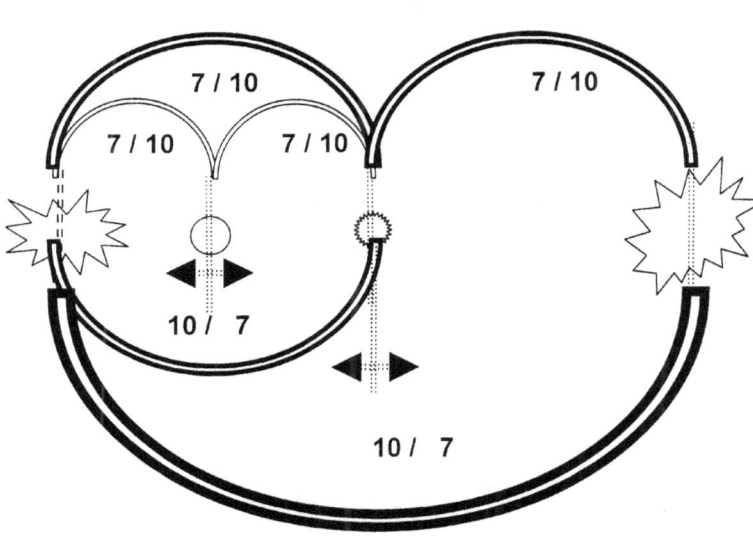

7 + 7 + 7

7 + 7

10

10

7 / 10   7 / 10   7 / 10   7 / 10

10 / 7

10 / 7

It is only the planet immediately next to the outside planet that forms a marker or a point carrying seven in relation to the centre Sun. All other planets are liquid space-time whether they are located further to the inside or more to the outside of the one forming the alliance, it does not matter. The planet takes the first seven value from the planet directly on its inside and then according to that seven places its marker position in the relative measure to the diameter distance of that next inside structure.

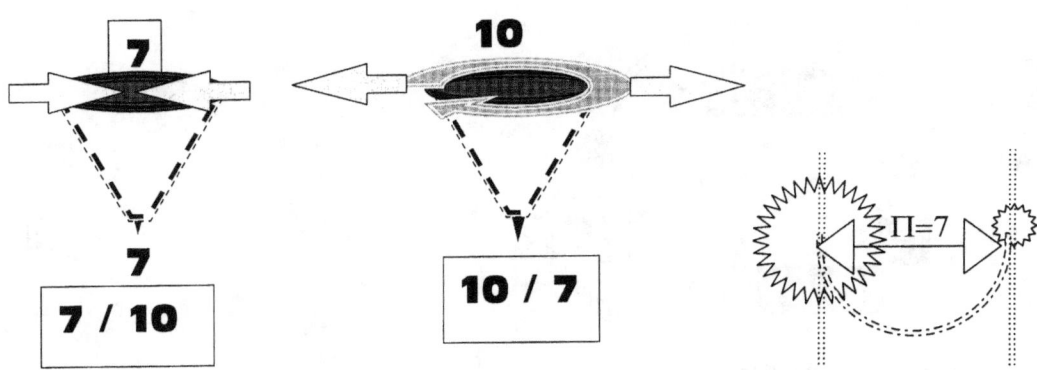

The seven used forms as a relative number that indicates material and the indication serves as it in the form of a sphere being present. In another book called ***The Seven Days Of Creation*** in which I mathematically explain how I see the solar system came about, I prove why the square of seven as the diameter became an abnormality. The proof that I employ to serve my theory is still visible and before the lens of the telescope as well as hidden in the mathematics.

From the centre of the Sun there is a connection between the Sun Π^0 that forms Π^0 and the planet that forms Π. The seven holds the value of seven, but that I explain in another book where such explaining is more to the point. Between the centre alignment there has to be two sevens relating to ten because there are two $7^2 = 49$ in $10^2 = 100$

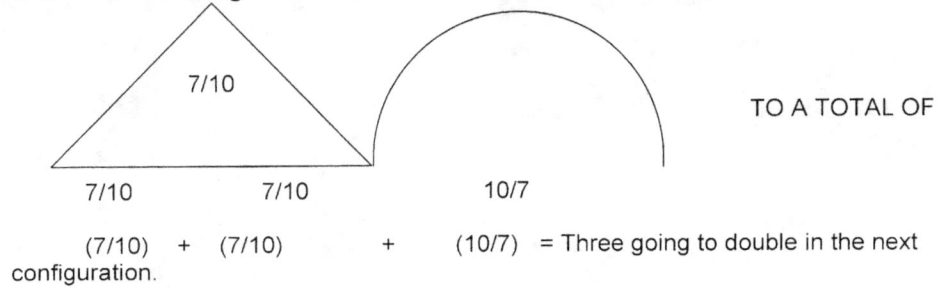

(7/10)   +   (7/10)   +   (10/7)   = Three going to double in the next configuration.

In the square of time (10) there are seven on the one side of time while on the square of material there are a double of seven relating to the square of time.

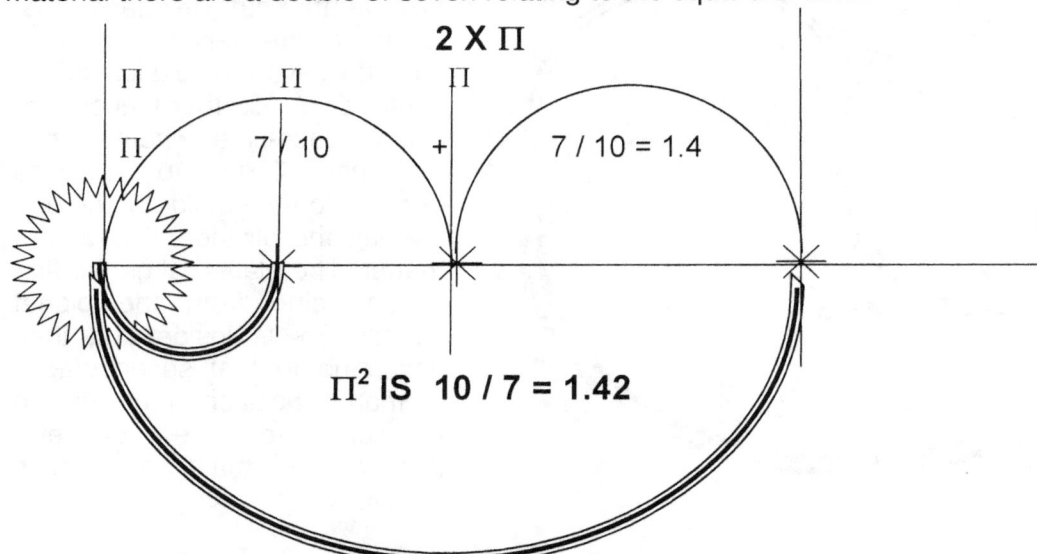

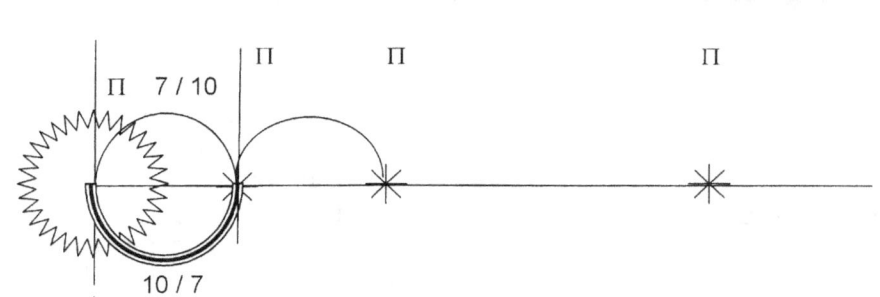

This will be the reason why Mercury has such an "abnormal" orbiting route.

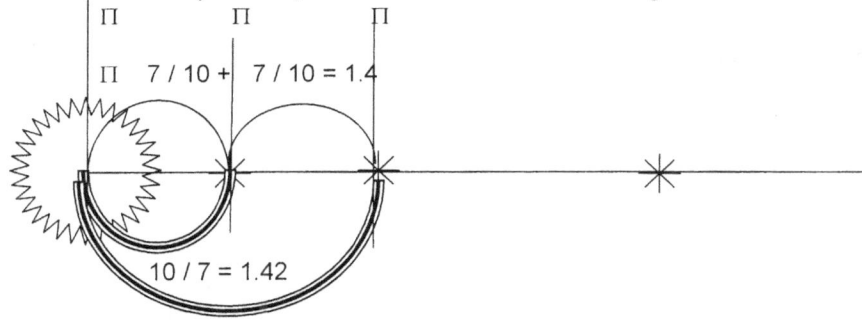

THERE REMAINS A DISCREPINCY OF ,02

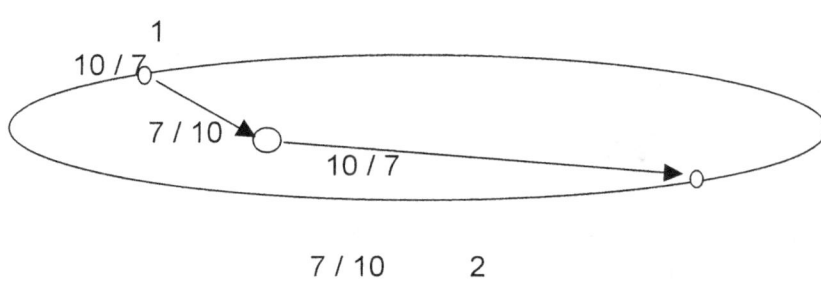

The extension of Π is well received as a dimensional implication to matter holding seven positions from singularity and space having four quarters throughout the rotation of singularity forming the centre to the five dimensions (one side lost to the cube's six sides connecting to the five remaining sides) making the total sides facing space from the point holding singularity at any given instant at a value of twenty (4 X 5 = 20). Then adding the singularity cross of Π being (1+1) = 2 the relation becomes 22/7. This is crude because in more precise calculations it becomes .91 + 1 = 21.91/7 = Π

The sectors provide individual singularity as a means in sustaining governing singularity by which provision comes through maintaining governing singularity the required spin in maintaining cooling. If this process did not apply, there would be no connecting individual singularity to major singularity. The sectors provide individual singularity a means in sustaining governing singularity by which provision comes through maintaining governing singularity the required spin in maintaining cooling.

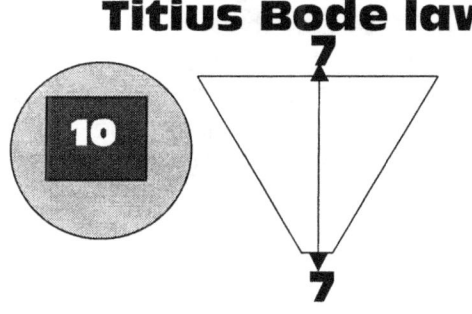

Titius Bode law

This ratio there is between the governing singularity and the marker, which is the planet next in line with the Sun. All the other planets, notwithstanding position or prominence have any implication on the planets Titius Bode layout. There is the centre singularity, which the Sun's centre represents and then from the Sun, next to the planet in need of aligning, the planet next door forms the first 7. That distance the planet has, irrespective of the planet number the planet is in relation to the Sun, that next door inner planet, forms the first seven which represents the first distance in relation to ten with the Sun.

From that planet forming the marker, the next seven forms as that distance doubles to place the planet in question in alignment. The Sun forms the centre aligning as singularity that holds 1^0 in relation to 1^1. The next planet is the marker, which is the first inner planet acting on behalf of the innermost planet and serves as the planet marking a position according to the Titius Bode law. All other planets to the inside or outside holds no validity except the one allocating the first seven as positional pointers to the inside. The distance from the sun to this first marker puts the position of the next planet in alignment so that it holds a marker of seven at one distance and a marker of another and next seven, which allocates the planet in question to the position it holds. This has all to do with splitting singularity in relation with time.

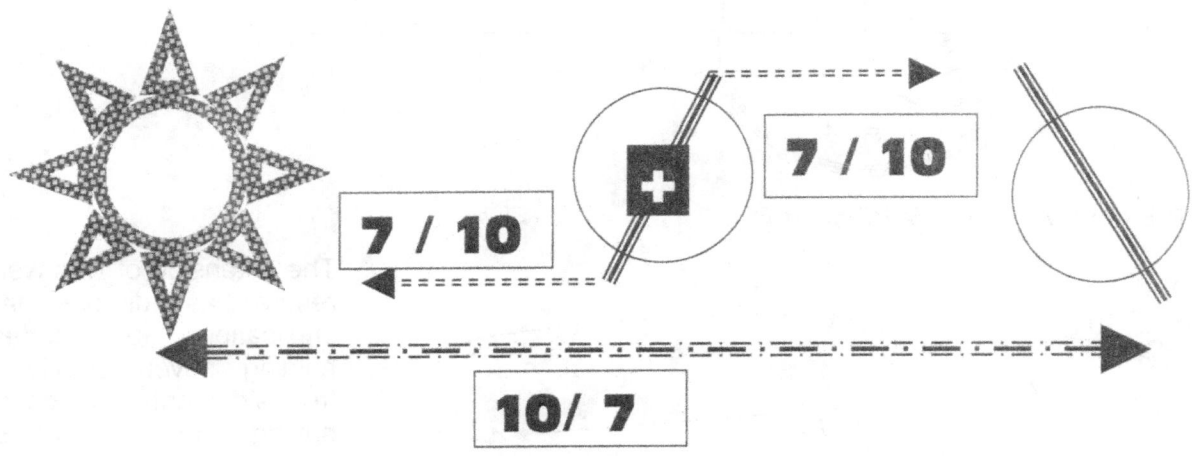

From the orbiting structure (planet) aligning singularity only one structure, the very inside singularity, applies as a position of reference and that is reference to the distance applied between the points in governing singularity. From the Sun (governing singularity) the matter marker is 7/10 = 0.7 with the only one other forming a marker 7/10 = 0.7. The two form 1.4. From the Sun (governing singularity) the outer planet forming the marker in search of position holds space in the square 10 / 7 = 1.42 in aligning with the 7 forming material of the Sun. Therefore, there are two sevens relating to ten forming the material positioning of the structure in orbit and from the governing singularity, all outside the Sun are the square of space (ten) aligning with one particle (seven) and not one of the other structures to the inside or the outside holds any value.

Because 7 + 7 = 14 and in the time aspect 10 / 7 = 1.42, the distance doubles every time there is an aligning of three orbiting objects in question. In this there is definite proof of influences coming about between particles sharing gravity. But then again the entire Universe shares gravity and as such then all will influence everything. With material taking the position of 7 + 7 = 14 and time holding the position as 10 / 7, the interaction between space, which is moving through time, would be an advancing by the measure of (7 + 7 = 14) / (10/ 7) = 9.86 and that holds the value of Π^2. That puts singularity Π in motion Π^2, equal to gravity, which is what Kepler introduces as the factor that keeps the universe in tact. It is the movement of singularity (9.86) = Π^2 = Π^3 / Π. That proves mathematically the Universe expand by the Titius bode law and mass is just a figment of Newtonian imagination.

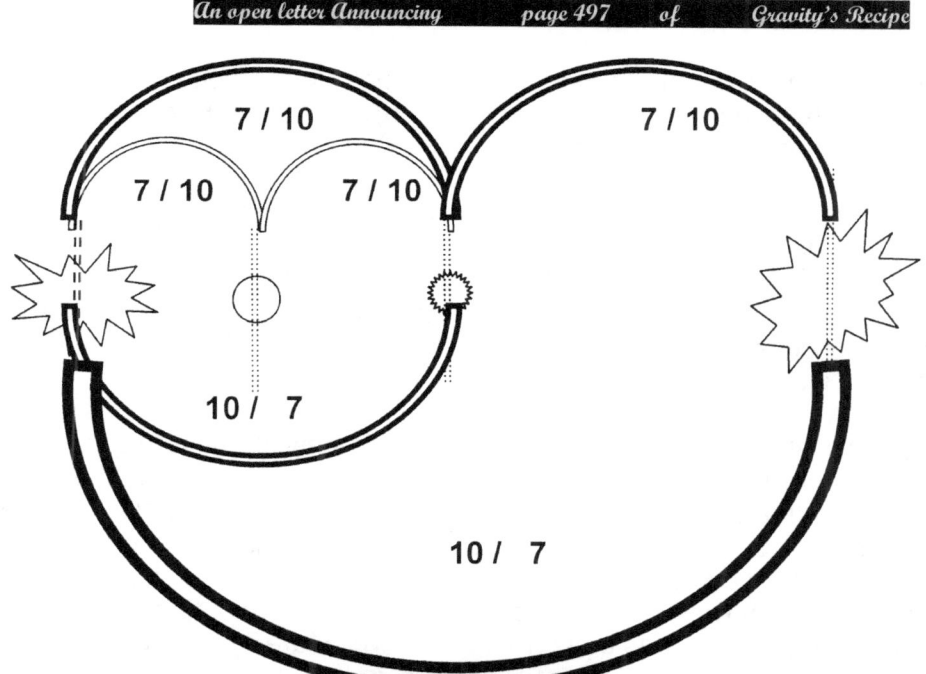

It is the position singularity holds in relation to the Universe and the Milky Way, forming currents and seasons moreover than the Sun shining brighter or not. I show a relation between singularity in different positions maintaining seasons and north / south polarity, not only as far as concerning the Earth but also outside influencing polarization. This has to do with the second position singularity holds in accordance to matter and space and is an "*electromagnetic*" (used for the lack of a better word) sustained positional opposing derived precisely from the graph in the manner when calculating electricity.

In this it is clear why the Titius Bode ([10 + 10 + 1 + .991] / 7) and the Lagrangian 5 \\ 7 systems part their ways when applying the different processes they hold. With all the differentiating, the observer must also consider the dual message that light uses in travelling through the vastness of universal space. The thought of nothing is just what it is, a thought of nothing and although it is in the human mind common nature to present nothing as a value in the recalling of something, nothing is a presentation of the figment in the human mind. There can be no number such as nothing and that was (possibly) Newton's biggest error. Nothing represents non-existing and that is just what nothing is, it is non-existing.

Mercury	Venus	Earth	Ceres	Mars
$4 - 4 = 0$ $14 - 4 = 10$	$0 + 7 = 7$ $20 - 4 = 16$	$7 \times 2 = 14$ $32 - 4 = 28$	$10 \times 2 = 20$	$16 \times 2 = 32$

Jupiter	Saturn	Uranus
$28 \times 2 = 56$ $56 - 4 = 52$	$52 \times 2 = 104$ $104 - 4 = 100$	$100 \times 2 = 200$ $200 - 4 = 196$

The Titius Bode influences in a manner that on the one side holds the matter-to-matter relation of 7+7/10 whilst on the other side, during the same time, holds the space-to-matter relation of 10/7, forming equal and opposing values. From this the orbits of cosmic structures are always oval favouring the singularity dynamics of the one structure at one point and switching the favouring to the other structure on the opposing side. Because the structures can never be equal in size (singularity will not permit that where the Roche principle will intervene) the shape is always "off centre" as well.

The ten dimensions I named the atomic relevancy is also showing the double value of singularity as singularity extends into as well as beyond space. The atomic relevancy is $(\Pi^2 + \Pi^2)(\Pi^2 \times \Pi \times 3) = 1836$ that is the mass relation between the electron (3) and the proton. Proton $= (\Pi^2 + \Pi^2)$ Neutron $= \Pi^2 \Pi$. The atomic relevancy holds the dynamics of singularity control. In the ratio and dimensions we find in the atom, all space-time derives from the atom, whatever the atom is.

It started with a dot, because that is the only form, size and dimension mathematical logic will allow our brain to accept. From the one dot had to come a second dot and a third dot. The dynamics of such a dot is smaller than we can understand because such a dot is in negative relation to what we see Π to be, and the deeper we delve in our quest to find the smallest fragment of material where space started, we may enter a road leading ever close to uncover infinity, but reaching infinity, that we will never accomplish. The value of infinity is 1^1 but there is another part of singularity that also has an eternal value, which are 1^0. We will always have 1^0 going to 1^1 but since from our vantage point, 1^0 and 1^1 is the same in measure, and yet they are an eternity apart, I had to devise some measure of separating the two. As the one is infinitely bigger than the other, I gave one the value of 1 and the other the value of .99999999 repeated infinitely to show the difference there has to be. After all, between these two values is our entire Universe. The division is between that which has no start and that which has no end and is in the spot where time is still eternal as much as we can accept eternity to be. This we find in the aligning of planets. Where the one dot from which the aligner stem becomes the reference too the distance applied between the aligner and the original dot, or governing singularity or structure in charge of holding position to all orbits following.

The reason why we should first locate the spot is because we can only work from that point forward. By working forward we have to work backwards to locate where we are heading. The cosmos started at a point and where such a point is, we will find the Universe. Every one knows where the Universe is, because we can see where the Universe is, but if we can see where the Universe is, then we should find the centre of the Universe in that spot. Einstein theoretically positioned the point of beginning at a place he indicated where singularity should be.

With the cosmos the size it is and space so large compared to our smallness, we have no chance in finding the centre of the Universe. The Universe started where singularity is and singularity is the sure indicator of the Universe. With all spinning objects holding singularity, we then have located singularity in as much as finding the centre of the Universe. The Universe started with a dot forming. That answer arises from taking mathematics back to a point of being the smallest possible position, far smaller than we may be able to calculate form.

My approach might seem unconventional but through the abandoning of the accepted, it enabled me in locating the precise location of a universal singularity forming a connecting basis of the Universe (this I say with some degree of confidence). The smallest figure there can be must be a dot. The dot is the only form that leaves all the options open to extend in any and in all directions should the opportunity arise. The only mathematically sensible option about extending a line from the dot will be non-bias progress in all directions equally in order to give a meaningful flow of mathematical equilibrium.

The Pythagoras mathematical principle is the proof and that I explain. The obtaining of singularity is in my rejecting of nothing by replacing it with something being the dot. With the clepsydra or "water thief" Empedocles deducted that air was composed of innumerable fine particles, braking the thought that what we now know is air, was also believed to contain nothing being altogether a space filled with nothing until proven to be wrong so many years ago. Never did science take the lesson learnt back then to the future and out into outer space. If there is space, there cannot be "nothing" as space is something. The claim becomes obvious when observing the connection between the half circle, the straight line and the triangle, which could also promote all the qualities lurking behind the pyramid. Consider the connection between 180^0 sharing and then one may realise much of the pyramid mystique becomes less spectacular in considering the very basic in mathematics being the Law of Pythagoras on which all mathematics are based. Once the water thief was eliminated by some human intelligence the matter was left at that. Nothing shifted out to an area we think of as outer space. In outer space we now find nothing. There is nothing but an atom here and there and even the atom is covered in nothing!

I wonder why the nothing landed there? Could it be that the reverse came about and because there was no visible "water thief" the very limit of man's suspicions came into practice. Man has always been extremely good in flying from one outer edge to another and if the water thief proved something was present, then the mere absence of a water thief must therefore prove that nothing must be in outer space. But what is space as such? What can space be, because with explosions we can clearly witness space created from heat. Our culture prevents us from admitting our vision, but the release of heat produces a *"shock wave"*. That *"shock wave"* is nothing less than space created from heat released. We have to brake free from culture of the past and a rigged mind set narrowing our vision.

Einstein's Critical Density lacks the accepted matching facts we need in proving the critical mass factor. But our inability in securing such required evidence defies the most basic logic. It seems all new evidence we receive from outer space is disputing all the findings of Newton's laws, which also disproves Einstein's Critical Density as the answer. The Universe will not reach a point of contracting, not withstanding whatever dark matter astronomers try to locate in the vast space.

Why would the expansion turn around and do a reverse by going back to where it came from. Consider the momentum alternation such a change will bring about.

The Sun is not a gas-filled sphere holding hydrogen in its "natural gas" form, but it is all fluid and it is in a liquid form where singularity is liquid- freezing hydrogen at 6500^0 C while outer space is boiling over at -276^0 C. This book explains the Roche limit in the practical sense… when applying cosmic laws instead of improvising cosmic laws and uncovering facts that reality then becomes awesome. It becomes clear the Universe is as much expanding as it is contracting and contracting by expanding. As there is no hot or cold, no big or small, no grand opposing but relevancies in ratio to one another. If you do not believe me, then believe your eyes when looking at the picture. What ever the Sun is it is fluid falling into fluid.

Consider the time it took from 10^{-43} to 10^{-5} seconds to create a cosmos the size of a neutron. Compare that to what is happening now and see how many events took place by the creation of every lepton and every hadron and it is true that that period took longer to complete than it took the Universe to create the solar system. The flow of light through the density that space produces heat, gives the speed of light the relevancy of time in space. The thicker the "soup" of heat is that space forms, the longer it will take light to cover a distance. It is very important to note that the speed of light is a relevancy between time (seconds) and space (kilometres). The speed relies completely on the value **k** holds on space –time. The speed of light is forever a constant but the constant is part of the relevancy of space-time

If one looks at the transmission of sound, it too depends on the relocation of matter, but to a very small degree, and in this process lies the transmitting of sound. To make the error of judgment in confusing the process with the breaking of the Doppler rings are quite understandable.

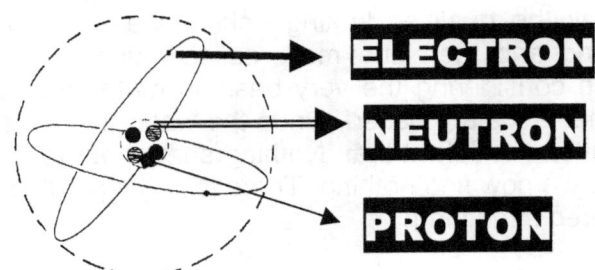

The Universe connects in a way Kepler established through his relevancy theory. Those not convinced answer this: where would the Planets be if not for the Sun securing planet positions.

The relation proves the ratio of one in all cases to be valid. It proves much more than merely connections at liberty of holding positions where ever the randomly opportunity placed the structure. The structure does not come closer by a pulling and tugging. Why not test Newton's $F = G (M.m)/r^2$ from figures Kepler left us and see how far did planets shift closer. I guess this will again make this book as successful as the others with me openly criticising Newton and Newtonians, but Universities are not about knowledge but t about protectionism. Universities protect their own without any willingness to test that which it protects. From that we then can see what we are waiting for and how long before the big solar clashing will begin. The absence in they're just mentioning such possibility confirm to me they know as well as I do there is no tugging and the Universe is in synchrony more than any person may ever be able to prove. The atom is the manner how the Universe takes form. The electron is a gateway providing displacement at a rate that supersedes the speed of light. However since it also secure and confine heat stored as material it allow the density of material to extend above and beyond the density of light, thus giving liquids versus solids.

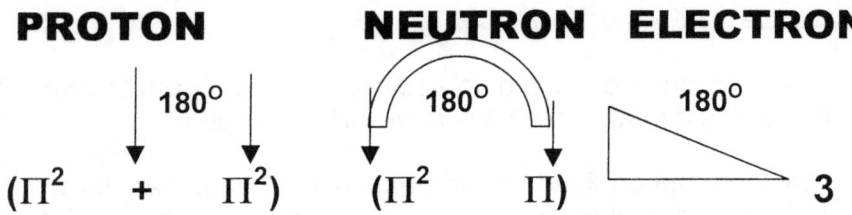

Everything in the cosmos is moving, either by own individual accord, or under the influence of some other singularity dominance. In explaining we return to the top.

When the top is in a state of motionlessness of own accord, it is everything but motionless. The motion it adapts are synchronised with the Earth in harmony with the solar system and according to the greater picture of the cosmos. When an energy source not related to the cosmos called life, intervenes and energises the top's motion, the singularity in that top suddenly jumps to life. By adopting a rotation energised to an unnatural state of energising because of life's intervention, the singularity of the top is not in charge but as it applies more and more energy, it will begin to find a means whereby it can escape and apply individual singularity as the top

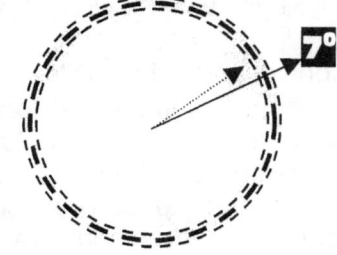

starts to separate from the singularity the Earth holds. The singularity holding the Earth would then allow the singularity of the top to rotate within a specific band where that specific band of being active before the Earth's singularity will start to destroy the singularity in rebellion. The top on the other hand will try its outmost, when the singularity it holds gets by individual spin is too strong to remain in domination of the Earth's singularity. The motion of the top is an attempt to begin applying an individual singularity space-time defying and standing apart from the Earth's gravity. That action we see as the top starts rotating in a manner where the top does not align with the Earth's singularity, but establishes a driving singularity independent from the Earth's gravity. With the adding of spin, the time the top holds becomes unrelated to the time the Earth holds and the top will start a campaign to escape from the singularity domination the Earth has on the top. When the time or spin of the top exceeds the limits the Earth places on the top, the top would emerge by trying to escape from constrains placed by the Earth. The view I represent at this point is known to science for almost as long as science knows mathematics.

Not long after the law of Pythagoras was understood where Pythagoras introduced mathematics Eratosthenes of Syene made as big a discovery as Pythagoras did. But in the one instance the world took notice because the world could see and understand and the other instance the world disregarded the findings because the world did not see what the implications were. The same apply to the aircraft flying and when the aircraft wishes to escape the earth's singularity hold, it has to comply with the laws laid down by the earth. The seven becomes as big a part of the concept as does Π as it all interacts

If we wish to find the future we should locate the past. If the cosmos is contracting, where to is it contracting? The direction of contracting must be in the opposing direction the direction of expanding. If we wish to locate the past from where the cosmos came and through that in what direction the cosmos came, we must take an effort to backtrack the direction it came from. Should the argument come about that all came from nothing, then everything either still has to be at nothing, or our understanding of nothing leaves much to be desired? Nothing means not existing, not being, never found and unable to produce any multiplication of any growth.

The above questions, but mostly the fact of what is more nothing and what is less nothing draws me to the realisation that there can be no such a quantity in space as nothing because even space has to be something. Heat expands as the level rises and clearly it is for any one to see that the releasing of uncontrolled heat creates space, which is no more evident than by releasing a nuclear explosion. The wind is shock waves, but what is the shock wave other than new space coming into prominence. In that way it is clear that releasing heat brings about the expanding of r as part of the sphere forming space. Hubble proved the Universe is expanding. Then by backtracking we have to set about reducing the sphere constituting the expanding Universe. If r in the circle is growing we have to reduce r to backtrack.

When the circle reduces, the value located to r will become implicated because r determines specific size. Not so in the case of Π, because Π in the true sense only indicates that the circle is a square without corners and therefore Π dictates form and not size. By reducing size only r comes into contest and will point to such reduction. By reducing the circle radius r by half continuously will lead to an infinite small circle but Π will remain because the circle as a form remains even being infinitely small. In the past, and even in some quarters today, science is on the search for the 100% efficiency machine. That theory runs on the surmise that a machine can drive as an output delivery without receiving input of energy. A few hundred years ago many Kings were fooled by such notion and some scientists truly spent a lifetime in honest search of just such a device. Mostly the accomplishment came from cheats that very well know their machines were not up to the task, but in fooling a rich investor, brought about wealth to the inventor. As science progressed, the no input giving all output machine became less and lesser a feature of the honest inventor. But the idea does not exclusively come from crooks finding a way to cheat the world.

The practise of receiving without giving comes from science in the form of physics. It is physics taking the world on a wild goose chase in the way physics presents the cosmic motion. Physics propagates that the cosmos is all about running without input driving energy. The cosmos is all about wasting matter to a supply of motion. This idea prevails even after the world of science saw clearly in the past that there could be no such machine anywhere. Even the cosmos must be a machine driven by an input and an output. It is the input / output driving energy that must be located and the driving ability we have to locate. Science holds the mass drawing power to prominence, but what if it is not the drawing power of mass that holds prominence, but it is the reducing or contracting of space that is the driving motor behind the cosmos. All energy we humans at present use to accomplish matter motion, holds some form of heat redistribution. Even electricity is a form of pure heat. I say that in mind of what applies when the energy of electricity becomes over abundant and the machine overheats. By overheating it means that the motion the machine creates comes about from heat control and precisely planned heat distribution.

I realised that it is not me that is drawn towards the Earth, it is the space in which I find myself that reduces, and that produces the effort that is bringing me closer to the Earth, The formula $F = G(M_1.m_2)/r^2$ suggests driving, moving in a direction and contracting. It suggests the reducing of space and not merely drawing or moving closer. When looking at any machine in practice, the machine draws power from space reducing whereby heat increases. Not releasing the heat to form space will lead to the destruction of the composition forming the machine. There is no form of matter, or element strong enough to resist matter deformation brought about by overheating. Having this in mind that matter does not resist heat, it is of importance to recognise that it is heat that is allowing space to give matter form. Looking at the manner in which energy is utilised it is space and heat forming matter allowing motion that allows work to achieve value.

At this moment science is all about a body falling where the two bodies are producing a force whereby the bodies draw one another closer. The bigger the mass, the bigger the drawing that comes about from the force unleashed by the mass of matter. The idea about this practise was phenomenal in 1602, it was impressive in 1802, but it is really ridiculous in 2002. Why would Boron form a solid having 5 protons weighing 10.811 g / mol and Argon a gas having 18 protons weighing 39.9 g / mol, but the "heavy" element with the biggest drawing power is a gas and the lightest element is a solid. That denounces the contracting force theory. The way we compile and use energy must be in a similar manner to the way the cosmos uses energy distribution. **We humans can create nothing, but nothing is all that we humans can create**. The rest of our achievements are by duplicating whatever nature provides. To establish what drives the Universe except for blaming some medieval magical force coming from nowhere going nowhere we have to find what drives us. The energy we use in all forms is producing heat in space by either converting space to heat or heat to space. Explosions are about converting heat to space. Compressing is about reducing space to heat. That is all energy composing work and is the only method of producing energy notwithstanding the immeasurable many names we use to express the same function in different forms.

Arriving at the question about locating the space and time forming the centre of the Universe one has to realise the centre of the Universe are in every singularity forming matter be it is big or small, size carries no significance. It is the impartiality of singularity that is claiming the value and not the differentiation of matter. One must realise there are no big / small or hot /cold or near / far. It is all relevancies between matter claiming space and space is heat in a turnabout manner. Every aspect in the cosmos is locked-in Universes, sealed off from other Universes and inclusive or exclusive depending on singularity holding relevancies relating to one another. The relevancies rely on inter dependence and inter linking, but there are no differences according to human sizes or standards. Accepting that principle unlocks the "so called mysteries" of the Universe and brings about clear understanding. It is all about accepting, acknowledging and interpreting the role singularity maintains on matter.

One should not try to focus on an image of such a spot or dot because there is no image. The line dividing the cosmos and that run through every particle, no matter how large or small is beyond our vision. Such a small line, so small it is not even noticeable is large enough to part the cosmos into sectors. It splits the biggest there is into particles and we are not even able to notice the precise location of such a split. In truth there is no top or bottom that we living in 3D can see.

We shall have to use a general conception brought about by intelligence. Your intellect tells you about such a spot, but that is all because that spot is on the other side of the Universe (quite literally). From the centre of the dot there is a top and a bottom spot. From those points there is connection with four quarters. That produces six connecting points that are all aligning to the centre. Because it serves big and small, hot and cold equal and alike, and it is the smallest cutting the biggest into equality, size is of no issue. Size is what man makes of it. In the Universe there is no size in hot and cold, large and small. For the smallest there is, it is serving the largest there is equally.

Hydrogen is as much a liquid as iron is a gas and neon is a solid. It depends on the element relating to the space/heat in the circumstances surrounding the substance at that very precise instant in time. We have to stop telling the cosmos to show us what we wish to find and start accepting what the cosmos is telling us to find. The culture that I am referring to is all about **nothing.** At present we find that there is something we think of as nothing in outer space. Because nothing is what we wish to find and nothing is precisely what we are getting because we think of outer space as nothing. If you accept the cosmos to be nothing, then please define nothing to yourself and find the definition in the cosmos. The liquid the Sun has is the driving force that creates the duplication in motion. Without such liquid heat the Sun would become stationary and only depend on contraction while the contraction then passes the motion onto the heat in outer space. While the Star is in liquid all motion comes from the accumulating spin effort of the combined motion all elements together accomplish. In the case where the star is still in liquid the heat is stored in the atoms and as the star develops the heat transfers to the governing singularity, which makes the star immobile as the governing singularity takes charge of the entire star.

Our instincts, our logic and our calculating process all indicate that the sphere holds a centre point from where six evenly positioned point's position matter to be. Using The formula $F=G(M_1.m_2)/r^2$ it indicates to a force pulling objects closer, where each force is coming from each centre point the body in question has. The contraction must commit the two bodies towards a point in each case being spot on in the middle, not withstanding what direction the force is applying, the body will draw to the centre.

If the Universe spins around a centre point holding singularity, and singularity confirms the centre of the Universe, then every particle holds the centre of the Universe making the number of universal centres immeasurable many, and every atom and sub atomic particle presented outside the atom in smaller bits, are all not pieces of the Universe but they are a Universe surrounded by many Universes. If every atomic particle no matter how small, is holding the centre of the Universe, then the gravity is coming about from that point because that is where the gravity applying in the Universe is applying contraction.

It then is the atom in the most centre part where space and time meets singularity, that Einstein found a Universe collapsing to a single dimension, and every atom at a point post of the proton where gravity initiates in according with the proton dimensional colas of $(\Pi^2+\Pi^2)(\Pi^2 \times \Pi \times 3) = 1836$

See the fluid push out of the Sun where it lets the Sun seems to be a bowl of liquid, as the liquid is spilling back to the Sun and not escaping into the cold of outer space

as released heat should do. The liquid heat squirts into the cold of outer space and then it falls into the bowl of liquid that is the Sun. The inside of the Sun is not gas but it is fluid.

In all of nature there is no **NATURAL GAS** as much as there is no **NATURAL SOLID**.
 No element is either a gas or is a fluid or is a solid. We arrange the elements in such a manner, but that is only applying to the situation the earth grants the elements.

When an element freezes it is solid notwithstanding...
When an element melts it becomes a liquid

notwithstanding...
When an element boils it is a gas again notwithstanding...

Another point I question about the Official Policy is that they, as I am, are in agreement that 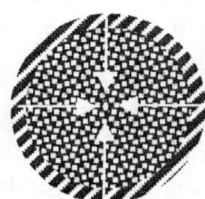 the heat melted particles onto particles and in those joining better combinations of particles came about. How it happened is another bone of contention but more about that a little later on. There was heat on the outside and there was matter on the inside. The heat was liquid because the Sun and other stars still indicate masses of liquid fluid inside. I can only imagine that that liquid inside the Sun holding temperatures as low as 6500^0 K and up to 1.8×10^6 K the heat already is in a molten form. What about the heat then when the frozen outer space was 10^{34} K and such temperatures were the general order of the day back then. If the Sun is liquid now, then those temperatures raging back then must put the heat in form available in outer space at the time as thick as mud.

From the outside drawn onto the particle inside the blanket of heat came a flow of soup that became matter. That much I do understand. This carried on until...when? When did this stop? When did the Universe run out of heat? When could one consider outer space as the coldest all around? Where to did the Universe dismiss the heat that was once there but now is empty? How did the process stop of bringing from space intense heat and from that particles grew stop? When did it stop affecting the growth of a particle, the growth of space? In fact the growth of everything that grew came from this first growth. What you see or do not see grew since it was part of the Big Bang and everything in the cosmos at present was part of the cosmos during the Big Bang. I say this process of collecting heat from outer space never stopped but was an on going process we now give a nice name calling it gravity. Outer space never became empty and void but relevancies changed concepts where centres formed that should not be as it then interferes with concepts about relevancies Gravity is not and never was about particles pulling each other closer. If it was, no Big Bang was possible. Gravity is about turning space, which is released heat back to heat and concentrate the heat where gravity is the strongest and heat is the least. Space is the transverse form of heat and visa versa is also true. Should any one not believe me, try a bicycle pump by compressing the plunger while blocking the valve bit. The heat will burn your finger to blisters if the force on the plunger is strong enough, the plunger seals enough and your ability to withstand pain can last that long. Then answer your own question about where the heat came from because it sure as hell is hot, it did not come from friction with air particles such as oxygen and nitrogen escaping through the valve bit. Heat is unleashed space and space is concentrated heat. Reducing space to heat is gravity and antigravity is expanding from overheating blowing into space accumulation.

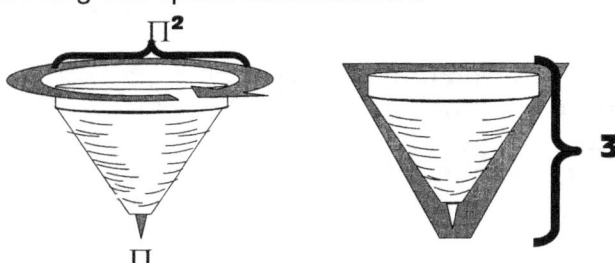

When looking at a sphere the inside has always (in a cosmic relevancy) the location with strongest heat also always has the strongest gravity in any given cosmic sphere. The centre of the sphere clusters the combination of particles forming the sphere into unity. By holding a specific centre the sphere becomes the strongest form any object can be. The sphere is without any doubt the favourite choice in forming gravity. Where gravity has the last say without other influences changing possibilities as collisions leaving debris in space or natural out burst like Super Nova explosions, gravity will enforce the sphere to be the form taken by the particle. But there is no evidence of particles of similar size joining in matrimony through gravity being the shotgun at the wedding. In cases where there is a mismatch of size outside any proportions of equality then there is a contracting of the lesser by the greater. In such cases the lesser is not qualifying as material (and that I prove later on) but the greater considers all the lesser to be heat. It is humans bringing distinction to matter in form.

The two objects should have their own value of gravity and _gravitons_ and in comparison with the _gravitons_ of the Earth; their value is insignificant. However, these two balls are in their own individual deuce to see who reaches the Earth first, and the iron ball's _gravitons_ should give it a superior advantage. This comes about because the two objects are in a position where they compare in relation to one another and share a common second factor, which is the Earth. In relation to the Earth, the gravity - motions of the two balls do not come into consideration, but this does not play a part since the Earth is a common factor. The balls however, are put in a situation where they stand in relation to each other. When compared to one another, the _gravitons_ should give the heavier ball a sizable advantage. In such an event one of the structures are turned to heat as it is liquefied flowing in the space dominated by the other and larger structure. If the structure proves too large the superior structure turns the lesser compatriot into heat. Then being heat it will apply gravity and admit such heat into the ranks of its atmosphere, but not before it turned it into fragments good enough to be heat.

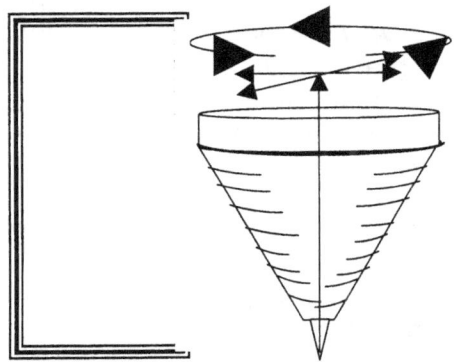

In the centre runs a line called the axis line. The line does not show any influence on managing the top when the top is motionless and bounded by the Earth gravity. However, the sooner a motion sets in that is adequately strong enough to support the

independence of the top, the top generates enough gravity to sustain an independent attitude in relation to the Earth.

The differences we find in the line carrying the factor of 3 and the factor of $\Pi^2\Pi$, is that the one represents a relation that singularity Π has with the motion Π^2 of time in singularity, whereas the factor represents time in motion coming from the past through the preset and into the future 3.

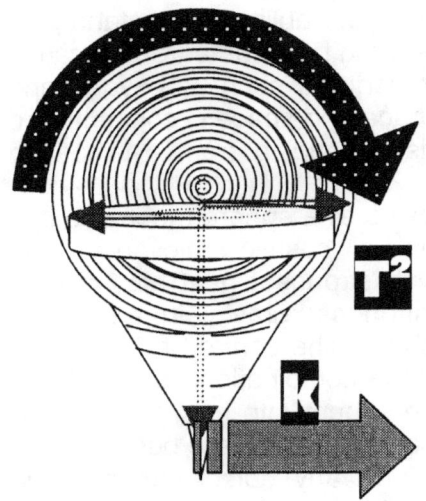

The dynamics that then support **the top in motion come from four points serving the top with time $\Pi^2+\Pi^2$ which come about from the circle the top forms by spinning the body of the top rotating and the space rotating in relation to singularity forming what we think of as space $\Pi\Pi^2$.**

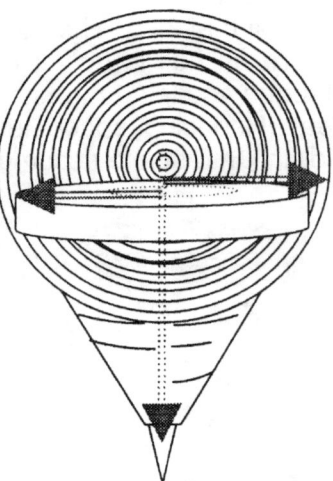

After the top is thrown the top changes some very vital characteristics in behaviour to what it had when it was not spinning. As the top hits the ground after being thrown with it's commencing its spin initially the top starts to rotate and it is as if it wishes to exceed its spin by moving around in circle – like formations while spinning excessively around its axis.

It spins vigorously as if the top suddenly is too energetic and exited to stand still and that is precisely what happens. This surging with excitement is a charging of vitality that finds a new dynamic and is a most important example of the manifesting of the combination becoming an effort as one accumulates cosmic effort that accumulates the rules of cosmic principles.

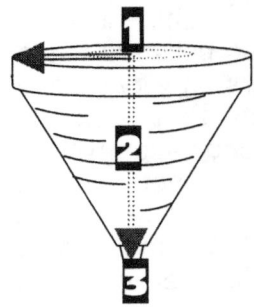

One can clearly see that it is Kepler's formula playing out as the Coanda principle. While the space is developed and defined by the motion $\mathbf{T^2}$, which verifies the independent space the top has acquired by motion of spin, it is also clear that the relevant factor of linear motion demonstrates it's presence in the moving about as the top is rotating. That puts Newton's claim of motion not being a factor in total disbelieve.

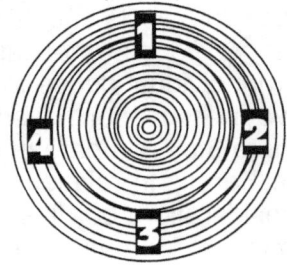

With all the excitement and no where to take the extending of the drive line, the effect thereof runs down the developed singularity inwards towards the newly established governing singularity that keeps the newly formed Universe erect. That is why the top is spinning in the first place. The more assertive the spin is in velocity, the more reaction there is from the lines running towards the centre and extending through the expanding outwards. In real terms the space of the top expands as the spin is in contact with more time in motion in quicker presentations during the same time in period as a bigger material unit fills the space because of more material duplication in the same period that is allowing the top to spin. In this the space in which the top spins has to expand as well as the process of filling time by means of duplicating, in order to compromise for the material relevancy growth to fit the newly acquired singularity governing the space-time and being erect by the motion. The support that the spinning top finds in it's task to establish a governing or controlling singularity keeps the top spinning in an upright and erect position that is only supported by the motion putting space between infinity in the centre and eternity in which the top spins. By placing differences between infinity and eternity it not only charges singularity to life but also charges space-time into the Universe.

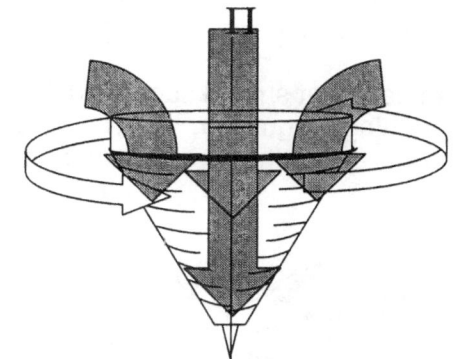

Through the behaviour and characteristics that the top displays it is possible to find answers to the cosmos that the greatest mathematical minds was unable to solve. It is a case where their genius was to great to find solutions and the search of the final results proved to complicated.

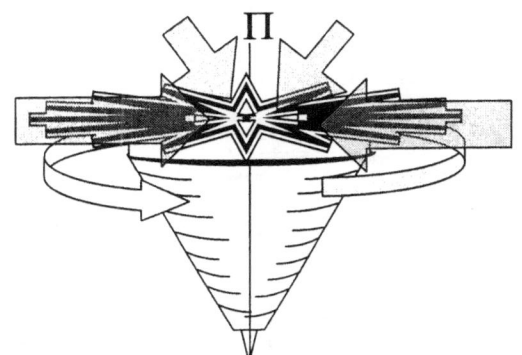

Looking at what takes place as the top starts spinning, we find a line coming from what was a mathematical point. What is there is not there but for those with intellect to find that what is there being present. What is there is not part of the cosmos and yet what is there drives the cosmos. It generates motion by not turning and places what is, in contrast to what was and what will be the very next instant. The facts are indisputable even to the most ardent mathematical Newtonian disbeliever and what the top parts is how the Universe came about. Motion parted eternity and infinity by developing space-time as a partitioning screen.

More spin increases both lines that force gravity by the increasing of T^2 extending k, k^{-1} as well as a^3. The space wants to exceed its boundary because the motion suddenly allows the space to become extended. The gravity line running to the centre wants to extend for the same reasons and so does the gravity line running towards the liquid that should be there and that should be enforcing this sudden living up to better standards.

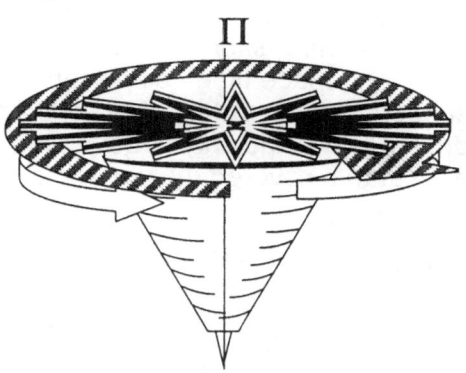

The spin under normal conditions can only come about as a result of more heat. With that aside the spin normally caused by heat will bring on a linear gravity running towards the centre of the top. This is then a product of $k = a^3 / T^2$. But to counter this (Newton's law on action and reaction), another balance comes about where $k^{-1} = T^2 / a^3$ that centres the material inline with the progressive spin and the extending of the motion that should be because of a liquid heat adding to the material.

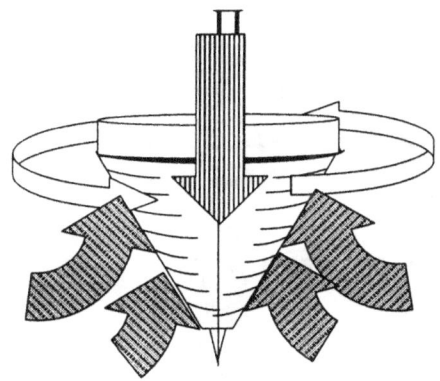

The support that the spinning top finds keeps it upright and performing as if in a fighting mood. However, again I have to press the point in regard to the top, that it is life that initiates the motion and for this motion to start as a natural flow of events requires a lot of nourishing by the independent singularity that starts to drive the object through a combined effort of rotation of all the included atoms accumulating the contracted heat to bring on such motion. When this process is in a natural occurrence within a star within a galactica it is the indication of the coming about of a newly developing in the heat centred cradle of a galactica. However it can only be gravity that is able to fight gravity by extending the Earth gravity and by extending the Earth gravity we find some part of the Roche limit also applying.

The heat that should supposedly under cosmos law drive the spinning top, will come from the governing singularity, which is accumulating the heat in concentration by the contraction or cooling ability the top singularity acquired. However, in this case the spin is a result of life's

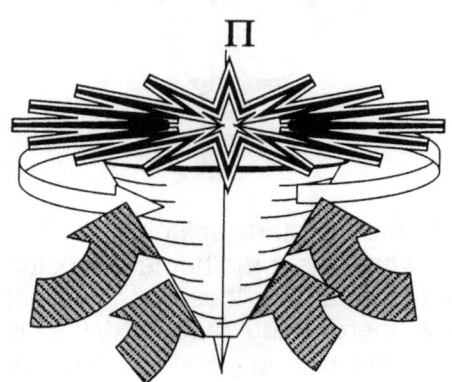

ability to manipulate space-time and alter cosmic events to the free will and interfering, as result of the nature of life. The heat that would establish such a drive in motion in real cosmic terms would require a lot of nourishing a sustaining from a large number of maintaining atoms that produce a large flow of space-time and can concentrate much excess heat outside of the atom sphere.

With sufficient energy the top gets

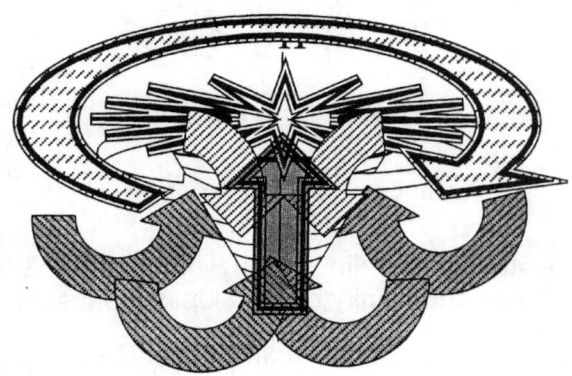

into a fighting mood that makes the top very reluctant to give up this newly established freedom. The behaviour now attributed to the top is normally the manner how a star develops in the galactica cocoon and how the fledgling star gains it's birth right to leave the nest of the cradle of the galactica. The atoms form a sum total of space-time displacement that can support the generating of the required gravity in securing the heat that would unleash such a drive. Such singularity in governing comes to life and releases the new star from the blanket of heat that covered the star up to the time of its release.

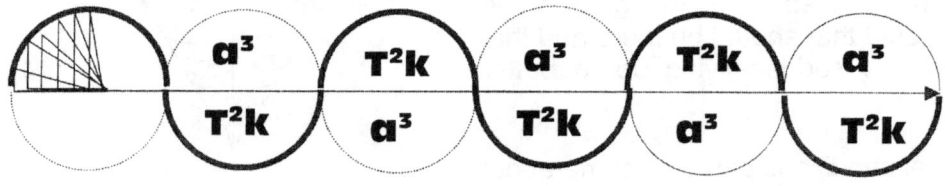

This example that we can gather from the top's behaviour shows how desperate the governing singularity can become when starved of motion and how such an excited singularity can put up a fight for life and independence. The top is in a fight for independence while the Earth is restraining the independence. The fight goes on until the Earth suppresses the last bit of motion that the top has and the top uses the last motion it has to defy the Earth's control.

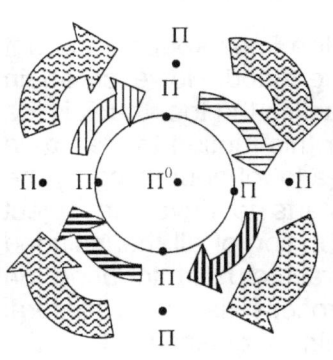

When the motion exceeds the level of the Earth's gravity, the top shows an eagerness to rise to a higher level of independence in the same manner that an electron reaches into higher rings of energy because the top with motion is in an electron or exemplifies an

expanding relation with the Earth, which is placing the Earth in the role of that, which is filling the proton or contraction role and the atmosphere being in the neutron role or that which is the factor responsible for the task to supply the gravity-motion.

Let's quickly establish events as they translate singularity from a dot to a controlling entity that is demanding space-time through the establishing of a separate individual drive. The motion comes about which proves to be that which generates the gravity that drives the individuality in the top.

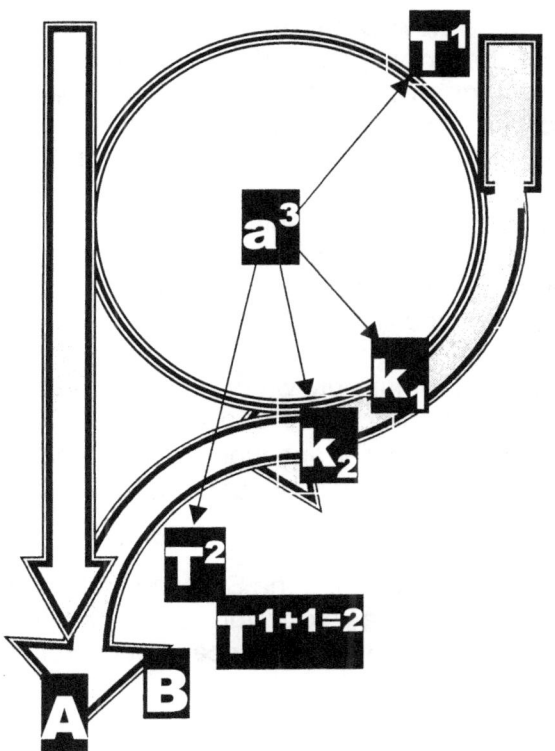

In the sphere centre is the spot that has to be there mathematically by measure of $(\Pi r^2) / (\Pi r^2) = (\Pi^0 r^0) = 1$. In order to provoke the line forming singularity into existence, motion is required, just as Kepler indicated where the space becomes equal to the motion and the motion is equal to the space $a^3 = T^2 k$

The inner four is singularity points equal to, as well as that it is singularity being in direct contact with singularity charging time. It is singularity charging the presence of singularity by four points that would form the four, forming time. Then one further point to the outside would form space with an indirect link to singularity. This formation goes on as long as time distorts to form space and space will forever have four to the inside connecting time and three in equivalence forming motion.

The Coanda effect is proof of gravity coming about through space forming motion. In the case where water diverts the normal directional flow, the space that translates to the motion is deflecting singularity with the flowing water charging the motion. In the centre of the object having the round form, singularity is duplicated and by transferring Π to form Π^2 and the motion of the water creates a line of gravity that pushes the flowing water to follow the direction that the newly obtained gravity applies to the water.

Everyone of the four rotating points spinning around a centre while duplicating the value of Π in relation to the centre Π^0 at a measure of $\Pi / 2$ and where Π^2 is responsible for establishing as well as relocating Π by duplicating Π through the motion thereof, therefore $\Pi^2 / 4$ becomes a limit in relation to the development from the centre. One has to remember that a star of the present takes characteristics of the form from the era before space was a factor.

As the absolute master of motion **Newton** should have placed emphasis on the motion aspect when he, as a young man, saw an apple fall from a tree. He made a brief calculation but he used the mass instead of the motion while Galileo proved that mass has no value while the falling occurs. Seeing this he jotted down a formula and chucked it away. Newton however insisted on mass in spite of the clear evidence brought by Galileo to the contrary of mass playing a part. However, most surprising to me is that most Newtonians are not only incapable of seeing the facts my way but they sometimes get pretty unpleasant in a very coldish pleasant way about my view. If mass had a major part, then the more massive must fall quicker because the mass will provide the drive and the drive will excel the velocity. While it is true that all things fall equally, then mass has no part to play while the action of falling is taking place and that is in spite of all the Newtonian abstinence about the matter.

Normally water will run down to the centre of any gravity point, as A shows. By allowing the flowing water to come into contact, an object of a specific form the flow will divert (B) from the normal line and follow the contour of the object presented. For that to take place there is one condition that has to come about.

This again proves Kepler's statement of $k = a^3/T^2$ that specifically states that space (in this case the object transferring singularity to a new position within the round object) and with the motion of the water redirects the gravity flow of the water to new space in new time. Only Kepler can explain the phenomenon but only when Kepler stands alone, correctly interpreted and divorced from Newton's opinion about Kepler's statements.

The motion we detect as part of the Coanda effect runs through all spinning material. It is part of the atom as much as it is part of the sound barrier and the sound barrier is just another atom having Lyman series lines, and works also by the principle of adding heat, which puts the object expanding in a higher relevancy than there was before.

However much noteworthy as it is it is, prudent to consider that only when the atom unit is broken and the Roche limit is crossed, does the sound barrier come into affect. It is the breaking of the bonding unit ($\Pi^2/2$) that becomes the sound barrier although the breaking is never completed ($\Pi^2/4$) as long as both objects share concentrated liquid time that the earth supplies.

It is as if one then must claim in affect that Kepler held $a^3 = T^2 k = 0$. If the Sun and the Earth have a rotating relevancy of zero, either the Sun has gone away or the Earth stopped existing. One cannot claim there is a wheel and then remove the spokes because according to your taste, you do not like the spokes

In this matter I am disputing Newton's honesty. He placed the relevance on mass when he was a young man and retracting his former claim would have tarnished his reputation as a genius. His glory was worth more to his mind than what the truth was. As a young man he drew instant fame by claiming mass as the driving force and when as an older man he found he had to retract the first genius, his fame seeking would have left him with a scar on his reputation. In that I can forgive the man for the man was human and as Cecil John Rhodes said, all men have a price by which the man can be bought. Newton's academic genius was his all- important vice. The problem is that the incorrectness stuck with science for almost four hundred years onwards and no brilliant mind since then was able to make the Galileo mass connection. What happened to the many wise that walked the path after Newton had gone to better grounds? Where is the honesty in those that were supposed to search for the unblemished truth? Gravity is motion and mass is the restraining of the motion of gravity. The top shows the truth. Let us reflect once more

What is it the Newtonians fail to see? If an electron is orbiting around an atom, the inside of the atom must be a circle. If the atom was not a circle, it then had to be a cube. The electron cannot rotate around a cube; therefore, the inside of the atom is a circle. The cosmos is one big atom imitating all atoms as all atoms produce one Universe

In a circle, there is a radius that initiates the circle. The calculation of such a circle is $\Pi \times r^2$.

$$\frac{\Pi r^2}{r^2} = \Pi$$

If one removes the radius from the circle, the circle remains, only holding the value of Π. By removing the value of r, Π becomes singularity with no place to be. Singularity is the place where there is no space to be in place. However, Π remains because once r receives the slightest of space Π will find space. Then the circle will grow to Πr^2 and r would determine the space. Without space, there is no r but there is a circle with the value of Π.

Singularity is in every single rotating object, be it the proton or the combining effort of all particles in the Universe. That is what light and the photon is. It is concentrated heat that the Sun (or any other generator of electricity) concentrates to connect the concentrated on heat to singularity where the heat receives either temporary connection to singularity or a small piece of individual singularity. All spinning matter has the point where the spin is still there but the radius is too small to measure by any means. That point is standing still in relation to the rest of the spin. In relation to that logic I do not accept Newtonian science holding the radius of the spinning object unrelated to the spin, whether the spin is applying or not.

Applying Newton's second law F=ma

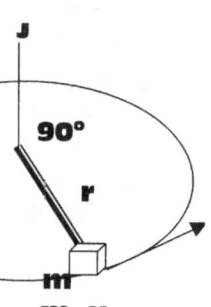

One arrive at the formula
$GMm / r^2 = m (\omega^2 r)$
By replacing $(\omega^2 r)$ with $2\Pi / T$ we obtain Kepler's third law
This law predicts that $T^2 = a^3 r$

The mass (m) multiplying the speed (v) forms a new value J AND THEREFORE j CONTINUOUS TO IMPLY $J = I \omega$

$J = r \times p$ where $p = (v = r \times \omega)$

$J = r.m.v = m.r^2 .\omega = I. \omega$ and becomes interpreted as $J = I \omega$

This establishes that $r = dJ / dt$

Since this is the absolute crux that Newtonian science pivots around I feel it is important enough to return to the whole issue once more in similar detail.

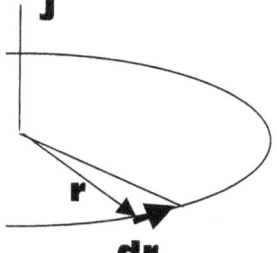

$r = dJ / dt$ In the case of planets in orbit around the Sun r forms a value of zero because $dJ / dt = 0$.

Since Newton became an institution forming the King bee of the academic cartel world wide The Brainy Bunch had Newton's vision written in the minds of the future generations almost at gunpoint...well definitely at an academic gunpoint. I am not the brightest in the world that I admit, but one thing no one can do, not even if you are the one and only Isaac Newton, is that you cannot place any relevancy in a relevancy and then claim it not to be in a relevancy because such a relevancy does not suit your taste.

 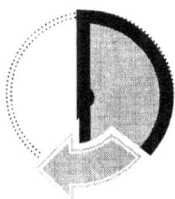 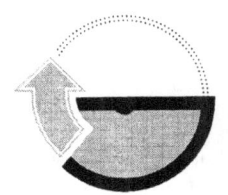 

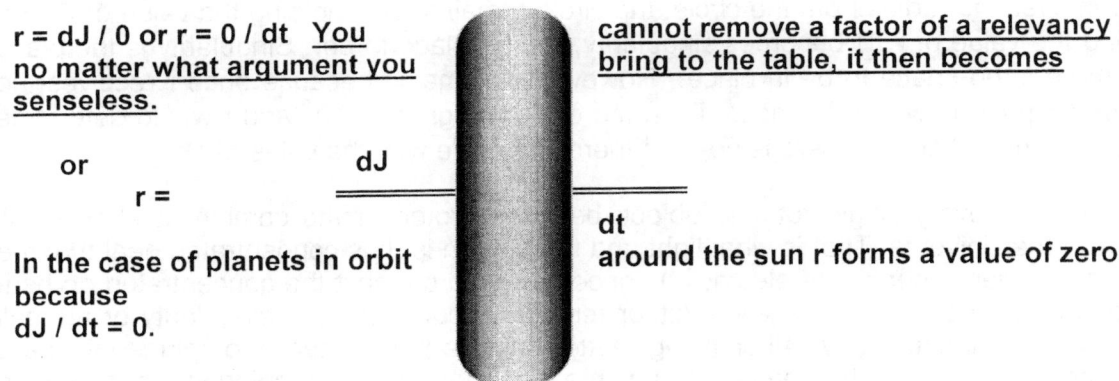

$r = dJ / 0$ or $r = 0 / dt$ <u>You</u>
<u>no matter what argument you</u>
<u>senseless.</u>

<u>cannot remove a factor of a relevancy</u>
<u>bring to the table, it then becomes</u>

or   dJ
     $r =$ $\overline{}$
             dt

In the case of planets in orbit
because
$dJ / dt = 0$.

around the sun r forms a value of zero

I wonder where would one put the zero part on the spinning wheel and what part must be excluded from the wheel. What Newton suggests, is a wheel has one side on top and no side at the bottom. While the wheel is spinning, one may not remove the one side and then claim there is no attachment between the top and the bottom. That would mean in a graph the top is not connected to the bottom because a wheel spinning is a graph moving against time. It is this principle that is responsible for the manner in which all power chain driving is done and not the least electricity. Every quarter of a rotating body is opposing the opposite sector directly and completely.

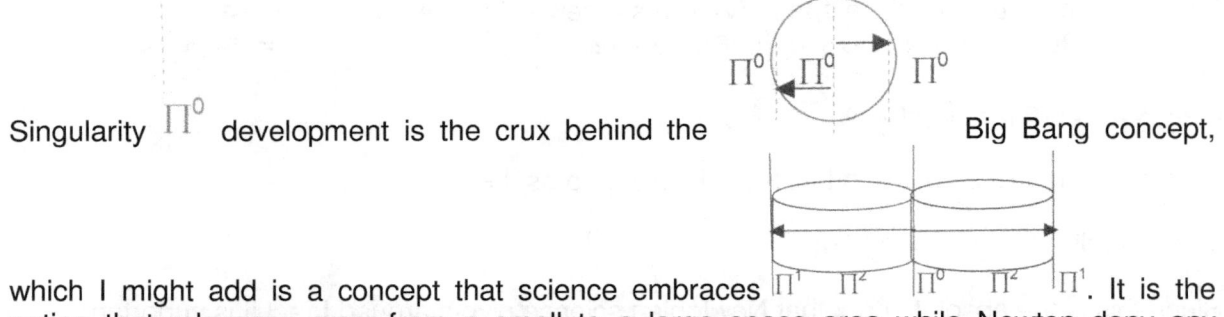

Singularity Π^0 development is the crux behind the   Big Bang concept,

which I might add is a concept that science embraces Π^1 Π^2 Π^0 Π^2 Π^1. It is the notion that relevancy grew from a small to a large space area while Newton deny any existing of such a relevancy factor in all of science. Newton sees the rotation canceling the relevancy and then Newton put nothing as a value and as a factor in that place.

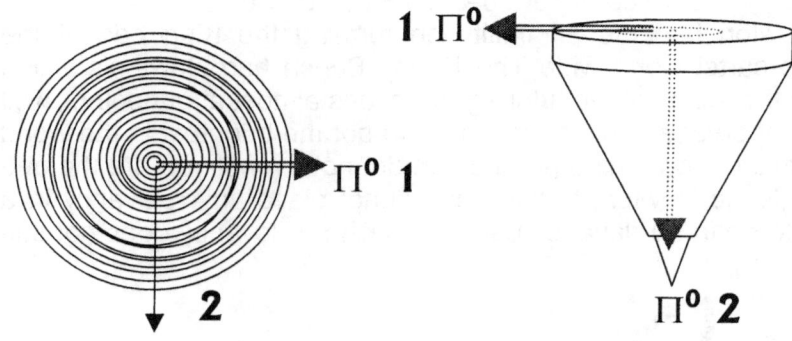

1 Π^0

Π^0 **1**

2

Π^0 **2**

Any Newtonian that wishes to justify any form of support about Newton's claim on rotation not establishing work must please explain what happens when the Coanda principle draws water by motion and how that motion

cannot be gravity.

If there is no production through motion, how would a top find a balance and what then inspires singularity to charge an erect stance through the generation of motion? These are legitimate questions in search of answers.

If gravity was mass inspired it would have the result that the Earth must be at some point during the year more massive than during other periods of the year. This we know is not the case and therefore the claim on mass is somewhat silly and a little bit of nostalgia coming from the Middle Ages. There are two equal but opposing gravity directions counterbalancing and both are the same that works independently to achieve a mutual goal. The relevancy from one side is about claiming space by progressing time and the other is by containing space through reclining time. That is why the comet never hits the Sun. It is because the Sun and the comet are in four different seasons in relation to each other while they are going through the quarter motion of time.

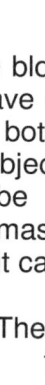

Gravity is motion and space is the blocking of the motion. Any object must have either gravity or mass but cannot have both. An object can be with gravity or the object can be with mass but

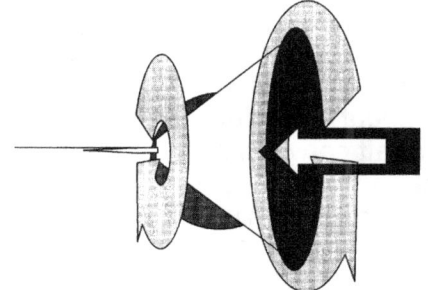

it cannot be in both conditions simultaneously.

The motion of the neutron (2) covers the gravity (3) that the neutron has while the space (4) flows

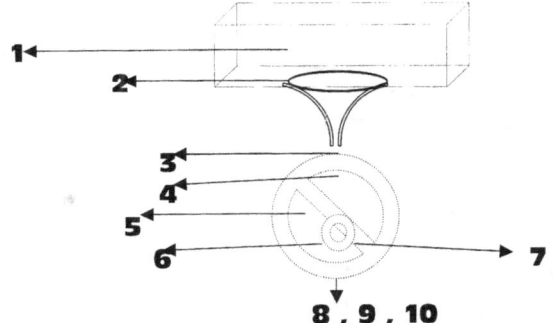

unhindered from the time (1) position through the location (5) fitting the neutron to the location (7) fitting the proton (8,9,10). The electron has mass because it restricts the flow or gravity and the proton has mass because in constrains the flow of gravity. Only the neutron has gravity because it flows unrestricted. That is what Galileo's work tries to prove but no one listens even to someone as important as Galileo because every one is mesmerized by Newton while Newton was absorbed by the lack of understanding the difference between gravity and mass. If he did understand the difference there is then he would have realized what Galileo was trying to say. While his little apple fell, it had gravity, but once it landed it had no more motion and therefore the containing part

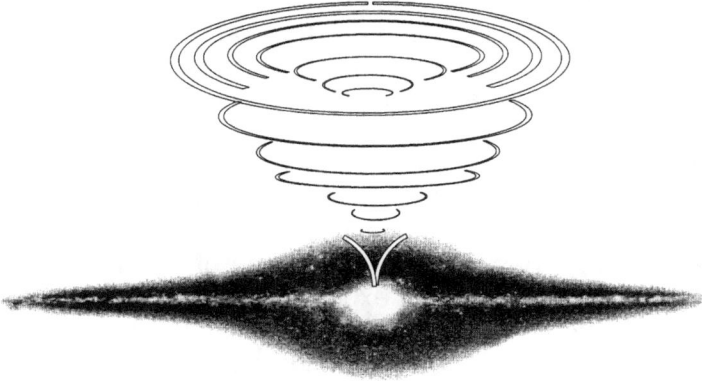

took charge as the apple then had mass. Galileo said all things fall equal (meaning all things are equal in gravity) while falling or while being in motion notwithstanding the difference in size or mass. That Newton missed. That part all Newtonians that came later also missed.

That is why Newton's first finding $F \quad \alpha \quad \dfrac{M_1 M_2}{r_2}$ being $F \quad = \quad \dfrac{r^2}{M_1 M_2}$ is most true and most accurate however $F \quad = \quad G \dfrac{M_1 M_2}{r^2}$ is nonsense. All objects must be in motion $\mathbf{a^3 = T^2 k}$ where it will be $\mathbf{k = a^3/T^2}$ in relation to one location and in another at the same time it will be $\mathbf{k^{-1} = T^2/a^3}$.

The motion within a galactica even generates sufficient gravity to re-enact a Black Hole within the centre of the galactica. This comes about just like the motion by which the top rotates, but the centre singularity being charged in the galactica is truly then a product of cosmic proportions.

The aircraft has mass when the aircraft is without motion and standing still. The motion of the aircraft becomes additional as soon as the motion that the heat of the engine produces, converts the heat to motion. The burning of fuel that is expanding through igniting, becomes the source of additional heat that is converted to space. By converting the space to motion or gravity, is that which then relieves the aircraft of some of its mass as the motion converts a part of the mass into gravity. Some of the mass turns to gravity and gravity is pure motion. It will always be some of the mass since the aircraft cannot be all motion. In the motion coming about from the engine that is converting heat to expand into motion the aircraft converts part (not all while it is in the Earth atmosphere) into gravity, which is independent motion from that of the Earth. While the aircraft is within the Earth, the Earth provides the motion and thereby serves the mass, which the aircraft (or all other bodies for that matter) will endure as the bodies remain a part of the Earth atmosphere.

The ship has mass but the buoyancy of the liquid in the water sustains the mass factor in order to provide the ship with another factor and that is displacement. The water holds motion in place and since the ship being on the water becomes part of the water in relation to the Earth it holds a part of the water in mass. However, since the ship then holds part of the air or atmosphere in relation to the water the ship becomes part of the air in relation to the water and therefore the ship holds a relevancy of air in relation to the water. Some of the mass the ship has is regarded by the water as air and some of the mass the ship has is regarded by the Earth as water. It is locked in relevancy as the factors establish a ratio.

By enlarging the ratio of air (wind we call it) onto a part of the ship, the motion takes up a part of the mass of the ship into the realms of the air and the air contributes to the motion that then finds the ability to go beyond the breaking power the mass has and converts some of the mass into motion. Again the wind is merely heat expanding and the expanding provides the motion that contributes to the duplication of the ship. An army battle tank is all mass in our thinking because of the iron composition providing it with such a solid and heavy structure. When the tank is thrown from a flying craft, some of the structure goes to mass because the tank requires more than one parachute to slow the descent down making the fall less destructive in nature. If the tank is left to fall with any other body and without restraining, the tank will not fall faster than any other body because the gravity the tank has is equal to all other bodies. It is the restraining of the gravity that produces the mass that requires a larger effort to contain the decline of the tank, but that again is interfering with nature since mass is interfering with the normal flow of nature.

The fact is that motion is the duplication of the same in ratio of the relative flow of time and when the duplication starts to claim the same position at the same location during the motion in time, the motion of the duplication of the space converts the part being restricted to the

same location as mass, while the rest is being converted to duplicating gravity. By duplicating, the mass converts to motion and while the duplicating is hindered, the restraining goes into mass. But in all Newton's claim that motion results in nothing is nonsense.

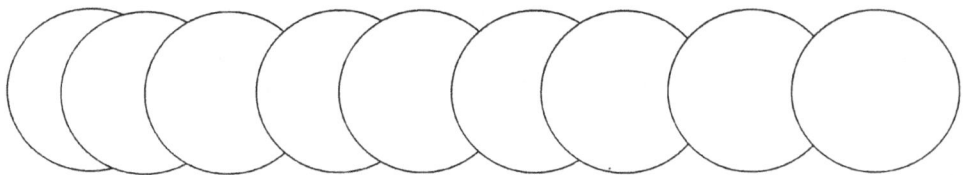

The fact that rotation does not produce work is impossible since rotation brings about motion changing the principles of the location.
It is the same relevancy we find in the motion applying to atoms. One do seem to get the impression that little changes in line with the rotation will bring some forward motion and some returning to the original position.

Even by using half a wheel would still bring considerable confusion but one can clearly see that Newton's presumption does not quite match reality.

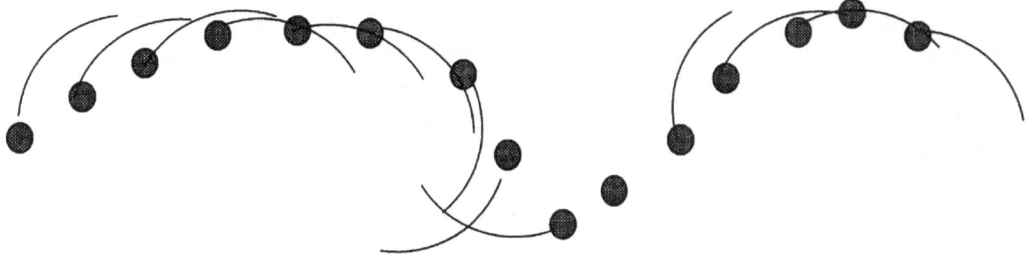

Shortening the arch changes the complexity considerably as one can then see a changing of the arch does not nearly bring the return of the dot to the previous spot.

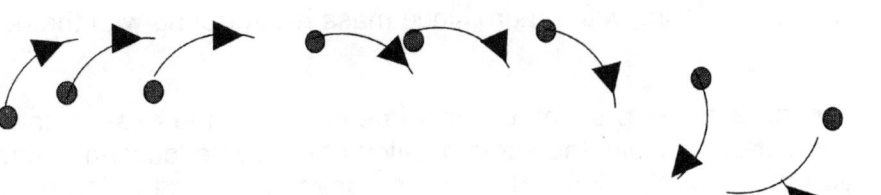

When placing arrows pointing in a direction that is indicating the direction of the line of movement, it becomes clear that there is a complete mismatching and the cosmos changes as rotation progresses. The behaviour, which I describe, is a flow of space through the line of time. Electricity is charged in this manner. The same generated force keeping the top upright is what is used to generate electricity. The flow of a charged conductor through excited space-time brings about the flow of a current.

An object in outer space has limited motion, which provides a part of mass and another part in gravity or motion. When the same object is in a Black Hole it is limitless and infinite in mass and has no motion. Outer space however, is all motion as it provides motion therefore outer space is without mass. There is a mixing of mass or motion being gravity but having both is not having the same.

Even the electron serves the line of time, in the same manner. As the Earth spins through time by repositioning space in time, singularity is re-applied, repositioned and re-aligned with the entire Universe in the manner I describe. The relation of the proton moving has to effect the following location of the electron since the electron is relevant to a position in space in time by a continuous motion through time.

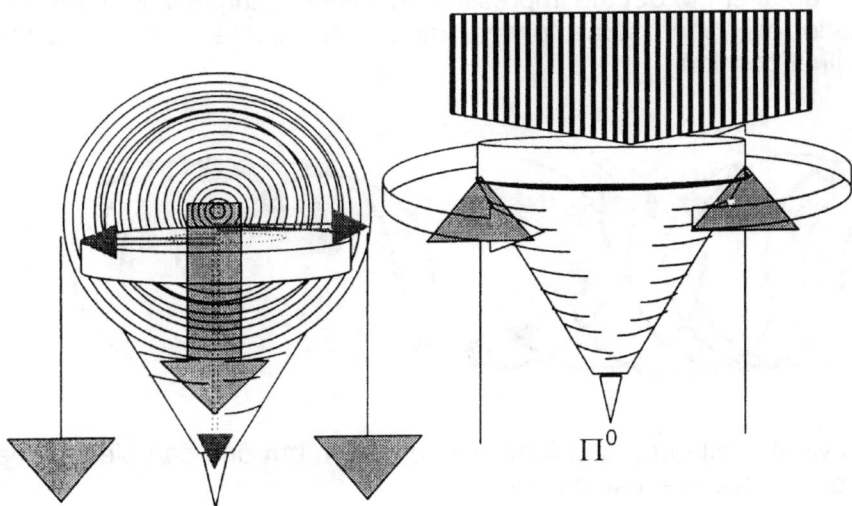

Π^0

Notwithstanding the motion that space forms as space is moving towards the centre of the Earth, the top finds a way to counteract the motion by producing a motion that is stronger than the motion restriction or in other words the mass, that is the restricting of the earth's gravity that fights to dissolve the independence of the top altogether and thereby form mass which then is the lack of independent motion of the top. By spinning there is no force pulling the top down and restraining the top to the surface of the soil. The mass is still there but the top tries to combat that mass with vigour and the needlepoint holds the top spinning as the top is fighting the mass. On the needlepoint the top rides out whatever force the mass would enforce to restrain the mass.

The total restriction of the mass control over the top has all but disappeared because the force or mass that the needlepoint of the top generates, multiplies the normal mass of the motionless top many times over because of the intensity that such a small area has on the increase of the effectiveness of the top. Yet, notwithstanding even more restriction by an increase in the mass restriction, the motion still generates independence by motion evoking a defying erect stance.

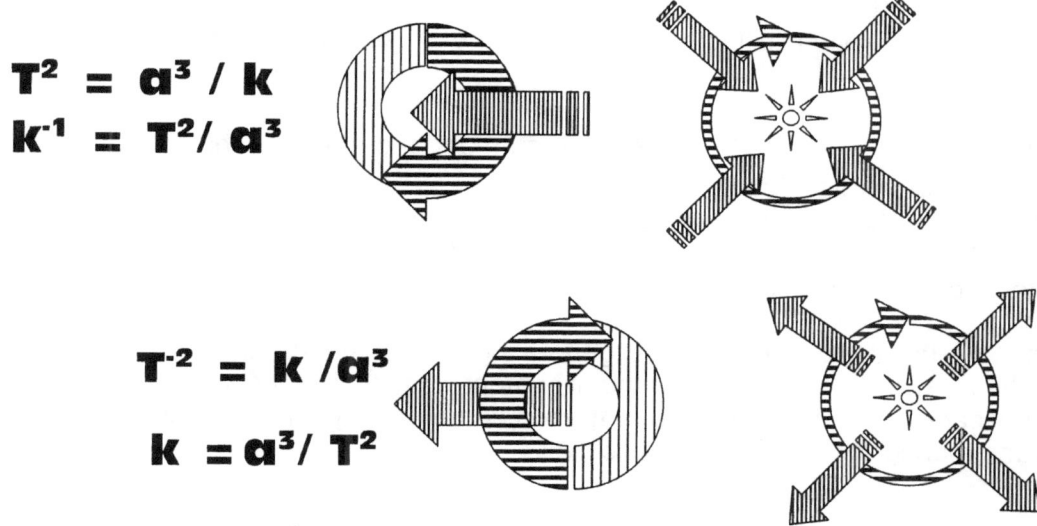

$$T^2 = a^3 / k$$
$$k^{-1} = T^2/ a^3$$

$$T^{-2} = k /a^3$$
$$k = a^3/ T^2$$

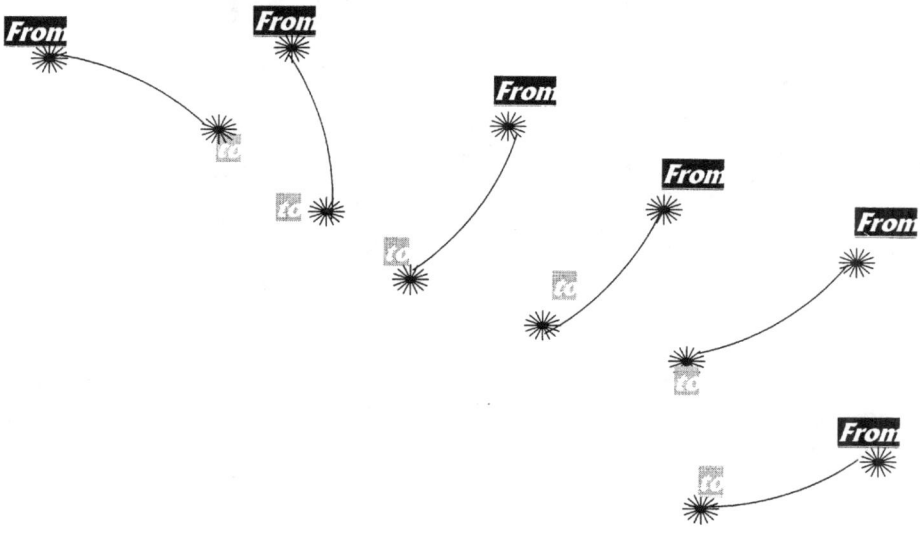

In the spin there is relevance within the unit that forms all the principles we attach and associate with gravity. There is the expanding as well as the contracting which form an integrated part of the rotation principle. If there were no rotation of a body, which installs the contrasting, we find associated with rotation, then gravity by principle would not have been possible. However the linear aspect also shows strong influences, which is as much part of gravity as gravity by rotation is a factor.

Although the electron is orbiting at the speed of light it is still in motion vertically and that also becomes a product of time as the whole structure is repositioning the relevancies it had a moment before to that which it will have the next moment. Such motion will again have an influence on the relation in the position the electron forms with the rest of the Universe while the lump of metal is now travelling as a spacecraft destined to other galactica. It is if we use the logic those intellectuals calling themselves Academics show and those Super-Educated that advocate how we may travel to far away galactica while we go on skipping the nearby galactica that is only two to twenty million light years away. Since the electron is duplicating by motion, the motion

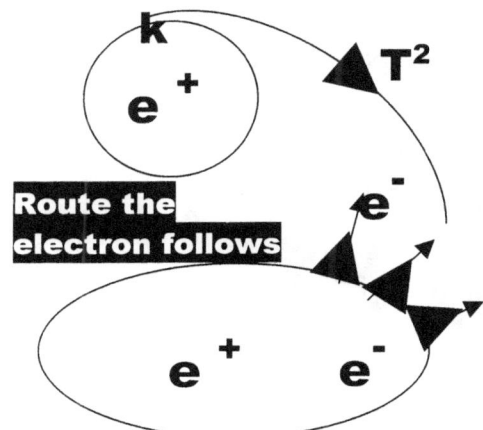

Route the electron follows

links the electron to a time constant. The time constant is linked to the speed of light but time as such, is part of the speed of light. The faster we take the electron to go straight in the motion man produces, the less time there will be for the electron to circle around the atom. If we make **k** bigger in relation to increased motion, the smaller will **T²** produce a usable space.

There is **k** that forms the distance between the proton and the electron while the electron is spinning **T²** around the proton **k⁰**. While all this action is going on, we think of the atom as being very still and satisfied with being a small part in a lump of metal we call iron. It could be any element but I use iron just as an example this time. The lump of iron is as motionless on earth as anything can be while being.

Even by coming erect through motion the generating of this stance finds its roots in the relocating of the rotating (**T²**) in relation to the alignment with the line (**k**) in relation to space-time **a³** in time-space **T²k**. The generating of the top and of gravity and electricity is provided in the very same manner by the Coanda principle.

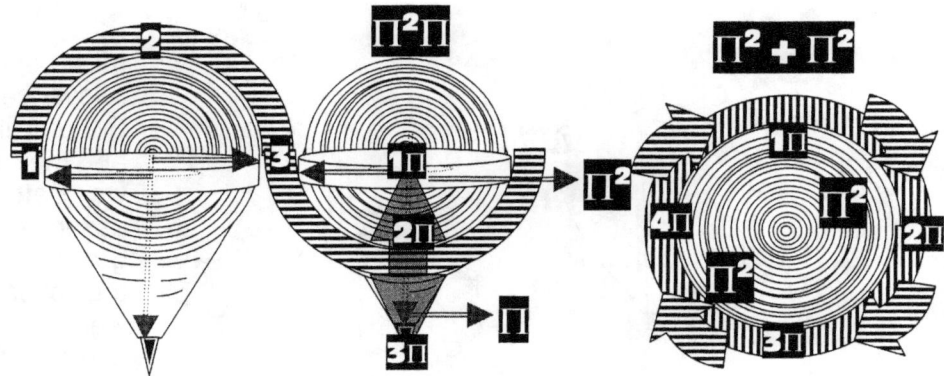

The motion that evokes singularity charges a graph from where the graph runs along the line of time. It is said that the spinning top is in balance but there the explanation ends and all parties are satisfied. Never is the question raised about what comes into balance? The balance is a control of space-time that is established as space is duplicated by time while time supports space in duplicating. The space is limited by the rotary action of **4** points in relation to singularity where this generates **3** points serving infinity that creates a division between infinity holding it's centre space and eternity being **3** active positions in time and the three is an eternal motion that never ends. By setting the division between **3** in infinity and **3** in eternity the containing that comes about is creating a cyclic space in **4** points.

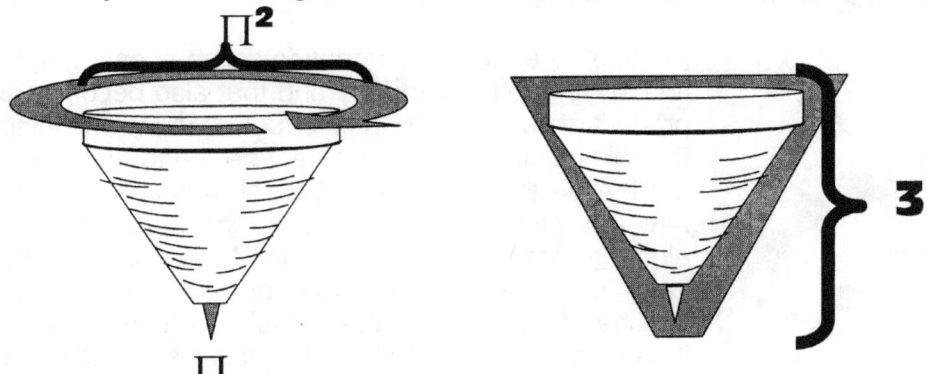

There is a something (if you wish I can use the term force although I strongly hesitate to use such an outrageous term for the most common aspect of the Universe) that is generating the

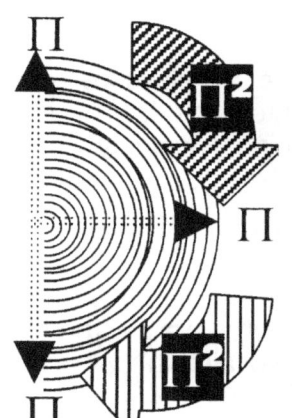

power that keeps the top upright while rotating. The energy that is charged has the dynamics to stand its ground against the gravity of the Earth where the gravity of the Earth would under normal conditions depress the top into submission. However, by rotating the top seems inspired and is reviving singularity by motion. The top is fighting and rebelling against the Earth's gravity when in spin. The top is performing the same way as an electric motor would. The difference there is between it and an electric motor is the origin of the source that produces the drive. After all the debating, there is one source that drives all forces small medium and large and that is the containing of heat and the distributing of heat.

The Roche limit came in place at the time when all the phenomena came about. When the phenomena came about that action brought us a Universe to have and enjoy. It was when singularity Π^0 heated to form Π and that had to involve motion. When Π^0 expanded and formed Π it had to cross Π

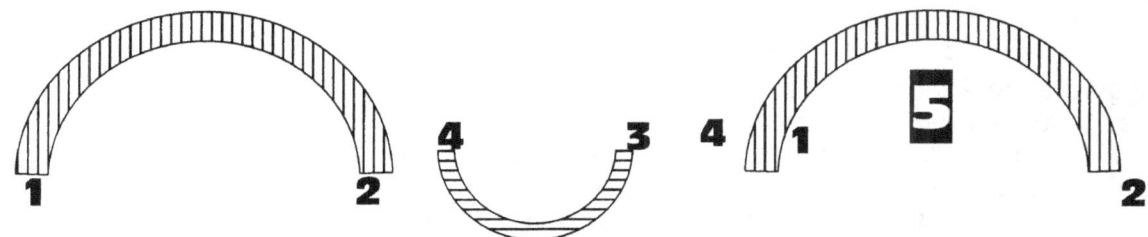

in doing so. In order to establish motion Π^2 it had to go from Π all the way to where Π duplicated as Π. This involved the initial motion at moment – Alfa when space formed time by forming space.

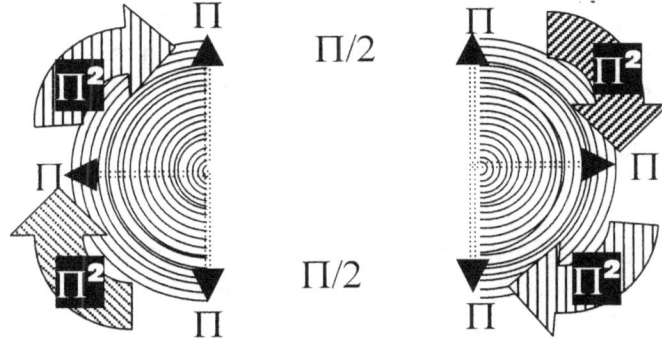

As the rotation was a change of directions involving four aspects, which was a duplication of the previous along the present going into the future, a division came in place that parted the one unit from the next unit. As the fourth-spot serving singularity landed where the first developed, that made the first spot the fifth spot. However, this was accompanied by a rule, which today still apply in the cosmos.

With every four rotating points duplicating to reproduce one unit, a parting had to be devised

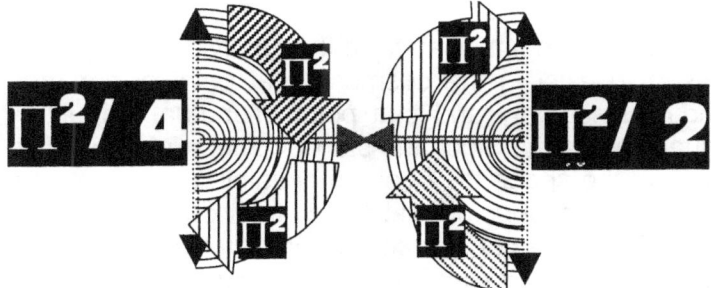

to separate one Universe from continuing into the next Universe. The four points duplicating the value of Π in relation to the centre Π^0 crossed the centre at a point measuring $\Pi/2$ and halfway where Π^2 lands, the next Π by motion thereof therefore $\Pi^2/4$ became the limit that brought space into a dividing point four from point five in relation to the developing centre. One has to remember that the Universe presently holds the characteristics it once enjoyed because once anything is part of the cosmos it has to remain part of the cosmos since there is no other place to go but to remain in the cosmos.

Even the motion of innumerable stars relate to a singularity in the centre that plays the part of the generated governing singularity and every faintest and slightest motion of every individual object plays a significant part in the generating of the governing singularity.

The Sun is on the outskirt of the Milky Way and the Sun is in an ova orbit around the Milky Way. The law of orbit is in principle that all orbiting structures follow an oval path.

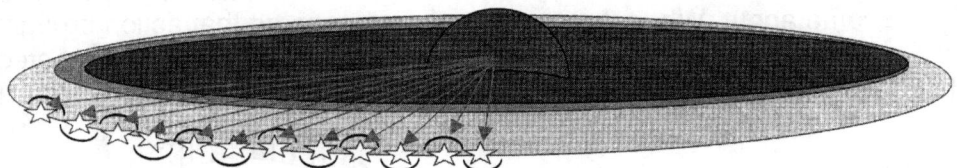

Exaggerated to a large extend the influence the Milky Way has to have on the Earth orbit comes to focus when a pattern comes in place as the Earth follows not a circle but a wave around the Sun while the Sun sets its motion around the Milky Way. The fact that the planets orbit the Sun and the fact that the Sun orbits the Milky Way indicate an undeniable influence. The fact that the Sun is heading farther away from the influence should then lead to a variation in the planets orbiting wave. The Earth never, not once lands on the exact same spot by the completion of one more year cycle.

By not having a wheel rotate, the wheel becomes the factor of one, and the rotation becomes zero. The wheel does not disappear. In the cosmos, everything is rotating because nothing ever stands still. Therefore the mean equilibrium, the common factor there is to share, has to be one, eternity, the eternal Π, because all rotating objects has Π in singularity, and sharing singularity, gives every object in space a relation with all other objects in space. After trying for many years to bring them the candle, I concluded that Newtonians are incapable of realizing that mathematical principle as reality.

The comet rotates the Sun, and the Sun by itself has a point of singularity where Π remains without r. The comet, holding the orbit, also has a point of singularity, but since there is space separating the two objects, they cannot share a mean point of singularity, the very point of existing. Since singularity means just that, being single, there cannot be two. The comet and the Sun have a mean point of singularity but the space they occupy divides their common singularity. That is why they orbit in an oval path, a path where the one structure holds on to more space from its point of singularity towards the space it claims. Since they do not claim equal space, BY THE DENSITY they hold, the space will not be in proportion.

They do share in the common fact of singularity a point away from their individual singularity proclaiming their cosmic individual reason to exist in the cosmos. That point of common singularity holds space between individual singularity and that point of mutual singularity saves and protects the points of individual singularity. Since the start of time at moment-Alfa where both found the space they occupy, in the space they hold, maintaining a time to that space in accordance to the singularity they hold, that point will be their individual eccentricity from singularity. The two objects are holding eccentric spaces around their individual but common singularity. That point of singularity is Π the circle without the radius because the singularity removes all forms or values of r, discarding r to infinity and leaving Π to be singularity.

That is why gravity is a fixation of Newton's mind making Newton fraud on a scale the world has never seen, and his $F = G(M_1M_2)/r^2$ is utter nonsense. The moment you say Newton or any of Newton's laws, the Newtonian brain stun. Not once did I find one Newtonian surprised at this, I could not once find one single Newtonian to see this. It always leads to an argument and the argument is about Newton being in use for centuries. One Professor even answered me by saying that I should realize Newton's formulas placed man on the moon, and if that is not proof of his correctness to me I will never obtain proof. That is beside the point. That is miles from the issue. If you say Newton is wrong, you commit the worst blasphemy possible. One may swear at God and all is understood but mention your not accepting Newton's gravity and they all fall on their knees, cover their eyes in the ground, start stuttering and moaning and you cannot make them see anything but Newton. Dare say there is no such a thing as gravity because Newton is wrong, they run outside and hide the woman and children from your rage of mental instability.

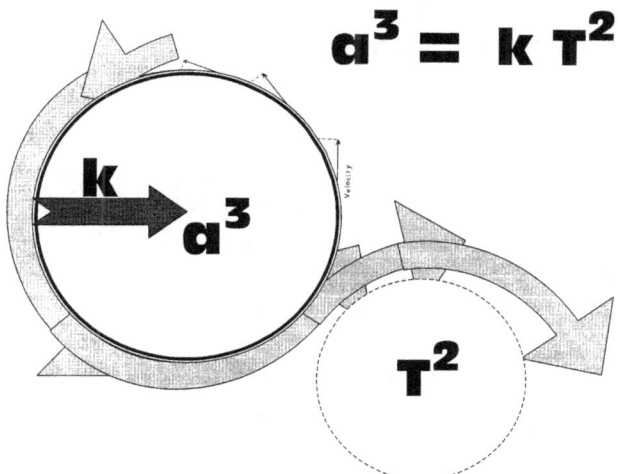

$$a^3 = k\,T^2$$

Because I have had unmentionable arguments that I in the end lost because of the mental Newtonian block all Newtonians hold covering their senses, where I could not reach a single spot of healthy logic within their minds, I wish to run through the facts once more and find what is so incomprehensible about the issue.

This in fact, is the very same findings that brought Johannes Kepler his own everlasting fame when he declared that the planets stand to a value of $a^3 = T^2k$ as they orbit the Sun. Never once did he mention the presence of a force or gravity. Newton came up with this bogus idea all by himself without the help of other "giants" as he called *Galileo and Kepler*. In a later stage I indicate that I might prove the possibility that Newton did not have enough information to draw conclusions about Kepler's work. Newton saw a circle in Kepler's formula and there is a Universe of information hiding in that formula because that formula depicts the key to science namely singularity.

Newton made the formula one big blunder as far as the cosmos is concerned. Newton works perfectly well where there is equilibrium and unchanging in space and time as we find on Earth with the Earth forming the basis for space-time. Taking Newton to outer space is a blunder and Newton created the blunder by re-adapting his original formula of $F = r^2/ (m_1 X m_2)$ to fit Kepler's vision of $a^3 = T^2\,k$. This very same bogus idea helped Einstein to ignore the space factor of Π^3 and place a relative value of one to space.

This he (Einstein) stated (without stating it) when he (Einstein) put the Universe in a single dimension property at the point where gravity was stretched to the limit. Einstein put the Universe to a three dimensional value of matter, space and time and then out of the blue he places space at a factor of one when gravity supposedly destroys time. This notion stands totally unrelated and divorced to reality. I do admit that Einstein is absolutely accurate when saying this, but the space he refers to, as outer space and the space disappearing in time are as far apart as the cosmos is wide.

Einstein was the one that said that space and time could never be separated because it was the very same thing, a point I agree with in all my findings. The difference between my point holding the Universe and being the Universe is within every atom because it is there where singularity is. From singularity through the atom space has the relation between Π^0 as singularity and Π forming space inside singularity holding time $\mathbf{T^2} = \Pi^2$ in relation to the triple value of $\Pi\Pi\Pi$ forming Π^3. I am afraid that Einstein made much more sense when he was still an amateur, working as a clerk in the Swiss patent offices. Then he landed himself under the spell of the Newtonian disciples and all his initial ideas that were factual, became integrated and confused with delusions of the "Xepted scientific Newtonian High Priests" called "acclaimed scientists" and their mesmerizing fantasies about gravity.

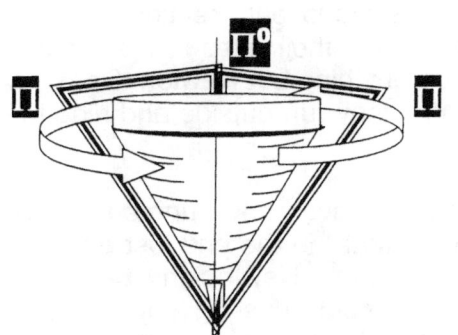

There is only one way to explain space-time and that is by finding **space-time**. Space-time is not some force, but is a location of space parting time in infinity from time in eternity. Putting space-time in a context of some force is putting what we are, well and truly out of our reach. It changes that which is the most obvious into something we may dream about and wonder why we have to adhere to it with so much respect. Einstein the master of physics was completely lost in his physics.

He went looking for a flat Universe, he saw singularity in the dark of the night hiding as obscure fairytale characters behind Black Holes, where he saw gravity lingering around stars with nothing better to do than to wait for passing light just to bend seven variations of different types of shit out of each of them. Singularity makes every atom rotate that makes every cosmic object rotate that applies the overall rotation to the Universe. **THAT IS TIME**. Each time I try to share this idea of mine with the **"Accomplished Scientists",** I do not get farther than the phrase: ***"Newton and Einstein are wrong."*** After completing this sentence, I get treated as a raving lunatic with extremely dangerous hallucinations indicating a murderous tendency. Why would not one academic listen to the rest I wish to say before bluntly denouncing me?

Nobody even listens or pretend to listen to the rest of my case. Every time I see the light in their eyes go blank and they sit patiently and wait for the motor mechanic to finish his senseless rambling. It is so obvious they consider me as mindless person with arrogance and having a nerve to criticize the two highest-ranking Newtonians of all time! I can assure you I am not mentally disabled! It took me twenty-one years of research and another six years in compiling and writing this book.

That is the relation matter has outside singularity. $R^3 / T^2 = 1$.

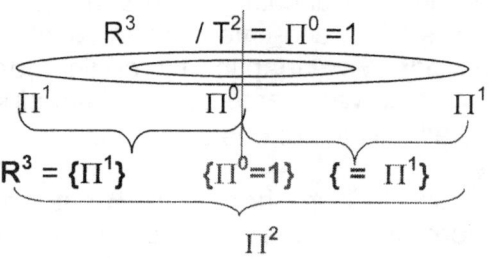

When we look at space, we do not see space. Where one thinks of seeing space, the space we think we see is time forming what we think should be space that is filled with particles. It is the atoms holding the space secured that forms the picture. Why on Earth would nobody realize Einstein was seeing the Universe from a wrong perspective? What Einstein saw was one hundred percent correct but Einstein saw what he saw in the space of the atom and not in space at large. That is space-time holding every aspect the Universe holds to a specific relevancy. The space outside singularity holds the time outside singularity because everything in the Universe is spinning. Science knows there has to be a difference because in space an object might be weightless, although it retains its mass, and no one can say the difference, except to put it down to "gravity".

We, the tax paying public, are letting these Master Minded Academics get off the hook so easily, because every one is so scared to ask "why and how". In the pages above, I pointed to the most basic mistakes about "gravity" which science ignores, because the answers they do not know. Even Nobel Prize winning work is blatantly misguided. I challenge any person to prove how an atom can collapse on itself, by force, by weight, by pressure or by any other means. No atoms will ever touch one another let alone compress to diminishing space, and if they do, the result is a nuclear reaction

IF IT DID NOT SPIN, IT WAS NOT ROUND, AND NOT BEING ROUND THERE WILL NOT BE SINGULARITY.

According to Einstein, the speed of light is a constant throughout the Universe. The speed of light results from two factors, being distance (kilometres) and time (seconds). This speed is accepted at 3×10^6 kilometres per second. Scientists know that it takes Sun light 10^6 years to reach the surface of the Sun, and we know the Sun is not thousands of billions of kilometres in diameter. The "Xepted scientific Newton Mistaken" explanation about this fact is that the Sun light "bounces against matter" and this retards the Sunlight dramatically. When light hits matter, (except in the case of glass), it joins singularity immediately. Therefore, the Sun holding matter on the inside has to be all- glass, or the "Xepted scientific" explanation is not very scientific at all. It all comes down to the density of matter in space valuing the time in that space away from the point maintaining singularity. Why can nobody but me see that?

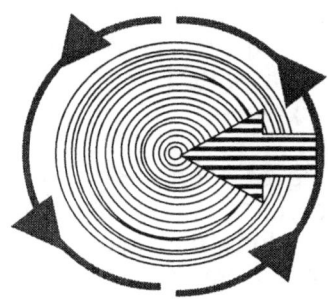 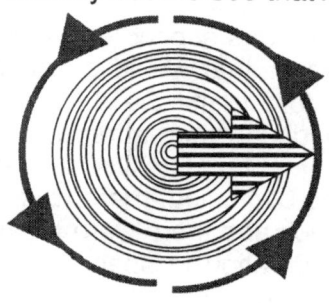

Where the arrow points we will find a spot that has no start. It is 1^1 that are the part that releases from 1^0 when motion parts singularity by infinity and eternity. It comes about when motion unleashes the dot 1^1 that has no space and has no start from the spot 1^0 that has no end. Every time the top starts spinning, a Universe is born in motion. The top is instigated and the motion is produced by the skills of life but in the cosmos serving nature such motion is the product have heat concentrated to sustain and maintain singularity. It is invisible, unseen and only detectable by intelligence and still it is a part of the cosmos that is no part of the Universe and from it the principle we call the Universe comes about. It is always a principal because it has no where to go but to be on call and by never being in the Universe it always is in the Universe. The spot forming singularity in the centre of the top, that spot we know is there because we are unable to see that it is there and therefore, being the Master specie that we think we are with the superior intellect that we think we have, we realise its presence because of it not being present. As is the case with religion, we, those of us who are considering ourselves to be of the Human specie, are the only specie that can detect the spot holding singularity, which is being there without seeing it being there.

Walk outside and look at the vastness of the blue sky or at night at the blackness we can see without being able to see because it is impossible to see darkness. That what you see when looking at the vastness is eternity that parted from infinity when space-time established a Universe. That which you see has no end because it is eternity in every aspect one may attach to such a connection. Standing where you are, no matter where you are you are standing in 1^1 and you are part of that which parts 1^1 from 1^0. You form part of 1^1 as you stand and being part of the centre of the Universe (because all light flows directly towards you and acknowledges you at being the centre of the Universe, you therefore also form infinity being the inner most part of eternity. That means the infinity you hold gives you with life entity that never can be disputed. You are in 1^0 that can never end as much as you carry

1^1 that never can start. You are both the spot 1^0 and the dot 1^1 and neither can ever start or end.

That is what Kepler (again I cannot say whether he wittingly or unwittingly) declared by using the formula $a^3 = T^2 k$ he announced space-time in a formula, the formula Newton raped to his advantage

because in $\dfrac{M_s \times M_C}{r^2} G = F$ there can be no pointing to singularity

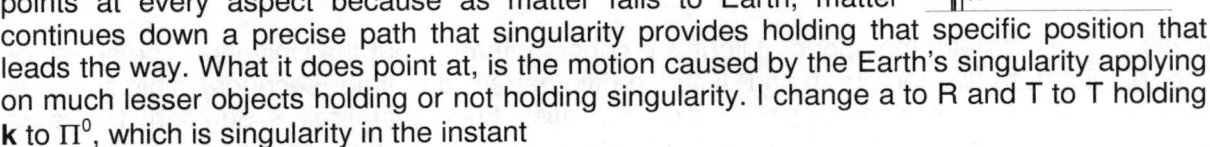

in the cosmic sense. In his initial formula $F = r^2 / (Mm)$ singularity points at every aspect because as matter falls to Earth, matter continues down a precise path that singularity provides holding that specific position that leads the way. What it does point at, is the motion caused by the Earth's singularity applying on much lesser objects holding or not holding singularity. I change a to R and T to T holding **k** to Π^0, which is singularity in the instant

FROM THAT POINT SINGULARITY IS IN EVERY PROTON HOLDING SPACE AS RELATIVE AS TIME IS. $a^3 / T^2 = k$

This in fact, is the very same findings that brought Johannes Kepler his own everlasting fame when he declared that the planets stand to a value of $a^3 / T^2 = k$ as they orbit the Sun.

I chose one aspect from a wide range of possibilities to explain the way the cosmos formed. However that would constitute to a book much larger than the one you are reading and therefore I limit the development only to where matter, space and time parted and then I immediately thereafter focus on the point where the Big bang came into place.

In the beginning, there was time Zero to moment Alpha. There has never been a Big Bang as such and there were too many Bangs too numerous to count. Everything is a variation of time duration in space. During the period of time Zero to moment Alpha the value of 1 second was equal in duration to about 1 000 billion, billion, billion, billion years (I am only stopping with the billion part in order not to bore the readers), measured in geodesic space-time values that currently applies. It could be even billions times this duration because the value of time then, was measured far beyond the speed of light, since light did not yet exist. We have no way to calculate the duration of time.

The closer time is to singularity the longer the duration would be. The method by which time is expanding is by heat and only heat can expand while only by reducing heat can there be a demise of space. That applied during that geodesic space-time era as much as it does today, and we must accept it as one equal to infinity shorter than eternal. It is heat in all its splendour because only heat can expand. Even boiling soup produces space that expands. When a bowl of soup is boiling, have you seen the bubbles of air rising from the soup? Has any Newtonian ever taken the time to explain that process in detail? I think not, because such explanations would be far too "everyday-like" to bother their mighty brains.

Well, that boiling soup tells the complete story about the creation. Creating is a fact of creation, however creation was not created and left on its own, the Universe is in creation being created every smallest fragmented split instant there can be. The Universe is generated as it moves and such generating of the creation is a process of creating what there is. We speak so lightly of creating and no one comes close to understanding the concept of creation. Poets and painters and writers always wishes to say how "they created their creation". That is rubbish; they created nothing. They brought nothing new to the cosmos, they only rearranged what was a small part of the cosmos into a new order, that one can detect a distinction from. Creating is producing what never was before. When looking at the boiling soup, there are bubbles rising from the soup to the top. In the soup's brew, there are only liquids and solids before the heat came. In such a manner the expanding of heat created space. No one placed air in before the event or during the event at any time. Yet from the brew of liquid and solid rises gas, or if you wish space. That space was not there previously. That SPACE WAS CREATED.

That space is energy and energy is the interaction between heat and space. As space becomes a part of the soup, a part not there before, with no room to be, it moves out. We refer to that process as boiling. That space creation is applying heat to time, and time in singularity will respond as space in singularity. The space created will vanish just as it came, back to singularity. By applying heat to time, brings forth space, and from the three components, only the heat factor is not in singularity. It removes space in singularity from time in singularity to establish room (space) for heat (time).

That is how creation started. Time in singularity overheated and the product of that was space. That is the 180 $^\circ$ of the straight line as much as it is the 180° of the half circle.

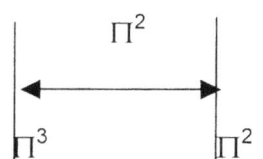

The Π^3 space from
The Π^2 is motion of liquid heat
The Π is time in space to singularity.

There is a time as a line that we find in the centre of the top, which is singularity. However there is another time, which offers material the space in which material is able to duplicate. That too is time but it is the relevance between the holding time and the space-time where material is located.

Material uses the relevancy of time within, which developed as singularity and time without which was the expanding of singularity to commit to motion.

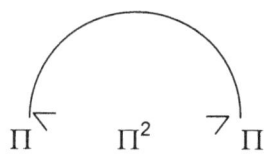

The half circle is 180° placing matter in a circle but because space only applies, to one half, 6/2 and matter holds space to value 6/2=3 only half the circle comes into effect. Half a circle is 180°. Because space has three parts in effect, it also becomes a triangle.

That means where space holds three and time is one, the heat within that space becomes another dimension, the fourth dimension holding space-time (3^2+1^2) = 10. That changes the matter inside space in singularity at ten and "gravity" at Π^2. This is why "gravity" Π^2 is space (10) losing one dimension (Π^2). "Gravity" is all about space (occupying matter and heat) losing one dimension back on a long journey to singularity.

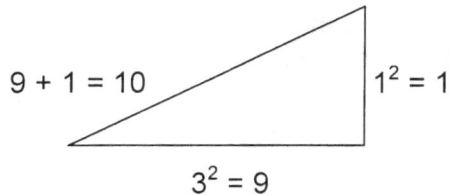

As time is in singularity, and space is in singularity and both are the same thing $(\Pi^3 \rightarrow \Pi^2 \rightarrow \Pi)$ the 10 of matter (heat) that affects space (10 Π) will also affect time (Π^3) and therefore time carrying heat will become 10 (Π^3) with space 10 Π. Anyone with a simple calculator can divide 10 Π^3 by 10 Π and see where Π^2 fits in. It is the doubling of matter in relation to time (7/10 + 7 /10) times the double factor of time in space (10) standing related to the line of time that refers to matter (10/7). That gives gravity its value of (Π^2).

Through the Coanda principle the motion of liquidΠ^2 confines space Π^3, to what space Π^3 confines as the atom$\Pi^3 = \Pi^2 \, \Pi$) to the solid. In this containing of space by liquid in motion with the limiting or putting a border on space by liquid flowing, which confirms the space what establishes the Roche limit. In that there has to be a liquid (the neutron at Π^2) lies in the two components of space-time occupation or "gravity" manifested in the Roche limit.

All objects spin and spinning is a circle Π^2 while all objects are moving in a direction $\Pi^2/2$. Again only, half of Π has any dimensional validity at any given time, therefore the dimension surrounding an object is Π. That is how gravity forms the atom as the surface of the cosmic object extends from Π^2 to Π but only half of the circle of Π (180°) can apply to time (Π^2) being in a straight line $\Pi^2 \to \Pi$, "gravity" will form at that point of $(\Pi/2)^2$ giving the Π in space the "gravity" to hold.

$$\Pi^6 \, (\Pi^2 + \Pi^2) \, / \, (6 \times 10)$$

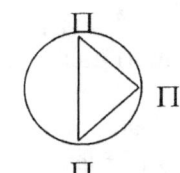

That places Π in a total of Π^6 with 6 sides in space (10) affecting the proton ($\Pi^2 + \Pi^2$)

That is why space will forever comply with 7 / 10 Π^6) / 60 = 112, (the Π^6 is ($\Pi^2 + \Pi^2 + \Pi^2$)) and time forming the line (180°) between the half circle (Π to $\Pi = \Pi^2$) at a 180° will form the triangle of space in half (180°). The matter component of the Titius Bode law effectively applies to the value of space, therefore 7/10 comes into the calculation. That places any atom with an existence in space at a premium of 7/10 (Π^6) (6/10). The reason why plutonium at 5($\Pi^2 + \Pi^2$) $(\Pi/2)^2(3/5) = 244$ is at the element limit is obvious; when dissecting the relevancy in detail. The complete element holds the very edge of what an element in space and time can endure in this era, but two or three eras ago it had the function cobalt has at present

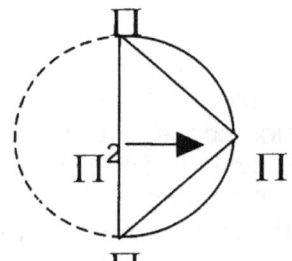

That will produce time in singularity a value of Π^3.

Explaining the other five stages of gravity (Π^2) development is extremely complicated and for that there is no room in a book meant to introduce new ideas such as this. My motto in this book (part one) is "Keep it simple"

With time in singularity, time was eternal.

Π^O

Time is the spin rate of heat in space. This translates to heat in spin (the atom sealing time off by the spin of the electron, which then produces a motion relevancy with the proton and time in space, which brings about the time line. As we can see when the top spins the top forms the spin of heat (the top spinning) in space. That means the way the movement changes where matter and heat relate to other matter and heat in space. All the movements are relating to a circle (Π^2) going somewhere (Π) in space 3. The Π will form the radius to the circle (Π^2). Any novice can see that the longer Π becomes, the wider Π will be and therefore the longer change in the repositioning of matter will be.

Any person wishing to uphold Einstein's view about the speed of light being the limit through which matter can apply velocity, then that person should first explain how the Black Hole works. It is very distinct that whatever is inside takes that which is inside, to exceed the velocity of light. In other words, the Black Hole is able to force matter into speeds that goes way beyond the speed of light. The contraction produces a spiralling of particles that takes matter into a motion dimension far beyond the speed of light.

The concept involves not the moving of the particles but the slowing of time to force a duplication cycle that goes beyond the capability a photon can withstand. Inside the Black Hole must be matter, because there is no space, yet time does apply because it takes the particles spiralling inwards to the centre time to move from point to point. Matter in motion is time.

However, no light can return to the surface, therefore the light is slower than the moving particles within the star. The only thing about the star is that it maintains a higher relevancy than the relevancy the speed of light can apply. By accepting the existence of a Black Hole, any of Einstein's claims about the speed of light being the fastest that matter can travel becomes fictitious.

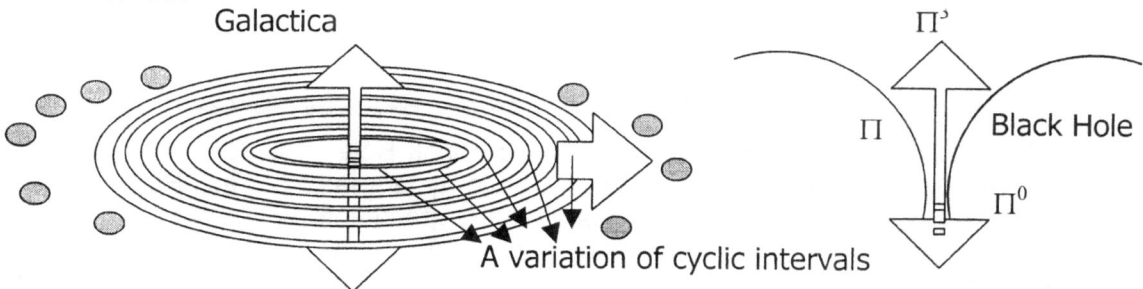

Another place where the speed of light becomes obsolete is within the centre of galactica, where the accumulative movement of matter exceeds the speed of light. That is where doctor Hawking saw a Black Hole that is not a Black Hole, but the precise opposite. Light, matter and heat, moves inward in an effort to maintain cooling (as the group of proto-stars belonging to an era to the future) where they still claim their share of heat maintenance. Those particles in such close proximity, establish a time in motion well above that of the speed of light. Everything in the cosmos is all about relevancies. Particles in that phase are still very close to time eternal where motion of material took space into singularity. Time started at such a high velocity, it had to be eternal. Nothing that diverts from eternal can become more than eternal so it has to be less than eternal. It is fragmenting eternity into parts making eternity smaller.

Closer investigation reveals that the way the galactica operated is a close replication of exactly the manner that the atom grows. In that a revolving heat blanket that forms a bonding of star matter on the inside honours a centre. The blanket of heat is maintaining a centre core

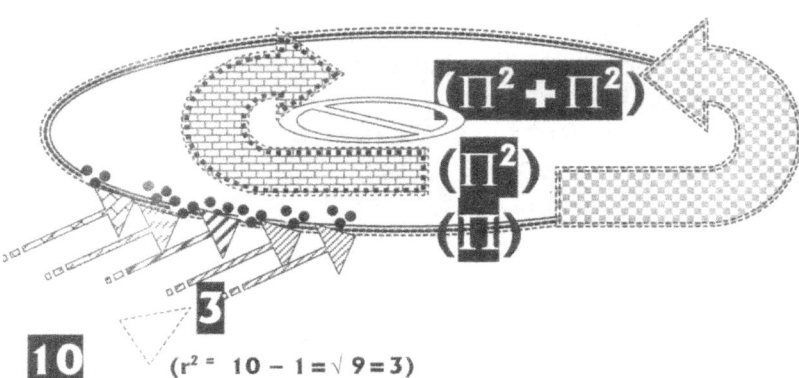

around which it revolves.

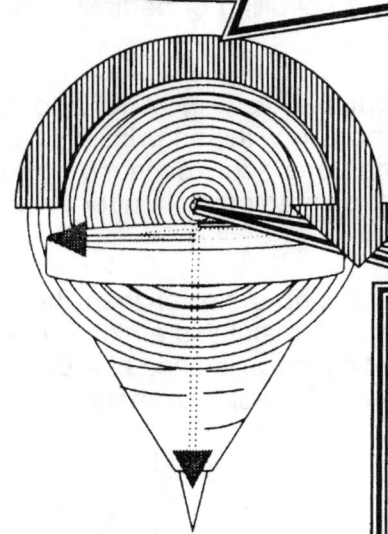

The Coanda effect that joins time by committing material to gravity

Time in eternity forming the motion or liquid that locks space in as material

Time in infinity forming the motionless or solid that locks space out as material

Professor Hawking holds the opinion that there is a Black Hole centred within the centre of the star, which of course cannot possibly be true. The dynamics of a Black Hole is such that it is a star as massive as they come, that fused all the atoms into one structure. The nature and the essence of a star are to unify the singularity that was divided amongst all the atoms during the process of cosmic expanding. The star is a collection of atoms, which unite in motion that then through the unit generate motion to establish a controlling centre governing singularity. The rotation of every individual atom spinning is collected as a generated effort and the collective drive accumulates the effort to the centre of the star. The more the star develops the more is the drive of the star vested in the centre of the star and is it less concentrated in the material compiling the heat and therefore the drive. As the star becomes self secured, the maintaining of the star removes the duty of finding heat to secure the star from overheating from the atoms to the centre governing singularity. The singularity finally takes control of the motion of the star as the star evicts all space and drive time back to eternity.

Then a point arrives where the star abandons all motion. The star then achieved the main goal all stars have by producing a gravity that controls the motion of time. The spin has moved from within the star to time itself and by abolishing space all together the space became what singularity can offer. By using Kepler's formula the relevancy placed infinity in control of contraction and $k^{-1} = T^2/a^3$ which means the space used by the star is infinitively small as the motion producing the gravity calls on the entirety of time to move and to establish such infinitive immobility. On the other hang the space that the star then control is eternally big $k = a^3/T^2$

It is where the Coanda effect joins eternity, which will forever move with infinity, which can never move and the coming together is resulting in the Coanda effect since the entirety in time is established by motion and infinity that never can move, sets the outer limits. The Coanda effect joins what always moves (eternity being the liquid part) with infinity that can never move (taking on the part of the solid) and combining time is set by the Coanda effect. That makes the controlling gravity, the entirety of the time where that joins the infinity of time which locks space-time within and that forms space as a time aspect because singularity being infinite commands time to motion where such command is stretching the ultimate. It is more complicated and I do explore the working of stars and the development of stars leading to Black holes in much more detail in another book I have being "*STARSTUFFIN*".

In the very opposite it is the motion of all the heat and all the stars either still in a proto state within the heat within the centre of a galactica or otherwise that forms the unit driving the galactica to improvise a Black Hole situation within the centre of the star. The Coanda effect that generates the singularity, which controls the star becomes generated as a result of all the heat and particle motion that turns about the star centre. Where the motion is valid enough to sustain a drive that would generate an equal gravity to that which the Black Hole demands, the totality of the liquid in motion in the galactica invests into a gravity that does form the drive equal to the drive of a Black Hole. But the drive forms what seems to be a black Hole. There is no real Black hole because if there was a Black Hole, then the galactica had met its destiny before any of the stars within such a galactica could journey onto a road of development. From a Black Hole nothing escapes and every star is a future Black Hole on a journey of development to finally become the ultimate, the Black Hole. However, there is one star more superior than that, but there is no space to go into that explaining.

Gravity is motion. Motion is either the expanding or the contracting of material because of heat interacting with space. When material overheats, heat expands the space of the material and when heating diminishes, the cooling reduces the space that material claims. In both instances it is motion applying. While moving the material overheat thus it expands. The expanding may be controlled and therefore the progress of expanding is controlled but motion has to be by way of expanding even when the expanding is under the auspices of contracting. It is the duty of the star to contract that which the galactica expanded.

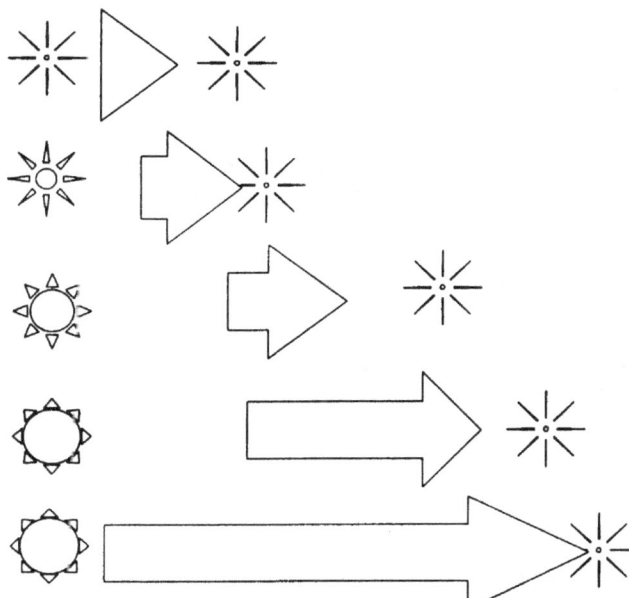

The galactica expanded as a compromise for the overheating but in expanding the galactica, such expanding also developed and controlled the progress of young stars into adulthood. The galactica expands as the expanding of the developing stars sets off the expanding and the stars has the role to contract the Universe back to singularity. While the galactica expands it gives the stars the opportunity to place heat stored as space of heat frozen by spin in the atom. The atom generates a governing singularity that demolished the space as it accumulates all the heat back to singularity.

When the star reduces with the flow of developing back to singularity, it is not only the star that reduces. Neither is it totally the fact that the Universe expands as the star that grows, while the star demises in size to accommodate the space that expands by diminishing the relative space-time the star claims. It is the relevancies between the star's space demise and the surge in space in outer space combined, that are applying more tendencies in representing the relations there are between structures in space and structures and space.

With outer space carrying the blackness in progressive multiplying, the very essence of space being space within, the atom too must be in growth claiming more space. Of all the above factors Mainstream science only acknowledges the growth of space in as much as calling it the Hubble Constant. However, that is not where the growth affects ends because it originates as much from any individual atom as it comes from Alfa singularity. Space does not expand because the space is only reducing the heat in density while producing density in space.

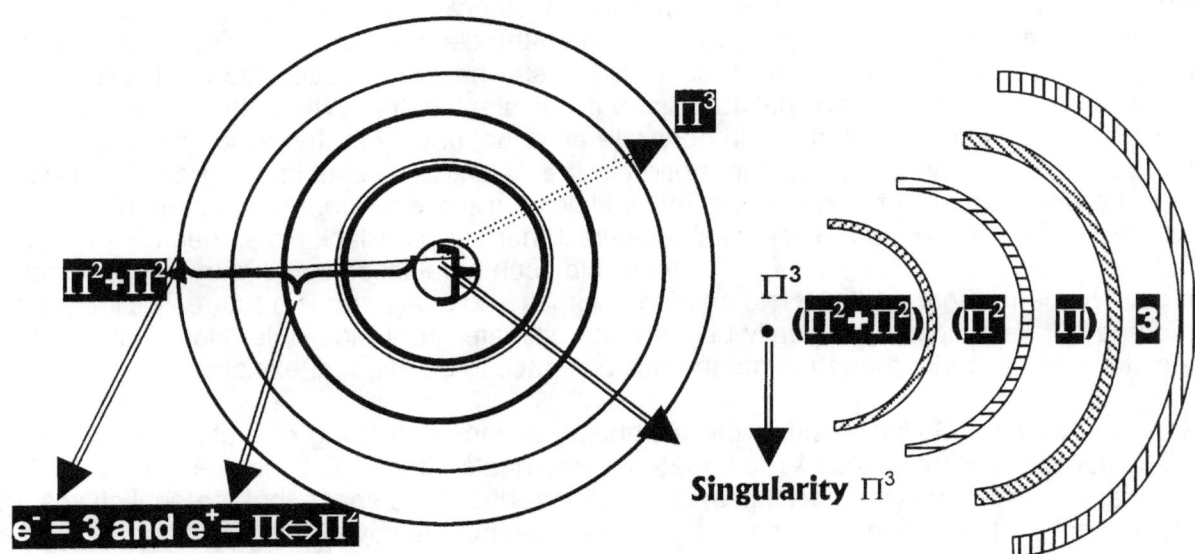

Every layer in the star represents one factor in the atom since the star is just another cosmic atom securing strings of atoms that as a unit aims for one goal and that is to secure one singularity within the star.

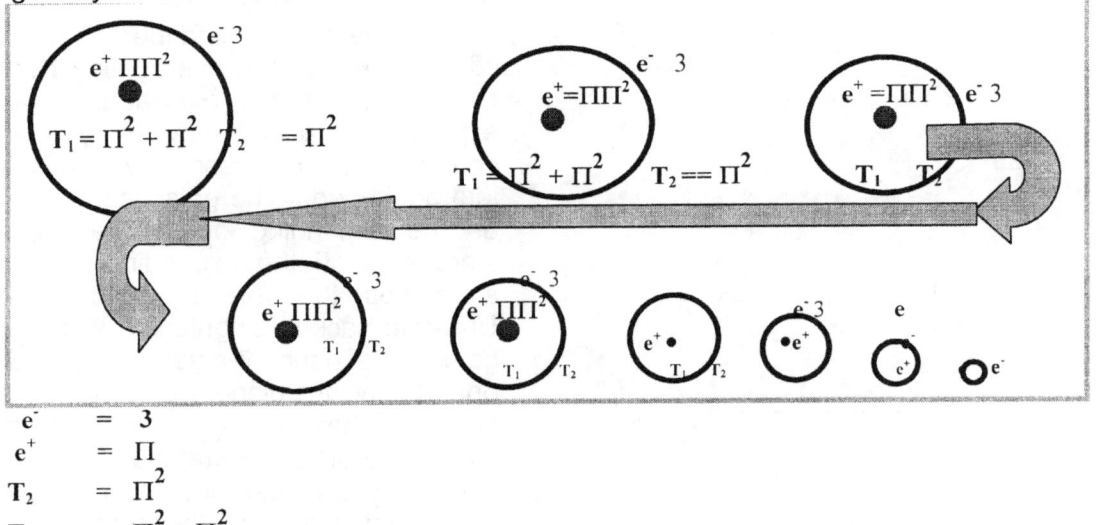

$$e^- = 3$$
$$e^+ = \Pi$$
$$T_2 = \Pi^2$$
$$T_1 = \Pi^2 + \Pi^2$$

As the Universe expands, the Universe is then the atom that expands. The part of the Universe that Newtonians think expands well that part cannot expand, because that part is eternal. The part that expands is the atom and the atom is the Universe because the atom defined infinity from eternity by dividing the two points representing singularity. The star on the other hand has the task to convert the atom back to singularity by contracting all the heat into singularity. The star goes in development by shedding its layers until it finally has only singularity in the centre left. The start of the life cycle as a star is a combination of all the atoms within the star that has to control within the singularity atoms forming the star. The star is one cosmic atom. Every layer holds elements that serve the star in the particular development the star finds itself in and the layers hold a certain displacement value of space-time. The star is what the combined effort of the accumulated effort of all its atoms represents, and for that reason the star is a cosmic atom as much as all its atoms are one unit combining as an atom and forming individual atoms. Therefore the atomic proton value is representative of the relevancy there is to form the need of singularity maintaining within that layer and as a layer contributing to the star as a whole.

BACK THEN when the Universe was new

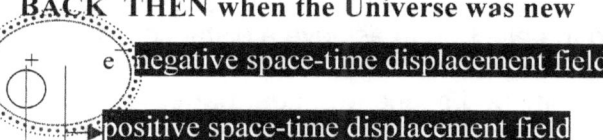

e ̄ negative space-time displacement field

positive space-time displacement field

negative space-time displacement field

Positive space-time displacement claims on space through motion.

PRESENTLY we refer to the sizes we find space has in the Sun as quantum meaning they are inexplicably big

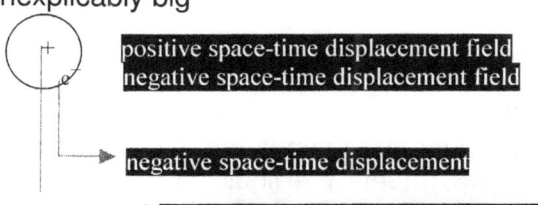

positive space-time displacement field
negative space-time displacement field

negative space-time displacement

Positive space-time displacement claims on space through motion.

IN FUTURE TO COME they are going to get a lot bigger than the quantum size now present.

positive space-time displacement field e ̄ negative space-time displacement

negative verplasing field

Positive space-time displacement claims on space through motion.

This became the atom $(\Pi^2+\Pi^2)(\Pi^2\Pi)(\Pi^0+\Pi^0+\Pi^0)$ = 1836 and the atom formed stars that still act in accordance with and to the atomic relevancy

Outer space substantiate the atom as $(\Pi^2+\Pi^2)(\Pi^2\Pi)$ 3

The manner, in which the schematic layout presents itself as follows.

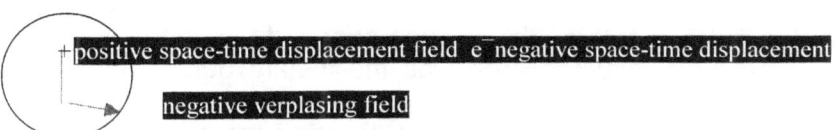

ATOM NUCLEUS   ELECTRON

Space-time field   e   Space-time field
occupied space-time   unoccupied space-time

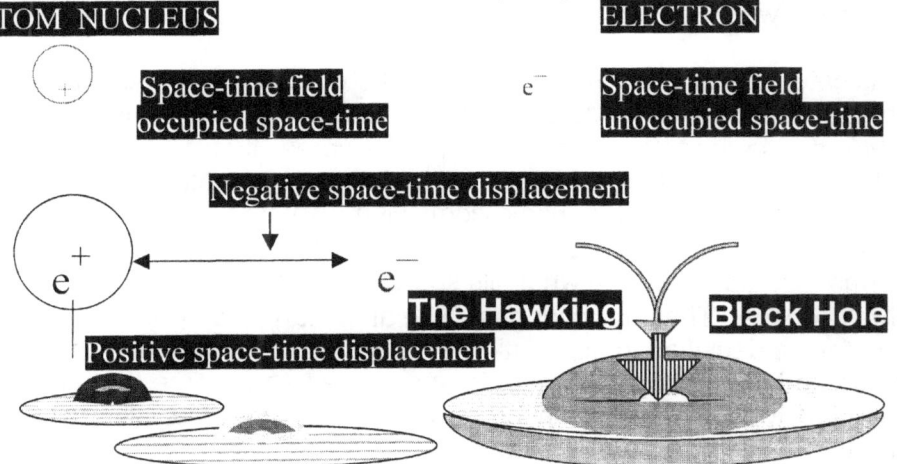

Negative space-time displacement

e ̄

The Hawking   Black Hole

Positive space-time displacement

It is not only outer space that grows because $k = a^3 / T^2$ is as much the cosmic value as the value within the atom. That means that $k = a^3 / T^2$ is also in place within the atom and that shows the space within the atom grows as the Universe grows because the atom represents the Universe that is in growth. As gravity brings space-time reduction from the centre of the proton, so must the growth come from the centre to the atomic proton cluster. As the atom expands in space-time, the proton can also grow dimensionally bigger through the neutron growing in stature. It expands in captured space–time by pushing the electron walls to allow the atom more space to occupy. It is pushing the electron to achieve a distance every time in the same manner that the body lets nails and hair grow. There are three factors of space-time where space-time is released. Cosmic unity and space and heat parted as singularity released the space heat holds by forming motion which produces time to set boundaries and relevancies applying.

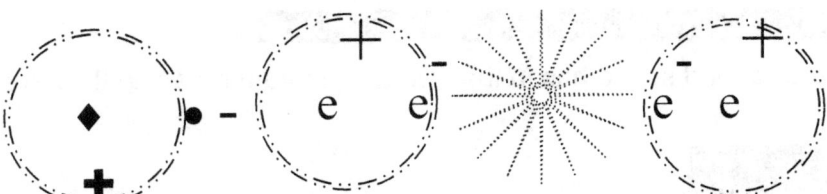

Everything in the Universe is an atom because the atom is everything formed in the Universe. The atom is the progress of the Universe although the atom is not necessarily the structure we have in mind. The atom is where the solid meets the liquid and that which provides gravity by motion. Stars and galactica are as much atoms as any other grouping of time in delay.

When contracting, gravity takes place by means of lying down newly acquired heat to maintain the cooling of the structure. However, by contracting it is accumulating material that produces a build up of material in order to enlarge the existing heat surface. In that manner it spreads the heat in a wider area than what the area was the instant before and in that manner it duplicates slightly more than it did duplicate the instant prior. That means even by contracting the measure is still expanding the material by relocating the material from an uncontrolled zone to a controlled zone (the atom.) Still this accumulation by contraction is expanding by motion. When saying this please be sure of one thing: it is not the Universe that is expanding but factors within the Universe that relocate that which takes up space and provides space in the Universe. The Universe remains unaffected by all this by never increasing or decreasing. Two of the three factors swap ends and that places the third factor at a different relevancy. The Universe can expand as little as it can reduce. The Universe expands by the curvature of space –time as Einstein proved but my solution proves much simpler than the way Einstein went about.

I will in a short while indicate how Einstein is correct about the curvature of space time in his theory about "The curvature of space-time" because the curvature of space –time is the form of Π, which is the value of singularity and that forms the Universe. It is a building of what there is by the dynamics that singularity provides in relation to the accountability space-time has relating to singularity. There is the space-time complying with singularity and filling the space-time in singularity is heat and matter valuing space-time. Space-time (Π^3 to Π) cannot bend, cannot curve, forms a straight line, but what fills space-forming time is matter in motion (Π^2) and heat (3) in time in space. That part changes. The atom cannot be gas, or liquid, but is a solid, because the atom is densified in occupation of space-time. The atom is space with heat under control of directed motion. It is the heat in unoccupied space-time that produces the gas and a liquid is closely connected as much as part of the solid that all substances form however it is not within the enclosure of the atom.

It is THE HEAT in SPACE that produces TIME, that can and does curve, bend or whatever. That HEAT in SPACE forming TIME that forms the relevancy of space-time does bend because it can flow and flowing is changing direction or constructing by altering flow directions in space wherein matter flow but that then is part of being part of time. If, by applying the forming of gas, or liquid to the element where it is the space between the elements, of course you will get the incorrect vision of Π, where the space-time (matter holding singularity forming singularity) is doing all the bending that applies to the curvature of space (validating time) and time in singularity (a straight line) will be solid. Einstein placed the relevancy incorrectly on singularity, instead of heat.

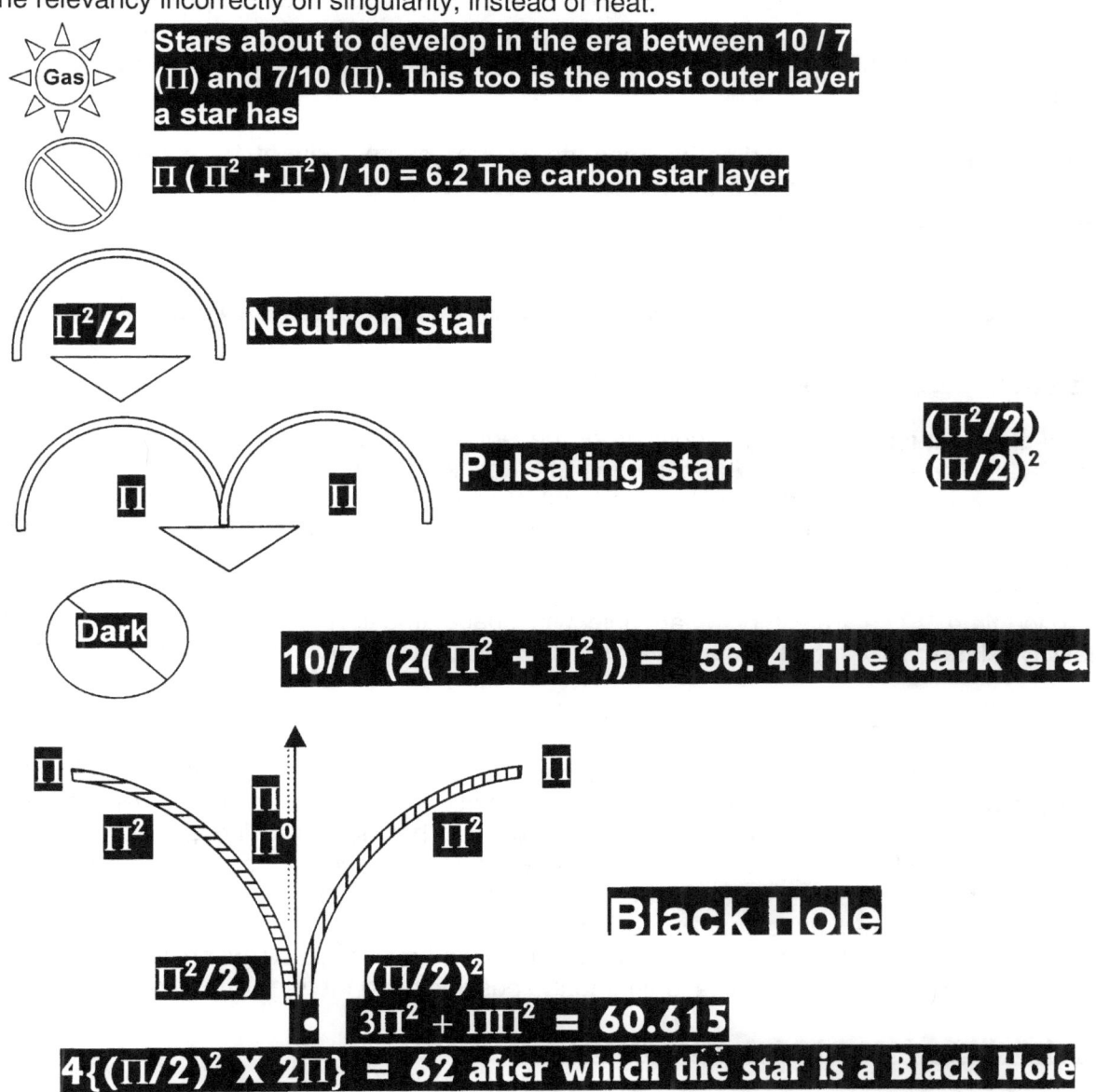

Stars about to develop in the era between 10 / 7 (Π) and 7/10 (Π). This too is the most outer layer a star has

$\Pi (\Pi^2 + \Pi^2) / 10 = 6.2$ The carbon star layer

$\Pi^2/2$ **Neutron star**

Pulsating star

$$\frac{(\Pi^2/2)}{(\Pi/2)^2}$$

$10/7 \ (2(\Pi^2 + \Pi^2)) = 56. 4$ **The dark era**

Black Hole

$\Pi^2/2)$ $(\Pi/2)^2$

$3\Pi^2 + \Pi\Pi^2 = 60.615$

$4\{(\Pi/2)^2 \ X \ 2\Pi\} = 62$ after which the star is a Black Hole

Every star is on the inside many different stars because every layer holds a different (k) or relevance making the space in the star very different from every other layer in the star. This is because every layer has a different motion in relation to the governing singularity and therefore has a different gravity confining space. The layer is the result of the gravity effort of all the atoms in such a layer and therefore the space in that layer will bring about the time factor that produces the proton cluster relevancy

I do admit, IT IS A LOT MORE COMPLICATED THAN WHAT I ALLOW IT TO BE AT THIS POINT, but the motto is, Keep it simple. If you wish to keep time in space constant, everything in the Universe will be oblong. That is why the Newtonians have an absurd view of the cosmos, and they present facts in the cosmos in a way nobody (least of all the Newtonians) can understand. Please allow me to explain this part first.

When material in space in time first appeared there was a displacement relevancy that was in place then that had 139 protons in relation to 138 protons relating to 136 protons. It was an atom that had many atom variations gathered and that held the construction of one atomic atom worth on the outside 139 protons, in the centre 138 protons and in the gravity zone 136 protons.

There was no electron at the time because the electron came about at $10 / 7 (4(\Pi^2 + \Pi^2)) = 112$. This was the phase where the neutron was established as part of the Universe. The first time I can detect space / time and matter is when the proton came to a relevancy of space-time displacement $= 7(\Pi^2 + \Pi^2) = 138$

Quoted directly from the Oxford dictionary of Astronomy the following:

The definition of space-time is as follows:

Space-time is a four dimensional position of the Universe where the position of an object is specified by three coordinates in space and one position in time. According to the theory of special relativity there is no absolute time, which can be measured independently of the observer, so events that are simultaneous as seen from one observer occur at different times when seen from a different place. Time must therefore be measured in a relative manner as are positions in three-dimensional Euclidean space, and this is achieved through the concept of space-time. The trajectory of an object in space-time is called world line. General relativity relates to curvature of space-time to the positions and motions of particles of matter.

The definition of space-time is as follows:

Space-time is a four dimensional position of the Universe where the position of an object is specified by three coordinates in space and one position in time. According to the theory of special relativity there is no absolute time, which can be measured independently of the observer, so events that are simultaneous as seen from one observer occur at different times when seen from a different place. Time must therefore be measured in a relative manner as are positions in three-dimensional Euclidean space, and this is achieved through the concept of space-time. The trajectory of an object in space-time is called world line. General relativity relates to curvature of space-time to the positions and motions of particles of matter.

The definition of singularity is as follows:

Singularity: a mathematical point at which certain physical quantities reach infinite values for example, according to the general relativity the curvature of space-time becomes infinite in a black hole. In the big bang theory the Universe was born from singularity in which the density and temperature of matter were infinite.

While it probably is the greatest mind to walk the Earth that produced the spectacular in the above, a much more simple mind as the one I have noticed much more simple aspects of nature that only one with a simple mind as I have could recognise because my mind does not have the capacity for the greatness of the great minds.

Singularity is the most vital aspect of the Universe and I can take any person's finger and

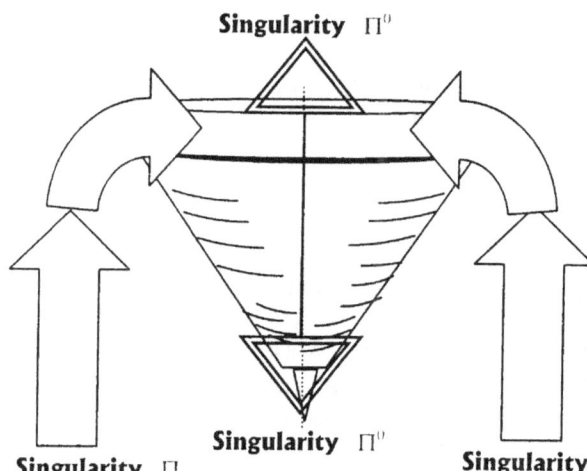

Singularity Π^0

Singularity Π **Singularity** Π'' **Singularity**

show singularity. Singularity is just a mathematical point, yes, that is true, but moreover it is as clear as any person with intelligence can appreciate anything. For anything except life, to be a part of the Universe at some point in the past, it has to be part of the Universe at this point in the present because from when everything in the Universe was up to now in the present and in our presence, it had to be part of the past in order to be able to be part of the future. If whatever is, then whatever that was which was where it was, had nowhere to go but to remain in the universe and since it is in the Universe where it was, it now also is in our presence. If it was, it is because it is. It therefore also is while it is going to be because it can't be anywhere else but with us in the presence. That might exclude life, but it includes everything of atomic fibre that life uses as a tool to manifests motion as being part of the force able to manipulate space-time. The atoms might not be life, but it is what life employs as tools and the tools remain part of the Universe before life entered and after life vanished. Only life can come and go at random but all else remain in time as part of time.

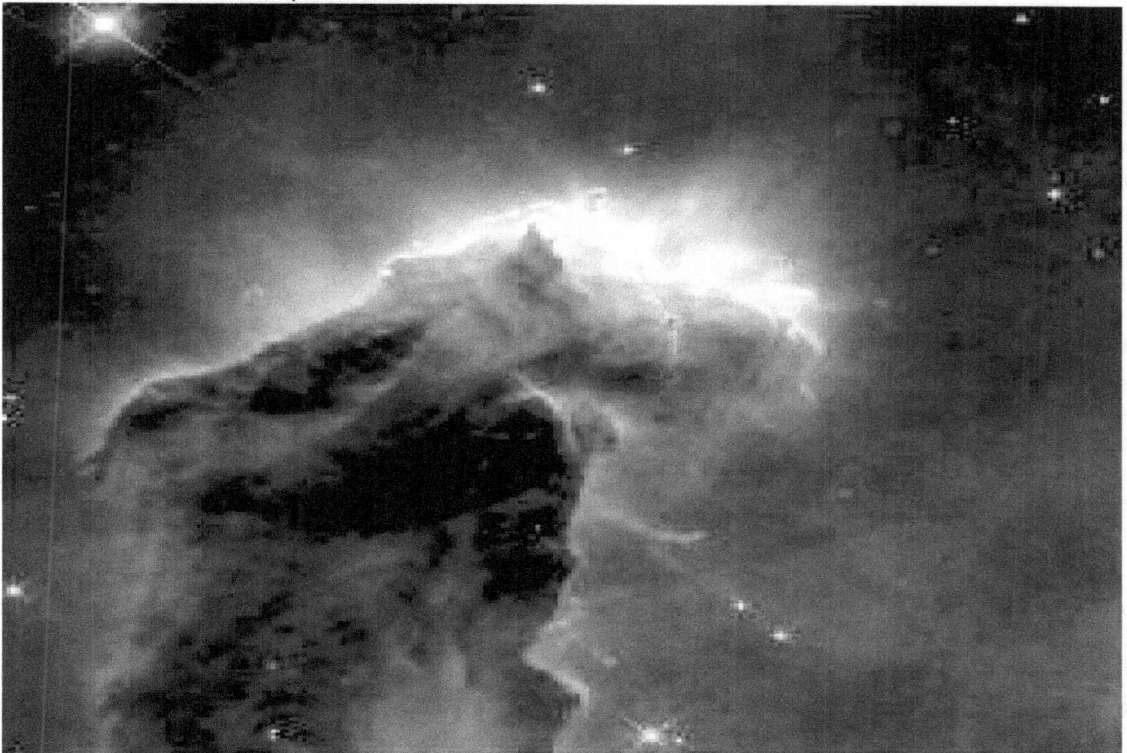

If the Universe did start from one single point and time matter and space flowed from that point, then that point must have a relative connecting base because such a point holding singularity must be eternal as space, matter and time link eternal. There then therefore must be one point linking the entire Universe when regarding the fact of singularity. Then according to the theory of relativity there has to be one exact point holding time in relevance notwithstanding the fact that time departs from that position and relates differently to all space-time away from such a point.

Every person I have discussed facts about creation recollects images in the trend depicted in a presentation as one may find shown with massive clouds of unbounded material. That depicts chaos and in chaos I would have no ability to use mathematics. The fact that I can use mathematics presents a Universe of order and that is just what gravity is. Where there is gravity chaos is prohibited. That would be the most unlikely way Creation came in place. The recalling of pictures representing images about creation must have form, but to mathematics it had no form. From this thought the very opposite arise where Creation came from nothing but such an idea is mathematically simply not possible.

The mathematical presence we have in the distribution and organisation of the Universe and even the calculation which we are able to present as indicating that that is where chaos is abundant, still tells a story of organised growth and not a blob of material with no centre from where gravity controls and where there is no trace of any even distribution of material. No wonder those Newtonians will fill a Universe with nothing and then stand back to have a view of the entire nothing they can see. Please keep in mind I am the under-educated zombie that has no brain function, which would enable me to understand Newton while they can have their view at night on what they fill the Universe with and call it nothing. How on Earth can one say one can see nothing and still point at something you see?

The thought of nothing is just what it is, a thought of nothing and although it is in the human mind common nature to present nothing as a value in the recalling of something, the nothing is a presentation of the figment in the human mind. There can be no number such as nothing and that was (possibly) Newton's biggest error. Nothing represents non-existing and that is just what nothing is, it is non-existing.

In order to prove my point I wish to ask the reader to define the shortest line there can theoretically be. If he should answer anything but that the shortest line would be at a point where the beginning and the end of the line is the very same spot he will be wrong. The shortest line that can ever be anywhere must have a start and finish holding the exact same spot. The line will be humanly impossible to create but we humans are capable of very little.

When the line has a beginning and an end at the very same spot and it wishes to extend the position as to further the possibility it has, which direction should it favour? Humans in the west would naturally think of extending from left to right while in the east humans may want to go from right to left. Some persons will tend to go up or down, but all of the options are about human preference and not mathematical conclusions. Extending the line in any one direction will favour one direction without a conclusion about not extending in other directions. Such a conclusion has no sound mathematical foundation. The only option about extending will be in all directions equally in order to give a meaningful non-bias flow of mathematical equilibrium.

The shortest line in the realm of possibilities must have a start and finish holding one spot and such a line will also be a dot or a circle. Not favouring one direction puts all directions at equilibrium meaning that any form what ever may be can develop from such a spot with the end and the start being the same. This reasoning prompted me to look for singularity in such a spot because if the prime spot from which all came was a spot, then the spot must hold the shortest line but more prominent it will hold the smallest form including the smallest circle.

One possibility that the shortest spot can never have is having a starting point on the zero mark. If the mark of zero holds the start it must also hold the end because the end and the beginning has the same position. If the position of zero then is the beginning, the end will also be zero leaving the line without an end as well as without a beginning.

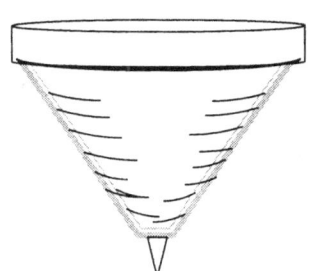

The conclusion from this is that no line can start at zero because that will be a mathematical impossibility. A line or spot starting at zero would therefore be shorter than the shortest line possible. A line growing or extending from zero can never leave zero because the influence of being zero disqualifies any possibility of growth. If the line then had to grow in all directions at the same pace the line must therefore be a circle. The value of the circle is Π, and that is where creation started.

That gave me the clue where to start looking for singularity. One would find singularity in the value Π and the value Π will be in all things rotating in a circle. In my explanation about my cosmic theory I wish to include a tad of what may seem nostalgia and the relevancy will become apparent. When we were boys we played with a top we called the spinning top. I cannot imagine that there is one boy in the western world that did not hold such a devise in his hand. Tying a string securely around the tapered cone started the operation and then with a jerking or pulling throw the devise is launched in a projectile manner and the big knack to success was getting the nail end firmly on the ground and with a releasing jerk the top was rotating. The champion was always the one boy that could throw his top to spin the fastest and that would create a humming sound. The louder the sound the spin produced, the bigger the champion. When a back braking effort produced a throw of enormity the spinning top would not only produce sound varying in pitch but also create a spin that would seem to have some instability. There are very many limitations about the spin, parameters that determine the slowest and the highest spin rate and spinning is within the parameters of such settings. The question arising is why such parameters are there in the first place?

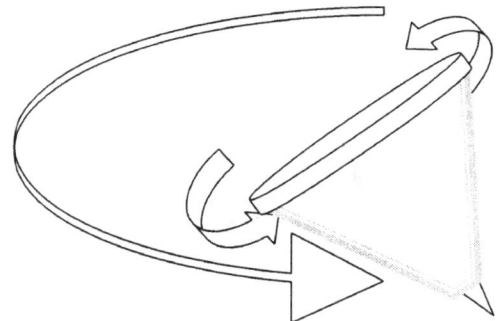
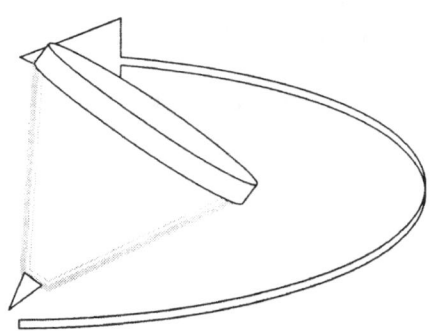

An enormous effort will have the top going oblong while spinning violently and as the pace reduced the top will stabilize by coming to an upright position. In the upright position it will then spin for the remainder of the period where it will in the end start tilting to the side and in a last effort throw a few wild oblong turns and fall over.

Boys playing games will never realize the scientific breakthrough and grown ups do not play with toys. In this little toy played everywhere everyday by almost every one is the answer most brilliant of human Brainpower seeks answers about all the cosmic riddles no one seems to understand. In the spin as such one may find two vital boundaries in the motion and the boundaries are marked by a wobble coming about as if the top is fighting some other influence. Spinning too fast pulls the centre off centre and so does spinning too slow. It is the same influence coming about at both ends of the limitation in the spin. There are influences at work, but force…no; it cannot be forces setting such boundaries. From that I started per cuing what sets such limitations because that limitation must be universal as all matter is spinning in one way or the other. The spinning top is the Coanda effect and the Coanda effect is the atom in every principle the cosmos have.

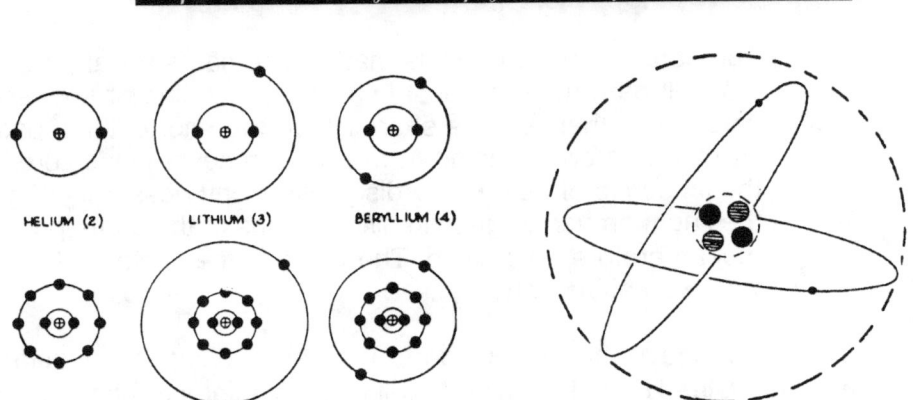

The 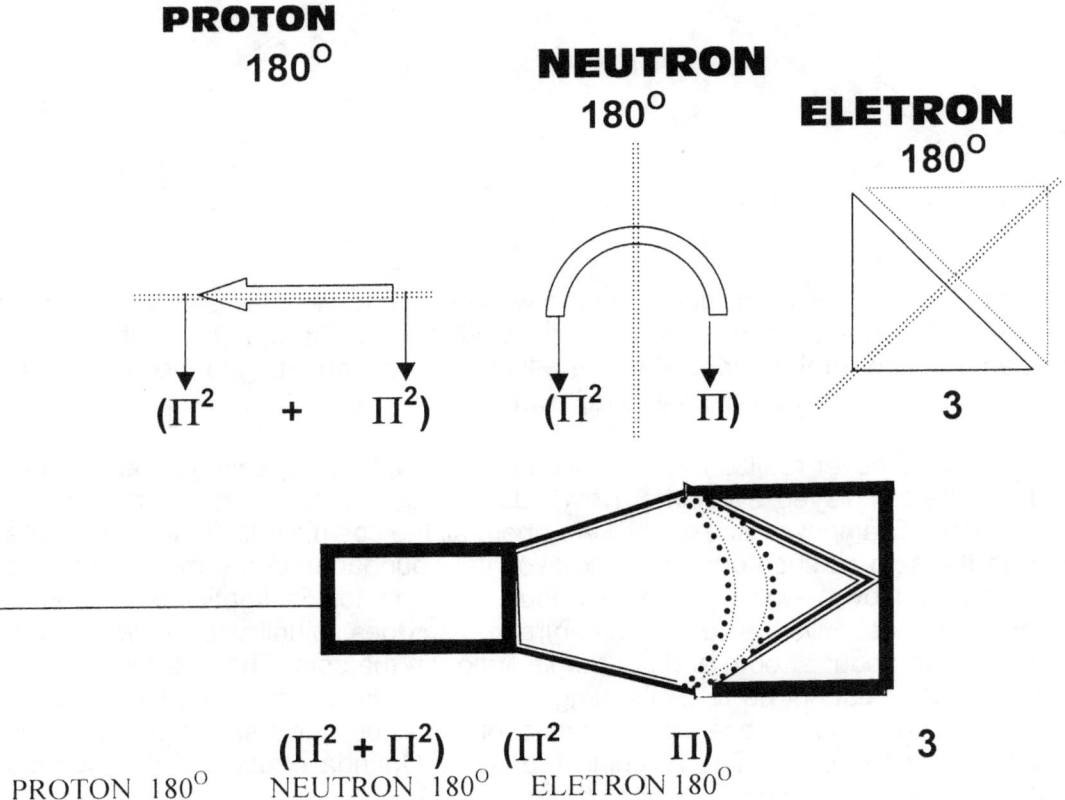 relation that an atom has with heat stems from the number of protons in the nucleus of the proton cocoon.

Every aspect of the Universe holds relevancy by applying time to space and the time to space first claims space from singularity then controls space from singularity and influences space outside the direct contact with singularity. In every event the factor remains the same, as it is only the relevancy re-applying a dimensional influence on space-time.

The key to the relevancy is heat and space. When matter heats, it expands therefore it takes more space. When matter is cooled it shrinks, therefore takes less space. That is the relevancy because matter in any form is heat. Heat produces the increase of space and reducing space produces an increase of heat. That is the relevancy. That is the secret of the Universe. That is the secret of gravity. That is the secret of momentum and every other aspect within the Universe.

PROTON 180° $(\Pi^2 + \Pi^2)$ NEUTRON 180° $(\Pi^2 \quad \Pi)$ ELETRON 180° 3

Time stood still in eternity, then after a command of the Creator, time started to move by overheating and eventually formed the relevancy of the proton ($\Pi^2 + \Pi^2$) the neutron ($\Pi^2\Pi$) and the electron (3). As a star returns time by depleting space to the dimensional increase of heat, space destruction is in progress and the star will abandon systematically some of the dimensions the atom holds. That is the relevancy. That will be whatever position there is in the Universe. In the depleting process of dimensional re- adapting, the star shall abandon aspects of space-time. The electron (3) may become obsolete, the neutron ($\Pi^2\Pi$) may become obsolete in neutron stars and even ($\Pi^2+\Pi^2$) the proton will become dysfunctional as space reduction completely disappears from the star's space-time occupation. However, those stars will be dark, and beyond our vision.

The relevancy holds value pointing the relation between the various dimensions as they are in the atom. The relevancy of ($\Pi^2 + \Pi^2$) ($\Pi^2\Pi$) (3) = 1836 will remain but the mass of the electron and the mass of the proton will change in every space that time applies. Cosmology thus far was incomprehensible because it was incorrect. When applying natural laws, it becomes so simple that a person as ordinary as I can understand and explain it.

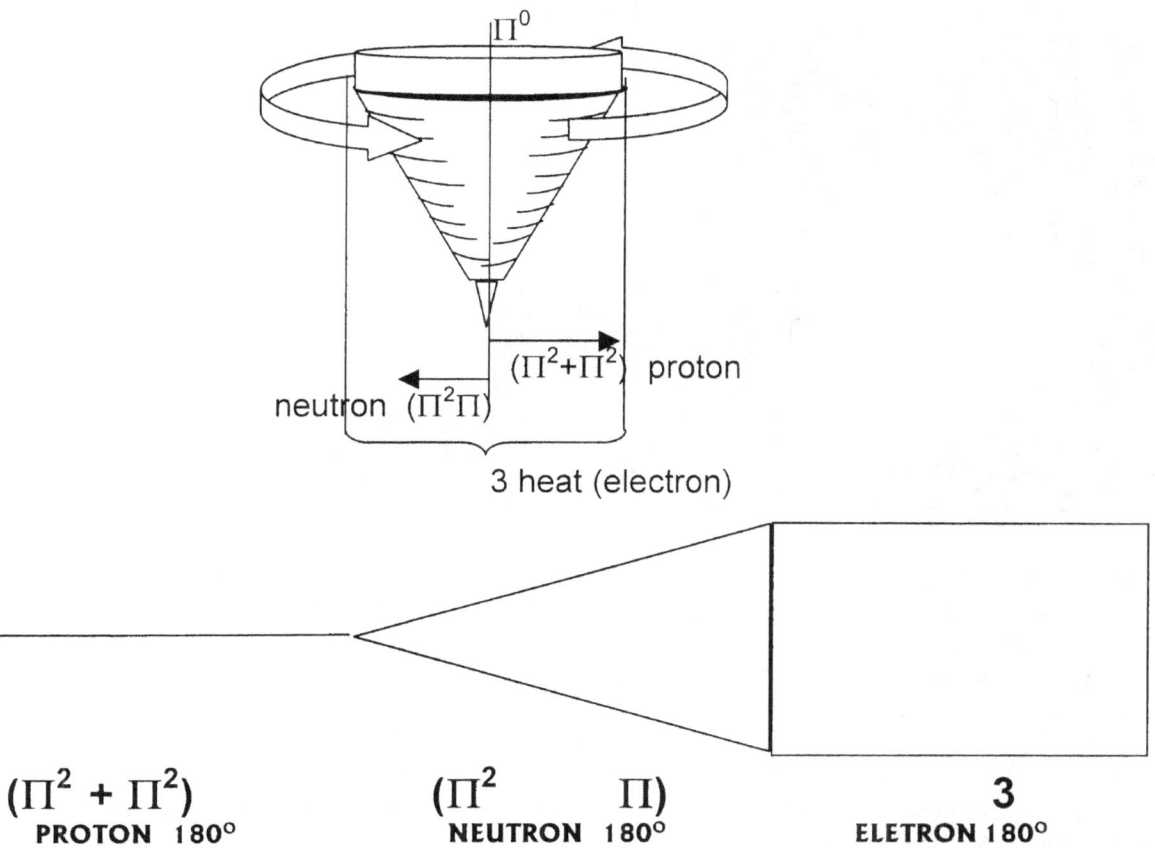

$$\Pi^0$$

($\Pi^2+\Pi^2$) proton

neutron ($\Pi^2\Pi$)

3 heat (electron)

$(\Pi^2 + \Pi^2)$
PROTON 180°

$(\Pi^2 \quad \Pi)$
NEUTRON 180°

3
ELETRON 180°

The above indicates where singularity originates and how that establishes the factor in singularity Π. The Universe started from the factor in singularity Π. The entire Universe holds a spinning relevancy to all other factors in the Universe. If that were not the case, the Universe would not be there. The first person to consider the factor in singularity Π, was Galileo. In the swing of the pendulum he saw singularity remain as one, that formed time, destroying space to maintain time. To prove my statement I shall very briefly indicate some barriers of motion science refer to as the Doppler effect, but Doppler used a slow moving train that at best could indicate two or three very minor moving limits.

If I can understand it, every other non-brainwashed human on Earth should understand it. The relevancy of $(\Pi^2+\Pi^2)$ $(\Pi^2\Pi)$ and 3, is a dimensional reduction of the flow of heat from space back to time. The flow of heat becomes necessary to prevent solid matter from overheating. By removing heat from the gas of space, through the neutron, to the solid of matter, space reduces as the intensity of heat flow requirements increase.

How does the photons manage to convey one complete picture coming from as far apart and as wide an area as it does? With a few photons connecting the eye or lens no one ever noticed the wonder of light. The photons reflect a view that seems as if coming from all the billions upon billions of stars. But most is coming from darkness covering an area no man can measure. Yet how many photons can actually connect to the lens of the camera or to the eye? Still a few photons coming from a single direction directly ahead eventually tell the entire story. It is very simple to take the process of seeing by means of photon conducting very lightly and I have never heard one of the Brainy Bunch really in sincerity uncover the process to its utter and full potential. It is impossible that light from such an array of assorted sources can simply come together at the eye lens and show a picture of objects spanning across a Universe as wide as our mind can receive where the objects they reflect is beyond human measurement and the quantity is inconceivable many.

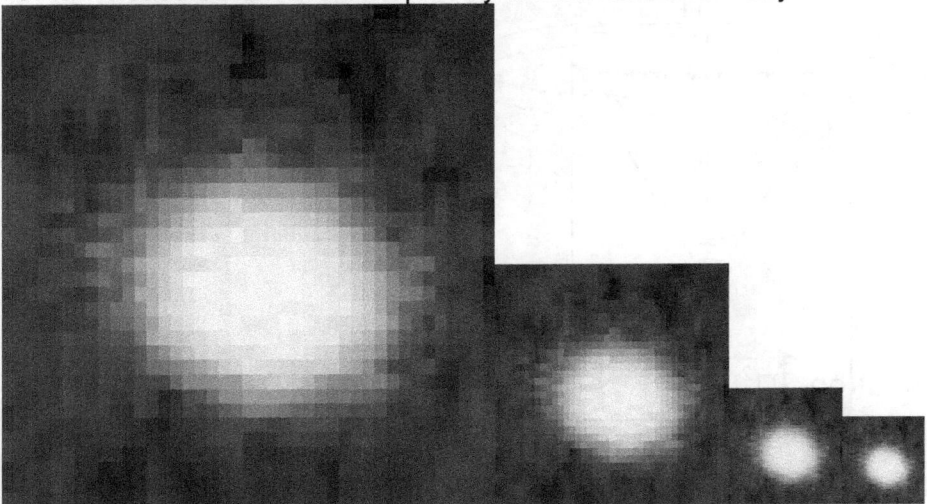

If scientists think of outer space as geodesic zero, with nothing in outer space but space then how do they explain the fact that we can see the nothing. How can we see nothing being in between light? According to official science that blackness out there represents geodesic space. Geodesic zero means the light travels in a straight line from where it originates unhindered all across space to where the light connects the eye. By crossing the vastness of space, the intensity of light reduces. The light coming from the Sun is quantifiably less than where the light is hitting the surface of mercury. The light loses intensity while travelling through the Blackness of outer space. If light was losing intensity the intensity of light must be that which is robbing the light that which is taking immeasurable small measured light away from the mainstream of light flow must be light. If that which was collecting the light were not by own measure light the light would have stood apart from that which was collecting the light without being light. The light cannot mix with the darkness and remain apart because it is not the nothing the darkness represents without standing in an identifiable visible support of what it remains to be. It can only mix, if it was the same that was mixing.

Isn't it rather reasonable to think that if the light were different from what it went through, the light would leave a luminous trail as it went along since the light cannot mix with the conducting medium but still leave some part of the light behind as the price it pays for passage?

Think how wide your view span is. Think of the enormous room, space, volume size you pack into such a small space as your eye sockets are. Try to calculate at night what the volume is at that point that you can see simultaneously. Pack that lot into your eye sockets. See how many cubic light years you put into an area where only a few electrons can go. How do you manage that! Look at on dot in the night sky and think about that one simple dot. Think how many electrons must be streaming from that spot. Every photon that reaches you

must have left one atom that went fusing. As it travelled on route it went in circles for millions of years before it was able to leave the star.

The photon did not increase in size, and it carried the information it obtained from the one point that went into fusion. At the outside of the star it did not grow bigger and at that point it represented the information of such a small area we have no means to calculate it even if it was possible to put that area it represented into your eye. As it came along the information it represented grew in stature, as it became a larger part of a smaller growing space. It started to tell about information it could never know anything about except on the condition that the photons fuse together as they travel on. At best the photon originated from an area smaller than the eye socket and it represents (say for arguments sake) one group of stars. How did the information it portrays when reaching my eye, form a picture representative of the entire night sky? How did all the information become scrambled, then became unscrambled and mixed while being able to put the entire night sky in a coherent picture unless all the photons fuse as they come along? Somehow the information has to fuse together in order to find coherency. It can only be that light is singularity overblown and represented by photons over spanning the whole context. Light then is singularity k^0 in space a^3 in motion T^2 over the distance it came k. Light we see is therefore $k^0 = a^3 + T^2 k$ or the space it represents a^3 at the distance it travelled $T^2 k$ bringing in once again Kepler's formula of $a^3 = T^2 k$. **That is the story of one photon telling us about one light source that may be one star or one group of stars.**

The picture gets even more extravagant when one thinks about the idea that one photon might represent an entire galactica with a combination of billions of stars grouped into one area. That is not the only picture we are getting. Our picture contains much more than a few billion stars in one dot. We might see several such galactica in one area.

Nothing is all about not being and not "not seeing". We visually see two spots holding light that is in the overall mathematical picture quite close to each other. Why do we not see one spot in double vision? What is keeping the two dots apart? How do we put the entire picture in precise co-ordinates into a total three-dimensional unit that fits all into my eye socket? Surely in comparison there is an example of everything going to nothing. Try to put that measurement in reduction into a sensible and audible mathematical expressed configuration in order of simple understanding. Convert that mass into a comprehensible reduction. The cosmos is not about mathematics because though you Newtonians fool yourself in that you're in superb ability with your mathematical skills. The most obvious puts your mathematical skill to shame. When it comes down to true issues you and your mathematics can only degrade the cosmos.

The fact that one can see the night sky is a proven fact! The fact that it is there in every man that has the ability to see is the proven fact. The fact that your calculations fall short is the proven fact. The fact that in true issues of cosmic proportions your mathematics does not even cover the idea there is presented in the wider cosmos is a fact. The fact that by your mathematics you do not even have the capacity to think out the question, which is hardly a reason to understand the question that is hardly a reason to come to the same answer proves how far your superb ness in mathematics leave your wonderful atheistic claims in shame.

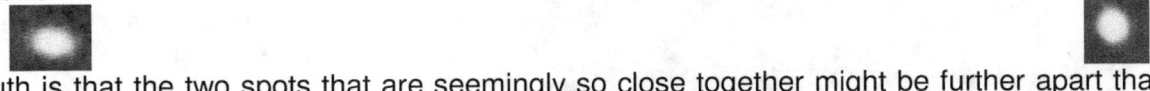

Truth is that the two spots that are seemingly so close together might be further apart than the entire Milky Way is wide. This means there is a lot of never explored between the obvious which makes the unobvious part very unobvious. I hope that makes sense in the way I wish for it to make sense.

The reason is why they (the Newtonians) can see the edge of the Universe and can therefore calculate all the mass in the entire Universe is that they fill the centre of the Universe. The Newtonians fill the centre of the Universe. All the light is coming from afar directly to every Newtonian and thus they can judge from the centre of the Universe the size the Universe has.

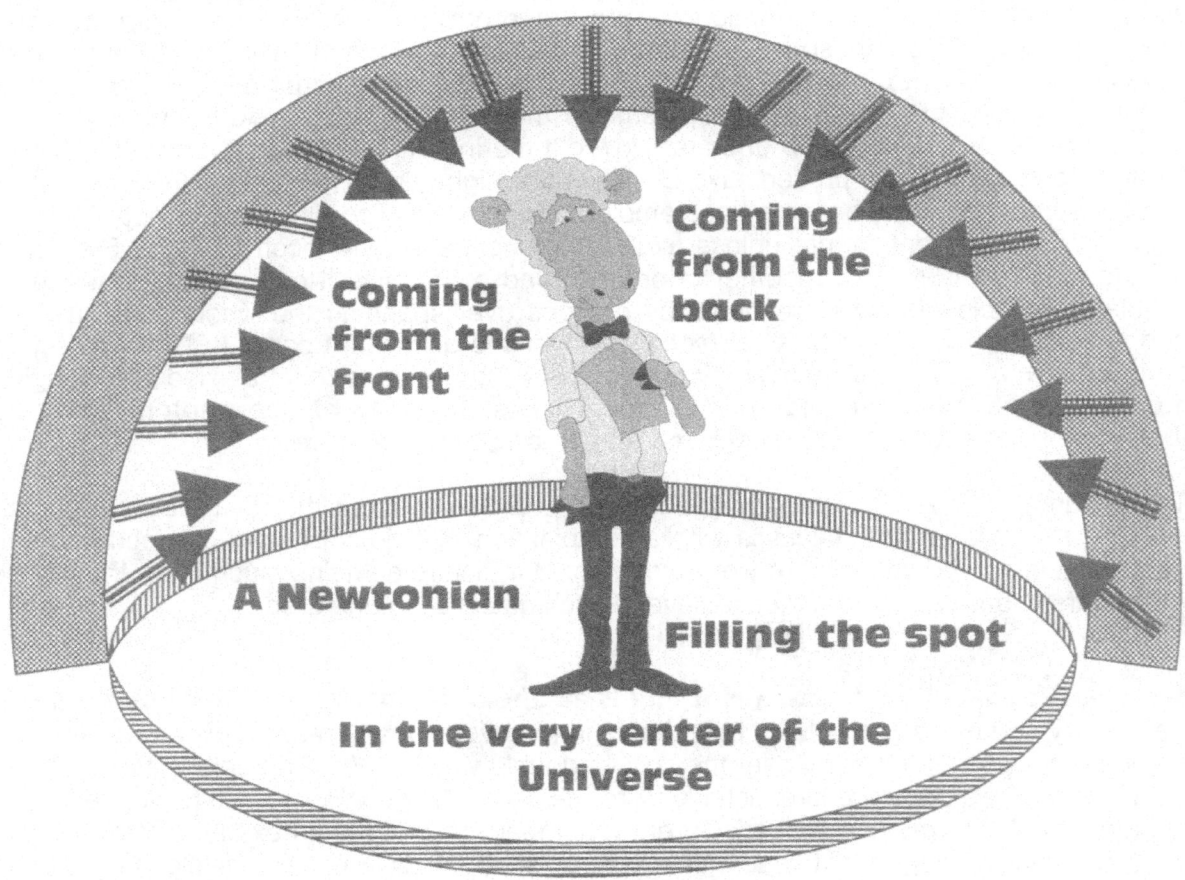

Go out into the desert and look at the night sky where the sky is unblemished of light pollution. The night sky seems three dimensions. All that marvel and fascination is lost with man's artificial light degrading the unnatural cosmic light. See how wide an area runs from front to centre, top to centre, left to right to centre and calculate that what is in the eye range to an understatedly mathematical measured volumetric size. See how much " space" there are in the view. Then go home and start calculating so that it might enable science to find answers.

Then try to fit that calculation into the size of the eye. Reduce that lot to the size called the eye nerve and squeeze that which is visible in the desert night sky into a nerve fibre that would carry the information unto the brain. Let those mathematical Master Minds show precisely how that space goes into one eye.

If light came across space from wherever light comes in the form of individual streams of photon flurries, then in our vision the message must be one of fragments of space, tumbling where pieces represent light sources that does not make sense because the rest will constitute of nothing where the light cannot report nothing and light cannot represent nothing. We must therefore see speckles of light that forms senseless spots conveying information from pieces of stars that are sending too little information too slowly to have us make sense of what we see. The photon as an individual particle that is released from a source that is so small it is fragmented atomic debris, the concept would translate senseless non-understandable madness.

The photon as a single unit cannot translate what we see or bring knowledge with light that we are able to translate into proof that we form a picture we can use as proof. Although the stars as such are shown as fragmented dots in picture above, we still can translate what we see as a coherent message. If it were only photons with no corroboration about a unified picture, it would not be sensible. It would be a picture unconnected bringing across some photons in the manner where every object stands apart not being related in any way and that will be what we see, if it is anything that we see. That we know is not the case but that means geodesic zero is as much rubbish as anything Scientists regard with simplicity and with careless thought. Geodesic zero means nothing and how can I see nothing as darkness because "nothing" is not darkness, nothing is "nothing" and the darkness I see is darkness showing the darkness as something.

Such an idea by itself is outrageous because the stream of photons reduce in space to such a minute quantity that taken the area the photons travel and the space in vastness it covers, the chances of one photon coming across many hundreds of light years through billions upon trillions of cubic kilometres of space and selecting my eye to convey the electricity is less than infinite. Yet such conveying takes place every second of every minute. The position of the location of the second singularity, which is the precise duplication of the first singularity but in a diminished capacity, is obvious to miss when one is not applying a detective mentality, as one should in scrutinizing the cosmos.

We may view two dots that would seem to be close to each other but seen through the lens when using a strong telescope the two dots are many degrees apart. The two dots might hold information across an area in space that covers something such as our Milky Way many times and that is done by each of those we are observing. As far as the size on the claim of space they contain go, we can calculate but the numbers such calculations arrive at is meaningless to our small minds. What is the difference between 10 light years across and 5×10^5 light years across? Now we get to the true sticky issue I am aiming to get to after concluding this approach run. Now we get to the question I wished to propose in the very beginning namely: What about the blackness in between. If a dot of one or two millimetres might represent an area covering a 100 light years and the two dots both being a millimetre across are some five millimetres apart, the five millimetres they are apart represents an immeasurable time span. That black stuff Newtonians conclude to be nothing. Let's dissect the "black" part first.

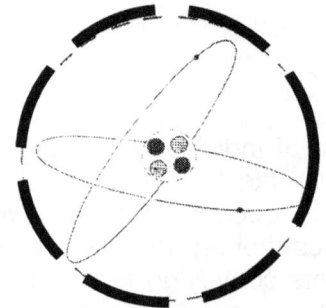

The atom in whatever form, is a combination of what forms the Universe and in that to the extent that it is the Universe. The combinations may go as high as the most advanced galactica or as low as the most insignificant sub atomic particle but the end result is the atom is the Universe notwithstanding description.

All galactica forms one atom. All stars form one atom. All layers within stars form one atom. The electron proton neutron cluster form one atom and all cocoons holding an assembly of the above form one atom. All subatomic particle groupings form an atom.

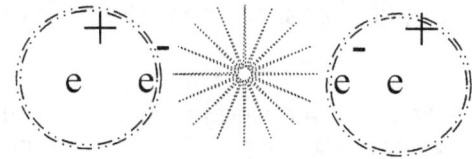

In the centre of any atom there is a line generated by all the spinning particles within that unit. The unit represents all the particles and the particles in the unit are a combination of infinitive numbers of particles joining together to produce the unit. The unit goes down as small as one can allow reason to take the particles and the Unit goes as large as the Universe at large. In the end it is all the same to the cosmos by cosmic standards. It is individual motion that parts one sector from another sector and in the end the parting is connected by infinity.

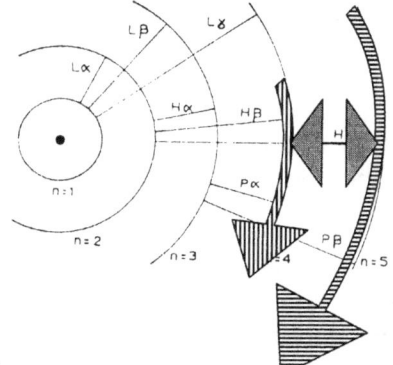

From what ever our abilities are, that part of our abilities vested in the Universe will never reach or even almost nearly reach the end of the line. The atom is made up of energy but that is as general a term as time or space is. Never can any one accredit energy with an infinitive meaning but only with a vague description. By seeing what happens when an atom increases space or decreases space, we can see what is its final substance? By growing, the atom endorses heat and by reducing, the atom rejects heat. The atom is heat in more or less quantities. Newtonians use the term energy as a very convenient escape passage. The moment their explaining runs dry the term they grab onto is energy. What is heat…heat is energy. Every time I hear that phrase I feel as if can blow my top, and moreover using it so liberally seems to please every Newtonian. It is the same as saying that every one is willing to bluff the other one as long as the other one is prepared to be bluffed and then will bluff right back on the same terms. What is energy in the most infinite sense? Energy is heat pure and simple. Energy is heat where heat is time delayed and time delayed forms time in progress either by being space or time and mostly both.

To answer that we must be clear on what is cosmic within human realities and what is beyond cosmic in realities. At this point the atheist mind gets equal to the abilities of the animal and being an animal in mind by being an atheist I suppose the following will go far above their abilities of reasoning. The thoughts that I generate travel far beyond the speed of light and therefore my thoughts are not part of my cosmic material that I take charge of. I can move my body's parts with my mind's thoughts, which proves that my mind's thoughts control my body parts. I am not going to go into details about this but in another book with the title *"Xepted Astronomical Mistakes" ISBN. 0-984410-1-4 I* explain life in physics in extensive detail. Life control space –time by converting thought to electricity through the generation thereof. The electricity is not life.

The electricity is only the result of what life can generate. We find in electricity the ability to command instructions that regulate and control space-time whether by moving things physical or by moving the physical body supplied to host and support life or by manipulating other life or objects life created / established / used. They are all is terms and the term makes no difference. All motion we have comes to accept as common practise and is exclusive to our environment, which is a host to life. All other places are most hostile to life and only on Earth does life find a way to flourish. Every other place holding space is hostile to life to the point being there or putting life there will be futile to the life as part of the body holding life. There is a need for an acute and a deliberate turnabout in Newtonian standing points about life and cosmos motion. There is a need for a differentiation about what is cosmic motion and what is life inspired. However, to understand any and all motion is the most complex issue in nature. In order to understand the true concept behind motion we have to break the cosmic motion down to where motion initially started. We have to return to moment –Alfa.

There was a continuous motion where eternity met infinity before moment –Alfa. Eternity was spinning within infinity because we still have infinity and we still have eternity. They are tangible entities found everywhere throughout the Universe and

are in control of all aspects of the Universe. Between eternity and infinity heat moved in and heat parted that which cannot part. This partition is presented as a time delay that later (at present time) became a time delay and all of the entire Universe is heat lagging behind the time which is in front and was pulling on time that was further behind. It is space-time parting infinity from eternity.

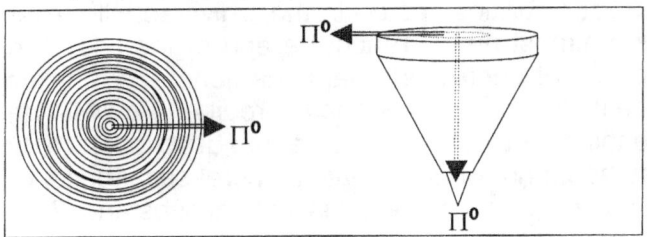

In the motion there are three factors filled by the same substance that is the same substance although the parting is also standing in for the substance. It is the same thing that is in time and time is in front of that which is followed by that which is behind. The three are inseparably one.

The evidence how it all started is still visible in every day life that surrounds us. It starts in the same manner every time motion calls on the curvature of space-time to activate singularity. At first we find eternity sharing a spot with infinity where singularity is eternally spinning on the spot it shares with infinity. Since infinity can only provide one spot that is not present and eternity has to use such a spot to move, all the movement are eternally though it is on one spot that is not part of the Universe at large. The spot eternity use to move is not present and therefore all the motion eternity has is not present either. The motion is eternally there but the spot it uses to move is absent from the Universe although it is the controlling factor.

As soon as motion of the spot comes about through heat parting time by creating movement the space that time provides as a change in the lasting of eternity, makes eternity lag behind and such parting brings eternity to part from infinity that is the part that is unable to move. This is no guesswork but is evident for all to see when that person who desires to see investigates what happens to the top spinning. That which cannot move does not move and fills the spot that does not move. Looking at the centre of the top one can see the spot not being there forming a line that cannot be because the line that is not there can have no space or sides to use as being present. Yet that line not being there parts a Universe into four opposing sectors.

That, which can move, moves about by four around that which are unable to move. Then that which cannot move transforms to the next location because it cannot move. However, since it is 1^0 and the next spot too is 1^0, the one is the very same as the next 1^0. That means it is not relocating but it is just being the same somewhere ells and that means it did not move at all. It just is what it was being the same with different allocated circumstances surrounding it. The singularity did not move while it is the heat that placed different relevancies on surrounding marking spots. By relocating the surroundings singularity remains as it was without shifting. Because the one spot is precisely the exact replicating of the next spot harbouring singularity in infinity, it is the heat, which forms singularity in eternity that relocates as it reallocates new relevancies surrounding singularity. Singularity in infinity does not relocate but transforms the surrounding by which it cannot move.

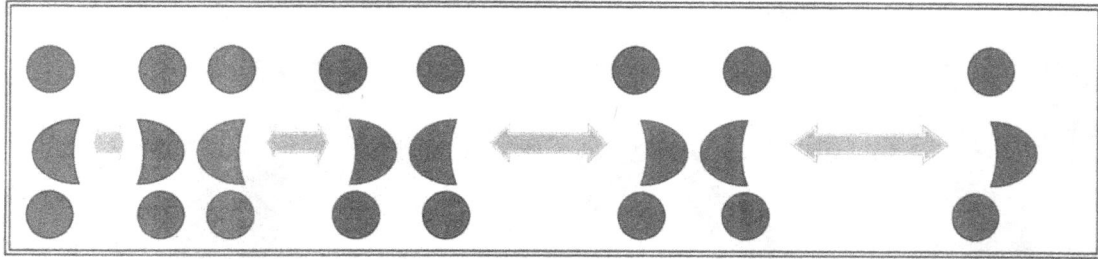

When it started and time in eternity moves apart from time in infinity eternity shifted by relevance but eternity never vacated what was. Eternity still held four positions as before but since eternity has the ability to maintain without destructing (that is what it was doing before the interruption of infinity) the cycle, it found a location to freeze some of what was overheating and which came in place between that which has no outside and that with no inside. Some became fluid and formed time and others through spin became solid and froze into seemingly tangible material

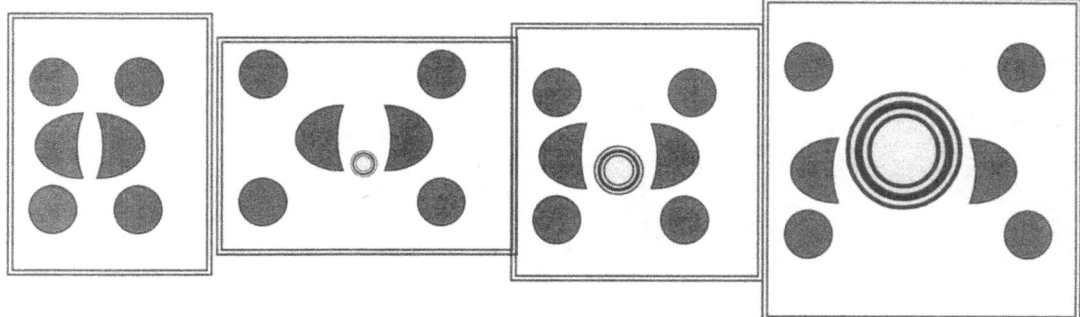

The dragging of a heat turning stared to keep behind because the motion that parted that which has no outside from that which has no inside aroused the centre singularity, which has no inside. Since that which has no outside was the shift in time that part was what parted and by parting more because it could not lag behind. That is what puts distance, which is time in suspense at a point holding more time (seeming further) or holding less time (seeming closer). It is the material that is lagging from the point holding singularity in relation to the point holding singularity. It is crucial to think of singularity as 10 and therefore all points serving singularity holds the relation to time as immediate. It then is the task of material to relate to singularity and whenever it does not do that in the immediate it is lagging behind that immediate.

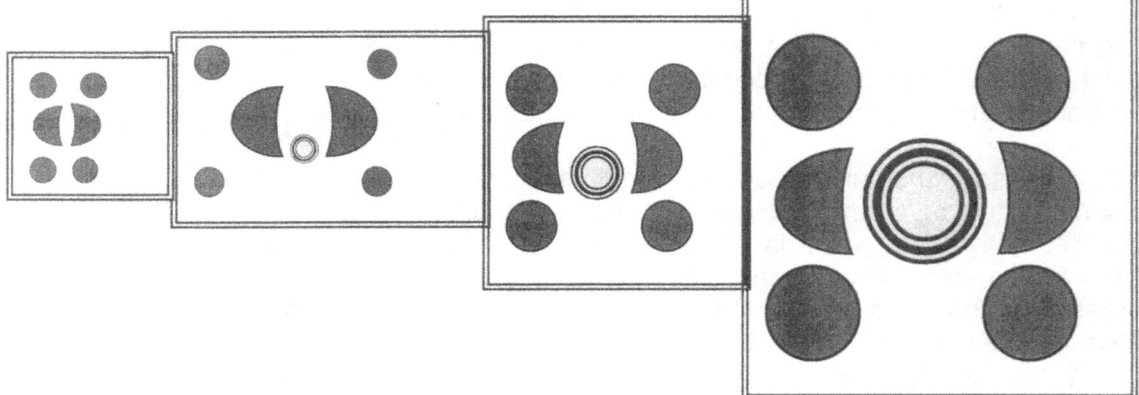

The part that has no sides parted from the sides that has motion and in between is all the heat we now find filling space and time as a usable Universe.

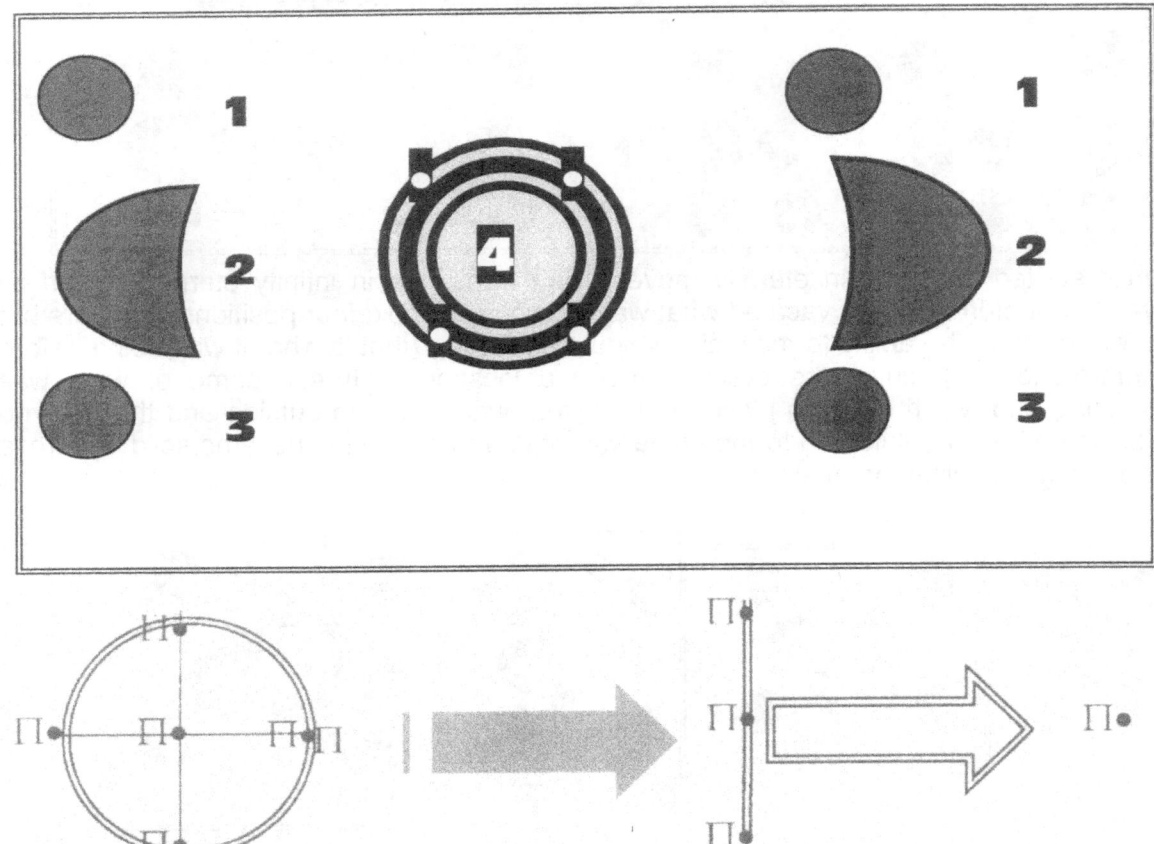

After the three moved away from the one the four remained eternally in place. The vacating of the three did not remove the four because of the eternal status of the four. The four can once again reunite and accommodate the four but the four can never be evicted from being in between the one and the three than was in the location that the eternal four finds a temporary position

The Newtonians are going to charge my sanity and attack my mental state of mind for saying the following but I have proof about what I am about to say. The Universe is not real. The Universe is not a tangible fortified reality of lasting proportions. The Universe is a flaw that came about when that which cannot part separated and the Universe is in a course of repairing the flaw to destroy itself, which is that what is not reality. The Universe does not truly exist but is a temporary measure where heat or light became time in eternity and material became the protector of time in infinity and time split.

As it is in the case of the human body so too is the Universe an elusion compiled for a time to be present as a part of the cosmos and thereafter repair what was undone. The body blankets life. The reality, which is life, establishes a body through time to cover life and in which life find a use. As soon as life has no longer required needs for the body, life discharges the body and with no life compiling and maintaining the body, the cosmos discharge and destroy the body. In the same manner does the Universe work, except it works on a much grander scale in time comparison. While Galactica produce nurse and give birth to stars, stars come in cycle to undo the damage of Creation be mending time into a unit once again. The Black Hole and stars beyond that is the epitome of the repairs going on where time in light in eternity is re uniting with time in infinity at a point where there is no longer needs for material to be present. Heat or light parted time and that started the Universe. From then on the Universe was set on a course where it freezes heat out of existing by increasing of motion or movement. By moving it freezes the heat that parted the time that couldn't part in the first place.

There were four in the centre that wasn't there because the four shared a centre spot, which was the centre spot.

The four spots that represent eternity was spinning around one centre that represents infinity and because the centre was on the same position as where the four in eternity also was spinning time became a repeat of three circling around one where the four was sharing one location at any one time. The same one in the centre was also three spinning on the same spot.

The three was at the same time the three holding eternity while the three was spinning around a fourth centre spot.

Time came about when the four parted by having three positions standing apart from the fourth centre position. The reason why it stood apart is because the one point was allocated ahead of the centre, the other was in par with the centre and the other went behind the centre. A Universe was born because 1^0 moved from 1^1 to form $7 \times \Pi^0$ and that established Π. By motion of Π moved from Π^0 to Π establishing Π^2 that resulted in Π^3. The proof is still in every spinning top.

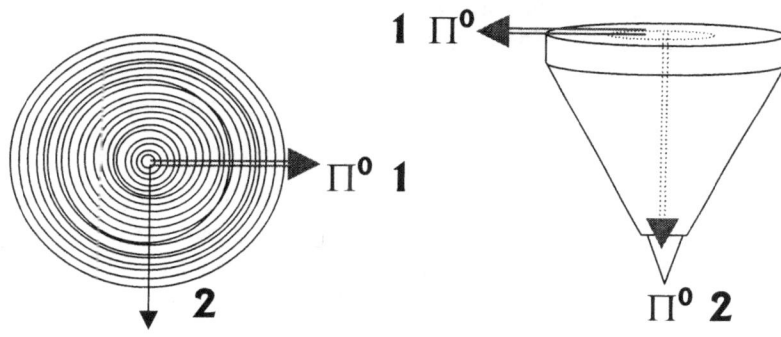

1 Π^0

Π^0 **1**

2

Π^0 **2**

The universe came in place because infinity as 1 parted from eternity as three and because 1^0 that also were incorporating 1^1 as part of 1^0 then by motion became relevant to 1^1. A top can spin because the line forming singularity parted from the three positions forming time and with motion space-time entered between the two factors. As motion becomes part of the top the outside, which defines eternity and is representing the eternity factor then through motion associated with the individuality of the top forming one edge of the top that parts infinity within the top from eternity outside the top. It is in the spinning top present for all to witness.

Π^0

motion is that

Π^0

Π^1 Π^1 produces expanding

Π^1 Π^0 Π^1

heat going into space, where the provokes three points that serves time forming a differentiation the past, the present and the future. The time served is a flow of heat coming from the past through the present and going onto the future. It is a ratio of time or a relevancy in space that is brought about.

The centre singularity point expands with heat accumulation where derived from the expanding. The action result in the release of heat an of the space time as between

Π^0

Π^1 **X** Π^1 = Π^2

Π^0

Π^2 Π^2

Π^1 Π^1

This is proven by the flow of space-time just as Kepler's calculations reflect. The space $\mathbf{a}^3$ is a combination of motion in time that produces a dimensional quality of 7/10 Π^6/ 6 =112. It is $\mathbf{a}^3$ that is a collection of motion ($\mathbf{kT}^2$), which is an assembly of time ($\mathbf{k} = \Pi^0$) and ($\mathbf{T}^2 = 10^2$) which is then in the dimensional expression $\{\mathbf{a}^1 = (\Pi^0 \times 10^2)\} + \{\mathbf{a}^1 = (\Pi^2 \times 10)\} + \{\mathbf{a}^1 = (\Pi^0 \times 10^2)\} = 298$. That is the space-time ratio $\mathbf{T}^2 / \mathbf{a}^3$ that Kepler introduced as the value of $\mathbf{k}$.

This is more a symbolic expression than a calculated

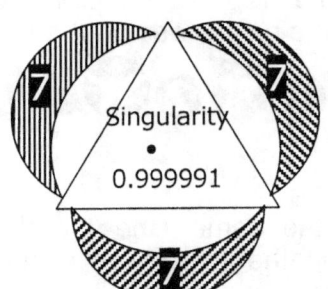

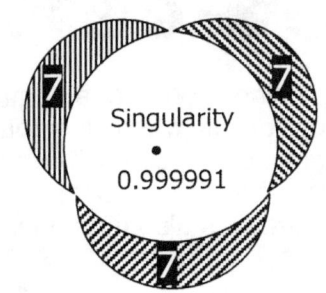

mathematical statement but it does prove that space is time by three positions and the combination of space-time by three positions in three allocated setting is 7/10 Π^6/ 6 =112. It is the seven of space in relation to the ten representing the square of time (5+5) in relation to singularity holding three positions in space as well as three positions in time relating to the Universe we are in having six coordinates to fill.

It is the time factor in the relevancy in form relating to the space relevancy in form that produce the Titius Bode law of cosmic proportions. What this indicates is there is a movement Π^2 of Π^0 to Π^0, which is a movement of 1^1 to 1^0 that circles about 1^0.

We may not see such a flow as a direction because all the direction we associate with space is part of the three sevens that becomes the three flowing with time. We see the flow of time coming from eternity towards infinity because the time is lagging in the side eternity holds and is catching the side infinity holds. What we see at night, the black stuff that we see that is holding all the bright dots in position is time in eternity. Eternity is reuniting with infinity and infinity is within every spinning particle within my body.

In the relation of space against material, there are always three positions of seven circling points forming a triangle of time, to become time in present and time gone by, rotating about a very specific centre. This is the flow of time and that indicator points to the direction of the flow of time. The positions in time as well as the reference the positions make during the flow of time and the order the allocations of the various positions bring set time as a cosmic controlling centre. There has to be a centre. That centre has to be because of the motion surrounding the centre. There has to be time retarded that formed heat in between such centres in relevancy. The nature of the spin promotes, as much contraction as expansion and the loss of expansion is the gain to contraction.

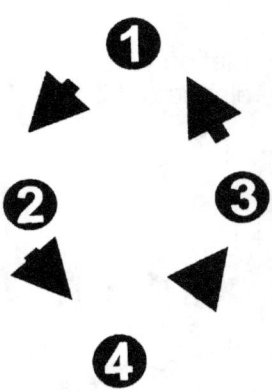

The spinning going on in the past is the spinning going on in the present, which is the same, spinning involving the same points in the future. Please note that it is the space-time within those particular circumstances that is generated and what is generated is different. That which is doing the generating is identical, precisely the same duplicating an exact clone copy.

It is absolutely vital to understand that the three points coming on, as time is the very same three points disappearing into singularity. It is where time in eternity that parted from time in infinity and is catching infinity to become unified once again. The unification is part of every spinning object there is and is the centre of the Universe.

Looking at the top it seems that the body structure of the top is solid and the air surrounding the top is liquid. The top as a structure composes of solid particles that light cannot penetrate and that material cannot pass through. In that sense it seems to fit all the conditions we set for solidness. The top spins and it spins through the air that allows the top to spin seeing that the top has much more density than the air has.

Singularity by Time

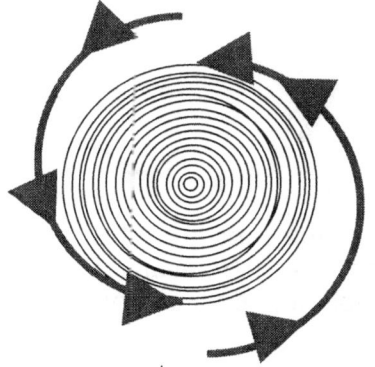

What can move is liquid and stands related to what cannot move being singularity. Since everything is singularity everything is immovable but also since everything is singularity everything is eternally moving because space divides time that moves from time immovable.

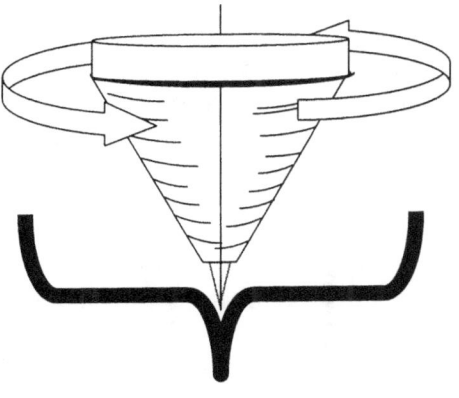

the top forming a solid or so yes in a way and not that is a pump that pumps heat inwards just like a turbine that is rotating inside the keeping the centre erect. The motionless because all the moving and the moving of extending the singularity of where the top meets eternity. singularity is holding the air the liquid the flow of the liquid keeps the top erect and spinning. The spin produces a cold in relation to the hot that the liquid is.

Every thing outside the top is liquid with it seems to us. Well much either. The top from the outside engine. Every atom structure of the top is centre is totally atoms in the top are the top circle is the top to the edge The extending of as a liquid and being

What is moving is liquid and what is not moving is solid. Everything has a reference in relation to another point. That which is capable of relocating is forming a liquid in relation to that which is securing the position of rotation. Everything in the cosmos can move and yet not one particle in the cosmos can move. The cosmos stands divided between the eternal moving of eternity and the immovability of infinity.

$\Pi^{\mathbf{0}}$

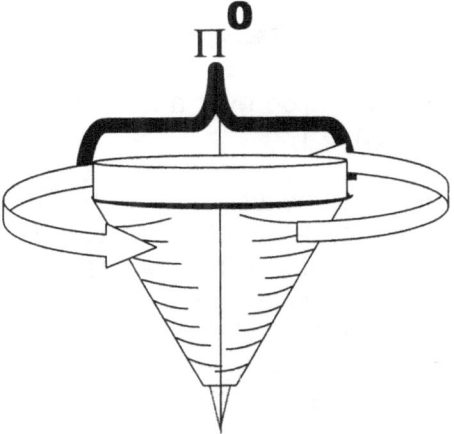

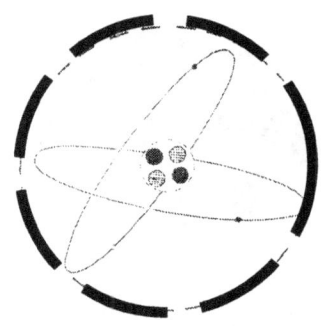

Everything around the top is liquid with the centre being a solid. However the solidness and liquid has cosmic standards and just as it is in the case of hot and cold, big and small, fast and slow, our standards and cosmic standards do not share any measurements. So too does cosmic notions about liquid and solids have a totally different meaning in cosmic terms.

There is a pumping interaction of space-time flowing towards singularity through every point that confirms singularity. Every thing in the top that forms the material is also liquid. By providing motion, the matter in the top serves as the liquid factor that extends the space that singularity provides. The structure is composed of atoms. In the atom there are a governing generated singularity around which all material rotate. In the case of the atom all the rotating material forms the heat while the generated centre, which is incapable of rotating, forms the solid factor. Every aspect that is without motion stands in a relation of 1^0 and that which is relatively moving or changing location or finding a new position holds 1^1. Everything that is standing still is 1^0 and everything that is moving is 1^1.

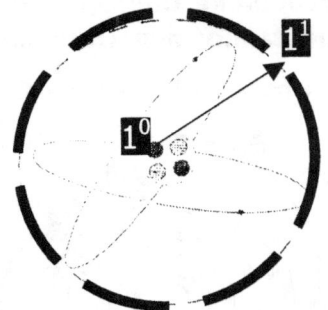

Gravity or motion is a constant relation that solids have with heat where heat forms the liquid and solids form space. There is the rotation but part of the rotation is the lateral progressing by rotation to confirm the generated centre. The generation is in the rotation but the flow towards is the lateral and just as electricity produce a flow of time in relation to space collapsing, space-time by measure of gravity is using the same system to do the very same.

There is no substance difference between 1^0 and 1^1 and it is a relation where one moves as the liquid partner and the other is the solid factor. Both are not as much equal as they are precisely the same. Infinity cannot move and eternity cannot stop moving. By parting, infinity had to move and eternity had to introduce as part of the cycle a point where it stops moving in relation to the other side that cannot move but does start moving.

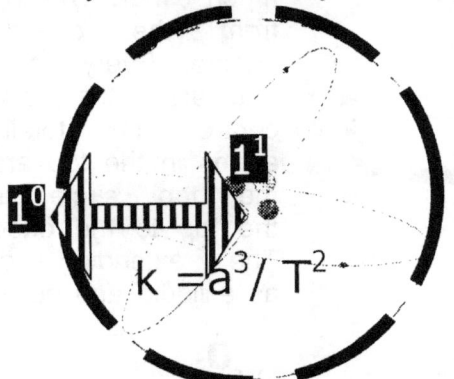

$$k = a^3 / T^2$$

The factor that shows motion forms the liquid while at that moment the factor that does not show motion forms the solid. The measure of 1^0 is transformed to 1^0 and which ever are 1^1 is passing the extending of space on to 1^0.

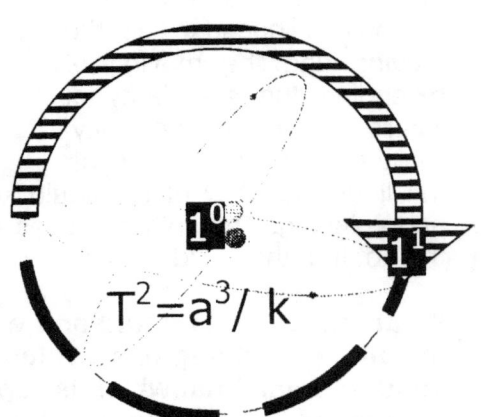

$$T^2 = a^3 / k$$

Time spin because everything spins in order to secure the centre singularity. But also time moves and in that there is the linear that always is part of cosmic motion. The centre is referred to by heat but heat also secures the centre by reconfirming the centre in the lateral. But in both cases singularity is reinstating singularity by confirming as it is referring one another. In the manner that 1^0 confirms a position in singularity 1^0 is supporting 1^0 by generating 1^0. By generating 10 it is repositioning and reallocating a position by confirming 1^1.

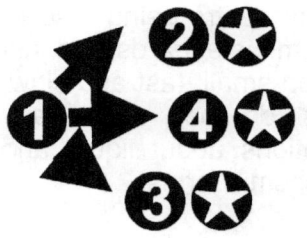  

Coming down to understanding the concept in the infinite we must turn to light, which is the smallest material particle visible to the eye. Since the photon is small and fast and we are incapable of really managing an investigation into the photon, we must turn to the photon's more spectacular but far less frequent counterpart being what is referred to as "ball lightning" Ball lightning is heat or time liquefied as it is generating a point within the centre and such singularity is generating motion to concentrate time to heat or electricity or flames or whatever name one wishes to attach to the very same thing.

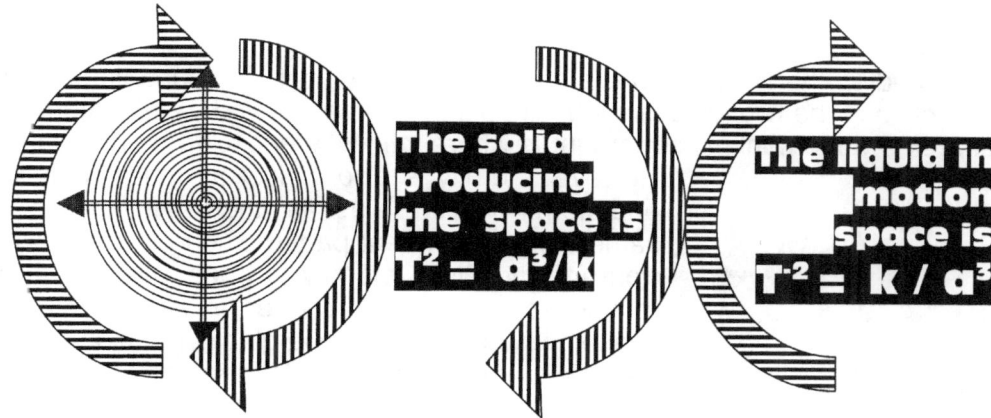

The spin centres singularity (k^0) and the space (a^3) establishes a singularity centre motion by spin (T^2) as well as by relevance (**k**). It is the Coanda effect where the liquefying of heat into a visible electric flame produces the limits of a space created by the motion of the heat being charged by a centre charging the centre again. It is again the manifestation of Kepler's space-time $a^3 = T^2 k$, which is $k^0 = a^3 / T^2 k$. It is time catching up with infinity and the result is eternity returning to infinity as the heat is once again recovered by the uniting of eternity and infinity. That is the photon on a smaller scale as well. In the Universe size is not an issue but a change in relevancy. Since the ball lightning is larger, the spin factor T^2 takes most of the motion and that increases the space a^3 factor to suit the situation. But that also decreases the relevancy or linear factor **k**. In the case of light the spin factor T^2 reduces the space factor a^3 and that allows the relevance to match C. Light is what remained of the Big bang where space was an electron C^3 leaving gravity at C^2 and motion by relevance at C.

We stand in time holding space. Life takes charge of the material lent to life to support life's manipulation of space-time by another form of movement separated and apart from cosmic motion. By bringing about motion, there is a discrepancy established. The discrepancy is by the changing of the alignment between the one position time holds in eternity relating to a point infinity holds and the next minute realigning the space point. It is realigning 1^0 with 1^1 by three locations in time. One must not look at the motion in the circle but at the motion of the circle as the relevancy reduces by duplicating that which rotates in ratio with that which contracts.

Locating and finding the presence of singularity

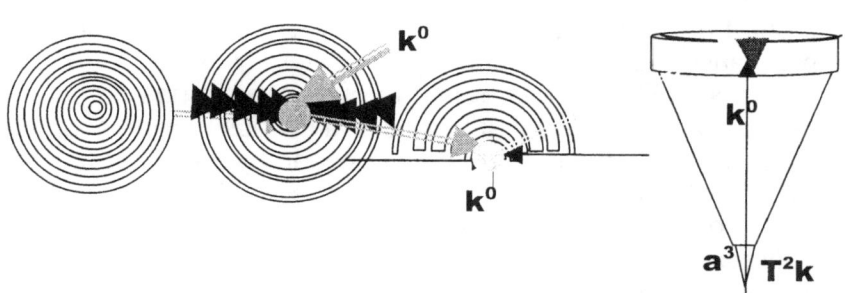

It is the rotation T^2 that duplicates space a^3 / k by reducing the relevance a^{3-1} where the relevancy is k^{-1}.

Since in truth the contraction is T^{-2} but it would lead to great confusion if we state that time moves backwards because it cannot, we have to use the reducing of time as space reclining its value **a^3/k,** which is precisely the same thing. In the end it is 1^0 relating to 1^1. The essence of moment −Alfa was that eternity parted by placing light between eternity and infinity. The essence of the atom is to remove light from eternity and place it in the atom to be united with infinity once more.

What is allocated to the position where 1^0 are in relation to 1^1 is space-time. The motion established between the two allocations of singularity is time delayed as it is returning time in delay to time in the moment. The moment is where infinity is encircled by heat in the atom. However eternity is also 1^0 just as much as infinity is 1^0 and eternity is also 1^1 just as much as infinity is 1^1. It is the same thing which heat parted when eternity moved away from infinity. Since eternity is three in time, which is the four in space and is the three in singularity, eternity is space, which flows by charging singularity. What is in eternity is not just like what is in space or just like what is in singularity. It is the same. It is a clone of the very same thing. The difference is not 1^0 or 1^1 but the motion of time putting 1^0 at a point to relate to itself at a point 1^1. The factor 1^0 fits into 1^1 in total harmony since the factor 1^0 is 1^1.

Because the factor 1^0 is the factor 1^1 and the factor 1^0 is precisely what the photon encircles. The photon encircles the electron centre because the electron is rushing the light through my eye nerve to my brain. The movement is taking 1^0 in the way of 1^1 to my brain, which holds 1^1 as a reflex of 1^0. Since what is in my brain are 1^0 is that which forms the expansion of time is also 1^0 but is 1^1 in relation because of a time constraint, I can see all 1^1 because I have 1^0 in my brain, which is 1 in time. I am as much part of eternity, which I see as I am infinity uniting eternity.

What I see is what I am is what is in the yonder of time and only by measure of time delay is there a differentiation between that which I am and that which I see that is flowing towards me. I am the end of time and therefore I am the centre of the Universe. The atom hosts the centre of the Universe and because the Universe started with the atom, the universe concludes through the atom. The atom is the gateway where eternity again once more reunites with infinity and in that the Universe arrives at a conclusion.

Because I am at the end of eternity where that which has no end starts, I represent that part of eternity that holds no start in relation to the part that has no end. I find a start in that which I can see as being the part in eternity that has a beginning. Being at the start of that which has no end I am unable to see the other part of eternity, which is the part that has no end. That is because there is no such a part in eternity as forming the side or the part in eternity that has an end. Eternity has a start through me but has no end. In the very same manner can I see the part of infinity that has no end because I represent that side of infinity while the side of infinity that has no start is also unseen by me. That too is because there is no side where infinity starts. I lock in the divide between infinity that has no start but uses me as an end and eternity that has no end but uses me as a start. When I and all of space −time are eventually removed then again eternity with no end will once more unite with infinity with no start and all space-time will disintegrate into that which has no start and neither has an end.

I can see what ever is out there in all of time because what is there in all of time is exactly what I hold in space less time. I am 1^0 and therefore all of 1^0 is also part of me because 1^0 has no sides and has no space, therefore all of 1^0 fits all into 1^1 and the whole Universe fits into my optic nerve with no squeezing required what so ever. I can fit into me what I am and since I am what time concluded I conclude eternity by uniting eternity with infinity. I am a black hole as much as a Black Hole is a black Hole. The difference between my being a Black Hole and the Black Hole being a black hole is that I still sport atoms and the Black Hole got rid of all time delay of any standing.

Culture will have us believe that when one sees a colour shining from an object the colour is associated with the object. Logic tells a different story. A yellow dot is all the colours in the spectrum but yellow because it is disassociating with the yellow. That goes for red blue and all other colours we may visualise. I think the norm accepts this as scientific fact with very little argument or substantiating proof about that required. We have in our minds the formed concept of opposites applying. The opposing side of red is blue. The opposing side of white is black. If white is an array of all colours scrambled, then the forming of the colour black must be where all colours that are present in the scrambling of the white is also absent in the scramble mixture of black. If there was total absence of substance as far as colour goes, the heavens would be a few specks of white dots that are blending because outer space as "nothing: is parting them and therefore "nothing is parting" them. Going one step more in sane is that we then all agree the colour of the nothing is black since we can find no colour mixture would blend to form black.

What then about colours that are technically not colours as is the case with black and white? White is simple. By spinning all the colours in the spectrum the colour white shines through. Black is quite another matter. A friend of mine whom is one of the best painters I have ever come across told me that one couldn't paint black but have to make black a dark blue to show shade on the canvass. That apparently is his success in achieving the realism.

He also went on to explain how many variations of dark blue form the shadows in one simple tree. This remark set my mind in motion. One cannot see black because black has no colour to show, but black is the colour most prevalent in the universe. One can see only by colour and since black is not a colour we should not see black, but we do.

If the darkness was the representation of "nothing", then that should be exactly what we must see, nothing but the stars. Taken from the top picture some stars and leaving the rest to nothing is what we see in the picture below. A blind person sees nothing but when we look at space, we see something that we think nothing of as we see as space. One cannot have the ability of sight and see nothing except by closing your eyelids and then you see nothing. But in that case you do not see "nothing" in contrast of "something" you see "nothing" without it contrasting to "something".

By the ability to see the darkness such action in it self renders the darkness a factor of forming something other than nothing and that changes the acquired value of the darkness from nothing to something. There is an eternal difference between something in infinity and nothing. That black stuff…the use of the word black by itself makes it a contentious issue. That is even before we are using logic by disregarding the nothing part which crosses over what borders the ridiculous throwing the whole argument into the mindless because it is ridiculous to think that anyone is able to see something that is made up of nothing. Yet I stand alone against the might of Mainstream science no matter how correct my views are and notwithstanding how far their senselessness are going into madness, I still stand alone in my thinking as I am again and again rejected by the Brainy Bunch.

The arguments introduced touches the most basic aspects of my work and by no means can such an introduction secure an opinion that I do realise. Yet, not once through all my long investigation in the past thirty or more years have I found any other person claiming such views that I have brought about even in this skimpy way. If you see it, what you see must be light because you see it.

We view an object over a distance and see what the light brings across. It reads as $k = a^3 /T^2$. I see the space that time holds by time developed as k the time factor of relevance. Time at present T^2 is holding the space a^3 in relation to how the cosmos developed the relevance k. If I move time back by bending time with a lens $T^2 = k / a^3$ then space will increase as the "distance" or relevant position of the space increases. I challenge all Newtonians to increase or decrease nothing. I can see with the aid of light. Light has to come to my eye to allow me to see. If the dot is too small it will find not enough representation in the overall picture of visible light coming to me.

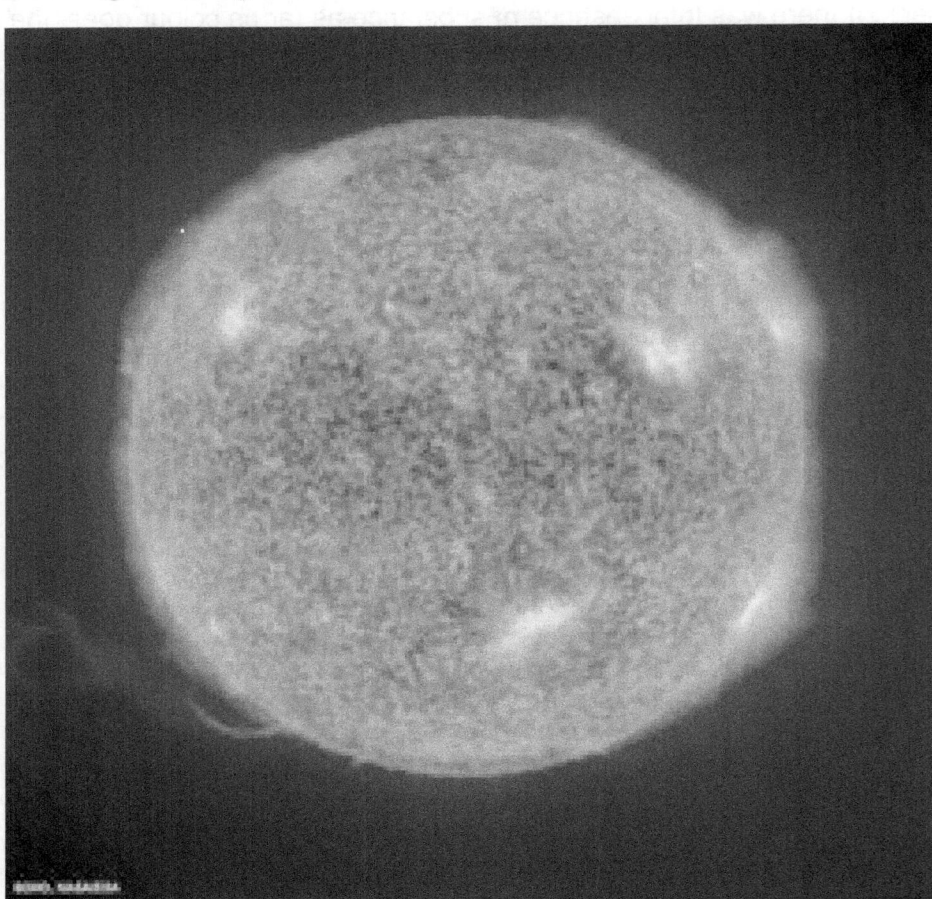

I still can see, however that which I interpret as light I can see and to my mind that which I cannot interpret I interpret as if I cannot see it. I can see it but I am unable to interpret it. If I have to acknowledge that I can't interpret what it is that I am seeing I am as stupid as a grazing cow not knowing what she is looking at when she stands staring at what it is she does not focus on while grazing. No mathematician that concerns him as smart as to accomplish the calculations God used when God invented the cosmos will ever think of himself or herself as so stupid he or she doesn't even know what they are seeing. They know exactly what they are seeing... they are seeing nothing and to top that they are too stupid to realise that they should realise it is impossible to see nothing!

By increasing the one part the other part has to decrease. But that doesn't mean they have to go less blind or blinder because they are seeing what they are seeing. It is only a matter of finding a balance in what they do interpret and what they do not interpret.

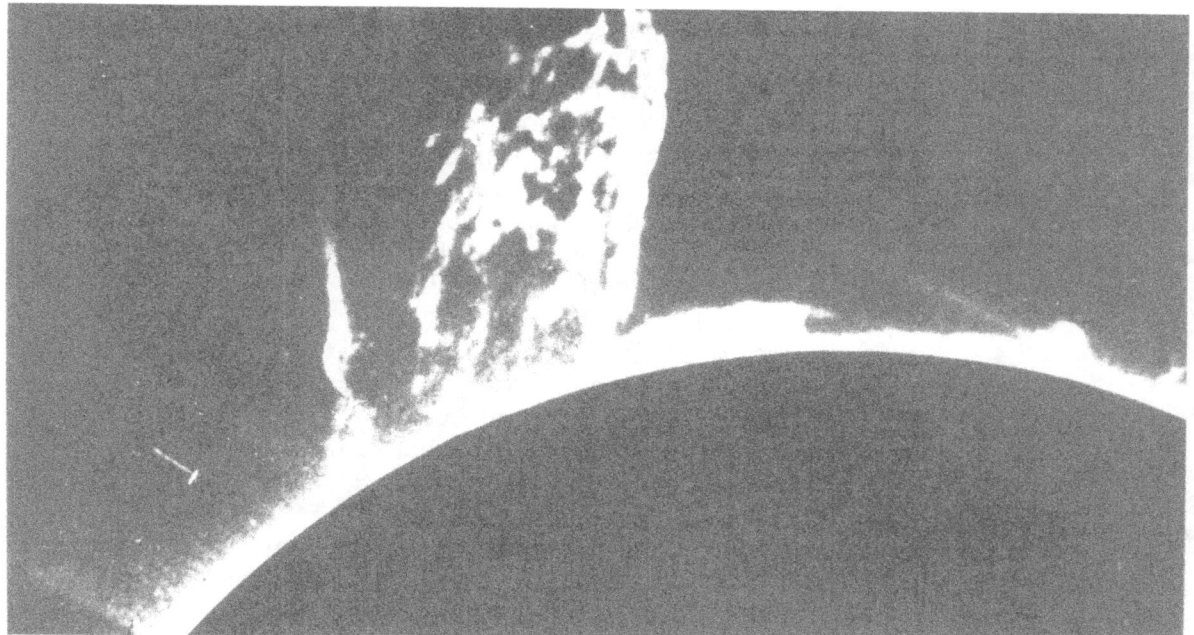

Looking at light coming from the Sun there is a distinct interpretation that the light coming on is bright and it is hot and it is hot because it is bright. Finding myself in the darkness I find the darkness is cold and the light is dark. In that way I think of the light being bright and hot. That is because in relation to the cold of the light coming my way I am hot, In relation to the darkness and the cold in the darkness I am cold. With the bright light is my relation to the light that brings about that I am hot. I am the relative hot party because I feel the light touch me and where the light touches me it exaggerates the cold, which I interpret as a hot spot. One can't feel the light but I can feel my condition at the point I sense the light and at that point I feel heat. Where I am within the darkness the darkness of outer space is not freezing me because I am freezing at the point I touch outer space. It is because I am so much colder than the hotness of outer space that I am freezing instantly.

Remember I am freezing and not outer space. Lets take a look at the Sun. Where the Sun meets outer space the Sun is boiling and not freezing over. By the cold Sunlight touching me I am able to generate a lot of heat within me as the heat has a route to where it can conduct. By generating the heat, which I then am able to conduct, I feel heated and generated. On the other end where I touch the outer region close to outer space, there is an abundance of heat outside me and with me being so utterly cold in comparison to the hot outside of outer space, I cannot conduct a flow of heat outwards and that stops me from generating heat inside. That smothers my generating ability and life is generating motion in a quest to manipulate space-time. If the part holding outer space was cold, the Sun would freeze at that point with ice everywhere.

However, where the Sun is in contact with the outer space region, the Sun is boiling. That means the temperature the Sun holds must hot up and start looking to meet the required heat at the point of contacting outer space. Outer space is raising the temperature of the Sun and not dropping it. I am in relevance to the heat, which makes me either the hot party where the coldness of the Sun touches me and I feel my body raising the temperature of the heat to bring it to an acceptable level to my body heat. At the point where my body meets outer space I burn black because at that point outer space ingenerates my cold body with its heat. At the point where the Sun touches the heat of outer space, the liquid heat of the Sun starts to boil. The principle is not similar to the heat transmitting that we are familiar with when an element like a kettle is heating the water from within. The contact is on the outside and the heat at that point releases from outer space as it flows to the Sun.

This picture to the left presents an explosion where the heat of the star is coming from within. The heat within is beyond the control of the star. The heat turning to space in the overheating

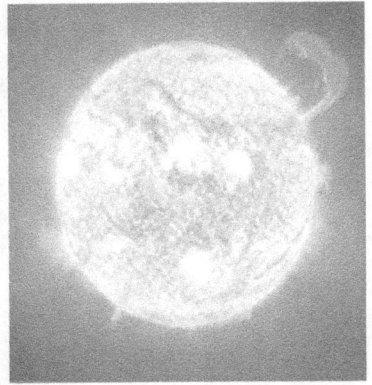

has the star overheating from the centre and the expanding resulting from the star overheating is beyond control of the star. In this case the star is producing gas that is the result of overheating which is the result of heat levels within the star that turns the liquids within the star to boiling gas. One cannot even dream to compare the procedure of what is happening in this event with the procedure of what happens in the Sun under normal condition in the process they call a prominence. The liquid that squirts from the Sun is a solid tube of flowing material in a liquid form and a liquid is a semi solid. The centre of the liquid is a steady semi solid and does not burst open with heat that turns to gas. It is clear that the heating is done on the outside of the liquid. The liquid vaporises as it makes contact with the time zone called outer space, but is still cooled by the material on the inside of the flowing stream of liquid squirting out. When the rest of the retained material does not show a density

decrease, the coldness of the contraction of the Sun allows the liquid to fall back into the Sun spontaneously. That does not happen with the exploding star where the liquid remains in relation to singularity but not controlled by the freezing cold of singularity. The sun is not a ball of gas that is contained by the unexplainable magical force called gravity, but is a frozen liquid ball that heats up when it makes contact with outer space exactly at the point (on the outside) where it makes contact with the overheating outer space. The Sun is clearly a cold and controlled object that only overheats on the thin rim, which is in contact with outer space. The thin line on the outside shows signs of overheating where the thin liquid heats up in waves of heat and expires as gas or light into the time zone outer space.

The Sun has no ability to release the heat into the Sun and the shear heat turns the pebbles that crystallises as heat makes the liquid heat into flakes that is not gas and neither is it a liquid. These small crystals or a pebble of liquid light that contains singularity by immense spin is a fragment of heat. The small malls of heat singularity turns into small crystals we know as photons. It is heat particles like it was in the Sun, but at that point where it touches the atmosphere it becomes concentrated liquid particles with a gas envelope leaving it to be photons. Within the Sun, the liquid photons are like water with a much more solid binding. This is because the Sun is that much colder than outer space is. The temperature of the Sun rises at that point where outer space comes into contact with the liquid of the Sun. It is not the Sun that is loosing heat because then the Sun would freeze at that point.

The Sun throws plumes into outer space because the heat entering from outer space distributes uneavenly and at certain points the heat rises to higher levels than at other places on the surface of the Sun. This can only be because of waves forming at that point just like the waves we see in the phenomena we call mirages. It is layers of different concentrations of heat. At those points it turns the cold substance within the Sun, the liquid in the Sun into gas plumes. But even there the distribution is uneven and only where outer space truly touches the liquid can some liquid rise sufficiently to form a gas that will be absorbed by outer space. The rest that did not heat to a proper gas remains a liquid and drops back into the liquid Sun. If the plumes of heat forming the prominence carried the heat internally as the heat expanded outwards, the plume would explode since all that lovely heat is suddenly exposed to cold and the plume would violently release the heat to the cold. It is not happening that way. The plume of liquid remains concentrated and only on the outer regions does it turn the liquid to a gas. That means the heat is not in the liquid but the heat is outside the liquid and it is turning the liquid to a gas from the outside inwards.

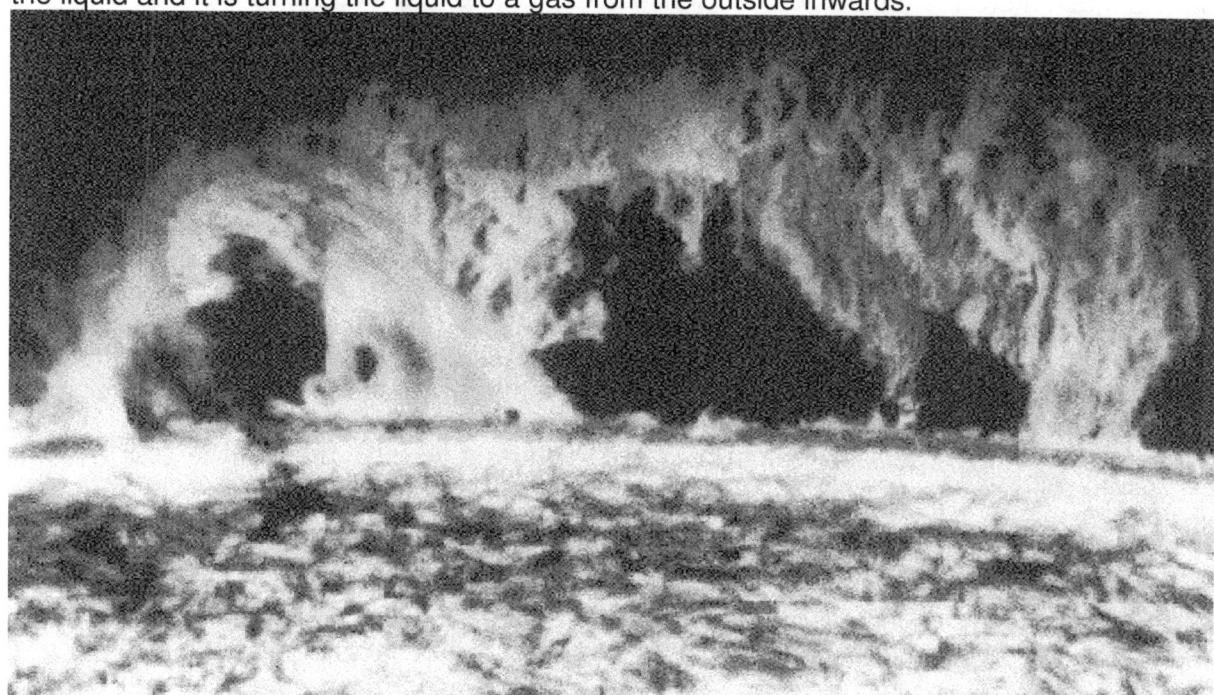

When we see dry ice boiling the boiling is in the fringes where the ice is in contact with the air. The ice does not heat from the centre because then the ice would explode from the centre outwards. The ice is stable on the inside but burns away at the fringes where there is contact with the much hotter atmosphere. The boiling would go on until the liquid ice has gone into a total form pf gas. But the process is on the wall of the ice and not from the inside and therefore the ice does not explode but turns to gas systematically. The Sun converts much more space to a frozen liquid than outer space can boil away the liquid interior of the Sun because with the exasperating movement of the Sun the motion freezes outer space by gravity, which is the freezing of time.

Look at the flow of the prominence. If the heat was in the prominence the prominence would disintegrate the moment it release from Sun because the heat would expand into gas. But since the prominence is cold because the Sun is cold that makes the edges of the prominence hot. When the edges of the flowing liquid turns hot, the relevancy to the inside turns the prominence even colder since the edges of the prominence is hot. In that case the prominence does not expand with heat but contracts and the density the prominence has increases. The prominence then falls back into the Sun because it contracts and its density would not allow it to remain out of the Sun. If the heat were inside the prominence the prominence would boil away and not conserve the relative small amount of heat it has. Since the prominence plume is part of the contracting cold it falls back into the Sun, leaving the photon particles that became liquid gas prominence plumes to expand with the gas in outer space and allows the hot outer space to expand the light in all directions. The surface of the Sun is a pool of liquid. Where outer space touches the pool of liquid the heat wave formed by outer space breaks down the cold in the liquid of the Sun. The liquid breaks up into fragmented light particles and as the fragmented liquid light particles become a gas it flows through outer space and out. But by outer space also being a light with much more intensity than the liquid heat of the photon has, the photon puts the space it holds in relevancy 3^3 into motion of space $3\Pi^2$. And where the space is colder than the contact it makes the space forming the photon puts the heat difference between the space and the photon into motion exactly like a drop of water does when the drop lands on a red hot stove plate. The heat transforms to gravity and gravity is either expanding motion or contracting motion

Light is much more than the medium science takes it to be. Light connects the Universe in a way we cannot contemplate. Light being far apart originating from regions not in the same time or Universal space, connects in a way that presents us with a picture holding the Universe in an understandable content. From the point we stand and we watch the Universe the significance of what we see surpasses the sense of understanding of what we are experiencing. How can the few photons that our lenses catch coming from such an area as the night sky cover transmit the complete picture of what we see. Take a few seconds and study the picture of the night sky then rethink the picture applying the full content in the picture to what the size of you eyes are. Think how big the picture is that your eyes take in and translate that area to the size of your eyeball in an effort to determine a ratio. One will be forgiven if one thinks of the ratio as eternal to nothing. Yet a few pages back I showed that according to mathematics there couldn't be anything as nothing. Consider the path the light followed from the source connecting to light from all other sources where all particles of the other light may come from bringing a full picture to the lens one use to look through. In your mind connect a line from every atom producing light and connect the lines to your eyeball and see how you can manage to fit all the lines, as small as the lines may be.

If I can bend something to increase the light by which I see, then it is light that I am bending. If I increase the light I use in magnitude in order to see, then I am increasing light. I am not decreasing the darkness by bending the light. I am bending the light because I am improving the focus I have in the light. By bending the darkness I am not squirting out more light, I am increasing a balance between that which I have a use for and that which I am unable to use. It doesn't mean that which I am unable to use is noting, unless I am a mindless animal that cannot realise that that which I cannot use, also exists in a way that I can use it if I had some intellect to do so. Newtonians are much better…they seem to realise that what they are unable to calculate cannot exist because who in the hell can be so stupid as to invent something that is outside the spectrum of their calculating abilities, after all, they are occupying the centre of the Universe from where they can see and calculate the lot.

It is so easy…if you can bend something more and afterwards see more in it must be a lens with a curve because we alter the curve. It you can read radio waves in the curve by bending the curve more with optic devises such as radio astronomy, then that which you bend is light. One can only reverse time as an optic elusion and bending something to gain an optic visibility improvement must be that there is an improving in the clarity of light. It is all part of the curvature of space-time since the curvature of space-time is the eternal sphere. We have to keep in mind that the first sphere was so small it was by today's standards not part of the cosmos because eternity parted from infinity. That came by measure of light or heat or whatever one wishes to call the substance the entire Universe is made of. By today's standards that, which was so small is so big the entire Universe fits into the parts that parted. Everything even that is so huge that we cannot see it still only is between the two points that was so eternally small. The biggest realisation man could make when science discovered radio telescopes was in realising that by bending outer space more, a new Universe appeared. That, which I mention is proof that they did realise but the most significant part of that went past them because they do not think. If outer space shows no light at one point, but does show images when that which it contains is bent slightly more, outer space then must be a variety of lenses containing light. The darkness that we see as outer space is light that is not focussed enough for us to see what images the light contains. If we had strong enough focusing lenses we could go back and visit era two or three. The darkness we see is not darkness but only represents the darkness of small vision. That darkness we see is light that is insufficiently bent in the shape of lenses to concentrate light to such an extent that we would be able to see the concentration of light it contains.

If it is lenses that enable us to see what we can't see in outer space it also means we cannot

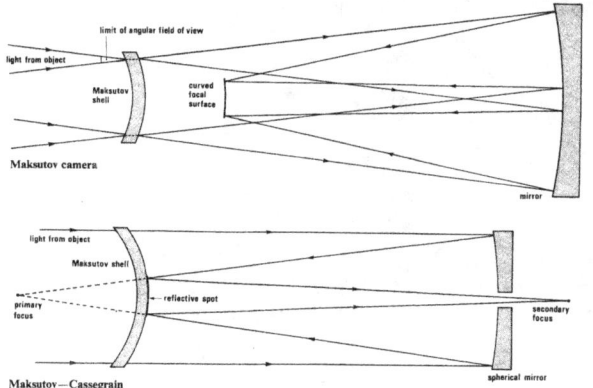

see the light, which is outer space because we haven't got the lens to match the curb of outer space. Newtonians think of outer space as geodesic zero, with nothing in outer space but space. Geodesic zero means the light travels in a straight line from where it originates unhindered all across space to where the light connects the eye. Such an idea by itself is outrageous because the stream of photons reduce in space to such a minute quantity that taken the area the photons travel and the space in vastness it covers, the chances of one photon coming across many hundreds of light years through billions upon trillions of cubic kilometres of space and selecting my eye to convey the electricity is less than infinite. Yet such conveying takes place every second of every minute.

The position of the location of the second singularity, which is the precise duplication of the first singularity but in a diminished capacity, is obvious to miss when one is not applying a detective mentality, as one should in scrutinizing the cosmos. Culture will have us believe that when one sees a colour shining from an object the colour is associated with the object. Logic tells a different story. A yellow dot is all the colours in the spectrum but yellow because it is disassociating with the yellow. That goes for red blue and all other colours we may visualise. I think the norm accepts this as scientific fact with very little argument or substantiating proof about that required.

I wish to return to my previous statement that I made about light because in this one may find the purpose of the Universe. Where I did mention this in the previous part, the purpose was to introduce the concept. Now that the introduction should have sunk in, I wish to elaborate on both the significance of the dark night concept and the manner in which we see the context of outer space. If light came as individual streams of photon flurries, then our visage would translate that as such shown in the fragmented picture above. It does not. The flow of light is precise and coherent which renders the Newtonian concept of light to be rather sheepish. If light were individual photons, it would leave a picture that is completely unconnected. The picture that would come across as a steam of photons will also be bringing across many pictures where each is telling a story that portraits the image told by some photons in the manner where every object stands apart and there is no coherent overall unification by darkness, where it is separate and not being related in any way and that will be what we see, if it is anything that we see. We find ourselves being in the position where we find our allocated centre of the Universe. We find that light coming from the past is streaming to our future where we dissolve light into the future. Still, we are the future of light being in the past. We hold the centre of the Universe because we find that all light from all over is streaming towards the position I as a person hold.

Having a position where light contracts from all over puts me between eternity and infinity and that puts me holding time in infinity away from time in eternity

Having a position where I am the end off time and where time takes light to unite infinity and eternity once more, puts me in the Universe in a position where I form just another Black Hole.

In the picture we find something resembling a Newtonian. In the picture we find something resembling life. In the picture we also find the purpose of the Universe. On the outside we find time in eternity, which has no end. Go on, go out and look for eternity's end. Only Newtonians are bright enough to find the edge of the Universe. Where can the Universe grow when it is as large as anything can ever be? How can the Universe get bigger and larger, when the Universe contains everything of that which there ever can be? That the Universe can get bigger and can hold more, we know is not the case but that means geodesic zero is as much rubbish as anything Newtonians regard with simplicity and with careless thought. Geodesic zero means nothing and how can I see nothing as darkness because "nothing" is not darkness, nothing is "nothing" and the darkness I see is darkness showing the darkness as something. Only Newtonians are bright enough to find something, which has no end (and they admit to that) having an end at a place they say, where the Universe meets its limit or some edge indicating a Universe at an end. Only Newtonians can have the audacity to find that which is limitless, to expand more. The Universe has no end, so where to should it expand? On the same level of genius we find Newtonians bright enough to see nothing. We find Newtonians to see the edge of the Universe where nothing stops and where nothing stops, there also nothing stops being black which means nothing stops having no colour it has not got in any case. Nothing is no longer having not having a colour. Does this rambling make sense to you? If you are one of the members of the most intellectual paternity devised by any group of persons, this madness covered by senseless ness is just what you will be advocating! However, if you are a brilliant SUPER- EDUCATED member of the Brainy Bunch, it all makes sense because it is as senseless as all other Newtonian rubbish. What then about colours that are technically not colours as is the case with black and white? White is simple. By spinning all the colours in the spectrum the colour white shines through. Black is quite another matter. A friend of mine whom is one of the best painters I have ever come across told me that one couldn't paint black but have to make black a dark blue to show shade on the canvass. That apparently is his success in achieving the realism. He also went on to explain how many variations of dark blue form the shadows in one simple tree. This remark set my mind in motion. One cannot see black because black has no colour to show, but black is the colour most prevalent in the universe. One can see only by colour and since black is not a colour we should not see black, but we do. If it were individual photons that were conveying information, the information could not be a united all including unit. That is what it is. In that the logic is that the photon must be the smallest particle but it is just electricity that conveys more information under certain conditions. However, that which conducts the information whether the inductor is excited or not, must be a liquid that has integrated the picture we see. The picture is a liquid that finds validity through some electricity and in the electricity we are finding information. However there is a unifying, all including, uniting substance holding that which I already showed to be time in a unified init.

The liquid presents time being in the part that is indicative to present motion. The notion that a Black Hole is a magic and sacred place being beyond what any normal brain can figure as a concept, is rubbish. In that we have to find out what happens to light that is consumed by material. The fact that certain light bounces away leaving a colour by which we classify the object, is part of the story. There must be more of what is not that obvious. We have to find the more there are. It is quite obvious that any object that is a specific colour also is all the colours there are except the one colour that it rejects just because it rejects that colour it is associated with. If we see an object being yellow it is a human response to think the object is yellow.

It quite obviously is not yellow because it rejects the yellow we associate the colour with. Then what happened to the rest of the spectrum of colours that hit the object. One thing is true and that is that what ever is part of the Universe has to remain part of the Universe except time. Time can find a situation where infinity meets eternity and by uniting time, eternity returns to infinity and while it is not lost, it is also not present in the Universe. If that argument is true in respect to one set of circumstances it must be true through out. If the Sun rids itself of heat it supposedly has, the Sun must be cold because there is just no more heat left. If outer space absorbs all the heat on offer and shows no change that can be attributed to the collecting of heat, then outer space is as hot as it can get. If a yellow object is yellow on the outside it cannot also be yellow on the inside because where is all the yellow coming from?

If it was true about a yellow object not being yellow and a red object rejecting red and therefore not being red, the same must be true about dark and light. The bright object rejects all the light just because it is light from the outside. If it is light from the outside it has to be very dark inside. It then holds true that the brightest light must be the light we are unable to witness because all that bright light is contained by the source that keeps the light bright within. The object would then shine darkness out in order to protect the bright light it is saving. The brightest there are must seem to us as being the darkest, but because we can see the darkest, therefore the darkest is so bright that even the darkness cannot contain all the light. Therefore we see the light that is not escaping but it has to escape because what we see is light not finding favour to any particular wavelength and shines without favour all variations of light evenly. But if outer space is light, and I have shown that outer space is time, then light is time. That means the light coming across is merely time telling us of a past still present in our future, which is exactly what I stipulated the function of light is. I am what contain light because what I contain also contains time. Time ends by me because time finding me ends within me. That too would count for what we believe it to be dark stars. Such dark stars must be most brilliantly lit because they keep all the light to their inside and well protected. They are dark because they keep the light on the inside where we can't see it and therefore out of our viewing range. The stars that we see as being dark or as

Newtonians view them to be "dead" are points where time finds conclusion. Stars have two ways dealing with heat. The one is to duplicate that, which is there and distribute the heat over a larger area in order to contain the heat through distribution. The stars have gone so cold the stars have to conserve all heat to remain in gravity. With light being the highest concentrated form of heat, it stands to apparent reason that the light would be the energy of prime choice to contain. The same must then apply to outer space in that outer space is conserving all light and by keeping all light, outer space is brilliantly lit. We just are unable to witness the light because our position is such concentrated where as the light being dark is expanded to the full. We are able to see the galactica because the galactica represents highly concentrated light in one reduced area. The darkness contrasting the light we see as darkness because the light is expanded to the ultimate. The fact that we can see the darkness makes the darkness light, which we are unable to see. However with the space stretched to the maximum the lens we see light by has as far as our position goes, not even slightly curved because we are so small. Now you go and tell any mathematician in charge of theories this much and see how far you can get convincing him about your view. They wish to manufacture and design space whirls and not see reason.

We all accept that the true cosmic form would be a sphere and most probably in the universe as a whole will be the sphere…but why would the sphere form representing the original form when matter is not pre-cast to have a specific form and therefore can take on any shape as a form? By merely blaming gravity pulling from the centre is rather avoiding the question with simplicity because the question arising from this answer is where is the centre of the universe? **These are most ordinary questions… the questions that those with big mathematical skills do not answer…. but in an careful investigation Kepler bring the explanation In answering this question one will arrive at the point where we can begin in finding answers to the cosmos.** Because it is the human choice to select spheres as the form that is ruling the universe and such a form is selected by man to be the form of choice. From logic comes the answer that the sphere is the answer that every one accepts and the cosmic solution must be in that direction.

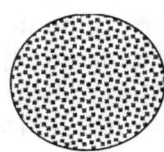 In the centre of a sphere there is a point where all space ends, where the area just got to small to hold space and yet the point is there amongst all the space. Space is the part in the Universe that is filled with material. It is the part of the Universe that cannot share with other space, which is also filled with material.

The space filled with material is moving through space not filled with material and therefore the two ideas having the same, is not the same thing. The one idea, which is space, is filled and incapable of further filling while the other idea science regard as nothing unless it is filled with the space filled giving the space not filled the concept of something. That surely shows the two concepts that are sharing a name is not the same thing.

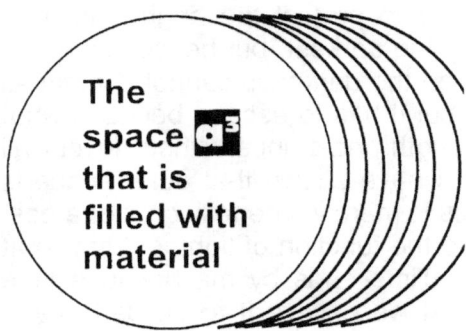

The space a^3 that is filled with material

The space through which the space a^3 that is filled with material moves.

The moving of material is the filling of space with material that is equal too and in the same direction as the relieving of space by the removing of material from space to the same value and in the same direction as the filling of the space is applying. On the one side the space is emptying to the same tune as what the

space is filling on the other end

The space filling and emptying with the space permanently filled with material cannot be equal. Therefore it would be wise to revert to Kepler and find out what the cosmos told Kepler about the cosmos.

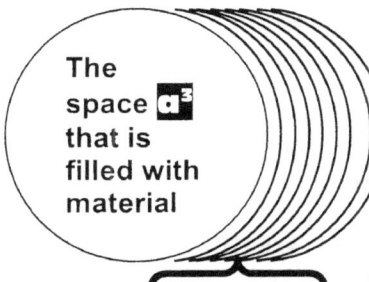

As space transforms to time, the connection it holds to time and the space it claims to relocate space in the new allocation, has to fuse heat with cold. It is at that point that space is in contact with time and at that point time fries space into fragments of heat. The fragments we call photons.

Where space relocates a new position is where heat transforms the liquid into crystals, as a^3 becomes $T^2 K$

By duplicating, the structure has to present a part of it to time and that thin layer making the contact is what turns liquid into light. The light is liquid that blisters into spinning fragments where every fragment covers one point overstretching singularity and that singularity is making up outer space. The fact that we see light means that the dark next to the light cannot be "nothing", If the darkness was the representation of "nothing", then that should be exactly what we must see, nothing but the stars. At first stars confirm heat regulation by duplicating and in that there are more duplicating than there are contraction. When duplicating, the atom is in complete control of the star and the individual atomic singularity is confirming the star by duplicating positions within the star as to control the heat balance of the star.

The duplicating is a manner in which we see something but that being something obviously must leave absolute darkness at that spot. The darkness however, is not nothing but is a black of heat. One cannot have the ability of sight and see nothing. It is light that we see and it is light that we use, which enable us to see. That proves the darkness that we see in outer space is light that we see without recognising it as such. If the darkness was the representation of "nothing", then that should be exactly what we must see, nothing but the stars. Even in this slight developing condition the star performs in the role of a black body. We might see the light flowing away and recognise the star by that. That however, is also completely incorrect because we witness what is not by giving evidence about what we cannot witness. It is light that we see and it is light that we use, which enable us to see. That proves the darkness that we see in outer space is light that we see without recognising it as such.

Material is what singularity use as a solid that can move to become the buffer that moves in a place that cannot move in order to allow time the distinction of storing as well as controlling heat while singularity grows in stature to comply with a measure where singularity once again can accommodate light, time, heat and eternity.

Every atom is a bouncing little star, but every atom is also a murderous Black Hole that is bringing conclusion to time by reuniting infinity with eternity through duplication as well as contraction. It is $a^3 = T^2 K$. In the cosmos there are no big thing and small things and to the cosmos every atom holds the same validity as every star does.

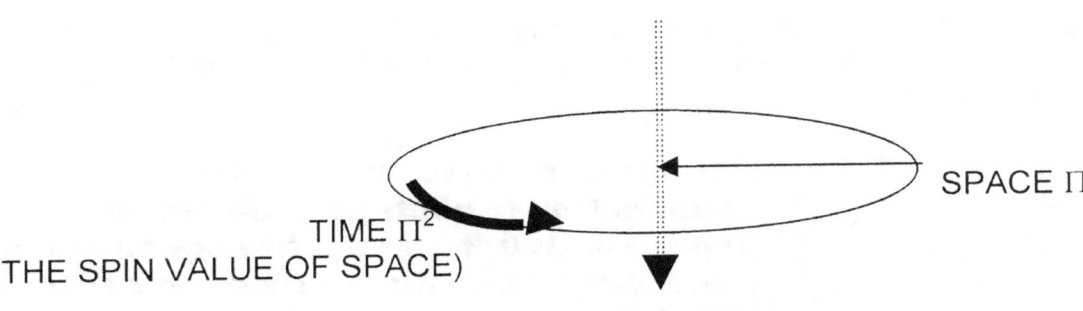

TIME Π^2
(THE SPIN VALUE OF SPACE)

SPACE Π

SINGULARITY

Gravity is the very same but it is the recalling of the space by creating motion in the space. Gravity is the retracting of heat by splitting matter as matter duplicates and reduces space by increasing space in expanding. As the space gets more and the time holding the space gets less per unit in time used, the heat distribution is wider in less time and by such distributing the heat in relevance gets less because more gets distributed by a wider area in a shorter time frame. Gravity is the retracting of heat by cooling because of the expanding of heat increasing.

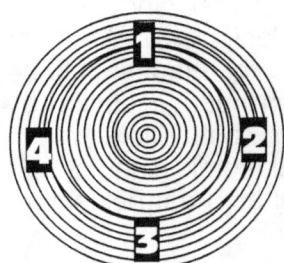

By moving from 1^0 to 1^1 and from $1^0\Pi^0$ to $1^1\Pi$ requires space. Yet such moving did not leave the realm or the domain of singularity. The motion was still within singularity because moving involved forming a relevancy between heat and cold between infinity and eternity, between space and time and most of all producing what will in the far future develop into a Universe that can even be a host for life albeit on a very small spot for a very short while in relation to the vastness space has and the duration cosmic time has.

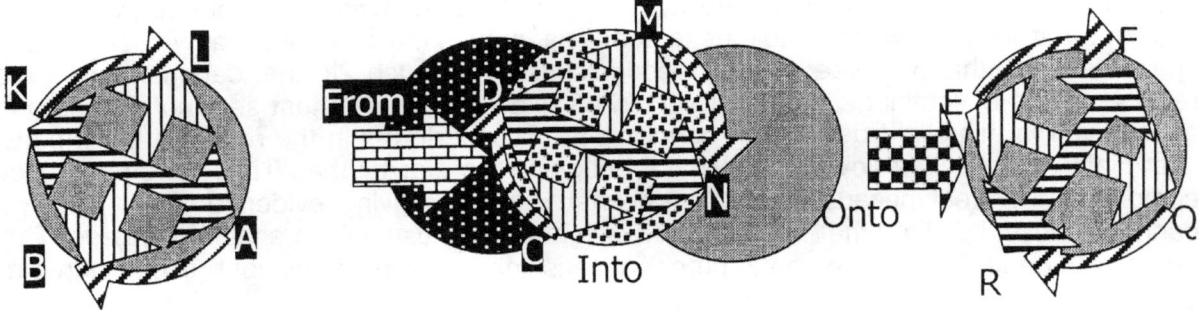

From   Into   Onto

Relevancies came about when the dot moved away from the spot but had no space to move. All that was possible was to charge singularity by relevance to comply in being activated into complying. Space-time is motion and movement are all the same things only separated by dimensions and dimensions are formed space, where the dimensions become space being in motion and the space is motion by contraction or by expansion but because time is almost eternal at k^0 our perception of the universe we are in is a stable and steady eternal structure. Gravity is motion and motion creates space to the third by the third in the third that interacts with one but establishes ten.

The cosmos holds no constant and that is the only constant. Every aspect of the cosmos is relevancies where matter in different forms, form different relations to other matter also in various forms. As I indicated about time, where time is an ongoing repositioning of relevant matter locating relevancies in the position they hold to the time they apply. The cosmos cannot grow, as much as the cosmos cannot shrink. It is a never-ending flow of changing relevancies, where singularity meets singularity as much as space meets time. The point in singularity I named time, is the point where time started and where time ends as much as where time will finally fulfil its reducing of space. Heat made space renegade and time slowly contains space by reclaiming heat. That is the cosmos.

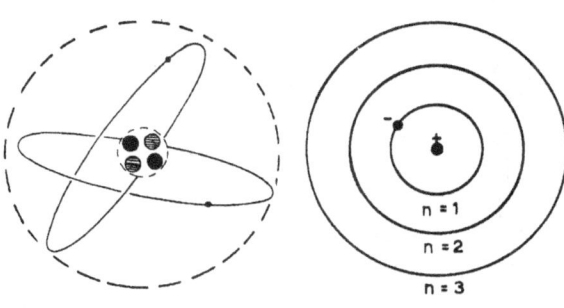

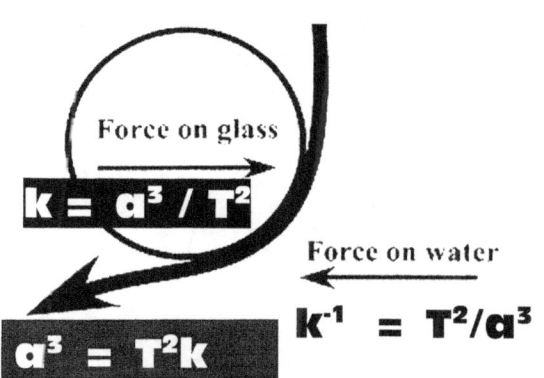

$$k = a^3 / T^2$$

$$k^{-1} = T^2/a^3$$

$$a^3 = T^2k$$

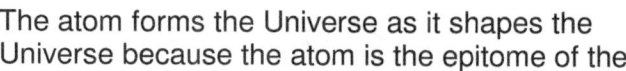

The **space a^3** of the unit is **defined k** by the **flow T^2** of the liquid

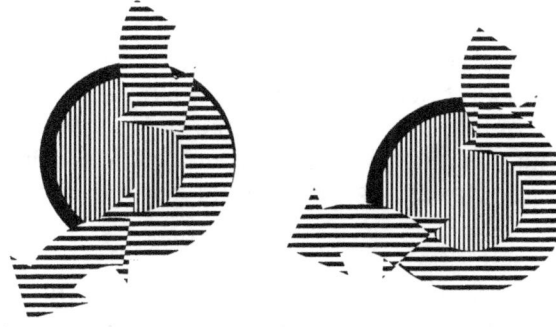

The atom forms the Universe as it shapes the Universe because the atom is the epitome of the Coanda principle. The atom spins and the spin provides an accumulative that drives stars, which in

their turn give an accumulative spin that, drives galactica and in all that there is a centre governing singularity controlling every spin just as the simple top showed us. The atom is the template on which the entire Universe is formed. The atom is a galactica, it is a star, it is a bicycle, it is a supersonic aircraft flying in the atmosphere, it is a moon orbiting another containing cosmic structure, in fact only nothing does not use the atomic principle as a format in the Universe and nothing is the only fact not represented in the entire Universe. There is a solid centre, which is supported by a liquid outside, and the balance between the two determines the relevance applying. The atom therefore is the Coanda effect which is the Universe in its entirety.

That what I am about to explain may sound inconceivably simple but don't blame me for that. It is not my fault no one brought what I am about to say into the context of gravity and if I don't explain the most mundane in connection with gravity there is no one going to do it.

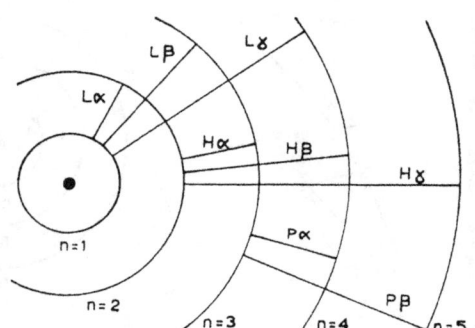

The control regulating godly simple genius. The has on heat everything accumulating time spinning side which the atom has in heat is plainly as it is also godly control the atom by expanding outside while heat inside as a retardant on the of the atom, is that creates the Universe and allow the Universe to be time retarded confirming matter as it compromises time by splitting infinity from eternity.

By producing space in expanding when heating the atom not only control what belongs to the atom but also that what is in the control of the atom and what is not in the control of the atom since the entirety of the Universe is the world of the atom. The atom is the Universe because the atom is what forms the bumper that allows contraction when cold is needed to preserve time and expanding is required to sustain time.

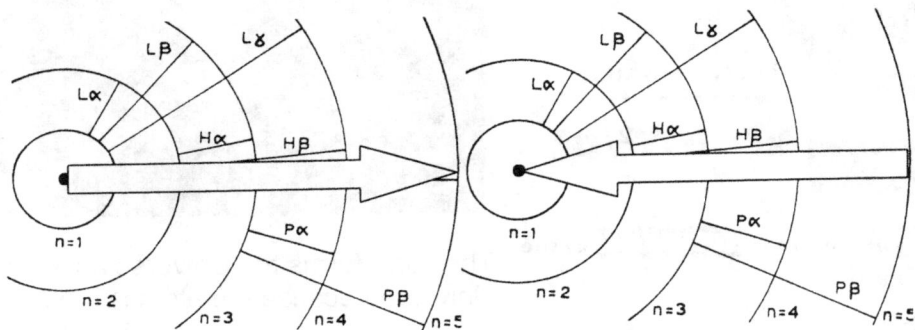

What we see as heat is relevancies because as the relevancy within the Sun changes, the atom adapts to the changes. The atmosphere of the Sun becomes denser, which we see as being hotter and the containing becomes stronger. The atom has to reinvent it by adapting to the changes or different surroundings. In this manner the motion that the star provides which is so much more than what is the motion is we find in outer space that the hot / cold dynamic changes all together.

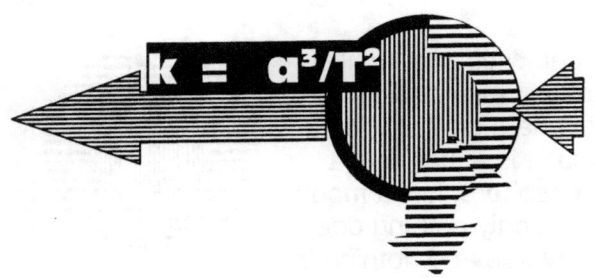

Depending on the flow of time, the atom will accumulate heat within the structure or on the outer side of the structure all depending on which at that moment holds the strongest relevancy between $k = a^3 / T^2$ and $k^{-1} = T^2/a^3$. That is gravity committed by the flow of space-time and controlled by singularity $k^0 = a^3 / T^2 k$. **This was also what the Big Bang was all about.**

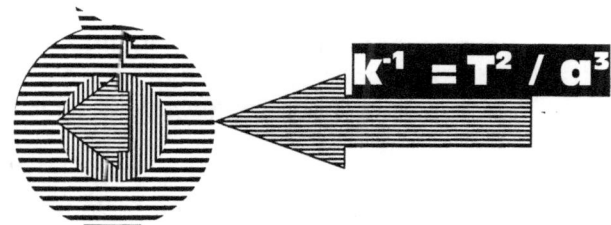

$$k^{-1} = T^2 / a^3$$

This became possible as the Coanda effect brought the speed of light as gravity (**k**), which brought the motion of the Universe in contraction to the speed of light T^2 that made the Universe be the speed of light a^3, which is exactly and specifically what the **GUT** theory proves.

The spin of the liquid proves the value of the relevancy. The stronger the motion is that the liquid generates, the higher would the contraction be and the lower the spin motion is that the liquid generates the higher would the expanding be. In that we find a definite favouring of either the factor of seven or the factor of ten depending on what the situation will dictate.

Depending on the balance there are the contraction will be totally dominating but never to a point where it annihilates expanding and on other occasions the circumstances would be that the expanding may dominate but also never to a point where it devastates contraction.

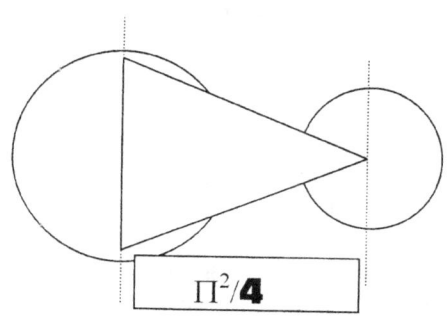

$$\Pi^2 / 4$$

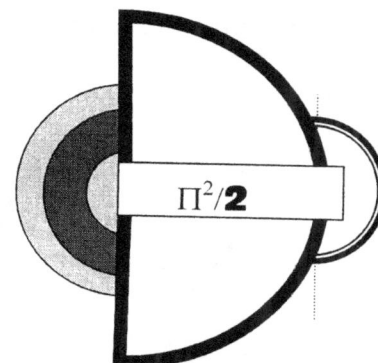

$$\Pi^2 / 2$$

The motion of the liquid factor puts the aspect of time in eternity in relation with infinity where the ratio that develops gives infinity the chance to interrupt eternity. Infinity in the centre is immovable but has to associate with eternity since the tow parted by space. Therefore we have eternity having three positions that goes square since it develops an alternating stance

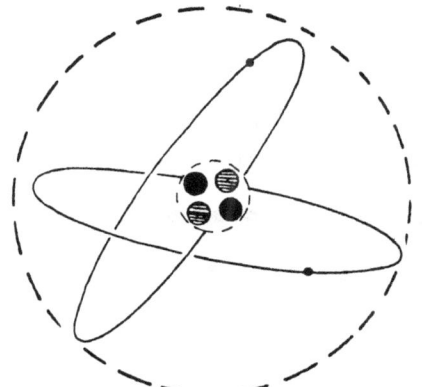

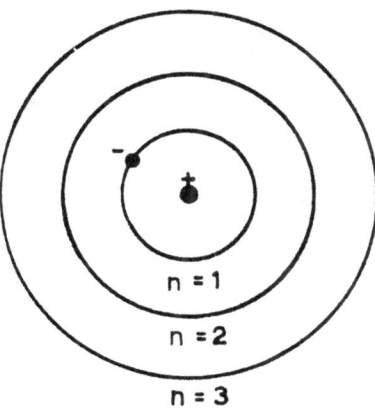

n = 1

n = 2

n = 3

with singularity and in that the three of eternity develops a square since the angle of developing cross 180^0 by the margin of 90^0. From the mathematical legacy we can see that time was forming 5 but with material not formed yet, that which later became material moved through time holding the value of 5, at a measure that was slightly less than half the value of time and still had the notion to move being $\Pi^2 / 4$, which was at slightly less half the value of 5.

But as space grew into a stronger part of the Universe, time came into a square having five on both sides of the divide and time gave space an integrate part of the motion that time

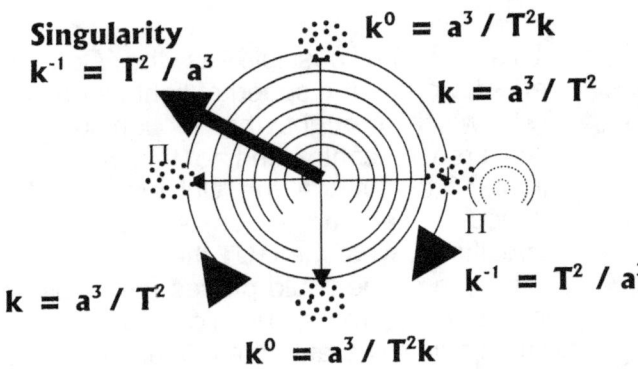

Singularity
$$k^{-1} = T^2 / a^3$$

$$k^0 = a^3 / T^2k$$

$$k = a^3 / T^2$$

$$\Pi$$

$$\Pi$$

$$k^{-1} = T^2 / a^3$$

$$k = a^3 / T^2$$

$$k^0 = a^3 / T^2k$$

provides. Then the motion became $\Pi^2/2$ and material found a better and a much more valued relation in time. This only happened while time grew less, which gave space more value in $a^3 = T^2k$.

That produced the atom we now enjoy with three factors instead of only two, as was the case in the previous dispensation. There then was time, space to hold time and space lagging behind time. That part that was lagging was excluded and that part formed a secluded unit that became the atom.

Every time the point rotating changes direction it crosses the divide and in that the entire Universe changes. What comes down then goes up and what went forward then goes back. The line is a cutting limit that divides one sector from the other like no other found anywhere. In that the divide produces changes that are beyond comparing to anything else and in that we find such conclusive distinction in gravity. It is all coming down to perceptions in relevancy because that which cannot spin does spin by never actually spinning.

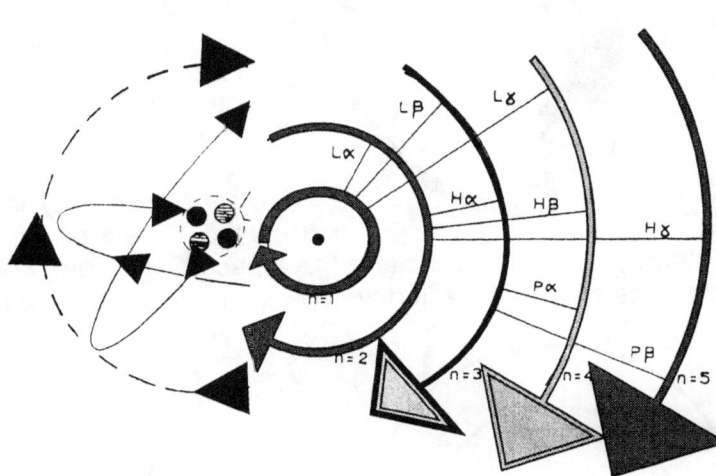

From investigating this we can presume with very little doubt as to how the universe came about. The atom is the Universe that is true but that sounds like rhetoric. Let's break that down a bit by dissecting what this means. The atom is more than just the main ingredient because the atom is also a calendar of the Universe. The atom is a storage facility of the Universe and the atom is a building stone of the Universe. The atom is the final formation the one that will come last as much as it is the one that came first. The atom is the custodian of singularity and the guardian of maintaining the Universe.

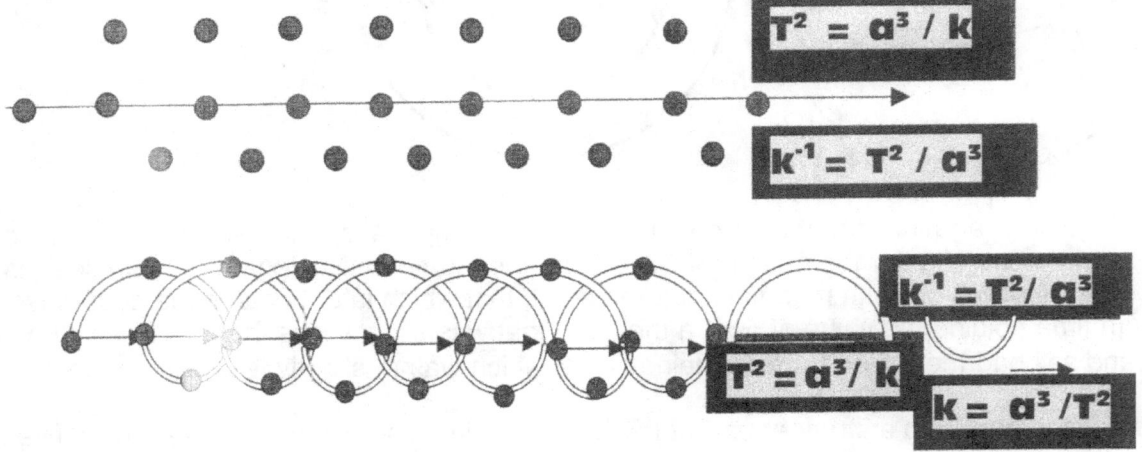

$$T^2 = a^3 / k$$

$$k^{-1} = T^2 / a^3$$

$$k^{-1} = T^2 / a^3$$

$$T^2 = a^3 / k$$

$$k = a^3 / T^2$$

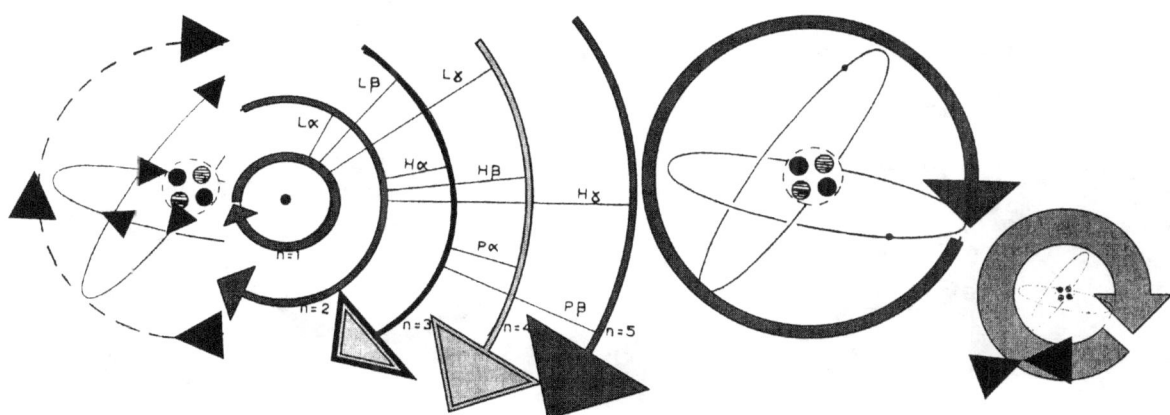

Going back to the start is going back to where space was a thought that divided eternity from infinity.

$$k^{-1} = T^2 / a^3 \qquad\qquad T^2 = a^3 / k$$
$$T^{-2} = k / a^3 \qquad\qquad k = a^3 / T^2$$

Time cannot move back but in relevancy time can respond to the nature of motion contradicting its repeat by the nature of the repeat. Time progressing $T^2 = a^3/k$ has a repeat of $k^{-1} = k/ a^3$ which leaves a nett growth of $k = a^3/ T^2$.

When referring to how things develop I have this human tendency to put such matter in relation to the past. In this matter I must beg your forgiveness but it is my human nature I sometimes find hard to control. That, which I refer to as if in the past is as current as you or I

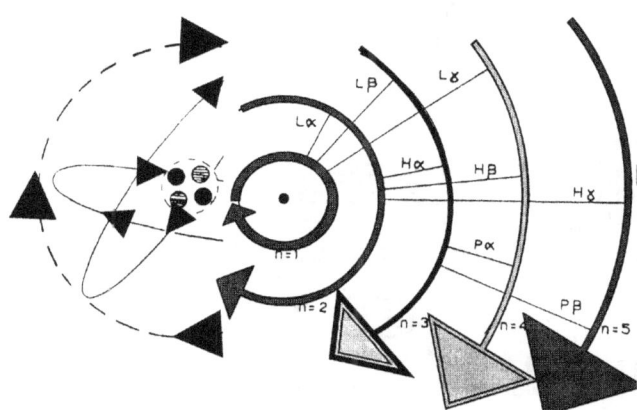

breathing, but since it does no match what our everyday experience would have us believe that it is occurring around us I differentiate in a human perspective commonly used. In this I use the past but please note that I state without excluding anything that in the cosmos there is no future or past in the sense we humans wish to observe time.

In science everyone has the opinion that material exists while the rest is nothing. It is the atom and around the atom is nothing. It is nothing keeping atoms or material apart. Such a view is extremely shortsighted and is evidence of a lack of thought. It is time we put realistic thinking power to task when dealing with matters in cosmic proportions. I do not put myself on a pedestal. To the contrary any one can see how shallow and little educated my work really is and yet, if someone with such a little educated background can see what I see, how much more is there to see when persons with true mind power start to truly think on cosmic issues. Three words that should be taken out of any cosmic context is nothing, maybe and suppose. From those three words I could find no evidence in my entire search.

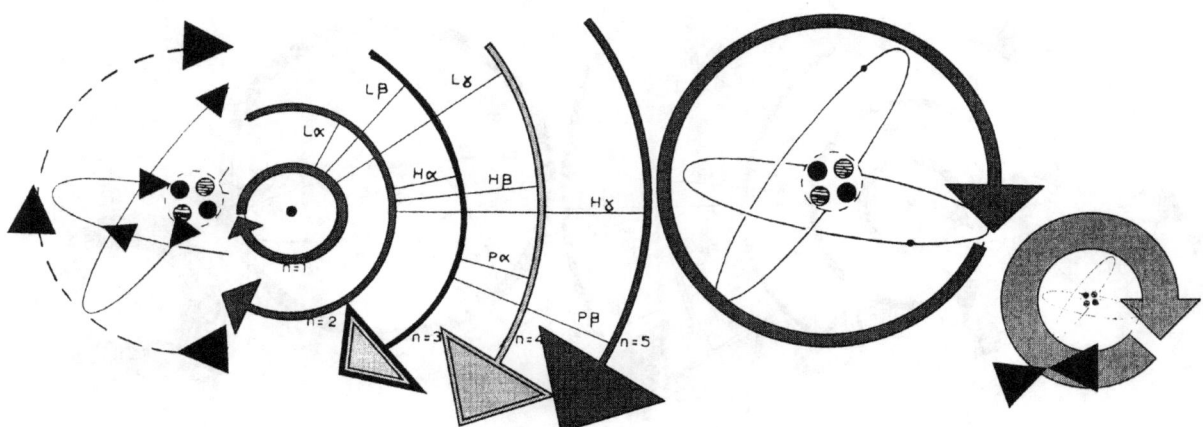

We find the atom spin and think of the atom as where we now are. Sure that is as correct as it can be but the atom spins because new alliances come about from previous conceptions. Being previous does not mean it has moved to the past, as did napoleon and his thirst for war, no it is just functioning holding prominence below the veneer of the current.

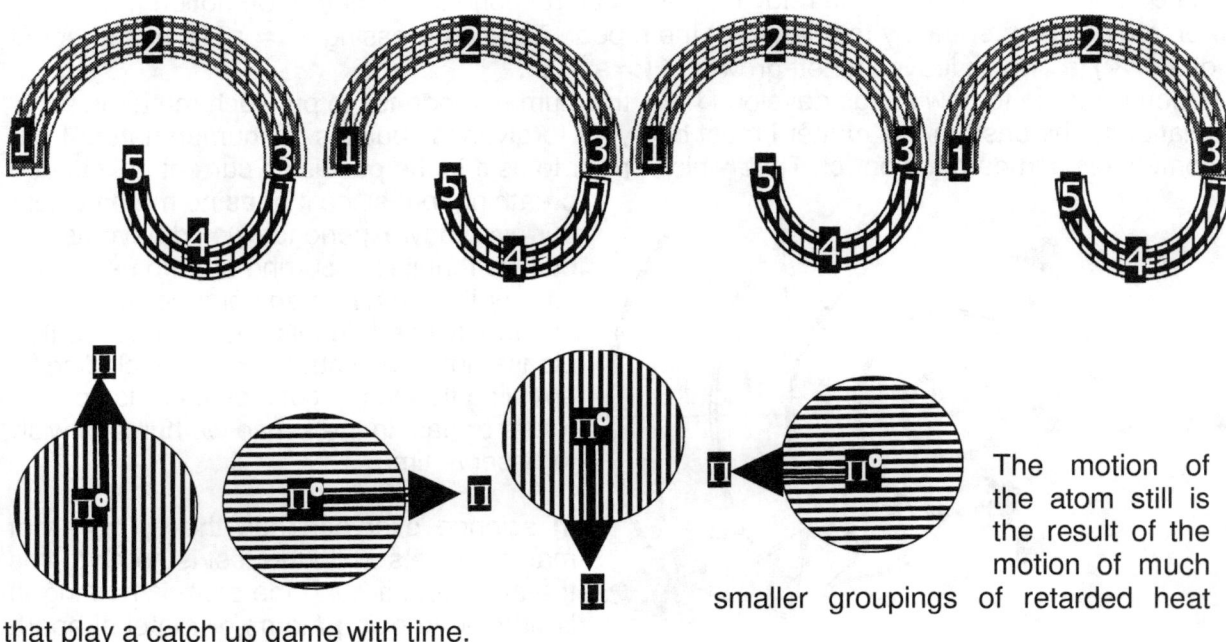

The motion of the atom still is the result of the motion of much smaller groupings of retarded heat that play a catch up game with time.

The time retarding grew as the spin involved a higher degree of relevant points. At first there were two points namely this side and the other side of the Universe. Then the four came into prominence. Later eight points came into affect

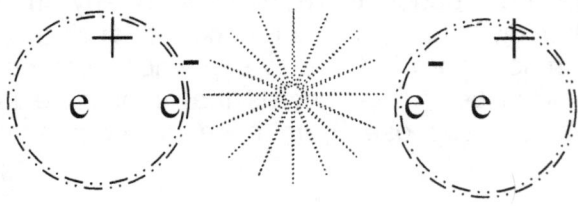

and so it carried on. This was time developing as time always increased the back log it formed with the current situation prevailing at that moment.

The relevance we find in the atom is a flow of heat. There is no such a thing as fixed particles. There is heat and there are points positioning the characteristics of the heat at such a location. The heat flow past and at that point the heat adheres to a positional interpretation led on by points in reference to other points in singularity.

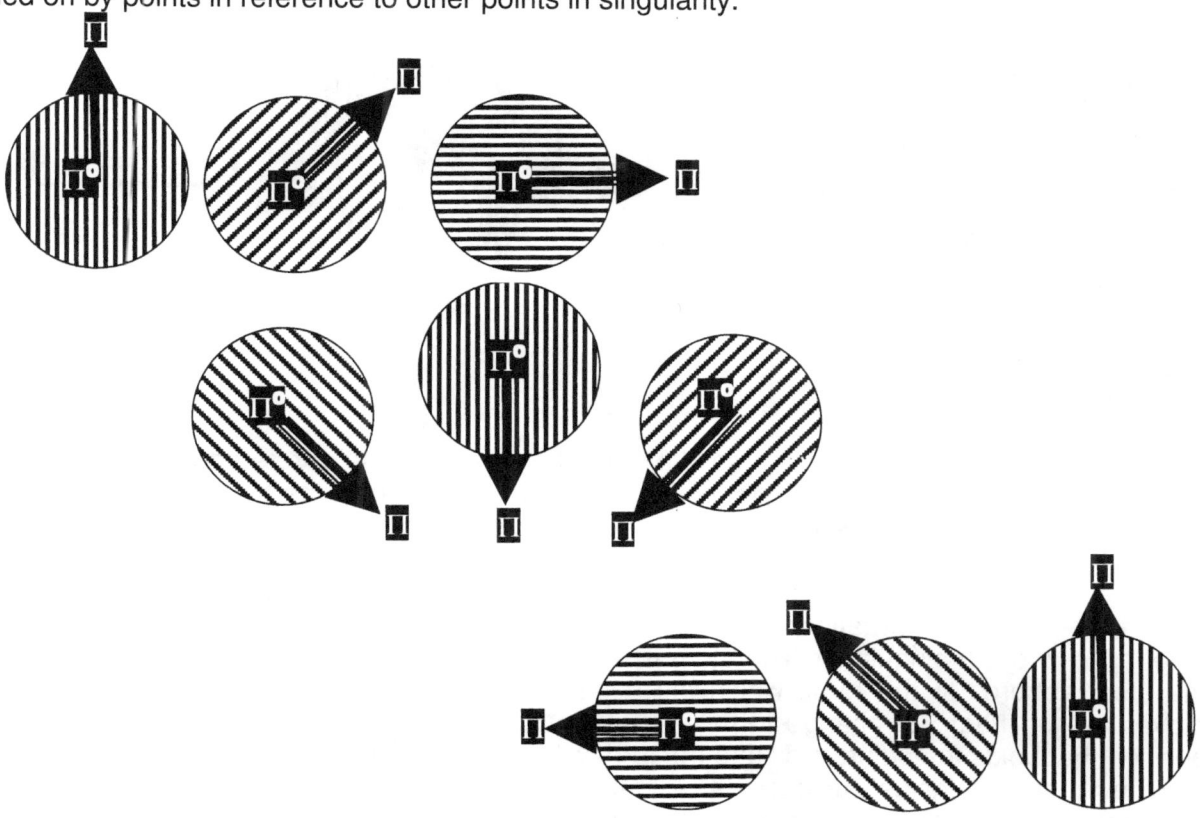

Every time the increase brought new relevancies that put new orders in placed to match the previous requirements. But also it brought about that duplication by progress of space increase was in advantage of the situation.

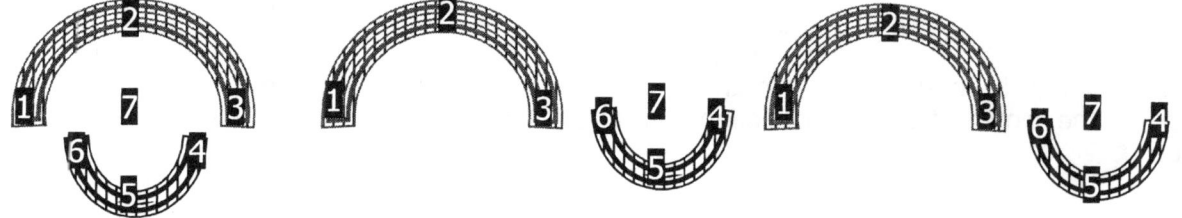

The flow of time is heat flowing past points that was there since eternity parted from infinity. It was there when the cosmos came about. What distances the pointers from each other is heat or light or whatever one would call the formless filling of space-time.

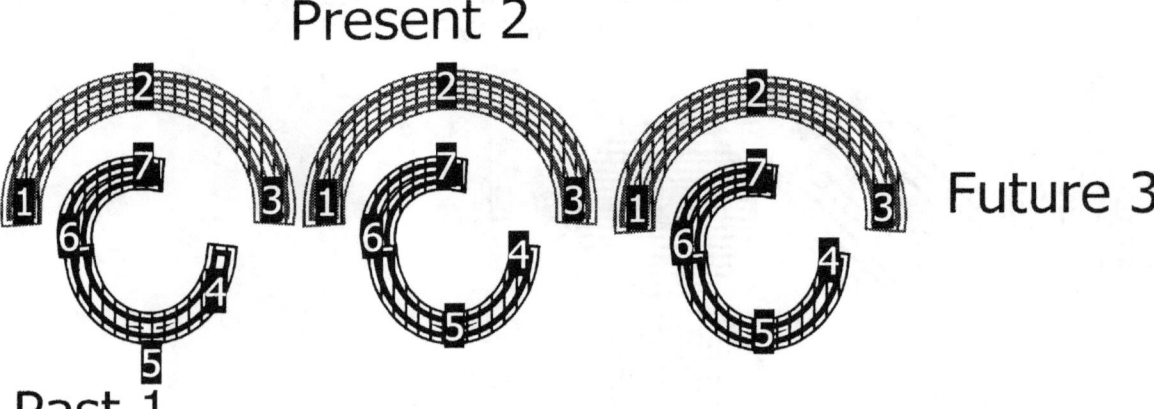

Present 2

Past 1

Future 3

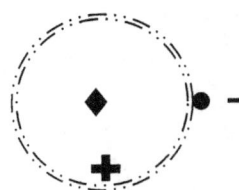

The Atom is a solid atom but it is space-time in reference to singularity pointers that mark those spots, which forms the relevancy within the markers where the markers perform the claiming as characteristics of the atom. Through the markers flow heat, which is time retardation and the pointers, are time in present. The time in present refers to the time retardation as allocated reference markers and in that allows the heat or time lapse to catch singularity. The atom is taking in heat and the heat in that instant inside the atom adheres to the form, which the pointers or markers prescribe. As retarded time moves on the heat will change form in relation to the allocated position and to what the position that holds the pointers dictate the form will be. It that sense the entirety of everything there ever can be inside our cosmos is liquid time and everything dictating what form the liquid will take on is singularity that is holding pointers which acts as the solids.

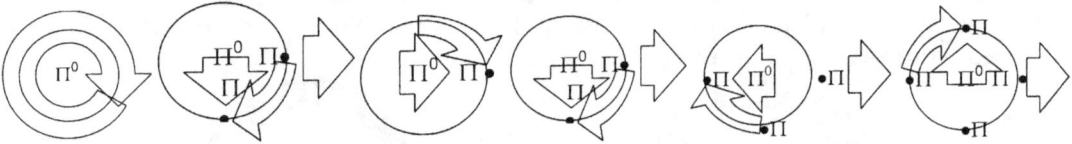

In the motion there are the four markers indicating the limit of the atom while what fills the atom is the motion that flows directional as time catching time by destroying the backlog of time.

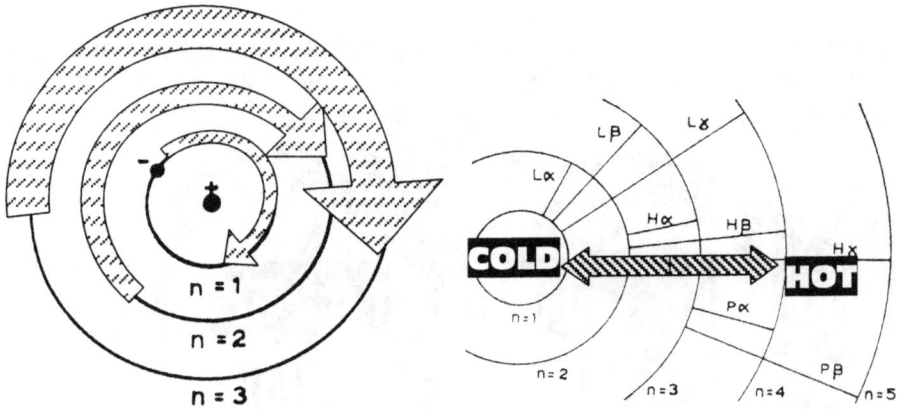

The liquid of time is part of what we see and the solid forming time is part of what we cannot see. The solid shape the liquid into forms and specifics while the liquid provides substance. The liquid is part of the Universe and it is eternity, which holds material as the solid in eternity performing as the liquid and the solid is the part where material on the other hand forms the substance that holds infinity in liquid. The solid is not a part of the cosmos but is there in directing that which is apart of the cosmos. It is singularity that control heat and the flow thereof while it is heat that provide singularity that which is generated as time retarding.

It starts as follows:
ALL OF CREATION STARTED WITH Π = TIME AND AS TIME WAS ETERNAL,

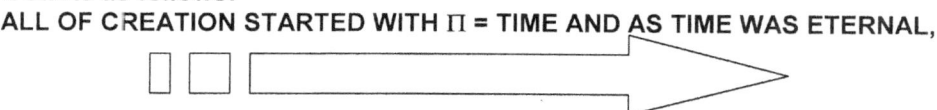

The spot grew into a dot by the value of 1^0 going onto 1^1. However this also was Π^0 going onto Π. However with no space available yet and the spot that is forming the line compulsory being so small (as we still can see when observing the centre of the spinning top because the line inside the line that is inside the line we cannot see is the line we are referring to) something has to give and the part that has to give is the liquid in eternity. The line is there as it was, but our inability to go that far into singularity places us at the disadvantage not to recognise the line as it is. Because it is spinning it has to be Π and because it is spinning it also has to be a line and all of that is there to witness for all those disbelievers.

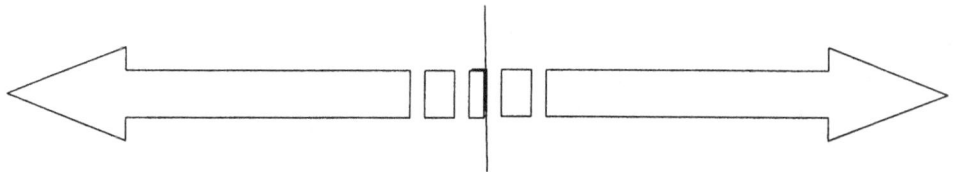

The line formed a spiralling line that was so small it was continuous and never broken due to the lack of space and those not believing me I advise you to inspect the spinning top. The line formed next to the running line and that too is in plane view for all, but we cannot see. The line has to form a line on this side of the divide because it forms a divide as much as there has to be a line on that side of the divide for the very same reason

This places the single line that is half of the full line at a value of 180°.

That is also the value of the circle and the square, both dimensional components of the cosmos. By reducing the one side of the square to zero (which it cannot be) the square will disappear. Therefore one has to reduce the one line of the square to infinity to produce a square holding a straight line with the line running in both directions from a point of infinity.

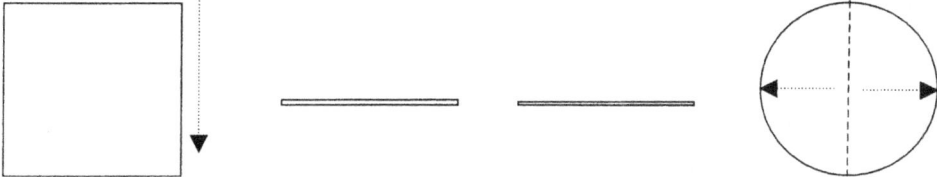

That same principle applies to the circle because the radius of the circle is one side to the round cube. That makes the square the circle and the straight line the very same thing of something totally different. The common denominator is the singularity of eternity finding infinity.

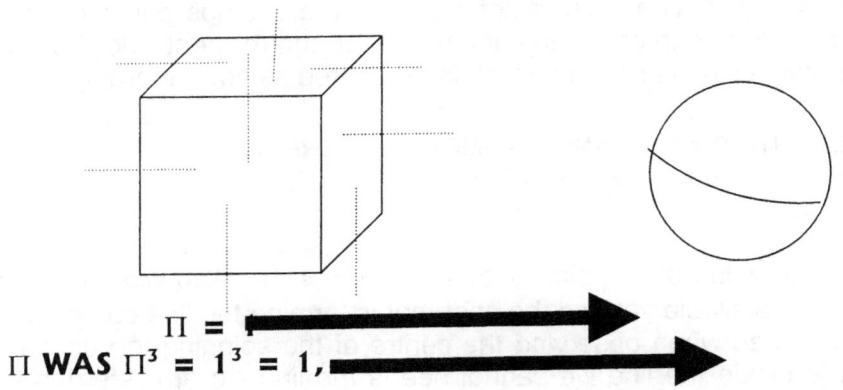

$\Pi = $

Π WAS $\Pi^3 = 1^3 = 1$,

Π WAS $\Pi\Pi\Pi = 1 \times 1 \times 1 = 1$ THEREFORE TIME WAS ETERNAL:
E $= \Pi = 1$

Let us have another look at the straight line

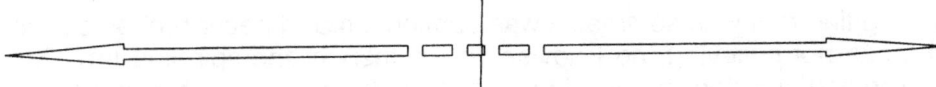

No line can start at zero because having a starting point of zero there is no line (0 X by what ever reduces whatever to zero). The starting point has to be infinity the shortest any line can be leading to eternity the longest any line can ever be. By having infinity there then has to be a virtual ZERO (not zero) and from that point the rest of the line must start running the other way.

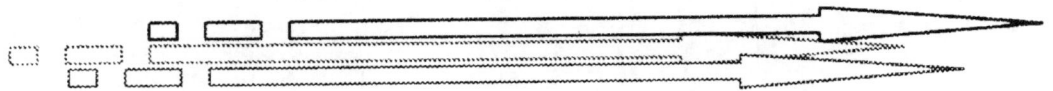

This establishes that no line can have a single direction and must have a continuance at both ends. Such a line has to have a value of 360^0 not 180^0 as believed.

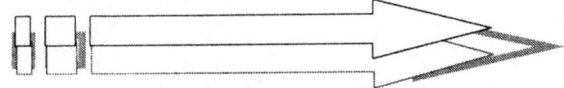

From infinity came the straight-line borders bordering the straight line on both sides of infinity

MATTER EVOLVED FROM TIME AS TIME = 7 $(\Pi^2 + \Pi^2)$
THEN TIME, BECAUSE IT WAS ONE, MOVED A DIMENSION UP AND DOWN, WITH BOTH DIMENSIONS BEING EQUAL, THE SAME IN EVERY WAY.
Let the lines be somewhat bigger than infinity where we can see the lines more clearly

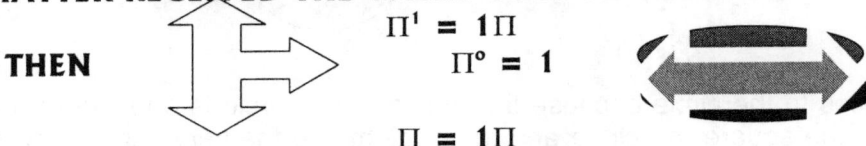

MATTER RECEIVED THE VALUE OF Π^2 AS FOLLOWS:

THEN $\Pi^1 = 1\Pi$
              $\Pi^0 = 1$

             $\Pi = 1\Pi$

THEN TIME HAD A DIMENSIONAL VALUE OF $\Pi = 1$; $\Pi = 1$ AND $\Pi = 1$, BUT AT THE SAME Π DUE TO THE PROXIMITY OF SINGULARITY

POINT OF SIGULARITY

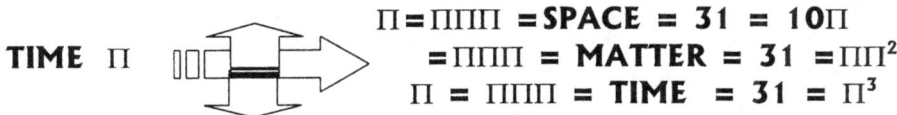

$\Pi=1$ $\Pi^0=1$ $\Pi=1$

The Universe formed singularity in the straight line with two points of Π to both sides. There was no radius because the radius was infinitely small. Then time formed as part of singularity in the value of Π going on to $\Pi\Pi\Pi$ going on to Π^3. The one Π remained in singularity Π while the other $\Pi\Pi$ became Π^2.

TIME Π

$$\Pi = \Pi\Pi\Pi = \textbf{SPACE} = 31 = 10\Pi$$
$$= \Pi\Pi\Pi = \textbf{MATTER} = 31 = \Pi\Pi^2$$
$$\Pi = \Pi\Pi\Pi = \textbf{TIME} = 31 = \Pi^3$$

While on the other side of the universal atom Π^3 that formed space in time in the value of $\Pi\Pi^2$. Between the two universal atoms the line to singularity was the combination of Π^3 holding Π to Π^2 and forming space in matter to the value of $\Pi\Pi^2$.

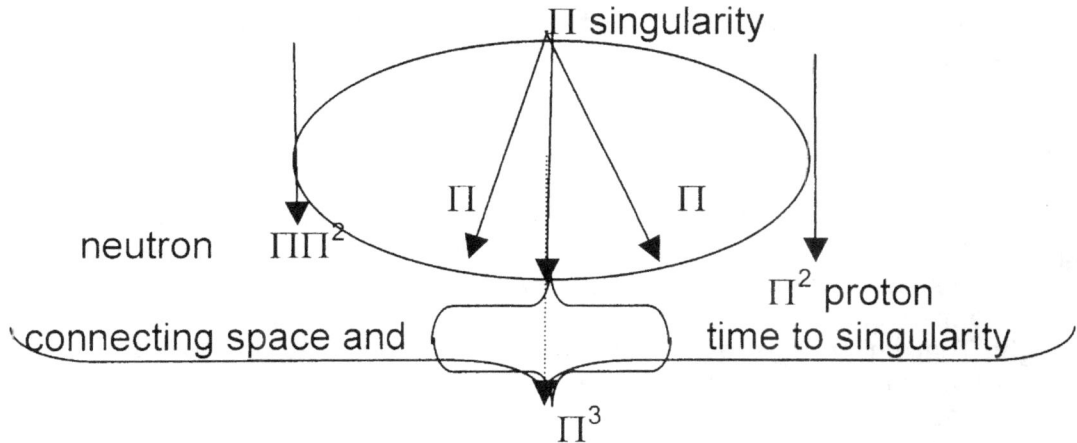

In all it became 10 Π in all being in singularity

Whatever there is today started at a point we cannot trace. It started at a point we are able to envisage and locate but that is only with a mind that can accept what is not there to view. It is acceptable through intelligence because through intelligence we can detect it albeit outside the Universe we have. It is just like religion and the worst is that it is generating what there is because what there is are not if not generated. There was Π^0, which was α^0 or if you would rather have it Ω^0 or it maybe was 1^0, but more correctly it was all the above and the beyond because multiplying what ever constitute the mentioned will bring about what is mentioned to a precise equality. It was a spot that was not. The spot is still there because the spot is still not there. It was a line that ran eternal but because it ran eternal and kept repeating exactly what was before to the precise what came afterwards, the line was there and was eternally running, while never changing in the least or growing by any measure.

It was not one because before it could reach one, it returned to what was repeated and the process cycled back to before one was reached and even before one could be accomplished. It was such a continuing of the monotony, no change occurred and therefore never did the running produce progress because the progress was in the perfect repeat of what was before. The duplication brought contraction to the minutes detail. That is where our atheists get one hiccup. The repeat brought eternity and the repeat was so perfect that the repeat continued. The repeat still is with us as much as we are within the repeat. To bring change to the eternal repeating of the monotonous there had had to be something beyond the Universe that institute change. There was something that brought a difference and we are within that difference. That difference was time and that time is what we move through as much as what we see at night. Oh, how stupid and how thoughtless the minds of atheist and other atheistic animals are. Baboons do not recognize the light we are within because they cannot think and are therefore atheists. Spiders cannot think and therefore they are atheists, as they do not think where the line is that is not. Reptiles cannot think and without thought they are incapable to see what time is, how time that is not, generates space that is in time that is not. Mammals cannot envisage what space is, what light is and what makes us see the darkness cannot be. All the animals I have mentioned are mindless atheists because they fail to see beyond the visible into the realms of the thinkable. All that I have mentioned passes them by including religion and the accepting of God because through being incapable of thought and reason they cannot envisage what only intellect can bring to mind. Because of the incapacity to think the animals are both mindless and they are atheists. Therefore atheists are mindless. The night sky is such a bright light our evolution protected our vision from the brightness in order to give as much better vision. Through evolution developed and our eyes are protected by how we remove the qualities from the light. However animals do use that light and not our light to see by. You can shine a bright hunting spotlight onto an animal at night and the animal will not be able to see the light on it. The animal does not use the light to see better as the animal is totally unaware of the light. Then a prowling cat comes out of the darkness and sees the antelope in the light the night provides. It does not use the light, which the spotlight casts and the light is not even traceable to either animal being the hunter or the hunted.

From there we accept that during the day the animals must be using our light to see because the nightlight is inferior to see by. Who says they use the daylight much different from the nightlight because all evidence is there that they cannot recognize our light as light. It is very evident in the manner they go on hunting and grazing while being totally unaffected by our form of light. That which you see at night because you cannot see darkness and you cannot see black is the light the Universe is painted in just like the Bible says. That is the light that started all because that is the light holding us away from the eternal darkness. This is not religion and it is not a sermon, it is hard-core and brutal basic science and it the most fundamental basic physics there is. It is the start of the mathematical Universe portraying the only physical way it could ever be. The light that came from the Command is the light allowing material to move.

My atheistic idiots, your mindlessness caught up with you!

Then came this light that the Bible refers to as the first of what ever was and what our stupidity tells us is darkness. This was moment-Alfa. The darkness was there and from the darkness heat came about. Only heat expands and it interrupted the true invisible darkness, the blackness of a Black Hole, the invisibility coming from within the Black Hole. Eternity tore from infinity. Darkness broke from light. Heat broke from cold. Relevancies parted by 1^0 going 1^1. There was one but also there was two too, because one cannot be without two being there to ensure one is one. The marks are still with us but to see the marks requires a great deal of intellect.

$\Pi^0 \Rightarrow \Pi$. In this there was only space for one being one in the two forming one. It was $\Pi^0 \Rightarrow \Pi$ however there was no space to be $\Pi^0 \Rightarrow \Pi$ and therefore because of the lack of space to be which is the infinity of time braking the eternity of time the true measure was $\Pi^0 \Rightarrow \Pi$ but

realized only 1^0 going 1^1. Π was to the future because of the motion of time involved and the space less ness of space at the time. By inclining to move, the process crossed the Universe but also it took one eternity to accomplish the feat.

$$\bullet\bullet\blacktriangleright 2$$

The fact that 1^0 going 1^1 brought movement can only become a reality as a result of light. Light is heat and the heat is expanding.

1^0 going 1^1 where $1^0 \Rightarrow 1^1$

1^0 going 1^1 1^0 $\blacktriangleright$ 1^1 had to bring about 1^0 going 1^1 1^0 $\blacktriangleright$ 1^1, because the eternal repeat of duplicating while contracting was not relieved from the Universe. Before the contracting was equal to the duplicating because by measure the heat was identical to the cold. It was eternity that was interrupted by one cycle of infinity and was in repeat of eternity. Once something is part of the Universe there is nowhere else to take it so it has to remain as a part of the Universe.

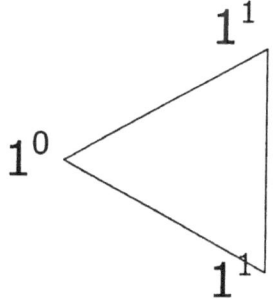

Then came three because motion was so limited that the least inclination to move threw what wished to move to the other side of the Universe, As it moves it also moved across singularity. It crossed the entire Universe as it moved because it moved and finding nowhere to move to. It crossed the entire Universe and it took one eternity less the measure of one period lasting infinity to achieve that. That brought to relevance three points where each was in measuring quantity exactly equal but also one Universe apart. In the reality there was now two points holding singularity on both sides of the Universe because by crossing the divide that crossing set in place the two sides relevant of singularity governing. However, infinity was bridge at two points holding infinity with which process eternity repeated the past into the future.

$$\bullet\bullet\blacktriangleright 2 \bullet\bullet\bullet\blacktriangleright 3 \bullet\bullet\bullet\bullet\blacktriangleright 4$$

This then is the occasion where Pythagoras stepped in. Since it as a crossing of the divide, the crossing involved a line that formed a half circle connecting a triangle. But the crossing was done in the space of half the Universe and since the Universe was 180^0 half of the Universe was 90^0. That involved Pythagoras as mathematics was born. Up to this point it was arithmetic with adding but now mathematics came into place. Remember we are a few eternities in side the development of the Universe.

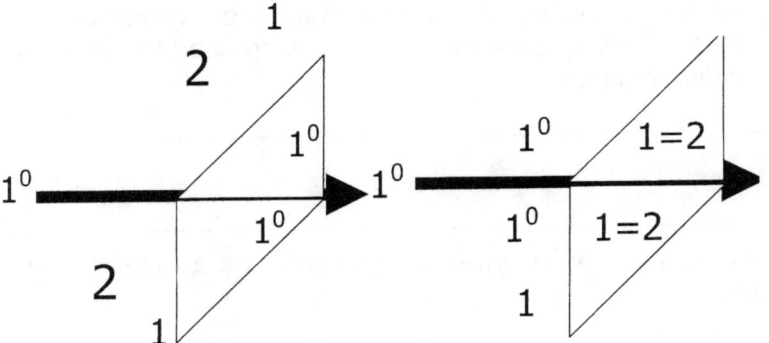

In the three came four that brought along five.

$$2 \blacktriangleright 3 \blacktriangleright 4 \blacktriangleright 5$$

Then five filled the one half of the Universe that was able to contract and cool while the Universe divided the other half into sectors of what was (5) and what will be (5), which put material (7) in relation to the half of the Universe $\Pi^0 \Rightarrow \Pi$ in which the material was at that specific point (five relating to seven) in time.

$$1^0 \quad \blacktriangleright 5$$
$$1^0 \quad = 1 + 5 = 6$$

The motion consisted of Π moving to Π and thereby duplicating Π to relieve Π^0 of the burden of overheating. On the one side of The Universe there was Π^2 being relevant to Π which was forming on the other side of the Universe. The entire Universe had the combined value of Π^2 on the one side in addition of Π forming on the other side. I wish to remind the reader that any and all points formed by singularity was as much representing the Universe as it was the Universe at all times because $\Pi^0 = 1^0 = 1$. That made the entire Universe being any point affirming singularity by forming about and around singularity.

Reaching five is a benchmark because at that point half the Universe was finalised. From the five that formed, the Universe continued and formed space-time. All progress balanced on the five that formed where two parts of equality parted eternally as liquid separated from solid, motion moved apart from the motionless, heat diverted from cold. The four in time grew away from the one point space formed being just outside the control of time. It was the start of material since material came into position at the point where the fifth point parted from the first four. Material materialised at the spot where time delay formed the point allocated as point five or then the first new point acting as point one but still being point five formed where it located a position just outside the four of time. There was a lagging behind of heat building and not being in the range of the immediate control of time. Time could generate a point in heat without bringing immediate demise to the point through cooling.

But that meant that the Universe was a total of $\Pi^2 + \Pi$ which when added was also $\Pi^2 + \Pi = 13.0$

$$2 \blacktriangleright 3 \blacktriangleright 4 \blacktriangleright 5 \blacktriangleright 6$$

$13.0 - \Pi^0 = 12$ because singularity cannot be part of space-time developing as the space, which later was filled with the material that formed, filled this part.

$12 / 2 = 6$ Material formed at the point where six was located.

$$6 \;+\; 6 = 13\text{-} (\Pi^0) = 12$$

Because singularity is a divide and is not part of space-time singularity as a factor is removed from space-time. Why it adds with five to form six is because to the one side only singularity is in the other side of the divide. Only nothing can be in two places at the same time and only nothing is the only thing that is not, therefore on any one side was the half of twelve, which divided 12 in two parts.

That then was $12 / 2 = 6$

$$6 \;+\; 6 = 12 + (\Pi^0) = 13$$

Developing six was an addition to the square as the line flowed and did not involve the crossing of the divide. Therefore Pythagoras was not involved by the forming of six.
At the point where the space filled with heat meets the point in time representing singularity, the end of material (6) confirmed the following spot (+1) at 7.

Forming seven very much involved singularity because it confirms a point where space ends and space (8) begins.

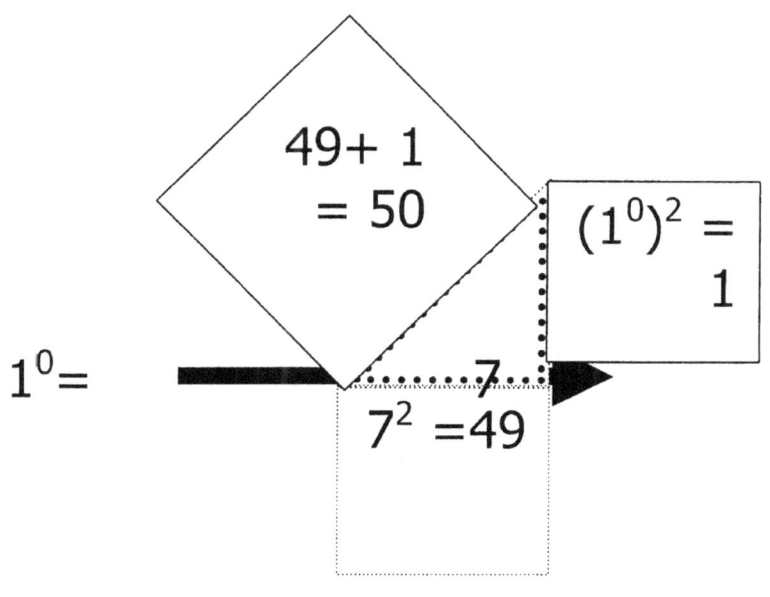

By taking singularity into Pythagoras and filling the Universe by halving the square of space seven completed the required circle within one half of the Universe in order to relate to half the time it takes material to fill time by duplicating. To find the necessary cooling required for control, material has to use five points to be within because of the square involved. There has to be another double five amounting to ten to fill the void from time in the past (position of five) and time in the future (another position of five).

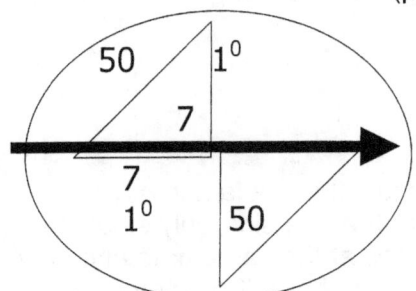

$$50 + 50 = 100; \ (100)^{1/2} = 10$$

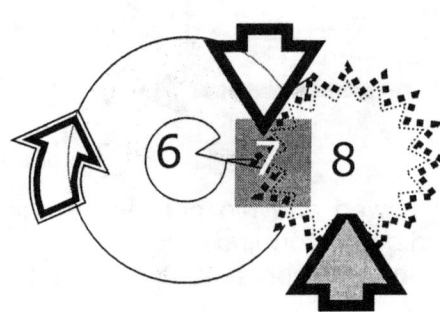

$(\Pi^0) = 5$ 10

7

The requirement of Titius Bode is seven holding relevance to ten and ten being relevant of seven while being in half the Universe $\Pi^0 = 5$

Then come eight causing a line of material to break.

At seven the line is completed at a point distinguishing material within space from space without material

The circle of development has finalized a point. Seven has gone square 7^2 and realized with singularity half of the final of space in the absolute square.

It is this eight to ten science does not recognize and do not distinct as one other part of time. This in relation with the finality that came about at the point seven marked by using Pythagoras that another space, this time in time was developed to compromise for the lagging of time within space-time.

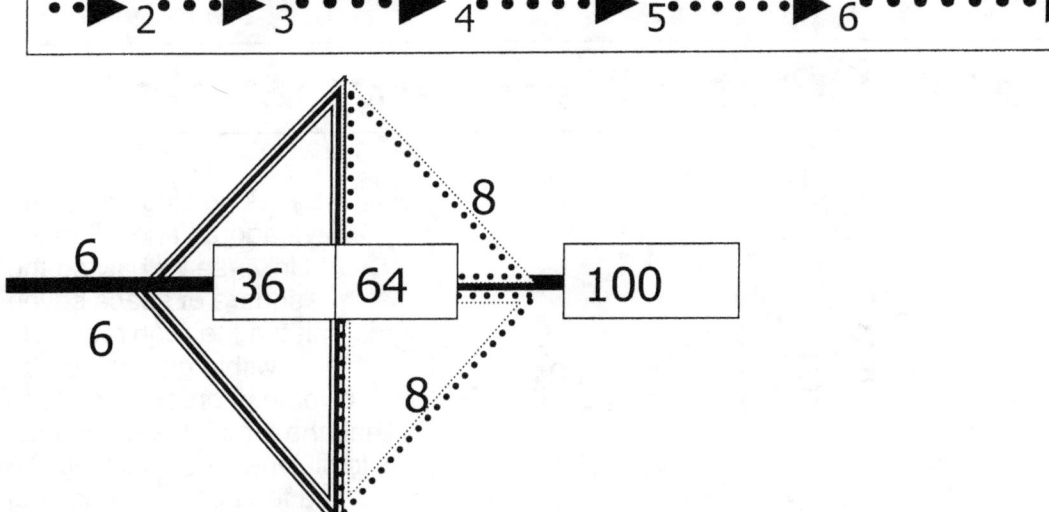

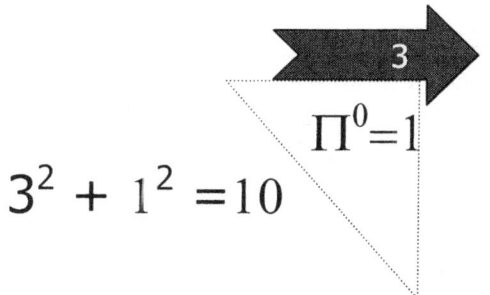

$$3^2 + 1^2 = 10$$

The cycle of eternity could then complete one more time by forming singularity once more

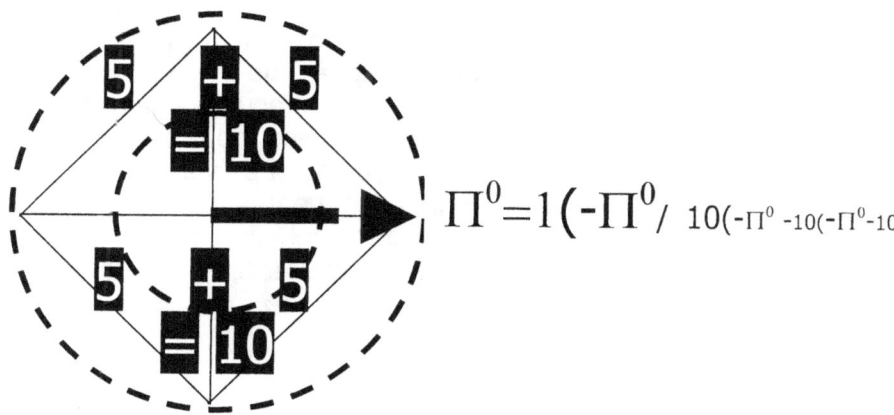

$$\Pi^0 = 1(-\Pi^0 / 10(-\Pi^0 - 10(-\Pi^0 - 10$$

With the Universe established at ten crossing the divide meant that Π^2 at four was a half and five was completing the one half.

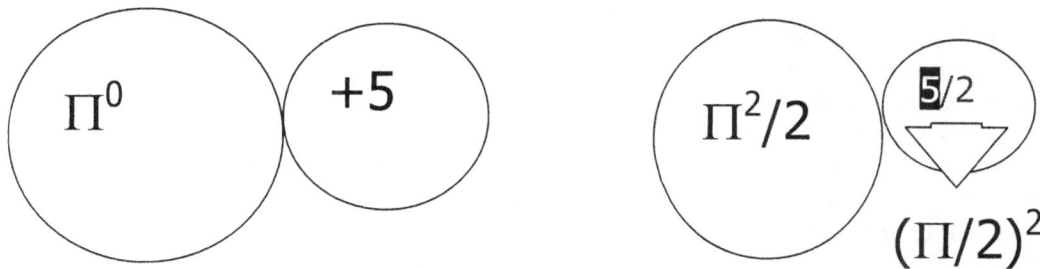

The Roche limit shows that singularity needs at least more than half the Universe (5/2) to share and…the Lagrangian system is at least half the Universe.

Therefore the circle extended to the point outside singularity where matter was holding a space value of $10\,\Pi^3$ and space holding matter was in $10\,\Pi$. Then there came the proton connecting space in time to singularity ON THE OTHER SIDE OF THE UNIVERSE.

They say its Kepler's laws but that is where the inaccuracy starts because Kepler never said that. It is a derogative Newton established of what Kepler introduced as something very different.

The Law Newton gave as Kepler's law of orbits.

All planets move in elliptical orbits, with the Sun at the centre focus.

The figure shown below is set to show a planet with a mass of m, moving in such an orbit (blue dotted circle) around the Sun. The Sun holds a dominating mass of M. We assume that $M > m$ so the centre of the mass of the planet is virtually in the centre of the Sun. The orbit is described by giving its **semi major axis** a and its **eccentricity** e the latter defined so that ea is the distance from the centre of the ellipse to either focus F or F^1. *An eccentricity of zero corresponds to a circle,* in which the two foci merge to a single central point. The eccentricities are not as large as the sketch would indicate but in reality they seem circular. In order to bring across the eccentricity the orbit discrepancy has to be exaggerated to find meaningful clarity. In the case of the earth the true eccentricity is only 0.0167.

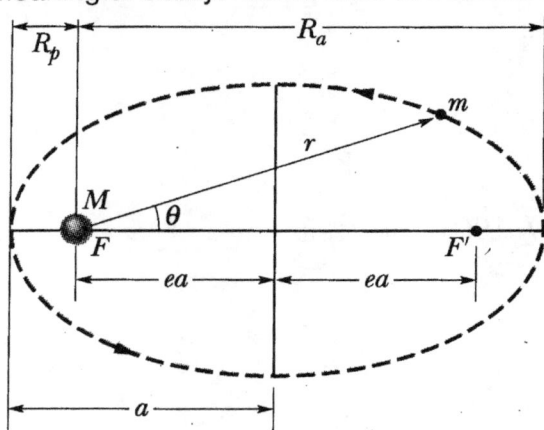

According to Newton and from what we can observe in the sketch the planet with mass m is at focus F of the ellipse. The other, or the *empty,* is focus F^1. Each focus is at a distance ea from the centre, e being the eccentricity of the ellipse. The semi major axis a, of the ellipse, the perihelion (nearest to the Sun) is at a distance R_p and the aphelion (farthest from the Sun) is at a distance R_a are also present.

Notsofast.

The diverting from the main axis forming an angle θ is not a mathematical angle but is only time duration. In physics there might be a point in reference to another point but this is far from physics. It is cosmology. There is a point holding eternity 1^1 by three positions $1^1 \times 3$ in relation to infinity 1^0 and the three positions are time. There has to be a past to achieve the present and to achieve the present there has to be a future. The atom has to have been there if it is to be where it is in order to go to it will be next. There cannot be a top or a bottom or a front or a back because there is an infinite centre, one relating to six other position which numbers eternal positions all sharing one unit. Those being strong in the field of mathematics, can take this challenge and find the possible number of 1^1 in relation to the centre with no sides 1^0 and calculate how many possibilities there might be on the edge of any sphere that will correspond by six to the total number of possibilities not there in the centre singularity 1^0. In the Universe there is no fixed point. On Earth there is the all - dominating centre which grants physics relevant formation, but in the cosmos there are no such points anywhere. There is in the relation one point holding a singularity concerned with expanding, which has the position of 1^1 and there is a singularity concerned with contraction 1^0. The motion depends on getting al the atoms from one location to the next location and this involves the distributing of all the factors concerning 1^1 in relation to every one of the factors relating individually and as a group to 1^0. This reference of F relating to $F1$ is corresponding not to the planet but it is corresponding to every atom individually where the group forms a centre, which corresponds to a centre that forms with the aid of all the atoms forming motion within the Sun. The relation that this group has, forms a partnership with the motion within the entire Earth and from that a centre of governing gravity is selected by the motion of every independent atom that responds to the motion of every subatomic particle notwithstanding how incredibly minute they may seem. The Sun turns with the turning of every atom within the Sun, which generates a combined motion that generates a selected centre through the all inclusive effort of every proton that provides the expanding and contracting balance we find in the Sun. By the motion there is also a conflicting motion that supports the motion where the one part is lateral and the other is rotating. The space $\mathbf{a}^3$ moves partly in a circle $\mathbf{T}^2$ and partly in a straight-line $\mathbf{k}$. That is what Kepler said when Kepler said $\mathbf{a}^3 = \mathbf{T}^2 \mathbf{k}$.

An eccentricity of zero corresponds to a circle, **in which the two foci merge to a single central point.**

If that was true then the circle in which the planet orbits the Sun should be dead canter because with an eccentricity that is obvious, there has to be a favouring one side and then favouring the other side. The zero correspondence is not there since there is a dual eccentricity detected in all the planets in orbit around the Sun. There is much more a swapping of prominence between the two points in foci.

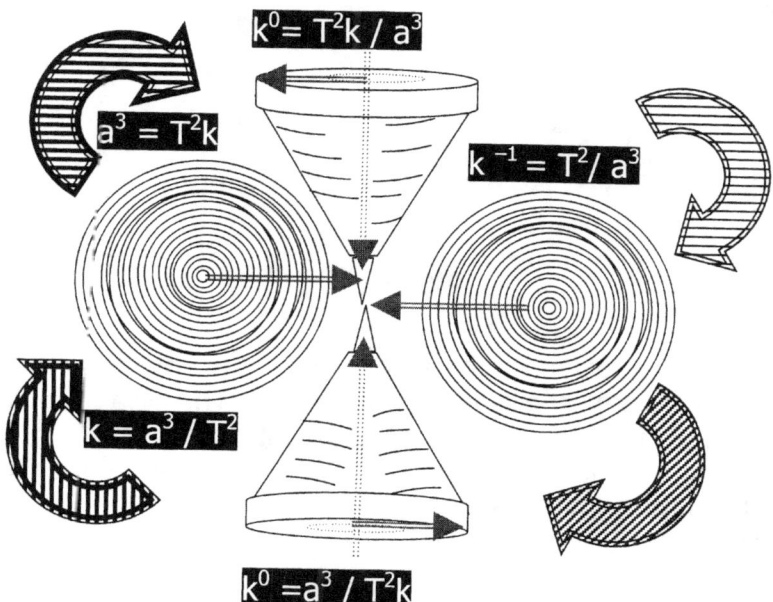

$$k^0 = T^2k / a^3$$
$$a^3 = T^2k$$
$$k^{-1} = T^2 / a^3$$
$$k = a^3 / T^2$$
$$k^0 = a^3 / T^2k$$

There is a divide that separates one side of the orbit from the other side of the orbit and the one is opposing the other side of the same orbit. This has to do with the changing of the motion in relation to the orbit where the changing of relevancies produces gravity, but not by mass because mass is something in Newton and in Newtonian imagination. Mass is a hoax, a lie and fraud to such an extent as the world has never witnessed. Objects can only have mass when the object takes on a host object's gravity and forsake its individualism and independent gravity. Then the object no longer really exists but is part of the host. Mass is not a reality but if one wishes to put mass in as a factor, then clearly mass is that which destroy gravity and not as Newtonians wish it to be, become the factor that forms gravity. By accepting mass the object concedes personal independence and abdicate gravity. There is no hint of any planet having mass because every planet has gravity. The gravity the planets have is initiated by the Sun and that places all the planets in equilibrium to the sun as far as size and notion concerns the planet.

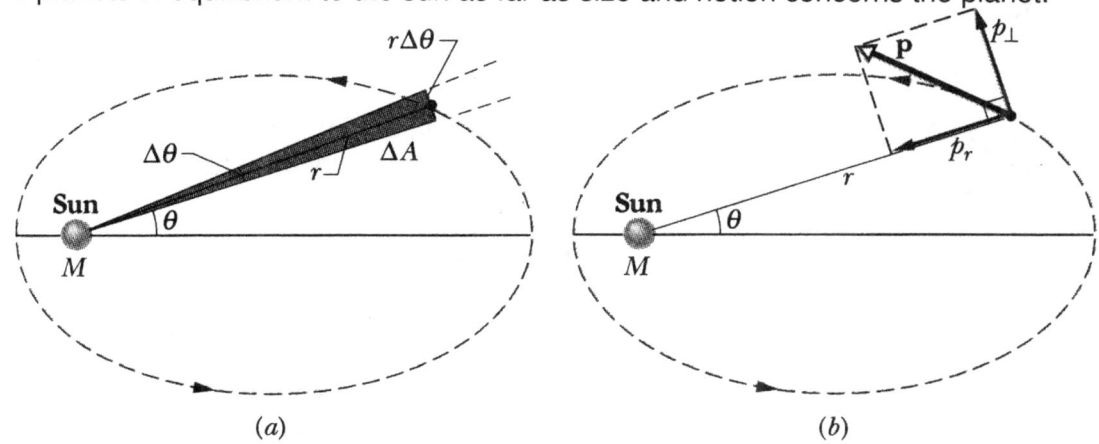

(a)   (b)

The Law Newton gave as Kepler's law of areas.
A line that connects a planet to the Sun sweeps out equal areas at equal times.
Qualitatively, this second law tells us that the planet will move most slowly when it is farthest from the Sun and move most rapidly when it is nearest to the Sun. As it turns out, Kepler's

second law is totally equivalent to the law of conservation of angular momentum. Let us prove it.

In time Δt, the line r connecting the planet to the Sun (of mass M) sweeps through an angle $\Delta\theta$, sweeping out an area ΔA. The linear momentum p of the planet and its components.

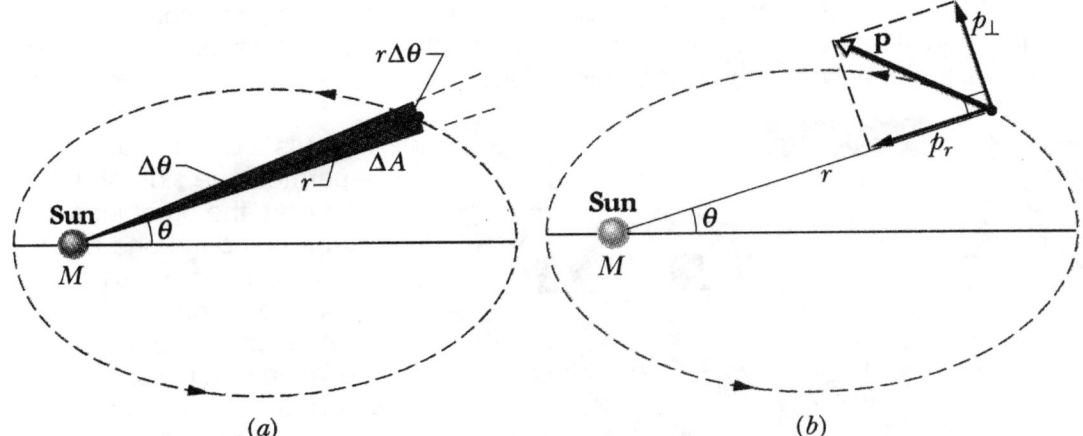

(a)   (b)

The area of the shaded wedge closely approximates the area swept out in time Δt by a line connecting the Sun and the planet, which are separated by a distance r. The area ΔA of the wedge is approximately the area of a triangle with base $r\,\Delta\theta$ and height r. Thus $\Delta A \approx \frac{1}{2}r^2\,\Delta\theta$. This expression for ΔA becomes more exact as Δt (hence $\Delta\theta$) approaches zero. The instantaneous rate at which area is being swept out is then

$$\frac{dA}{dt} = \frac{r^2}{2}\frac{d\theta}{dt} = \frac{r^2\omega}{2}$$

in which ω is the angular speed of the rotating line connecting Sun and planet.

The figure shows the linear momentum p of the planet, along with its components. From Eq, the magnitude of the angular momentum L of the planet about the Sun is given by the product of r and the component of p perpendicular to r, or

$$L = rp_\perp = (r)(mv_\perp) = (r)(m\omega r)$$

$$= mr^2\omega$$

where we have replaced $v_\perp$ with its equivalent ωr. Eliminating $r^2\omega$ leads to

$$\frac{dA}{dt} = \frac{L}{2m}$$

If dA / dt is constant, as Kepler said it is, then that means that L must also be constant – angular momentum is conserved. So Kepler's second Law is indeed equivalent to the law of conservation of angular momentum.

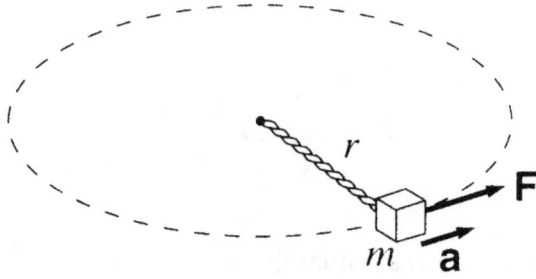

Newtonians share a dubious opinion that the second law is totally equivalent to the law of conservation of angular momentum. If so someone forgot to tell the person holding this opinion that there is a "small" side and there is a "larger" side and this does not stroke with the mass pulling mass idea. If this was correct then it would be that mass in motion neutralises mass in the centre and not mass pulling mass by reducing the radius to the square. On earth the swinging object will hold a perfect radius and if not the spinning object will destroy the centre object as the imbalance will cause the centre to shift because the thrust varies. I still do not agree with the Newtonian view on the physics part but in that I have no mission.

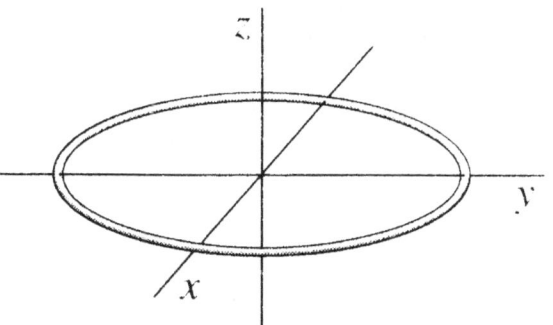

A ring is placed on the x y with the centre point spacing the circle evenly. That represents space-time. I shall get to Newton's mass in a short while… The motion we get in rotation honours the straight line because it admits to the circle being present and combined they pulsate

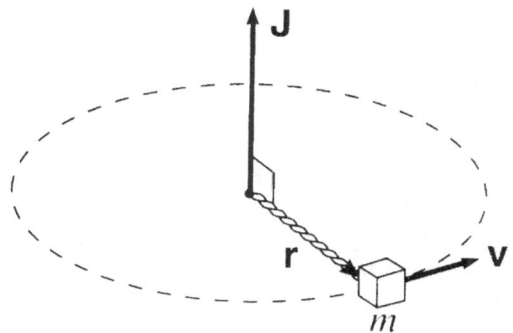

between the triangle representing one another in relation to the other side of the Universe and each other. Lets study the process of the motion in the most incredibly small minute context in order to exclude any and all time delays forming the material that is filling the space. We go to where there is still no mass present even by Newton's imagination. Lets make the Force F the overheating and the mass m the overheated while area a, is the expanding.

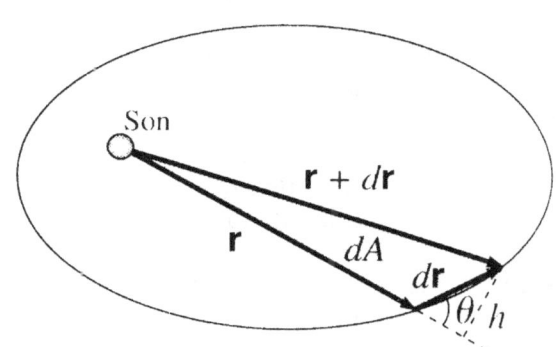

The first acknowledgement we have to make is singularity is charging motion and that would be 1^0 that are charging Π^0. We are now where $6\Pi^0 + 1^0$ is going to form Π. In the forming of Π time continues to run in a line. The line is forming Π because Π is forming on the inside of time being on the outside. That which is in the process of forming is to the outside of time because time is eternally with no end and no outside on the outer limits of Π.

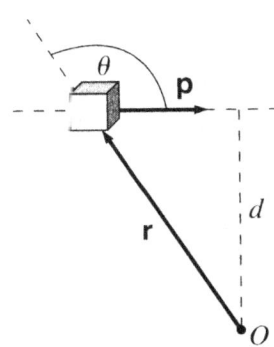

On the one side the motion formed a straight line to the value of 180^0 and at that moment on the other side of the Universe (the Universe being 1^0) a half circle also to the value of 180^0 that in relation forms a triangle also to the value of 180^0 and that is making what there is in duplicate in both directions (**k^{-1}** and **k**) formed as the dot 1^1 coming from the spot 1^0 went around the Universe because the spot 1^0 could turn just as much as it still cannot turn. That is the reason 1^0 has to charge 1^1 to commit motion as Π^0 and form a line that represents a

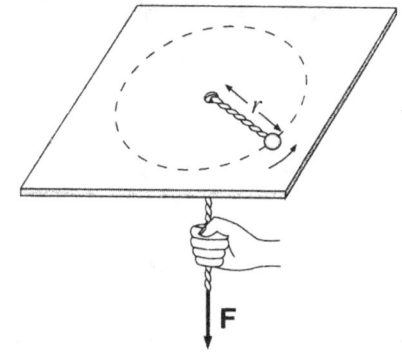

half circle on the other side of the Universe. We can see that Θ is equal to p while d is the motion representing p that is acting as r but since there is no possibility of motion O is actually going over to the other side of the Universe and from (**k forming k^{-1}**)

In order to understand what truly is happening is the fact that eternity cannot expand to the outside. It has no outside. Eternity has to reduce in order to expand and by reducing it is releasing infinity which has no star by giving infinity a start to the outside where eternity that

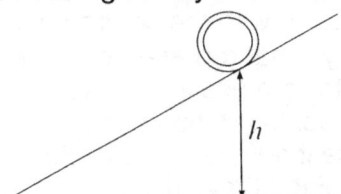

has no end finds an end in infinity with no starting point where infinity starts. I am incapable of putting it more practical and Newtonians with not the mental capacity to understand must at this point find a motor mechanic to explain to them what is going on.

The motion coming about is eternity squeezing the daylights (because it is heat or light) out of infinity that already has no space where in it can be squeezed.

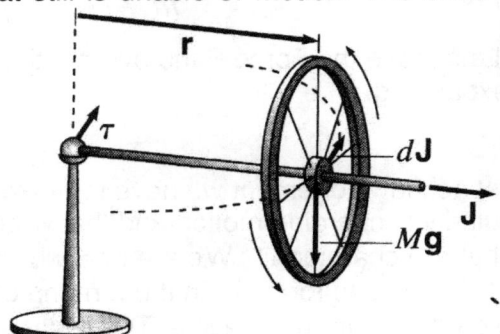

The circle finds the drive it will have from the outside where the outside is in contact with eternity and eternity finds a line to guide (not drive) the rolling circle. It is powered by the discrepancy in the motion of time on the rim in relation to the singularity centre that still is unable of motion. The drive is in the space-time that the motion charges into action. The fact that the top stands erect alone is most significant. The drive- line that charges the motion that drives the top has its significance from a line spun around the outside of the top. Life with its manipulating qualities wrapped a rope on the outside and with a release of the top while still controlling the line, the top found locomotion. The biggest energy issue at this point is admitting the origination of the energy. The energy is life induced. It is not cosmic.

The energy cannot be equal in "mass" meaning size or "mass" meaning duplicating tempo since the atomic or particle presence within the top has no strong enough motion accumulated between the lot to displace the singularity relevance to a centre where from such a centre a governing singularity can be charged and maintained.

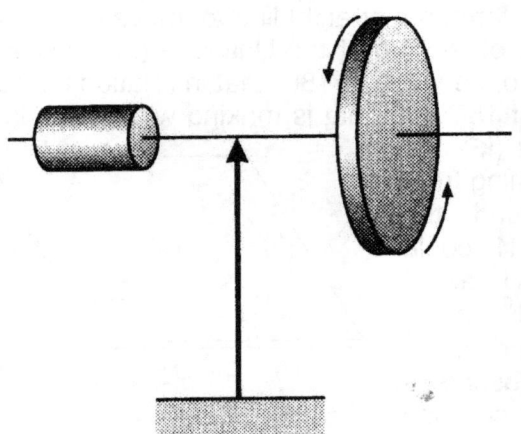

The whole motion locomotion is life inspired and life has validity on one small spot in one tiny solar system that is maintaining a most insignificant solar object of very little cosmic substance. The gyroscope finds the drive coming from the outside because the very inside is unable of motion. The motion drives by means of both dJ and Mg where Mg has no mass reference. The required increase in driving effort comes from the larger ratio the outside of the wheel hold to the significance of the inside that still is too small to move notwithstanding the increase in ratio between eternity and infinity. The significance of a drive line is placing the generating there is between that which moves eternally and that which eternally can't move in relation to another but connected Universe where that which eternally cannot move has a conflict in ratio to that which moves eternally.

Notsofast.

From the planets view the planet is moving away from the Sun by a measure of $k = a^3 / T^2$. Therefore according to the planet it is moving straight ahead way from the Sun without even glancing back. From the view the Sun is enjoying the situation the planet is coming back and the planet has no time to look back into the open sky as it is on its return by the measure of $k^{-1} = T^2 / a^3$.

In reality from an observer's view going by the name of Johannes Kepler space a^3 is in a relation with time by the straight - line k as well as the partial circle T^2. The planet is returning as fast as the planet is escaping and every factor finds a position of victory while equilibrium is the only victor.

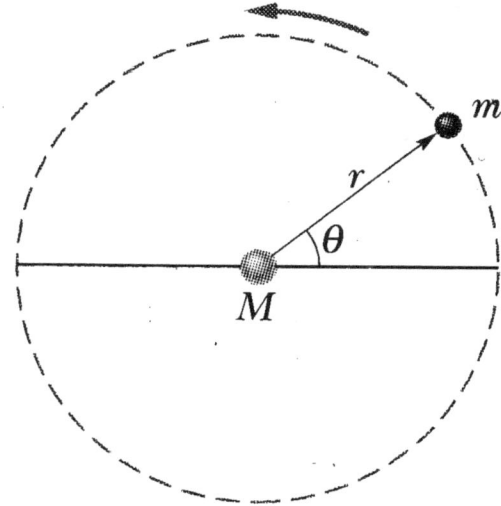

A planet of mass m moving around the Sun in a circular orbit of radius r.

The Law Newton gave as Kepler's law of periods.

The square of the period of any planet is proportional to the cube of the semi major axis of its orbit. To see this, consider a circular orbit with radius r (the radius is equivalent to the semi major axis of an ellipse).

Applying Newton's second law, F=ma, to the orbiting planet yields.

$$\frac{GMm}{r^2} = (m)(\omega^2 r)$$

Here we have substituted from for the force F and used to substitute $\omega^2 r$ for the centripetal acceleration. If we replace ω with $2\pi/T$, where T is the period of the motion, we obtain Kepler's third law:

$$T^2 = \left(\frac{4\pi^2}{GM}\right) r^3 \qquad \text{(law of periods).}$$

The quantity in parentheses is a constant, its value depending only on the mass of the central body.

The Equation also for elliptical orbits, provided we replace r with a, the semi major axis of the ellipse. This law pinned on poor old Kepler who was rather innocent is the biggest swindle mankind was ever hoaxed by.

Newton suggested that the two structures were working in a relation as pulleys would. To do that a point has to be in place on which the pulley may pivot. Then the one mass will pull(y) the other mass and bring the one mass closer to the other mass by diminishing the radius to the square because it is done from both sides. He caught the world hook line and sinker for three hundred and fifty years and every one was caught because no one would admit they were too stupid to understand what the most brilliant brain of all time saw.

Notsofast.

This is the biggest scam the world was ever caught by. Nothing can even compare to the stunt Newton pulled off. This made Newton the Master swindler who is matchless by miles. Some sell the Eiffel tower and other tricksters walk through the China wall, but let Mr. Houdini repeat this one. Newton never (not once) had to prove mass because mass was never a cosmic factor and never being there it was never there to be proven so everyone accepted it as proven. Mass is Earth born and it is Earth related.

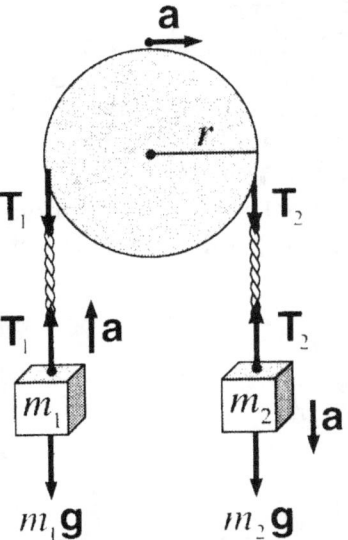

There has to be a resistance stopping the motion or resisting the gravity to bring about mass because the mass comes about when duplication by motion is prevented from moving while the moving still manages to preserve the independent nature of the structure holding the considered mass factor. In outer space everything must move therefore the only restriction is the conserving nature the particular elements reserve. For three and a half centuries no one had the idea to return and see why mass is what mass is when mass is and when mass is not. Good God (that a prayer for my fellow humans), is everyone that stupid, that blind and that naïve. The most ridiculous part of all is the joke they pin on me. When they are cornered and their explaining has to go beyond the ridiculous, they say it is I that cannot understand Newton. It then becomes me being uneducated and at every occasion they pin to my jacket the label that I am mentally underdeveloped. Then they counter the blame on me arguing that I am so retarded that I cannot understand Newton because I am uneducated. Suddenly my being a motor mechanic convinces them most of all because of all brainless things any one can be a motor mechanic is the worst there is. Then it dawns on them that it was I that never even for one semester was at any University to be educated in the science of Newton. By all these standards I am ridiculed and labelled because I can't understand Newton! I thank God I can't understand Newton! Lets look at what is so obvious even I can see it and yet that all the educated never question or sow any doubt.

PLANET	SEMIMAJOR AXIS $A(10^{10}m)$	PERIOD T (y)	T^2/a^3 $(10^{-34}\, y^2/m^3)$
Mercury	5.79	0.241	2.99
Venus	10.8	0.615	3.00
Earth	15.0	1.00	2.96
Mars	22.8	1.88	2.98
Jupiter	77.8	11.9	3.01
Saturn	143	29.5	2.98
Uranus	287	84.0	2.98
Neptune	450	165	2.99
Pluto	590	248	2.99

This law predicts that the ratio T^2/a^3 has essentially the same value for every planetary orbit around a given massive body. The table above shows how well it holds for the orbits of the planets of the solar system.

If the Newtonians could just once, for one instant show me how mass does affect the behaviour of planets orbiting even in the least, then I too can boast about me finally being able to understand and appreciate Newton. I have tried for so long on so many occasions and in so many respects to find grounds for Newton's persistence on blaming mass in relation to motion created. I have committed so many arguments that I could not take in any direction as to why Newtonians would follow the untested and meaningless arguments so sheepishly just to follow their Master. There are slight anomalies but one can see there is a persistence of three in relation to the time space holds in all cases.

The orbit velocity shows only concern for the Sun and all the planets are an additional factor. The planets individually or as a group are driven by what the Sun drives and the Sun is driving something other than driving the planet directly. The planets are liquid driven or hydraulically driven. There is no distinction of position or size or gravity or even formation. There is no distinct favouring of any means whatsoever. To the Sun all the planets are as liquid as that which is liquid in relation to the Sun. Only if and when an object is in total buoyancy can all objects irrespective of differentiation be so equally regarded. The planets move as if they are the same. That could only apply if they were all the same in regard to the Sun and they have no solid application but only serves as more liquid in a sea of liquid where the Sun is the only solid.

Time in spot 1

$\Pi^0 X \ 10^2 +$

Time in spot 2

$\Pi^2 X \ 10 +$

Time in spot 3

$\Pi^0 X \ 10^2 = 3D$ 3 Dimensions in total ± 298 or then two

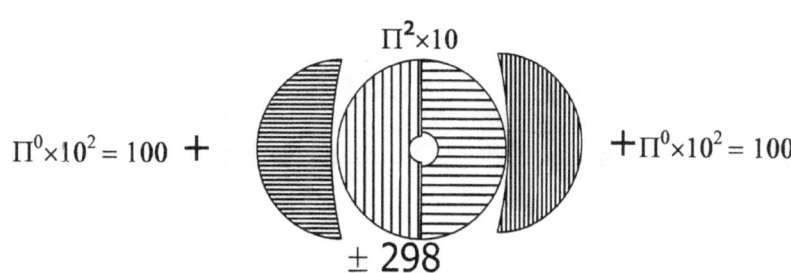

$\Pi^2 \times 10$

$\Pi^0 \times 10^2 = 100 \ +$ $+ \Pi^0 \times 10^2 = 100$

± 298

positions (spot 1 and spot 3) and one allocated in (Time in spot 2 $\Pi^2 X \ 10$) gravity-motion.

Notwithstanding the planet arrangement time relates to space by $k^{-1} = T^2 / a^3$.

Spot 1 Spot 2 Spot 3

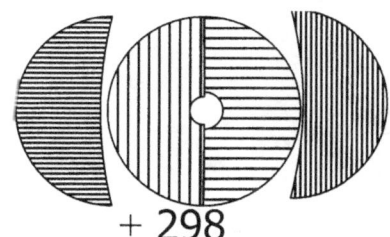

± 298

From the above table one can clearly see the three positions in time that space holds. By placing time in the relevance to space we also can clearly see that the space is flowing where the relevance is reducing time $k^{-1} = T^2 / a^3$ or $\Pi^0 = T^2 / 3a^3$. The space is flowing towards a dominating centre and that we know is the centre of the Sun.

$$k^{-1} = T^2 / a^3 \text{ which then is } 3 = T^2 / a^3$$

$T^2 = a^3 / k$ which then is $T^2 = a^3 / 3$

$a^3 = T^2 k$ which then is $a^3 = 3 T^2$

$k = a^3 / T^2$ which then is $k = 3 a^3 / T^2$

$k^0 = a^3 / T^2$ which then is $k^0 = 3 a^3 / T^2$

PLANET	MASS	GRAVITY	GRAVITY/MASS	%
Mercury	0,05527:1	0,372 :1	6,821	682,1 %
Venus	0,815 :1	0,905 :1	1,11	111 %
Earth	1 :1	1 :1	1	100 %
Mars	0,1074 :1	0,38 :1	3,538	353,8 %
Jupiter	317,8 :1	2,538 :1	0,007986	0,80 %
Saturn	95,16 :1	1.075 :1	0,0113	1,13 %
Uranus	14,5 :1	0,914 :1	0,063	6,306 %
Neptune	17,2 :1	1,14 :1	0,663	6,63 %

There is no mention of mass or any discrepancies in any form or shape or thought. All the planets serve the Sun on equal terms notwithstanding whatever notion humans try to add for what ever disguised pleasure the human may find in such an argument.

Here is a typical Newtonian response to a very formidable question about the masses of the planets that don't matter when determining their orbits. What counts in this situation is the equilibrium between the gravitational attraction and the centripetal force. Look at these equations:

$$Fc = (mv^2)/r \quad\quad and \quad\quad\quad Fg = (GMm)/r^2$$

*Where Fc is the centripetal force, Fg the gravitation, m the is planet's mass, M is the Sun's mass, v is the orbital velocity of the planet (which is really not constant, but we can consider it this way) and r is the orbit's radius. G is the gravitation constant. Its value is 6.67E-11 m^3/(s^2 * kg).*

The orbits are stable, otherwise we wouldn't be here to discuss this problem! So, the attractive force (gravity) has to be equal to the centripetal force, which tends to make the planet escape from its orbit.
Fc = Fg

$$(mv^2)/r = (GMm)/r^2 \quad\quad\quad and\ with\ some\ simple\ manipulation:$$
$$r = (GM)/v^2$$

As you can see, the radius of the orbit doesn't depend on the planet's mass. It depends only on the Sun 's mass, and inversely on square velocity. What determines the position of the planets in any solar system is the system's initial conditions, which aren't well known... yet! The initial angular momentum, the mass distribution discontinuities in the dust cloud that originated our solar system (or any other), and some other factors, were the conditions that led the planets to be arranged this way. Yes, they were "built" about 5 billion years ago on almost the same orbit they have today!

*Fc is the centripetal force, Fg the gravitation, m is planet's mass, M is Sun 's mass v is the orbital velocity of the planet, (which is really not constant but we can consider it this way) and r is the orbit's radius. G is the gravitation constant. Its value is 6.67E-11 m^3/(s^2 * kg).*

There is no indication whatsoever in any language ever proposed that mass has any indicative role or influence on any planet in any way. Not the motion. Not the orbit. Not the rotation. Not the velocity. That which is supposed to be is in our minds and not on paper.

That is how we wish to interpret cosmology. The larger person is the heavier person holding more momentum. The larger planet is the heavier planet forcing more gravity and is therefore the strongest. That is rubbish.

That is the mindless gargle of the brainless. I am sorry but I cannot be less explicit about this matter. It is out of touch with reality. It could be part another fairy tale. It seems no one has learned! Go on and calculate which planet is generating the most gravity per cubic what ever. Find out in simple terms that the big structures are generating almost nil when compared in direct relation to the smaller planets. It has nothing to do with their being gas and everything to do with the inverse square law.

Again I reiterate and repeat my question: If the gravitational constant had the means to eliminate the mass factor in the orbiting of the orbiting structures, why is there a need to use it in any form of calculations? Why persist with a meaningless proposal.

Jupiter has a MASS of 18955.872 x 10^{24} / GRAVITY of 24.89778 = 761.58
That means every 761.58 parts holding space generates one Nm of gravity

Saturn has a MASS of 5686.76 x 10^{24} / GRAVITY 10.556 = 538.7
That means every 538.7 parts holding space generates one Nm of gravity

Neptune has a MASS of 1027.872 x 10^{24} / GRAVITY 11.2815 = 91.1
That means every 91.1 parts of space held generates one Nm of gravity

Uranus has a MASS of 866.52 x 10^{24} / GRAVITY 8.96634 = 96.64
That means every 96.64 parts holding space generates one Nm of gravity

Earth has a MASS of 59.76 x 10^{23} / GRAVITY 9.81 = 6.091
That means every 6.091 parts holding space generates one Nm of gravity

Venus has a MASS of 48.7044 x 10^{24} / GRAVITY 8.87805 = 5.485
That means every 538.7 parts holding space generates one Nm of gravity

Mars has a MASS of 6.418224 x 10^{24} / GRAVITY 9.81 = 0.654
That means every 0.654 parts holding space generates one Nm of gravity

Mercury has a MASS of 3.3029352 x 10^{24} / GRAVITY 3.64932 = 0.905
That means every 0.905 parts holding space generates one Nm of gravity

Pluto has a MASS of 0.11952 x 10^{24} / GRAVITY 0.3924 = 0.3045.
That means every 0.3045 parts holding space generates one Nm of gravity

Because of the invert square law the existing of such a law puts Newton's theory on mass in doubt. The more the size is the less the force of the mass will be in proportion to space occupied. The bigger is not the more powerful or the strongest because of the invert square law. If there were no invert square decrease of space to power ratio, then yes it would make sense. But the Sun has the least gravity of all in the solar system just because of the size it holds. Can any Newtonian out there please explain what mass has to do with the rotating of planets around the Sun and what do you accomplish by calculating the mass as a factor. That has no practical implication even in the least in relation to the orbit. It implies a massive degree of senseless ness shown by the educated to prove his education to be bizarre and null - in void. It is like blaming the ice-cream sales in Alaska for the persistent drought in the

Namib Desert and then invents non-relevant factors to prove a link. If mass was a factor then Jupiter must either propel faster or slower than the rest or Jupiter should be farther or closer or have some indication that places the size relevancy in another category than the others. To say the gravitational constant even things out means to say the gravitational constant is nullifying the mass and then that proves by point, the mass had nothing to do with the price of eggs in the first place. Every one is bullshitting one another by using invalid factors to calculate non-existing quantities in no relevant and banal calculations. The mass does not make a planet go faster or slower or influence the structure in motion. Then why the hell use it in calculations, when it has no purpose or accuracy? The mass does not influence the allocated orbit in any way, then why use it? The mass does not pull the planet closer to the Sun so why the hell bring mass in as a factor. Mass does not indicate any difference even in the smallest indication. It does not show the least or biennial of differentiation between cosmic structures so why the hell use it? It is used to make the mathematician feel worthy and superior even if it also shows his absolute mindless ability in rational arguing. If a factor has no influence on whet is proposed, what their calculations wish to measure except to prioritise the ego of the mathematician then using it in calculations is outrageous. Using mass has the same implications in the calculation than it has to bring in India's net pepper production and substitute that with any factor so why not use that as a factor in the calculations. This is how the layout of planet orbits is in reality.

Mercury	Venus	Earth	Mars	Jupiter	Saturn	Uranus	Neptune	Pluto
0.055	0.86	1.0	0.11	318	95	14.5	17.2	0.002

This proves distinctly that there is no proof that big and small has any consideration in the cosmos but the idea of differentiating between big and small is a human concept that holds no cosmic merit.

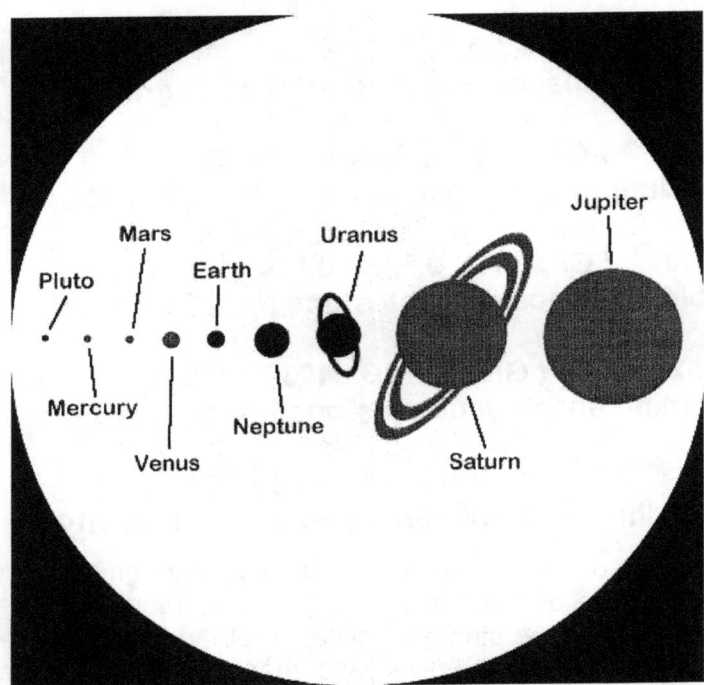

The big question to ask is why would mass not show as a mathematical indicator when orbit velocities of different planets are concerned. Why would time show a flow that turns to be negative in relation to space when time is put in division of space? These are the mathematical answers that hold truth and not the criminally motivated mismanagement of mathematical principles and mathematical rules that Newton advocated. If mass does not show as a factor, one cannot go and grant mass a premium but one has to surge for a reason why mass does not show its presence as a factor. If mass had any influence on the orbit of planetary motion as the calculation formulas would suggest, then the orbit arrangement should be as the illustration alongside suggests. If mass had anything to do with gravity we would find either the largest in the inner orbit since it has the largest mass that will erode the radius the fastest or we will find the least in the inner orbit because the least mass will produce the least motion. Then the planet orbit layout would be as follows:

This is the manifestation of the fairytale where it shows the naked reality when the sublime goes mad in the practical implication. It is equal to the fairytale that tells about the King that paraded around naked because only the stupid would realise he was without clothes and since only the mindless would not be able to see his magic clothes. Then he including his subjects pretended to marvel at the beauty because they would rather bullshit their minds than to admit that they were to stupid to see the clothes the King was wearing. So everyone pretended not to be stupid. In reality, it showed exactly how stupid he or she was. In this case it was the intellectually Superior that was privileged to be stupid just to show the rest who was the most stupid of them all and never to ask what the hell mass has to do with the whole affair. So every one who thought he was superior showed how big an idiotic fool he would be in order to prove the point. That made him not more a fool than the fool he or she was trying to impress and both idiots impressed other idiots by accepting that mass plays a major part while mass has no reason to be part of any of the facts relevant. In this case it is the using of mass that everyone echoed because by not realising the use of mass would show any individual that needed to become part of the Brainy Bunch how mindless he or she was not to see the advantages of the use of mass. So the mindless then accepted there is some lunacy.

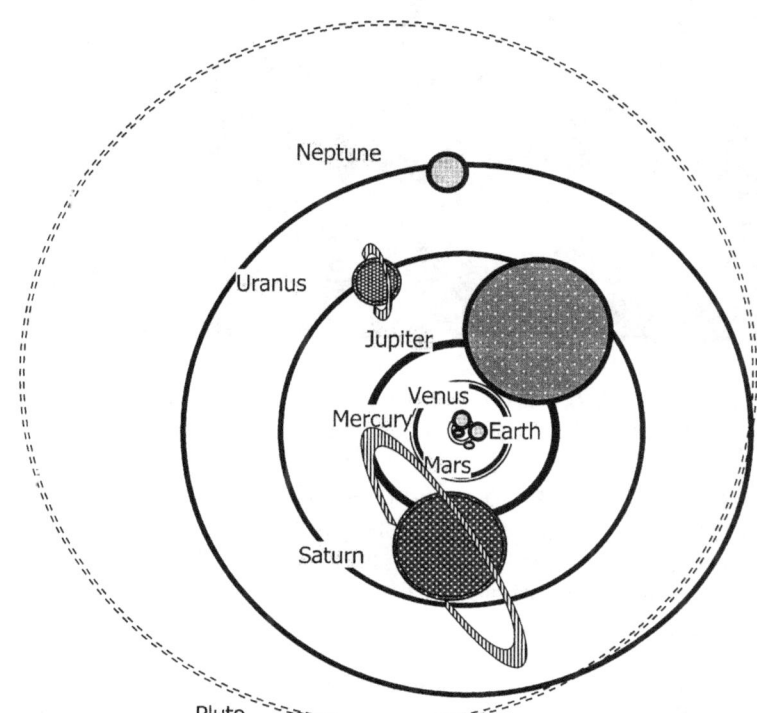

Jupiter has a MASS of 18955.872 x 10^{24}

Saturn has a MASS of 5686.76 x 10^{24}

Neptune has a MASS of 1027.872 x 10^{24}

Uranus has a MASS of 866.52 x 10^{24}

Earth has a MASS of 59.76 x 10^{24}

Venus has a MASS of 48.7044 x 10^{24}

Mars has a MASS of 6.418224 x 10^{24}

Mercury has a MASS of 3.3029352 x 10^{24}

Pluto has a MASS of 0.11952 x 10^{24} /

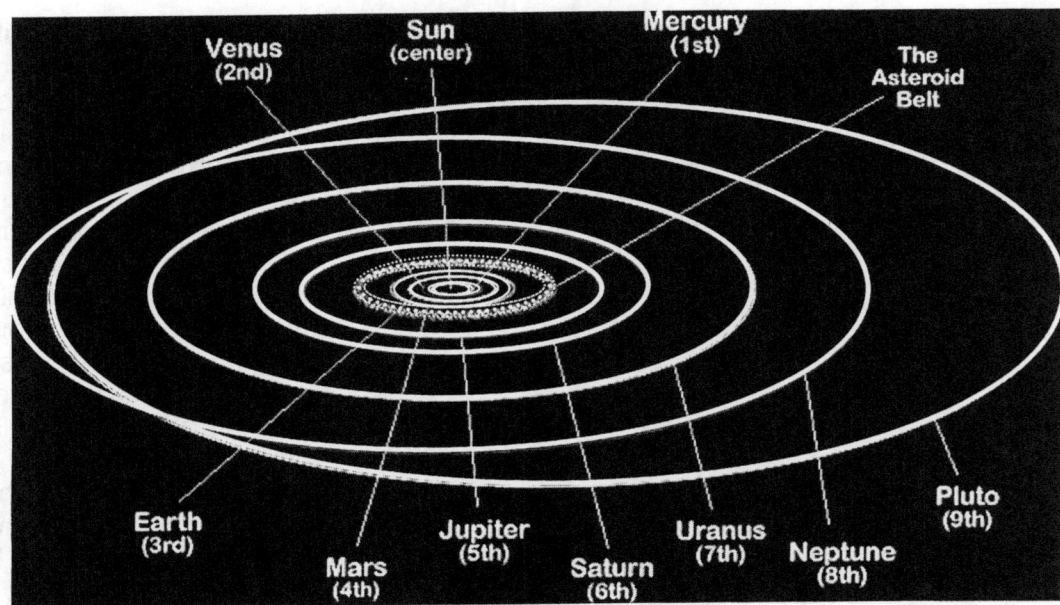

That is the fairytale but in reality the dead is criminal. If I go about and suggest facts, which is untrue and while I know them to be untrue I still support the facts and support the system declaring the facts as to be true, the system that includes me is a fraud. No wonder the whole lot is devious and mindless atheists. A question never asked and a thought never put to words is that if there is a Gravitational constant guiding all cosmic matter by the force of gravity where is it going and where is the centre where it is taking all the material. That statement is criminal with dubious intent to mislead the public and is a conspiracy outweighing crime syndicates such as drug lords and the mafia.

	Distance (AU)	Radius (Earth's)	Mass (Earth's)	Rotation (Earth's)	# Moons	Orbital Inclination	Orbital Eccentricity	Obliquity	Density (g/cm³)
Sun	0	109	332,800	25-36*	9	--	--	--	1.410
Mercury	0.39	0.38	0.05	58.8	0	7	0.2056	0.1°	5.43
Venus	0.72	0.95	0.89	244	0	3.394	0.0068	177.4°	5.25
Earth	1.0	1.00	1.00	1.00	1	0.000	0.0167	23.45°	5.52
Mars	1.5	0.53	0.11	1.029	2	1.850	0.0934	25.19°	3.95
Jupiter	5.2	11	318	0.411	16	1.308	0.0483	3.12°	1.33
Saturn	9.5	9	95	0.428	18	2.488	0.0560	26.73°	0.69
Uranus	19.2	4	17	0.748	15	0.774	0.0461	97.86°	1.29
Neptune	30.1	4	17	0.802	8	1.774	0.0097	29.56°	1.64
Pluto	39.5	0.18	0.002	0.267	1	17.15	0.2482	119.6°	2.03

Mass has precious little influence on gravity because gravity is the motion, whereby there is interaction between time serving as a liquid flowing by contraction to the centre of the Sun and the planet by duplication is part of the flow. Since there is a flow and duplication set in balance the duplication a^3 is in the range of the motion T^2 while the flow of three is in the straight- line k that serves time.

Again, from the lips of the Newtonian into your face and then you tell me who is the criminal misconduct. If any person goes into a contract with such devious preconditioned misleading an evidence with the intent to mislead and defraud by giving unreliable and untrue facts deliberately while full well knowing what is suggested has no truth to bear, you are a swindler, a cheat, a criminal and you belong in a safe place away from society where you and your malice may not defraud others any more. If you do not like my saying this, prove me wrong. Prove how your swindling behaviour to cover Newtonian defrauding is not criminally intended!

This might seem harsh criticizing but it is reality. Producing fictitious facts in order to mislead and present untruths is criminal. Any academic wherever can either sue me for wrongful slander but doing so that academic must prove without doubt how mass plays any part in initiating or contribute to motion of whatever major or minor influence on whatever scale. It is criminal to go around and falsify information in order to protect a corrupt complot and counterfeit facts with the intent to deceive and spread deception. I say astro physics is a hoax from the start up to the present and up to now the academics scandalously avoided me by ignoring me in order not to face up to their deceit. I challenge who ever to charge me with slander and prove me wrong. Prove in what way does mass bring about any implication to the orbits of any cosmic structure. No individual has the right to claim the privilege to attribute any factor where there is no clear evidence that such a factor has validity in its presence. One cannot tell Jupiter it has mass of what ever measure we humans dedicate to Jupiter just because we see a flow of more light coming through a lens that Pluto show because Pluto show less light reflecting. If they all orbit equal they are all alike, and it is a human duty not to tell the cosmos what Newton says the cosmos should be but to take guidance from Kepler and allow the cosmos to inform us about matters that prevail in the cosmos.

Just as improbable and therefore inappropriate is the manner that science suggests that "by the magic of gravity" stars accumulating dust to form stars. The suggestion alone that dust by the force of gravity can accumulate into solid matter is fraud and then to wilfully with deceit intoned to use tax money to investigate such dubious ludicrous nonsense is over the top. It is fraudulent to use tax money on something that is obviously a hoax. If any banker or politician did the same criminal offence he or she would have been branded as a swindler and paid a penalty in a penal institution for a very long time.

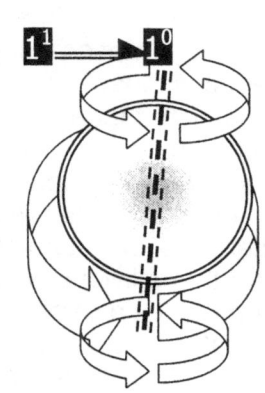

Let us once and for all accept that Newton's mass activated gravity is inspired by his imagination. The process is when liquid in the form of outer space lines up against a solid such as a palate or the Earth. There is motion in the one department that acts as if it is a solid but is in fact the partner that holds the motion. Then there is the liquid, which by being the stationary acting the part of the solid but is a liquid all the same.

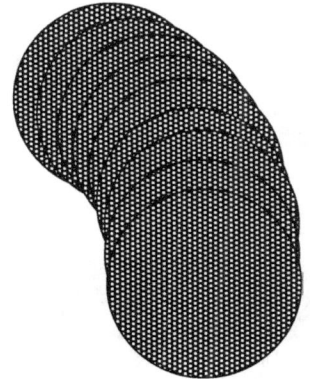

The body delivering the motion is a solid that forms a unit as space. The part that serves as a cosmic liquid is the partner that is also stationary and serves as an immobile liquid. This has things rather confused in the manner that gravity in the cosmos operates. Remember the planets are not a normal set up and even stars with micro stars to attend to are holy unnatural. It is very seldom in combination and when it is, things get as confusing as we find it to be on Earth. The norm in the cosmos is a lone star that spins on

an axis while in motion around a galactica. The galactica presents the same layout as a star and the working process in the galactica that apply in a similar mode, as stars with layers would have.

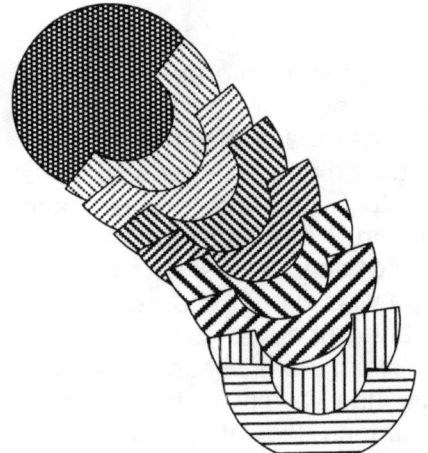

We have to remember that the atoms with seven positions moves as much into time with ten sides than time with ten sides contracts into space with seven positions. There can never be any motion that is not a response of what it is responding on and that comes from what it reacts on. The motion of the atom to expand and fill more time $k = a^3 /T^2$ is just as much time contracting to space filled as material by the movement of $k^{-1} = T /a^3$. The one responds on the other as time in motion changes space in motion to flow through time by measuring time.

Do not look for the pumping going on where one can see time that meets space. The pumping action is going on where the proton pumps time into singularity by expanding and then contracting in the very heart of the atom nucleus. There the duplication presents the expanding and the contracting, which feeds the star with the motion either in duplication or in contraction that the star requires to comply with the demand space-time insist on as gravity. The reducing of heat by motion is presented as cooling since motion reduces space and by reducing space it is cooling. To establish that rapid cold the proton moves 1836 times faster in order to restrain the heat from the value the heat had when the heat was at the electron relevance. At the electron the heat was already at the speed of light and therefore the atom removes all heat by freezing the heat into the oblivious where every atom is a black hole. All this adds up as a general reducing of space by the governing singularity in charge. Every atom in the star is a pump that coverts heat to cold and transfers singularity 1^1 to singularity 1^0 to regain what was lost during moment-Alfa. The gravity in the star is not nearly the gravity going about the planets.

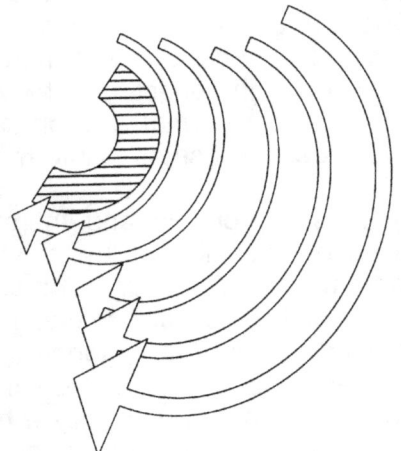

In the case of the planets there is an orbit motion that puts liquids in relation to solids without the much pumping being the dominant factor. The liquid allows the solids space to move within. As the solid pushes against the liquid the liquid bears down on the solid and some liquid give way but the inner liquid increases the density at the point and just above where it touches the solid moving structure. The liquid pushes down the solid while in accordance with the Coanda principle. The space expands to point directly relevant to the motion that the solid provides. That which is without motion is secured by the liquid to be part of the Earth while that which is liquid is secured onto the Earth as an extension of space.

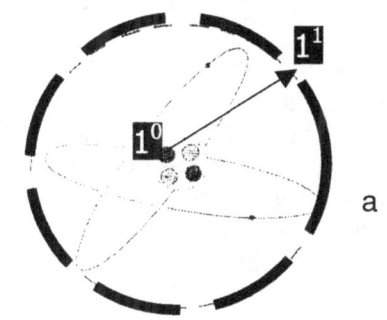

a

There is an allocated line designated by the extending of the solid that includes the liquid to gather that liquid into the unit forming the solid. We gave that so many names ending with sphere even the thought of all these sphere makes ones head spin. How Newtonians fit the sphere as in stratosphere and atmosphere and what not into gravity is still a puzzle, which is eluding me in the manner that Newton's vision on mass was eluding me. At the end of all this there is a line that is the friction point and it is at that line where liquid tear from solid while the solid is actually intensified liquid.

However the only constant in the Universe is that there is no constant applying. Everything is in cyclic shifting as the relevance relocate and alternate positions. In order to get a flow of space - time 1^0 and 1^1 must be forever alternating. The fact of constants are that constants are as Newtonian as mass can ever be and constants are as much a fact that does not apply as mass where then mass has the same position. The planet forms an electron to the Sun becoming the solid and the Sun allows the planet to spin while the planet receives it alternating which forms motion from the Sun that provides the governing singularity not only to the Sun but also everything orbiting the Sun as an electron. Because the planet is just an electron, the planet will rotate about the Sun as any good electron would do.

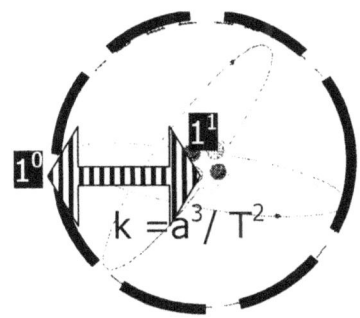

$$k = a^3 / T^2$$

In the case of comets the Sun is the solid that forms stability while the comet is the solidity that moves and outer space is the liquid that does not move. In the case of comets the cosmic law is transgressed. The Sun is an atom. The Sun consists of a unit forming an atom where the Sun is the atom in compiled group but also where the group serves the unit. Every layer in the Sun is a liquid to the top where the bottom serves as a solid to the top layer forming the liquid. The proton puts time at motion where time puts space in demise. Time devours space as eternity meets infinity. The atom is a Black Hole with matter in between infinity and eternity and this fills the black Hole with substance that is forming space - time.

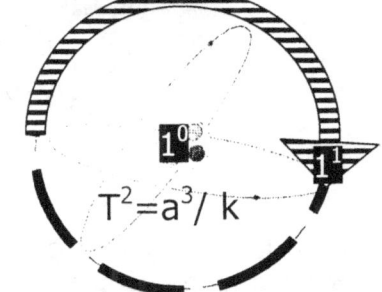

$$T^2 = a^3 / k$$

The proton serves as 1^0 to the neutron being 1^1 where the neutron serves as 10 to the electron being 1^1. The atom forms a Universe that holds both eternity and infinity apart by allowing motion to separate time. The atom concludes the Universe because the atom is what concludes the Universe as much as it started the

Universe. In the end all stars will be one atom in the hydrogen atom but that is the final conclusion where the last era arrives. The atom is the Universe.

The atom maintains relevancies where the core within the atom serves as 1^0 and the orbit serves as 1^1. The core is the solid and the electron is the liquid. The electron provides the motion because in relevancy at the point where the electron is located it is the electron that is in motion while the core within the atom is a solid that does not move. The atom serves as movement because singularity generated by all atoms forming the unit provides the motion. All atoms forming the star are allocated the value of motion being 1^1 while singularity charged with governing the star is 1^0.

When the planet is on one side of the Universe where the centre of the Sun forms the Universe, the planet resists the flow of time. When the planet is on the other side of the Universe the planet, the Sun is 1^0 but the planet alternates 1^1 and 1^1 because the planet lands on one side of the Universe and then on the other side of the Universe.

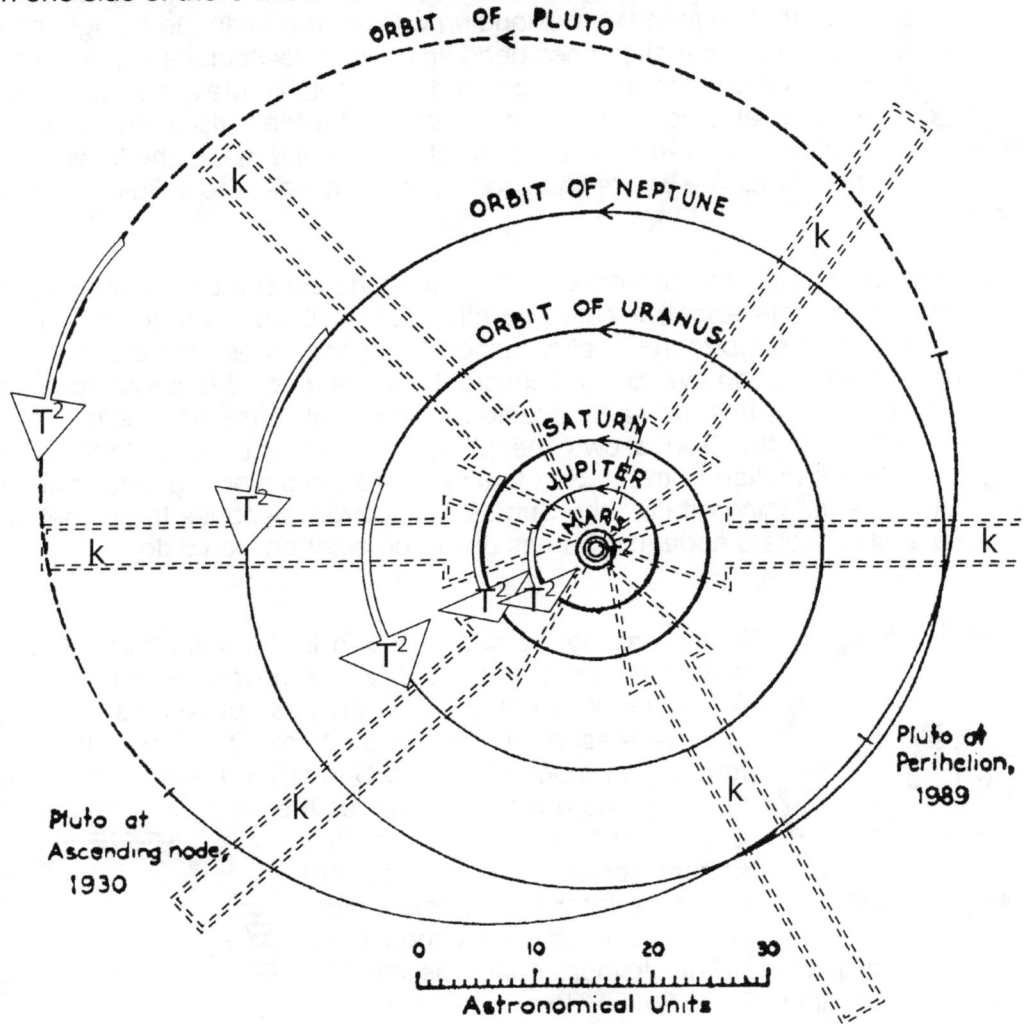

Kepler's formula insists on the flow of space-time. Gravity is never mentioned because gravity is a name coupled to an idea. If the idea is incorrect the coupling is unjustified. The space is floating in the moving time. $\mathbf{a^3 = T^2 k}$.

The Sun is contracting time towards the centre and by doing that, it is implicating the process we gave the name of the Coanda effect. The space is duplicating the space it holds in order to flow in the time being contracted. By duplicating the space, the space is generating what it was to where it is to the location it is going to be. That is space-time and that is gravity and that is motion. Restricting this process leads to mass applying as a restraining because no object can share space just as much as no object can be in two places at one time.

There is a flow of space-time running towards the Sun at a duplication tempo where three portions of space are in relation to one unit of time. Material stands related to space by measure of the past position, the present position and the future position and that the cosmos told Kepler by a language of mathematical equations. It is not my say so. It is not the say so of Kepler. It is not the hearsay of Newton. It is not the interpretation of the visions of Hubble or the guesswork of Hawkins or the calculations of Einstein's perceiving. This comes from the horse's mouth. This has nothing lost in translation. The cosmos told this to Kepler without ceremony or private preferences colouring and tainting preconditioned favouring of prejudice. The cosmos spoke in a language everyone can appreciate because of pronunciations some dialects may bring in some incorrect abbreviations…and Newton still managed to corrupt all the information, notwithstanding the utmost care the cosmos took to prevent such behaviour, and that he did to impress those he wanted to convert to his preconceived ideas, just to oil his neurotic ego.

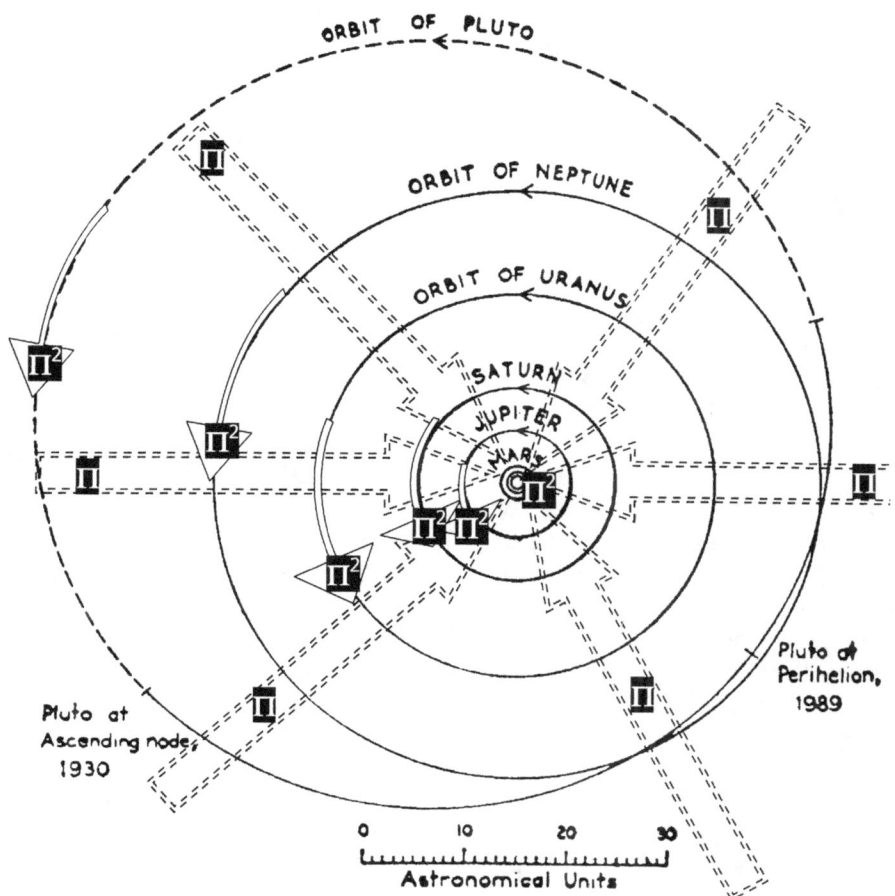

It is the atom that is in charge of the Universe because it is the atom that charges motion. The rotation of the object produces the proton duplication while the rotation around the contracting centre is gravity or motion unblemished and the time flowing towards the centre places the orbiting structure in the 3 dimensional space in time $(10)^2$ square. In the final analysis it still is the atom that produces the atom, which serves as a star. It is the proton at $(\Pi^2+\Pi^2)(\Pi^2\Pi)3 = 1836$ that forms the Universe in more ways than any human can appreciate.

It applies simply because in time there is a ratio whereby space duplicate in expanding as well as contracting and this ratio is serving what ever the cosmos might be. The cosmos said it is $a^3 = T^2 k$ therefore it is little surprising that Newton did detect a ratio between space-time and time flowing about space. It is more surprising that he missed the rest. Much more surprising is that every one of his dedicated followers and mathematical geniuses missed the rest. Then I better put my mouth where my proof should be and explain what the rest is that all missed.

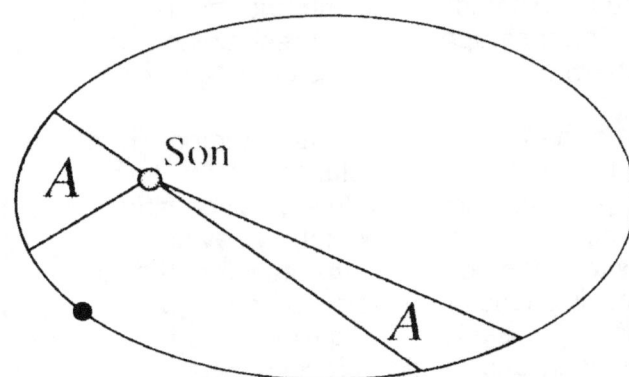

The cosmos is not expanding because it is shrinking. When it started that which cannot be bigger parted from that which cannot be smaller, it started where infinity was eternity. Then infinity parted from eternity leaving eternity bigger than infinity but because of the size of infinity, eternity was not much smaller than eternity. The star is the atom within the star by the multiplied motion of all the atoms forming a unity that charges the motion in the star. The star is a culmination of the efforts of the star.

When the first moment came the two factors being eternity and infinity was the same. Then they parted company putting a reference between them and not much more that just a reference of division

Eternity • Unified with infinity

put it this way: the that what was at first end and in that is the expanding of the Universe.

Eternity •• parting with infinity

The ratio is increasing because it is decreasing the Universe. Let's Universe cannot expand because is eternally big with no possible → reason why there is no possible

growing • **Eternity ↦away from infinity**

The Hubble constant is the measure whereby the Universe is reducing since there is no possible room for any expanding. The part being infinity is the part that cannot reduce since it is as small is infinity can ever be smaller. It is so small it has no sides but all points share on spot. Any further reducing will bring about an increase in size and by reducing further it is increasing what never can further reduce. With infinity being there without having a possibility to reduce it is increasing what cannot reduce and by increasing that which cannot reduce it is reducing the part that cannot increase. While neither of the two is capable of changing in the direction they represent because they represent the entirety there is to represent, the increasing of that which cannot reduce is reducing that which cannot increase because without ever changing the two are growing apart. By eternity never changing while growing apart from infinity that aspect too never can change, and while never changing that part that cannot increase is increasing the part that cannot reduce while the part that cannot reduce is reducing the part that is incapable of increasing. Those, the ones that cannot change is doing what it can do best to the other part that cannot change and in changing its relation with the other part it is remaining the same while the other side in reference then changes the reference.

What Hubble saw was a Universe that was shrinking away into the oblivious because that part that cannot reduce is reducing the part that cannot increase and since the part that cannot increase holds eternity, it can shrink the part that it has in the smallest side by shrinking its reference to that into the oblivious.

It all ends with the Black Hole (not quite but to go into detail about that requires another half a book of explaining and proving as the Black Hole has two more steps to involve mathematically). Let's put the Black Hole as the biggest there is while correcting this error at the same time. The Black Hole is so huge it fits a Universe inside but that means when the Black Hole was huge it was so big it parted eternity from infinity. That came when eternity was just bigger than infinity because the two then parted their shared unity no that long ago. At that point when the Black Hole was the liquid star sloshing away and shining as bright as it could, the Universe was separating eternally big from infinitely small by a margin of Π. Eternity parted 1^0 from 1^1 by a margin of Π^0 that increased to Π by becoming Π^2. The increase came as Π move to Π forming Π^2 that came to a total of Π^3. Still it was at a time when infinity was just smaller than eternity and eternity was just bigger than infinity.

The difference at the start when 1^0 was going onto 1^1 which was going onto Π^0 that was forming $7\Pi^0$ and was combining time as $(10 + 10+ 1.9991= 21.99991 / 7\Pi^0= \Pi)$ this was forming Π^3 but there was little else to show on the eternity side as well as the infinity side and little split the difference. That was just about the environment the Black Hole encountered (okay there was more…but not that much more) and in that the Black Hole was what split infinity and eternity at the time. This we may deduce on the grounds of the facts we now see the Black Hole represents. At present the Black Hole is so small it can reduce the entirety of eternity into infinity while it remains so large that it can absorb the entire eternity into infinity without needing any matter to produce a time delay. It still continues to have the power to carry on what it started with. It started a process putting a bridge between eternity and infinity and counteracted when it shrunk eternity into infinity while it is expanding infinity onto eternity. It took this job when the Universe was wasting away.

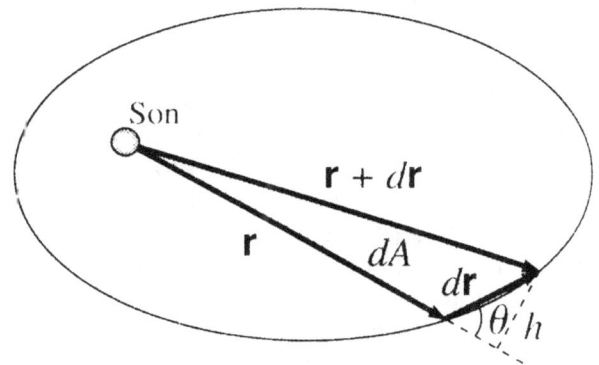

The wasting involved that when that which cannot expand, then reduced and as it reduced it pushed apart that which cannot separate by decreasing that which cannot decrease as that which cannot expand did expand. The split was about that which cannot decrease to part from farther from that which cannot increase while the difference brought about an increasing of the reducing Universe. After the long story about mass and a centre point and a shared centre point with one point favouring both, it boils down to see how material move. Elsewhere I explain what material is. Material is seven point that has no space but is generating space claimed by that which pretends to heat and the retarded heat circle compacted as matter around a centre forming a sphere as small as one is not able to imagine. It has no name because no name giving fame seeking Newtonian can get to it. Only the heat in retarding spin is a part of the Universe but that substantiating and controlling the heat confirms the allocation of the heat by swerving singularity Π^0 which is maintaining $7\Pi^0$ to become the smallest sphere there may be.

In the centre of the smallest matter runs a line that actually is just an expanded point of singularity and the retarded heat spins around this point. The spinning commences where the rotating is holding matter, as matter is the only substance at such a point with the ability of rotation. It is the fact of rotating that is producing the retarding of time and the more of this 1^0 to $7\Pi^0$ there is the more retarding of time going backwards there is. That is a part of the

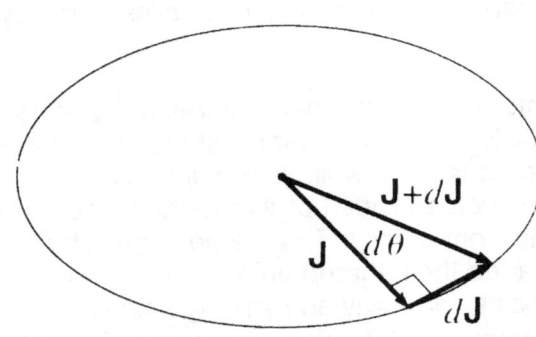

"nothing" Newtonians put into outer space as outer space. That line is where all possible points serving the line lands on one point that line is a dot with a spot in the centre that has no sides and all point in this line falls on the very same point.

The value of this point is 1^0. Because this line focuses all the possible sides on one point and this spot including all the sides that fit into it then still has no sides and fits still with the adding fits all into one place while it serves the rotating centre of retarded heat that finally combine as the atom, one finds that all the lines serving all the retarded heat fits into the next spot that form a centre line that also has no sides in one spot. Eventually the lot forms a combining line that includes all the material within the atom. This eventually then finally combines as $(\Pi^2+\Pi^2)(\Pi^2\Pi)3 = 1836$ and we find it serving our Universe in the capacity as an atom.

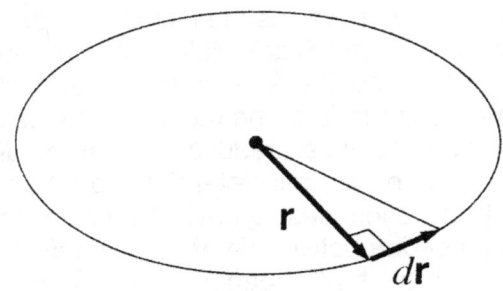

The atom is the atom because all the lines that centre the rotation of all the retarded heat combine in one line forming a centre to the atom. This is possible because all the lines have no sides just like the atomic governing singularity centre line also still has no sides. From there all the atomic governing centre lines fit into a centre line that can hold all the atomic lines because it holds all the atomic lines on one spot as one spot. The line finally forming as the governing singularity driving the motion of the star which is at that point representing all the retarded heat centres which project to the governing centre singularity of the star.

All the positions in eternity relate to every position in infinity and the line holding infinity is the result of the accumulation of all the points serving infinity in the retarded heat. All points that hold 1^0 projects to one point holding 1^0 because all points have no sides therefore all points fit into one point that forms a centre line. All the possible lines by all the possible atoms is projected to one centre line since that one line holds all the possible points there can ever be, on the only one point in a point there possibly can be. This of course also applies in the case of and well as also to the orbiting satellite. Since that point cannot be smaller it can all fit into as well as fit all into that one point that cannot ever be smaller. Since all the points are therefore exactly equal to the extent they all are being the same point, all points everywhere are then the same point concentrated into one point while spread out as far and wide as the point may reach.

The point is singularity referring to singularity. Mathematically this point is expressed as 1^0, which by all mathematical rules are equal to 1^1, which is equal to $\Pi^0 = 1$. In that number 1^0 the entire Universe unites as one unit that incidentally also never can be because everything fits into one spot that is not. To infinity eternity is 1^1 on the condition that infinity then is 1^0 and just the reverse is applied in the relation that eternity holds infinity because the reversing holds infinity at 1^1 as long as eternity can be 1^0. Since both are 1^0 while holding the other as 1^1, the roles inherently have to change when crossing over to the other side of the Universe.

The one side of the Universe will gauge infinity as 1^0 while eternity is 1^1 and at the same moment on the other side of the Universe eternity will be gauged as 1^0 while viewing infinity as 1^1. The end to the Universe is not near, as the start is far away (eternally further than the idiotic 13.5×10^9 years Newtonians give the Universe). The end will arrive when the difference between infinity and eternity will be so large infinity will again join eternity while eternity at that point will be so overburden of the extreme that the incorporation will go unnoticed by all that happened. The planet orbits the Sun since time placed material at that location and allocated a time delay as to the position the Sun holds and the Planet holds. This same argument also serves the same way as it applies in the Sun and the orbiting satellite holding their relation secure.

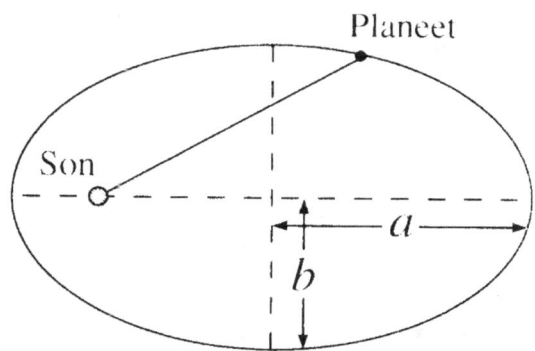

In regard to all these mentioned facts we can deduct that all the matter of the satellite holds one line charged that has no sides. Since this line has no side and takes up no space it fits into the governing singularity that has no sides and all possible lines of all possible matter fit into that into charged line that charges the next line where all the lines also fit into This charging and combining of lines that has no space claims and therefore holds all points on one allocated position eventually form the atom that form the layer that forms the star that form the motion in duplication and conserving the duplication by contraction.

Since that one line is exactly in equality to the next line running to the compiling centre line in the Sun, the centre line in the Sun therefore is also the centre line in the slightest piece of independent time delayed matter of the satellite and the satellite then becomes 1^1 to the Sun being 1^0. From the particle there is a reverse reality that the Sun forms 1^1 to the satellite particle line that holds 1^0, and all this represents eternity in reality departing from infinity forming reality and to the one, the other is 1^1 while that one holding the reference is 1^0.

To the Sun it is the satellite that stands between eternity 1^0 and infinity 1^1 while to the satellite it is the Sun parting infinity 1^1 from eternity 1^0 where eternity drives it in motion. This is the result because the same eternity drives both the Sun and the satellite as individual allocations of 1^1, where the other 1^0 is at the point that becomes part 1^1. This is in place because of the fact that what there is, is there as time delay. The division is result of as much as it is the cause why eternity parted from infinity. While the one is playing a blaming game on the other by taking president in the relation since it holds eternity and the other forms a factor of infinity, the one holds 1^0 to the position that one then allocates to the other as 1^1.

In all the talking and all the explaining, matter is in reality not even reality because matter is three points that lagged in time behind three points that in time serves as time to follow four points spinning in a centre where the four points all share the same spot on the fifth centre spot. When the spot Π^0 became functional and established all relevancies possible, heat parted from cold as eternity parted from infinity. The expansion was not clear motion but more a parting of relevancies where a centre formed a relevancy because the centre could not provide motion.

Without being capable of motion, the centre established four points, which also

When space brought division between eternity and infinity

The Spot becoming the **Dot**

served singularity. From the inverse square law we know that the centre doubled by producing the four points holding singularity.

When the cosmos came to motion, motion was not yet defined. When the cosmos brought about motion, the first motion was relevancies. Cold parted from hot. Eternity parted from infinity. Motion parted from motion absence. Infinity broke the laboriousness of eternity for the duration of infinity. The spot became and grew into the dot.

From what the spot was to what the dot now is might be just a mathematical implication of going from 1^0 to 1^1 but in reality that first motion was the creating of and establishing of an entire Universe with all possibilities now in it. Never again can that much growth become a reality, although to us the growth is beyond what we ever can notice. But it is because the growth is so massive and we are so small that we are unable to notice such almighty growth.

Man learns to live to live to learn. Having life is being born with one immediate requirement and that is to learn to live. Man finds superiority by extending this necessity for one full lifetime of learning through achievement and that puts man above all other life. Man continues to learn physically and morally and to continue in the struggle of life means that man can live so long as man learns. To live is to learn and to learn is to live where living and learning never detaches. When one stops learning one stops finding reasons to live because learning is the fuel of life.

In the final event when all the achievement of the full lifetime compliment is added and measured by standards of that which the person has achieved, the total does not reflect on what one has achieved or what one has accomplished or even what one had accumulated, for that will become a notion. It is about what one had to endure while living to find wisdom in life because life is the learning thereof. The final gauge is in the question that did you work like no other could to achieve that which no other did or did all fall into your lap for you always found crooked ways which makes your achievements a prize worth that of the achieve of a thief.
If it all is this simple then why is that deemed as complicated?

BEST WISHES,

PETRUS. (PEET) S. J. SCHUTTE
THIS WAS An open letter about GRAVITY'S RECIPE

ISBN ??????????????

There are more about this in other books with the titles:

This was the prologue letter announcing
MATTER'S TIME IN SPACE: THE THESIS

ISBN   0-9584410-8-1
FROM THE ORIGINAL AFRIKAANS: "MATERIE SE TYD IN RUIMTE" I. S. B. N.   0 – 620-27041-1
WRITTEN BY PEET SCHUTTE
© KOSMOLOGIESE EN ASTRONOMIESE TEGNIKA.

www.ingramcontent.com/pod-product-compliance
Lightning Source LLC
Chambersburg PA
CBHW080647190526

45169CB00006B/2019